Problems in
Real and Functional
Analysis

Problems in Real and Functional Analysis

Alberto Torchinsky

Graduate Studies
in Mathematics

Volume 166

American Mathematical Society
Providence, Rhode Island

2010 *Mathematics Subject Classification.* Primary 26-01, 28-01, 46-01, 47-01.

For additional information and updates on this book, visit
www.ams.org/bookpages/gsm-166

Library of Congress Cataloging-in-Publication Data
Torchinsky, Alberto.
   Problems in real and functional analysis / Alberto Torchinsky.
      pages cm. — (Graduate studies in mathematics ; volume 166)
   Includes index.
   ISBN 978-1-4704-2057-4 (alk. paper)
      1. Mathematical analysis—Textbooks.   2. Functional analysis—Textbooks.   3. Set theory—
Textbooks.   I. Title.

QA300.T65   2015
515′.7–dc23

                                                                              2015022653

To Massi

Contents

Preface

Students tell me that they learn mathematics primarily from doing problems. They say that a good course is one that motivates the material discussed, building on basic concepts and ideas leading to abstract generality, one that presents the "big picture" rather than isolated theorems and results. And, they say that problems are the most important part of the learning process, because the problems force them to truly understand the definitions, comb through the proofs and theorems, and think at length about the mathematics.

Exercises that require basic application of the theorems highlight the power of the theorems. They also offer an opportunity to encourage students to construct examples for themselves. Problems can also be used to explore counterexamples to conjectures. Supplying a counterexample helps the student gain insight into theorems, including an understanding of the necessity of the assumptions. Well-crafted problems review and expand on the material and give students a chance to participate in the mathematical process. Open-ended problems ("Discuss the validity of . . . ") afford the students the opportunity to adjust to researching and discovering mathematics for themselves.

The purpose of this book is to complement the existing literature in introductory real and functional analysis at the graduate level with a variety of conceptual problems, ranging from readily accessible to thought provoking, mixing the practical and the theoretical. Students can expect the solutions to be written in a direct language, one they can understand; always the most "natural" rather than the most elegant solution is presented.

The book consists of twenty chapters: Chapters 1 through 10 contain the Problems, and Chapters 11 to 20 contain (selected) Solutions. Chapters

1 to 7 cover topics in real analysis, from set theory and metric spaces to Fubini's Theorem, and Chapters 8 to 10 cover topics in functional analysis, from functionals and linear operators on normed linear spaces to Hilbert spaces. Each of the Problem chapters opens with a brief reader's guide stating the needed definitions and basic results in the area and follows with a short description of the problems. There are 1,457 problems.

The notation used throughout the book is standard or else is explained as it is introduced. "Problem 2" means that the result alluded to appears as the second item of the chapter in question, and "Problem 3.2" means that it appears as the second item in Chapter 3.

It is always a pleasure to acknowledge the contributions of those who make a project of this nature possible. Mary Letourneau was the best editor and Arlene O'Sean the best project manager this ambitious project could have had. My largest debt is to the students who attended the real variables courses I taught through the years and kept a keen interest in learning throughout the ordeal. Many examples, counterexamples, problems, and solutions are due to them. They also proofread the text and made valuable suggestions. I owe them much. I assume full responsibility for any typos that the text may have and apologize for any confusion they may cause. To quote from John Dryden: "Errors, like straws, upon the surface flow; He who would search for pearls must dive below."

A. Torchinsky

Part 1

Problems

Chapter 1

Set Theory and Metric Spaces

In this chapter we revisit basic notions of set theory and metric spaces. We consider sets in a naive fashion. As Cantor said, "A *set* is a collection into a whole of definite, distinct objects of our intuition or our thought." We say that the sets A and B are *equivalent* if there is a 1-1, onto function $f : A \to B$. A *finite* set is one that is empty, denoted $\emptyset$, or equivalent to $\{1, \ldots, n\}$ for some $n \in \mathbb{N}$; any set that is not finite is called *infinite*. Infinite sets equivalent to $\mathbb{N}$ are called *countable*, all other infinite sets are *uncountable*. Often the term countable is applied to a set that is equivalent to any subset of $\mathbb{N}$.

Equivalent sets cannot essentially be told apart, which motivates the following informal definition. We associate with a set A its *cardinal number*, denoted $\operatorname{card}(A)$ or a, with the property that any two equivalent sets have the same cardinality. 0 is the cardinal number of the class of sets equivalent to $\emptyset$, n that of $\{1, \ldots, n\}$, $\aleph_0$ that of $\mathbb{N}$, and c that of $[0,1]$ or $\mathbb{R}$.

The inclusion relation for sets translates into a comparison relation for cardinal numbers. More precisely, given cardinals a, b, we say that $a \le b$ if there are sets A, B with $\operatorname{card}(A) = a$ and $\operatorname{card}(B) = b$ such that A is equivalent to a subset of B. The Cantor-Bernstein-Schröder theorem asserts that if there exist injective functions $f : A \to B$ and $g : B \to A$, then there exists a bijection $h : A \to B$. Thus if $\operatorname{card}(A) \le \operatorname{card}(B)$ and $\operatorname{card}(B) \le \operatorname{card}(A)$, then $\operatorname{card}(A) = \operatorname{card}(B)$.

As for the arithmetic operations, given cardinals a, b and disjoint sets A, B with $\operatorname{card}(A) = a$ and $\operatorname{card}(B) = b$, $a + b$ is defined as the cardinal of

$A \cup B$; the product $a \cdot b$ is similarly defined as the cardinal of $A \times B$. And b^a is defined as the cardinal of B^A, the collection of maps from A into B.

We say that $(M, \prec)$ is an *ordered set* if the relation $\prec$ on $M \times M$ is a partial order on M, i.e., it satisfies the following three properties: (i) $m \prec m$ for every $m \in M$. (ii) If $m_1 \prec m_2$ and $m_2 \prec m_1$, then $m_1 = m_2$. (iii) If $m_1 \prec m_2$ and $m_2 \prec m_3$, then $m_1 \prec m_3$.

Given an ordered set $(M, \prec)$, we say that $m \in M$ is the *first element* of M if m precedes any other element of M. We say that an ordered set $(M, \prec)$ is *well-ordered* if it has a first element and any of its subsets ordered with the restriction order has a first element. Zermelo proved that every set can be well-ordered provided the axiom of choice is assumed. In one of its equivalent formulations the axiom of choice states that given an arbitrary family $\mathcal{A} = \{A_i : i \in I\}$ of nonempty sets indexed by a (nonempty) set I, there exists a function $f : I \to \bigcup_{i \in I} A_i$, called the choice or selection function, such that $f(i) \in A_i$ for each $i \in I$.

In particular, the axiom of choice is equivalent to Zorn's lemma — or Zorn's dilemma as Zorn used to say — which can be stated as follows. M is said to be *totally ordered* if for any $m \neq m' \in M$, $m \prec m'$ or $m' \prec m$. And $M' \subset M$ is said to have an *upper bound* $m \in M$ if $m' \prec m$ for all $m' \in M'$; note that m need not be an element of M'. An element $m \in M$ is said to be *maximal* if there is no $m' \in M$ so that $m \prec m'$. Finally, we say that $A \subset M$ is a *chain* in M if A, equipped with the induced order relation $\prec |_A$, is totally ordered. Zorn's lemma asserts that if every chain in a partially ordered set $(M, \prec)$ has an upper bound, then $(M, \prec)$ has a maximal element.

As an immediate application of Zorn's lemma it follows that a linear space over a field contains a maximal linearly independent set, i.e., a basis. A Hamel basis is a basis of $\mathbb{R}$ as a linear space over $\mathbb{Q}$.

Finally, we need *ordinals*, in particular one, Ω. By Zermelo's theorem there exist uncountable well-ordered sets and there is one with the property that all of its initial segments are countable. The ordinal of this set is denoted Ω.

Recall that a *metric space* (X, d) is a nonempty set X together with a nonnegative real-valued function d on $X \times X$, called a metric, such that for all $x, y, z \in X$ the following three properties hold: (i) $d(x, y) = 0$ iff $x = y$. (ii) $d(x, y) = d(y, x)$. (iii) $d(x, y) \leq d(x, z) + d(z, y)$.

In a metric space the *balls* $B(x, r) = \{y \in X : d(x, y) < r\}$, $r > 0$, induce a natural topology on X where the *open sets* O are those sets such that if $x \in O$, there exists $B(y, r) \subset O$ with $x \in B(y, r)$; *closed sets* are the complements of open sets. We say that $x \in A \subset X$ is an *interior point* of

A if $x \in B(y, r)$ where $B(y, r) \subset A$; int(A), the *interior* of A, denotes the collection of interior points of A and is the largest open subset of A. The *closure* $\overline{A}$ of A is the smallest closed subset of X that contains A, i.e., the intersection of all closed sets containing A.

We say that a metric space (X, d) is *complete* if all Cauchy sequences of (X, d) converge. Cantor's nested theorem asserts that the intersection of any nested sequence of nonempty compact subsets of a metric space X is nonempty iff (X, d) is complete.

We say that G is a G_δ set in X if G is the intersection of a countable family of open sets in X; similarly, F is an F_σ set in X if F is the countable union of closed sets in X.

We say that $D \subset X$ is *dense* if $D \cap O \neq \emptyset$ for every open set O in X. A set in X is said to be *nowhere dense* if its closure has empty interior. The sets of *first category* in X are those that are countable unions of nowhere dense sets; all other sets are said to be of *second category* in X.

We say that a metric space (X, d) is a *Baire space* if every set of first category in X has empty interior. The Baire category theorem asserts that a complete metric space (X, d) is of second category in itself, i.e., X cannot be represented as a countable union of nowhere dense sets.

The problems in this chapter cover the various areas described above. They include observations dealing with the basic set-theoretical nature of sets, including the fact that the countable intersection of dense G_δ sets is a dense G_δ set, Problem 4, and the construction and properties of the Cantor set, or Cantor discontinuum, Problems 18–19, as well as the Cantor-Lebesgue function Problem 20. In the area of the Baire category problems include the existence (and abundance) of functions satisfying various properties, Problems 37–38, as well as the nature of the set of discontinuities of a continuous function, Problems 35–37. In the area of limits of continuous functions, we consider if $\chi_\mathbb{Q}$, the characteristic function of the rationals, is the limit of continuous functions, Problem 41, and whether pointwise convergence corresponds to metric convergence for an appropriate metric, Problem 42. The properties of the Baire class $\mathcal{B}_1$, i.e., those functions that are pointwise limits of continuous functions, are covered in Problems 43–51. Cardinality and cardinal arithmetic are discussed in Problems 54–70, and the Hamel basis is discussed in Problems 72–74.

The interested reader can further consult, for instance, K. Devlin, *The Joy of Sets: Fundamentals of Contemporary Set Theory*, Springer-Verlag, 2000; W. Brito, *El Teorema de Categoría de Baire. Sus Aplicaciones*, Editorial Académica Española, 2011; R.-L. Baire, *Sur les fonctions de variables réelles*, Annali di Mat. Ser. **3** (1899), no. 3, 1–123.

Problems

1. Let (X, d) be a metric space. Prove that the following statements are equivalent: (a) (X, d) is a Baire space. (b) The countable intersection of open dense sets is dense. (c) If A is of first category, A^c contains a dense G_δ subset.

2. Let (X, d) be a Baire space. Discuss the validity of the following statement: $A \subset X$ is nowhere dense iff $\overline{A^c} = X$.

3. Let (X, d) be a Baire space. Prove: (a) An open subset O of X is a Baire space in the induced metric. (b) If $\{F_n\}$ are closed subsets of X with $X = \bigcup_n F_n$, then $\bigcup_n \mathrm{int}(F_n)$ is dense in X.

4. Let (X, d) be a complete metric space. Prove: (a) The countable intersection of dense G_δ sets in X is a dense G_δ set in X. (b) If a set and its complement are dense subsets of X, at most one can be G_δ. (c) A countable dense subset of X cannot be G_δ.

5. Give an example of: (a) A sequence of open dense subsets of $[0, 1]$ whose intersection is a countable subset of $[0, 1]$. (b) A G_δ subset of $\mathbb{R}$ that is neither open nor closed. (c) A subset of $\mathbb{R}$ that is neither G_δ nor F_σ. (d) $A \subset \mathbb{R}$ such that $A \in F_{\sigma\delta} \setminus F_\sigma$ and $B \subset \mathbb{R}$ such that $B \in G_{\delta\sigma} \setminus G_\delta$.

6. Let (X, d) be a complete metric space. Prove: (a) A nonempty countable closed subset A of X has isolated points. (b) If X is perfect, X is uncountable.

7. We say that $A \subset \mathbb{R}$ has the Baire property if $A = G \triangle P$ with G open and P of first category. Prove that if A, B are sets of second category that have the Baire property, then $A + B$ and $A - B$ contain an interval.

8. Let (X, d) be a metric space, $\varphi : X \to X$ a homeomorphism, and $A \subset X$. Prove that A and $\varphi(A)$ are of the same category.

9. Let $d(x) = d(x, \mathbb{Z})$ denote the distance from $x \in \mathbb{R}$ to $\mathbb{Z}$. For $q \in \mathbb{N}$ and $\alpha > 0$, let $U_\alpha(q) = \{x \in \mathbb{R} : d(qx) < q^{-\alpha}\}$ and $Y_\alpha = \{x \in \mathbb{R} : x$ belongs to infinitely many $U_\alpha(q)\}$. Prove: (a) Y_α is a G_δ subset of $\mathbb{R}$ and $X = \bigcap_{\alpha \in \mathbb{R}^+} Y_\alpha$ is a dense G_δ subset of $\mathbb{R}$. (b) For $x \in \mathbb{R}$, $x \notin X$ iff there exists a polynomial P with real coefficients such that $P(n)d(nx) > 1$ for all $n \in \mathbb{N}$.

10. We say that a property in a metric space is generic if it holds except possibly in a set of first category. Prove that a generic point in $\mathbb{R}^2$ has both coordinates irrational.

11. We say that a real number x is Diophantine of exponent $\alpha > 0$ if there exists a constant $c > 0$ such that $|x - p/q| > cq^{-\alpha}$ for all rationals p/q. We denote by $\mathcal{D}(\alpha)$ the set of Diophantine numbers of exponent α and by $\mathcal{D} = \bigcup_\alpha \mathcal{D}(\alpha)$ the collection of Diophantine numbers. A real number x that is neither rational nor Diophantine is said to be a Liouville number; $\mathcal{L}$ denotes the collection of Liouville numbers. Prove: (a) If an irrational number x is algebraic of degree $d > 1$, $x \in \mathcal{D}(d)$. (b) $\mathcal{D}$ is of first category and, therefore, generic real numbers are Liouville.

12. Prove that if $A \subset \mathbb{R}$ is of first category, then $A^c - A^c = \mathbb{R}$.

13. Let $\mathcal{A} = \{x \in \mathbb{R} : \text{the decimal expansion of } x \text{ contains every possible finite pattern of digits}\}$. Prove that $\mathcal{A}$ is a dense G_δ subset of $\mathbb{R}$.

14. Let M be a closed subset of $\mathbb{R}$. Prove that M cannot be written as $M = \bigcup_n M_n$ with $M_n \subset \overline{M \setminus M_n}$ for all n.

15. Let ξ be an irrational number. Prove that $X = \{x \in \mathbb{R} : x = m + n\xi, m, n \text{ integers}\}$ is dense in $\mathbb{R}$.

16. Let $G \subset \mathbb{R}$ be an open set unbounded above and let $\{\lambda_n\}$ be such that $\lambda_n \to \infty$ and $d = \limsup_n (\lambda_{n+1} - \lambda_n) = 0$. Prove that every open interval of $\mathbb{R}$ contains a point x with the property that $x + \lambda_n \in G$ for infinitely many n.

17. Let G be an unbounded open subset of $(0, \infty)$ and $D = \{x \in (0, \infty) : nx \in G \text{ for infinitely many integers } n\}$. Prove that D is dense in $(0, \infty)$.

18. Let $\{a_n\}$ be a fixed sequence in $[0, 1]$ such that $a_0 = 1$ and $0 < 2a_n < a_{n-1}$ for $n \geq 1$. Let $P_0 = [0, 1]$ and let P_1 be the set obtained by removing the middle open interval of P_0 of length $a_0 - 2a_1$, i.e., $P_1 = [0, a_1] \cup [1 - a_1, 1]$; note that each interval of P_1 is closed and has length a_1. Next, having constructed P_n let P_{n+1} be the subset of P_n obtained by removing the middle open interval of length $a_n - 2a_{n+1}$ of each of the 2^n disjoint closed intervals, each of length a_n, that comprise P_n. Thus P_{n+1} consists of 2^{n+1} closed intervals each of length a_{n+1}. Finally, let $P = \bigcap_{n=0}^\infty P_n$.

(a) Give an explicit description of the P_n. (b) Let $S = \{x : x_n = 0 \text{ or } x_n = 1\}$ be the space of sequences with terms 0 or 1. Set $r_n = a_{n-1} - a_n$ and let $\varphi : S \to [0, 1]$ be given by $\varphi(x) = \sum_n x_n r_n$. Prove that $\varphi : S \to P$ is 1-1 and onto. That is, P consists precisely of those numbers in $[0, 1]$ of the form $\sum_n x_n r_n$ with $x_n = 0$ or $= 1$.

Also, prove: (c) P has empty interior. (d) Every point in P is an accumulation point of P and, consequently, P is uncountable. (e) Each P_n is closed and has Lebesgue measure $|P_n| = 2^n a_n$. Thus P is a Borel set with Lebesgue measure $|P| = \lim_n 2^n a_n$. (f) If $0 \leq \beta < 1$, $\{a_n\}$ can be chosen so that P has Lebesgue measure $|P| = \beta$.

19. Let p be an integer greater than or equal to 3 and λ a real number such that $0 < \lambda \le p - 2$. Proceeding as in the construction of the P_n in Problem 18 remove open intervals of length λp^{-n} at the n-th stage and let $P_\lambda = \bigcap_n P_n$. (a) Find the Lebesgue measure of P_λ.

Now set $p = 3$ and $\lambda = 1$. The resulting set is called the Cantor set, or Cantor discontinuum, and is denoted C. Prove: (b) C is an uncountable perfect set that contains no intervals and has Lebesgue measure 0. Also, C consists of those $x \in [0, 1]$ with ternary expansion $x = \sum_n x_n/3^n$ where $x_n \ne 1$ for all n. (c) C is symmetric about the point $1/2$. Moreover, if $x \in C$, $x/3 \in C$ and, if $x < 1/3$, $3x \in C$. (d) Each point of $[0, 1]$ is the midpoint of not necessarily distinct points of C. (e) $C - C = [-1, 1]$ and $C + C = [0, 2]$. Also: (f) Give examples of rational and irrational $x \in [0, 1]$ that are, and are not, in C. (g) Characterize the set of left endpoints of the removed open intervals in the construction of C.

20. The Cantor-Lebesgue function $f(x)$ is defined on $[0, 1]$ in two steps, first on C and then on $[0, 1] \setminus C$. If $x \in C$, let

$$f(x) = \frac{1}{2}\sum_n \frac{x_n}{2^n} \quad \text{where} \quad x = \sum_n \frac{x_n}{3^n} \, , \ x_n = 0, 2 \, .$$

Prove: (a) $f : C \to [0, 1]$ is onto. (b) If a, b are the endpoints of any of the open intervals removed in the construction of C, then $f(a) = f(b)$.

Next, if $x \in [0, 1] \setminus C$, x is in exactly one of the intervals (a, b) removed from $[0, 1]$. By (b), $f(a) = f(b)$ and we define $f(x) = f(a)$ for all $x \in (a, b)$. Prove: (c) f is monotone and continuous. (d) f satisfies $f(x) = 2f(x/3)$ for all $x \in [0, 1]$. (e) Determine where f fails to be differentiable.

21. Discuss the validity of the following statement: There exist an interval $J = [a, b]$ and a strictly increasing function g on J such that $g'(x) = 0$ a.e. on J.

22. Let $X = \{\text{sequences } x : x_n = 0 \text{ or } = 1 \text{ for all } n\}$. Prove that equipped with the metric $d(x, y) = \sum_n 2^{-n}|x_n - y_n|$, (X, d) is homeomorphic to the Cantor discontinuum, i.e., there is a continuous bijection $\varphi : X \to C$ with φ^{-1} continuous.

23. Prove that $I = [0, 1]$ is not the union of a countable family of: (a) Pairwise disjoint nonempty closed sets. (b) Cantor sets of positive Lebesgue measure.

24. Suppose that to every $x \in \mathbb{R}$ there corresponds a set $P(x) \subset \mathbb{R}$ such that $x \notin P(x)$ and x is not a limit point of $P(x)$. We say that two points $x, y \in \mathbb{R}$ are independent if $x \notin P(y)$ and $y \notin P(x)$; $A \subset \mathbb{R}$ is said to be independent if every pair of points of A is independent. Prove that there is an independent subset of $\mathbb{R}$ which is of second category in $\mathbb{R}$.

25. Discuss the validity of the following statements: There is an ordering $r_1, r_2, \ldots$ of the rationals in $[0,1]$ such that: (a) $\lim_n r_n$ exists. (b) $\sum_n r_n^k < \infty$ for some integer k. (c) $\sum_n r_n^n < \infty$.

26. For $f : [0,1] \to \mathbb{R}$, let

$$D^+ f(a) = \limsup_{x \to a^+} \frac{f(x) - f(a)}{x - a}.$$

Prove that for each $a \in [0,1)$, $G_a = \{f \in C(I) : D^+ f(a) = \infty\}$ is a dense G_δ subset of $C(I)$.

27. Let (X, d) be a complete metric space and $\mathcal{F} = \{f_\lambda\}_{\lambda \in \Lambda}$ a pointwise bounded family of continuous functions on X, i.e., for each $x \in X$ there is $M(x) < \infty$ such that $|f_\lambda(x)| \leq M(x)$ for all $f_\lambda \in \mathcal{F}$. Prove that $\mathcal{F}$ is uniformly bounded in a ball B of X, i.e., there exist a constant $M > 0$ and a ball B of X such that $|f_\lambda(x)| \leq M$ for all $f_\lambda \in \mathcal{F}$ and $x \in B$.

28. Let (X, d) be a Baire space and $f : X \to \mathbb{R}$ a lower semicontinuous function on X. Prove that f is bounded on a nonempty open subset O of X.

29. Let $\{f_n\}$ be continuously differentiable functions on $\mathbb{R}$. Prove: (a) If $|f_n(x)|, |f_n'(x)| \leq M$ for each $x \in \mathbb{R}$ and all n, $\{f_n\}$ has a uniformly convergent subsequence. (b) If for each $x \in \mathbb{R}$ there are finite numbers $M(x), N(x)$ such that $|f_n(x)| \leq M(x), |f_n'(x)| \leq N(x)$ for all n, a subsequence $\{f_{n_k}\}$ of $\{f_n\}$ converges uniformly in an interval $J \subset \mathbb{R}$.

30. Let $\{f_n\} \subset C(\mathbb{R})$ be such that for each $x \in \mathbb{R}$ there exists an integer n so that $f_n(x) = 0$. Let $O = \{x \in \mathbb{R} :$ there exist an integer n and a neighborhood V_x of x such that $f_n|_{V_x} = 0\}$. Prove that O is an open dense subset of $\mathbb{R}$.

31. Let $f : \mathbb{C} \to \mathbb{C}$ be an entire function. Prove that f is a polynomial or, given a positive integer n and a closed disk $D_r(z_0)$, $f^{(n)}(z) \neq 0$ for some $z \in D_r(z_0)$.

32. Let $f : \mathbb{R} \to \mathbb{R}$ be an infinitely differentiable function such that for all $x \in \mathbb{R}$ there exists an integer n (which may depend on x) with $f^{(n)}(x) = 0$. Prove that f is a polynomial.

33. Let (X, d) be a metric space and $A \subset X$. Prove that χ_A is lower semicontinuous iff A is open.

34. Let (X, d) be a Baire space and O an open subset of X. Prove: (a) The set of points of discontinuity $D(\chi_O)$ of the characteristic function of O is a nowhere dense subset of X. (b) Given open sets $\{O_n\}$, there exists $x \in X$ such that χ_{O_n} is continuous at x for each n.

35. Let (X, d) be a Baire space and f a function on X. Prove: (a) $D(f) = \{x \in X : f \text{ is discontinuous at } x\}$ is an F_σ set. (b) f is continuous on a dense subset of X iff $D(f)$ is of first category in X.

36. Discuss the validity of the following statements: There exists a real-valued function f on $[0, 1]$ such that $C(f)$, the points of continuity of f, are: (a) The rationals in $[0, 1]$. (b) The numbers with a finite decimal expansion. (c) The algebraic numbers in $[0, 1]$.

37. Recall that a function $f : \mathbb{R} \to \mathbb{R}$ is left-continuous at $x \in \mathbb{R}$ if given $\varepsilon > 0$, there exists $\delta > 0$ such that $|f(x) - f(y)| < \varepsilon$ for $x - \delta < t < x$; similarly for right-continuous. Discuss the validity of the following statement: There exists a function $f : \mathbb{R} \to \mathbb{R}$ that is left-continuous everywhere and right-continuous nowhere.

38. For f defined on an open interval $J \subset \mathbb{R}$, let $f'_g(x)$ and $f'_d(x)$ denote the left-hand side derivative and the right-hand side derivative of f at $x \in J$, respectively, whenever they exist. Prove that $A = \{x \in J : f'_g(x) \neq f'_d(x)\}$ is countable.

39. Let $A = \{a_n\}$ be a countable subset of $\mathbb{R}$. Construct a nondecreasing real-valued function f on $\mathbb{R}$ with $D(f) = A$.

40. Let (X, d) be a Baire space, (Y, d') a separable metric space, and $f : X \to Y$ with the property that the preimage of an open set in Y is F_σ in X. Prove that f is continuous in a dense G_δ subset of X.

41. Prove that $\chi_\mathbb{Q}$ is not the pointwise limit of a sequence of continuous functions. However, show that there exist continuous functions $\{f_{m,n}\}$ on $\mathbb{R}$ such that $\lim_n \lim_m f_{m,n}(x) = \chi_\mathbb{Q}(x)$ for all $x \in \mathbb{R}$.

42. Prove that $X = \{f : \mathbb{R} \to \mathbb{R}\}$ cannot be equipped with a metric d so that convergence in (X, d) is equivalent to pointwise convergence.

43. Let (X, d) be a metric space. We say that a function f on X is of Baire class 1, and this is denoted $f \in \mathcal{B}_1$, if f is the pointwise limit of a sequence of continuous functions on X. Prove that if X is a Baire space and $f \in \mathcal{B}_1$, the set $C(f)$ of points of continuity of f is a dense G_δ subset of X.

44. Prove that if f on $[0, 1]$ has a finite number of discontinuities, $f \in \mathcal{B}_1$. In particular, $\mathcal{B}_1$ contains the step functions.

45. Let $\{f_n\} \subset \mathcal{B}_1$ be defined on I and $\sum_n M_n$ a convergent series of positive real numbers. Prove that if $|f_n(x)| \leq M_n$ for $x \in I$ and all n, then $\sum_n f_n(x) = f(x) \in \mathcal{B}_1$.

46. Let $\{f_n\}$ be $\mathcal{B}_1$ functions on I. Discuss the validity of the following statement: If f_n converges uniformly to f in I, then $f \in \mathcal{B}_1$.

47. Let (X, d) be a metric space and f a semicontinuous function on X. Prove that $D(f)$ is of first category in X.

48. Prove that if a metric space (X, d) is of the first category in itself there exists an everywhere discontinuous bounded lower semicontinuous function on X.

49. Prove that if $f : [0, 1] \to \mathbb{R}$ is semicontinuous, then $f \in \mathcal{B}_1$.

50. Let (X, d) be a metric space and A a first category F_σ subset of X. Construct $f \in \mathcal{B}_1$ with $D(f) = A$.

51. Let (X, d) be a complete metric space and $f \in \mathcal{B}_1$. Prove that if $O \subset \mathbb{R}$ is open, then $f^{-1}(O)$ is an F_σ subset of X.

52. Let $f : (0, \infty) \to \mathbb{R}$ be continuous and suppose that for $\lambda_n \to \infty$ with $d_n = \lambda_{n+1} - \lambda_n \to 0$, $\lim_n f(x + \lambda_n)$ exists for all x in an open interval J. Prove that $\lim_{x \to \infty} f(x)$ exists.

53. Let $f : (0, \infty) \to \mathbb{R}$ be a continuous function such that $\lim_n f(nx) = 0$ for each $x > 0$. Prove that $f(x) \to 0$ as $x \to \infty$.

54. Prove that the set of real numbers in $[0, 1]$ which have two decimal expansions (one terminating in 9's and one in 0's) is countable.

55. Prove that $X = \{0, 1\}^{\mathbb{N}}$ is uncountable.

56. Discuss the validity of the following statement: There is a set $\mathcal{F} \subset \mathcal{P}(\mathbb{N})$ that satisfies the following two properties: $\mathrm{card}(\mathcal{F}) = c$ and, if $A, B \in \mathcal{F}$, $\mathrm{card}(A \cap B) < \infty$.

57. Let A be a countable subset of $\mathbb{R}$. Is it possible to translate A by a real number r into $r + A$ so that $A \cap (r + A) = \emptyset$?

58. Let $A \subset \mathbb{R}$ be uncountable. Prove that for some $t \in \mathbb{R}$, $A \cap (-\infty, t)$ and $A \cap (t, \infty)$ are both uncountable. Furthermore prove that there exist $a < b \in \mathbb{R}$ such that $A \cap (-\infty, a)$ and $A \cap (b, \infty)$ are both uncountable.

59. Let $\mathcal{A}$ be a set of real numbers with the property that $|a_1 + \cdots + a_n| \leq 1$ for every finite subset $a_1, \ldots, a_n$ of $\mathcal{A}$. Prove that $\mathcal{A}$ is at most countable.

60. Construct $A \subset \mathbb{R}$ with $\mathrm{card}(A) = c$ such that A intersects every closed nowhere dense subset of $\mathbb{R}$ in a countable set.

61. Prove: (a) $\aleph_0 + n = \aleph_0$ for all $n = 1, 2, \ldots$. (b) $\aleph_0 + \aleph_0 = \aleph_0$.

62. Let A be an infinite set. Prove: (a) A contains a countable subset. (b) If $\mathrm{card}(A) = a$, then $a + \aleph_0 = a$. (c) For infinite cardinals $a \leq b$, $a + b = b$.

63. Let A be an infinite set. Prove: (a) A can be expressed as a pairwise disjoint union of countable subsets of A. (b) If A is uncountable, A can be expressed as $A = X \cup X^c$ where X and X^c are uncountable.

64. Prove that if A is an infinite set, then $A \times A \sim A$. Therefore, if $\text{card}(A) = a$, then $a \cdot a = a$.

65. Given an infinite cardinal b, let $\{a_d\}_{d \leq b}$ be cardinals such that $a_d \leq b$ for all d. Prove that $\sum_{d \leq b} a_d \leq b$. In particular, prove that the countable union of countable sets is countable.

66. Discuss the validity of the following statements: (a) There exists an uncountable pairwise disjoint collection of balls in $\mathbb{R}^n$. (b) There exists an uncountable pairwise disjoint collection of spheres in $\mathbb{R}^n$. (c) There exists an uncountable pairwise disjoint collection of figure eights in the plane.

67. We say that a sequence $\{a_n\}$ is eventually constant if there is an integer N such that $a_n = a_N$ for all $n \geq N$. What is the cardinality of the set of all eventually constant sequences of rational numbers? Real numbers?

68. Find the cardinality of $C(I)$ and $\mathcal{B}_1$.

69. Prove that $\text{card}(\mathcal{B}(\mathbb{R})) = c$.

70. Compute the cardinality of the family $\mathcal{C}$ of the compact subsets of $\mathbb{R}$.

71. Construct a set $B \subset \mathbb{R}$ such that both B and B^c intersect every uncountable compact subset of the line.

72. Prove that if H is a Hamel basis of $\mathbb{R}$, $\text{card}(H) = c$.

73. Prove that there are 2^c different Hamel bases of $\mathbb{R}$.

74. Prove that the Cantor discontinuum contains a Hamel basis.

75. Prove that there exist pairwise disjoint $\{A_n\} \subset \mathbb{R}$ such that $\mathbb{R} = \bigcup_n A_n$ and for any open interval $J \subset \mathbb{R}$, $A_n \cap J$ is of second category for all n.

Chapter 2

Measures

This chapter is devoted to the notion of measure, a generalization of such elementary concepts as the length of a line segment, the area of a rectangle in the plane, and the volume of a parallelepiped in $\mathbb{R}^n$. The natural setting for measures is one that allows to operate freely with sets, including the taking of limits; this is achieved with the introduction of algebras and σ-algebras of sets.

A nonempty class $\mathcal{A}$ of subsets of a universal set X is called an algebra of subsets of X, or plainly an *algebra*, provided the following two properties hold: (i) If $A \in \mathcal{A}$, then $X \backslash A \in \mathcal{A}$. (ii) If $A_1, \ldots, A_n \in \mathcal{A}$, then $\bigcup_{k=1}^{n} A_k \in \mathcal{A}$. Given a family $\mathcal{C}$ of subsets of X, the intersection of all algebras containing $\mathcal{C}$ clearly contains $\mathcal{C}$ and is an algebra; this algebra is denoted $\mathcal{A}(\mathcal{C})$ and is called the algebra generated by $\mathcal{C}$.

As for the limit, given a sequence $\{A_n\}$ of subsets of X we consider the sets $\limsup_n A_n = \{x \in X : x$ belongs to infinitely many $A_n\}$ and $\liminf_n A_n = \{x \in X : x$ belongs to all but finitely many $A_n\}$. In concrete terms,

$$\limsup_n A_n = \bigcap_n \left(\bigcup_{k=n}^{\infty} A_k \right) \quad \text{and} \quad \liminf_n A_n = \bigcup_n \left(\bigcap_{k=n}^{\infty} A_k \right).$$

When the above limits are equal we say that $\{A_n\}$ converges to the common value, which is denoted $\lim_n A_n$.

It becomes quickly apparent that limiting operations are not necessarily closed in an algebra, and one is thus led to the concept of σ-algebra. We say that an algebra $\mathcal{M}$ of subsets of X is a σ-*algebra* of subsets of X if it satisfies the additional property: (iii) If $\{A_n\} \subset \mathcal{M}$, then $\bigcup_n A_n \in \mathcal{M}$.

Given a family $\mathcal{C}$ of subsets of X, there is the question of determining the smallest family of subsets of X that contains the limits of sequences in $\mathcal{C}$; the answer is, of course, the smallest σ-algebra of subsets of X that contains $\mathcal{C}$. Now, the intersection of all σ-algebras containing $\mathcal{C}$ clearly contains $\mathcal{C}$ and is a σ-algebra; this σ-algebra is denoted $\mathcal{M}(\mathcal{C})$ and is called the σ-*algebra generated by* $\mathcal{C}$. One of the most important examples in this setting is $\mathcal{M}(\mathcal{O}_n)$, the σ-algebra generated by the open subsets $\mathcal{O}_n$ of $\mathbb{R}^n$; these are the *Borel subsets of* $\mathbb{R}^n$ and the σ-algebra is denoted $\mathcal{B}(\mathbb{R}^n)$.

Other important concepts discussed below include monotone classes and λ-systems; their main use is to prove that if the class of sets generating a σ-algebra satisfies a certain property, the property is satisfied by all the sets in the σ-algebra.

A *set function* ψ on an algebra $\mathcal{A}$ of subsets of X is a function that assigns to each set in $\mathcal{A}$ a real value or $\pm\infty$; we require that if ψ assumes infinite values, they all be of the same sign. ψ is said to be additive provided the following property holds: if $A_1, \ldots, A_n \subset \mathcal{A}$ are pairwise disjoint, then $\psi(\bigcup_{k=1}^{n} A_k) = \sum_{k=1}^{n} \psi(A_k)$.

Finally, a set function μ on a σ-algebra $\mathcal{M}$ of subsets of X is said to be a *measure* provided the following three properties hold: (i) $\mu : \mathcal{M} \to [0, \infty]$. (ii) $\mu(\emptyset) = 0$. (iii) If $\{A_n\} \subset \mathcal{M}$ are pairwise disjoint, then $\mu(\bigcup_n A_n) = \sum_n \mu(A_n)$.

We then say that $(X, \mathcal{M}, \mu)$ is a *measure space* and the sets in $\mathcal{M}$ are said to be *measurable*. In particular, when $\mathcal{M} = \mathcal{B}(\mathbb{R}^n)$, μ is called a Borel measure. Borel measures on $\mathbb{R}$ that are finite on bounded sets are characterized in terms of distribution functions. A nondecreasing right-continuous function F on $\mathbb{R}$ is said to be a *distribution function* if it has a limit as $x \to \pm\infty$. If F is a distribution function, the set function μ_F given by

$$\mu_F((x, y]) = F(y) - F(x), \quad \text{all } x < y \in \mathbb{R},$$

can be extended to a Borel measure on $\mathbb{R}$, called the measure induced by F.

The basic continuity properties of measures are: (Continuity from below) If $A_1 \subset A_2 \subset \ldots$ is a nondecreasing sequence of measurable sets, then $\mu(\bigcup_n A_n) = \lim_n \mu(A_n)$. (Continuity from above): If $A_1 \supset A_2 \supset \ldots$ is a nonincreasing sequence of measurable sets and $\mu(A_n) < \infty$ for some n, then $\mu(\bigcap_n A_n) = \lim_n \mu(A_n)$.

A measure space $(X, \mathcal{M}, \mu)$ is said to be *complete* if given $N \in \mathcal{N} = \{N \in \mathcal{M} : \mu(N) = 0\}$ and $B \subset N$, then $B \in \mathcal{M}$ and $\mu(B) = 0$.

A measure space $(X, \mathcal{M}, \mu)$ is said to be *finite* if $\mu(X) < \infty$, a *probability measure space* if $\mu(X) = 1$, and σ-*finite* if there are pairwise disjoint finite measure measurable subsets $\{X_n\}$ of X such that $X = \bigcup_n X_n$. And, a measure is said to be *semifinite* if $\mu(X) = \infty$ and given $0 < M < \infty$, there exists $A \subset X$ with $M < \mu(A) < \infty$.

Let $(X, \mathcal{M}, \mu)$ be a measure space. We say that $A \in \mathcal{M}$ with $0 < \mu(A) < \infty$ is an *atom* if whenever $B \subset A$ and $\mu(B) < \mu(A)$, then $\mu(B) = 0$. We say that μ is *purely atomic* if every measurable set of positive measure contains an atom. A measure with no atoms is called *nonatomic*.

In this chapter we pose the questions as to whether there exist an infinite σ-algebra with countably many sets, Problem 14, and if a σ-algebra can be extended to include a given nonmeasurable set, Problem 18, and the original measure extended to a measure in this new setting, Problems 41–43. The uniqueness of measures is discussed in Problems 39–40, Problem 71, and Problem 95. The notion of induced σ-algebra is discussed in Problem 21 and that of induced measure in Problems 45–47. The completion of measures is covered in Problems 54–56. In Problem 64 we consider whether an arbitrary measure can be written as the sum of a purely atomic measure and a nonatomic measure. The support of a measure is discussed in Problems 75–76. Various properties of measures, such as regularity, semifiniteness, and saturation, are covered in Problems 70, Problems 79–83, and Problem 82, respectively. Metric properties of the space of measures are discussed in Problem 84 and Problems 89–90. π-systems and λ-systems, including Dynkin's $\pi - \lambda$ theorem, are covered in Problems 93–98. Independent sets, including the second Borel-Cantelli Lemma, are discussed in Problems 106–110, and measure preserving transformations in Problems 111–112. Distribution functions are covered in Problems 113–121, and finally, doubling measures, in Problem 127.

The interested reader can further consult P. Halmos, *Measure theory*, Springer-Verlag, New York, 1974; V. I. Bogachev, *Measure theory*, I and II, Springer-Verlag, New York, 2007; and P. Billingsley, *Probability and Measure*, J. Wiley, New York, 1986.

Problems

1. Let $\mathcal{A}$ be an algebra of subsets of X and $A, B \notin \mathcal{A}$. Is the same true of $A \cup B$?

2. Show that the union of algebras of subsets of X is not necessarily an algebra. How about the union of an increasing collection of algebras?

3. We say that $\mathcal{F} \subset \mathcal{P}(X)$ is a π-system on X if $\mathcal{F}$ is closed under finite intersections. Let $\mathcal{F}$ be a π-system on X such that $X \in \mathcal{F}$ and if

$A \in \mathcal{F}$, then $A^c = \bigcup_{k=1}^n A_k$ with $A_k \in \mathcal{F}$. Prove that $\mathcal{A} = \{A \subset X : A = \bigcup_{k=1}^n A_k, A_k \in \mathcal{F}\} = \mathcal{A}(\mathcal{F})$, the algebra of subsets of X generated by $\mathcal{F}$.

4. A left half-open interval in $\mathbb{R}$ is an interval of the form $(a, b]$ where $-\infty \le a < b \le \infty$; in particular, (a, ∞) is a left half-open interval. By a left half-open interval J in $\mathbb{R}^n$ we mean a set of the form $J = \prod_{k=1}^n J_k$ where J_k is a left half-open interval in $\mathbb{R}$, $1 \le k \le n$. Let $\mathcal{F}$ denote the collection of left half-open intervals in $\mathbb{R}^n$ and $\mathcal{A}$ the collection of finite unions of intervals in $\mathcal{F}$. Prove that $\mathcal{A}$ is an algebra of subsets of $\mathbb{R}^n$.

5. Let X be an infinite set. (a) Prove that $\mathcal{A} = \{A \subset X : A$ is finite or $X \setminus A$ is finite$\}$ is an algebra of subsets of X. (b) Given $\emptyset \ne A \subset X$, let $\mathcal{S}_A = \{B \subset X : B \subset A$ or $B^c \subset A\}$. Prove that $\mathcal{S}_A$ is a σ-algebra of subsets of X.

6. Let X be an infinite set. Describe the σ-algebra generated by: (a) the singletons of X, and (b) pairs of elements of X.

7. Is the union of an increasing family of a σ-algebra of subsets of X a σ-algebra of subsets of X?

8. Let (X, d) be a metric space and $\mathcal{S} = \{A \subset X : A$ is open$\}$. When is $\mathcal{S}$ a σ-algebra of subsets of X?

9. Given a nonempty class $\mathcal{C}$ of subsets of X, let $\mathcal{C}_1 = \{A \subset X : A \in \mathcal{C}$ or $A^c \in \mathcal{C}\}$, $\mathcal{C}_2 = \left\{\bigcap_n A_n : A_n \in \mathcal{C}_1\right\}$, and $\mathcal{C}_3 = \left\{\bigcup_n A_n : A_n \in \mathcal{C}_2\right\}$. Prove that $\mathcal{C}_3 = \mathcal{M}(\mathcal{C})$.

10. Let $\mathcal{R}$ be a nonempty collection of subsets of X closed under countable unions and relative complements (i.e., if $A, B \in \mathcal{R}$, then $A \setminus B$ in $\mathcal{R}$), $\mathcal{R}' = \{A^c : A \in \mathcal{R}\}$, and $\mathcal{S} = \mathcal{R} \cup \mathcal{R}'$. Prove that $\mathcal{S}$ is a σ-algebra of subsets of X.

11. Let $\mathcal{M}$ be a σ-algebra of subsets of X, $A \subset X$, not necessarily in $\mathcal{M}$, and $\mathcal{M}_A = \{M \cap A : M \in \mathcal{M}\}$. Prove that $\mathcal{M}_A$ is a σ-algebra of subsets of A.

12. Let $\mathcal{M}$ be a σ-algebra of subsets of X that contains countably many nonempty pairwise disjoint sets. Prove that $\mathcal{M}$ is uncountable.

13. Let $\mathcal{M}$ be a σ-algebra of subsets of X. Given $x \in X$, let $\mathcal{C}_x = \{A \in \mathcal{M} : x \in A\}$ and $B_x = \bigcap_{A \in \mathcal{C}_x} A$. Prove that for $x \ne y$, B_x, B_y are identical or disjoint. Are the B_x necessarily measurable? Show that an arbitrary union of B_x is not necessarily measurable.

14. Does there exist an infinite σ-algebra $\mathcal{M}$ with countably many sets?

15. Let $\mathcal{M}$ be a σ-algebra of subsets of X and $\mathcal{A} = \{A_n\}$ measurable sets. Prove that $B_k = \{x \in X : x$ belongs to exactly k sets in $\mathcal{A}\}$, $k \ge 1$, is measurable.

16. Let $\mathcal{C}_1$ and $\mathcal{C}_2$ be collections of subsets of X. Prove that if $\mathcal{C}_1 \subset \mathcal{M}(\mathcal{C}_2)$ and $\mathcal{C}_2 \subset \mathcal{M}(\mathcal{C}_1)$, then $\mathcal{M}(\mathcal{C}_1) = \mathcal{M}(\mathcal{C}_2)$.

17. Let $\mathcal{E}$ be a collection of subsets of X. Prove that $\mathcal{M}(\mathcal{E}) = \bigcup_{\mathcal{F}} \mathcal{M}(\mathcal{F})$ where $\mathcal{F}$ ranges over all countable families of $\mathcal{E}$.

18. Let $\mathcal{S}$ be a σ-algebra of subsets of X and $V \notin \mathcal{S}$. Prove: (a) $\mathcal{M}(\mathcal{S}, V)$, the σ-algebra of subsets of X generated by $\mathcal{S}$ and V, coincides with $\mathcal{F} = \{M \subset X : M = (A \cap V) \cup (B \cap V^c) \text{ with } A, B \in \mathcal{S}\}$. (b) If $\{M_n\}$ are pairwise disjoint sets in $\mathcal{M}(\mathcal{S}, V)$, there are $\{A_n\}$ and $\{B_n\}$, each consisting of pairwise disjoint sets in $\mathcal{S}$, such that $M_n = (A_n \cap V) \cup (B_n \cap V^c)$ for all n.

19. Let $\mathcal{M}$ be a σ-algebra of subsets of X, $\{A_n\} \subset \mathcal{M}$, and $\{B_n\}$ given by $B_1 = A_1$, $B_{n+1} = B_n \triangle A_{n+1}$, $n = 1, 2, \ldots$. Characterize those sequences $\{A_n\}$ for which $\lim_n B_n$ exists.

20. Let $\mathcal{S}(\mathbb{R})$ be a σ-algebra of subsets of $\mathbb{R}$, $\mathcal{A} = \{\emptyset, \{-\infty\}, \{\infty\}, \{-\infty, \infty\}\}$, and $\overline{\mathbb{R}} = \mathbb{R} \cup \{-\infty, \infty\}$. Prove: (a) $\mathcal{S}(\overline{\mathbb{R}}) = \{B \cup C : B \in \mathcal{S}(\mathbb{R}), C \in \mathcal{A}\}$ is a σ-algebra of subsets of $\overline{\mathbb{R}}$, specifically the σ-algebra generated by $\mathcal{S}(\mathbb{R})$ and $\mathcal{A}$. (b) $\mathcal{S}(\mathbb{R}) = \mathcal{S}(\overline{\mathbb{R}}) \cap \mathbb{R}$.

21. Given a mapping $T : X \to Y$ and a σ-algebra $\mathcal{N}$ of subsets of Y, let $\mathcal{M}_T = \{T^{-1}(B) \subset X : B \in \mathcal{N}\}$. Prove that $\mathcal{M}_T$ is a σ-algebra of subsets of X; $\mathcal{M}_T$ is called the σ-algebra induced by T and is the smallest σ-algebra of subsets of X that makes T measurable.

22. Given a mapping $T : X \to Y$ and a collection $\mathcal{C}$ of subsets of Y, characterize $\mathcal{M}(T^{-1}(\mathcal{C}))$, the σ-algebra generated by $T^{-1}(\mathcal{C}) = \{T^{-1}(C) : C \in \mathcal{C}\}$.

23. Let f be a function defined on $\mathbb{R}$ with values in $(\mathbb{R}, \mathcal{B}(\mathbb{R}))$. Prove that the singletons $\{x\} \in \mathcal{M}_f$ for all $x \in \mathbb{R}$ iff f is injective.

24. Let f be a nondecreasing real-valued function on $\mathbb{R}$. Discuss the validity of the following statement: $\mathcal{M}_f = \mathcal{B}(\mathbb{R})$.

25. Consider the function f on $\mathbb{R}$ with values in $(\mathbb{R}, \mathcal{B}(\mathbb{R}))$ given by $f(x) = x^2$. Characterize the σ-algebra $\mathcal{M}_f$ induced by f. Also, if $g(x) = x$, is $g : (\mathbb{R}, \mathcal{M}_f) \to (\mathbb{R}, \mathcal{B}(\mathbb{R}))$ measurable?

26. Consider the function F on $\mathbb{R}^2$ with values in $(\mathbb{R}, \mathcal{B}(\mathbb{R}))$ given by $F(x, y) = x - y$. Characterize the σ-algebra $\mathcal{M}_F$ induced by F.

27. Let $\mathcal{E}$ be an arbitrary collection of subsets of $\mathbb{R}^n$ that contains all the open and closed sets. Prove that if $\mathcal{E}$ is closed under countable unions and intersections, $\mathcal{E}$ contains $\mathcal{B}(\mathbb{R}^n)$.

28. Let $(X, \mathcal{M}, \mu)$ be a measure space and $\{A_n\} \subset \mathcal{M}$. Prove that there are pairwise disjoint sets $\{E_n\} \subset \mathcal{M}$ such that $E_n \subset A_n$ and $\bigcup_n E_n = \bigcup_n A_n$. Furthermore, $\mu(\bigcup_n A_n) = \mu(\bigcup_n E_n) = \sum_n \mu(E_n) \le \sum_n \mu(A_n)$.

29. Let $(X, \mathcal{M}, \mu)$ be a finite measure space and $\mathcal{S} = \{A \in \mathcal{M} : \mu(A) = 0$ or $\mu(A) = \mu(X)\}$. Prove that $\mathcal{S}$ is a σ-algebra of subsets of X.

30. Let $\mathcal{A}$ be an algebra of subsets of X, ψ a nonnegative set function on $\mathcal{A}$ such that $\psi(\emptyset) = 0$, and $\mathcal{C} = \{C \in \mathcal{A} : \psi(A) = \psi(C \cap A) + \psi(C^c \cap A)$ for all $A \in \mathcal{A}\}$. Prove: (a) $\mathcal{C}$ is an algebra of subsets of X. (b) The restriction of ψ to $\mathcal{C}$ is finitely additive. (c) If $\{C_1, \ldots, C_n\} \subset \mathcal{C}$ are pairwise disjoint and $A \in \mathcal{A}$, then $\psi(\bigcup_{k=1}^{n}(C_k \cap A)) = \sum_{k=1}^{n} \psi(C_k \cap A)$.

31. Let $\mathcal{M}$ be a σ-algebra of subsets of X and ψ a positive, finitely additive set function on $\mathcal{M}$. Prove that ψ is σ-additive iff the following condition holds: If $\{A_n\}$ is a decreasing sequence of sets in $\mathcal{M}$ with $\bigcap_n A_n = \emptyset$, then $\lim_n \psi(A_n) = 0$.

32. Let $(X, \mathcal{M}, \mu)$ be a finite measure space and ψ a positive, finitely additive set function on $\mathcal{M}$. Prove that if for any sequence $\{A_n\} \subset \mathcal{M}$ with $\mu(A_n) \to 0$ and also $\psi(A_n) \to 0$, ψ is a measure on $\mathcal{M}$.

33. Let $\mathcal{M}$ be a σ-algebra of subsets of X and ψ a positive, finite set function on $\mathcal{M}$. Prove that ψ is σ-additive iff ψ is additive and σ-subadditive.

34. Let (X, d) be a complete metric space and $\mathcal{S} = \{A \subset X : A$ is of first or second category$\}$. For $A \in \mathcal{S}$ define

$$\mu(A) = \begin{cases} 0, & \text{if } A \text{ is of first category,} \\ 1, & \text{if } A \text{ is of second category.} \end{cases}$$

Prove that $\mathcal{S}$ is a σ-algebra of subsets of X and that μ is a measure on $\mathcal{S}$.

35. Let (X, d) be a complete metric space. Prove that $\mathcal{C} = \{A \subset X : A$ has the Baire property$\}$ is the σ-algebra generated by the open sets together with the sets of first category.

36. Let $\mathcal{S} = \{A \in \mathcal{B}(\mathbb{R}^2) : $ if $(x, y) \in A$, then $(y, x) \in A\}$. Prove that $\mathcal{S}$ is a σ-algebra of subsets of $\mathbb{R}^2$. Furthermore, show that if μ is a measure on $\mathcal{S}$, it is possible to extend μ to $\mathcal{B}(\mathbb{R}^2)$ and that the extension need not be unique.

37. We say that a nonempty collection $\mathcal{M}_0$ of subsets of a universal set X is a monotone class if it is closed under countable increasing unions and countable decreasing intersections. Specifically, if $\{A_n\} \subset \mathcal{M}_0$ and (i) $A_1 \subset A_2 \subset \ldots$, then $\bigcup_k A_k \in \mathcal{M}_0$, and (ii) $A_1 \supset A_2 \supset \ldots$, then $\bigcap_k A_k \in \mathcal{M}_0$. Prove: (a) If $\mathcal{M}_0$ is monotone, $\mathcal{N} = \{A \in \mathcal{M}_0 : A^c \in \mathcal{M}_0\}$

is monotone. (b) Let $\mathcal{C}$ be a collection of subsets of X. If $\mathcal{M}_0$ is monotone, $\mathcal{N} = \{A \in \mathcal{M}_0 : A \cap C \in \mathcal{M}_0 \text{ for all } C \in \mathcal{C}\}$ is monotone. (c) Given a collection $\mathcal{C}$ of subsets of X, there is a smallest monotone class of subsets of X that contains $\mathcal{C}$; this class is denoted $\mathcal{M}_0(\mathcal{C})$ and is called the monotone class generated by $\mathcal{C}$. (d) If a monotone class $\mathcal{M}_0$ is an algebra of subsets of X, $\mathcal{M}_0$ is a σ-algebra of subsets of X.

38. Let $\mathcal{A}$ be an algebra of subsets of X. Prove that $\mathcal{M}_0(\mathcal{A}) = \mathcal{M}(\mathcal{A})$.

39. Give an example of a collection $\mathcal{C}$ of subsets of X and measures μ, ν on $\mathcal{M}(\mathcal{C})$ such that $\mu(A) = \nu(A)$ for all $A \in \mathcal{C}$ yet μ and ν do not coincide on $\mathcal{M}(\mathcal{C})$. On the other hand, prove that if $\mathcal{A}$ is an algebra of subsets of X and μ, ν are finite measures on $\mathcal{M}(\mathcal{A})$ such that $\mu(A) = \nu(A)$ for all $A \in \mathcal{A}$, then $\mu = \nu$.

40. Let $\mathcal{A}$ be an algebra of subsets of X and μ, ν measures on $\mathcal{M}(\mathcal{A})$ that are σ-finite relative to $\mathcal{A}$, i.e., $X = \bigcup_n A_n$ with the A_n pairwise disjoint sets in $\mathcal{A}$ and $\mu(A_n) = \nu(A_n) < \infty$ for all n. Prove that if $\mu(A) = \nu(A)$ for all $A \in \mathcal{A}$, then $\mu = \nu$. Is the result true if the measures are not σ-finite relative to $\mathcal{A}$?

41. Let $(X, \mathcal{S}, \mu)$ be a finite measure space, $V \notin \mathcal{S}$, and $\mathcal{M}(\mathcal{S}, V)$ as in Problem 18. Let μ_i and μ_e be defined on $\mathcal{M}(\mathcal{S}, V)$ by $\mu_i(M) = \sup_{A \subset M, A \in \mathcal{S}} \mu(A)$ and $\mu_e(M) = \inf_{A \supset M, A \in \mathcal{S}} \mu(A)$, respectively. Prove: (a) If $M, N \in \mathcal{M}(\mathcal{S}, V)$ with $M \cap N = \emptyset$ and $M \cup N \in \mathcal{S}$, then $\mu_i(M) + \mu_e(N) = \mu(M \cup N)$. (b) If $M_k \subset A_k$ with $\{A_k\} \subset \mathcal{S}$ and the A_k pairwise disjoint, then $\mu_i(\bigcup_k M_k) = \sum_k \mu_i(M_k)$ and $\mu_e(\bigcup_k M_k) = \sum_k \mu_e(M_k)$.

42. Let $(X, \mathcal{S}, \mu)$ be a finite measure space and $V \notin \mathcal{S}$. Prove that the set functions on $\mathcal{M}(\mathcal{S}, V)$ given by $\nu_*(M) = \mu_i(M \cap V)$ and $\nu^*(M) = \mu_e(M \cap V)$, $M \in \mathcal{M}(\mathcal{S}, V)$, are measures on $\mathcal{S}$.

43. Let $(X, \mathcal{S}, \mu)$ be a finite measure space and $V \notin \mathcal{S}$. Prove that the set functions μ_*, μ^* on $\mathcal{M}(\mathcal{S}, V)$ given by $\mu_*(M) = \mu_i(M \cap V) + \mu_e(M \cap V^c)$ and $\mu^*(M) = \mu_e(M \cap V) + \mu_i(M \cap V^c)$ are extensions of μ to $\mathcal{M}(\mathcal{S}, V)$ that satisfy $\mu_*(V) = \mu_i(V)$ and $\mu^*(V) = \mu_e(V)$. Furthermore, if ξ is a real number such that $\mu_i(V) \leq \xi \leq \mu_e(V)$, there exists a measure ν on $\mathcal{M}(\mathcal{S}, V)$ such that $\nu|_\mathcal{S} = \mu$ and $\nu(V) = \xi$.

44. Let $(X, \mathcal{M}, \mu)$ be a measure space and $T : (X, \mathcal{M}) \to (Y, \mathcal{N})$ a measurable map, i.e., $\mathcal{M}_T \subset \mathcal{M}$. Discuss the validity of the following statement: $T(\mathcal{M})$ is a σ-algebra of subsets of Y.

45. Let $(X, \mathcal{M}, \mu)$ be a measure space and $T : X \to Y$ a mapping from X onto Y. Prove that $\mathcal{N} = \{B \subset Y : T^{-1}(B) \in \mathcal{M}\}$ is a σ-algebra of subsets of Y and that the set function ν on $\mathcal{N}$ given by $\nu(B) = \mu(T^{-1}(B))$ is a measure, called the measure induced by T.

Furthermore, if μ is a probability measure, so is ν; similarly for σ-finiteness and completeness. Finally, if $S : Y \to Z$, characterize the measure induced by $S \circ T : X \to Z$.

46. Let $\mathcal{M} = \{A \subset \mathbb{R} : A \text{ is countable or } A^c \text{ is countable}\}$ and μ the measure on $\mathcal{M}$ given by $\mu(A) = 0$ for A countable and $= \infty$ otherwise. If $Y = \{0, 1\}$, prove that the mapping $T : (\mathbb{R}, \mathcal{B}(\mathbb{R})) \to (Y, \mathcal{P}(Y))$ defined by $T(x) = 0$ if $x \in \mathbb{Q}$ and $= 1$ if $x \in \mathbb{R} \setminus \mathbb{Q}$ is measurable and determine the measure induced by T.

47. Describe the measure μ induced by $T(x) = |x|$ on $(\mathbb{R}, \mathcal{B}(\mathbb{R}))$ endowed with the Lebesgue measure λ.

48. Let $(X, \mathcal{M}, \mu)$ be a measure space and $\mathcal{A}$ an algebra of subsets of X with $\mathcal{M} = \mathcal{M}(\mathcal{A})$ such that X is σ-finite relative to $\mathcal{A}$. Prove that $\mathcal{A}$ is dense in $\mathcal{M}$ in the sense that given $\varepsilon > 0$ and $M \in \mathcal{M}$ with $\mu(M) < \infty$, there is $A \in \mathcal{A}$ such that $\mu(M \triangle A) < \varepsilon$.

49. Let $(X, \mathcal{M}, \mu)$ be a measure space, $\{A_1, \ldots, A_n\} \subset \mathcal{M}$ finite measure sets, denote $\mathcal{I}_k^n = \{I \subset \{1, \ldots, n\} : \operatorname{card}(I) = k\}$, $1 \le k \le n$, and let $A_I = \bigcap_{\ell \in I} A_\ell$. Prove the inclusion-exclusion principle, i.e., $\mu(A_1 \cup \ldots \cup A_n) = \sum_{k=1}^n (-1)^{k+1} \sum_{I \in \mathcal{I}_k^n} \mu(A_I)$. Deduce from this Poincaré's formula

$$\operatorname{card}\left(\bigcup_{i=1}^n A_i \right) = \sum_{k=1}^n (-1)^{k+1} \sum_{I \in \mathcal{I}_k^n} \operatorname{card}(A_I) \,.$$

50. Let $(X, \mathcal{M}, \mu)$ be a measure space and $\{A_n\} \subset \mathcal{M}$ such that each A_n intersects at most one other A_m with $n \neq m$. Prove that $\mu(\bigcup_n A_n) \le \sum_n \mu(A_n) \le 2\,\mu(\bigcup_n A_n)$.

51. Let X denote the collection of permutations σ on $\{1, 2, \ldots, n\}$ and consider the probability measure on $\mathcal{P}(X)$ given by $\mu(\sigma) = 1/n!$ for $\sigma \in X$. Let $A_n = \{\sigma \in X : \sigma(m) \neq m \text{ for all } m = 1, \ldots, n\}$. Find $\mu(A_n)$ and compute $\lim_n \mu(A_n)$.

52. Let $\mathcal{M}$ be a σ-algebra of subsets of X, $Y \subset X$, and suppose that μ is a measure on $(Y, \mathcal{M}_Y)$. Does $\nu(M) = \mu(M \cap Y)$ define a measure on $(X, \mathcal{M})$?

53. Let $(X, \mathcal{M}, \mu)$ be a measure space where μ is not complete, $\mathcal{N} = \{N \in \mathcal{M} : \mu(N) = 0\}$, $\mathcal{M}_1 = \{A \in \mathcal{M} : M_1 \subset A \subset M_2 \text{ with } M_1, M_2 \in \mathcal{M}, M_2 \setminus M_1 \in \mathcal{N}\}$, $\mathcal{M}_2 = \{M \cup B : M \in \mathcal{M}, B \in \mathcal{P}(\mathcal{N})\}$, and $\mathcal{M}_3 = \{M \triangle B : M \in \mathcal{M}, B \in \mathcal{P}(\mathcal{N})\}$. Prove that $\mathcal{M}_1 = \mathcal{M}_2 = \mathcal{M}_3$ is a σ-algebra of subsets of X.

54. Let $(X, \mathcal{M}, \mu)$ be a measure space where μ is not complete, $\mathcal{N} = \{N \in \mathcal{M} : \mu(N) = 0\}$, and $\mathcal{S} = \mathcal{S}(\mathcal{M}, \mathcal{P}(\mathcal{N}))$ the σ-algebra generated by

$\mathcal{M}$ and $\mathcal{P}(\mathcal{N})$. Prove that there is a unique extension μ_1 of μ to $\mathcal{S}$ such that $(X, \mathcal{S}, \mu_1)$ is complete; $(X, \mathcal{S}, \mu_1)$ is called the completion of $(X, \mathcal{M}, \mu)$. Furthermore, μ is σ-finite iff μ_1 is σ-finite.

55. Let $(X, \mathcal{M}, \mu)$ be a complete measure space. Prove: (a) If $A \cup N \in \mathcal{M}$ where $\mu(N) = 0$, then $A \in \mathcal{M}$. (b) If $A, B \subset X$ are such that $A \in \mathcal{M}$ and $\mu(A \triangle B) = 0$, then $B \in \mathcal{M}$ and $\mu(B) = \mu(A)$.

56. Let $(X, \mathcal{M}, \mu)$ be a measure space and $(X, \mathcal{S}, \mu_1)$ its completion. For $A \subset X$ define $\mu^*(A) = \inf\{\mu(M) : M \in \mathcal{M}, A \subset M\}$ and $\mu_*(A) = \sup\{\mu(M) : M \in \mathcal{M}, M \subset A\}$. Prove: (a) If $A \in \mathcal{S}$, $\mu^*(A) = \mu_*(A) = \mu_1(A)$. (b) Conversely, if $A \subset X$ is such that $\mu^*(A) = \mu_*(A) < \infty$, then $A \in \mathcal{S}$.

57. Let $(X, \mathcal{M}, \mu)$ be a σ-finite measure space and $\mathcal{C}$ a collection of pairwise disjoint sets in $\mathcal{M}$ of positive measure. Prove that $\mathcal{C}$ is countable.

58. Let $(X_n, \mathcal{M}_n, \mu_n)$ be measure spaces where the X_n form a pairwise disjoint partition of $X = \bigcup_n X_n$. Prove that $\mathcal{M} = \{B \subset X : B \cap X_n \in \mathcal{M}_n$ for all $n\}$ is a σ-algebra of subsets of X and $\mu(B) = \sum_n \mu_n(B \cap X_n)$, $B \in \mathcal{M}$, a measure on $\mathcal{M}$. Also, prove that μ is σ-finite iff all the μ_n are σ-finite.

59. Let $\mathcal{M}$ be a σ-algebra of subsets of X and $\{\mu_k\}$ a monotone sequence of measures on $\mathcal{M}$. Discuss the validity of the following statement: The set function μ on $\mathcal{M}$ defined by $\mu(A) = \lim_k \mu_k(A)$, $A \in \mathcal{M}$, is a measure on $\mathcal{M}$.

60. Let $(X, \mathcal{M}, \mu)$ and $(X, \mathcal{M}, \nu)$ be finite measure spaces and let λ denote the set function on $\mathcal{M}$ given by $\lambda(A) = \mu(A) + \nu(A)$, $A \in \mathcal{M}$. Prove that λ is a measure.

61. Let $\mathcal{M}$ be a σ-algebra of subsets of X and μ, ν measures on $\mathcal{M}$ with $\nu(A) \le \mu(A)$ for $A \in \mathcal{M}$. Prove that there exists a measure λ on $\mathcal{M}$ such that $\mu(A) = \nu(A) + \lambda(A)$ for $A \in \mathcal{M}$. Show that in general λ is not unique and give a condition that ensures it is.

62. Let $(X, \mathcal{M}, \mu)$ be a finite measure space. Prove: (a) A is an atom iff $\mu(A \cap B) = 0$ or $\mu(A \setminus B) = 0$ for all $B \in \mathcal{M}$. (b) If μ is nonatomic, given A with $\mu(A) > 0$ and $\varepsilon > 0$, there exists $B \subset A$ with $0 < \mu(B) < \varepsilon$. (c) If μ is nonatomic, for every $\eta \in [0, \mu(X)]$ there is $A \in \mathcal{M}$ such that $\mu(A) = \eta$.

63. Let $\mathcal{M}$ be a σ-algebra of subsets of X and ν and λ purely atomic measures on $\mathcal{M}$. Prove that $\mu = \nu + \lambda$ is a purely atomic measure on $\mathcal{M}$.

64. Let $(X, \mathcal{M}, \mu)$ be a measure space. Prove that there exist a purely atomic measure μ_1 and a nonatomic measure μ_2 such that $\mu = \mu_1 + \mu_2$.

65. Let $(X, \mathcal{M}, \mu)$ be a measure space such that $\{\mu(A) : A \in \mathcal{M}\}$ consists of exactly N nonnegative real numbers together with ∞. Prove that there exist a positive integer $n \le N$ and pairwise disjoint measurable sets $A_1, \ldots, A_n, A_\infty$ such that $\mu(A_\infty) = \infty$, the A_k are atoms with $0 < \mu(A_k) < \infty$ for $k = 1, \ldots, n$, and $\mu(X \setminus (A_1 \cup \ldots \cup A_n \cup A_\infty)) = 0$.

66. Let $(X, \mathcal{M}, \mu)$ be a finite measure space and $\{A_r\}_{r \in (0,1]}$ measurable subsets of X satisfying $A_r \subset A_s$ for $r \leq s$. Prove that $A = \bigcap_{r>0} A_r$ is measurable and $\mu(A) = \lim_{r \to 0} \mu(A_r)$.

67. Let $(X, \mathcal{M}, \mu)$ be a finite measure space and suppose that for some $\eta > 0$, $X = \bigcup_n X_n$ with $\sum_n \mu(X_n) \leq \mu(X) + \eta$. Prove that for all $A \in \mathcal{M}$, $\sum_n \mu(X_n \cap A) \leq \mu(X \cap A) + \eta$.

68. Let $(X, \mathcal{M}, \mu)$ be a measure space and $\{A_n\}$ measurable sets with $\mu(\bigcup_n A_n) < \infty$. Discuss the validity of the following statement: $\mu(\bigcup_n A_n) = \sum_n \mu(A_n)$ iff $\mu(A_n \cap A_m) = 0$ for all $n \neq m$.

69. Let $(X, \mathcal{M}, \mu)$ be a probability measure space and $\{A_n\} \subset \mathcal{M}$ such that $\mu(A_n) = 1$ for all n. Prove that $\mu(\bigcap_n A_n) = 1$.

70. Let μ be a finite Borel measure on $\mathbb{R}^n$. Prove that μ is *regular*, i.e., for any $B \in \mathcal{B}(\mathbb{R}^n)$, $\mu(B) = \sup\{\mu(C) : C \subset B, C \text{ compact}\} = \inf\{\mu(A) : A \supset B, A \text{ open}\}$. In other words, for any $B \in \mathcal{B}(\mathbb{R}^n)$ both inner and outer regularity hold. Does the result hold if μ is σ-finite?

71. Let μ, ν be finite Borel measures on $\mathbb{R}^n$ such that $\mu(C) = \nu(C)$ for every compact C. Prove that $\mu = \nu$.

72. Let ψ be a nonnegative, finite, finitely additive set function on $\mathcal{B}(\mathbb{R}^n)$ that is inner regular. Prove that ψ is σ-additive and, hence, a measure.

73. Let μ be a Borel measure on $\mathbb{R}^n$. Prove: (a) If there is a compact set $K \subset \mathbb{R}^n$ with $\mu(K) = \infty$, given $\eta > 0$, there is an open cube Q of sidelength less than η such that $\mu(Q) = \infty$. (b) If μ is inner regular and $V = \{\bigcup O : O \text{ is open and } \mu(O) = 0\}$, then $\mu(V) = 0$.

74. Does there exist a probability Borel measure μ such that $\mu(\{x\}) = 0$ for all $x \in \mathbb{R}^n$ and $\mu(B)$ only assumes the values 0 or 1 for all $B \in \mathcal{B}(\mathbb{R}^n)$?

75. For a Borel measure μ on $\mathbb{R}^n$, the *support* $\text{supp}(\mu)$ of μ is defined as the set $\text{supp}(\mu) = \{x \in \mathbb{R}^n : \mu(O) > 0 \text{ for every open } O \text{ containing } x\}$. Let μ be a finite Borel measure on $\mathbb{R}^n$ with $\mu(\mathbb{R}^n) = \eta$. Prove: (a) F, the support of μ, satisfies $\mu(F) = \eta$ and $\mu(K) < \eta$ for every proper compact subset K of F. (b) $\mathbb{R}^n \setminus F$ is the largest open set in $\mathbb{R}^n$ with μ measure zero.

76. Given a compact K in $\mathbb{R}^n$ construct a Borel measure μ supported in K and a probability Borel measure ν supported in K.

77. Let μ be a Borel probability measure on $I = [0, 1]$, $m = \int_I x \, d\mu(x)$, and $a = \int_I x^2 \, d\mu(x) - m^2$. Prove that $a \leq 1/4$ and determine the unique measure μ for which $a = 1/4$.

78. Let X be an uncountable set and $\mathcal{M}$ the σ-algebra of subsets of X given by $\mathcal{M} = \{A \subset X : A$ is countable or $X \setminus A$ is countable$\}$. Prove: (a) If $\{A_n\}$ are pairwise disjoint measurable sets with union X, exactly one A_n is uncountable. (b) Let μ be the counting measure on $\mathcal{M}$, i.e., for $A \in \mathcal{M}$, $\mu(A)$ is the number of elements in A if $\operatorname{card}(A) < \infty$ and $\mu(A) = \infty$ otherwise. Then μ is a semifinite, but not σ-finite, measure on $\mathcal{M}$. (c) Let ν be the set function on $\mathcal{M}$ given by $\nu(A) = 0$ if A is at most countable and $= 1$ if A^c is at most countable. Then ν is a probability measure and those $A \in \mathcal{M}$ whose complement is at most countable are atoms of ν.

79. Let $(X, \mathcal{M}, \mu)$ be a measure space where μ is semifinite and $A \in \mathcal{M}$ with $\mu(A) = \infty$. Prove: (a) Given $c > 0$, there is a measurable $B \subset A$ with $c < \mu(B) < \infty$. (b) A contains a measurable subset B with $\mu(B) = \infty$ such that B is the countable union of sets of finite measure.

80. Let $(X, \mathcal{M}, \mu)$ be a measure space and ν the set function on $\mathcal{M}$ given by $\nu(A) = \sup\{\mu(B) : B \in \mathcal{M}, B \subset A, \mu(B) < \infty\}$, $A \in \mathcal{M}$. Prove: (a) ν is a semifinite measure. (b) If μ is semifinite, $\nu = \mu$. (c) $\mu = \nu + \lambda$ where ν is semifinite and the measure λ assumes only the values 0 and ∞.

81. Prove that a semifinite Borel measure μ on $\mathbb{R}^n$ is inner regular.

82. Let $(X, \mathcal{M}, \mu)$ be a measure space, $\mathcal{M}_f = \{M \in \mathcal{M} : \mu(M) < \infty\}$, and $\mathcal{M}_{loc} = \{A \subset X : A \cap M \in \mathcal{M}$ for all $M \in \mathcal{M}_f\}$; sets in $\mathcal{M}_{loc}$ are called locally measurable. Prove: (a) $\mathcal{M}_{loc}$ is a σ-algebra of subsets of X that contains $\mathcal{M}$. (b) We say that μ is saturated if $\mathcal{M} = \mathcal{M}_{loc}$. If μ is σ-finite, μ is saturated. (c) The set function ν on $\mathcal{M}_{loc}$ given by $\nu(A) = \mu(A)$ if $A \in \mathcal{M}$ and $= \infty$ otherwise is a saturated measure on $\mathcal{M}_{loc}$. (d) If $(X, \mathcal{M}, \mu)$ is complete, $(X, \mathcal{M}_{loc}, \nu)$ is complete.

(e) Also, we say that $N \in \mathcal{M}$ is locally null if $\mu(A \cap N) = 0$ for all $A \in \mathcal{M}_f$. Discuss the validity of the following statement: If N is locally null, then $\mu(N) = 0$.

83. Let $(X, \mathcal{M}, \mu)$ be a measure space where μ is semifinite and let λ be the set function on $\mathcal{M}_{loc}$ given by $\lambda(A) = \sup\{\mu(B) : B \subset A\}$, $A \in \mathcal{M}$. Prove that λ is a saturated measure on $\mathcal{M}_{loc}$ that extends μ. Discuss the validity of the following statement: With ν as in Problem 82, $\lambda = \nu$.

84. Let $\mathcal{M}$ be a σ-algebra of subsets of X and $\mathcal{P}$ the linear space of finite measures on $\mathcal{M}$. Prove that the function d in $\mathcal{P} \times \mathcal{P}$ given by $d(\mu, \nu) = \sup_{A \in \mathcal{M}} |\mu(A) - \nu(A)|$ is a distance and that $(\mathcal{P}, d)$ is a complete metric space.

85. Let $(X, \mathcal{M}, \mu)$ be a measure space and $\{A_n\} \subset \mathcal{M}$. Discuss the validity of the following statements: (a) $\mu(\liminf_n A_n) \leq \liminf_n \mu(A_n)$. (b) $\limsup_n \mu(A_n) \leq \mu(\limsup_n A_n)$. (c) If $\lim_n A_n$ exists, then $\mu(\lim_n A_n) = \lim_n \mu(A_n)$.

86. Let $(X, \mathcal{M}, \mu)$ be a measure space and $\{A_n\}$ measurable sets such that $\mu(\bigcup_n A_n) < \infty$. Compute $\eta = \limsup_n \mu(\liminf_k (A_n \cap A_k^c))$.

87. Let $(X, \mathcal{M}, \mu)$ be a finite measure space and $\{A_n\} \subset \mathcal{M}$ such that $\sum_n \mu(A_n \setminus A_{n+1}) < \infty$. Prove: (a) $\mu(\liminf_n A_n) = \mu(\limsup_n A_n) = L$. (b) $\lim_n \mu(A_n) = L$.

88. Let $(X, \mathcal{M}, \mu)$ be a probability measure space and $A_1, \ldots, A_{10}$ distinct subsets of X with $\mu(A_n) = 1/3$, all n. Prove that $A = \{x \in X : x$ belongs to at least four distinct $A_n\}$ is a measurable set of positive measure. Can the same conclusion be reached with only nine A_n?

89. Let $(X, \mathcal{M}, \mu)$ be a probability measure space. Given $A, B \in \mathcal{M}$, we say that $A \sim B$ iff $\mu(A \triangle B) = 0$. (a) Prove that $\sim$ is an equivalence relation on $\mathcal{M}$. (b) Consider the quotient space $\widetilde{\mathcal{M}} = \mathcal{M}/\sim$ and for $[A], [B] \in \widetilde{\mathcal{M}}$ let $d([A], [B]) = \mu(A \triangle B)$. Prove that $(\widetilde{\mathcal{M}}, d)$ is a complete metric space.

90. Let $(X, \mathcal{M}, \mu)$ be a measure space and $(\widetilde{\mathcal{M}}, d)$ as defined in Problem 89. Prove that the mappings from $(\widetilde{\mathcal{M}}, d) \times (\widetilde{\mathcal{M}}, d) \to (\widetilde{\mathcal{M}}, d)$ given by $\phi_1(A, B) = A \cup B$, $\phi_2(A, B) = A \cap B$, and $\phi_3(A, B) = A \triangle B$ are continuous and that the mapping $\phi : (\widetilde{\mathcal{M}}, d) \to (\widetilde{\mathcal{M}}, d)$ given by $\phi(A) = A^c$ is an isometry.

91. Let $(X, \mathcal{M}, \mu)$ be a probability measure space and $A_1, \ldots, A_N$ measurable sets such that $\sum_{n=1}^{N} \mu(A_n) > N - 1$. Prove that $\mu(\bigcap_{n=1}^{N} A_n) > 0$.

92. Let $(X, \mathcal{M}, \mu)$ be a measure space and $\{A_n\}$ a decreasing sequence of measurable sets such that $\mu(A_n \setminus A_{n+2}) = 2^{-n}$ and $\mu(A_1 \setminus A_2) = 1/3$. Find $\mu(\bigcup_n (A_n \setminus A_{n+1}))$.

93. Let $\mathcal{D} \subset \mathcal{P}(X)$. We say that $\mathcal{D}$ is a λ-*system*, or *Dynkin system*, on a set X if (i) $X \in \mathcal{D}$, (ii) $A, B \in \mathcal{D}$ and $A \subset B$, then $B \setminus A \in \mathcal{D}$, and (iii) $\{A_n\} \subset \mathcal{D}$ and $A_n \subset A_{n+1}$, $n \geq 1$, then $\bigcup_n A_n \in \mathcal{D}$. (a) Give an example of a λ-system that is not an algebra.

Let $\mathcal{D}$ be a λ-system on X. Prove: (b) Condition (ii) in the definition of $\mathcal{D}$ is equivalent to: If $A \in \mathcal{D}$, $A^c \in \mathcal{D}$. (c) Condition (iii) in the definition of $\mathcal{D}$ is equivalent to: If $\{B_n\}$ are pairwise disjoint sets in $\mathcal{D}$, $\bigcup_n B_n \in \mathcal{D}$. (d) If $\mathcal{F} \subset \mathcal{P}(X)$, there is a smallest λ-system $\mathcal{D}(\mathcal{F})$, say, containing $\mathcal{F}$; $\mathcal{D}(\mathcal{F})$ is called the λ-system generated by $\mathcal{F}$. (e) For $A \in \mathcal{D}$, $\mathcal{D}_A = \{B \subset X : A \cap B \in \mathcal{D}\}$ is a λ-system on X. (f) $\mathcal{D}$ is a σ-algebra iff $\mathcal{D}$ is closed under intersections.

94. (Dynkin's $\pi - \lambda$ theorem) Let $\mathcal{D}$ be a λ-system and $\mathcal{F} \subset \mathcal{D}$ a π-system. Prove that $\mathcal{M}(\mathcal{F}) \subset \mathcal{D}$.

95. Let $\mathcal{F}$ be a π-system in X and μ, ν measures on $\mathcal{M}(\mathcal{F})$ that are σ-finite relative to $\mathcal{F}$, i.e., $X = \bigcup_n A_n$ with the A_n pairwise disjoint sets in $\mathcal{F}$ and $\mu(A_n) = \nu(A_n) < \infty$ for all n. Prove that if $\mu(A) = \nu(A)$ for all $A \in \mathcal{F}$, then $\mu = \nu$. Is the result true if the measures are not σ-finite relative to $\mathcal{F}$?

96. Let $(X, \mathcal{M}, \mu)$ be a probability measure space and let $\mathcal{F}, \mathcal{F}_1 \subset \mathcal{M}$ be π-systems in X. Prove that $\mu(B \cap C) = \mu(B)\mu(C)$ for all $B \in \mathcal{M}(\mathcal{F}), C \in \mathcal{M}(\mathcal{F}_1)$ iff $\mu(B \cap C) = \mu(B)\mu(C)$ for all $B \in \mathcal{F}, C \in \mathcal{F}_1$.

97. Let $\mathcal{F}$ be a π-system on X, $(X, \mathcal{M}(\mathcal{F}), \mu)$ a probability measure space, and $\phi : X \to X$ such that $\mathcal{M}_\phi \subset \mathcal{M}(\mathcal{F})$. Prove that if $\mu(F) = \mu(\phi^{-1}(F))$ for $F \in \mathcal{F}$, then $\mu(A) = \mu(\phi^{-1}(A))$ for $A \in \mathcal{M}(\mathcal{F})$.

98. Let X, Y be nonempty sets, $\phi : X \to Y$, and $\mathcal{D}_X$ a λ-system in X. Prove that $\mathcal{D}_Y = \{F \in \mathcal{P}(Y) : \phi^{-1}(F) \in \mathcal{D}_X\}$ is a λ-system in Y.

99. Let $(X, \mathcal{M}, \mu)$ be a measure space, $\{A_n\} \subset \mathcal{M}$ with $\lim_n \mu(X \setminus A_n) = 0$, and $G = \{x \in X : x$ belongs to finitely many $A_n\}$. Prove that G is a measurable set with $\mu(G) = 0$.

100. Let $(X, \mathcal{M}, \mu)$ be a finite measure space and $\{A_n\} \subset \mathcal{M}$. Prove: (a) If $\sum_n \mu(A_n) < \infty$, then $\mu(\limsup_n A_n) = 0$. (b) If $\mu(\limsup_n A_n) = 0$, then $\lim_n \mu(A_n) = 0$. (c) If $\lim_n \mu(A_n) = 0$, there is a subsequence $\{A_{n_k}\}$ of $\{A_n\}$ such that $\mu(\limsup_{n_k} A_{n_k}) = 0$. Is it true that $\mu(\limsup_n A_n) = 0$?

101. Let $(X, \mathcal{M}, \mu)$ be a probability measure space and $\{A_n\} \subset \mathcal{M}$ such that $\limsup_n \mu(A_n) = 1$. Prove that given $0 < \eta < 1$, there is a subsequence $\{A_{n_k}\}$ of $\{A_n\}$ such that $\mu(\bigcap_{n_k} A_{n_k}) > 1 - \eta$.

102. Let μ be a nonatomic Borel measure on I and define for $n \geq 1$, $a_n = \max\{\mu([i/n, (i+1)/n]) : 0 \leq i \leq n - 1\}$. Prove that $\lim_n a_n = 0$.

103. Let $(X, \mathcal{M}, \mu)$ be a probability measure space, $\{A_n\} \subset \mathcal{M}$, and $0 < \eta \leq 1$. Prove that the following statements are equivalent: (a) $\mu(\limsup_n A_n) \geq \eta$. (b) If $\mu(B) > 1 - \eta$, then $\sum_n \mu(A_n \cap B) = \infty$.

104. Let $(X, \mathcal{M}, \mu)$ be a finite measure space, $M \in \mathcal{M}$, and $\{A_n\} \subset \mathcal{M}$ such that $\sum_n \mu(M \cap A_n \cap A_{n+1}^c) < \infty$ and $\mu(\liminf_n M \cap A_n) = 0$. Prove that $\mu(\limsup_n A_n) \leq \mu(X) - \mu(M)$.

105. Let $(X, \mathcal{M}, \mu)$ be a finite measure space and $\{A_n\} \subset \mathcal{M}$ such that $\mu(A_n) \geq \eta > 0$, all n. Prove that $\mu(\limsup_n A_n) \geq \eta$. Is the conclusion true if $\mu(X) = \infty$? On the other hand, show that there exists a sequence $\{A_n\}$ as above so that for any of its subsequences $\{A_{n_k}\}$, if the measurable set $B \subset A_{n_k}$ for all n_k, then $\mu(B) = 0$.

106. Let $(X, \mathcal{M}, \mu)$ be a probability measure space. Recall that we say that $\{A_n\} \subset \mathcal{M}$ is *independent* if for every finite subfamily $A_{n_1}, \ldots, A_{n_k}$ of $\{A_n\}$, $\mu(\bigcap_{k=1}^m A_{n_k}) = \prod_{k=1}^m \mu(A_{n_k})$. Prove that if $\{A_n\}$ are independent sets, $A_1^c, A_2, \ldots, A_n, \ldots$ are independent sets.

107. Let $(X, \mathcal{M}, \mu)$ be a probability measure space. Prove that $A \in \mathcal{M}$ is independent of all $B \in \mathcal{M}$ iff $\mu(A) = 0$ or $\mu(A) = 1$.

108. Let $(X, \mathcal{M}, \mu)$ be a probability measure space and $\{A_n\}$ independent measurable sets. Prove that if $\sum_n \mu(A_n) = \infty$, then $\mu(\limsup_n A_n) = 1$.

109. Let $(X, \mathcal{M}, \mu)$ be a probability measure space, $\{A_n\}$ independent measurable sets with $\mu(A_n) = \alpha_n$ where $\sum_n \alpha_{2n}\alpha_{2n+1} = \infty$, and $C_n = A_n \cap A_{n+1}$, $n = 1, 2, \ldots$. Compute $\mu(\limsup_n C_n)$.

110. Let $(X, \mathcal{M}, \mu)$ be a probability measure space and $\{A_n\}$ independent measurable subsets of X such that $\mu(A_n) < 1$ for all n. Prove that $\mu(\bigcup_n A_n) = 1$ iff $\mu(\limsup_n A_n) = 1$.

111. Let $(X, \mathcal{M}, \mu)$ be a measure space. We say that $T : X \to X$ is *measure preserving* if $T^{-1}(A) \in \mathcal{M}$ for all $A \in \mathcal{M}$ and $\mu(T^{-1}(A)) = \mu(A)$. Prove that if T is measure preserving and $\{A_n\}$ are measurable sets such that $\sum_n \mu(A_n) < \infty$, then for μ-a.e. $x \in X$ there exists an integer $N(x)$ such that $T^n(x) \notin A_n$ for all $n \geq N(x)$.

112. Let $(X, \mathcal{M}, \mu)$ be a probability measure space and $T : X \to X$ measure preserving. Prove that the following are equivalent: (a) If $A \in \mathcal{M}$ and $T^{-1}(A) = A$, then $\mu(A) = 0$ or 1. (b) If $A \in \mathcal{M}$ and $T^{-1}(A) \subset A$, then $\mu(A) = 0$ or 1. (c) If $A \in \mathcal{M}$ and $T^{-1}(A) \supset A$, then $\mu(A) = 0$ or 1.

113. Given nonnegative reals $\{p_n\}$, let $F(x) = \sum_{n \leq x} p_n$. Prove that F is a distribution function.

114. Let μ be a Borel measure on $\mathbb{R}$ such that $\mu(J) < \infty$ for bounded intervals $J \subset \mathbb{R}$. Prove that there exists a distribution function F such that $\mu = \mu_F$.

115. Let ψ be the set function on $\mathcal{B}(\mathbb{R})$ given by $\psi(A) = \sum_n \frac{1}{2^n} \chi_A(\frac{1}{n})$, $A \in \mathcal{B}(\mathbb{R})$. Prove that ψ is a Borel measure associated to a distribution F and find F.

116. Let μ_F be a Borel measure on $\mathbb{R}$ associated to the distribution function F such that $\mu_F((a-1, a]) < \infty$ for all $a \in \mathbb{R}$. Prove: (a) $\mu_F(\{a\}) = F(a) - F(a^-)$. Obtain the expression for $\mu_F(J)$ where J is an interval in $\mathbb{R}$. (b) Compute $\mu_F(B_j)$ for each of the following sets: $B_1 = \{2\}, B_2 = [-1/2, 3), B_3 = (-1, 0] \cup (1, 2)$, and $B_4 = [0, 1/2) \cup (1, 2]$, where

$$
F(x) = \begin{cases}
0, & x < -1, \\
1 + x, & -1 \leq x < 0, \\
2 + x^2, & 0 \leq x < 2, \\
9, & x \geq 2.
\end{cases}
$$

117. Let μ_F be the Borel measure on $\mathbb{R}$ induced by the distribution function F. Prove that $\mu_F(\{x\}) = 0$ iff F is continuous at x.

118. Let F be a continuous distribution function and μ_F the measure induced by F. Discuss the validity of the following statement: If B is a Borel set with $\mu_F(B) > 0$ and $\mu_F(\mathbb{R} \setminus B) = 0$, B is dense in $\mathbb{R}$.

119. Given

$$F(x) = \begin{cases} x, & x < 0, \\ x + n, & n \le x < n + 1, \ n = 0, 1, 2, \ldots, \end{cases}$$

find $\mu_F(A)$ for an arbitrary $A \in \mathcal{B}(\mathbb{R})$.

120. Let F be a distribution function of a probability measure μ such that $F(x) = 0$ or $= 1$ for all $x \in D$, where D is dense in $\mathbb{R}$. Prove that μ is a Dirac measure.

121. Given a right-continuous nondecreasing function $F : [a, b] \to \mathbb{R}$ with $F(a) = 0$ and $F(b) = L$, $-\infty < a < b < \infty$, let $F^{-1}(t) = \inf\{s \in [a, b] : F(s) \ge t\}$. Prove: (a) $\{t \in [0, L] : F^{-1}(t) \le s\} = [0, F(s)]$ for every $s \in [a, b]$ and F^{-1} is Borel measurable on $[0, L]$. (b) If μ is the set function on the Borel subsets of $[a, b]$ given by $\mu(A) = |\{t \in [0, L] : F^{-1}(t) \in A\}| = |F^{-1}(A)|$, then μ is a measure and $\mu([a, s]) = \mu((a, s]) = F(s)$ for all $s \in [a, b]$.

122. Given a right-continuous, nondecreasing function $F : \mathbb{R} \to [0, 1]$, find $F^{-1}(u)$ if F corresponds to a discrete function that takes the values $x_1, x_2, \ldots, x_k$ with $|\{F = x_i\}| = p_i$, $1 \le i \le k$.

123. Let F be a nondecreasing, right-continuous function on $\mathbb{R}$, $\Phi : \mathbb{R} \to \mathbb{R}$ continuous, increasing, with a continuous inverse, and μ_F and $\mu_{F \circ \Phi}$ the measures induced by F and $F \circ \Phi$, respectively. Prove that $\mu_{F \circ \Phi}(\Phi^{-1}(A)) = \mu_F(A)$ for all Borel $A \subset \mathbb{R}$.

124. Let μ be a finite Borel measure on $\mathbb{R}^2$, $S_x = \{y \in \mathbb{R}^2 : |x - y| = 1\}$, and $\varphi(x) = \mu(S_x)$. Prove that φ is continuous at x iff $\varphi(x) = 0$.

125. Given a finite Borel measure μ on $\mathbb{R}^n$ and an open subset O of $\mathbb{R}^n$, let $\varphi : \mathbb{R}^n \to \mathbb{R}$ be defined by $\varphi(x) = \mu(x + O)$. Is φ continuous? If not, is φ lower or upper semicontinuous?

126. Let μ be a Borel measure on $\mathbb{R}^n$. Prove that the upper k-density $D_k(x) = \limsup_{r \to 0+} r^{-k}\mu(B(x, r))$ is a Borel measurable function of x for each given $k > 0$.

127. Given a cube Q in $\mathbb{R}^n$ of sidelength $\ell(Q)$, let λQ denote the cube concentric with Q of sidelength $\ell(\lambda Q) = \lambda \ell(Q)$. A Borel measure μ on $\mathbb{R}^n$ is said to be *doubling* if $\mu(2Q) \le c\,\mu(Q)$ for all cubes Q. Prove that if μ is doubling, $\mu(\mathbb{R}^n) = 0$ or $\mu(\mathbb{R}^n) = \infty$.

Chapter 3

Lebesgue Measure

This chapter is devoted to the Lebesgue measure on $\mathbb{R}^n$. The σ-algebra $\mathcal{L}(\mathbb{R}^n)$ of Lebesgue measurable sets is required to contain the intervals I of $\mathbb{R}^n$, i.e., those sets of the form $I = \{x \in \mathbb{R}^n : a_k \leq x_k \leq b_k, k = 1, \ldots, n\}$, and the Lebesgue measure to be complete, translation invariant, and to agree with the volume on intervals. First one defines the Lebesgue outer measure, a nonnegative σ-subadditive set function on $\mathcal{P}(\mathbb{R}^n)$, and the Lebesgue measurable sets are selected as those subsets of $\mathbb{R}^n$ that can be approximated by open sets in the sense of the Lebesgue outer measure. The Lebesgue measure is then the restriction of the Lebesgue outer measure to $\mathcal{L}(\mathbb{R}^n)$. Specifically, given a subset A of $\mathbb{R}^n$, the *Lebesgue outer measure* $|A|_e$ of A is defined as the quantity $|A|_e = \inf \left\{ \sum_k v(I_k) : A \subset \bigcup_k I_k \right\}$, where $v(I_k)$ denotes the volume of the closed interval I_k and the infimum is taken over the family of all coverings of A by countable unions of closed intervals. The Lebesgue outer measure is additive for sets that are far apart, i.e., if $A, B \subset \mathbb{R}^n$ are such that $d(A, B) > 0$, then $|A \cup B|_e = |A|_e + |B|_e$, σ-subadditive, and $|I|_e = v(I)$ for all intervals in $\mathbb{R}^n$.

We say that $A \subset \mathbb{R}^n$ is *Lebesgue measurable*, or $A \in \mathcal{L}(\mathbb{R}^n)$, if, for any $\varepsilon > 0$, there is an open set $O \supset A$ such that $|O \setminus A|_e < \varepsilon$. $\mathcal{L}(\mathbb{R}^n)$ is then a σ-algebra of subsets of $\mathbb{R}^n$ and the restriction of the Lebesgue outer measure to $\mathcal{L}(\mathbb{R}^n)$ is a measure, which is called the *Lebesgue measure on $\mathbb{R}^n$* and is denoted $|\cdot|$.

Sets in $\mathcal{L}(\mathbb{R}^n)$ can be described in one of two ways, namely, $A \in \mathcal{L}(\mathbb{R}^n)$ iff $A = H \cup N$, where H is F_σ and $N \in \mathcal{L}(\mathbb{R}^n)$ is null, and $A \in \mathcal{L}(\mathbb{R}^n)$ iff $A = G \setminus N$, where G is G_δ and N is null. Also, for arbitrary sets $A \subset \mathbb{R}^n$, there exist $H \subset A \subset G$ such that H is F_σ, G is G_δ, and $|H| = |A|_e = |G|$.

Also, the Lebesgue measure is *regular*. That is, if $A \in \mathcal{L}(\mathbb{R}^n)$ and $\varepsilon > 0$, there are $F \subset A \subset O$ such that F is closed, O is open, and $|O \setminus F| < \varepsilon$. Moreover, if $|A| < \infty$, F can be chosen to be compact.

A characterization of $\mathcal{L}(\mathbb{R}^n)$ due to Carathéodory highlights the interplay between the Lebesgue measurable sets and the Lebesgue measure, and its interest lies in the fact that it can be used to define the Lebesgue measure, as well as other measures on more general σ-algebras of sets. Carathéodory's characterization states that $A \in \mathcal{L}(\mathbb{R}^n)$ iff $|E|_e = |E \cap A|_e + |E \setminus A|_e$ for every $E \subset \mathbb{R}^n$.

The problems in this chapter include the existence of Lebesgue non-measurable sets, Problem 1, and properties of the Lebesgue outer measure, including the existence of infinitely many pairwise disjoint subsets of $[0, 1]$ with Lebesgue outer measure 1, Problem 8, and the Lebesgue outer measure version of the Borel-Cantelli lemma, Problem 22. The fact that a translation invariant measure on the Borel subsets of $\mathbb{R}$, as well as one that satisfies appropriate "dilation" properties, is a multiple of the Lebesgue measure is addressed in Problem 23 and Problem 54, respectively. A couple of important properties of Lebesgue measurable sets of positive measure are that there exists a cube that contains an arbitrarily large proportion of the set in question, Problem 55, and the Steinhaus theorem that asserts that the difference set of a set of positive Lebesgue measure contains a neighborhood of the origin, Problem 67. Also, such sets contain arbitrarily long sequences, Problem 38.

The action of linear and Lipschitz maps on subsets of $\mathbb{R}^n$ is addressed in Problem 49 and Problem 43, respectively. The properties of points of density, and dispersion, of a Lebesgue measurable set are discussed in Problems 78–81. The Lebesgue measurability, and measure, of various sets defined in terms of general expansions, including dyadic, ternary, or decimal, is discussed in Problems 85–90. Properties of the Hamel basis are discussed in Problems 101–104. The relation between category and measure are discussed in Problems 111–113.

The interested reader can further consult R. L. Wheeden and A. Zygmund, *Measure and integral*, Marcel Dekker, 1977.

Problems

1. Construct a Lebesgue nonmeasurable set in $\mathbb{R}^n$.

2. How many Lebesgue nonmeasurable sets are there in $\mathbb{R}$?

3. Prove that $\mathcal{L}(\mathbb{R}^n)$ is the smallest σ-algebra of subsets of $\mathbb{R}^n$ that contains $\mathcal{B}(\mathbb{R}^n)$ and the sets of Lebesgue measure 0.

4. Let B be a Lebesgue nonmeasurable subset of $\mathbb{R}^n$. Prove that there exists $B_0 \subset B$ such that $B_0 \notin \mathcal{L}(\mathbb{R}^n)$ and if $A \subset B_0$ is Lebesgue measurable, $|A| = 0$.

5. Prove that $B \notin \mathcal{L}(\mathbb{R}^n)$ iff there exists $\varepsilon > 0$ such that for every Lebesgue measurable $A \subset B$, $|B \setminus A|_e \geq \varepsilon$.

6. Does there exist a Lebesgue nonmeasurable set $B \subset \mathbb{R}$ such that $A = \{x \in B : x \text{ is irrational}\}$ is measurable?

7. Prove that if $A \subset \mathbb{R}^n$ intersects every compact subset of $\mathbb{R}^n$ of positive measure, $|A|_e = \infty$.

8. Do there exist pairwise disjoint subsets $\{B_n\}$ of $[0,1]$ such that $|B_n|_e = 1$ for all n?

9. Let $\{I_n\}$ be a finite collection of open intervals that covers $[a,b] \cap \mathbb{Q}$. Prove that $\sum_n |I_n| \geq b - a$. Does it follow that $|[a,b] \cap \mathbb{Q}| \geq b - a$?

10. Prove that for a compact subset A of $\mathbb{R}^n$, $|A|$ can be computed using finite covers, i.e., $|A| = \inf\{\sum_{k=1}^{N} v(I_k) : A \subset \bigcup_{k=1}^{N} I_k\}$.

11. Let $B \subset \mathbb{R}^n$ be bounded and for an interval J consider the expression $|J| - |J \setminus B|_e$. Prove that as long as J contains B this expression is independent of J.

12. Let $\{A_k\} \subset \mathbb{R}^n$ be pairwise disjoint measurable sets and $A \subset \mathbb{R}^n$. Prove that $|A \cap \bigcup_k A_k|_e = \sum_k |A \cap A_k|_e$.

13. Let A, B be subsets of $\mathbb{R}^n$ with $|A|_e, |B|_e < \infty$ such that $|A \cup B|_e = |A|_e + |B|_e$. Prove: (a) $|A \cap B| = 0$. (b) If $A \cup B \in \mathcal{L}(\mathbb{R}^n)$, then $A, B \in \mathcal{L}(\mathbb{R}^n)$.

14. Let $A, B \subset \mathbb{R}^n$. Prove: (a) If $A \in \mathcal{L}(\mathbb{R}^n)$, $|A \cap B|_e + |A \cup B|_e = |A| + |B|_e$. (b) If there exists a measurable C such that $A \subset C$ and $|B \cap C| = 0$, then $|A \cup B|_e = |A|_e + |B|_e$.

15. Discuss the validity of the following statement: If $A \subset \mathbb{R}^n$ is such that $\inf\{|G| : A \subset G, G \text{ open in } \mathbb{R}^n\} = \sup\{|F| : F \subset A, F \text{ closed in } \mathbb{R}^n\}$, then A is Lebesgue measurable.

16. Discuss the validity of the following statement: $A \subset \mathbb{R}^n$ is Lebesgue measurable iff $|Q| = |Q \cap A|_e + |Q \setminus A|_e$ for all cubes Q in $\mathbb{R}^n$.

17. Let $A \subset \mathbb{R}^n$ with $|A|_e < \infty$. Prove that $A \in \mathcal{L}(\mathbb{R}^n)$ iff given $\varepsilon > 0$, there exists a finite collection of closed intervals $I_1, \ldots, I_N$, say, such that $|A \triangle \bigcup_{k=1}^{N} I_k|_e < \varepsilon$.

18. Let $A \in \mathcal{L}(\mathbb{R}^n)$ with $|A| < \infty$. Prove that there exists $\{A_k\}$, where each A_k is a finite union of intervals, with the following two properties: $\lim_k \chi_{A_k}(x) = \chi_A(x)$ a.e. and $|A \triangle (\liminf_k A_k)| = 0$.

19. Prove that the Lebesgue outer measure is continuous from below. Specifically, if $\{A_k\} \subset \mathbb{R}^n$ is an increasing sequence, then $|\bigcup_k A_k|_e = \lim_k |A_k|_e$.

20. Suppose $\{A_j\} \subset \mathbb{R}^n$ is such that for a strictly increasing sequence $n_1 < n_2 < \cdots$, $|\bigcup_{j=1}^{n_k} A_j|_e = \sum_{j=1}^{n_k} |A_j|_e$. Prove that $|\bigcup_j A_j|_e = \sum_j |A_j|_e$.

21. Give an example of: (a) Pairwise disjoint sets $\{A_k\}$ such that $|\bigcup_k A_k|_e < \sum_k |A_k|_e$. (b) A decreasing sequence $\{A_k\}$ with $|A_1|_e < \infty$ and $|\bigcap_k A_k|_e < \lim_k |A_k|_e$.

22. Let $\{A_k\} \subset \mathbb{R}^n$ be such that $\sum_k |A_k|_e < \infty$. Prove that $\limsup_k A_k$ is a measurable set of measure 0.

23. Let μ be a Borel measure on $\mathbb{R}$ such that $\mu((0,1]) < \infty$ and D a dense subset of $\mathbb{R}$ such that $\mu(d + (a,b]) = \mu((a,b])$ for $d \in D$ and $a < b \in \mathbb{R}$. Prove that μ is a multiple of the Lebesgue measure on the Borel sets, i.e., for every $A \in \mathcal{B}(\mathbb{R})$, $\mu(A) = c|A|$, where $c = \mu((0,1])$.

24. Prove that $A \subset \mathbb{R}^n$ has $|A| = 0$ iff given $\varepsilon > 0$, A can be covered infinitely often by intervals $\{I_k\}$ with $\sum_k v(I_k) < \varepsilon$.

25. Let $A \in \mathcal{L}(\mathbb{R}^n)$ with $|A| > 0$, $\{x_k\}$ a bounded sequence in $\mathbb{R}^n$, and $A_k = x_k + A$ for $k \geq 1$. Prove that $|A| \leq |\limsup_k A_k|$.

26. Discuss the validity of the following statements: (a) A measurable set in $\mathbb{R}^n$ has measure 0 iff its closure has measure 0. (b) There is an open set A of arbitrarily small measure whose boundary ∂A has arbitrarily large measure.

27. Let H be a hyperplane in $\mathbb{R}^n$ perpendicular to a coordinate x_k direction, $1 \leq k \leq n$. Prove that $|H| = 0$.

28. Prove that a line segment in $\mathbb{R}^2$ has measure 0.

29. Let $\mathcal{A}$ be the collection of finite unions of sets of the form $(a,b] \cap \mathbb{Q}$ where $-\infty \leq a < b \leq \infty$. (a) Prove that $\mathcal{A}$ is an algebra of subsets of $\mathbb{Q}$ and identify $\mathcal{M}(\mathcal{A})$. (b) Discuss the validity of the following statement: There exists a measure μ defined on $\mathcal{P}([0,1] \cap \mathbb{Q})$ such that $\mu((a,b]) = b - a$ for all rationals $a < b$.

30. Discuss the validity of the following statement: If A is a closed subset of $\mathbb{R}^n$ and $O_k = \{x \in \mathbb{R}^n : d(x, A) < 1/k\}$, then $|A| = \lim_k |O_k|$.

31. For $A \in \mathcal{L}(\mathbb{R}^n)$, let $\varphi_A(x) = |A \cap B(0, |x|)|$. Find $\lim_{|x| \to \infty} \varphi_A(x)$ and discuss the validity of the following statement: φ_A is uniformly continuous in $\mathbb{R}^n$.

32. Let $A \subset \mathbb{R}^n$ with $|A|_e > 0$. Prove that for $\eta \in (0, |A|_e)$, there is a compact $K \subset A$ with $|K| = \eta$.

33. Let $A \in \mathcal{L}(\mathbb{R})$ with $|A| < \infty$. Prove that $|A \cap (-\infty, x)| = |A \cap (x, \infty)|$ for some $x \in \mathbb{R}$.

34. Let $A \in \mathcal{L}(\mathbb{R}^n)$ have infinite measure and let $\{\lambda_k\}$ be positive real numbers. Prove that there are pairwise disjoint measurable subsets $\{A_k\}$ of A such that $|A_k| = \lambda_k$ for all k.

35. Let Q be a cube in $\mathbb{R}^n$. Prove that if $A \subset Q$ is such that $|A| = |Q|$, A is dense in Q.

36. Discuss the validity of the following statements: (a) There is a closed set $F \subset [0,1]$ consisting entirely of irrational numbers such that $|F| = \eta$, $0 < \eta \leq 1$. (b) Given $\varepsilon > 0$, there is an open dense $O \subset \mathbb{R}$ with $|O| \leq \varepsilon$. (c) For any $0 < \varepsilon < 1$ there is a closed nowhere dense subset F of $[0,1]$ with $|F| \geq 1 - \varepsilon$. (d) Every closed subset of $\mathbb{R}$ with empty interior has Lebesgue measure 0.

37. Let A be a measurable subset of $\mathbb{R}^n$ with positive finite measure. Prove that $\lim_{|x| \to 0} |A \cap (x + A)| = |A|$.

38. Let $A \in \mathcal{L}(\mathbb{R})$ have $|A| > 0$. Prove that A contains arbitrarily long arithmetic progressions, that is, for every positive integer n, there exist $a \in A$ and $h > 0$ such that the points $a, a+h, a+2h, \ldots, a+(n-1)h$ belong to A.

39. Let $A \in \mathcal{L}(\mathbb{R}^2)$ have $|A| > 0$. Discuss the validity of the following statements: (a) A contains the vertices of an equilateral triangle. (b) There exists $\varepsilon > 0$ such that, for all $0 < \eta < \varepsilon$, A contains the vertices of a square of sidelength η.

40. Let A be a bounded measurable set in $\mathbb{R}^n$ with $|A| > 0$. Prove that $\lim_{|x| \to 0} |A \triangle (x + A)| = 0$. Show by means of an example that the result may fail for sets of infinite measure.

41. Let $A \in \mathcal{L}(\mathbb{R}^n)$ with $|A| < \infty$ and put $\varphi(x) = |A \cap (x + A)|$ for $x \in \mathbb{R}^n$. Prove that $\lim_{|x| \to \infty} \varphi(x) = 0$.

42. Let $A \in \mathcal{L}(\mathbb{R}^n)$ and $B \in \mathcal{L}(\mathbb{R}^n)$ be bounded, and put $\varphi(x) = |A \cap (x + B)|$. Prove that $\varphi(x)$ is a continuous function of $x \in \mathbb{R}^n$.

43. Let $T : \mathbb{R}^n \to \mathbb{R}^n$ be a locally Lipschitz transformation, i.e., $|T(x) - T(y)| \leq M_K |x - y|$ for all x, y in a compact K of $\mathbb{R}^n$ and a constant M_K that may depend on K. Discuss the validity of the following statements: (a) If $A \subset \mathbb{R}^n$ is bounded, $|T(A)|_e < \infty$. (b) If $|A| = 0$, $|T(A)| = 0$. (c) T maps measurable sets into measurable sets. (d) If $A \in \mathcal{L}(\mathbb{R}^n)$ has $|A| < \infty$, then $|T(A)| < \infty$.

44. Suppose that $A \subset [0,1]$ is measurable. Prove that $B = \cos(A) = \{\cos(x) : x \in A\}$ is measurable and $|B| \leq .85 \, |A|$.

45. Let $A \subset \mathbb{R}^n$, $|A|_e < \infty$, and $r \in \mathbb{R}$. Prove: (a) The dilation of A by r, $rA = \{rx : x \in A\}$, satisfies $|rA|_e = |r|^n |A|_e$. (b) A is measurable iff rA is measurable for some $r \neq 0$, and then $|rA| = |r|^n |A|$.

46. Let $A \in \mathcal{L}(\mathbb{R}^n)$. Prove that $B = -A \cup A \in \mathcal{L}(\mathbb{R}^n)$ and $|A| \leq |B| \leq 2|A|$.

47. Let $A \in \mathcal{L}(\mathbb{R})$ be such that for each irrational x, exactly one of $-x, x$ is in A. Prove that $|A| = |A^c| = \infty$.

48. Let $A, B \subset I^n$ be Lebesgue measurable with $|A| + |B| > 1$. Prove that there exist $x \in A, y \in B$ such that $y = (1, \ldots, 1) - x$.

49. Let $T : \mathbb{R}^n \to \mathbb{R}^n$ be a linear map. Prove: (a) For a closed cube Q_r of sidelength r, $|T(Q_r)|_e = \eta \, |Q_r|$ where $\eta > 0$ is a constant independent of Q_r and compute η explicitly. (b) Let M be the $n \times n$ matrix such that $T(x) = Mx$ for all $x \in \mathbb{R}^n$. Then $|T(A)| = |\det(M)| \, |A|$ for all $A \in \mathcal{L}(\mathbb{R}^n)$.

50. Prove that the Lebesgue measure is rotation invariant.

51. Can $\mathbb{R}^n$ be written as a countable union of hyperplanes?

52. An ellipsoid $\mathcal{E}$ in $\mathbb{R}^n$ with center x_0 is a set of the form $\mathcal{E} = \{x \in \mathbb{R}^n : \langle A(x - x_0), x - x_0 \rangle \leq 1\}$ where A is an $n \times n$ positive definite matrix and $\langle \cdot, \cdot \rangle$ denotes the Euclidean inner product. Find $|\mathcal{E}|$.

53. An isometry $\phi : \mathbb{R}^n \to \mathbb{R}^n$ is a mapping such that $|\phi(x) - \phi(y)| = |x - y|$ for all $x, y \in \mathbb{R}^n$. Prove that if ϕ is an isometry, $|\phi(A)| = |A|$ for all $A \in \mathcal{L}(\mathbb{R}^n)$.

54. Let μ be a Borel measure on $\mathbb{R}$ that is finite on bounded subsets of $\mathbb{R}$ such that $\mu(tA) = |t| \, \mu(A)$ for all $t \in \mathbb{R}$, $A \in \mathcal{B}(\mathbb{R})$. Prove that $\mu(A) = \eta \, |A|$ where $\eta = \mu((0, 1])$.

55. Let $A \subset \mathbb{R}^n$ have $|A|_e > 0$ and $0 < \eta < 1$. Prove that there is a cube Q in $\mathbb{R}^n$ such that $|A \cap Q|_e / |Q| \geq \eta$. Furthermore, prove that Q may be assumed to have arbitrarily small measure.

56. Let $A \subset \mathbb{R}^n$ be such that $|A \cap Q|_e \geq \eta \, |Q|$ for some $\eta > 0$ and all cubes $Q \subset \mathbb{R}^n$. Prove that if $A \in \mathcal{L}(\mathbb{R}^n)$, $|A^c| = 0$, and show that if $A \notin \mathcal{L}(\mathbb{R}^n)$, $|A^c|_e$ may be positive.

57. Let $A \subset \mathbb{R}^n$ with $|A|_e > 0$ and $0 < \eta < 1$. Prove that there is a cube Q in $\mathbb{R}^n$ centered at a point of A such that $|A \cap Q|_e > \eta \, |Q|$.

58. Let $A \in \mathcal{L}(\mathbb{R}^n)$ be such that, given $x \neq y \in A$, $(x+y)/2 \notin A$. Prove that $|A| = 0$.

59. Prove that if $A \subset \mathbb{R}^n$ has $|A|_e < \infty$, then $A^c - A^c = \mathbb{R}^n$.

60. Discuss the validity of the following statement: If $A, B \in \mathcal{L}(\mathbb{R})$ and $|A| = |B| = 0$, then $|A + B| = 0$.

61. Let A, B be measurable sets in $\mathbb{R}$ of finite measure such that $A + B$ is measurable. Prove: (a) $|A| + |B| \leq |A + B|$. (b) If $0 < \eta < 1$ and $(1 - \eta)A + \eta B$ is measurable, then $|A|^{1-\eta}|B|^\eta \leq |(1 - \eta)A + \eta B|$.

62. Let $A \subset [-1, 1]$ be measurable with $|A| > 1$. Prove that $1 \in A - A$.

63. Prove that if $A \in \mathcal{L}(\mathbb{R})$ has $|A| > 0$, $A - A$ contains: (a) An irrational point x. (b) A rational point $0 \neq x$.

64. Prove that if $A \in \mathcal{L}(\mathbb{R}^n)$ has $|A| > 1$, $A - A$ contains a point with integer coordinates.

65. Let B be a convex set in $\mathbb{R}^n$ centrally symmetric with respect to 0 with $|B| > 2^n$. Prove that B contains a point $x \neq 0$ with integer coordinates.

66. Prove that if $A \in \mathcal{L}(\mathbb{R})$ has $|A| > \eta > 0$, $A - A$ and $A + A$ contain a measurable subset with measure $> 2\eta$.

67. Prove that if $A \in \mathcal{L}(\mathbb{R}^n)$ has $|A| > 0$, $A - A$ contains a neighborhood of the origin.

68. Prove that if $A, B \in \mathcal{L}(\mathbb{R}^n)$ have positive measure, $A - B$ and $A + B$ contain a nonempty open interval.

69. Prove that if $A \in \mathcal{L}(\mathbb{R}^n)$ has $|A| > 0$, then $\operatorname{card}(A) = c$. In other words, if $A \in \mathcal{L}(\mathbb{R}^n)$ and $\operatorname{card}(A) < c$, then $|A| = 0$.

70. Let $E = \{(x, y) \in \mathbb{R}^2 : x - y \notin \mathbb{Q}\}$. Prove that E does not contain a set of type $A \times B$ with $A, B \in \mathcal{L}(\mathbb{R})$ with positive measure.

71. Prove that if $A \in \mathcal{L}(\mathbb{R}^n)$ is a subset of a Vitali Lebesgue nonmeasurable set V, then $|A| = 0$.

72. Prove that if $A \subset \mathbb{R}^n$ has $|A|_e > 0$, A contains a Lebesgue nonmeasurable subset.

73. Let V be a Vitali set in $[0, 1)$, $\{q_k\}$ the rationals in $(0, 1)$, $V_k = q_k + V$, and $A_n = \bigcup_{k=1}^n V_k$. Prove that the A_n are Lebesgue nonmeasurable.

74. Let $f : [0, 1] \to [0, 1]$ be a continuous function. Discuss the validity of the following statements: (a) If f is Lipschitz and $A \in \mathcal{L}([0, 1])$, then $f^{-1}(A) \in \mathcal{L}([0, 1])$. (b) If f is a homeomorphism, i.e., f is invertible and has a continuous inverse, and $K \subset [0, 1]$ is a compact set of measure 0, then $|f(K)| = 0$.

75. Prove that if $\emptyset \neq O \subset \mathbb{R}^n$ is open, O can be expressed as the countable union of disjoint open balls and a set of measure 0.

76. Discuss the validity of the following statement: There exists a sequence of pairwise disjoint closed disks $\{D_n\}$ contained in the open unit square $Q = (0,1) \times (0,1)$ of $\mathbb{R}^2$ such that $\sum_n |D_n| = 1$.

77. Given a measurable set $A \subset (0,1)$ with $|A| > 1 - 1/N$, let $B = \bigcup_{n=0}^{\infty} (n + A)$. Prove that given a finite collection of points $x_1, \ldots, x_N$ in $\mathbb{R}$, there exists $x \in (0,1)$ such that $x - x_n \in B$ for $1 \leq n \leq N$.

78. We say that $A \in \mathcal{L}(\mathbb{R}^n)$ has density d at $x \in \mathbb{R}^n$ if

$$\lim_{r \to 0^+} \frac{|A \cap B(x,r)|}{|B(x,r)|} = d \,.$$

If $d = 1$ we say that x is a *point of density* of A and if $d = 0$ we say that x is a point of dispersion of A. (a) Given $x_0 \in \mathbb{R}$ and $d \in (0,1)$, construct a set $A \subset \mathbb{R}$ with density d at x_0. (b) Construct B with no density at 0.

79. Prove that if 0 is a point of density of $A \in \mathcal{L}(\mathbb{R}^n)$, then $\lambda A \cap A \neq \emptyset$ for $|\lambda| \geq 1$.

80. Prove that if 0 is a point of density of $A \in \mathcal{L}(\mathbb{R}^n)$, there exist $x_k \to 0$ such that (a) $x_k \in -A \cap A$ for all k, or (b) $x_k, 2x_k \in A$ for all k.

81. Let $A \in \mathcal{L}(\mathbb{R}^n)$. We say that A has a well-defined density $D(A)$ if the limit

$$D(A) = \lim_{r \to \infty} \frac{|A \cap B(0,r)|}{|B(0,r)|}$$

exists. Discuss the validity of the following statements: (a) $D(A)$ is well-defined for all $A \in \mathcal{L}(\mathbb{R})$. (b) If $D(A)$ and $D(B)$ are well-defined and $A \cap B = \emptyset$, $D(A \cup B)$ is well-defined and $D(A \cup B) = D(A) + D(B)$. (c) If $\{A_m\}$ are pairwise disjoint sets with well-defined density and $A = \bigcup_m A_m$, then A has a well-defined density and $D(A) = \sum_m D(A_m)$.

82. Prove that if $A \in \mathcal{L}(\mathbb{R}^n)$ has $|A| > 0$, there is a sequence $\{x_k\}$ in $\mathbb{R}^n$ such that $|\mathbb{R}^n \setminus \bigcup_k (x_k + A)| = 0$. For $\{x_k\}$ one may take any countable dense set D in $\mathbb{R}^n$.

83. Let $A = \bigcup_{c \in C} I_c$ where the I_c are intervals of length $1/10$ centered at the points c of the Cantor discontinuum. Find $|A|$.

84. Let $\{\eta_k\}$ be a sequence in $(0,1)$. Let $P_0 = [0,1]$, P_1 the set obtained by removing the middle open interval of P_0 of relative length η_1, and, having constructed P_k, let P_{k+1} be the subset of P_k obtained by removing the middle open interval of relative length η_k of each of the 2^k disjoint closed intervals that comprise P_k. Let $P = \bigcap_{k=0}^{\infty} P_k$. Prove that $|P| = 0$ iff $\sum_n \eta_n = \infty$.

85. Fix an integer $\ell > 1$. For $x \in [0,1]$ consider its expansion in base ℓ,

$$x = \sum_n \frac{x_n}{\ell^n} \,, \qquad 0 \leq x_n \leq \ell - 1 \text{ for all } n \,,$$

where in case of ambiguity we pick the nonterminating expansion. When $\ell = 2$ the expansion is called dyadic, when $\ell = 3$ ternary, and when $\ell = 10$ decimal. (a) Let $A_n(k) = \{x \in I : x_n = k\}$ where $0 \le k \le \ell - 1$ and $n \in \mathbb{N}$. Prove that $A_n(k)$ is Borel measurable and find its measure. (b) Let $A^k = \{x \in [0, 1] : x_n \ne k, \text{ all } n\}$, $0 \le k \le \ell - 1$. Prove that A^k is Borel measurable and find its measure. (c) Let $B_k = \{x \in [0, 1] : x_n = k$ for infinitely many $n\}$, $0 \le k \le \ell - 1$. Prove that B_k is Borel measurable and find its measure. (d) Describe $A^k + A^k$, $0 < k < \ell$, prove that it is measurable, and determine its measure.

86. Suppose that $\ell \ge 3$ and let $0 \le m < n < \ell$. (a) Let $A = \{x \in [0, 1] : x_k = m$ or $x_k = n$ for all $k\}$. Prove that A is an uncountable compact set of measure 0. (b) Let $B = \{x \in [0, 1] : m$ appears before n in the expansion of $x\}$. Prove that B is Borel and find $|B|$.

87. Suppose that $\ell \ge 3$ and let $0 \le \ell_1, \ell_2 < \ell$, $\ell_1 \ne \ell_2$. Let $A = \{x = \sum_k x_k \ell^{-k} \in [0, 1] : \text{if } x_m = \ell_1 \text{ there is } n < m \text{ such that } x_n = \ell_2\}$. Prove that A is measurable and compute its measure.

88. Let $A = \{x = \sum_k x_k 2^{-k} \in [0, 1] : x_k = 0, 1, \text{ and } x_{2k} = 0 \text{ for all } k\}$. Prove that A is a compact, uncountable, nowhere dense set of measure 0.

89. Let $A = \{x = \sum_k x_k 10^{-k} : x_k = 0, \ldots, 9, \text{ and, for each } n = 0, 1, \ldots, \text{there is } m \text{ with } 2^n \le m < 2^{n+1} \text{ and } x_m = 0\}$. Discuss the validity of the following statement: A is Lebesgue measurable and $|A| = 0$.

90. Given $x \in \mathbb{R}$, define the mantissa $M(x)$ of x as $M(x) = x - [x]$ where $[x]$ is the integer part of x; M assumes values in $[0, 1)$. Prove that if $A_n = [2^{-n} M(\ln(n + 1)), 2^{-n} M(\ln(n + 1)) + 1]$, then $|\limsup_n A_n| = 0$.

91. Construct a set $A \subset \mathbb{R}$ with $|A| = 0$ such that $A \cap J$ is uncountable for every interval J.

92. Let A be such that $|A \cap J| = \eta |J|$ where $0 \le \eta \le 1$ is fixed and J is a subinterval of $\mathbb{R}$. Prove that $|A \cap [0, 1]| = 0$ or $= 1$.

93. Let $A \in \mathcal{L}(\mathbb{R}^n)$ and $D \subset \mathbb{R}^n$ a dense subset of $\mathbb{R}^n$ such that $d + A = A$ for all $d \in D$. Prove that $|A| = 0$ or $|A^c| = 0$.

94. Let $A \in \mathcal{L}(\mathbb{R})$ have the following property: If $x \in A$ and if the decimal expansions of x and y differ in finitely many places, then $y \in A$. Give examples of such sets and prove that $|A| = 0$ or $|\mathbb{R} \setminus A| = 0$.

95. Let $A \in \mathcal{L}(\mathbb{R}^n)$ have $|A| > 0$, let $\mathbb{Q}^n = \{q\}$ be the rational n-tuples in $\mathbb{R}^n$, and let $B = \bigcup_{q \in \mathbb{Q}^n} (q + A)$. Prove that $|B^c| = 0$.

96. We say that $p \in \mathbb{R}$ is a period of A if $p + A = A$. Let $A \in \mathcal{L}(\mathbb{R})$ have arbitrarily small periods. Prove that $|A| = 0$ or $|A^c| = 0$.

97. Let $A \in \mathcal{L}(\mathbb{R}^+)$ be such that $rA = A$ for all rationals $r > 0$. Prove that $|A| = 0$ or $|A^c| = 0$.

98. Let $A \in \mathcal{L}(\mathbb{R}^n)$ be such that for all x in a dense subset D of $\mathbb{R}^n$, $|A \triangle (x + A)| = 0$. Prove that $|A| = 0$ or $|A^c| = 0$.

99. Construct a rotation invariant probability measure μ on the unit circle $S^1 = \{(x, y) \in \mathbb{R}^2 : x^2 + y^2 = 1\}$ in $\mathbb{R}^2$.

100. Let $\{a_n\}$ be a real sequence and $\{\lambda_n\}$ a strictly positive real sequence. Prove that if $\sum_n \sqrt{\lambda_n} < \infty$, $\sum_n \lambda_n |x - a_n|^{-1} < \infty$ for x a.e. in $\mathbb{R}$.

101. Prove that the existence of a Hamel basis of $\mathbb{R}$ as a linear space over $\mathbb{Q}$ implies the existence of a Lebesgue nonmeasurable set.

102. Prove that if a Hamel basis H of $\mathbb{R}$ is Lebesgue measurable, then $|H| = 0$.

103. Prove that the Cantor discontinuum contains a Hamel basis of $\mathbb{R}$.

104. Construct a Lebesgue nonmeasurable Hamel basis for $\mathbb{R}$.

105. Characterize the differentiable functions $f : [0, 1] \to [0, 1]$ such that $|f^{-1}(J)| = |J|$ for every interval $J \subset [0, 1]$.

106. Let $\mathcal{O} \subset \mathbb{R}$ be open and $\varphi : \mathcal{O} \to \mathbb{R}$ a continuous function so that $\varphi^{-1}(J) \in \mathcal{L}(\mathbb{R})$ for every open interval J of $\mathbb{R}$ and $|\varphi^{-1}(J)| = |J|$. Prove that if $A \in \mathcal{L}(\mathbb{R})$, $\varphi^{-1}(A) \in \mathcal{L}(\mathbb{R})$ and $|\varphi^{-1}(A)| = |A|$.

107. Discuss the validity of the following statement: The mapping $f : [0, 1] \to [0, 1]$ given by $f(x) = 2x$ if $0 \le x \le 1/2$ and $= -2x + 2$ otherwise, is measure preserving.

108. Does there exist a function $f : \mathbb{R} \to [0, 1]$ such that the set $D(f)$ of discontinuities of f has measure 0 yet $D(f) \cap I$ is uncountable for every I?

109. Let $(\widetilde{\mathcal{L}}, d)$ denote the metric space introduced in Problem 2.88 corresponding to the Lebesgue measure on I. Prove that $\mathcal{S} = \{[A] \in \widetilde{\mathcal{L}} : |A \cap J| > 0$ and $|A^c \cap J| > 0$ for all intervals $J \subset I\}$ is of second category in $(\widetilde{\mathcal{L}}, d)$.

110. Let $\{\lambda_n\}$ be positive reals such that $\sum_n \lambda_n < \infty$ and for $k > 0$ let

$$A_k = \left\{ x \in \mathbb{R} : \left| x - \frac{p}{q} \right| \ge \frac{\lambda_q}{k|q|} , \text{ for all integers } p, q, q \ne 0 \right\}.$$

Prove that $\left| \mathbb{R} \setminus \bigcup_k A_k \right| = 0$.

111. Give an example of a subset of $\mathbb{R}$: (a) Of first category and full measure. (b) Of second category and measure 0.

112. Discuss the validity of the following statement: There exists a set $A \subset \mathbb{R}$ of first category and measure 0 such that $A + A = \mathbb{R}$.

113. Discuss the validity of the following statement: If A, B are disjoint subsets of $\mathbb{R}$ with A of second category, B positive measure, and $A \cup B = \mathbb{R}$, then $A + B$ contains an interval.

114. Discuss the validity of the following statements: (a) Given an integer n, there are n pairwise disjoint Borel subsets $A_1, \ldots, A_n$ of $\mathbb{R}$, say, such that $|A_k \cap J| > 0$ for every nonempty open interval J and $1 \leq k \leq n$. (b) There are pairwise disjoint Borel sets $\{A_n\}$ in $\mathbb{R}$ such that $|A_n \cap J| > 0$ for every nonempty open interval J and all n.

115. Let $D = \{m + \sqrt{2}n : m, n \in \mathbb{Z}\}$ and for $x, y \in \mathbb{R}$ consider the equivalence relation $x \sim y$ iff $x - y \in D$. By the axiom of choice there is a set $\mathcal{V}$, say, consisting of one point from each equivalence class. Prove that $\mathcal{V} \notin \mathcal{L}(\mathbb{R})$.

Chapter 4

Measurable and Integrable Functions

This chapter is devoted to the notions of measurable and integrable functions. Given a σ-algebra $\mathcal{M}$ of subsets of X, we say that a real-valued function f on X is *measurable* if the level sets $\{x \in X : f(x) > \lambda\}$ of f are measurable, i.e., are in $\mathcal{M}$, for all $\lambda \in \mathbb{R}$. An extended real-valued function f on X is measurable if $\{f = -\infty\} \in \mathcal{M}$ and $\{\lambda < f < \infty\} \in \mathcal{M}$ for each real λ. In the case of a measure space $(X, \mathcal{M}, \mu)$ the measurability of various expressions involving f, such as $f^+(x) = \vee(0, f(x))$ and $|f(x)|^p$, $0 < p < \infty$, for instance, follow at once from the fact that if f is a measurable μ-a.e. finite function on X and ψ a real-valued continuous function on $\mathbb{R}$, then $\psi \circ f$ is measurable.

We say that a function φ on X is *simple* if $\varphi = \sum_{k=1}^{n} \lambda_k \chi_{A_k}$ where the λ_k are scalars and the A_k measurable subsets of X. Every measurable function f is the μ-a.e. pointwise limit of simple functions assuming rational values; when f is nonnegative the functions are given by

$$f_n(x) = \begin{cases} (k-1)/2^n, & (k-1)/2^n \le f(x) < k/2^n, \text{ for } k = 1, 2, \ldots, n2^n, \\ n, & f(x) \ge n. \end{cases}$$

For arbitrary functions $f = f^+ - f^-$ the simple functions are obtained by approximating f^+ and f^- separately and then subtracting to approximate f.

The *integral* of a nonnegative simple function φ over X with respect to μ is denoted $\int_X \varphi(x) \, d\mu(x)$, or $\int_X \varphi \, d\mu$, and is defined as the quantity $\int_X \varphi \, d\mu = \sum_{k=1}^{n} \lambda_k \mu(A_k)$. Now, given a nonnegative measurable function f

on X, let $\mathcal{F}_f = \{\varphi : \varphi \text{ is simple and } 0 \leq \varphi \leq f\}$. The *integral of f over X with respect to μ* is denoted $\int_X f(x)\,d\mu(x)$, or $\int_X f\,d\mu$, and is defined as $\int_X f\,d\mu = \sup\left\{\int_X \varphi\,d\mu : \varphi \in \mathcal{F}_f\right\}$. Arbitrary functions $f = f^+ - f^-$ have an integral defined by $\int_X f\,d\mu = \int_X f^+\,d\mu - \int_X f^-\,d\mu$ provided at most one of the integrals in the right-hand side of the definition is infinite.

We say that a measurable function f on X is *integrable* if $\int_X |f|\,d\mu < \infty$ and the linear space $L^1(X)$ is defined as $L^1(X) = \{f : f \text{ is measurable and } \int_X |f|\,d\mu < \infty\}$; endowed with the metric $d(f,g) = \int_X |f-g|\,d\mu$, $(L^1(X), d)$ becomes a complete metric space.

Integrable functions satisfy Chebychev's, or Markov's, inequality, i.e., $\lambda\mu(\{|f| > \lambda\}) \leq \int_{\{|f|>\lambda\}} |f|\,d\mu$ for all $\lambda > 0$.

In the case of the Lebesgue measure in $\mathbb{R}^n$, or $A \in \mathcal{L}(\mathbb{R}^n)$, the integral of f is denoted $\int_{\mathbb{R}^n} f(x)\,dx$, or $\int_A f(x)\,dx$, respectively, and if a function f defined on a finite interval of $\mathbb{R}^n$ is Riemann integrable, f has a Lebesgue integral and both integrals coincide. The Lebesgue integral is translation invariant and simple or compactly supported smooth functions are dense in $L^1(\mathbb{R}^n)$ in the sense that given $\varepsilon > 0$, there is a simple or compactly supported smooth function g such that $d(f,g) = \int_{\mathbb{R}^n} |f(x) - g(x)|\,dx < \varepsilon$. Also, $L^1(\mathbb{R}^n)$ functions are continuous in the metric, i.e., if $f \in L^1(\mathbb{R}^n)$, given $\varepsilon > 0$, there exists $\delta > 0$ such that $\int_{\mathbb{R}^n} |f(x+y) - f(x)|\,dx \leq \varepsilon$ whenever $|y| < \delta$.

An essential feature of the Lebesgue integral is the Lebesgue differentiation theorem. It states that if f is locally integrable, i.e., integrable on bounded subsets of $\mathbb{R}^n$, then with $Q(x,r)$ denoting the cube centered at x of sidelength $r > 0$,

$$\lim_{r \to 0^+} \frac{1}{|Q(x,r)|} \int_{Q(x,r)} f(y)\,dy = f(x) \quad \text{a.e. } x \in \mathbb{R}^n\,.$$

We say that a point $x \in \mathbb{R}^n$ in the domain of a locally integrable function f on $\mathbb{R}^n$ is a *Lebesgue point of f* if

$$\lim_{r \to 0^+} \frac{1}{|Q(x,r)|} \int_{Q(x,r)} |f(y) - f(x)|\,dy = 0\,.$$

Then, the strong version of the Lebesgue differentiation theorem holds, i.e., almost every point in the domain of a locally integrable function f on $\mathbb{R}^n$ is a Lebesgue point of f.

Some of the important questions in $L^1(X)$ concern the passing of the limit under the integral sign. A basic result is the Beppo Levi or monotone convergence theorem (MCT), which states that if $\{f_n\}$ is a nondecreasing sequence of nonnegative finite μ-a.e. measurable functions on X, then $f(x) = \lim_n f_n(x)$ exists everywhere on X, f is nonnegative and measurable, and $\lim_n \int_X f_n \, d\mu = \int_X f \, d\mu$. The possibility that the limit is infinite is allowed.

Now, limits are hard to find and so Fatou's lemma is an essential tool in dealing with convergence. It states that if $\{f_n\}$ are extended real-valued measurable functions on X so that $g \leq f_n$ μ-a.e. for an integrable function g, then $\int_X \liminf_n f_n \, d\mu \leq \liminf_n \int_X f_n \, d\mu$.

Then there is one of the most celebrated theorems in real variables, namely, the Lebesgue dominated convergence theorem (LDCT). It states that if $\{f_n\}$ are extended real-valued measurable functions on X such that $\lim_n f_n = f$ exists μ-a.e. and $|f|, |f_n| \leq g$ μ-a.e. for $g \in L^1(X)$ and all n, then $\lim_n \int_X f_n \, d\mu = \int_X f \, d\mu$.

Now to the problems. In Problem 13 we consider the relation of a measurable function and the nature of the underlying measure space, in particular when it is not complete. Also the general form of a measurable function: it is a linear combination of not necessarily pairwise disjoint characteristic functions of measurable sets with rational coefficients, Problems 17–18. In the particular case of $\mathbb{R}^n$, various classes of functions, such as monotone or semicontinuous, are Borel or Lebesgue measurable, Problems 21–29. On the other hand, for an arbitrary σ-algebra of subsets of $\mathbb{R}^n$, not all continuous functions are measurable, Problem 30. The composition of measurable functions can also be problematic, Problems 36–38. The properties of measurable, not necessarily continuous, functions on $\mathbb{R}$ that satisfy the relation $f(x+y) = f(x) + f(y)$ for all real x, y, are explored in Problems 45–46.

That integrable functions in Euclidean space can be approximated in the L^1 norm as well as μ-a.e., and continuous integrable functions everywhere, and often uniformly, by simply constructed classes of functions is done in Problems 54–55. The question as to whether two functions that have the same integral over all measurable sets are equal μ-a.e. is considered in Problem 69. And, the fact that if a Lebesgue integrable function on $\mathbb{R}$ has vanishing integral over all intervals of fixed length $c > 0$, it vanishes a.e., is given in Problem 72. That integrable functions have a higher order of integrability is done in Problem 73. The relation between the Riemann improper integral and the Lebesgue integral on $\mathbb{R}$ is explored in Problems 93–95. Various forms of the Lebesgue differentiation theorem are given in Problems 111–112. The question as to whether Lebesgue integrable functions on $\mathbb{R}^n$ tend to 0 as $|x| \to \infty$ is discussed in Problems 122–130. A general version of the Riemann-Lebesgue lemma is given in Problem 147.

Problems

1. Let f, g be Lebesgue nonmeasurable functions on I. What can one say about $f + g$, $|f|$, and fg?

2. Let $(X, \mathcal{M}, \mu)$ be a measure space, f a measurable function on X, and $g = f$ μ-a.e. Is g measurable?

3. Let $\mathcal{S}$ denote the σ-algebra of subsets of $\mathbb{R}$ given by $\mathcal{S} = \{A \subset \mathbb{R} : A = -A\}$. Characterize the measurable functions: (a) $f : (\mathbb{R}, \mathcal{S}) \to (\mathbb{R}, \mathcal{S})$. (b) $g : (\mathbb{R}, \mathcal{S}) \to (\mathbb{R}, \mathcal{B}(\mathbb{R}))$.

4. Let $X = (0, 1]$. Consider the σ-algebras of subsets of X, $\mathcal{S}_1 = \mathcal{M}\{(0, 1/2], (1/2, 3/4], (3/4, 1]\}$ and $\mathcal{S}_2 = \mathcal{M}\{(0, 1/4], (1/4, 1/2], (1/2, 1]\}$. Give the following examples of functions f on X: (a) f is $\mathcal{S}_1$ measurable but not $\mathcal{S}_2$ measurable. (b) f is $\mathcal{S}_2$ measurable but not $\mathcal{S}_1$ measurable. (c) f is both $\mathcal{S}_1$ and $\mathcal{S}_2$ measurable. (d) f is neither $\mathcal{S}_1$ nor $\mathcal{S}_2$ measurable.

5. Let $\mathcal{S}$ be the σ-algebra of subsets of $\mathbb{R}$ given by $\mathcal{S} = \{\emptyset, (-\infty, 0], (0, \infty), \mathbb{R}\}$. Describe the $\mathcal{S}$ measurable functions f on $\mathbb{R}$. If $|f|$ is measurable, does it follow that f is measurable?

6. Let $\mathcal{M}$ be a σ-algebra of subsets of X and f a function on X such that $|f|$ is measurable. Find an additional condition that ensures that f is measurable.

7. Given $a, b, c \in \mathbb{R}$, let $\mathrm{mid}\,(a, b, c)$ denote the middle value when a, b, c are arranged in an increasing (or decreasing) order. Let $\mathcal{M}$ be a σ-algebra of subsets of X and given measurable functions f_1, f_2, f_3, let $g(x) = \mathrm{mid}\,(f_1(x), f_2(x), f_3(x))$. Prove that g is measurable.

8. Let $\mathcal{M}$ be a σ-algebra of subsets of X and f_1, f_2, f_3, f_4 measurable functions on X. Let

$$f(x) = \begin{cases} f_1(x), & f_2(x) < f_3(x), \\ f_4(x), & \text{otherwise.} \end{cases}$$

Prove that f is measurable.

9. Let $\mathcal{M}$ be a σ-algebra of subsets of X, $\{A_n\} \subset \mathcal{M}$ with $\bigcup_n A_n = X$, $\{f_n\}$ measurable functions on X such that $f_n = f_m$ on $A_n \cap A_m$, and $f : X \to \mathbb{R}$ given by $f(x) = f_n(x)$ for $x \in A_n$. Prove that f is a well-defined measurable function on X.

10. Let $(X, \mathcal{M}, \mu)$ be a measure space, f a measurable function on X, and O an open subset of $\mathbb{R}$ containing the origin. Prove that there exists $A \in \mathcal{M}$ with $\mu(A) > 0$ such that $f(x) - f(y) \in O$ whenever $x, y \in A$.

11. Let $(X, \mathcal{M}, \mu)$ be a measure space and f a measurable function on X that is not constant μ-a.e. Prove that there exists $\lambda \in \mathbb{R}$ such that $\mu(\{f \leq \lambda\}) > 0$ and $\mu(\{f > \lambda\}) > 0$.

12. Let $\mathcal{S}$ denote the σ-algebra of subsets of $\mathbb{R}^2$ generated by the open balls $B(0, r)$ of $\mathbb{R}^2$ centered at the origin. Prove: (a) The function $u : (\mathbb{R}^2, \mathcal{S}) \to (\mathbb{R}^+, \mathcal{B}(\mathbb{R}^+))$ given by $u(x_1, x_2) = (x_1^2 + x_2^2)^{1/2}$ is measurable. (b) The function $v : (\mathbb{R}^+, \mathcal{B}(\mathbb{R}^+)) \to (\mathbb{R}^2, \mathcal{S})$ given by $v(x) = (x, 0)$ is measurable. (c) Let $a, b \in \mathbb{R}^2$ be such that $u(a) = u(b)$ and put $\mathcal{C} = \{A \in \mathcal{S} : a$ and b are simultaneously in A or in $A^c\}$. Prove that $\mathcal{C} = \mathcal{S}$. (d) Discuss the validity of the following statement: If $f : (\mathbb{R}^2, \mathcal{S}) \to (\mathbb{R}^+, \mathcal{B}(\mathbb{R}^+))$ is measurable, then $f(3, 4) = f(4, 3) = f(0, 5)$.

13. Let $(X, \mathcal{M}, \mu)$ be a measure space where μ is not complete and $(X, \mathcal{S}, \mu_1)$ is its completion. Prove that $f : (X, \mathcal{S}) \to (\mathbb{R}, \mathcal{B}(\mathbb{R}))$ is measurable iff there exists a measurable $g : (X, \mathcal{M}) \to (\mathbb{R}, \mathcal{B}(\mathbb{R}))$ such that $f = g$ μ_1-a.e.

14. Construct a Lebesgue nonmeasurable function $f : \mathbb{R} \to \mathbb{R}$ such that if $g : \mathbb{R} \to \mathbb{R}$ satisfies $|g(x) - f(x)| < 1$ for all $x \in \mathbb{R}$, g is Lebesgue nonmeasurable.

15. Let $A \in \mathcal{L}(\mathbb{R})$. Prove that given $\varepsilon > 0$, there is a continuous function f on $\mathbb{R}$ such that $|\{x \in \mathbb{R} : f(x) \neq \chi_A(x)\}| < \varepsilon$.

16. Let $f : [a, b] \to \mathbb{R}$ be measurable and finite a.e. Prove that f is the a.e. pointwise limit of polynomials.

17. Let $\mathcal{M}$ be a σ-algebra of subsets of X and f an extended real-valued measurable function on X. Prove that f can be expressed as $f = \sum_n \lambda_n \chi_{A_n}$ with the λ_n rational and the A_n in $\mathcal{M}$, not necessarily pairwise disjoint.

18. Let $\mathcal{M}$ be a σ-algebra of subsets of X and f a nonnegative measurable function on X. Prove that f can be expressed as $f = \sum_n n^{-1} \chi_{A_n}$ where the A_n are in $\mathcal{M}$.

19. Let $\mathcal{F}$ be a π-system in X with $X \in \mathcal{F}$ and $\mathcal{V}$ a linear space of functions on X such that: (1) $\chi_A \in \mathcal{V}$ for all $A \in \mathcal{F}$. (2) If $\{f_n\} \subset \mathcal{V}$ is an increasing sequence with $\lim_n f_n = f$, then $f \in \mathcal{V}$. Prove that $\mathcal{V}$ contains all measurable functions $f : (X, \mathcal{M}(\mathcal{F})) \to (\mathbb{R}, \mathcal{B}(\mathbb{R}))$.

20. Let $\mathcal{M}$ be a σ-algebra of subsets of X and $\overline{\mathbb{R}} = [-\infty, \infty]$ the extended real line. Prove that $f : (X, \mathcal{M}) \to (\overline{\mathbb{R}}, \mathcal{B}(\overline{\mathbb{R}}))$ is measurable iff $f|_{f^{-1}(\mathbb{R})} : (X, \mathcal{M}) \to (\mathbb{R}, \mathcal{B}(\mathbb{R}))$ is measurable and $f^{-1}(\{\infty\})$ and $f^{-1}(\{-\infty\}) \in \mathcal{M}$.

21. Let $f(x)$ be a continuous function on $[0, 1]$. Prove that $A = \{x \in [0, 1] : f(x) > f(y)$ for all $y \in [0, x)\}$ is Borel measurable.

22. Let f be a bounded function on $\mathbb{R}$ and $A = \{x \in \mathbb{R} : \lim_{y \to x} f(y) = f(x)\}$. Prove that $A \in \mathcal{B}(\mathbb{R})$.

23. Prove that a function f on $\mathbb{R}$ is measurable iff for every finite $a < b$ the restriction $f_{a,b} = f\chi_{[a,b]}$ of f to $[a,b]$ is measurable.

24. Prove that if $f : \mathbb{R}^n \to \mathbb{R}$ is continuous a.e., f is Lebesgue measurable.

25. Prove that if f on $\mathbb{R}^n$ is lower semicontinuous, f is Borel measurable.

26. Let f be a function on $\mathbb{R}$ that is right-continuous at each point of $\mathbb{R}$. Prove: (a) f is Borel measurable. (b) f is continuous a.e. in $\mathbb{R}$.

27. Let $f : \mathbb{R} \to \mathbb{R}$ be monotone. Prove that f is Borel measurable.

28. Suppose that f is differentiable everywhere on $\mathbb{R}$. Prove that f' is Borel measurable.

29. Discuss the validity of the following statement: If $f : [0,1] \to \mathbb{R}$ is bounded and Lebesgue measurable, then f agrees a.e. with a function in the Baire class $\mathcal{B}_1$. If the statement is false, consider the following alternative: f agrees a.e. with a function in the Baire class $\mathcal{B}_k$, for some k.

30. Let $\mathcal{M}$ be a σ-algebra of subsets of $\mathbb{R}^n$ that does not contain $\mathcal{B}(\mathbb{R}^n)$. Prove that there is a continuous real-valued function on $\mathbb{R}^n$ that is not measurable with respect to $\mathcal{M}$.

31. Let $f : \mathbb{R}^n \to \mathbb{R}^m$ be continuous. Prove that $f^{-1}(B) \in \mathcal{B}(\mathbb{R}^n)$ for all $B \in \mathcal{B}(\mathbb{R}^m)$.

32. Let $f : \mathbb{R}^n \to \mathbb{R}^n$ be continuous, 1-1, and onto. Prove that f maps Borel sets onto Borel sets.

33. Discuss the validity of the following statement: If f is a Lebesgue measurable function on $\mathbb{R}^n$ and $B \in \mathcal{B}(\mathbb{R})$, $f^{-1}(B) \in \mathcal{L}(\mathbb{R}^n)$.

34. Let $f : \mathbb{R}^n \to \mathbb{R}$ be measurable. Prove that $f^{-1}(A) \in \mathcal{L}(\mathbb{R}^n)$ for every $A \in \mathcal{L}(\mathbb{R})$ iff $f^{-1}(N)$ has measure 0 whenever $N \subset \mathbb{R}$ is a null set.

35. Let $B \notin \mathcal{L}(\mathbb{R})$ and $A = \{(x,x) \in \mathbb{R}^2 : x \in B\}$. Prove that $A \in \mathcal{L}(\mathbb{R}^2) \setminus \mathcal{B}(\mathbb{R}^2)$.

36. Let $f, g : \mathbb{R} \to \mathbb{R}$ be such that f is Lebesgue measurable, g continuous, and $f \circ g$ well-defined. Discuss the validity of the following statement: $f \circ g$ is measurable.

37. Let f be a measurable function on $\mathbb{R}^n$ and $\varphi : \mathbb{R}^n \to \mathbb{R}^n$ a continuous function with the property that for every null set $N \subset \mathbb{R}^n$, $\varphi^{-1}(N) \in \mathcal{L}(\mathbb{R}^n)$. Prove that $f \circ \varphi$ is Lebesgue measurable.

38. Let $\varphi : [-\infty, \infty] \to [-\infty, \infty]$ be monotone and suppose that $f : \mathbb{R} \to [-\infty, \infty]$ is measurable. Prove that $\varphi \circ f$ is measurable.

39. Let f be a measurable function on X and $g(x) = 0$ if $f(x)$ is rational and $= 1$ if $f(x)$ is irrational. Is g measurable?

40. Let $f : \mathbb{R} \to \mathbb{R}$ and $L_\lambda(f) = \{x \in \mathbb{R} : f(x) = \lambda\}$, λ real. Prove: (a) If f is measurable, $L_\lambda(f)$ is measurable for all λ. (b) If f is a simple function and $L_\lambda(f)$ is measurable for every $\lambda \in \mathbb{R}$, f is measurable. (c) Suppose $L_\lambda(f)$ is measurable for all $\lambda \in \mathbb{R}$. Does it follow that f is measurable?

41. Let $f : \mathbb{R}^n \to \mathbb{R}$ be Lebesgue measurable and define $g : \mathbb{R} \to [0, \infty]$ by $g(\lambda) = |\{x \in \mathbb{R}^n : f(x) = \lambda\}|$. Prove that g is Lebesgue measurable and compute $\int_{\mathbb{R}} g(\lambda)\, d\lambda$.

42. Let f be the function on $[0, 1]$ defined as follows: If $x = \sum_n x_n \ell^{-n}$, $0 \leq x_n \leq \ell - 1$, let $f(x) = \max_n x_n$. Prove that f is Lebesgue measurable.

43. Consider ternary expansions $x = \sum_n x_n 3^{-n}$ in $[0, 1)$ (possibly terminating in 0's) and, for $x, y \in [0, 1)$, let $g(x, y)$ denote the smallest n such that $x_n = y_n$ and $g(x, y) = \infty$ if no such n exists. Prove that g is Lebesgue measurable.

44. Consider the functions $\{f_n\}$ on $[0, 1]$ given by $f_n(x) = x_n$ where $x = \sum_n x_n 2^{-n}$, $x_n = 0, 1$. (a) Prove that f_n is measurable, $n \geq 1$. (b) If $\sigma : \mathbb{N} \to \mathbb{N}$ denotes a permutation, let $f(x) = \sum_n x_{\sigma(n)} 2^{-n}$. Prove that f is well-defined and measurable in $[0, 1]$, and calculate $|\{f < \lambda\}|$ for $\lambda \in [0, 1]$.

45. Let f be a measurable function on $\mathbb{R}$ that satisfies $f(x + y) = f(x) + f(y)$ for all real x, y. Prove that $f(x) = cx$ for some constant c provided: (a) f is integrable in some neighborhood of a point in $\mathbb{R}$, or (b) f assumes finite values in a set of positive measure.

46. Let $f : \mathbb{R} \to \mathbb{R}$ be a discontinuous function that satisfies $f(x + y) = f(x) + f(y)$ for all $x, y \in \mathbb{R}$. Prove: (a) For every irrational x there is a sequence $x_n \to x$ such that $f(x_n) \to 0$. (b) $f^{-1}(\{0\})$ is dense in $\mathbb{R}$ iff for every Hamel basis $\{e_\lambda\}_{\lambda \in \Lambda}$ of $\mathbb{R}$ over $\mathbb{Q}$, $\mathcal{F} = \{f(e_\lambda)\}$ is linearly dependent over $\mathbb{Q}$. (c) If $f(1) = 1$, the graph $G(f)$ of f is dense in $\mathbb{R}^2$.

47. Let f be a Lebesgue measurable function on $\mathbb{R}$ with periods s, t with s/t irrational. Prove that $f = c$ a.e.

48. Prove that if a continuous $f : \mathbb{R}^n \to \mathbb{R}$ vanishes a.e. in $\mathbb{R}^n$, $f(x) = 0$ for all $x \in \mathbb{R}^n$.

49. Given $f : [0, 1] \to \mathbb{R}$, which of the following statements implies the other? (a) f is continuous a.e. in $[0, 1]$. (b) There is a continuous function $g : [0, 1] \to \mathbb{R}$ such that $g = f$ a.e.

50. Let X be an uncountable set and $\mathcal{M}$ the σ-algebra of subsets of X given by $\mathcal{M} = \{A \subset X : A$ is countable or $X \setminus A$ is countable $\}$. For $A \in \mathcal{M}$, let μ be the counting measure on $\mathcal{M}$, i.e., $\mu(A) =$ the number of elements in A if $\mathrm{card}(A) < \infty$ and $= \infty$ otherwise. Characterize the measurable functions on X and describe $L^1(X)$.

51. Let $(X, \mathcal{M}, \mu)$ be a measure space. Discuss the validity of the following statement: $f \in L^1(X)$ iff $\lim_{\lambda \to \infty} \int_{\{|f| > \lambda\}} |f| \, d\mu = 0$.

52. Let $(X, \mathcal{M}, \mu)$ be a finite measure space and f a measurable function on X. For an integer n, let $A_n = \{|f| > n\}$ and $B_n = \{n \le |f| < n + 1\}$. Prove that the following statements are equivalent: (a) $f \in L^1(X)$. (b) $\sum_n n\mu(B_n) < \infty$. (c) $\sum_n \mu(A_n) < \infty$.

53. Show that $(C(I), \|\cdot\|_1)$ is not complete.

54. Consider a grid $\{Q_\ell^k\}$ of $\mathbb{R}^n$ consisting of nonoverlapping dyadic cubes with $|Q_\ell^k| = 2^{-nk}$, all ℓ. Given $f \in L(\mathbb{R}^n)$, let $\{s_k\}$ be given by $s_k(x) = |Q_\ell^k|^{-1} \int_{Q_\ell^k} f(y) \, dy$ for $x \in Q_\ell^k$, all ℓ. Prove that $s_k(f) \to f$ in $L^1(\mathbb{R}^n)$ and a.e.

55. Let f be a continuous function on $[-1/2, 1/2]$ with $f(-1/2) = f(1/2) = 0$ and $\{s_k\}$ nonnegative integrable functions on $[-1, 1]$ with integral 1 for all k such that $\lim_k \int_{\{\delta \le |x| \le 1\}} s_k(x) \, dx = 0$ for each $\delta > 0$. Prove that if $f_k(x) = \int_{-1/2}^{1/2} s_k(x - y) f(y) \, dy$, $x \in [-1/2, 1/2]$, then $f_k \to f$ uniformly in $[-1/2, 1/2]$.

56. Let $f \in L^1(\mathbb{R}^n)$ and $g \in L^\infty(\mathbb{R}^n)$. Prove that

$$\lim_{|h| \to 0} \int_{\mathbb{R}^n} |f(x + h) - f(x)| \, g(x) \, dx = 0.$$

57. Let $(X, \mathcal{M}, \mu)$ be a measure space, g a nonnegative measurable function on X, and ν the set function on $\mathcal{M}$ given by $\nu(A) = \int_X g\chi_A \, d\mu$, $A \in \mathcal{M}$. Prove: (a) ν is a measure. (b) If f is a nonnegative measurable function on X, $\int_X f \, d\nu = \int_X fg \, d\mu$. (c) If ν is σ-finite, g is finite μ-a.e.

58. Let $(X, \mathcal{M}, \mu)$ be a measure space, f a measurable function on X, and $\nu = f \circ \mu$ the set function on the Borel subsets of $\mathbb{R}$ given by $\nu(B) = \mu(f^{-1}(B))$, $B \in \mathcal{B}(\mathbb{R})$. Prove: (a) ν is a Borel measure on $\mathbb{R}$. (b) A Borel measurable function $\varphi : \mathbb{R} \to \mathbb{R}$ has an integral with respect to ν iff $\varphi \circ f$ has an integral with respect to μ and in that case $\int_X \varphi \circ f \, d\mu = \int_{\mathbb{R}} \varphi \, d\nu$.

59. Let $\mathcal{M}$ be a σ-algebra of subsets of X. We say that $F : (X, \mathcal{M}) \to (\mathbb{R}^2, \mathcal{B}(\mathbb{R}^2))$ is measurable if $F^{-1}(A \times B) \in \mathcal{M}$ for all open intervals $A, B \subset \mathbb{R}$. (a) Let f, g be real-valued functions on X and $F : X \to \mathbb{R}^2$ be given by $F(x) = (f(x), g(x))$, $x \in X$. Prove that F is measurable iff $f : (X, \mathcal{M}) \to (\mathbb{R}, \mathcal{B}(\mathbb{R}))$ and $g : (X, \mathcal{M}) \to (\mathbb{R}, \mathcal{B}(\mathbb{R}))$ are measurable. (b) Let $\Phi : (X, \mathcal{M}) \to (\mathbb{R}^2, \mathcal{B}(\mathbb{R}^2))$ be measurable. Prove that the set function μ_Φ given by $\mu_\Phi(E) = \mu(\Phi^{-1}(E))$ for $E \in \mathcal{B}(\mathbb{R}^2)$ is a measure. (c) Prove that $\int_{\mathbb{R}^2} f \, d\mu_\Phi = \int_X f \circ \Phi \, d\mu$ for all μ_Φ integrable functions $f : (\mathbb{R}^2, \mathcal{B}(\mathbb{R}^2)) \to (\mathbb{R}, \mathcal{B}(\mathbb{R}))$.

60. Let $\varphi : [0, 1] \to \mathbb{R}$ be a strictly monotone absolutely continuous function with continuous derivative and $J = \varphi(I)$. Prove: (a) $\int_{\varphi^{-1}(A)} |\varphi'(x)| \, dx$

$= |A|$ for every Borel $A \subset J$. (b) If f is integrable on J, $\int_{\varphi(A)} f(y)\, dy = \int_A f(\varphi(x))\, |\varphi'(x)|\, dx$ for every Borel $A \subset (a, b)$.

61. Let $f \in L^1(\mathbb{R})$ be a nonnegative function with integral 1, $F(x) = \int_{-\infty}^x f(y)\, dy$, and μ_F the Borel measure induced by F. Prove that for a Borel measurable function g, $\int_{\mathbb{R}} g(x) f(x)\, dx = \int_{\mathbb{R}} g(x)\, d\mu_F(x)$.

62. Let $g : \mathbb{R} \to [0, \infty)$ be a continuous function such that $g(0) > 0$ and $g(x) = 0$ if $x \notin [-1, 1]$. Prove that if h is a measurable function such that $g(x - n) \le h(x)$ for $n = 0, 1, 2, \ldots$, then $h \notin L^1(\mathbb{R})$.

63. Let $(X, \mathcal{M}, \mu)$ be a measure space and f a nonnegative integrable function on X. Prove that for every $\varepsilon > 0$, there is $A \in \mathcal{M}$ with $\mu(A) < \infty$ such that $\int_X f\, d\mu \le \int_A f\, d\mu + \varepsilon$.

64. Let $(X, \mathcal{M}, \mu)$ be a measure space and f a nonnegative measurable function on X. Discuss the validity of the following statement: $f \in L^1(X)$ iff given $\varepsilon > 0$, there exists $\delta > 0$ such that $\int_A f\, d\mu < \varepsilon$ whenever $\mu(A) < \delta$.

65. Let $(X, \mathcal{M}, \mu)$ be a measure space and $f \in L^1(X)$. Show that the parameter δ in the definition of absolute continuity of the integral cannot be chosen to depend on ε and $\|f\|_1$ alone.

66. Let $(X, \mathcal{M}, \mu)$ be a measure space and $A \subset X$ with $\mu(A) < \infty$. Prove that for some $x \in A$, $|f(x)| \le (1/\mu(A)) \int_A |f|\, d\mu$.

67. Let $(X, \mathcal{M}, \mu)$ be a measure space and f a measurable function on X such that $\int_A f\, d\mu \ge 0$ for all $A \in \mathcal{M}$. Prove that $f \ge 0$ μ-a.e. From this deduce that if $\int_A f\, d\mu = 0$ for all $A \in \mathcal{M}$, then $f = 0$ μ-a.e.

68. Let $(X, \mathcal{M}, \mu)$ be a measure space and $\mathcal{F}$ a π-system of subsets of X such that $\mathcal{M}(\mathcal{F}) = \mathcal{M}$. Suppose that f is integrable, or measurable and nonnegative μ-a.e. Prove that if $\int_A f\, d\mu = 0$ for all $A \in \mathcal{F}$ and $A = X$, then $f = 0$ μ-a.e.

69. Let $(X, \mathcal{M}, \mu)$ be a measure space and f, g measurable functions on X. Discuss the validity of the following statement: If

$$\int_A f\, d\mu = \int_A g\, d\mu, \quad \text{all } A \in \mathcal{M},$$

then $f = g$ μ-a.e.

70. Let $(X, \mathcal{M}, \mu)$ be a semifinite measure space, $F \subset \mathbb{R}$ a closed set, and f an integrable function on X such that

$$\frac{1}{\mu(A)} \int_A f\, d\mu \in F, \quad \text{all } A \in \mathcal{M} \text{ with } \mu(A) < \infty.$$

Prove that $f(x) \in F$ for μ-a.e. $x \in X$.

71. Let $(X, \mathcal{M}, \mu)$ be a measure space and $f \in L^1(X)$ such that $|\int_A f \, d\mu| \le c\mu(A)$ for all measurable A with $\mu(A) < \infty$ and a constant c. Prove that $|f| \le c$ μ-a.e.

72. Let $f \in L^1(\mathbb{R})$ and $c > 0$. Prove that if $\int_J f(y) \, dy = 0$ for all intervals J with $|J| = c$, then $f = 0$ a.e.

73. Let $f \in L^1(X)$. Prove that there is a nondecreasing Borel function $\varphi : \mathbb{R}^+ \to \mathbb{R}^+$ with $\lim_{t \to \infty} \varphi(t)/t = \infty$ such that $\int_X \varphi(|f|) \, d\mu < \infty$.

74. Let $(X, \mathcal{M}, \mu)$ be a measure space, f an integrable function on X, and g a bounded measurable function on X such that $a \le g(x) \le b$ μ-a.e. Prove that $\int_X g |f| \, d\mu = c \int_X |f| \, d\mu$ for some c with $a \le c \le b$.

75. Let $(X, \mathcal{M}, \mu)$ be a measure space and $f, g \in L^1(X)$. Prove that $\min(f, g)$ is integrable and $\int_X \min(f, g) \, d\mu \le \min(\int_X f \, d\mu, \int_X g \, d\mu)$. When does equality hold?

76. Let $(X, \mathcal{M}, \mu)$ be a measure space and f an integrable function on X such that $|\int_X f \, d\mu| = \int_X |f| \, d\mu$. Prove that f is of constant sign μ-a.e.

77. Let $f(x) = 0$ at each point of the Cantor discontinuum C and $f(x) = n$ in each interval of length $1/3^n$ in the complement of C. Prove that f is measurable and calculate $\int_I f(x) \, dx$.

78. Let f be the function on $[0, 1]$ given by $f(x) = 0$ if x is rational and $= n$ if there are exactly n zeros immediately after the period in the decimal representation of x. Prove that f is Borel measurable and compute $\int_I f(x) \, dx$.

79. Let

$$F(x) = \begin{cases} -2, & x < 1, \\ 1, & 1 \le x < 4, \\ 2, & x \ge 4 . \end{cases}$$

Compute $\int_\mathbb{R} x^2 \, d\mu_F(x)$ where μ_F denotes the Borel measure induced by F.

80. Let f be defined on $(0, 1)$ by

$$f(x) = \begin{cases} n^{-1}, & \text{if } n \text{ is the smallest integer with } x_n = 7, \\ 0, & \text{otherwise.} \end{cases}$$

Prove that f is measurable and compute $\int_I f(x) \, dx$.

81. Let $[t]$ denote the integral part of $t > 0$ and for $x \in [0, 1]$ put

$$f(x) = \begin{cases} 0, & x \text{ rational,} \\ [1/x], & \text{otherwise,} \end{cases} \quad \text{and} \quad g(x) = \begin{cases} 0, & x \text{ rational,} \\ \left[1/x\right]^{-1}, & \text{otherwise.} \end{cases}$$

Prove that f, g are measurable and compute their integrals.

82. Compute

$$\int_0^4 \left(\frac{1}{2} - \sqrt{x + \sqrt{x + \sqrt{x + \cdots}}} \right)^2 dx\,.$$

83. Given $x \in [0,1]$, let $x = \sum_n x_n \ell^{-n}$ denote the expansion of x in base ℓ, $0 \le x_n \le \ell - 1$, all n, let $x_n(x)$ denote the function on I given by $x_n(x) = x_n$, all n, and put for an integer $m > 0$,

$$f(x) = \sum_n \frac{x_n(x)}{m^n}\,, \quad g(x) = \sum_n (-1)^{x_n(x)} \frac{1}{m^n}\,.$$

Prove that f and g are well-defined and compute their integrals over I.

84. Discuss the validity of the following statement: If μ is a Borel measure on $\mathbb{R}$ such that $(\int_0^x t\, d\mu(t))^2 = \int_0^x t^3\, d\mu(t)$ for all $x > 0$, then μ is the Lebesgue measure on $\mathcal{B}(\mathbb{R})$.

85. Let $A \in \mathcal{L}(\mathbb{R}^n)$, M an invertible $n \times n$ matrix, and $f \in L^1(A)$. Prove that $f \circ M \in L^1(M^{-1}A)$ and $\int_A f(x)\, dx = |\det(M)| \int_{M^{-1}A} f \circ M(x)\, dx$.

86. Let $f \in L^1(\mathbb{R}^n)$. Prove that if $\lambda > 0$, $\int_{\mathbb{R}^n} f(\lambda x)\, dx = \lambda^n \int_{\mathbb{R}^n} f(x)\, dx$.

87. Let $T : \mathbb{R}^2 \to \mathbb{R}^2$ be the linear transformation given by $T(x,y) = (2x+y, x-y)$. Prove that if $f \in L^1(\mathbb{R}^2)$ is invariant under T, i.e., $f \circ T = f$, then $\int_{\mathbb{R}^2} f(x)\, dx = 0$.

88. Let A be a real $n \times n$ symmetric positive definite matrix, i.e., $\langle Ax, x \rangle > 0$ for $x \ne 0$. Compute $\int_{\mathbb{R}^n} e^{-\langle Ax, x \rangle}\, dx$.

89. Given a distribution function $F : \mathbb{R} \to \mathbb{R}$ and a continuous, increasing, invertible function $\Phi : \mathbb{R} \to \mathbb{R}$, let μ_F and $\mu_{F \circ \Phi}$ denote the measures induced by F and $F \circ \Phi$, respectively. Prove that if $f \in L^1(\mathbb{R}, \mu_F)$, then $f \circ \Phi \in L^1(\mathbb{R}, \mu_{F \circ \Phi})$ and $\int_{\mathbb{R}} f\, d\mu_F = \int_{\mathbb{R}} f \circ \Phi\, d\mu_{F \circ \Phi}$.

90. Let $f \in L^1(\mathbb{R})$. Prove: (a) $\sum_{n=-\infty}^{\infty} f(x + n)$ converges absolutely for x a.e. in $\mathbb{R}$. (b) If $\sum_n |f(x + n)|$ is integrable, $f = 0$ a.e. (c) If $\{\lambda_n\}$ is a real sequence, $\sum_n f(\lambda_n (x - x_n)) \in L^1(\mathbb{R})$ for any sequence $\{x_n\}$ in $\mathbb{R}$ iff $\sum_n |\lambda_n|^{-1} < \infty$.

91. Let $f(x)$ be a strictly positive continuous function on $[0,1]$ and $A = \{(x,y) \in \mathbb{R}^2 : 0 < x < 1, 0 < y < f(x)\}$. Compute $|A|$.

92. Discuss the validity of the following statement: A function f on $\mathbb{R}^+$ is Riemann integrable in the improper sense iff $f \in L^1(\mathbb{R}^+)$ in the Lebesgue sense.

93. Discuss the validity of the following statement: A function f on a finite interval $[a, b]$ is Riemann integrable in the improper sense iff $f \in L^1([a, b])$ in the Lebesgue sense.

94. Let $f : (0,1) \to \mathbb{R}$ be given by $f(x) = x^\eta \ln^n(1/x)$ for $-1 < \eta < 0$ and $n \geq 0$. Prove that f is Lebesgue integrable, that its integral coincides with an improper Riemann integral of f, and compute it.

95. Compute $\int_0^1 x^\eta (1-x)^{-1} \ln(x)\, dx$, $\eta > -1$.

96. Let f be given on $\mathbb{R}^+$ by $f(x) = (x \ln^2(x))^{-1}$ if $x \in (0, e^{-1})$ and $= 0$ if $x \notin (0, e^{-1})$, and $F(x) = x^{-1} \int_0^x f(t)\, dt$. Verify that $f \in L^1(\mathbb{R})$ and discuss the validity of the following statement: $\int_0^\eta F(x) dx < \infty$ for some $\eta > 0$.

97. Let f be the function on $\mathbb{R}^+$ given by

$$f(x) = \int_0^{\pi/2} \frac{\cos(t)}{t + x}\, dt\,, \quad x > 0\,.$$

Prove that f is well-defined and continuous, give the asymptotic expressions for f at 0^+ and ∞, and determine, if they exist, $\lim_{x \to 0+} f(x)$ and $\lim_{x \to \infty} f(x)$.

98. Let f_a and g_a denote the functions on $\mathbb{R}$ given by $f_a(0) = 0$ and $f_a(x) = 1/(|x|^a + |x|^{1/a})$ for $x \neq 0$, and $g_a(y) = \int_0^\infty f_a(x) \cos(xy)\, dx$, respectively. Find the values of $a > 0$ for which: (a) $f_a \in L^1(\mathbb{R})$. (b) $x f_a(x) \in L^1(\mathbb{R})$. (c) $g_a \in C^1(\mathbb{R})$.

99. Let $F(x)$ be given by

$$F(x) = \int_1^\infty \frac{t^x}{1 + t^2}\, dt\,, \quad -\infty < x < 1\,.$$

(a) Compute $\lim_{x \to -\infty} F(x)$ and $\lim_{x \to 1-} F(x)$. (b) Prove that for all k the derivative $F^{(k)}$ of order k of F is given by

$$F^{(k)}(x) = \int_1^\infty \frac{t^x}{1 + t^2} \ln^k(t)\, dt\,, \quad x \in (-\infty, 1)\,.$$

(c) Prove that $F'(x)^2 \leq F(x) F''(x)$, $x \in (-\infty, 1)$, and conclude that the function $\ln(F)(x)$ is convex in $(-\infty, 1)$.

100. Compute $I = \int_R e^{-x^2} \cos(ax)\, dx$, $a \in \mathbb{R}$.

101. Let $f \in L(\mathbb{R})$ be such that if φ is a compactly supported smooth function, $\int_\mathbb{R} f(x) \varphi'(x)\, dx = 0$. Prove that $\int_\mathbb{R} f(x) \psi(x)\, dx$ is independent of ψ where ψ is compactly supported and has integral 1.

102. Let $(X, \mathcal{M}, \mu)$ be a measure space and f, g complex-valued, integrable functions on X such that $\|f + g\|_1 = \|f\|_1 + \|g\|_1$. Prove that $\overline{f(x)}g(x) \geq 0$ μ-a.e.

103. Suppose that $f \in L^1(I)$ satisfies $\int_I x^k f(x)\, dx = 0$, $0 \leq k \leq n - 1$, and $\int_I x^n f(x)\, dx = 1$. Prove that $|f(x)| \geq 2^n(n+1)$ on a subset of I of positive measure.

104. Let f, g be continuous functions on $[0,1]$ such that f is nonincreasing and $0 \le g \le 1$. Prove that with $\eta = \int_0^1 g(x)dx$, $\int_0^1 f(x)g(x)\,dx \le \int_0^\eta f(x)\,dx$.

105. Let f, g be increasing functions on $[0,1]$. Prove: (a) $\int_0^1 f(x)g(x)\,dx \ge (\int_0^1 f(x)\,dx)(\int_0^1 g(x)\,dx)$. (b) $\int_0^1 f(x)g(1-x)\,dx \le (\int_0^1 f(x)\,dx)(\int_0^1 g(x)\,dx)$.

106. Let $(X, \mathcal{M}, \mu)$ be a probability measure space and f, g, h nonnegative integrable functions on X with integral 1. Prove: (a) $\mu(\{f, g \le 5\}) \ge 3/5$. (b) $\mu(\{f, g, h \le 5\}) \ge 1/5$.

107. Let f be a locally integrable function on $[0, \infty)$ such that for each $a > 0$, $\int_0^\infty |f(x)|\, e^{-ax}\, dx < \infty$. Does it follow that f is integrable? What if there exists M such that $\int_0^\infty |f(x)|e^{-ax}\, dx \le M < \infty$, all $a > 0$? What can one say about f if there exists M' such that $\int_0^\infty |f(x)|e^{ax}\, dx \le M' < \infty$, all $a > 0$?

108. Given f on $\mathbb{R}^+$, let $F(x) = \sum_n f(n^2 x)$, $x > 0$. Prove: (a) If f is integrable, the series defining $F(x)$ converges absolutely for a.e. $x > 0$, F is integrable, and $\int_0^\infty F(x)\, dx = A \int_0^\infty f(x)\, dx$ where A is a constant to be determined. (b) If f is continuous and $(1 + x)f(x)$ is bounded, $F(x)$ is defined for all $x > 0$, and $\lim_{x \to 0^+} \sqrt{x}F(x) = B$ exists where B is a constant to be determined.

109. Let $(X, \mathcal{M}, \mu)$ be a measure space with $X = \mathbb{R}^n$ and suppose that $B(0, 1/n) \in \mathcal{M}$ for all n. Discuss the validity of the following statement: If $f \in L^1(X)$, $\lim_n \int_{B(0,1/n)} f\, d\mu = 0$.

110. Let μ be a Borel measure on I so that $\int_I fg\, d\mu = (\int_I f\, d\mu)(\int_I g\, d\mu)$ for all continuous f, g on I. Prove: (a) There exists $x_0 \in I$ such that $\int_I f\, d\mu = 0$ implies $f(x_0) = 0$ for all $f \in C(I)$. (b) $\mu = \delta_{x_0}$.

111. Let $A \subset \mathbb{R}^n$ have $|A| = 0$. Prove that if $Q(x, \ell)$ denotes the closed cube centered at x of sidelength ℓ, there exists a nonnegative $f \in L^1(\mathbb{R}^n)$ such that

$$\liminf_{\ell \to 0} \frac{1}{|Q(x, \ell)|} \int_{Q(x,\ell)} f(y)\, dy = \infty, \quad \text{all } x \in A.$$

112. Let $f \in L(\mathbb{R}^n)$ and assume that for each $x \in \mathbb{R}^n$, there exist cubes $\{Q_{x,k}\}$ containing x such that $\int_{Q_{x,k}} f(y)\, dy = 0$, all k, and $\lim_k |Q_{x,k}| = 0$. Prove that $f = 0$ a.e.

113. Let $A \in \mathcal{L}(\mathbb{R}^n)$. Prove: (a) Almost every $x \in A$ is a point of density of A and almost every $x \in A^c$ is a point of dispersion of A. (b) If $0 < |A| < \infty$ there are a subset A_0 of A and $N > 0$ such that $|A_0| > |A|/2$ and $n|A \cap (x - 1/n, x + 1/n)| \ge 1$ for all $x \in A_0$ and $n \ge N$. (c) Let $\widehat{A} = \{x \in \mathbb{R}^n : |A \cap B(x, r)| > 0 \text{ for all } r > 0\}$. Compute $|A \setminus \widehat{A}|$.

114. Suppose $A \subset I$ satisfies the following property: There exists a dense set $D \subset I$ such that if $a, b \in D$, then $|A \cap (a, b)|/|(a, b)| = |A \cap I|/|I|$. Prove that $|A| = 1$ or $|A| = 0$.

115. Let $A \in \mathcal{L}(\mathbb{R}^k)$ and $\varepsilon > 0$. Prove that there exist $x \in \mathbb{R}^k$ and $r > 0$ such that $|A \cap B(x, r)|/|B(x, r)| < \varepsilon$ or $1 - \varepsilon < |A \cap B(x, r)|/|B(x, r)|$.

116. Discuss the validity of the following statement: There exists $A \in \mathcal{L}(\mathbb{R})$ such that if $f(x) = |A \cap (-1, x)|$, $f'(0) = 1/4$.

117. Given a closed subset F of $\mathbb{R}$, let $\delta(x) = d(x, F)$. Prove that $\delta(x + y)/|y| \to 0$ as $|y| \to 0$ for a.e. $x \in F$.

118. Let $f \in L(I)$. (a) Does there exist $x \in I$ such that $\int_0^x f(y)\, dy = \int_x^1 f(y)\, dy$? (b) What can one say about f if $\int_0^x f(y)\, dy = \int_x^1 f(y)\, dy$ for all $x \in I$?

119. Prove that if $f \in L^1([a, b])$ satisfies

$$\lim_{h \to 0} \frac{1}{h} \int_a^b |f(t + h) - f(t)|\, dt = 0\,,$$

then $f = $ constant a.e.

120. Let $f \in L^1([0, 1])$, $\int_I f(x)\, dx = L$. Does there exist a subset $A \subset [0, 1]$ such that $|A| = 1/2$ and $\int_A f(x)\, dx = L/2$?

121. Let $(X, \mathcal{M}, \mu)$ be a probability measure space and f a nonnegative measurable function on X with $\int_X f\, d\mu = 1$. Prove: (a) For all $0 < \eta < 1$, $\int_{\{f > \eta\}} f\, d\mu > 1 - \eta$. (b) If f is bounded above by 1, then $f = 1$ μ-a.e. in X.

122. Discuss the validity of the following statement: If f is a continuous integrable function on $\mathbb{R}$, $f(x) \to 0$ as $|x| \to \infty$.

123. Let $(X, \mathcal{M}, \mu)$ be a measure space and $f \in L^1(X)$. Prove that given $\varepsilon > 0$, there is a measurable set A with $\mu(A) < \infty$ such that $\sup_{x \in A} |f(x)| < \infty$ and $\int_{A^c} |f|\, d\mu \leq \varepsilon$.

124. Let $f \in L^1(\mathbb{R}^n)$. Prove that given $\varepsilon > 0$, there exist: (a) $R > 0$ and $B \subset \{|x| > R\}$ with $|B| < \varepsilon$ such that $|f| \leq \varepsilon$ for $x \in \{|x| > R\} \setminus B$. (b) $B \in \mathcal{L}(\mathbb{R}^n)$ such that $|B| < \varepsilon$ and $f(x)\, \chi_{B^c}(x) \to 0$ as $|x| \to \infty$.

125. Let $f \in L^1(\mathbb{R})$, $\{\beta_n\}$ a positive sequence and $\{\alpha_n\}$ such that $\sum_n \beta_n/|\alpha_n| < \infty$. Prove that $\lim_n \beta_n |f(\alpha_n x)| = 0$ a.e.

126. We say that a function φ on $\mathbb{R}^n$ is radial if $\varphi(x) = \varphi(|x|)$, $x \neq 0$. Prove that if φ is a nonnegative, nonincreasing, integrable radial function, then $\lim_{R \to \infty} R^n \varphi(R) = 0$.

127. Let f be a uniformly continuous, integrable function on $[0, \infty)$. Prove that $\lim_{x \to \infty} f(x) = 0$.

128. Suppose $f \in L^1(\mathbb{R}^+)$ is nonnegative and satisfies a Lipschitz condition $|f(x) - f(y)| \leq M\,|x - y|$. Prove that $\liminf_n \sqrt{n}\, f(n) = 0$.

129. Let $f \in L^1(\mathbb{R})$ and $\{a_n\}$ such that no more than N of the a_n lie in any interval of length 1. Prove that $\lim_n f(x + a_n) = 0$ a.e.

130. Let $f \in L^1([0,1])$. (a) Prove that there is a sequence $\{a_n\}$ decreasing to 0 such that $a_n |f(a_n)| \to 0$. (b) Let $\{f_k\}$ be integrable functions. Does there exist a sequence $\{b_n\}$ decreasing to 0 such that $\lim_n b_n |f_k(b_n)| \to 0$ for all k?

131. Suppose that $f \in L^1([0,1])$ and $\lim_{x \to 1^-} f(x) = \lambda$. Compute

$$\lim_n n \int_0^1 x^n f(x)\, dx\,.$$

132. Let f be a Lebesgue integrable function on finite intervals of $\mathbb{R}^n$ such that $\lim_{|x| \to \infty} f(x) = \lambda$. Compute

$$\lim_{\ell \to \infty} \frac{1}{\ell^n} \int_{Q(0,\ell)} f(x)\, dx\,.$$

133. Given $f \in L^1(\mathbb{R}^n)$ and $\lambda > 0$, compute

$$\lim_{\ell \to \infty} \frac{1}{\ell^{n\lambda}} \int_{Q(0,\ell)} |x|^\lambda f(x)\, dx\,.$$

134. Let $(X, \mathcal{M}, \mu)$ be a probability measure space. Does there exist a nonnegative integrable function f on X such that $\lim_n \int_X f^n\, d\mu = 2$?

135. Let $(X, \mathcal{M}, \mu)$ be a finite measure space, f a nonnegative measurable function on X, and g a nondecreasing function on $\mathbb{R}$ that is continuous at 0 and ∞. Find $\lim_n \int_X g(f^n)\, d\mu$.

136. Let $(X, \mathcal{M}, \mu)$ be a finite measure space and f a measurable function on X. Find $\lim_n \int_X |\cos(\pi f(x))|^n\, d\mu(x)$.

137. Let $(X, \mathcal{M}, \mu)$ be a finite measure space and f a nonnegative measurable function on X such that there exist a constant $c > 0$ and a sequence $n_k \to \infty$ such that $\int_X f^{n_k}\, d\mu = c$. Prove that f is the characteristic function of a measurable set.

138. Let $(X, \mathcal{M}, \mu)$ be a measure space and f a measurable function on X such that $\int_X f^n\, d\mu = c$ for $n = k, k+1, k+2$ where k is a positive integer. Prove that if k is even, $f = \chi_A$ μ-a.e. for some measurable set $A \subset X$ and that the same holds if k is odd provided that $f \geq 0$ μ-a.e. Can you think of a circumstance under which only two successive indices suffice?

139. Prove that given $\varepsilon > 0$, there is a nonnegative integrable function f on I such that $f(x) = 0$ on a set of measure $\geq 1 - \varepsilon$ and $\int_a^b f(x)\, dx > 0$ for any interval $(a, b) \subset I$.

140. Let $\{r_k\}$ denote an enumeration of $\mathbb{Q}^n$, J_k the interval centered at r_k of sidelength 1, $\{\alpha_k\}$ a nonnegative sequence such that $\sum_k \alpha_k < \infty$, and set

$$f(x) = \sum_k \alpha_k \frac{1}{|x - r_k|^{n/2}} \chi_{J_k}(x).$$

Prove: (a) $f \in L^1(\mathbb{R}^n)$. (b) f^2 is finite a.e. but is not integrable on any interval of $\mathbb{R}^n$. (c) f is discontinuous at every point of $\mathbb{R}^n$ and unbounded on every interval of $\mathbb{R}^n$ and it remains so after any modification on a Lebesgue set of measure 0.

141. Let $f \in L^1(\mathbb{R}^n)$, $B(x,r)$ the ball centered at x of radius r, and $g(x,r) = \int_{B(x,r)} f(y)\, dy$. Prove: (a) For each fixed $x \in \mathbb{R}^n$, $g(x,r)$ is a continuous function of r. (b) For each fixed $r > 0$, $g(x,r)$ is a uniformly continuous function of x and $\lim_{|x|\to\infty} g(x,r) = 0$.

142. Let $(X, \mathcal{M}, \mu)$ be a measure space and $f \in L^1(X)$. For $r > 0$ let $B_r = \{|x| \le r\}$ and define $g : [0, \infty) \to \mathbb{R}$ by $g(r) = \int_{B_r} f\, d\mu$. Prove that g is continuous at r iff $\mu(\partial B_r) = 0$.

143. Let $(X, \mathcal{M}, \mu)$ be a measure space, $\{A_n\} \subset \mathcal{M}$, and $f \in L^1(X)$ a μ-a.e. strictly positive function on X. Discuss the validity of the following statement: $\lim_n \int_{A_n} f\, d\mu = 0$ iff $\lim_n \mu(A_n) = 0$.

144. Let $\varphi : I \to I$ be a Borel measurable function. Prove that the following are equivalent: (a) For all Borel $A \subset I$, $|\varphi^{-1}(A)| = |A|$. (b) If f is a nonnegative Borel measurable function on I, $\int_I f(x)\, dx = \int_I f(\varphi(x))\, dx$. (c) If f is a continuous function on I, $\int_I f(x)\, dx = \int_I f(\varphi(x))\, dx$. (d) For all $x \in I$, $|\{t \in I : \varphi(t) \le x\}| = x$.

145. Prove that if $\varphi : \mathbb{R} \setminus \{0\} \to \mathbb{R}$ is given by $\varphi(x) = x - 1/x$ and $g = f \circ \varphi$, then $\int_{\mathbb{R}} f(x)\, dx = \int_{\mathbb{R}} g(x)\, dx$.

146. Let $(X, \mathcal{M}, \mu)$ be a probability measure space and $T : X \to X$ a measure preserving mapping. Prove that the following are equivalent: (a) If $A \in \mathcal{M}$ and $T^{-1}(A) = A$, then $\mu(A) = 0$ or 1. (b) If $A \in \mathcal{M}$ and $\mu(A \triangle T^{-1}(A)) = 0$, then $\mu(A) = 0$ or 1. (c) If $f \in L^1(X)$ and $f \circ T = f$ μ-a.e., then f is constant μ-a.e.

147. Let $f \in L^1(\mathbb{R})$ and ψ, φ bounded measurable functions on $\mathbb{R}$. Prove: (a) If $\psi(x + \beta) = -\psi(x)$ for some $\beta > 0$ and all $x \in \mathbb{R}$, then $\lim_n \int_{\mathbb{R}} f(x)\, \psi(nx)\, dx = 0$. (b) If $\varphi(x + \beta) = \varphi(x)$ for some $\beta > 0$ and all $x \in \mathbb{R}$, then $\lim_n \int_{\mathbb{R}} f(x)\, \varphi(nx)\, dx = (\beta^{-1} \int_0^\beta \varphi(x)\, dx) \int_{\mathbb{R}} f(x)\, dx$.

148. Let $a_n(x) = 2 + \sin(nx)$, $n = 1, 2, \ldots$, $x \in [0, 1]$. Prove: (a) If $f \in L^1([0, 1])$, $\lim_n \int_I f(x) a_n(x)\, dx = 2 \int_I f(x)\, dx$. (b) $\lim_n \int_I f(x) a_n(x)^{-1} dx$ exists and compute it.

149. Let $(X, \mathcal{M}, \mu)$ be a measure space and f, g independent measurable functions on X. Prove that $\cos(f)$ and $\sin(g)$ are independent.

150. Let $X = [0, 1]$ and for $x = \sum_n x_n 2^{-n} \in X$, let $f_n(x) = x_n$. Prove that if $n \neq m$, f_n and f_m are independent with respect to the Lebesgue measure.

151. Let $(X, \mathcal{M}, \mu)$ be a probability measure space and f, g bounded functions on X. Prove that if f and g are independent, then $\int_X fg \, d\mu = (\int_X f \, d\mu)(\int_X g \, d\mu)$.

152. Let $x = x_0.x_1 x_2 \ldots$ be the decimal expansion of $x \in \mathbb{R}$; the expansion is unique except for a set of measure 0, which can be disregarded. Define

$$\varphi_n(x) = \begin{cases} 1, & x_n \text{ even,} \\ -1, & x_n \text{ odd.} \end{cases}$$

Prove that if $f \in L^1(\mathbb{R})$, $\lim_n \int_\mathbb{R} f(x)\varphi_n(x) \, dx = 0$.

153. Let $g \in L^1(\mathbb{R}^n)$ be such that $\int_Q g(ry + s) \, dy = 0$ where Q is a fixed cube in $\mathbb{R}^n$ and $r \in \mathbb{Q}$ and $s = (s_1, \ldots, s_n) \in \mathbb{Q}^n$ are arbitrary. Prove that $g = 0$ a.e.

154. Let $r_1, r_2, \ldots$ be an enumeration of the rationals in $[0, 1]$, $f_n(x) = H(x - r_n)$ where H is the Heaviside function, and put $f(x) = \sum_k 2^{-k} f_k(x)$. Compute $\int_I f(x) \, dx$.

155. Let $(X, \mathcal{M}, \mu)$ be a measure space and $f \in L^1(X)$. Evaluate

$$L = \lim_{h \to 0} \int_X \frac{|1 + h\,f(x)| - 1}{h} \, d\mu(x) \, .$$

156. Let $(X, \mathcal{M}, \mu)$ be a finite measure space, $\{A_n\} \subset \mathcal{M}$ with $\mu(A_n) \geq \varepsilon > 0$ for all n, and let $n(x)$ denote the number of integers k such that $x \in A_k$. Prove: (a) $n(x) \geq 2$ for some $x \in X$. (b) $\sup_{x \in X} n(x) = \infty$. (c) $\mu(\{x \in X : n(x) = \infty\}) \geq \varepsilon > 0$.

157. Let $(X, \mathcal{M}, \mu)$ be a probability measure space and $A_1, \ldots, A_n$ measurable subsets of X such that μ almost every x in X belongs to at least k of these subsets. Prove that $\max_{1 \leq m \leq n} \mu(A_m) \geq k/n$.

158. Let $(X, \mathcal{M}, \mu)$ be a measure space, $\{A_n\} \subset \mathcal{M}$, and $A = \bigcup_n A_n$. Prove that if each $x \in A$ belongs to at most k different A_n's, $\sum_n \mu(A_n) \leq k\,\mu(A)$.

159. Let $(X, \mathcal{M}, \mu)$ be a probability measure space, $0 < \eta < 1$, and $A_1, \ldots, A_N$ in $\mathcal{M}$ with $\mu(A_n) \geq \eta$ for all n. Let $0 < a < \eta$ and $A = \{x \in X : x \in A_n \text{ for at least } aN \text{ values of } n\}$. Prove that $\mu(A) \geq (\eta - a)/(1 - a)$.

160. Let $\{p_n\}$ be real-valued polynomials with p_n of degree exactly n, all n, and $f \in L^1(I)$ such that $\int_I f(x)\, p_n(x)\, dx = 0$ for all n. Prove that $f = 0$ a.e.

161. Characterize those locally integrable functions φ on $\mathbb{R}$ which satisfy $\int_{\mathbb{R}} H_J(x)\varphi(x)\, dx = 0$ for all dyadic intervals $J \subset \mathbb{R}$. Here $\{H_J\}$ denotes the Haar system of functions on $\mathbb{R}$.

162. Let $f : \mathbb{R}^+ \to \mathbb{R}$ be a C^1 function such that $\varphi(t) = \sup\{|f'(x)| : x \geq t\}$, $t \geq 0$, satisfies $\int_0^\infty \varphi(t)\, dt < \infty$. Prove that for $a > 0$, $\Phi(a, x) = (f(ax) - f(x))/x$ is integrable and compute its integral.

163. Let $f : [0, \infty) \to [0, \infty)$ be continuous, nonincreasing, and integrable. Prove that

$$\lim_{x \to \infty} \frac{1}{f(x)} \int_x^\infty f(s)\, ds = 0$$

iff for each $t > 0$,

$$\lim_{x \to \infty} \frac{f(x + t)}{f(x)} = 0 \,.$$

164. Let $f \in L^1([0, 1])$, A_n measurable subsets of I, and $g_n(x) = \int_0^x \chi_{A_n}(t)\, f(t)\, dt$, all n. Prove that a subsequence $\{g_{n_k}\}$ of $\{g_n\}$ converges uniformly to a continuous function g in $[0, 1]$.

165. Let $(X, \mathcal{M}, \mu)$ be a measure space and f, g nonnegative integrable functions on X with integral 1. Let λ and ν denote the probability measures with density f and g with respect to μ, respectively. Prove that the total variation $\|\lambda - \nu\| = \sup_{A \in \mathcal{M}} |\lambda(A) - \nu(A)|$ is equal to $(1/2) \int_X |f - g|\, d\mu$.

166. Let μ be a Borel measure on $(\mathbb{R}, \mathcal{B}, \mu)$ and define $I : [0, \infty) \to \mathbb{R}$ by $I(x) = \int_{[0,x]} f(t)\, d\mu(t) = \int_{\mathbb{R}} \chi_{[0,x]}(t)\, f(t)\, d\mu(t)$, $f \in L^1(\mathbb{R})$. Prove that if μ has no atom at x, i.e., $\mu(\{x\}) = 0$, $I(x)$ is continuous at x.

167. Let f be a nonnegative absolutely continuous function on $I = [0, 1]$, $M = \max_{x \in I} f(x)$, and $A = \{x \in I : f(x) = M\}$. Which of the following statements is true? (a) $\lim_n \int_I (f(x)/M)^n\, dx = |A|$. (b) $\lim_n \int_I |f'(x)|\, (f(x)/M)^n\, dx = 0$.

168. Prove that if $\Lambda = \sum_n \lambda_n < \infty$ where the λ_n are positive, there is a sequence $\mu_n \to \infty$ such that $\sum_n \lambda_n \mu_n \leq 2\Lambda$.

169. Let $(X, \mathcal{M}, \mu)$ be a probability measure space and f a nonnegative integrable function on X. Prove that $\int_X \ln(f)\, d\mu \leq \ln(\int_X f\, d\mu)$.

170. Let $G(t) : \mathbb{R} \to \mathbb{R}^+$ be given by $G(t) = \int_{[0,1]} |\varphi(x) - t|\, dx$, where φ is integrable. Prove that G is continuous and find the condition that guarantees that G is differentiable at t.

Chapter 5

L^p **Spaces**

This chapter is devoted to the spaces of p-integrable functions on a measure space $(X, \mathcal{M}, \mu)$, $0 < p \leq \infty$. For $p < \infty$, the expression $\|f\|_p = (\int_X |f|^p \, d\mu)^{1/p}$ is called the L^p *norm* of f and $L^p(X) = \{f : f$ is defined on X, measurable, and $\|f\|_p < \infty\}$ is a linear class of functions which, endowed with the metric $d_p(f, g) = \|f - g\|_p$, $1 \leq p < \infty$, or $d_p^p(f, g) = \|f - g\|_p^p$ when $0 < p < 1$, becomes a complete metric space. When $p = \infty$, the expression $\|f\|_\infty = \inf\{\lambda > 0 : \mu(\{|f| > \lambda\}) = 0\}$ is called the μ-*essential supremum* of f and the linear space $L^\infty(X) = \{f : f$ is defined on X, measurable, and $\|f\|_\infty < \infty\}$, endowed with the metric $d_\infty(f, g) = \|f - g\|_\infty$, is a complete metric space. As is the case for $L^1(X)$, simple functions are dense in $L^p(X)$ and, for the Lebesgue measure, compactly supported smooth functions are dense in $L^p(\mathbb{R}^n)$, $1 \leq p < \infty$.

Functions in L^p satisfy Chebychev's inequality $\lambda^p \mu(\{|f| > \lambda\}) \leq \|f\|_p^p$, $0 < \lambda < \infty$, and an efficient way to evaluate the L^p norm is given by $\int_X |f|^p \, d\mu = p \int_0^\infty \mu(\{|f| > \lambda\}) \lambda^{p-1} \, d\lambda$, $0 < p < \infty$. The *weak L^p spaces* are closely related to the L^p spaces and are defined as Wk-$L^p(X) = \{f : f$ is measurable on X and $\lambda^p \mu(\{|f| > \lambda\}) \leq c$ for a finite constant c and all $\lambda > 0\}$, $0 < p < \infty$. The infimum of the constant c is called the Wk-$L^p(X)$ norm of f.

Essential tools in dealing with these spaces include various inequalities. Jensen's inequality states that, in a probability measure space, if f is integrable and φ is convex on $\mathbb{R}$, then $\varphi(\int_X f \, d\mu) \leq \int_X (\varphi \circ f) \, d\mu$. If φ is concave, the inequality is reversed

Hölder's inequality states that, if $1 < p, q < \infty$ are conjugate indices, i.e., $1/p + 1/q = 1$, then $\int_X |fg| \, d\mu \leq \|f\|_p \|g\|_q$, with equality when $|g(x)|$

$= c|f(x)|^{p-1}$ μ-a.e. Also, $\int_X |fg|\, d\mu \le \|f\|_1 \|g\|_\infty$, with equality when $|g| = \|g\|_\infty$ μ-a.e.

Next, Minkowski's inequality states that if $f, g \in L^p(X)$, $1 \le p \le \infty$, then $\|f + g\|_p \le \|f\|_p + \|g\|_p$. The case of equality here depends on p. For $1 < p < \infty$ the condition is: There exist nonnegative constants A, B not both 0 such that $Af = Bg$ μ-a.e. On the other hand, if $p = 1$ the condition is: There exists a nonnegative measurable function h such that $fh = g$ μ-a.e. on the set $\{fg \ne 0\}$.

Finally, Minkowski's integral inequality states

$$\left(\int_X \left| \int_Y h(x,y)\, d\nu(y) \right|^p d\mu(x) \right)^{1/p} \le \int_X \left(\int_Y |h(x,y)|^p\, d\mu(x) \right)^{1/p} d\nu(y).$$

Now, the problems. The central role of the distribution function of a function f, i.e., the measure of the level sets of f, in the computation and estimation of the L^p norm of f for $0 < p < \infty$ is covered in Problems 5–7 and Problems 30–32. The Marcinkiewicz Wk-$L^p(X)$ classes are discussed in Problem 35 and Problem 53.

The separability of $L^p(X)$ is considered in Problem 43. Hölder's inequality is exploited in Problems 45–46, Problems 93–94, and Problem 97. The version of Hölder's inequality for indices $0 < p < 1$ is given in Problem 119. The fact that an inclusion between L^p spaces reflects the nature of the measure in the underlying measure space is covered in Problems 57–58. And, that in a finite measure space it is possible to define a function whose L^p norm grows at an arbitrary rate is done in Problem 61.

$L^\infty(X)$ is discussed in Problems 65–68, and interesting questions, including when it is finite dimensional, are considered.

Jensen's inequality is explored in Problems 81–82, Hardy's inequality in Problems 84–87, and Schur's lemma in Problems 88–89.

Problems 100–104 examine $\lim_{p \to 0^+} \|f\|_p$, and Problems 122–124 examine the limit $\lim_{p \to \infty} \|f\|_p$. Properties of the sequence ℓ^p spaces, including compactness, are discussed in Problems 131–148. The local L^p spaces defined in open subsets of $\mathbb{R}^n$ are studied in Problems 155–156.

Problems

1. Can it happen that $f^2 \in L^1(X)$, yet $f \notin L^2(X)$?

2. Prove that for all $0 < p \le \infty$, the equivalence class of an $L^p(\mathbb{R}^n)$ function contains at most one continuous function.

3. Let $(X, \mathcal{M}, \mu)$ be a finite measure space, $0 < p < \infty$, and $f \notin L^p(X)$. Prove that $f\chi_{\{|f|>\lambda\}} \notin L^p(X)$ for any $\lambda > 0$.

4. Let $(X, \mathcal{M}, \mu)$ be a measure space and $0 < p < \infty$. Given a measurable function f on X, let

$$f_n(x) = \begin{cases} n, & f(x) > n, \\ f(x), & |f(x)| \le n, \\ -n, & f(x) < -n. \end{cases}$$

Prove: (a) If $f \in L^p(X)$, $\|f_n - f\|_p \to 0$. (b) If $f \notin L^p(X)$, $\{\|f_n\|_p\}$ is an increasing unbounded sequence.

5. Let $(X, \mathcal{M}, \mu)$ be a measure space, f a measurable function on X, and $A_n = \{2^n < |f| \le 2^{n+1}\}$, $B_n = \{|f| > 2^n\}$. Given $0 < p < \infty$, prove that the following statements are equivalent: (a) $f \in L^p(X)$. (b) $I_p = \sum_{n=-\infty}^{\infty} 2^{np}\mu(A_n) < \infty$. (c) $J_p = \sum_{n=-\infty}^{\infty} 2^{np}\mu(B_n) < \infty$. Furthermore, verify that $\|f\|_p^p \sim I_p \sim J_p$.

6. Let $(X, \mathcal{M}, \mu)$ be a measure space and $1 \le p < \infty$. Prove that $f \in L^p(X)$ iff $\sum_n n^{-(p+1)}\mu(\{|f| > 1/n\}) + \sum_n n^{p-1}\mu(\{|f| > n\}) < \infty$.

7. Let φ be a nonnegative, nonincreasing measurable radial function and $0 < p < \infty$. Prove that $\varphi \in L^p(\mathbb{R}^n)$ iff $\sum_{k=-\infty}^{\infty} 2^{kn}\,\varphi(2^k)^p < \infty$.

8. For $0 < \alpha < \beta < \infty$, consider

$$f(x) = \frac{1}{x^\alpha + x^\beta}, \quad x > 0.$$

For what values of p does $f \in L^p([0,\infty))$?

9. Let $f : \mathbb{R}^3 \to \mathbb{R}$ be given by $f(x_1, x_2, x_3) = (x_1^2 + x_2^4 + x_3^6)^{-p}$. For what positive values of p is f integrable: (a) at the origin, and (b) at infinity?

10. Let f be the function on $\mathbb{R}^n$ given by $f(x) = (1 - |x|)\,\chi_{\{|x|<1\}}(x)$. For what values of p, $-\infty < p \le \infty$, is f^p integrable?

11. Let $0 \le \alpha, \gamma < \infty$, and suppose that f, g are measurable functions on $\mathbb{R}^n$ that satisfy

$$|f(x)| \le \begin{cases} |x|^{-\alpha} \ln^\gamma(1/|x|), & |x| \le 1, \\ 0, & |x| > 1, \end{cases} \qquad |g(x)| \le \begin{cases} |x|^{-\alpha} \ln^{-\gamma}(1/|x|), & |x| \le 1, \\ 0, & |x| > 1. \end{cases}$$

For what values of p are f, g in $L^p(\mathbb{R}^n)$?

12. Let $0 \le \beta, \gamma < \infty$, and suppose that f, g are measurable functions on $\mathbb{R}^n$ that satisfy

$$|f(x)| \le \begin{cases} 0, & |x| \le 1, \\ |x|^{-\beta} \ln^\gamma(|x|), & |x| > 1, \end{cases} \qquad |g(x)| \le \begin{cases} 0, & |x| \le 1, \\ |x|^{-\beta} \ln^{-\gamma}(|x|), & |x| > 1. \end{cases}$$

For what values of p are f, g in $L^p(\mathbb{R}^n)$?

13. Let $0 < p < \infty$. Give an example of a measure space $(X, \mathcal{M}, \mu)$ and a function f on X so that $f \in L^p(X)$ and $f \notin L^q(X)$, $0 < q \leq \infty$, for all $q \neq p$.

14. Let $(X, \mathcal{M}, \mu)$ be a measure space and $f \in L^p(X)$. Prove that there exists a convex increasing function $\varphi : [0, \infty) \to \mathbb{R}$ such that $\varphi(0) = 0$, $\lim_{t \to \infty} \varphi(t)/t^p = \infty$, and $\int_X \varphi(|f|)\, d\mu < \infty$.

15. Let $(X, \mathcal{M}, \mu)$ be a finite measure space, $0 < q < \infty$, and f a nonnegative measurable function on X such that $\mu(\{f > t\}) \leq c/(1 + t^q)$, $t > 0$. For what values of p does $f \in L^p(X)$?

16. Let $(X, \mathcal{M}, \mu)$ be a probability measure space and $f \in L^2(X)$ such that $\int_X f\, d\mu = 0$ and $\int_X f^2\, d\mu = 1$. Prove that $\mu(\{f > t\}) \leq 1/(1 + t^2)$, $t > 0$.

17. Let $(X, \mathcal{M}, \mu)$ be a finite measure space, $1 \leq p < q < \infty$, and $\lambda > 0$. Prove that $\|f\|_p \leq \lambda\, \mu(\{|f| \leq \lambda\})^{1/p} + \|f\|_q\, \mu(\{|f| > \lambda\})^{1/p - 1/q}$.

18. Let $(X, \mathcal{M}, \mu)$ be a probability measure space and f a measurable function on X such that $\mu(f^{-1}([k/2^n, (k+1)/2^n))) = 2^{-n}$ for $n \geq 1$, $k = 0, \dots, 2^n - 1$. Compute $\int_X f^2\, d\mu$.

19. Let $(X, \mathcal{M}, \mu)$ be a probability measure space and f a measurable function on X such that $M_f(t) = \int_X e^{tf(x)}\, d\mu(x) < \infty$ for some $\varepsilon > 0$ and $|t| < \varepsilon$; M_f is called the moment generating function of f. Prove: (a) $\int_X |f|^r\, e^{tf}\, d\mu < \infty$, all $0 < r < \infty$, and $|t| < \varepsilon$. (b) $M_f^k(t)$, the k-th derivative of $M_f(t)$, exists and find $\lim_{t \to 0} M_f^k(t)$.

20. For $f \in L^p(I)$, $1 \leq p < \infty$, let $\pi_n(f) : [0, 1] \to \mathbb{R}$ be given by $\pi_n f(0) = 0$ and $\pi_n f(x) = 2^n \int_{k2^{-n}}^{(k+1)2^{-n}} f(y)\, dy$ if $k2^{-n} < x \leq (k+1)2^{-n}$ for $k \in \{0, \dots, 2n - 1\}$, $n \geq 1$. Prove: (a) $|\pi_n f(x)|^p \leq \pi_n |f(x)|^p$ for $x \in [0, 1]$, and, consequently, $\|\pi_n(f)\|_p \leq \|f\|_p$. (b) $\lim_n \pi_n(f) = f$ in $L^p(I)$.

21. Let $\{f_k\}$ be bounded measurable functions on $\mathbb{R}^n$ with each f_k vanishing outside $B(0, 1/k)$ and having integral equal to η, such that $\|f_k\|_1 \leq c$ for all k. For $g \in L^p(\mathbb{R}^n)$ define

$$g_k(x) = \int_{\mathbb{R}^n} f_k(y)\, g(x - y)\, dy.$$

Prove that $g_k \in L^p(\mathbb{R}^n)$ and $g_k \to \eta g$ in $L^p(\mathbb{R}^n)$.

22. Let $f \in L^p(I)$ be nonnegative. Find

$$\lim_n \left(n^{p-1} \sum_{k=0}^{n-1} \left(\int_{[k/n, (k+1)/n)} f(y)\, dy \right)^p \right)^{1/p}.$$

23. Let $(X, \mathcal{M}, \mu)$ be a measure space and $0 < p < \infty$. Prove that μ is σ-finite iff $L^p(X)$ contains a strictly positive function.

24. Let $(X, \mathcal{M}, \mu)$ be the measure space where $X = \{a, b\}$, $\mathcal{M} = \mathcal{P}(X)$, and $\mu(\{a\}) = 1$, $\mu(\{b\}) = \infty$. Discuss the duality relation between $L^p(X)$ and $L^q(X)$, $1/p + 1/q = 1$, for $1 \leq p < \infty$.

25. Let $(X, \mathcal{M}, \mu)$ be a measure space and g a measurable function on X such that $fg \in L^p(X)$ whenever $f \in L^p(X)$. Does it follow that $g \in L^\infty(X)$? If so, what is $\|g\|_\infty$ equal to?

26. Let $(X, \mathcal{M}, \mu)$ be a measure space, $1 < p < \infty$, and consider the following statement: If a measurable function f on X is such that fg is integrable for every $g \in L^q(X)$, $1/p + 1/q = 1$, then $f \in L^p(X)$. Prove that the statement is true iff μ is semifinite.

27. Prove: (a) If φ is a nonnegative function on $\mathbb{R}^n$, then $A = \{f \in L^p(\mathbb{R}^n) : |f(x)| \leq \varphi(x) \text{ a.e.}\}$ is a closed subset of $L^p(\mathbb{R}^n)$. (b) $B = \{f \in C(I) : \int_I f^2(x)\, dx > 1\}$ is open in $C(I)$.

28. Let $1 \leq p \leq \infty$. Discuss the validity of the following statement: $\mathcal{F} = \{f \in L^p(\mathbb{R}) : f \geq 0 \text{ a.e.}\}$ has empty interior.

29. Let $(X, \mathcal{M}, \mu)$ be a nonatomic finite measure space, φ a continuous function on $\mathbb{R}$, and $0 < p, q < \infty$. Prove that $\varphi(f) \in L^q(X)$ for all $f \in L^p(X)$ iff $\limsup_{|t| \to \infty} |\varphi(t)|^q / |t|^p < \infty$.

30. Let $(X, \mathcal{M}, \mu)$ be a measure space, f a measurable function on X, and $0 < p < \infty$. Prove: (a) If $f \in L^p(X)$, then $\lim_{\lambda \to \infty} \lambda^p \mu(\{|f| > \lambda\}) = 0$. (b) $\lim_{\lambda \to \infty} \lambda^p \mu(\{|f| > \lambda\}) = 0$ does not imply that $f \in L^p(X)$. Nevertheless, if $\mu(X) < \infty$, $f \in L^r(X)$, $0 < r < p$. (c) If $\mu(X) < \infty$ and $\mu(\{|f| > \lambda\}) \leq c\lambda^{-p} \ln(\lambda)^{-(1+\varepsilon)}$ for λ large, then $f \in L^p(X)$.

31. Let $(X, \mathcal{M}, \mu)$ be a measure space, f a measurable function on X, $1 \leq p < \infty$, and $q - p = r > 0$. Discuss the validity of the following statement: $I = \sum_n 2^{-nr} \int_{\{|f| \leq 2^n\}} |f|^q \, d\mu < \infty$ iff $f \in L^p(X)$.

32. Let $(X, \mathcal{M}, \mu)$ be a finite measure space, f a measurable function on X, and $0 < p < \infty$. Prove that the following statements are equivalent: (a) $\int_X |f|^p \, d\mu < \infty$. (b) $\sum_n \mu(\{|f|^p > n\}) < \infty$. (c) If $\alpha > 1$, $\gamma, \beta > 0$ are such that $\gamma\beta - \alpha + 1 = p\beta$, then $\sum_n n^{-\alpha} \int_{\{|f| \leq n^\beta\}} |f|^\gamma \, d\mu < \infty$. (d) If $0 < \alpha < 1$, $0 < \gamma \leq p$, $\beta > 0$ are such that $\gamma\beta - \alpha + 1 = p\beta$, then $\sum_n n^{-\alpha} \int_{\{|f| > n^\beta\}} |f|^\gamma \, d\mu < \infty$.

33. Let τ_y denote the translation operator $\tau_y f(x) = f(x + y)$ and $1 \leq p < \infty$. Discuss the validity of the following statements: (a) τ_y is bounded in $L^p(\mathbb{R}^n)$. (b) For $\varphi \in L^\infty(\mathbb{R}^n)$, $\lim_{|h| \to 0} \int_{\mathbb{R}^n} |\tau_h f(x) - f(x)|^p \varphi(x)\, dx = 0$. (c) $\tau_y f$ converges to f in $L^p(\mathbb{R}^n)$ as $|y| \to \infty$. If not, does it converge weakly?

34. Let $\mathcal{F}$ be a relatively compact subset of $L^p(\mathbb{R}^n)$, $1 \leq p < \infty$. Prove that $\mathcal{F}$ is bounded and that, given $\varepsilon > 0$, there exists $\delta > 0$ such that $\|\tau_h f - f\|_p < \varepsilon$ for all h with $|h| \leq \delta$ and $f \in \mathcal{F}$.

35. Let $(X, \mathcal{M}, \mu)$ be a σ-finite measure space, f a measurable function on X, and $0 < r < p < \infty$. Prove that $f \in \text{Wk-}L^p(X)$ iff there exists a positive finite constant c such that $\int_A |f|^r \, d\mu \leq c\,\mu(A)^{1-r/p}$ for every measurable A with $\mu(A) < \infty$.

36. Let $(X, \mathcal{M}, \mu)$ be a finite measure space and, for $1 < p < \infty$, put $A_p(f) = \int_0^\infty \mu(\{|f| > \lambda\})^{1/p} \, d\lambda$. Prove: (a) $\|f\|_p \leq c_p A_p(f)$. (b) For $r > p$, $A_p(f) \leq c_{p,r} \|f\|_r$ with $c_{p,r} \to \infty$ as $r \to p$.

37. Let $S_p = \{g \in L^p([0,1]) : \int_0^1 g^2(t) \, dt = 1\}$ and for $f \in L^p([0,1])$ let $d_p(f, S_p)$ denote the distance from f to S_p in $L^p([0,1])$. If $f(x) = x - 1/2$, find $d_1(f, S_1)$, $d_2(f, S_2)$, and $d_\infty(f, S_\infty)$.

38. Let f be a continuous function on I with $f(0) = 0$ that has a right-hand derivative at 0. Prove that $x^{-3/2} f \in L^p(I)$, $1 \leq p < 2$. Furthermore show that without the differentiability assumption at the origin the result is false.

39. Let f be a continuously differentiable function on $[0,1]$ with $f(0) = 0$, $1/p + 1/q = 1$, and $0 < r < \infty$. Prove: (a) $|f(x)| \leq x^{1/q} \|f'\|_p$, $x \in (0,1)$. (b) $\|f\|_r \leq (q/(q+r))^{1/r} \|f'\|_p$.

40. Let f be a differentiable function such that $f, f' \in L^2(\mathbb{R})$. Prove that f is bounded and $|f(x)| \leq \|f\|_2 \|f'\|_2$, all $x \in \mathbb{R}$.

41. Let $\mathcal{F} = \{f \in C^1([-1,1]) : \int_{-1}^1 |f'(x)|^2 \, dx \leq 1\}$. Discuss the validity of the following statement: If $0 \leq \alpha \leq 1$, then

$$\eta = \sup_{f \in \mathcal{F}} \sup_{x,y \in [-1,1]} \frac{|f(x) - f(y)|}{|x - y|^\alpha} < \infty.$$

42. A step function is one that can be written as a finite linear combination of pairwise disjoint intervals of $\mathbb{R}^n$. Prove that step functions are dense in $L^p(\mathbb{R}^n)$, $1 \leq p < \infty$.

43. Let $(X, \mathcal{M}, \mu)$ be a σ-finite measure space and $1 \leq p < \infty$. Prove that if $\mathcal{M}$ is countably generated, $L^p(X)$ is separable.

44. Let $(X, \mathcal{M}, \mu)$ be a measure space. Discuss the validity of the following statement: If $f, g \in L^2(X)$, then $fg \in L^2(X)$.

45. Prove that if $\|fg\|_r \leq c\|f\|_p \|g\|_q$ for all $f \in L^p(\mathbb{R}^n)$ and $g \in L^q(\mathbb{R}^n)$, then $1/p + 1/q = 1/r$.

46. Let $(X, \mathcal{M}, \mu)$ be a measure space and $0 < p, q < \infty$. Prove that if $f \in L^p(X)$ and $g \in L^q(X)$, then $fg \in L^r(X)$ where $1/r = 1/p + 1/q$ and $\|fg\|_r \leq \|f\|_p \|g\|_q$.

47. Let $(X, \mathcal{M}, \mu)$ be a measure space, $1 \leq r < p$, and g a measurable function on X such that $fg \in L^r(X)$ for all $f \in L^p(X)$. Prove: (a) The mapping T given by $T(f) = fg$ is bounded from $L^p(X)$ into $L^r(X)$. (b) If μ is σ-finite, then $g \in L^q(X)$ where $1/r = 1/p + 1/q$.

48. Let T be a linear mapping from real $L^p(\mathbb{R})$ into real $L^q(\mathbb{R})$, $0 < p, q < \infty$, such that $f \geq 0$ implies $T(f) \geq 0$. Prove that T is bounded.

49. Given $f \in L^p(I)$, $1 < p \leq \infty$, let $g(x) = f(x^\eta)\cos(x)$ and $h(x) = f(x^\eta)\sin(x)$, $1 < \eta < \infty$. For what values of p are g and h integrable?

50. Let $f \in L^p(\mathbb{R}^n), 0 < s < p < \infty$, and $\alpha > n(1 - s/p)$. Prove:

(a) $\lim_{\eta \to 0+} \eta^{n(s/p-1)} \int_{B(0,\eta)} |f(x)|^s \, dx = 0.$

(b) $\lim_{\eta \to \infty} \eta^{\alpha + n(s/p-1)} \int_{B(0,\eta)^c} |f(x)|^s \, |x|^{-\alpha} \, dx = 0\,.$

51. Let $(X, \mathcal{M}, \mu)$ be a probability measure space, $1 < p < \infty$, and $f \in L^p(X)$ a nonnegative function on X such that $\int_X f \, d\mu > \lambda$. Prove that $\int_X (f - \lambda) \, d\mu \leq \mu(\{f > \lambda\})^{1/q}(\int_X f^p \, d\mu)^{1/p}$ where $1/p + 1/q = 1$.

52. Let $(X, \mathcal{M}, \mu)$ be a measure space, f, g measurable functions on X, and $0 < p < \infty$. Prove that if $0 < \alpha < p$, $0 < \beta, r$, and $\mu(\{|g| > \lambda\}) \leq \lambda^{-\alpha} \int_{\{|f| > \lambda^\beta\}} |f|^r \, d\mu$ for all $\lambda > 0$, there is a constant $c_{\alpha, p}$ such that $\int_X |g|^p \, d\mu \leq c_{\alpha, p} \int_X |f|^{r + (p-\alpha)/\beta} \, d\mu.$

53. Let $(X, \mathcal{M}, \mu)$ be a measure space and $0 < p < q < \infty$. Prove that if $f \in L^p(X) \cap L^q(X)$, then $f \in L^r(X)$ for $p < r < q$ and $\|f\|_r \leq \|f\|_p^\eta \|f\|_q^{1-\eta}$ where $0 < \eta < 1$ is given by the relation $r = \eta p + (1 - \eta)q$. What if $f \in \text{Wk-}L^p(X) \cap \text{Wk-}L^q(X)$ instead?

54. Let $Q(0, \ell) \subset \mathbb{R}^n$ denote the cube centered at the origin of side-length ℓ and $f \in L^p(\mathbb{R}^n)$, $1 \leq p \leq \infty$. Discuss the validity of the following statement: $\lim_{\ell \to \infty} \int_{Q(0, 2^{\ell+1}) \setminus Q(0, 2^\ell)} f(x) \, dx = 0.$

55. Let $f \in L^p(I)$ and $0 < p < q < \infty$. Prove that given $\varepsilon > 0$ and $K > 0$, there exists a function $g \in L^p(I)$ such that $\|f - g\|_p < \varepsilon$ and $\|g\|_q > K$.

56. We say that a locally integrable function f on $\mathbb{R}$ has weak derivative g if $\int_{\mathbb{R}} f(x)\varphi'(x) \, dx = -\int_{\mathbb{R}} g(x)\varphi(x) \, dx$ for all compactly supported smooth functions φ. Prove: (a) f has weak derivative 0 iff $f = c$ a.e. (b) If $f \in L^p(\mathbb{R})$, $1 \leq p < \infty$, f has weak derivative 0 iff $f = 0$ a.e.

57. Let $(X, \mathcal{M}, \mu)$ be a measure space and $0 < p < q \leq \infty$. Prove that $L^p(X) \not\subset L^q(X)$ iff X contains sets of arbitrarily small positive measure.

58. Let $(X, \mathcal{M}, \mu)$ be a measure space and $0 < p < q < \infty$. Prove that $L^q(X) \not\subset L^p(X)$ iff X contains sets of arbitrarily large finite measure. How about $q = \infty$?

59. Let $(X, \mathcal{M}, \mu)$ be a probability measure space and suppose that f is such that $\|f\|_p = \|f\|_q$ for some $0 < p < q < \infty$. Prove that $|f|$ is constant μ-a.e.

60. Let $(X, \mathcal{M}, \mu)$ be a measure space and $0 < p, q < \infty$. When are $L^p(X)$ and $L^q(X)$ equal?

61. Let $\lim_p \Psi(p) = \infty$. Construct a function $f \in \bigcap_{p<\infty} L^p(I) \setminus L^\infty(I)$ such that $\lim_p \|f\|_p = \infty$ and $\|f\|_p \leq \Psi(p)$ for all p large.

62. Let μ be a nonatomic Borel measure on $\mathbb{R}^n$ and $f \in L^1(\mathbb{R}^n)$ with $\mu(\{|f| > 0\}) > 0$. Given $1 < p < \infty$, construct a measurable function g such that $\int_{\mathbb{R}^n} |f\,g|\,d\mu < \infty$ and $\int_{\mathbb{R}^n} |f|\,|g|^p\,d\mu = \infty$.

63. Discuss the validity of the following statements: (a) Simple functions are dense in $L^\infty(I)$. (b) $C([0,1])$ is dense in $L^p([0,1])$, $1 \leq p \leq \infty$.

64. Let $(X, \mathcal{M}, \mu)$ be a measure space, $\mathcal{N} = \{A \in \mathcal{M} : \mu(A) = 0\}$, and for measurable f let $A(f) = \{\lambda \in \mathbb{R} : \{|f| > \lambda\} \in \mathcal{N}\}$. Prove: (a) If $A(f) = \emptyset$, $\|f\|_\infty = \infty$. (b) If $A(f) \neq \emptyset$, $|f| \leq \|f\|_\infty$ μ-a.e. and $\|f\|_\infty \in A(f)$.

65. Let $(X, \mathcal{M}, \mu)$ be a measure space and $f \in L^\infty(X)$. Prove that $f \geq 0$ μ-a.e. iff $\|\lambda - f\|_\infty \leq \lambda$ for all $\lambda \geq \|f\|_\infty$.

66. Let $(X, \mathcal{M}, \mu)$ be a measure space and f a real-valued measurable function on X. The μ-essential infimum of f, or plainly the essential infimum of f, is defined as ess inf $f = \sup\{\lambda : \mu(\{f < \lambda\}) = 0\}$. Prove that if $f \neq 0$ is nonnegative, ess inf $f = 1/\text{ess sup}(1/f)$.

67. Let $(X, \mathcal{M}, \mu)$ be a measure space. Prove that the following statements are equivalent: (a) No collection of pairwise disjoint measurable subsets of X with positive measure is infinite. (b) $L^\infty(X)$ is finite dimensional. (c) $L^\infty(X)$ is separable. (d) If $f : (X, \mathcal{M}) \to (\mathbb{R}, \mathcal{B}(\mathbb{R}))$ is measurable, $f \in L^\infty(X)$.

68. Let $(X, \mathcal{M}, \mu)$ be a measure space, $\mathcal{M}_f = \{M \in \mathcal{M} : \mu(M) < \infty\}$, and $\mathcal{N}_f = \{A \in \mathcal{M} : \mu(A \cap M) = 0 \text{ for all } M \in \mathcal{M}_f\}$. We say that a measurable function f on X is locally μ-essentially bounded if $\|f\|_{\infty,loc} = \inf\{\lambda > 0 : \{|f| > \lambda\} \in \mathcal{N}_f\} < \infty$; $\|f\|_{\infty,loc}$ is called the local μ-essential supremum of f and we let $L^\infty_{loc}(X) = \{f : f \text{ is measurable and locally } \mu\text{-essentially bounded}\}$. Let $A(g)_f = \{\lambda \in \mathbb{R} : \{|g| > \lambda\} \in \mathcal{N}_f\}$. Prove: (a) If $A(g)_f = \emptyset$, then $\|g\|_{\infty,loc} = \infty$. (b) If $A(g)_f \neq \emptyset$, then $\{|g| > \|g\|_{\infty,loc}\} \in \mathcal{N}_f$, or $\|g\|_{\infty,loc} \in A(g)_f$. (c) Modulo functions that coincide except on sets in $\mathcal{N}_f$, $(L^\infty_{loc}(X), \|\cdot\|_{\infty,loc})$ is a Banach space. (d) $L^\infty(X) \subset L^\infty_{loc}(X)$ and $\|f\|_{\infty,loc} \leq \|f\|_\infty$ for all $f \in L^\infty(X)$.

69. Let $(X, \mathcal{M}, \mu)$ be a measure space and f a measurable function on X such that $fg \in L^p(X)$ whenever $g \in L^p(X)$. Discuss the validity of the following statement: $f \in L^\infty_{loc}(X)$.

70. Let $f \in L^\infty([a,b])$ be such that $\lim_{x \to a^+} f(x) = \gamma$ exists. Discuss the validity of the following statement:

$$\lim_{t \to a^+} \int_a^t \frac{f(x)}{\sqrt{(x-a)(t-x)}} \, dx$$

exists.

71. Let $\mathcal{P} = \{\text{polynomials } p : p(1) = 0\}$. Prove that $\mathcal{P}$ is dense in $L^q(I)$ for $1 \le q < \infty$.

72. Prove that $\mathcal{D} = \{e^{-|x|^2} p(x) : p \text{ is a polynomial}\}$ is dense in $L^q(\mathbb{R}^n)$, $1 \le q < \infty$.

73. Let $\{p_m\}$ be a sequence of polynomials that converge uniformly to a function f on $\mathbb{R}^n$. Prove that f is a polynomial.

74. Let $(X, \mathcal{M}, \mu)$ be a finite measure space where $X = [0,1]$. Prove: (a) $x^n \to 0$ in $L^p(X)$, $1 \le p < \infty$, iff $\mu(\{1\}) = 0$. (b) $x^n \to 0$ in $L^\infty(X)$ iff $\mu([1-\eta, 1]) = 0$ for some $\eta > 0$.

75. Let $(X, \mathcal{M}, \mu)$ be a measure space, $0 < r \le p < \infty$, and suppose that $\{A_n\}$ are measurable sets such that $\lim_n \mu(A_n) = 0$. Prove that if $f \in L^p(X)$, then $\lim_n \mu(A_n)^{r/p-1} \int_{A_n} |f|^r \, d\mu = 0$.

76. Given $f \in L^1([0,1])$ and $0 < p < \infty$, let $f_p(x) = px^{p-1} f(x^p)$. Discuss the validity of the following statement: $\lim_{p \to 1} \int_{[0,1]} |f_p(x) - f(x)| \, dx = 0$.

77. Let μ be a probability Borel measure on $\mathbb{R}$ and φ, ψ monotone functions on $\mathbb{R}$. If φ, ψ are monotone in the same sense (i.e., both decreasing or increasing), prove: (a) $\left(\int_\mathbb{R} \psi(f) \, d\mu\right) \left(\int_\mathbb{R} \varphi(f) \, d\mu\right) \le \int_\mathbb{R} \psi(f)\varphi(f) \, d\mu$. (b)

$$\int_\mathbb{R} \psi(f)\, \varphi(f) \, d\mu + \varphi\left(\int_\mathbb{R} f \, d\mu\right) \psi\left(\int_\mathbb{R} f \, d\mu\right)$$
$$\ge \psi\left(\int_\mathbb{R} f \, d\mu\right) \int_\mathbb{R} \varphi(f) \, d\mu + \varphi\left(\int_\mathbb{R} f \, d\mu\right) \int_\mathbb{R} \psi(f) \, d\mu.$$

On the other hand, if φ, ψ are monotone in the opposite sense, the opposite inequalities hold.

78. Let $(X, \mathcal{M}, \mu)$ be a measure space and f on X such that $|f| < 1$ μ-a.e. Prove that

$$\int_X \frac{1}{1-f^2} \, d\mu \le \left(\int_X \frac{1}{1-f} \, d\mu\right) \left(\int_X \frac{1}{1+f} \, d\mu\right).$$

79. Let $(X, \mathcal{M}, \mu)$ be a measure space and $1 \le p < \infty$. Suppose that $f \in L^p(X)$ satisfies $\int_X fg \, d\mu = 0$ for all $g \in \mathcal{D}$, a dense class in $L^q(X)$ where $1/p + 1/q = 1$. Prove that $f = 0$ μ-a.e.

80. Let $(X, \mathcal{M}, \mu)$ be a measure space, f a measurable function on X, and φ, ψ monotone functions on $\mathbb{R}^+$ such that $\varphi(\lambda |f|)$ and $\psi(\lambda |f|)$ are integrable for all $\lambda > 0$. Prove: (a) If φ is increasing, then $\mu(\{|f| > t\}) \le \inf_{\lambda>0} \varphi(\lambda t)^{-1} \int_{\{|f|>t\}} \varphi(\lambda |f|) \, d\mu$, $t > 0$. (b) If ψ is decreasing, then $\mu(\{|f| < t\}) \ge \sup_{\lambda>0} \psi(\lambda t)^{-1} \int_{\{|f|<t\}} \psi(\lambda |f|) \, d\mu$, $t > 0$.

81. Suppose φ is a real continuous function such that $\varphi(\int_0^1 f(x) \, dx) \le \int_0^1 \varphi(f(x)) \, dx$ for all bounded measurable functions f. Prove that φ is convex. Similarly, if $\int_0^1 \psi(f(x)) \, dx \le \psi(\int_0^1 f(x) \, dx)$, ψ is concave.

82. Let $(X, \mathcal{M}, \mu)$ be a probability measure space and φ a strictly monotone function on $\mathbb{R}$ with a strictly monotone derivative. Prove that $\int_X \varphi(f) \, d\mu = \varphi(\int_X f \, d\mu)$ iff $f = c$ μ-a.e. on X.

83. Let $(X, \mathcal{M}, \mu)$ be a measure space and $f, g \in L^p(X)$, $1 \le p < \infty$, with $\|f\|_p, \|g\|_p \le M$. Prove that $\int_X \big| |f|^p - |g|^p \big| \, d\mu \le 2p \, M^{p-1} \|f - g\|_p$.

84. Prove Hardy's inequality: If f is a nonnegative $L^p([0, \infty))$ function, $1 < p < \infty$, and

$$F(x) = \frac{1}{x} \int_0^x f(t) \, dt, \quad x > 0,$$

then $\|F\|_p \le p' \|f\|_p$ where p' is the conjugate to p. Furthermore, prove that p' is the best possible constant. What if $p = 1$? $p = \infty$?

85. Let $\{f_1, \ldots, f_n\} \subset L^p([0, \infty))$, $1 < p < \infty$, be nonnegative functions. Prove that if

$$F_k(x) = \frac{1}{x} \int_0^x f_k(t) \, dt, \quad 1 \le k \le n, x > 0,$$

then $\|(\prod_{k=1}^n F_k)^{1/n}\|_p \le (p'/n) \| \sum_{k=1}^n f_k \|_p$.

86. Prove the following variants of Hardy's inequality: (a) Let f be a nonnegative locally integrable function such that $\int_0^\infty x^{-r}(xf(x))^p \, dx < \infty$, $1 < p, r < \infty$. If $F(x) = \int_0^x f(t) \, dt$, $x > 0$, then

$$\int_0^\infty x^{-r} F(x)^p \, dx \le \left(\frac{p}{r-1}\right)^p \int_0^\infty x^{-r}(xf(x))^p \, dx.$$

(b) Let f be a nonnegative locally integrable function such that for $0 < r < 1$ and $1 < p < \infty$, $\int_0^\infty x^{-r}(xf(x))^p \, dx < \infty$. If $F(x) = \int_x^\infty f(t) \, dt$, $x > 0$, then

$$\int_0^\infty x^{-r} F(x)^p \, dx \le \left(\frac{p}{1-r}\right)^p \int_0^\infty x^{-r}(xf(x))^p \, dx.$$

87. Let $f_1, \ldots, f_n$ be nonnegative $L^p([0, \infty))$ functions, $1 < p < \infty$. Prove that if $F_k(x) = \int_0^x f_k(t) \, dt$, $1 \le k \le n, x > 0$, and $1 < r < \infty$, then $\|(\prod_{k=1}^n F_k)^{1/n}\|_{L_\mu^p} \le (p/n(r-1)) \| \sum_{k=1}^n f_k \|_{L_\mu^p}$.

88. Let $K(s,t)$ be a nonnegative integrable function defined for $s,t \geq 0$ which is homogenous of degree -1, i.e., $K(\lambda s, \lambda t) = \lambda^{-1} K(s,t)$, $\lambda > 0$. Also, suppose that $\int_0^\infty t^{-1/p} K(1,t)\, dt = \gamma < \infty$ for some $1 < p < \infty$. Let $Tf(s) = \int_0^\infty K(s,t) f(t)\, dt$. Prove that $\|T(f)\|_p \leq \gamma \|f\|_p$.

89. Verify that $K(x,y) = 1/(x+y)$ and $K_1(x,y) = 1/\max(x,y)$ satisfy the assumptions of Problem 88 and calculate the norm of the associated integral mappings.

90. Let f, g be nonnegative lower semicontinuous functions on $\mathbb{R}$ and h a function on $\mathbb{R}$ such that $f(x)^{1-\eta} g(y)^\eta \leq h((1-\eta)x + \eta y)$ for all $x, y \in \mathbb{R}$ and some $\eta \in (0,1)$. Prove that if Φ is increasing and $\Phi(t) = \int_0^t \Phi'(s)\, ds$, then

$$\left(\int_{\mathbb{R}} \Phi(f(x))\, dx \right)^{1-\eta} \left(\int_{\mathbb{R}} \Phi(g(x))\, dx \right)^{\eta} \leq \int_{\mathbb{R}} \Phi(h(x))\, dx .$$

91. Let $(X, \mathcal{M}, \mu)$ be a measure space, f a nonnegative integrable function on X, g a nonnegative measurable function on X, and $1 \leq p < \infty$. Prove that $(\int_X gf\, d\mu)^p \leq \|f\|_1^{p-1} \int_X g^p f\, d\mu$.

92. Let $(X, \mathcal{M}, \mu)$ be a measure space and $1 \leq p \leq \infty$. Discuss the validity of the following statement: Given $f \in L^p([0,1])$, there is a unique $g \in L^q([0,1])$ with $\|g\|_q = 1$ such that $\int_I f(x) g(x)\, dx = \|f\|_p$.

93. Characterize the nonnegative functions $g(x) \in L^4(I)$ that satisfy $(\int_I x g(x)\, dx)^4 = (27/125) \int_I g(x)^4\, dx$.

94. Prove that

$$I = \left(\int_0^1 \frac{x^{1/2}}{(1-x)^{1/3}}\, dx \right)^3 < \frac{8}{5} .$$

95. Let $(X, \mathcal{M}, \mu)$ be a measure space and $f, g \in L^3(X)$ such that $\|f\|_3 = \|g\|_3 = 1$. Prove: (a) $f^2 g \in L^1(X)$. (b) If $\int_X f^2 g\, d\mu = 1$, then $g = |f|$ μ-a.e.

96. Let $(X, \mathcal{M}, \mu)$ be a probability measure space and f, g measurable functions on X. Prove that if $|\mathrm{Cov}(f,g)| = \sqrt{\mathrm{Var}(f)}\sqrt{\mathrm{Var}(g)}$, there exist constants a and b such that $g = af + b$ μ-a.e.

97. Prove that $\int_0^\pi x^{-1/3} \sin^{2/3}(x)\, dx \leq 2^{1/3} \pi^{2/3}$.

98. Let $(X, \mathcal{M}, \mu)$ be a measure space, $0 < p < q < \infty$, $A \subset X$ with $0 < \mu(A) < \infty$, and f a nonnegative function on X such that

$$\left(\frac{1}{\mu(A)} \int_A f^p\, d\mu \right)^{1/p} \geq \alpha, \quad \left(\frac{1}{\mu(A)} \int_A f^q\, d\mu \right)^{1/q} \leq \beta .$$

For $0 < \eta < 1$, let $A_\eta = \{f \geq \alpha\eta\}$. Find a constant $c > 0$ depending on η, α, β, such that $\mu(A_\eta) \geq c\,\mu(A)$ holds.

99. Let $f \in L^1(\mathbb{R}) \cap L^2(\mathbb{R})$ and put $f_0(x) = xf(x)$. Prove that $\|f\|_1^2 \leq 8\|f\|_2\|f_0\|_2$.

100. Let $(X, \mathcal{M}, \mu)$ be a probability measure space and f an integrable function on X such that $\mu(\{f \neq 0\}) < 1$. Prove that $\lim_{p \to 0+} \|f\|_p = 0$.

101. Let $(X, \mathcal{M}, \mu)$ be a finite measure space and $f \in L^r(X)$ for some $0 < r \leq \infty$. Does $\lim_{p \to 0+} \int_X |f|^p \, d\mu$ exist? If so, what is it equal to?

102. Let $(X, \mathcal{M}, \mu)$ be a probability measure space, f a nonnegative function in $L^r(X)$ for some $r > 0$ such that $\ln(f) \in L^1(X)$, and F be given by $F(p) = \int_X f^p \, d\mu$, $p \in [0, r]$, $F(0) = 1$. Prove that F is right-differentiable in $[0, r)$ and compute the value of the derivative.

103. Let $(X, \mathcal{M}, \mu)$ be a probability measure space and f a simple function on X. Prove that $\lim_{p \to 0+} \|f\|_p$ exists and compute it.

104. Let $(X, \mathcal{M}, \mu)$ be a probability measure space and $f \in L^r(X)$ for some $0 < r < \infty$ and $\int_X \ln(|f|) \, d\mu > -\infty$. Prove that $\lim_{p \to 0+} \|f\|_p = \exp(\int_X \ln(|f|) \, d\mu)$.

105. Let $(X, \mathcal{M}, \mu)$ be a probability measure space and $0 < p < \infty$. Prove that $(\int_X |f|^p \, d\mu) \ln(\int_X |f|^p \, d\mu) \leq \int_X |f|^p \ln(|f|^p) \, d\mu$.

106. Let $(X, \mathcal{M}, \mu)$ be a measure space and f a nonnegative function in $L^r(X) \cap L^s(X)$ where $1 \leq r < p < s < \infty$. Prove that $(\int_X f^p \ln(f) \, d\mu)^2 \leq (\int_X f^p \, d\mu)(\int_X f^p \ln^2(f) \, d\mu)$ and deduce that $\phi(p) = \ln(\|f\|^p)$ is convex.

107. Let $a_1, \ldots, a_n$ be nonnegative reals and $p_1, \ldots, p_n > 1$ such that $\sum_{k=1}^n 1/p_k = 1$. Prove that $\prod_{k=1}^n a_k \leq \sum_{k=1}^n a_k^{p_k}/p_k$.

108. Let $(X, \mathcal{M}, \mu)$ be a measure space and f a nonnegative function on X with $\int_X f \, d\mu = 1$. Prove that for every $A \subset X$ with $\mu(A) < \infty$ we have $\int_A \ln(f) \, d\mu \leq \mu(A) \ln(1/\mu(A))$, and $\int_A f^p \, d\mu \leq \mu(A)^{1-p}$ for $0 < p < 1$.

109. Prove that if $f, g \in L^p(X)$, $0 < p < 1$,

$$\|f + g\|_p \leq 2^{(1/p)-1} \left(\|f\|_p + \|g\|_p \right).$$

110. Let $J = [0, \pi]$ and $f \in L^2(J)$. Is it possible to have simultaneously $\int_J (f(x) - \sin(x))^2 \, dx \leq 4/9$ and $\int_J (f(x) - \cos(x))^2 \, dx \leq 5/9$?

111. Let f be a nonnegative integrable function on $\mathbb{R}^n$. Prove that for all real λ,

$$\left(\int_{\mathbb{R}^n} f(x) \cos(\lambda|x|) \, dx \right)^2 + \left(\int_{\mathbb{R}^n} f(x) \sin(\lambda|x|) \, dx \right)^2 \leq \left(\int_{\mathbb{R}^n} f(x) \, dx \right)^2.$$

112. Given that $\int_0^\infty e^{-x} \sin x \, dx = 1/2$, estimate $\int_0^\infty e^{-x} (3 + 2\sin x)^\eta \, dx$ where $0 < \eta < 1$.

113. Let $(X, \mathcal{M}, \mu)$ be a finite measure space and $1/p + 1/q = 1$, $1 < p < \infty$. Prove that if $f \in L^p(X)$ has integral 0,

$$\left| \int_X fg \, d\mu \right| \leq \left(\int_X |f|^p \, d\mu \right)^{1/p} \left(\int_X |g - \tfrac{1}{\mu(X)} \int_X g \, d\mu|^q \, d\mu \right)^{1/q}.$$

114. Let $(X, \mathcal{M}, \mu)$ be a probability measure space, $f \in L^p(X)$, $1 < p < \infty$, and $\lambda \in (0, 1)$. Prove: (a) $(1 - \lambda) \int_X f \, d\mu \leq \int_X f \chi_{\{f \geq \lambda \int_X f \, d\mu\}} \, d\mu$. (b) $\mu(\{f \geq \lambda \int_X f \, d\mu\})^{p-1} \geq (1 - \lambda)^p (\int_X f \, d\mu)^p / \int_X f^p \, d\mu$.

115. Let $(X, \mathcal{M}, \mu)$ be a measure space, f a measurable function on X, and $r > s > t > 0$. Prove: (a) If $f \in L^r(X) \cap L^t(X)$, $\|f\|_s^s \leq \|f\|_r^r + \|f\|_t^t$. (b) $(\int_X |f|^s \, d\mu)^{r-t} \leq (\int_X |f|^r \, d\mu)^{s-t} (\int_X |f|^t \, d\mu)^{r-s}$.

116. Let $(X, \mathcal{M}, \mu)$ be a finite measure space and f, g nonnegative measurable functions on X such that $f(x) g(x) \geq 1$. For what values of p and q, $0 < p, q \leq \infty$, is it true that

$$\left(\frac{1}{\mu(X)} \int_X f^p \, d\mu \right)^{1/p} \left(\frac{1}{\mu(X)} \int_X g^q \, d\mu \right)^{1/q} \geq 1 \, ?$$

117. Let $(X, \mathcal{M}, \mu)$ be a finite measure space and f a measurable, finite μ-a.e. function defined on X. Prove that for any $0 < p, q < \infty$,

$$\left(\frac{1}{\mu(X)} \int_X e^{pf(x)} \, d\mu(x) \right)^{1/p} \left(\frac{1}{\mu(X)} \int_X e^{-qf(x)} \, d\mu(x) \right)^{1/q} \geq 1.$$

118. Let $(X, \mathcal{M}, \mu)$ be a measure space, $0 < p < \infty$, and suppose there is a measurable function f on X such that f and $1/f$ are in $L^p(X)$. What can one say about μ?

119. Let $(X, \mathcal{M}, \mu)$ be a measure space and $0 < p < 1$, $-\infty < q < 0$ be conjugate indices, i.e., $1/p + 1/q = 1$. Suppose that f, g are positive measurable functions on X such that f^p, g^q are integrable, and fg is integrable. Prove: (a) $(\int_X f^p \, d\mu)^{1/p} (\int_X g^q \, d\mu)^{1/q} \leq \int_X fg \, d\mu$. (b) $\left(\int_X g^q \, d\mu \right)^{1/q} = \inf_f \left[\int_X fg \, d\mu / \left(\int_X f^p \, d\mu \right)^{1/p} \right]$, and if $0 < \|fg\|_1 < \infty$ equality in (a) holds iff $A|f|^p = B|g|^q$ μ-a.e. on X for some $A, B > 0$.

120. Let $(X, \mathcal{M}, \mu)$ be a finite measure space and $1 \leq p \leq r \leq \infty$. Compute $M = \sup\{\|f\|_p / \|f\|_r : f \in L^r(X), f \neq 0\}$.

121. Let $(X, \mathcal{M}, \mu)$ be a measure space and f a function on X such that $f \in L^r(X)$ for all $r < p$. Discuss the validity of the following statement: $f \in L^p(X)$ and $\|f\|_p = \lim_{r \to p^-} \|f\|_r$.

122. Let $(X, \mathcal{M}, \mu)$ be a finite measure space. Discuss the validity of the following statement: $\bigcap_{p < \infty} L^p(X) = L^\infty(X)$.

123. Let $(X, \mathcal{M}, \mu)$ be a measure space and $f \in L^r(X)$ for some $r < \infty$. Prove that $\lim_{p \to \infty} \|f\|_p = \|f\|_\infty$. Does the conclusion hold if the assumption that f belongs to some $L^r(X)$ is dropped?

124. Let $(X, \mathcal{M}, \mu)$ be a measure space, $f \in L^1(X) \cap L^\infty(X)$ with $\mu(\{|f| > 0\}) > 0$, and for each positive integer n, put $\alpha_n = \int_X |f|^n \, d\mu$. Prove that $\lim_n \alpha_{n+1}/\alpha_n$ exists and calculate it.

125. Let $(X, \mathcal{M}, \mu)$ be a finite measure space. Prove: (a) For a nonnegative function f on X,

$$\lim_n \frac{1}{n} \ln \left(\int_X e^{nf} \, d\mu \right) = \|f\|_\infty \,.$$

(b) If h is a bounded continuous function on X,

$$\lim_n \frac{1}{n} \ln \left(\int_X e^{-nh} \, d\mu \right) = - \min_{x \in X} h(x) \,.$$

126. Let $\{a_n\}$ be a sequence of positive numbers such that $\sum_{n=1}^\infty a_n^\varepsilon < \infty$ where $0 < \varepsilon < 1$. Prove that if $\{x_n\} \subset [0, 1]$,

$$\sum_n \frac{a_n}{x - x_n}$$

converges absolutely a.e., and that the statement is not true for $\varepsilon = 1$.

127. Let $(X, \mathcal{M}, \mu)$ be a measure space, f a function on X such that $\|f\|_p \leq cp^\eta$, where c is a constant independent on p, and $0 < \eta < 1$. Prove that f is exponentially integrable and $\int_X (\exp((\gamma|f(x)|)^{1/\eta}) - 1) \, d\mu(x) \leq c_1$ where c_1 is a constant independent of f and $\gamma = \eta^{1/\eta}/6c$.

128. Let $(X, \mathcal{M}, \mu)$ be a measure space, $\{f_n\}$ nonnegative measurable functions on X, and $1 < p < \infty$. Prove that

$$\left(\sum_n \left(\int_X f_n \, d\mu \right)^p \right)^{1/p} \leq \int_X \left(\sum_n f_n^p \right)^{1/p} d\mu \leq \sum_n \int_X f_n \, d\mu \,.$$

129. Let $1 \leq p \leq \infty$. We say that a mapping $T : L^p(I) \to L^p(I)$ is a contraction if $\|T(f)\|_p \leq c\|f\|_p$ for some constant $0 < c < 1$. Let T be given by $Tf(t) = \int_0^t f(s) \, ds$. Prove: (a) If $1 < p < \infty$, T is a contraction on $L^p(I)$. (b) T^2 is a contraction on $L^\infty(I)$ but T is not. (c) T is not a contraction on $L^1(I)$ but $T^n(f) \to 0$ in $L^1(I)$ for all $f \in L^1(I)$.

130. Discuss the validity of the following statement: The linear space of sequences with finitely many nonzero terms is dense in ℓ^p for $1 \leq p \leq \infty$.

131. Let $0 < p < q \leq \infty$. Prove: (a) $\ell^p \subset \ell^q$, $0 < p < q \leq \infty$. (b) $\bigcup_{p<q} \ell^p \subsetneq \ell^q$. (c) For $x \in \bigcup_{p<\infty} \ell^p$, $\lim_{p \to \infty} \|x\|_p = \|x\|_\infty$.

132. Let $1 \leq p, q < \infty$, $q \neq p$. Prove that $(\ell^p, \| \cdot \|_q)$ is not complete.

133. Let $\varphi : \mathbb{N} \to \mathbb{N}$ be a bijection. Prove that $\sum_n \varphi(n)/n^2 = \infty$.

134. Let $A = \{x \in \ell^2 : \sum_n |\sin(x_n)| < \infty\}$. Prove that A is a dense subspace of ℓ^2 and characterize it.

135. Let $M = \{x \in \ell^p : \sum_n n^p |x_n|^p \leq 1\}$, $1 \leq p < \infty$. Prove that M is convex.

136. Let $x = \{x_n\}$ be a sequence such that $\sum_n n|x_n|^p < \infty$. Prove that $x \in \ell^r$ for $0 < r < p < 2r$.

137. Discuss the validity of the following statements: (a) $A = \{x \in \ell^2 : |x_n| \leq 1/n$ for all $n\}$ is closed in ℓ^2. (b) $B = \{x \in \ell^2 : |x_n| < 1/n$ for all $n\}$ is open in ℓ^2.

138. Characterize those positive sequences $\{\lambda_n\}$ such that $A = \{x \in \ell^2 : |x_n| < \lambda_n$ for all $n\}$ is open in ℓ^2.

139. Let λ be a nonnegative sequence and for $1 \leq p < \infty$, let $C_\lambda = \{x \in \ell^p : |x_n| \leq \lambda_n$ for all $n\}$. Prove that C_λ is compact iff $\{\lambda_n\} \in \ell^p$.

140. Given a nonnegative real sequence $\lambda \notin \ell^p$, $1 \leq p < \infty$, let $C_\lambda = \{x \in \ell^p : |x_k| \leq \lambda_k$ for all $k\}$ and $M_\lambda = B \cap C_\lambda$ where B is the closed unit ball in ℓ^p. Discuss the validity of the following statement: M_λ is compact.

141. Let $\{b_n\}$ be an unbounded sequence of positive numbers. Show that there is a positive sequence $\{a_n\}$ such that $\sum_n a_n < \infty$ but $\sum_n a_n b_n = \infty$. Along similar lines, let $\{c_n\} \notin \ell^1$. Show there is a positive sequence $\{d_n\}$ with $\lim_n d_n = 0$ such that $\sum_n c_n d_n = \infty$.

142. Let $a = \{a_n\}$ be a real sequence such that $\sum_n a_n b_n$ converges for every $b = \{b_n\} \in c_0$. Prove that $a \in \ell^1$.

143. Let $a = \{a_n\}$ be a real sequence such that $\sum_n a_n b_n$ converges for every $b = \{b_n\} \in \ell^p$. Prove that $a \in \ell^q$, where $1/p + 1/q = 1$.

144. Prove that for summable sequences $\{a_n\}$, $\{b_n\}$ of nonnegative reals, $(\sum_n a_n) \ln((\sum_n a_n)/(\sum_n b_n)) \leq \sum_n a_n \ln(a_n/b_n)$.

145. Let $\{a_n\}$, $\{b_n\}$ be sequences of positive real numbers such that $0 < a_n < 1$, all n, and $\sum_n b_n = 1$. Prove that $\sum_n (\ln(a_n)) b_n \leq \ln(\sum_n a_n b_n)$ and $\lim_k \prod_{n=1}^k a_n^{b_n} \leq \sum_n a_n b_n$.

146. Construct sequences $\{x^n\}$ such that, simultaneously, $\lim_n \|x^n\|_1 = \infty$, $\lim_n \|x^n\|_2 = 1$, and $\lim_n \|x^n\|_\infty = 0$.

147. Let $A_p = \{x \in \ell^p : \sum_n x_n = 0\}$. For what values of p is A_p closed in ℓ^p, $1 \leq p \leq \infty$?

148. Prove that ℓ_0^p, $1 \leq p < \infty$, is of first category in itself.

149. Prove that for $1 < p \leq \infty$, $L^p(I)$ is of first category in $L^1(I)$.

150. Let $(X, \mathcal{M}, \mu)$ be a finite measure space and $Y \subset L^1(X)$ a closed subspace of $L^1(X)$ with the property that if $f \in Y$, then $f \in L^p(X)$ for some $p > 1$. Prove that there is a fixed $p > 1$ so that $Y \subset L^p(X)$.

151. Let $f \in L^1(\mathbb{R}^n) \cap L^\infty(\mathbb{R}^n)$ satisfy $\|f\|_\infty < \pi/2$. Prove that $\lim_k \int_{\mathbb{R}^n} \sin^k(f(x))\, dx = 0$. What if $f \in L^p(\mathbb{R}^n)$ instead, $1 < p < \infty$?

152. Show that $L^1(\mathbb{R}) \cap L^2(\mathbb{R})$ is not closed in $(L^1(\mathbb{R}), \| \cdot \|_1)$ or in $(L^2(\mathbb{R}), \| \cdot \|_2)$.

153. Let $(X, \mathcal{M}, \mu)$ be a finite measure space and $1 < p < \infty$. Explain the following situation: $L^p(X)$ is separable but the subspace $L^p(X) \cap L^\infty(X)$, which is dense in both $L^p(X)$ and $L^\infty(X)$, is not separable.

154. Let $(X, \mathcal{M}, \mu)$ be a measure space and $1 \leq p < r \leq \infty$. Prove that $\|f\| = \|f\|_p + \|f\|_r$ is a norm in $L^p(X) \cap L^r(X)$ and that $(L^p(X) \cap L^r(X), \| \cdot \|)$ is a Banach space. Furthermore, if $p < q < r$, the inclusion map $L^p(X) \cap L^r(X) \hookrightarrow L^q(X)$ is continuous.

155. Let O be an open set of finite measure in $\mathbb{R}^N$ and $1 \leq p < \infty$. We denote $L^p_c(O) = \{f \in L^p(O) : \operatorname{supp}(f) \subset O \text{ is compact}\}$ and $L^p_{loc}(O) = \{f : O \to \mathbb{R} : f \text{ is measurable and } f\chi_K \in L^p(O) \text{ for all compact } K \subset O\}$. (a) Equip $L^p_{loc}(O)$ with a complete metric d such that for each sequence $\{f_n\} \subset L^p_{loc}(O)$ and $f \in L^p_{loc}(O)$, $\lim_n d(f_n, f) = 0$ iff $\lim_n \|(f_n - f)\chi_K\|_p = 0$ for all compact $K \subset O$. (b) Prove that $L^p_c(O)$ is dense in $(L^p_{loc}(O), d)$.

156. Let O be an open set in $\mathbb{R}^N$ of finite measure, $1 \leq p < \infty$, and q the conjugate exponent of p. (a) Given $g \in L^q_c(O)$, let L_g denote the linear functional on $L^p_{loc}(O)$ given by $L_g(f) = \int_O f(x)g(x)\, dx$. Prove that L_g is a well-defined continuous linear functional on $L^p_{loc}(O)$ and that the mapping $\Phi : L^q_c(O) \to L^p_{loc}(O)^*$ given by $\Phi(g) = L_g$ is a linear isomorphism ($L^p_{loc}(X)$ endowed with the distance d, or the norm q).

(b) We say that a linear functional ℓ on $L^p_c(O)$ is continuous if for every compact $K \subset O$, the restriction of ℓ to the space $\{f \in L^p(O) : \operatorname{supp}(f) \subset K\}$, endowed with the norm of $\| \cdot \|_p$, is continuous. Denote by $L^p_c(O)^*$ the collection of continuous linear functionals on $L^p_c(O)$. If $g \in L^q_{loc}(O)$, let ℓ_g denote the linear functional on $L^p_c(O)$ given by $\ell_g(f) = \int_O f(x)g(x)\, dx$. Prove that the mapping $\Psi(g) = \ell_g$ establishes a linear isomorphism from $L^q_{loc}(O)$ onto $L^p_c(O)^*$.

Chapter 6

Sequences of Functions

This chapter is devoted to some of the various ways sequences of functions converge to a limit. The general setting is that of a real-valued sequence $\{f_n\}$ defined on a measure space $(X, \mathcal{M}, \mu)$; when the functions are complex-valued the results often follow by considering the real and imaginary parts separately.

First, there is pointwise convergence. We say that $\{f_n\}$ *converges pointwise to f at $x \in X$* if $\lim_n f_n(x) = f(x)$; here the term convergence means to a finite limit. Of course, there is also the familiar notion of uniform convergence.

Now, the notion of μ-a.e. convergence is relevant in the presence of a measure. We say $\{f_n\}$ *converges to f μ-a.e. on $A \subset X$* if $\mu(\{x \in A : \lim_n f_n(x) \neq f(x)\}) = 0$. Egorov's theorem states that, if $\{f_n\}, f$ are measurable functions on a finite measure space $(X, \mathcal{M}, \mu)$ and $f_n \to f$ pointwise μ-a.e. in X, then, for every $\varepsilon > 0$, there exists a measurable $B \subset X$ such that $\mu(B) < \varepsilon$ and f_n converges uniformly to f on $X \setminus B$.

Another way to incorporate the measure is convergence in measure. We say that $\{f_n\}$ *converges to f in measure* if

$$\lim_n \mu(\{x \in X : |f_n(x) - f(x)| > \varepsilon\}) = 0 \quad \text{for each } \varepsilon > 0.$$

When $\mu(X) = 1$ we say that the convergence is in the sense of probability. Finally, we say that $\{f_n\}$ *converges to f completely* if $\sum_n \mu(\{|f_n - f| > \eta\}) < \infty$ for all $\eta > 0$.

The monotone convergence theorem (MCT), Fatou's lemma, and the Lebesgue dominated convergence theorem (LDCT) deal with conditions that allow for the limit to be taken under the integral sign, including norm convergence. We say that $\{f_n\}$ *converges to* f *in* $L^p(X)$, $0 < p \le \infty$, if $\lim_n \|f_n - f\|_p = 0$. And, we say that $\{f_n\}$ *converges weakly to* f *in* $L^p(X)$, $1 \le p < \infty$, and write $f_n \rightharpoonup f$ in $L^p(X)$, if $\lim_n \int_X f_n g \, d\mu = \int_X f g \, d\mu$ for all $g \in L^q(X)$, where q is the conjugate Hölder index of p given by the relation $1/p + 1/q = 1$.

The problems in this chapter include the question of whether measurability is preserved under the operations of sup and inf and the taking of pointwise limits, addressed in Problems 1–5. The behavior of the various types of convergence with respect to usual concepts is covered in the next few problems; for instance, convergence in measure is highlighted in Problems 27–34 and Problem 36. As for μ-a.e. convergence, the question as to whether it is equivalent to a metric convergence is considered in Problem 35.

The version of Fatou's lemma for convergence in measure is discussed in Problem 38 and that of the LDCT in Problem 42. And, the fact that the existence of a majorant is not needed for the conclusion of the LDCT to hold is covered in Problems 53–54. As for convergence in L^p, deep and useful results are covered in Problems 84–85.

The concept of a *uniformly absolutely continuous* family of functions is introduced in Problem 90, and its central role in the theory of L^1, as well as L^p, convergence is explored in Problems 91–101.

The role of uniform convergence as it relates to L^p convergence is explored in Problems 117–119. Weak convergence is discussed in Problems 145–156.

The interested reader can further consult his or her favorite real variables textbook or the classics, including I. P. Natanson, *Theory of functions of a real variable*, Ungar Publishing Co, 1955; and E. Hewitt and K. Stromberg, *Real and abstract analysis*, Springer-Verlag, New York, 1965.

Problems

1. Discuss the validity of the following statement: If $\mathcal{M}$ is a σ-algebra of subsets of X, $\Lambda = [0, 1]$, and $\{f_\alpha\}_{\alpha \in \Lambda}$ measurable functions on X, then $\sup_{\alpha \in \Lambda} f_\alpha(x)$ is measurable.

2. Let $(X, \mathcal{M}, \mu)$ be a measure space and $\{f_\alpha\}_{\alpha \in \Lambda}$, $\{g_\alpha\}_{\alpha \in \Lambda}$ measurable functions on X. Consider the following statement: If $f_\alpha \le g_\beta$ μ-a.e. for $\alpha, \beta \in \Lambda$, then $\sup_{\alpha \in \Lambda} f_\alpha \le \inf_{\beta \in \Lambda} g_\beta$ μ-a.e. (a) Show that the statement is not necessarily true for arbitrary Λ. (b) What condition on Λ ensures that the statement is true?

3. Let $\{f_\alpha\}_{\alpha \in \Lambda}$ be measurable functions on $\mathbb{R}$ and $A = \{x \in \mathbb{R} : \lim_{\alpha \to \beta} f_\alpha(x) \text{ exists}\}$. (a) Show that A need not be measurable. (b) What condition on $\{f_\alpha\}$ ensures the measurability of A? (c) What condition on Λ ensures the measurability of A?

4. Let $\mathcal{M}$ be a σ-algebra of subsets of X, $\{f_n\}$ measurable functions on X, and $f = \liminf_n f_n$. Discuss the validity of the following statement: If $A_n = \{f_n > 0\} \in \mathcal{M}$ for all n, then $A = \{f > 0\} \in \mathcal{M}$.

5. Let $(X, \mathcal{M}, \mu)$ be a measure space and $\{f_n\}$ measurable functions on X such that $f_n \to f$ μ-a.e. Is f measurable?

6. Let $(X, \mathcal{M}, \mu)$ be a measure space and $\{f_n\}$, f measurable functions on X. Prove: (a) $B = \{x \in X : \lim_n f_n(x) \neq f(x)\} = \bigcup_k \bigcap_m \{x \in X : \sup_{n \geq m} |f_n(x) - f(x)| \geq 1/k\}$. (b) If $f_n \to f$ μ-a.e., then

$$\mu\left(\bigcap_m \{\sup_{n \geq m} |f_n - f| \geq \varepsilon\}\right) = 0 \quad \text{for every } \varepsilon > 0.$$

7. Let $(X, \mathcal{M}, \mu)$ be a probability measure space and $\{f_n\}$ measurable functions on X. Prove that the following statements are equivalent: (a) There is a subsequence $\{f_{n_k}\}$ of $\{f_n\}$ that converges to 0 μ-a.e. (b) There is a sequence $\{c_n\}$ with $\limsup_n |c_n| > 0$ such that $\sum_n c_n f_n(x)$ converges μ-a.e. (c) There is a sequence $\{c_n\}$ with $\sum_n |c_n| = \infty$ such that $\sum_n c_n f_n(x)$ converges absolutely μ-a.e.

8. Let $\mathcal{M}$ be a σ-algebra of subsets of X and $\{f_n\}$ measurable functions on X. Prove that $A = \{x \in X : f_n(x) = 1 \text{ for exactly one } n\}$ is measurable.

9. Let $(X, \mathcal{M}, \mu)$ be a finite measure space and $\{f_n\}$ measurable functions on X such that $|f_n(x)| \leq M(x)$ for each $x \in X$ and all n, where $M(x)$ is finite everywhere. Discuss under what conditions on the f_n and $M(x)$ the following holds: Given $\varepsilon > 0$, there exist a measurable set A and a constant M such that $\mu(X \setminus A) < \varepsilon$ and $|f_n(x)| \leq M$ for all n and $x \in A$.

10. Let $(X, \mathcal{M}, \mu)$ be a probability measure space and $\{f_n\}$ measurable functions on X. Discuss the validity of the following statement: $f_n \to f$ μ-a.e. where f is finite μ-a.e. iff given $\varepsilon > 0$, there is a finite constant M_ε such that $\mu(\{\sup_n |f_n| \leq M_\varepsilon\}) \geq 1 - \varepsilon$.

11. Let $\{f_n\}$ be the measurable functions on $I = [0, 1]$ given by $f_n(x) = n$-th digit in the binary expansion of $x \in I$; since there are countably many binary rationals the choice at those points is irrelevant. Suppose $\{g_n\}$ are Lebesgue measurable functions such that $\sup_n |\{x \in I : g_n(x) \neq f_n(x)\}| =$

$\eta < 1/4$. Prove that no subsequence $\{g_{n_k}\}$ of $\{g_n\}$ converges in a subset of I of measure 1.

12. (Borel-Cantelli) Let $(X, \mathcal{M}, \mu)$ be a measure space and $\{A_n\} \subset \mathcal{M}$. Prove that if $\sum_n \mu(A_n) < \infty$, then $\mu(\limsup_n A_n) = 0$.

13. Let $(X, \mathcal{M}, \mu)$ be a probability measure space and $\{f_n\}$ measurable functions on X such that $\mu(\{|f_n| \leq 2^{-n}\}) > 1 - 3^{-n}$. Does it follow that $\sum_n |f_n(x)|$ converges μ-a.e.?

14. Let $(X, \mathcal{M}, \mu)$ be a probability measure space, $\{f_n\}$ measurable functions on X, and $\{\lambda_n\}$ such that $\sum_n \lambda_n$ and $\sum_n \mu(\{f_n \neq \lambda_n\})$ converge. Prove that $\sum_n f_n$ converges μ-a.e.

15. Let $(X, \mathcal{M}, \mu)$ be a probability measure space, $\{f_n\}$ integrable functions on X with $\int_X |f_n| \, d\mu \leq c < \infty$, and for a positive sequence $\{\lambda_n\}$ such that $\sum_n \lambda_n^{-1} < \infty$, let $A_n = \{|f_n - \int_X f_n \, d\mu| \geq \lambda_n\}$. Prove that $\mu(\limsup_n A_n) = 0$.

16. Let $(X, \mathcal{M}, \mu)$ be a probability measure space and $\{f_n\}$ nonnegative independent functions on X. Prove that $\sup_n f_n(x) < \infty$ μ-a.e. iff $\sum_n \mu(\{f_n > M\}) < \infty$ for some $M < \infty$.

17. Let $(X, \mathcal{M}, \mu)$ be a probability measure space and $\{f_n\}$ independent functions on X such that $\mu(\{f_n > \lambda\}) = \lambda^{-5}$ for all $\lambda > 1$ and all n. Prove that $\limsup_n \ln(f_n(x))/\ln(n) = c$ μ-a.e. for some number c, and find c.

18. Let $(X, \mathcal{M}, \mu)$ be a probability measure space and for independent measurable functions $\{f_n\}$ on X let $\mu(\{f_n = 1\}) = p_n$ and $\mu(\{f_n = 0\}) = 1 - p_n$. Prove: (a) $f_n \to 0$ in probability iff $p_n \to 0$. (b) $f_n \to 0$ μ-a.e. iff $\sum_n p_n < \infty$.

19. Let $(X, \mathcal{M}, \mu)$ be a probability measure space and $\{f_n\}$ and $\{g_n\}$ measurable functions on X such that $f_n \to f$ μ-a.e. and $\sum_n \mu(\{f_n \neq g_n\}) < \infty$. Prove that $g_n \to f$ μ-a.e.

20. Let $(X, \mathcal{M}, \mu)$ be a finite measure space and $\{f_n\}$ finite μ-a.e. measurable functions on X. Prove that there is a positive sequence $\{\lambda_n\}$ such that $\lim_n \lambda_n f_n(x) = 0$ μ-a.e.

21. Let $(X, \mathcal{M}, \mu)$ be a finite measure space and $\{f_n\}$ nonnegative measurable functions on X such that $\lim_n f_n = 0$ μ-a.e. Prove that there exists a nondecreasing integer sequence λ with $\lambda_n \to \infty$ such that $\lim_n \lambda_n f_n(x) = 0$ μ-a.e.

22. Let $(X, \mathcal{M}, \mu)$ be a measure space and $\{f_n\}$ measurable functions on X such that $f_n \to 0$ μ-a.e. Discuss the validity of the following statement: $f_n \to 0$ in measure.

23. Let $(X, \mathcal{M}, \mu)$ be a measure space, $\{\lambda_n\}$ a positive sequence, and $\{f_n\}$ measurable functions on X such that $\sum_n \mu(\{|f_n|/\lambda_n > 1\}) < \infty$. Prove: (a) $-1 \le \liminf_n (f_n(x)/\lambda_n) \le \limsup_n (f_n(x)/\lambda_n) \le 1$ μ-a.e. (b) If $\{\lambda_n\}$ tends to 0, then $\lim_n f_n = 0$ μ-a.e. and $\lim_n f_n = 0$ in measure.

24. Let $(X, \mathcal{M}, \mu)$ be a measure space, $\{f_n\}$ measurable functions on X, and $\sum_n \lambda_n$ a convergent series with $\lambda_n > 0$ for all n. Prove that if $\sum_{n=1}^{\infty} \mu(\{|f_{n+1} - f_n| \ge \lambda_n\}) < \infty$, there exists f on X such that $f_n \to f$ μ-a.e.

25. Let $(X, \mathcal{M}, \mu)$ be a measure space and $\{f_n\}, f$ measurable functions on X. Prove that if $\lim_n f_n = f$ completely, then $f_n \to f$ μ-a.e.

26. Let $(X, \mathcal{M}, \mu)$ be a probability measure space and $\{f_n\}, f$ measurable functions on X. Prove that if $f_n \to f$ μ-a.e., a subsequence $\{f_{n_k}\}$ of $\{f_n\}$ converges to f completely.

27. Let $(X, \mathcal{M}, \mu)$ be a measure space and $\{f_n\}, f$ measurable functions on X such that $f_n \to f$ in measure. (a) Prove that a subsequence $\{f_{n_k}\}$ of $\{f_n\}$ converges to f μ-a.e. (b) Must $\lim_n \int_X |f_n - f|\, d\mu = 0$? (c) Assume that $\{f_n\}$ is an increasing sequence that converges to f in measure. Prove that $\lim_n \int_X f_n\, d\mu = \int_X f\, d\mu$. (d) Prove that if a sequence $\{f_n\}$ of nonnegative measurable functions converges to f in measure, then $f \ge 0$ μ-a.e.

28. Let $(X, \mathcal{M}, \mu)$ be a measure space and $\{f_n\}$ a monotone sequence of measurable functions on X that converges to f in measure. Prove that $f_n \to f$ μ-a.e.

29. Let $\{f_n\}$ be a sequence of nondecreasing functions on $[0, 1]$ that converges to f in measure. Prove that if x is a point of continuity of f, then $\lim_n f_n(x) = f(x)$.

30. Let $(X, \mathcal{M}, \mu)$ be a finite measure space and $\{f_n\}$ a sequence of measurable functions on X that is Cauchy in measure. Prove that $\{f_n\}$ converges to a measurable function f in measure.

31. Let $(X, \mathcal{M}, \mu)$ be a finite measure space and $\{f_n\}$ measurable functions on X. Prove: (a) If $f_n \to f$ in measure and $f_n \to g$ μ-a.e., then $f = g$ μ-a.e. (b) If $\{f_n\}$ converges to f and g in measure, then $f = g$ μ-a.e.

32. Let $(X, \mathcal{M}, \mu)$ be a finite measure space and $\{f_n\}$ measurable functions on X. Prove that $f_n \to f$ in measure iff every subsequence $\{f_{n_k}\}$ of $\{f_n\}$ in turn has a subsequence $\{f_{n_{k_\ell}}\}$ that converges to f μ-a.e.

33. Let $(X, \mathcal{M}, \mu)$ be a finite measure space and $\{f_n\}, \{g_n\}$ measurable functions on X that converge in measure to f and g, respectively. Prove that if $\psi : \mathbb{R} \times \mathbb{R} \to \mathbb{R}$ is continuous, $\psi(f_n, g_n)$ converges to $\psi(f, g)$ in measure. Also, show that the finiteness of $\mu(X)$ and the continuity of ψ are necessary.

34. Let $(X, \mathcal{M}, \mu)$ be a probability measure space and for measurable functions $\{f_n\}, f$ on X, let $F_n(t) = \mu(\{f_n \leq t\})$ and $F(t) = \mu(\{f \leq t\})$, $t \in \mathbb{R}$. Prove that if $f_n \to f$ in measure, then $\lim_n F_n(t) = F(t)$ at each point of continuity t of F.

35. Let $(X, \mathcal{M}, \mu)$ be a finite measure space with no atoms and $\mathcal{F}$ the class of measurable functions on X where functions that agree μ-a.e. are identified. Discuss the validity of the following statement: There exists a metric $\rho : \mathcal{F} \times \mathcal{F} \to \mathbb{R}^+$ such that convergence in the metric ρ is equivalent to μ-a.e. convergence.

36. Let $(X, \mathcal{M}, \mu)$ be a finite measure space and $\mathcal{F}$ the class of measurable functions on X where functions that agree μ-a.e. are identified. (a) Prove that $d, d_1 : \mathcal{F} \times \mathcal{F} \to [0, \mu(X)]$ given by

$$d(f,g) = \int_X \frac{|f - g|}{1 + |f - g|}\, d\mu\,, \quad d_1(f,g) = \int_X (|f - g| \wedge 1)\, d\mu\,, \quad f, g \in \mathcal{F}\,,$$

respectively, are metrics on $\mathcal{F}$. (b) Discuss the validity of the following statement: $(C(I), d)$ is a complete metric space. (c) Prove that $\lim_n d(f_n, f) = 0$ iff $\lim_n d_1(f_n, f) = 0$ iff $f_n \to f$ in measure.

37. Let $(X, \mathcal{M}, \mu)$ be a probability measure space and φ a nonnegative continuous function on X with $\varphi(0) = 0$. We say that $\{f_n\}$ converges to 0 in φ if $\lim_n \int_X \varphi(f_n)\, d\mu = 0$. Prove that the following statements are equivalent: (a) There exists φ such that $\{f_n\}$ converges to 0 in φ iff $f_n \to 0$ μ-a.e. (b) If $\lim_n f_n = 0$ in probability, then $\lim_n f_n = 0$ μ-a.e. (c) X is a countable union of atoms.

38. Let $(X, \mathcal{M}, \mu)$ be a measure space and $\{f_n\}$ nonnegative measurable functions on X such that $\lim_n f_n = f$ in measure. Prove that $\int_X f\, d\mu \leq \liminf_n \int_X f_n\, d\mu$.

39. Let $(X, \mathcal{M}, \mu)$ be a measure space and $\{f_n\}, f$ measurable functions on X such that $f_n \to f$ in measure and $\varphi = \sup_n |f_n| \in L^1(X)$. Prove that $\lim_n \int_X |f_n - f|\, d\mu = 0$ and show that the conclusion may not hold if φ is not integrable.

40. Let $(X, \mathcal{M}, \mu)$ be a measure space and $\{f_n\}, f$ nonnegative measurable functions on X such that $f_n \leq f$ μ-a.e., all n, and $f_n \to f$ μ-a.e. Prove that $\int_X f\, d\mu = \lim_n \int_X f_n\, d\mu$.

41. Let $(X, \mathcal{M}, \mu)$ be a measure space and $\{f_n\}, f$ nonnegative integrable functions on X such that $f_n \to f$ μ-a.e. and $\int_X f_n\, d\mu \to \int_X f\, d\mu$. (a) Prove that for every measurable $A \subset X$, $\lim_n \int_A f_n\, d\mu = \int_A f\, d\mu$. (b) What if $f_n \to f$ in measure instead? (c) What if the functions are allowed to have variable sign? If $f \notin L^1(X)$?

42. Let $(X, \mathcal{M}, \mu)$ be a measure space and $\{f_n\}, g$ integrable functions on X such that $|f_n| \leq g$ μ-a.e. and $f_n \to f$ in measure. Prove that $\lim_n \int_X f_n \, d\mu = \int_X f \, d\mu$.

43. Let $(X, \mathcal{M}, \mu)$ be a measure space and $\{f_n\}, \{g_n\}$ measurable functions on X such that g_n, g are integrable, $g_n \to g$ in $L^1(X)$, $f_n \to f$ in measure, and $|f_n| \leq g_n$ μ-a.e., all n. Prove that $\lim_n \int_X |f_n - f| \, d\mu = 0$.

44. Let $(X, \mathcal{M}, \mu)$ be a measure space, $\{f_n\}$, f measurable functions on X, and $g \in L^1(X)$ such that $f_n \to f$ μ-a.e. and $|f_n| \leq g$ μ-a.e. for all n. Prove that given $\varepsilon > 0$, $f_n \to f$ uniformly on $X \setminus B$ where $B \in \mathcal{M}$ has $\mu(B) < \varepsilon$.

45. Let $0 < \eta \leq 1$. Prove that, for no bounded sequences $\{a_n\}$ and $\{b_n\}$, $f_n(x) = a_n \sin(2\pi n x) + b_n \cos(2\pi n x)$ tends to η a.e. in $[0, 1]$.

46. Let $A \subset [0, 2\pi]$ have positive measure and $\{r_n\}$ a real sequence. Find $\lim_n \int_A \cos^2(nx + r_n) \, dx$ and $\lim_n \int_A \sin^2(nx + r_n) \, dx$.

47. Let $(X, \mathcal{M}, \mu)$ be a measure space, $\{f_n\}$ measurable functions on X with $\|f_n\|_\infty \leq c$, all n, and $\lim_n \int_A f_n \, d\mu = 0$ for all A with $\mu(A) < \infty$. Prove: (a) If a subsequence $\{f_{n_k}\}$ of $\{f_n\}$ converges for x in A with $0 < \mu(A) < \infty$, it converges necessarily to 0. (b) If $f_n(x) = \sin(nx)$, $n = 1, 2, \ldots$, no subsequence $\{f_{n_k}\}$ of $\{f_n\}$ converges for all x in a set $A \subset \mathbb{R}$ with $|A| > 0$.

48. Let $(X, \mathcal{M}, \mu)$ be a measure space, $\{A_n\} \subset \mathcal{M}$ such that $\mu(A_n) \geq \eta > 0$ for all n, and $\{\lambda_n\}$ nonnegative reals such that $\varphi(x) = \sum_n \lambda_n \chi_{A_n}(x) < \infty$ μ-a.e. Discuss the validity of the following statement: $\sum_n \lambda_n < \infty$.

49. Let $(X, \mathcal{M}, \mu)$ be a measure space and $\{f_n\}$, f integrable functions on X. Consider the following statements: (a) $f_n \to f$ μ-a.e. and $\int_X f_n \, d\mu \to \int_X f \, d\mu$. (b) $\int_X |f_n - f| \, d\mu \to 0$. Which statement implies the other?

50. Let $(X, \mathcal{M}, \mu)$ be a measure space and $f \in L^1(X)$. Discuss under what conditions on a sequence $\{A_n\}$ of measurable subsets of X does it follow that $\lim_n \int_{A_n} f \, d\mu = 0$.

51. Let $\varphi(x)$ be a positive function on $[0, \infty)$ such that, together with x, $\varphi(x)$ decreases to 0 as $x \to \infty$, and $f_n = \lambda_n \chi_{A_n}$ where $\{\lambda_n\}$ is a nonnegative sequence and $A_n \subset [0, \infty)$ are measurable sets such that $0 \leq f_n \leq \varphi$ and $f_n \to 0$ a.e. Prove that $\lim_n \int_0^\infty f_n(x) \, dx = 0$.

52. Let $(X, \mathcal{M}, \mu)$ be a measure space, $\{f_n\}$ integrable functions on X, and f a measurable function on X such that $\lim_n \int_X |f_n - f| \, d\mu = 0$. Prove that $f \in L^1(X)$ and if $\mu(A_n \triangle A) \to 0$, then $\int_{A_n} f_n \, d\mu \to \int_A f \, d\mu$.

53. Let $\{f_n\}$ be given by $f_n(x) = \chi_{[n, n+1/n]}(x)$, $n \geq 1$. Prove: (a) $f_n \to 0$ a.e. and in Lebesgue measure. (b) $\lim_n \int_{[0, \infty)} f_n(x) \, dx = 0$. (c) If

$\varphi_m(x) = \sup_{n \geq m} |f_n(x)|$, $\varphi_m \notin L^1([0, \infty))$. (d) Given $\varepsilon > 0$, it is not true that $\lim_m |\{\varphi_m \geq \varepsilon\}| = 0$.

54. Let $\{f_n\}$ be given on $[0, 1]$ by $f_n(x) = (n/\ln(n))\chi_{(0,1/n)}(x)$. Verify that $f_n \to 0$ a.e. and prove that $\int_I f_n(x)\, dx \to 0$ although for no $\varphi \in L(I)$ does it hold that $f_n \leq \varphi$ a.e. for all n and, consequently, the LDCT does not apply.

55. Let $a \in \mathbb{R}$. Compute

$$\lim_n n \int_a^\infty \frac{1}{(1 + n^2 x^2)}\, dx\,.$$

56. Let $f_n(x) = n^2 x (1 - x)^n$ on $[0, 1]$. Find $f(x) = \lim_n f_n(x)$ and decide whether $\lim_n \int_0^1 f_n(x)\, dx = \int_0^1 f(x)\, dx$.

57. Find conditions on a nonnegative measurable function φ and $\lambda \in \mathbb{R}$ for the following limit to exist:

$$\lim_n \int_0^n \left(1 + \frac{x}{n}\right)^n e^{-\lambda x}\, \varphi(x)\, dx\,.$$

58. For integers $n \geq 1$, consider the integrals

$$I_n = \int_0^1 \frac{1 + x^2 + \cdots + x^{2n}}{1 + x + \cdots + x^n}\, dx\,, \qquad J_n = \int_{1/2}^1 \frac{x + x^2 + \cdots + x^n}{x + x^2/2 + \cdots + x^n/n}\, dx\,.$$

Discuss the validity of the following statements: (a) $\{I_n\}$ has a finite limit as $n \to \infty$. (b) $\{J_n\}$ has a finite limit as $n \to \infty$.

59. Find

$$\lim_n \int_1^{2n} \left(1 - \frac{x}{n}\right)^n dx\,.$$

60. Let φ in $L^\infty(\mathbb{R}^+)$. Find $\lim_n \int_0^n x^n e^{-nx} \varphi(x)\, dx$.

61. Prove that $G(x) = \sum_{n=1}^\infty n^{-2} \ln(1 + n^2 x)$, $x > 0$, is continuously differentiable and compute $G'(x)$.

62. Let $F(x) = \sum_n e^{-nx} a_n$, $a_n > 0$, $x > 0$. Prove that $\sum_n a_n < \infty$ iff the right-hand derivative of F exists at the origin.

63. Let $a \geq 0$, $b > 0$, and

$$f(x) = \frac{x e^{-ax}}{1 - e^{-bx}}, \qquad x \in \mathbb{R}^+\,.$$

Compute $\int_{\mathbb{R}^+} f(x)\, dx$.

64. By differentiating the relation $\int_0^\infty e^{-tx}\, dx = 1/t$ for $t > 0$, prove that $\int_0^\infty x^n\, e^{-x}\, dx = n!$, for all $n \in \mathbb{N}$.

65. Prove that $\sum_{n=0}^\infty \int_0^{\pi/2} (1 - \sqrt{\sin(x)})^n \cos(x)\, dx$ is finite and find its value.

66. Compute $\int_0^1 x^{-x}\,dx$.

67. Find sequences $\{A_n\}, \{B_n\} \subset \mathbb{R}$ such that $\lim_n A_n = \lim_n B_n = \mathbb{R}$,

$$\lim_N \int_{A_N} \frac{x}{1+x^2}\,dx \neq \lim_N \int_{B_N} \frac{x}{1+x^2}\,dx,$$

and both of the limits are finite.

68. Let $p(x) = \sum_{j=1}^N a_j x^j$. Find $\lim_k \int_0^k p(x) x^n (1 - x/k)^k\,dx$.

69. Find

$$\lim_k \int_0^{k^2} \frac{2x}{k^2}\left(1 - \frac{x}{k^2}\right)^k dx.$$

70. Prove that if $0 < a < 1$ and $b > 0$,

$$\lim_{r \to \infty} \int_0^r \frac{x^{a-1}}{1+x}\left(1 - \frac{x}{r}\right)^{br} e^{bx}\,dx = \frac{\pi}{\sin(\pi a)}.$$

71. Prove that

$$\lim_n \int_0^\infty \frac{x}{x^4 + x^2 + 1}\, n \sin(x/n)\,dx = \frac{\pi}{2\sqrt{3}}.$$

72. In $\mathbb{R}^n$, let $|f(x)| \leq (1 + |x|)^\alpha$, $\alpha < 1$. Find

$$\lim_{\lambda \to \infty} \int_{\mathbb{R}^n} f(x) \frac{1}{|x|(1 + |x|)^{n+1}} \lambda \sin(|x|/\lambda)\,dx.$$

73. Calculate

$$\lim_n \int_0^\infty \frac{1}{1 + (x/2\pi)^n + \frac{1}{4n}\sin(x^{-1}e^x)} \frac{1}{\cos(x) - 2}\,dx.$$

74. Let f be a positive, monotone, integrable function on $[0,1]$, and $g_n(x) = f(x^n)$, $n \geq 1$. Find $\lim_n \int_0^1 g_n(x)\,dx$.

75. Find

$$\lim_n \int_0^\infty \frac{\sin(x^n)}{x^n}\,dx.$$

76. Let $0 < \eta < 1$ and consider $\{f_n\}$ on I given by $f_n(x) = x^{-\eta}\sin^n(1/x)$ for $x \neq 0$ and $f_n(x) = 0$ for $x = 0$. Does $\lim_n \int_I f_n(x)\,dx$ exist?

77. Given $\varphi(x)$ so that $x^\alpha |\varphi(x)| \leq c$, $\alpha < 1/2$, find

$$\lim_n \int_0^1 e^{1/x} \frac{1}{1 + n^2 x}\sin(ne^{-1/x})\,\varphi(x)\,dx.$$

78. Let $\{r_n\}$ be a sequence of real numbers. Prove that

$$\lim_n \int_0^\infty e^{-x}\sin^n(x + r_n)\,dx = 0.$$

79. Calculate

$$\lim_n \int_I \frac{n(1-x)^2}{(1+nx)\ln^2(x)} \cos(nx)\, dx.$$

80. Let $(X, \mathcal{M}, \mu)$ be a measure space and $\{f_n\}$ functions on X such that $\sup_n \|f_n\|_2 \leq c$ and $\int_X f_n f_m\, d\mu = 0$ for $n \neq m$. Prove that given $\varepsilon > 0$ and A with $\mu(A) > 0$, $f_n(x) > \varepsilon$ for x in A for at most finitely many n.

81. Let $f \in L^1(\mathbb{R}^n)$ and $\{y_k\} \subset \mathbb{R}^n$ a bounded sequence. Prove that $\{f(\cdot + y_k)\}$ has a subsequence that converges in $L^1(\mathbb{R}^n)$ and a.e.

82. Let $(X, \mathcal{M}, \mu)$ be a measure space, $0 < p < 1$, and suppose that $f_n \to f$ in $L^p(X)$. Prove that $|f_n|^p \to |f|^p$ in $L^1(X)$.

83. Let $(X, \mathcal{M}, \mu)$ be a measure space and $\{f_n\}, f$ nonnegative integrable functions on X with integral 1 such that $f_n \to f$ μ-a.e. Compute $\lim_n \int_X (f_n^{1/2} - f^{1/2})^2\, d\mu$.

84. Let $(X, \mathcal{M}, \mu)$ be a measure space, $0 < p < \infty$, and $\{f_n\}, f \in L^p(X)$ such that $f_n \to f$ μ-a.e. Prove: (a)

$$2^p \int_X |f|^p\, d\mu \leq \liminf_n \left(2^p \int_X |f_n|^p\, d\mu - \int_X |f_n - f|^p\, d\mu\right).$$

(b) $\lim_n \|f_n - f\|_p = 0$ iff $\|f_n\|_p \to \|f\|_p$. (c) What can be said about the case $p = \infty$?

85. Let $(X, \mathcal{M}, \mu)$ be a measure space, $0 < p < \infty$, and $\{f_n\}, f$ measurable functions on X. Prove that if $\|f_n\|_p \leq c$ and $f_n \to f$ μ-a.e., then $\lim_n \int_X |\,|f_n|^p - |f - f_n|^p - |f|^p|\, d\mu = 0$.

86. Let $(X, \mathcal{M}, \mu)$ be a measure space and $\{f_n\}, f$ measurable functions on X. Discuss the validity of the following statements: (a) $f_n \to f$ in $L^\infty(X)$ iff $f_n \to f$ almost uniformly. (b) $f_n \to f$ in $L^\infty(X)$ iff there exists a measurable set B with $\mu(B) = 0$ and $\lim_n f_n(x) = f(x)$ uniformly for $x \in B^c$.

87. Let $(X, \mathcal{M}, \mu)$ be a measure space and $\{A_n\}$ measurable subsets of X. Prove that if $\chi_{A_n} \to f$ in $L^p(X)$, $1 \leq p \leq \infty$, f is equal μ-a.e. to the characteristic function of a measurable set.

88. Let $(X, \mathcal{M}, \mu)$ be a measure space, $1 \leq p < \infty$, and $\{f_n\} \subset L^p(X)$ such that $\sum_n \|f_n\|_p < \infty$. Prove that $f_n \to 0$ μ-a.e.

89. Let $(X, \mathcal{M}, \mu)$ be a probability measure space, $\{f_n\}$ independent square integrable functions on X with $\|f_n\|_2 \leq M$ for all n, and

$$\varphi_n(x) = \frac{1}{n} \sum_{k=1}^n \left(f_k(x) - \int_X f_k\, d\mu\right), \quad n = 1, 2, \ldots.$$

Prove: (a) $\|\varphi_n\|_2 \to 0$. (b) $n^\alpha \varphi_n \to 0$ in measure for $\alpha < 1/2$.

90. Let $(X, \mathcal{M}, \mu)$ be a measure space. We say that $\mathcal{F} = \{f\} \subset L^1(X)$ is uniformly absolutely continuous (or uniformly integrable) if given $\varepsilon > 0$, there exists $\delta > 0$ such that $\sup_{f \in \mathcal{F}} |\int_A f \, d\mu| < \varepsilon$ whenever $\mu(A) < \delta$. Prove that $\mathcal{F}$ is uniformly absolutely continuous iff $\{|f| : f \in \mathcal{F}\}$ is uniformly absolutely continuous.

91. Let $(X, \mathcal{M}, \mu)$ be a finite measure space and $\{f_n\}, f$ integrable functions on X. Discuss the validity of the following statements: $\{f_n\}$ is uniformly absolutely continuous provided that: (a) The f_n are nonnegative, $f_n \to f$ μ-a.e., and $\int_X f_n \, d\mu \to \int_X f \, d\mu$. (b) $\lim_n \int_X |f_n - f| \, d\mu = 0$.

92. Let $(X, \mathcal{M}, \mu)$ be a measure space and $\mathcal{F}$ a bounded set in $L^1(X)$. Prove that the following statements are equivalent: (a) $\mathcal{F}$ is uniformly absolutely continuous. (b) If the modulus of uniform continuity $\omega(\mathcal{F})$ of $\mathcal{F}$ is given by

$$\omega(\mathcal{F}, \varepsilon) = \sup_{f \in \mathcal{F}} \left\{ \int_A |f| \, d\mu : \mu(A) \leq \varepsilon \right\}, \quad \omega(\mathcal{F}) = \lim_{\varepsilon \to 0^+} \omega(\mathcal{F}, \varepsilon),$$

then $\omega(\mathcal{F}) = 0$. (c) Let $B_{f,R} = \{|f| > R\}$. If $\widetilde{\omega}(\mathcal{F})$ is given by

$$\widetilde{\omega}(\mathcal{F}, R) = \sup_{f \in \mathcal{F}} \int_{B_{f,R}} |f| \, d\mu \quad \text{and} \quad \widetilde{\omega}(\mathcal{F}) = \lim_{R \to \infty} \widetilde{\omega}(\mathcal{F}, R),$$

then $\widetilde{\omega}(\mathcal{F}) = 0$. (d) There exists a nonnegative Borel measurable function φ on $\mathbb{R}^+$ such that $\lim_{t \to \infty} \varphi(t)/t = \infty$ and $\sup_{f \in \mathcal{F}} \int_X \varphi(|f|) \, d\mu < \infty$.

93. Let $(X, \mathcal{M}, \mu)$ be a probability measure space. Discuss the validity of the following statement: $\mathcal{F} \subset L^1(X)$ is bounded iff $\mathcal{F}$ is uniformly absolutely continuous.

94. Let $(X, \mathcal{M}, \mu)$ be a measure space, $\{f_n\}$ integrable functions on X, and f a finite μ-a.e. function on X such that $\lim_n f_n = f$ μ-a.e. Prove: $f \in L^1(X)$ and $\lim_n \int_X |f_n - f| \, d\mu = 0$ iff (a) $\{f_n\}$ is uniformly absolutely continuous, and (b) given $\varepsilon > 0$, there exists $A \in \mathcal{M}$ with $\mu(A) < \infty$ such that $\int_{A^c} |f_n| \, d\mu \leq \varepsilon$ for all n.

Prove that this result, known as Vitali's theorem, implies the LDCT.

95. Let $(X, \mathcal{M}, \mu)$ be a probability measure space, $\{f_n\} \subset L^1(X)$, and f a μ-a.e. finite function on X. Prove: $f \in L^1(X)$ and $f_n \to f$ in $L^1(X)$ iff (a) $f_n \to f$ in probability, and (b) $\{f_n\}$ is uniformly absolutely continuous.

96. Let $(X, \mathcal{M}, \mu)$ be a measure space and $\mathcal{F} = \{f_n\}$ a bounded sequence in $L^1(X)$. Prove that there exist a subsequence $\{f_{n_k}\}$ of $\{f_n\}$ and a pairwise disjoint sequence $\{A_k\} \subset \mathcal{M}$ with $\mu(A_k) \to 0$ such that $\{\chi_{A_k^c} f_{n_k}\}$ is uniformly absolutely continuous.

97. Let $(X, \mathcal{M}, \mu)$ be a probability measure space, $\{f_n\} \subset L^p(X)$, $1 \leq p < \infty$, such that $f_n \to f$ in probability, and f μ-a.e. finite on X. Prove that the following are equivalent: (a) $\{|f_n|^p\}$ is uniformly absolutely continuous. (b) $f \in L^p(X)$ and $f_n \to f$ in $L^p(X)$. (c) $\|f_n\|_p \to \|f\|_p < \infty$.

98. Let $(X, \mathcal{M}, \mu)$ be a measure space and $\mathcal{F} = \{f_n\}$ a bounded sequence of nonnegative $L^1(X)$ functions that converges to g in measure. Prove that g is integrable and the following statements are equivalent: (a) $\{\int_X f_n \, d\mu\}$ is a convergent numerical sequence. (b) For every subsequence $\mathcal{F}' = \{f_{n_k}\}$ of $\{f_n\}$, $\lim_{n_k} \int_X f_{n_k} \, d\mu = \omega(\mathcal{F}') + \int_X g \, d\mu$ and $\omega(\mathcal{F}) = \omega(\mathcal{F}')$. (c) There exist subsequences $\mathcal{F}' = \{f_{n_k}\}$ and $\mathcal{F}'' = \{f_{n_\ell}\}$ of $\{f_n\}$ such that $\lim_{n_k} \int_X f_{n_k} \, d\mu = \liminf_n \int_X f_n \, d\mu$, $\lim_{n_\ell} \int_X f_{n_\ell} \, d\mu = \limsup_n \int_X f_n \, d\mu$, and $\omega(\mathcal{F}') = \omega(\mathcal{F}'')$.

99. Let $(X, \mathcal{M}, \mu)$ be a measure space and $\mathcal{F} = \{f_n\}$ a bounded sequence of $L^1(X)$ functions that converges to g in measure. (a) Prove that $\limsup_n \int_X |f_n - g| \, d\mu \leq \omega(\mathcal{F})$. (b) Assuming that $\lim_n \int_X |f_n - g| \, d\mu$ exists, find it. (c) Prove that if $\mu(X) < \infty$, $f_n \to g$ in $L^1(X)$ iff $\limsup_n \|f_n\|_1 \leq \|g\|_1$.

100. Let $(X, \mathcal{M}, \mu)$ be a finite measure space and $\mathcal{F} = \{f_n\}$ a bounded sequence in $L^1(X)$ such that $f_n \to f$ μ-a.e. Prove that $\{\|f_n\|_1\}$ converges iff $\{\|f_n - f\|_1\}$ converges, and then $\lim_n \|f_n - f\|_1 = \omega(\mathcal{F}) = \lim_n(\|f_n\|_1 - \|f\|_1)$.

101. Let $(X, \mathcal{M}, \mu)$ be a measure space and $\mathcal{F} = \{f_n\}$ a bounded sequence of nonnegative $L^1(X)$ functions that converges in measure to an integrable function g. Prove that the following statements are equivalent: (a) $\mathcal{F}' = \{f_{n_k}\}$ is a subsequence of $\{f_n\}$ and $\lim_{n_k} \int_X f_{n_k} \, d\mu = \liminf_n \int_X f_n \, d\mu$. (b) $\omega(\mathcal{F}') = \min\{\omega(\mathcal{F}'') : \mathcal{F}'' = \{f_{n_\ell}\}$ is a subsequence of $\{f_n\}\}$.

102. Let $(X, \mathcal{M}, \mu)$ be a measure space, $\{f_n\}$ measurable functions on X such that $f_1 \geq f_2 \geq \ldots \geq 0$, and $f(x) = \lim_n f_n(x)$, $x \in X$. Discuss the validity of the following statements: (a) $\lim_n \int_X f_n \, d\mu = \int_X f \, d\mu$. (b) If $\lim_n \int_X f_n \, d\mu = 0$, then $f = 0$ μ-a.e. on X.

103. Let $(X, \mathcal{M}, \mu)$ be a probability measure space, $\{f_n\}$ nonnegative integrable functions on X such that $\lim_n \int_X f_n \, d\mu = 1$, and $\{B_n\}$ with $\mu(B_n) \to 1$ such that $\int_{B_n} f_n \, d\mu \to 0$. Prove that $f(x) = \sup_n f_n(x) \notin L^1(X)$.

104. Let $(X, \mathcal{M}, \mu)$ be a measure space, $\{f_n\}$ nonnegative measurable functions on X that converge to 0 μ-a.e., and $\varphi_n = \sup\{f_1, \ldots, f_n\}$. Prove that if $\int_X \varphi_n \, d\mu \leq c$ for all n, then $\lim_n \int_X f_n \, d\mu = 0$.

105. Let $(X, \mathcal{M}, \mu)$ be a measure space and $\{f_n\}$ measurable functions on X such that $\mu(\{|f_n| > \lambda\}) \leq c\,\mu(\{|f_1| > \lambda\})$, $\lambda > 0$, for all n. Furthermore, suppose that $\|f_1\|_1 = 1$ and let $\varphi_n(x) = \sup_{1 \leq k \leq n} |f_k(x)|$. Discuss the validity of the following statement: $\lim_n n^{-1} \int_X \varphi_n \, d\mu = 0$.

106. Let $(X, \mathcal{M}, \mu)$ be a measure space and $\{f_n\}$ nonnegative measurable functions on X. Prove that if $\sup_n f_n$ is integrable, then

$$\limsup_n \int_X f_n \, d\mu \leq \int_X \limsup_n f_n \, d\mu \, .$$

Verify that the inequality may be strict and that it may fail unless $\sup_n f_n \in L^1(X)$.

107. Let $(X, \mathcal{M}, \mu)$ be a measure space and $1 \leq p < \infty$. Prove that $\{f_n\}$ on X is Cauchy in $L^p(X)$ iff the following conditions hold: (a) $\{f_n\}$ is Cauchy in measure. (b) $\{|f_n|^p\}$ is uniformly absolutely continuous. (c) Given $\varepsilon > 0$, there is a set A with $\mu(A) < \infty$ such that $\int_{A^c} |f_n|^p \, d\mu < \varepsilon$ for all n.

108. Let $(X, \mathcal{M}, \mu)$ be a measure space, $0 < p < \infty$, and $\{f_n\}$, f in $L^p(X)$ such that $f_n \to f$ in $L^p(X)$. Prove that there exist a subsequence $\{f_{n_k}\}$ of $\{f_n\}$ and $h \in L^p(X)$ such that $f_{n_k} \to f$ μ-a.e. and $|f_{n_k}| \leq h$ μ-a.e.

109. Let $(X, \mathcal{M}, \mu)$ be a measure space, $1 \leq p < \infty$, $\{f_n\}$ measurable functions on X, and $S_n = f_1 + \cdots + f_n$. Prove: (a) If $f_n \to 0$ μ-a.e., $S_n/n \to 0$ μ-a.e. (b) If $f_n \to 0$ in $L^p(X)$, $S_n/n \to 0$ in $L^p(X)$. (c) If $f_n \to 0$ in measure, it does not necessarily follow that $S_n/n \to 0$ in measure.

110. Let $(X, \mathcal{M}, \mu)$ be a measure space with $\mu(X) = 2$ and $\{f_n\}$ nonnegative $L^2(X)$ functions such that $\|f_n\|_1 = 2$ and $\big|\, \|f_n\|_2 - \sqrt{2}\,\big| \leq 2^{-n}$, all n. (a) Prove that $f_n \to 1$ μ-a.e. on X. (b) If the assumption that $f_n \geq 0$ is dropped, what conclusion follows?

111. Let $(X, \mathcal{M}, \mu)$ be a finite measure space and $\{f_n\} \subset L^2(X)$ with $\|f_n\|_2 \leq 1$ for all n. Prove that given $\varepsilon > 0$, there is an integer N such that $\mu(\{|f_n| \leq n^{2/3} \text{ for } n > N\}) \geq (1 - \varepsilon)$.

112. Let $(X, \mathcal{M}, \mu)$ be a finite measure space, $\{f_n\}$, f integrable functions on X such that $f_n \geq \eta > 0$ μ-a.e. and $f_n \to f$ in $L^1(X)$, and $\varphi : [\eta, \infty) \to [0, \infty)$ a Lipschitz function. Prove that the $\varphi(f_n)$ and $\varphi(f)$ are integrable and $\varphi(f_n) \to \varphi(f)$ in $L^1(X)$.

113. Let $(X, \mathcal{M}, \mu)$ be a measure space, $0 < p, q < \infty$, and $\{f_n\}$ defined on X such that $\|f_n\|_p \leq c$ for all n and $\|f_n\|_q \to 0$ as $n \to \infty$. Prove that for any r between p and q, $\|f_n\|_r \to 0$ as $n \to \infty$.

114. Let $(X, \mathcal{M}, \mu)$ be a measure space and $0 < p < \infty$. Suppose that $0 < r < p$, and let $q = rp/(p - r)$. Prove that if $\|f_n - f\|_p \to 0$ and $\|g_n - g\|_q \to 0$, then $\|f_n g_n - fg\|_r \to 0$.

115. Let $(X, \mathcal{M}, \mu)$ be a measure space, $1 < p < \infty$, and $\{f_n\}$, f nonnegative functions on X such that $\lim_n f_n = f$ in $L^p(X)$. Prove that $f_n^r \to f^r$ in $L^{p/r}(X)$ for all $r \in [1, p)$.

116. Let $(X, \mathcal{M}, \mu)$ be a measure space and $1 \leq p \leq \infty$. Prove: (a) If $f, g \in L^p(X)$, then $f \vee g = \max(f, g) \in L^p(X)$. (b) If $f_n \to f$ and $g_n \to g$ in $L^p(X)$, then $f_n \vee g_n \to f \vee g \in L^p(X)$.

117. Let $\{f_n\}$ be measurable functions on I such that $f_n \to f$ uniformly in $[\varepsilon, 1]$ for each $0 < \varepsilon < 1$, and $\int_I |f_n(x)|^p \, dx \leq c$ for $c > 0$, some $1 \leq p < \infty$, and all n. Discuss the validity of the following statement: $\lim_n \int_I |f_n(x) - f(x)| \, dx = 0$.

118. Let $\{f_n\}$ be continuous functions on $\mathbb{R}^+$ such that $\{f_n\}$ converges uniformly to $f \in C(\mathbb{R}^+)$. (a) Suppose that $\lim_{a \to \infty} \int_0^a f_n(x) \, dx$ exists and denote it $\int_0^\infty f_n(x) \, dx$. Show that $\lim_{a \to \infty} \int_0^a f(x) dx$ does not necessarily exist in $\mathbb{R}$. (b) Suppose that in addition $\lim_n \int_0^\infty f_n(x) \, dx = \lim_{a \to \infty} \int_0^a f(x) \, dx$ exists. Does it follow that $\lim_n \int_0^\infty f_n(x) \, dx = \int_0^\infty f(x) \, dx$?

119. Let $(X, \mathcal{M}, \mu)$ be a measure space, $0 < p < \infty$, and $\{f_n\} \subset L^p(X)$ such that $f_n(x) \to f(x)$ uniformly on X. Discuss the validity of the following statement: $f \in L^p(X)$ and $\lim_n \int_X |f_n - f|^p \, d\mu = 0$.

120. Let $(X, \mathcal{M}, \mu)$ be a measure space, $0 < p \leq \infty$, and $\{f_n\}, f$ in $L^p(X)$ such that $f_n \to f$ μ-a.e. Prove that $\|f\|_p \leq \liminf_n \|f_n\|_p$.

121. Let $(X, \mathcal{M}, \mu)$ be a finite measure space and $\{f_n\}$ measurable functions on X such that $f_1 \in L^2(X)$ and $\mu(\{|f_n| > \lambda\}) \leq c\mu(\{|f_1| > \lambda\})$ for all $\lambda > 0$. Prove: (a) For all $\varepsilon > 0$, $n\mu(\{|f_1| \geq \varepsilon\sqrt{n}\}) \to 0$ as $n \to \infty$. (b) $n^{-1/2} \sup_{1 \leq k \leq n} |f_k(x)| \to 0$ in measure.

122. Let $(X, \mathcal{M}, \mu)$ be a measure space and $\{f_n\}, \{g_n\}$ measurable functions on X. Further suppose that g_n, g are integrable, $g_n \to g$ in $L^1(X)$, and $f_n \leq g_n$ for all n. Prove that $\limsup_n \int_X f_n \, d\mu \leq \int_X \limsup_n f_n \, d\mu$.

123. Let $(X, \mathcal{M}, \mu)$ be a measure space and $\{f_n\}$ integrable functions on X that converge to f in $L^1(X)$. Show that if $\{g_n\}, g \in L^\infty(X)$ are such that $g_n \to g$ μ-a.e. it does not necessarily follow that $\lim_n \int_X f_n g_n \, d\mu = \int_X fg \, d\mu$. What additional condition on $\{g_n\}$ ensures that it does?

124. Let $(X, \mathcal{M}, \mu)$ be a measure space and $\{f_n\}, f$ nonnegative measurable functions on X such that $\sup_n \int_X f_n \, d\mu = K < \infty$ and $f_n \to f$ μ-a.e. (a) Prove that $\int_X f \, d\mu \leq K$. (b) Does it follow that $\lim_n \int_X f_n \, d\mu = \int_X f \, d\mu$? If not, what additional condition on $\{f_n\}$ ensures that it does?

125. Let $(X, \mathcal{M}, \mu)$ be a probability measure space. Give an example of integrable functions $\{f_n\}$ and an integrable function f on X such that $f_n \to f$ μ-a.e., $\int_X |f_n| \, d\mu = 2$, all n, and $\int_X |f| \, d\mu = 1$. Furthermore, prove that for any such sequence, $\lim_n \int_X |f_n - f| \, d\mu = 1$.

126. Let $(X, \mathcal{M}, \mu)$ be a measure space and $\{f_n\}, \{g_n\}, \{h_n\}$ and f, g, h in $L^1(X)$ such that $f_n \to f, g_n \to g, h_n \to h$ μ-a.e. Assume further that

$h_n \leq f_n \leq g_n$ μ-a.e. and $\int_X g_n \, d\mu \to \int_X g \, d\mu$, $\int_X h_n \, d\mu \to \int_X h \, d\mu$. Prove that $\int_X f_n \, d\mu \to \int_X f \, d\mu$.

127. Let $(X, \mathcal{M}, \mu)$ be a measure space and $\{f_n\}, f, g, \{f_n g\}, fg$ integrable functions on X such that $f_n \to f$ μ-a.e. What can one say about $\int_X f_n g \, d\mu$?

128. Let $(X, \mathcal{M}, \mu)$ be a measure space, $1 \leq p < \infty$, and $\{f_n\}, f, \{g_k\}$ functions on X such that $f_n \to f$ in $L^p(X)$ and $\|g_k\|_q \leq c$ for all k, where q is the conjugate to p. Suppose in addition that $\lim_k \int_X f_n g_k \, d\mu = 0$ for each n. Prove that $\lim_k \int_X f g_k \, d\mu = 0$.

129. Let $(X, \mathcal{M}, \mu)$ be a measure space and $\{f_n\}, f$ nonnegative integrable functions on X such that $f_n(x) \to f(x)$ μ-a.e. Let $A_n = \{f_n(x) < 2f(x)\}$. Prove: (a) $\int_{A_n} f_n \, d\mu \to \int_X f \, d\mu$. (b) If in addition $\int_X f_n \, d\mu \to \int_X f \, d\mu$, then $\int_X |f_n - f| \, d\mu \to 0$.

130. Construct a sequence of nonnegative functions $\{f_n\}$ on $[0, 1]$ such that $\lim_n f_n(x) = 0$ a.e., $\int_0^1 f_n(x) dx = 1$ for all n, and for each $g \in C([0, 1])$, $\lim_n \int_0^1 f_n(x) \, g(x) \, dx = (g(1/5) + g(2/5) + g(3/5))/3$.

131. Let $\{f_n\} \subset L^1(I)$ and $h \in L^1(I)$ be such that $|f_n| \leq h$ a.e., all n. Further suppose that $\lim_n \int_I f_n(x) g(x) \, dx = 0$ for each $g \in C(I)$. Discuss the validity of the following statements: (a) $\int_A f_n(x) \, dx \to 0$ for each measurable $A \subset I$. (b) $\lim_n \int_I |f_n(x)| \, dx = 0$.

132. Let $\{f_k\}$ be integrable functions on $[0, 1]$. Discuss the validity of the following statement: There exists a sequence $\{a_n\}$ decreasing to 0 such that $\lim_n a_n |f_k(a_n)| = 0$ for all k.

133. Let $(X, \mathcal{M}, \mu)$ be a measure space, f a nonnegative integrable function on X with $\int_X f \, d\mu = c > 0$, and $\varphi : [0, \infty) \to [0, \infty)$ a nondecreasing bounded function that satisfies the following properties: $\varphi(0) = 0$, $\varphi'(0) = 1$, and $\varphi'(x)$ or $\varphi(x)/x$ bounded. Prove that

$$\lim_n \int_X n \, \varphi((f(x)/n)^\alpha) \, d\mu(x) = \begin{cases} \infty, & 0 < \alpha < 1, \\ c, & \alpha = 1, \\ 0, & 1 < \alpha < \infty. \end{cases}$$

Finally, give examples of such functions φ and prove that the conclusion is true for $\varphi(x) = \ln(1 + x)$, which is unbounded.

134. Let $(X, \mathcal{M}, \mu)$ be a measure space and $\{f_n\}$ integrable functions on X such that $\|f_n\|_1 \geq \eta > 0$ for all n and if $A_n = \{|f_n| > 0\}$, $\lim_n \mu(A_n) = 0$. Prove that $\lim_n \|f_n\|_p = \infty$, $1 < p < \infty$.

135. Let $\{\varphi_k\}$ be an approximate identity in $L(\mathbb{R}^n)$, i.e., the φ_k are nonnegative, have integral 1, and their integral outside a neighbourhood of the origin tends to 0. Prove that $\|\varphi_k\|_p \to \infty$, $1 < p < \infty$.

136. Let $(X, \mathcal{M}, \mu)$ be a probability measure space and $\{A_n\}$ measurable sets with $\mu(A_n) = 1/2$ for all n and $\mu(A_n \cap A_m) = 1/4$ for $n \neq m$. Prove that if $f_n(x) = n^{-1} \sum_{j=1}^n \chi_{A_j}(x)$, then $\lim_n \int_X |f_n - 1/2|^p \, d\mu = 0$, $0 < p \leq 2$.

137. Let $x = \sum_k x_k 2^{-k}$ denote the dyadic expansion of $x \in [0,1]$. Prove that

$$S_n(x) = \frac{x_1 + \cdots + x_n}{n} \to \frac{1}{2} \quad \text{as } n \to \infty$$

in $L^2([0,1])$.

138. Let $(X, \mathcal{M}, \mu)$ be a measure space, $0 < p < \infty$, and $\{f_n\}$ functions on X such that $\|f_n\|_p \leq c$. Prove that for $\varepsilon > 1/p$, $\lim_n n^{-\varepsilon} f_n(x) = 0$ μ-a.e. on X.

139. Let $(X, \mathcal{M}, \mu)$ be a finite measure space and $\{f_n\}$ nonnegative functions on X such that $\int_X f_n^r \, d\mu \leq n^{-(r+\delta)}$, $\delta > 0$. Prove: (a) $\lim_n f_n(x) = 0$ μ-a.e. (b) $\sum_n f_n$ converges in $L^p(X)$, $1 \leq p < r$.

140. Construct nonnegative functions $\{f_n\}$ on $[0,1]$ such that

$$\lim_n \int_I f_n(x) \, dx = \infty \quad \text{and} \quad \lim_n \int_I \frac{f_n(x)}{1 + f_n(x)} \, dx = 0.$$

141. Let $X = \{\text{sequences } x : x_n = 0 \text{ for all } n \geq n_x\}$ and for $x, y \in X$ set

$$\rho(x, y) = \sum_n 2^{-n} \frac{|x_n - y_n|}{1 + |x_n - y_n|}.$$

(a) Show that (X, ρ) is not a complete metric space. (b) Describe a complete sequence space (Y, ρ) such that X is dense in Y. (c) What if X is endowed with $\|\cdot\|_\infty$?

142. Let $\{a_{n,k}\} \subset \mathbb{R}$ with $|a_{n,k}| \leq 1$ for $n, k = 1, 2, \ldots$, and $\lim_k a_{n,k} = 0$ for each n. Prove that $\lim_k \sum_n a_{n,k}/n^p = 0$ for $p > 1$.

143. Let $(X, \mathcal{M}, \mu)$ be a finite measure space, $0 < \alpha \leq \beta$, and $\{f_n\}, g$ integrable functions on X such that $\|f_n\|_1 \leq c$ and $n^{-\alpha}|f_n|^2 \leq g$ μ-a.e. Compute $\lim_n n^{-\beta} \int_X |f_n|^2 \, d\mu$.

144. Let $(X, \mathcal{M}, \mu)$ be a measure space and $\{f_n\}$ nonnegative measurable functions on X such that there exist $M > 0$ and positive convergent series $\sum_n \alpha_n, \sum_n \beta_n < \infty$ such that $\int_{\{f_n \leq M\}} f_n \, d\mu \leq \alpha_n$ and $\mu(\{f_n > M\}) \leq \beta_n$ for all n. Prove that $\sum_n f_n(x) < \infty$ μ-a.e.

145. Let $f \in L^1(\mathbb{R})$ have integral 1 and put $f_n(x) = nf(nx)$, all n. Discuss the validity of the following statement: $f_n \rightharpoonup f$ in $L^1(\mathbb{R})$.

146. Let f in $L^p(\mathbb{R}^N)$, $1 < p < \infty$, and with $\alpha > 0$ real, define $f_n(x) = n^\alpha f(nx)$, $n = 1, 2, \ldots$. Does $\{f_n\}$ converge strongly in $L^p(\mathbb{R}^N)$? Weakly in $L^p(\mathbb{R}^N)$?

147. Let $(X, \mathcal{M}, \mu)$ be a σ-finite measure space and $\{f_n\}, f$ in $L^p(X)$, $1 < p < \infty$. Prove that $f_n \rightharpoonup f$ in $L^p(X)$ iff $\|f_n\|_p \le c$ for all n and $\int_A f_n \, d\mu \to \int_A f \, d\mu$ for all A with $\mu(A) < \infty$.

148. Let $(X, \mathcal{M}, \mu)$ be a σ-finite measure space, $1 \le r < p < \infty$, and $\{f_n\}$ measurable functions on X such that $\|f_n\|_p \le c$ and $\|f_n\|_r \to 0$. Prove that $f_n \rightharpoonup 0$ in $L^p(X)$.

149. Let $(X, \mathcal{M}, \mu)$ be a measure space, $1 < p < \infty$, and $\{f_n\}$ measurable functions on X such that $\|f_n\|_p \le c$, and $f_n(x) = 0$ or $|f_n(x)| \ge n$ μ-a.e. for all n. Prove that $f_n \rightharpoonup 0$ in $L^p(X)$.

150. Let $(X, \mathcal{M}, \mu)$ be a measure space and $\{f_n\}$ a decreasing sequence of nonnegative functions on X such that $f_n \rightharpoonup 0$ in $L^p(X)$. Prove that $f_n \to 0$ in measure.

151. Let $(X, \mathcal{M}, \mu)$ be a σ-finite measure space and $\{f_n\}$ measurable functions on X such that $f_n \rightharpoonup f$ in $L^p(X)$ and $f_n \to g$ μ-a.e. Prove that $f = g$ μ-a.e.

152. Let $(X, \mathcal{M}, \mu)$ be a finite measure space and $\{f_n\}, f$ functions in $L^p(X)$, $1 < p < \infty$, such that $f_n \rightharpoonup f$ in $L^p(X)$ and $\lim_n \mu(\{|f_n| > \lambda\}) = 0$ for some $\lambda > 0$. Prove that $f \in L^\infty(X)$ and $\|f\|_\infty \le \lambda$.

153. Let $(X, \mathcal{M}, \mu)$ be a measure space, $1 < p < \infty$, and $\{f_n\} \subset L^p(X)$. Discuss the validity of the following statement: If $f_n \rightharpoonup f$ in $L^p(X)$, a subsequence $\{f_{n_k}\}$ of $\{f_n\}$ converges to f μ-a.e.

154. Let $(X, \mathcal{M}, \mu)$ be a measure space, $1 < p < \infty$, and $\{f_n\}$ defined on X with $\|f_n\|_p \le c$, all n, such that $f_n \to f$ μ-a.e. (a) Prove that $f \in L^p(X)$ and $f_n \rightharpoonup f$ in $L^p(X)$. Is the result true for $p = 1$? (b) Is the conclusion in (a) true if $f_n \to f$ in measure? (c) Discuss the validity of the following statement: If $1 \le r < p$, then $\lim_n \|f_n - f\|_r = 0$. But not for $r = p$.

155. With $\alpha \in (1, \infty)$, let f_n be the functions on $[-1, 1]$ given by

$$f_n(x) = \left(\frac{n}{1 + n^2 x^2}\right)^{1/\alpha}, \quad x \in [-1, 1], \ n = 1, 2, \ldots.$$

(a) Let $1 \le p < \alpha$. Prove that $\lim_n \|f_n\|_p = 0$. (b) Prove that $\{f_n\}$ converges weakly to 0 in L^p for $p = \alpha$. Does $\{f_n\}$ have a strongly convergent subsequence? (c) Let $p > \alpha$. Does $\{f_n\}$ converge strongly in $L^p([-1, 1])$? If not, does it have a weakly convergent subsequence?

156. Let $(X, \mathcal{M}, \mu)$ be a probability measure space and $\{f_n\}$ functions on X such that $\lim_n \int_A f_n \, d\mu$ exists for all $A \in \mathcal{M}$. Prove: (a) $\{f_n\}$ is uniformly absolutely continuous. (b) There exists $f \in L^1(X)$ such that $f_n \rightharpoonup f$ in $L^1(X)$.

157. Let $\{f_k\}, f \in L^1([0,1])$ be such that $f_k \rightharpoonup f$ in $L^1([0,1])$, $\mathcal{F} = \{\chi_A : A \in \mathcal{L}([0,1])\}$, and $\mathcal{F}_n = \{\chi_A \in \mathcal{F} : |\int_A (f_k(x) - f(x))\, dx| \leq \varepsilon$ for all $k \geq n\}$. Prove: (a) $\mathcal{F}$ and $\mathcal{F}_n$ are closed in $L^1(I)$. (b) $\{f_k\}$ is uniformly absolutely continuous.

158. Let $\{f_n\}$ denote the Borel measurable functions on $[0,1]$ given by $f_n = 1$ a.e. on $(2k/n, (2k+1)/n)$ for $(2k+1)/n \leq 1$, and $f_n = -1$ a.e. on $(2k - 1/n, 2k/n)$ for $2k/n \leq 1$. Prove that $f_n \rightharpoonup 0$ in $L^p(I)$ for $1 \leq p < \infty$.

159. For $t \in [0,1)$, let $v^t(x)$ denote the 1-periodic extension of $\chi_{[0,t)}(x)$ to $\mathbb{R}$, and define the sequence $\{v_n\}$ on $[0,1]$ by $v_n^t(x) = v^t(nx)$. Prove that $v_n^t \rightharpoonup t\chi_{[0,1]}$ in $L^p([0,1])$ for all $1 \leq p < \infty$.

160. Let $(X, \mathcal{M}, \mu)$ be a finite measure space, $1 \leq p, q < \infty$, $\varphi : \mathbb{R} \to \mathbb{R}$ a continuous function such that $|\varphi(t)| \leq c(1 + |t|^{p/q})$ for all $t \in \mathbb{R}$, and consider the mapping $T : L^p(X) \to L^q(X)$ given by $T(u)(x) = \varphi(u(x))$. Prove: (a) T is continuous from $L^p(X)$ into $L^q(X)$. (b) In $X = [0,1]$, if $u_n \rightharpoonup u$ in $L^p(X)$ implies that $T(u_n) \rightharpoonup T(u)$ in $L^q(X)$, then φ is linear.

161. Let $(X, \mathcal{M}, \mu)$ be a measure space, $2 \leq p < \infty$, and $\{f_n\}, f$ defined on X with $\|f_n\|_p \to \|f\|_p$, such that $f_n \rightharpoonup f$ in $L^p(X)$. Prove that $f_n \to f$ in $L^p(X)$.

Product Measures

Let $\mathcal{M}$ be a σ-algebra of subsets of X and $\mathcal{N}$ a σ-algebra of subsets of Y. The *product σ-algebra* $\mathcal{M} \otimes \mathcal{N}$ of subsets of $X \times Y$ is defined as follows: With $\mathcal{R} = \{A \times B \subset X \times Y : A \in \mathcal{M}, B \in \mathcal{N}\}$ the rectangles in $\mathcal{M} \times \mathcal{N}$, $\mathcal{M} \otimes \mathcal{N} = \mathcal{M}(\mathcal{R})$ is the σ-algebra generated by $\mathcal{R}$. $\mathcal{M} \otimes \mathcal{N}$ satisfies the following property: Given $E \subset X \times Y$, consider the sections $E_x = \{y \in Y : (x, y) \in E\}$ and $E^y = \{x \in X : (x, y) \in E\}$. Then, if $E \in \mathcal{M} \otimes \mathcal{N}$, $E_x \in \mathcal{N}$ for all $x \in X$ and $E^y \in \mathcal{M}$ for all $y \in Y$.

This result serves as a starting point towards the definition of the *product measure* $\mu \otimes \nu$ on $\mathcal{M} \otimes \mathcal{N}$. Specifically, if $(X, \mathcal{M}, \mu)$ and $(Y, \mathcal{N}, \nu)$ are σ-finite measure spaces, $\mu \otimes \nu$ is the unique σ-finite measure on $\mathcal{M} \otimes \mathcal{N}$ such that $\mu \otimes \nu(A \times B) = \mu(A)\,\nu(B)$ for all $A \times B \in \mathcal{R}$. Moreover, for $E \in \mathcal{M} \otimes \mathcal{N}$, $\mu \otimes \nu(E)$ can be computed in one of three ways, namely,

$$\int_{X \times Y} \chi_E\, d(\mu \otimes \nu) = \int_X \nu(E_x)\, d\mu(x) = \int_Y \mu(E^y)\, d\nu(y).$$

Along similar lines, if $f : X \times Y \to \mathbb{R}$, the *$x$-section* $f_x : Y \to \mathbb{R}$ of f is defined for each $x \in X$ by $f_x(y) = f(x, y)$ and the *y-section* $f^y : X \to \mathbb{R}$ by $f^y(x) = f(x, y)$ for each $y \in Y$. Now, if $f : (X \times Y, \mathcal{M} \otimes \mathcal{N}) \to \mathbb{R}$ is measurable, then $f_x : (Y, \mathcal{N}) \to \mathbb{R}$ and $f^y : (X, \mathcal{M}) \to \mathbb{R}$ are measurable for every $x \in X$ and $y \in Y$, respectively.

The main results concerning integration are Tonelli's theorem and Fubini's theorem. They establish that, if $f(x, y) : (X \times Y, \mathcal{M} \otimes \mathcal{N}) \to \mathbb{R}$ is measurable, then $g : (X, \mathcal{M}) \to \mathbb{R}$ and $h : (Y, \mathcal{N}) \to \mathbb{R}$ given by $g(x) = \int_Y f_x(y)d\nu(y)$ and $h(y) = \int_X f^y(x)\, d\mu(x)$, respectively, are measurable and that, under appropriate conditions, the integral of f over $X \times Y$ is equal to the iterated integrals of f, i.e.,

$$\int_{X\times Y} f\, d(\mu\otimes\nu) = \int_Y \int_X f^y(x)\, d\mu(x)\, d\nu(y) = \int_X \int_Y f_x(y)\, d\nu(y)\, d\mu(x)\,.$$

For Tonelli's theorem f is assumed to be nonnegative and then the integrals, whether finite or infinite, are equal. Fubini's theorem establishes that, if at least one of the three integrals $\int_{X\times Y} |f|\, d(\mu\otimes\nu)$, $\int_Y \int_X |f^y(x)|\, d\mu(x)\, d\nu(y)$, or $\int_X \int_Y |f_x(y)|\, d\nu(y)\, d\mu(x)$ is finite, then $f_x \in L^1(Y)$ for μ-a.e. $x \in X$, $f^y \in L^1(X)$ for ν-a.e. $y \in Y$, $g(x) \in L^1(X)$, $h(y) \in L^1(Y)$, f is integrable, and the integral of f over $X \times Y$ is equal to the iterated integrals of f.

Since the product of complete measure spaces is not in general complete, it is of interest to consider the version of this theorem in this case. Specifically, let $(X, \mathcal{M}, \mu)$, $(Y, \mathcal{N}, \nu)$ be complete σ-finite measure spaces and let $\mathcal{S} = \overline{\mathcal{M}\otimes\mathcal{N}}$ and $\eta = (\mu\otimes\nu)_1$ denote the σ-algebra of subsets of $X \times Y$ and the measure on $\mathcal{S}$, respectively, constructed in Problem 2.54 such that $(X \times Y, \mathcal{S}, \eta)$ is the completion of $(X \times Y, \mathcal{M}\otimes\mathcal{N}, \mu\otimes\nu)$.

Then, if $f : (X \times Y, \mathcal{S}) \to \mathbb{R}$ is nonnegative and measurable, $f_x(\cdot) : (Y, \mathcal{N}) \to \mathbb{R}$ is measurable for μ-a.e. $x \in X$, $f^y(\cdot) : (X, \mathcal{M}) \to \mathbb{R}$ is measurable for ν-a.e. $y \in Y$, $\int_X f(x, y)\, d\mu(x)$ is $(Y, \mathcal{N})$ measurable, $\int_Y f(x, y)\, d\nu(y)$ is $(X, \mathcal{M})$ measurable, and

$$\int_{X\times Y} f\, d(\mu\otimes\nu) = \int_Y \int_X f^y(x)\, d\mu(x)\, d\nu(y) = \int_X \int_Y f_x(y)\, d\nu(y)\, d\mu(x)\,.$$

The possibility that the integrals are infinite is allowed.

Along similar lines, if $f : (X \times Y, \mathcal{S}) \to \mathbb{R}$ is measurable, and $f \in L^1(X \times Y)$ with respect to the measure η, then $f_x \in L^1(\nu)$, $f^y \in L^1(\mu)$ for μ-a.e. x and ν-a.e. y, respectively. Moreover, $\int_X f(x, y)\, d\mu(x)$ is $(Y, \mathcal{N})$ integrable, $\int_Y f(x, y)\, d\nu(y)$ is $(X, \mathcal{M})$ integrable, and if one of the three integrals considered in Fubini's theorem is finite, the three integrals in the conclusion above are finite and equal.

We adopt the notation λ_n for the Lebesgue measure on $\mathbb{R}^n$, $\lambda_1 = \lambda$. Also, $\{x \in X : g(x) > t\} = \{g > t\}$.

Convolution next. Given $f, g \in L^1(\mathbb{R}^n)$, $f(x - y)g(y)$ is measurable on $\mathbb{R}^n \times \mathbb{R}^n = \mathbb{R}^{2n}$, and the *convolution* $f * g$ of f and g is defined as $f * g(x) = \int_{\mathbb{R}^n} f(x - y)g(y)\, d\lambda_n(y)$, $x \in \mathbb{R}^n$. $f * g$ is well-defined and finite for x λ_n-a.e. in $\mathbb{R}^n$ and is translation invariant, i.e., if τ_h represents the translation by h given by $\tau_h f(x) = f(x - h)$, then $\tau_h(f * g)(x) = (\tau_h f) * g(x)$.

Moreover, the convolution operation is commutative and associative. Furthermore, if $g \in L^1(\mathbb{R}^n)$ and $f \in L^p(\mathbb{R}^n)$, $1 \le p \le \infty$, $f * g(x)$ is a well-defined $L^p(\mathbb{R}^n)$ function that satisfies $\|f * g\|_p \le \|g\|_1 \|f\|_p$.

The problems in this chapter include the natural questions as to whether $\mathcal{B}(\mathbb{R}^m) \otimes \mathcal{B}(\mathbb{R}^n) = \mathcal{B}(\mathbb{R}^{m+n})$ and $\mathcal{L}(\mathbb{R}^m) \otimes \mathcal{L}(\mathbb{R}^n) = \mathcal{L}(\mathbb{R}^{m+n})$, addressed in Problem 3 and Problems 5–6, respectively.

The notion of product measure allows us to consider whether the integral of a nonnegative function can be evaluated as the area under its graph, Problem 25. And, on a different note, if the Lebesgue measure is the only measure on $\mathbb{R}^n$ which is translation invariant, Problem 29.

Sections of measurable functions are discussed in Problems 31–34 and Problems 38–39. The assumptions of Fubini's theorem are elucidated in Problem 40. Problem 60 is the prototype of results that deal with the μ-a.e. finiteness of a function. Integrals of independent functions are considered in Problems 82–84. And, properties of convolutions, including the existence of a unit, are addressed in Problems 86–90, Problems 92–95, Problems 97–100, and Problem 102.

Problems

1. Let $\mathcal{M}$ be an algebra of subsets of X, $\mathcal{N}$ an algebra of subsets of Y, and $\mathcal{R} = \{A \times B \subset X \times Y : A \in \mathcal{M}, B \in \mathcal{N}\}$ the rectangles in $\mathcal{M} \times \mathcal{N}$. Prove that the collection of finite unions of rectangles in $\mathcal{R}$ form an algebra of subsets of $X \times Y$.

2. Let $\mathcal{M}$ be a σ-algebra of subsets of X and $\mathcal{N}$ a σ-algebra of subsets of Y. Discuss the validity of the following statement: $A \times B \in \mathcal{M} \otimes \mathcal{N}$ iff $A \in \mathcal{M}$ and $B \in \mathcal{N}$.

3. Discuss the validity of the following statements: (a) $\mathcal{B}(\mathbb{R}^m) \times \mathcal{B}(\mathbb{R}^n) = \mathcal{B}(\mathbb{R}^{m+n})$. (b) $\mathcal{B}(\mathbb{R}^m) \otimes \mathcal{B}(\mathbb{R}^n) = \mathcal{B}(\mathbb{R}^{m+n})$.

4. Discuss the validity of the following statement: If $B \subset \mathbb{R}^2$ is such that $B_x \in \mathcal{B}(\mathbb{R})$ for all $x \in \mathbb{R}$ and $B^y \in \mathcal{B}(\mathbb{R})$ for all $y \in \mathbb{R}$, then $B \in \mathcal{B}(\mathbb{R}^2)$.

5. Prove that $\mathcal{B}(\mathbb{R}^{m+n}) \subsetneq \mathcal{L}(\mathbb{R}^m) \otimes \mathcal{L}(\mathbb{R}^n) \subsetneq \mathcal{L}(\mathbb{R}^{m+n})$.

6. Prove that $\mathcal{L}(\mathbb{R}^{m+n})$ is the completion of $\mathcal{L}(\mathbb{R}^m) \otimes \mathcal{L}(\mathbb{R}^n)$ with respect to the Lebesgue measure.

7. Let $A \in \mathcal{L}(\mathbb{R}^m)$ and $B \notin \mathcal{L}(\mathbb{R}^n)$. Prove that $A \times B \in \mathcal{L}(\mathbb{R}^{m+n})$ iff $\lambda_m(A) = 0$.

8. Let $A \subset \mathbb{R}^m$. Prove that the following are equivalent: (a) $A \in \mathcal{L}(\mathbb{R}^m)$. (b) $A \times B \in \mathcal{L}(\mathbb{R}^{m+n})$ for every $B \in \mathcal{L}(\mathbb{R}^n)$. (c) $A \times \mathbb{R}^n \in \mathcal{L}(\mathbb{R}^{m+n})$.

9. Let $(X_k, \mathcal{M}_k, \mu_k)$ be σ-finite measure spaces, $k = 1, 2, 3$. Prove: (a) $\mathcal{M}_1 \otimes (\mathcal{M}_2 \otimes \mathcal{M}_3) = (\mathcal{M}_1 \otimes \mathcal{M}_2) \otimes \mathcal{M}_3$. (b) $(\mu_1 \otimes \mu_2) \otimes \mu_3 = \mu_1 \otimes (\mu_2 \otimes \mu_3)$

is the unique σ-finite measure on the σ-algebra defined in (a) that assigns to $A_1 \times A_2 \times A_3 \subset X_1 \times X_2 \times X_3$ the value $(\mu_1 \otimes \mu_2)(A_1 \times A_2)\mu_3(A_3) = \mu_1(A_1)\mu_2(A_2)\mu_3(A_3)$.

10. Let $(X, \mathcal{M}, \mu)$ and $(Y, \mathcal{N}, \nu)$ be σ-finite measure spaces such that $(X \times Y, \mathcal{M} \otimes \mathcal{N}, \mu \times \nu)$ is complete. Prove that if $\mathcal{N}$ contains a nonempty null set, then $\mathcal{M} = \mathcal{P}(X)$.

11. Characterize: (a) $\mathcal{P}(\mathbb{N}) \otimes \mathcal{P}(\mathbb{N})$. (b) $\mu \otimes \mu$, where μ denotes the counting measure on $\mathbb{N}$.

12. Let $(X, \mathcal{M}, \mu)$ and $(Y, \mathcal{N}, \nu)$ be measure spaces and $E \subset X \times Y$ such that $\nu(E_x) = 0$ for μ-a.e. $x \in X$. Prove that $\mu \otimes \nu(E) = 0$ and that $\mu(E^y) = 0$ for ν-a.e. $y \in Y$.

13. Let $(X, \mathcal{M}, \mu)$ and $(Y, \mathcal{N}, \nu)$ be probability measure spaces and $E \subset X \times Y$ such that $\nu(E_x) = 0$ for μ-a.e. $x \in X$ and if $B = \{y \in Y : \mu(X \setminus E^y) < 1\}$, then $\nu(B) > 0$. Prove that $E \notin \mathcal{M} \otimes \mathcal{N}$.

14. Let $A \subset \mathbb{R}^{m+n}$. (a) If $\lambda_m \otimes \lambda_n(A) = 0$, prove that $\lambda_n(A_x) = 0$ λ_m-a.e. $x \in \mathbb{R}^m$ and $\lambda_m(A^y) = 0$ λ_n-a.e. $y \in \mathbb{R}^n$. (b) If A is a Lebesgue measurable subset of $\mathbb{R}^{m+n}$ such that for almost every $x \in \mathbb{R}^m$, $\lambda_n(A_x) = 0$, prove that A has measure zero and that for almost every $y \in \mathbb{R}^n$, $\lambda_m(A^y) = 0$.

15. Let $A \in \mathcal{B}(\mathbb{R}^2)$. Suppose that for each $t \in \mathbb{R}$, $\{(x, y) \in A : x - y = t\}$ is a finite set. Prove that $\lambda_2(A) = 0$.

16. For each $x \in \mathbb{R}$, let L_x denote the closed line segment in the plane joining $(x, 0)$ and $(0, 2)$. For each $A \subset \mathbb{R}$, the triangle $T(A)$ over A is defined as $T(A) = \{L_x : x \in A\}$. Prove that for $A \in \mathcal{B}(\mathbb{R})$, $T(A)$ is measurable in the plane and $\lambda_2(T(A)) = \lambda(A)$.

17. Let $(X, \mathcal{M}, \mu)$ and $(Y, \mathcal{N}, \nu)$ be probability measure spaces and $E \subset X \times Y$ such that $\mu \otimes \nu(E) = \eta^2 < 1$. Let $A = \{x \in X : \nu(E_x) \geq \eta\}$. Find the sharp upper bound for $\mu(A)$.

18. Let $(X, \mathcal{M}, \mu)$ and $(Y, \mathcal{N}, \nu)$ be probability measure spaces and $E \subset X \otimes Y$ such that $\mu(E^y) \geq \alpha$ for all $y \in Y$. Prove that if $\beta < \alpha$ and $A = \{x \in X : \nu(E_x) \geq \beta\}$, then $\mu(A) \geq (\alpha - \beta)/(1 - \beta)$.

19. Let $(X, \mathcal{M}, \mu)$, $(Y, \mathcal{N}, \nu)$ be probability measure spaces and $E \in \mathcal{M} \otimes \mathcal{N}$. Prove that if $\nu(E_x) \leq 1/2$ μ-a.e. in X, then $\nu(\{y \in Y : \mu(E^y) = 1\}) \leq 1/2$.

20. Prove that $\mathbb{R}^2$ cannot be written as a countable union of zero sets of nontrivial polynomial $p(x, y)$ in two variables.

21. Let μ be a finite Borel measure on $[0, \infty)$. (a) Characterize $\mathcal{B}(\mathbb{R}^+) \otimes \mathcal{P}(\mathbb{N})$. (b) Prove that there exists a unique measure π on $\mathcal{B}(\mathbb{R}^+) \otimes \mathcal{P}(\mathbb{N})$ such that

$$\pi(B \times \{n\}) = \int_B e^{-t} \frac{t^n}{n!} \, d\mu(t), \quad \text{all } n.$$

(c) Give an example where π is and is not given by a product of measures on $\mathcal{B}(\mathbb{R}^+)$ and $\mathcal{P}(\mathbb{N})$.

22. Let $(X, \mathcal{M}, \mu)$ be a measure space and for $A \subset X$ and $0 \le a \le \infty$, put $A_a = \{(x, y) : x \in A, 0 \le y \le a\}$ for finite a and $A_\infty = \{(x, y) : x \in A, 0 \le y < \infty\}$. Prove that if $A \in \mathcal{M}$, then $A_a \in \mathcal{M} \otimes \mathcal{B}(\mathbb{R})$ and compute $\mu \otimes \lambda(A_a)$.

23. Let $(X, \mathcal{M}, \mu)$ be a σ-finite measure space and $f : X \to \mathbb{R}$ a non-negative function with graph $\Gamma(f) = \{(x, f(x)) \in X \times \mathbb{R}^+ : x \in X\}$. Prove that if $f : (X, \mathcal{M}) \to \mathbb{R}$ is measurable, $\mu \otimes \lambda(\Gamma(f)) = 0$.

24. Let f be a Borel measurable function on $\mathbb{R}$ and $E = \{(x + f(x), x - f(x)) : x \in \mathbb{R}\}$. Discuss the validity of the following statement: $\lambda_2(E) = 0$.

25. Let $(X, \mathcal{M}, \mu)$ be a measure space and f a nonnegative measurable function on X. We denote by $R(f, A) = \{(x, y) \in X \times \mathbb{R}^+ : x \in A, 0 \le y \le f(x) \text{ if } f(x) < \infty, \text{ and } 0 \le y < \infty \text{ if } f(x) = \infty\}$ the region of f over A. Prove that $R(f, A) \in \mathcal{M} \otimes \mathcal{B}(\mathbb{R})$ and compute $\mu \otimes \lambda(R(f, A))$.

26. Let δ denote the Dirac measure concentrated at the origin of $\mathbb{R}$ and λ the Lebesgue measure on $\mathbb{R}$. Compute: (a) $\lambda \otimes \delta(E)$ where $E = \{(x, y) \in \mathbb{R}^2 : |x - y| < 1\}$. (b) $\int_{\mathbb{R}^2} f \, d(\lambda \otimes \delta)$ where $f(x, y) = 1/(x^2 + (y - 1)^2)$.

27. Let μ, ν be σ-finite Borel measures on $\mathbb{R}$ and $\Delta = \{(x, y) \in \mathbb{R} \times \mathbb{R} : x = y\}$ the diagonal of $\mathbb{R}^2$. Compute $\mu \otimes \nu(\Delta)$.

28. Let $(X, \mathcal{M}, \mu)$ and $(X, \mathcal{M}, \nu)$ be nontrivial σ-finite measures such that $\mu \otimes \nu((X \times X) \setminus \Delta) = 0$ where $\Delta = \{(x, y) \in X \times X : x = y\}$. Prove that there exist $a \in X$ and $\alpha, \beta \in \mathbb{R}$ such that $\mu = \alpha\delta_a$ and $\nu = \beta\delta_a$, with δ_a the Dirac measure concentrated at $a \in X$.

29. Discuss the validity of the following statement: If μ is a translation invariant measure on $\mathcal{L}(\mathbb{R}^n)$, i.e., $\mu(x + A) = \mu(A)$ for all $A \in \mathcal{L}(\mathbb{R}^n)$ and $x \in \mathbb{R}^n$, then μ is a multiple of the Lebesgue measure λ_n.

30. Let $f : \mathbb{R}^N \times \mathbb{R} \to \mathbb{R}$. Discuss the validity of the following statements: (a) If $f_x : \mathbb{R} \to \mathbb{R}$ and $f^y : \mathbb{R}^N \to \mathbb{R}$ are Borel measurable for all x in $\mathbb{R}$ and $y \in \mathbb{R}^n$, respectively, f is Borel measurable. (b) If $f_x : \mathbb{R} \to \mathbb{R}$ is left-continuous for all $x \in \mathbb{R}^N$ and $f^y : (\mathbb{R}^N, \mathcal{B}(\mathbb{R}^N)) \to (\mathbb{R}, \mathcal{B}(\mathbb{R}))$ measurable for all $y \in \mathbb{R}$, f is Borel measurable.

31. Let f be defined on $[a, b] \times \mathbb{R}^n$ such that $f_t(x)$ is continuous for each $t \in [a, b]$ and $f^x(t)$ is Lebesgue measurable for each $x \in \mathbb{R}^n$. Prove that $f \in \mathcal{B}([a, b]) \otimes \mathcal{L}(\mathbb{R}^n)$.

32. Let $f : \mathbb{R}^{m+n} \to \mathbb{R}$ be Lebesgue measurable such that if $A = \{(x,y) \in \mathbb{R}^{m+n} : f(x,y) \neq 0\}$, then $\lambda_m \otimes \lambda_n(A) = 0$. Prove that $\lambda_n(A_x) = 0$ λ_m-a.e. $x \in \mathbb{R}^m$ and $\lambda_m(A^y) = 0$ λ_n-a.e. $y \in \mathbb{R}^n$.

33. Let $f : \mathbb{R}^m \times \mathbb{R}^n \to \mathbb{R}$ be a Lebesgue measurable function such that f_x is continuous for λ_m-a.e. $x \in \mathbb{R}^m$ and $f^y = 0$ for λ_n-a.e. in $y \in \mathbb{R}^n$. Prove that for λ_m-a.e. $x \in \mathbb{R}^m$, $f(x,y) = 0$ for all $y \in \mathbb{R}^n$.

34. Let $f(x,y)$ be Lebesgue measurable in $\mathbb{R}^{m+n}$. Suppose that for λ_m-a.e. $x \in \mathbb{R}^m$, $f(x,y)$ is finite for λ_n-a.e. $y \in \mathbb{R}^n$. Prove that for λ_n-a.e. $y \in \mathbb{R}^n$, $f(x,y)$ is finite for λ_m-a.e. $x \in \mathbb{R}^m$.

35. Let $Q_m \subset \mathbb{R}^m, Q_n \subset \mathbb{R}^n$ be cubes and $f \in L^1(Q_m \times Q_n)$. Prove that for λ_{n+m}-a.e. $(x,y) \in Q_m \times Q_m$,

$$\lim_{\varepsilon \to 0} \frac{1}{|Q(y,\varepsilon)|} \int_{Q(y,\varepsilon)} f(x,z)\, d\lambda_n(z) = f(x,y).$$

36. Let $Q_m \subset \mathbb{R}^m, Q_n \subset \mathbb{R}^n$ be cubes and $f : Q_m \times Q_n \to \mathbb{R}$ a bounded measurable function such that f_x is continuous for each $x \in Q_m$ and f^y is Lebesgue measurable for $y \in D$, a dense subset of Q_n. Prove that $g(y) = \int_{Q_m} f^y(x)\, d\lambda_m(x)$ is a well-defined continuous function on Q_n.

37. Let $Q_m \subset \mathbb{R}^m, Q_n \subset \mathbb{R}^n$ be cubes of measure 1, $0 < \beta < \alpha < 1$, and $f : Q_m \times Q_n \to \mathbb{R}^+$ a nonnegative measurable function such that $\int_{Q_n} f(x,y)\, d\lambda(y) > \alpha$ for each $x \in Q_m$. Prove that if $A = \{y \in Q_n : \int_{Q_m} f(x,y)\, d\lambda(x) \geq \beta\}$, then $\lambda_m(A) \geq \alpha - \beta$.

38. Let $f : [a,b] \times [a,b] \to \mathbb{R}$ be a measurable function such that f_x is measurable and integrable with a continuous integral for λ-a.e. $x \in [a,b]$ and f^y is measurable and integrable with a continuous integral for λ-a.e. $y \in [a,b]$, and such that the iterated integrals of f are finite and equal. Discuss the validity of the following statement: $f \in L^1([a,b] \times [a,b])$.

39. Discuss the validity of the following statement: There exists $f : I \times I \to \mathbb{R}$ such that $\int_I f(x,y)\, d\lambda(x) = \infty$ for all rationals $y \in [0,1]$ yet $\int_{I \times I} f\, d\lambda_2 = 0$.

40. Let $([0,1], \prec)$ be a well-ordering of $[0,1]$ with ordinal Ω. Prove that if $S = \{(x,y) \in [0,1] \times [0,1] : x \prec y\}$ and $f = \chi_S$, the iterated integrals of f exist but are not equal.

41. Let $\Delta = \{(x,y) \in I \times I : x = y\}$ denote the diagonal of the unit square $I \times I$ in $\mathbb{R}^2$, λ the Lebesgue measure on $\mathcal{M} = \mathcal{B}([0,1])$, and μ the counting measure on $\mathcal{N} = \mathcal{P}([0,1])$. Prove: (a) $\Delta \in \mathcal{M} \otimes \mathcal{N}$. (b) $\lambda \otimes \mu(D)$, $\int_I \int_I \chi_\Delta(x,y)\, d\lambda(y)\, d\mu(x)$, and $\int_I \int_I \chi_\Delta(x,y)\, d\mu(x)\, d\lambda(y)$ are all different. (c) There is more than one measure ν on $I \times I$ such that $\nu(A \times B) = \lambda(A)\, \mu(B)$ for all $A, B \in \mathcal{B}(I)$.

42. $(X, \mathcal{M}, \mu), (Y, \mathcal{N}, \nu)$ be σ-finite measure spaces and $f \in L^1(X), g \in L^1(Y)$. Prove that $h(x, y) = f(x)g(y) \in L^1(X \times Y)$ and $\int_{X \times Y} h \, d(\mu \otimes \nu) = (\int_X f \, d\mu)(\int_Y g \, d\nu)$.

43. Let $E = \mathbb{R}^+ \times \mathbb{R}^+$ and $f(x, y) : E \to \mathbb{R}$ such that $f_x \in L^1(\mathbb{R}^+)$ for every $x \in \mathbb{R}^+$ and $\int_{\mathbb{R}^+} f_x(y) \, d\lambda(y) = g(x)$ is a continuous function of x. Discuss the validity of the following statement: $f \in L^1(E)$ and $\int_E f \, d\lambda \otimes \lambda = \int_{\mathbb{R}^+} g(x) \, d\lambda(x)$.

44. Let $v_n(r)$ denote the volume of the ball centered at the origin with radius r in $\mathbb{R}^n$; $v_n(1) = v_n$. (a) Identify the function $f(t)$ such that $v_{n+1} = v_n \int_{-1}^1 f(t)^n \, d\lambda(t)$, all n, and prove that $\lim_n f(t)^n = 0$ for $0 < |t| < 1$. (b) Prove that for any $A > 0$, $\lim_n A^n v_n = 0$.

45. Let $(X, \mathcal{M}, \mu)$ be a finite measure space and g a measurable function on X. Discuss the validity of the following statement: If $f(x, y) = 2g(x) - 3g(y) \in L(X \times X)$, then $g \in L^1(X)$.

46. Let f be a measurable periodic function with period 1, i.e., $f(t+1) = f(t)$, $t \in \mathbb{R}$, such that $\int_0^1 |f(a+t) - f(b+t)| \, d\lambda(t) \le c$ for all a, b and a finite constant $c > 0$. Prove that $f \in L^1([0, 1])$.

47. Let $f : \mathbb{R}^+ \to \mathbb{R}$ be an absolutely continuous function, $f(x) = f(0) + \int_{[0,x]} g(t) \, d\lambda(t)$, $x \ge 0$, where $g \in L^1(\mathbb{R}^+)$ and $0 < a < b < \infty$. Compute

$$J = \int_0^\infty \frac{f(bx) - f(ax)}{x} \, d\lambda(x) \, .$$

48. Let f be a nonnegative measurable function on $[0, 1]$. Prove that

$$\int_0^1 \int_x^1 f(x)f(y) \, d\lambda(y) \, d\lambda(x) = \frac{1}{2}\left(\int_0^1 f(x) \, d\lambda(x) \right)^2 .$$

49. Let $g \in L^1(\mathbb{R})$. Find those values of α for which the function $\phi(x, t) = |x|^{-\alpha} t^{\alpha-1} g(x) \chi_{[-t,t]}(x)$ is integrable on $\mathbb{R} \times \mathbb{R}^+$.

50. Let $f \in L^1(\mathbb{R})$ satisfy $\int_{\mathbb{R}} \int_{\mathbb{R}} f(4x)f(x+y) \, d\lambda(x) \, d\lambda(y) = 1$. Compute $\int_{\mathbb{R}} f \, d\lambda$.

51. Let $f : I \times I \to \mathbb{R}$ be given by $f(x, y) = (1 - xy)^{-a}$ where $a > 0$. (a) Discuss the validity of the following statement: The following three quantities are equal: $\int_I \int_I f(x, y) \, d\lambda(x) \, d\lambda(y)$, $\int_I \int_I f(x, y) \, d\lambda(y) \, d\lambda(x)$, and $\int_{I \times I} f \, d\lambda_2$. (b) Establish for what values of a, $\int_{I \times I} f \, d\lambda_2 < \infty$.

52. Let $\alpha \in \mathbb{R}$ and f on $\mathbb{R}^+ \times \mathbb{R}^+$ given by $f(x, y) = (1 + x + y)^{-\alpha}$. Determine the values of α for which f is integrable and compute the integral.

53. Let $C = \{(x, y, z) \in \mathbb{R}^3 : 0 < x, y, z < 1\}$. Decide whether $f : C \to \mathbb{R}$ given by $f(x, y, z) = (1 - xyz)^{-1}$ is integrable and compute its integral.

54. Let $J = (0,1) \times (0,1) \times (0,1)$, a, b, c positive real numbers, and $f(x,x,z) = (x^a + y^b + z^c)^{-1}$ for $(x,y,z) \neq (0,0,0)$ and $f(x,x,z) = 0$ otherwise. Characterize those values of a, b, c, so that: (a) $f \in L^1(J)$. (b) $f \in L^1(\mathbb{R}^3 \setminus J)$.

55. Given $f \in L^1(\mathbb{R})$, let $S_n(x) = n^{-1} \sum_{j=0}^{n-1} f(x + j/n)$ and $S(x) = \int_x^{x+1} f(y)\, d\lambda(y)$. Prove that $\lim_n S_n = S$ in $L^1(\mathbb{R})$.

56. Let μ_1, μ_2 be finite Borel measures on $\mathbb{R}$. Prove that there exists a finite Borel measure μ such that $\int_X f\, d\mu = \int_X \int_X f(x+y)\, d\mu_1(x)\, d\mu_2(y)$, all $f \in C_c(\mathbb{R})$. Furthermore, μ is unique.

57. Let $1 \leq p \leq \infty$ and define

$$Tf(x) = \int_{\mathbb{R}} \frac{f(y)}{1 + (1 + y^2)^2 x^2}\, d\lambda(y).$$

For which p does $f \in L^p(\mathbb{R})$ imply that $Tf \in L^1(\mathbb{R})$?

58. Let $f \in L^1(\mathbb{R}^{m+n}) \cap L^p(\mathbb{R}^{m+n})$, $1 < p < \infty$, and set $u(y) = \int_{\mathbb{R}^m} |f(x,y)|\, d\lambda_m(x)$ for $y \in \mathbb{R}^n$ and $v(x) = (\int_{\mathbb{R}^n} |f(x,y)|^p\, d\lambda_n(y))^{1/p}$ for $x \in \mathbb{R}^m$. (a) Prove that $u \in L^1(\mathbb{R}^n)$ and $v \in L^p(\mathbb{R}^m)$. (b) $(\int_{\mathbb{R}^n} u(y)^p\, d\lambda_n(y))^{1/p} \leq \int_{\mathbb{R}^m} v(x)\, d\lambda_m(x)$ if $u \in L^p(\mathbb{R}^n)$ and $v \notin L^1(\mathbb{R}^m)$ if $u \notin L^p(\mathbb{R}^n)$.

59. Let μ be a finite Borel measure on $\mathbb{R}$ and put $F(x) = \mu((-\infty, x])$. For $c > 0$ evaluate $J = \int_{\mathbb{R}} [F(x + c) - F(x)]\, d\lambda(x)$.

60. Let μ be a finite measure on $\mathbb{R}$ and define

$$f(x) = \int_{\mathbb{R}} \frac{\ln(|x - t|)}{|x - t|^{1/2}}\, d\mu(t), \quad x \in \mathbb{R}.$$

Prove that $f(x)$ is finite λ-a.e. for $x \in \mathbb{R}$.

61. Let $F \subset \mathbb{R}$ be a closed set with $\lambda(F^c) < \infty$ and denote by $\delta_F(x) = d(x, F)$ the distance of x to F. Prove: (a) δ_F is Lipschitz continuous, in fact, $|\delta_F(x) - \delta_F(y)| \leq |x - y|$. (b) If $M(x) = \int_{\mathbb{R}} \delta_F(y)/|x - y|^2\, d\lambda(y)$, then $M(x) = \infty$ for $x \in F^c$ and $M(x) < \infty$ for λ-a.e. $x \in F$.

62. Let $f \in L^1(I)$ and, for $x \in [0,1]$, define $g(x) = \int_x^1 t^{-1} f(t)\, d\lambda(t)$. Prove that g is Lebesgue integrable on $(0,1)$, and $\int_I g\, d\lambda = \int_I f\, d\lambda$.

63. Calculate

$$I = \int_{[0,\infty)} \int_{[0,\sqrt{\pi}]} \frac{x^3 y^3 \cos(y^2)}{(x^4 + y^4)^{3/2}}\, d\lambda(y)\, d\lambda(x).$$

64. Let $f : \mathbb{R}^2 \to \mathbb{R}$ be twice continuously differentiable in a neighborhood of $(a, b) \in \mathbb{R}^2$. Prove that the mixed derivatives f_{xy}, f_{yx} are equal, i.e., $f_{xy}(a, b) = f_{yx}(a, b)$.

65. Let $A \subset I$ be measurable. Compute $\int_I \int_I \sin(\pi t \chi_A(x))\, d\lambda(x)\, d\lambda(t)$.

66. Calculate

$$J = \int_0^\infty \int_0^\infty \frac{1}{(1+y)(1+x^2 y)}\, d\lambda(x)\, d\lambda(y)$$

and deduce the value of

$$K = \int_0^\infty \frac{\ln(x)}{x^2 - 1}\, d\lambda(x) \quad \text{and} \quad L = \int_0^1 \frac{\ln(x)}{x^2 - 1}\, d\lambda(x)\,.$$

67. Let $f : \mathbb{N} \times \mathbb{N} \to \mathbb{Z}$ be given by

$$f(m,n) = \begin{cases} 1, & m = n, \\ -1, & m = n + 1, \\ 0, & \text{otherwise.} \end{cases}$$

Discuss the validity of the following statement: $\sum_{m,n} |f(m,n)| < \infty$.

68. Compute

$$\int_0^\infty \frac{\sin(x)}{x}\, d\lambda(x)\,.$$

69. Let $E = \{(x,y) \in \mathbb{R}^2 : 0 < x < \infty, 0 < y < 1\}$. Compute

$$\int_E y \sin(x) e^{-xy}\, d(\lambda \otimes \lambda)(x,y)\,.$$

70. Let $(X, \mathcal{M}, \mu)$ be a measure space, $f \in L^p(X)$, $0 < p < \infty$, and $g : \mathbb{R}^+ \to \mathbb{R}^+$ be given by $g(t) = p\, t^{p-1} \mu(\{|f| > t\})$. Prove that g is Lebesgue measurable and $\int_X |f|^p\, d\mu = \int_0^\infty g(t)\, d\lambda(t)$.

71. Let $(X, \mathcal{M}, \mu)$ be a σ-finite measure space and $f \in L^1(X)$. Prove that $\int_X f\, d\mu = \int_0^\infty \mu(\{f > t\})\, d\lambda(t) - \int_0^\infty \mu(\{f < t\})\, d\lambda(t)$.

72. Let $(X, \mathcal{M}, \mu)$ be a σ-finite measure space and f, g nonnegative integrable functions on X. Prove that $\int_X fg\, d\mu = \int_0^\infty \varphi(t)\, d\lambda(t)$ where $\varphi(t) = \int_{F_t} g(x)\, d\mu(x)$ and $F_t = \{x \in X : f(x) > t\}$.

73. Let $(X, \mathcal{M}, \mu)$ be a σ-finite measure space and for nonnegative measurable functions f, g on X, let $F_t = \{f > t\}$ and $G_s = \{g > s\}$. Prove that $\int_X fg\, d\mu = \int_0^\infty \int_0^\infty \mu(F_t \cap G_s)\, d\lambda(s)\, d\lambda(t)$.

74. Let $(X, \mathcal{M}, \mu)$ be a σ-finite measure space and f, g integrable functions on X. Prove that

$$\int_X (f - g)\, d\mu = \int_{-\infty}^\infty \mu(\{f \geq t > g\})\, d\lambda(t) - \int_{-\infty}^\infty \mu(\{f < t \leq g\})\, d\lambda(t)\,.$$

75. Let $(X, \mathcal{M}, \mu)$ be a σ-finite measure space, f, g real-valued integrable functions on X, and $F_t = \{f > t\}, G_t = \{g > t\}$. Prove that $\int_X |f - g|\, d\mu = \int_{-\infty}^\infty \mu((F_t \setminus G_t) \cup (G_t \setminus F_t))\, d\lambda(t)$.

76. Let $(X, \mathcal{M}, \mu)$ be a measure space and f a positive measurable function on X. Prove that $\int_X \ln(1 + f)\, d\mu = \int_0^\infty (1 + t)^{-1}\, \mu(\{f > t\})\, d\lambda(t)$.

77. Let f be a nonnegative Lebesgue integrable function on $C = [0, 1] \times [0, 1]$ with integral 1. Prove that given $0 < \eta < 1$, there exists $x_0 \in [0, 1]$ and $R_{x_0} = \{(x, y) \in C : 0 \le x \le x_0, 0 \le y \le 1\}$ such that $\int_{R_{x_0}} f\, d(\lambda \otimes \lambda) = \eta$.

78. Let $a_1, a_2, a_3 > 1$, f a nonnegative measurable function on $\mathbb{R}^+$, and $J = \int_{\mathbb{R}^+ \times \mathbb{R}^+ \times \mathbb{R}^+} f(x_1 + x_2 + x_3)\, x_1^{a_1 - 1} x_2^{a_2 - 1} x_3^{a_3 - 1}\, d\lambda_3(x_1, x_2, x_3)$. Express J as a one-dimensional integral over $\mathbb{R}^+$.

79. Let $(X, \mathcal{M}, \mu), (Y, \mathcal{N}, \nu)$ be measure spaces, k a measurable function on $X \times Y$ such that $\int_X |k(x, y)|\, d\mu(x) \le A$ and $\int_Y |k(x, y)|\, d\nu(y) \le B$, and $Tf(x) = \int_Y k(x, y) f(y)\, d\nu(y)$. Prove that if $1 < p < \infty$, T is well-defined in $L^p(X)$ and with q the conjugate to p, $\|Tf\|_p \le A^{1/p} B^{1/q} \|f\|_p$.

80. Let $k : \mathbb{R}^n \times \mathbb{R}^n \to \mathbb{R}$ and $w : \mathbb{R}^n \to \mathbb{R}^+$ be measurable functions such that for all $x, y \in \mathbb{R}^n$,

$$\int_{\mathbb{R}^n} |k(x, y)| w(x)\, d\lambda_n(x) \le A w(y), \qquad \int_{\mathbb{R}^n} |k(x, y)| w(y)\, d\lambda_n(y) \le A w(x).$$

Prove that $Tf(x) = \int_{\mathbb{R}^n} k(x, y) f(y)\, d\lambda_n(y)$ maps $L^2(\mathbb{R}^n)$ boundedly into itself with norm $\le A$.

81. Let $(X, \mathcal{M}, \mu)$ be a finite measure space, f, g measurable functions on X, μ_f, μ_g the image measures defined in Problem 4.58, $\Phi(x) = (f(x), g(x))$, $x \in X$, and μ_Φ the image measure defined in Problem 4.58. Prove that f, g are independent iff $\mu_\Phi = \mu_f \otimes \mu_g$.

82. Let $(X, \mathcal{M}, \mu)$ be a finite measure space and f, g independent integrable functions on X. Prove that $\int_X fg\, d\mu = (\int_X f\, d\mu)(\int_X g\, d\mu)$.

83. Let $(X, \mathcal{M}, \mu)$ be a finite measure space, f, g independent measurable functions on X, and $0 < p < \infty$. Prove that $f + g \in L^p(X)$ iff $f \in L^p(X)$ and $g \in L^p(X)$.

84. Let $(X, \mathcal{M}, \mu)$ be a finite measure space, f, g independent measurable functions on X with $\int_X g\, d\mu = 0$, and $1 \le p < \infty$. Prove that $\int_X |f|^p\, d\mu \le \int_X |f + g|^p\, d\mu$.

85. Let f_a, $a > 0$, be the function on $\mathbb{R}^n$ given by $f_a(x) = \exp(-a|x|^2/2)$. Compute $f_a * f_b(x)$, $a, b > 0$, $x \in \mathbb{R}^n$.

86. Let $f \in L^p(\mathbb{R}^n)$, $g \in L^q(\mathbb{R}^n)$, where $p \ge 1$ and q is the conjugate exponent to p. Prove: (a) $f * g$ is defined on $\mathbb{R}^n$, is uniformly continuous, and verifies $\|f * g\|_\infty \le \|f\|_p \|g\|_q$. (b) If $p > 1$, $f * g(x) \to 0$ as $|x| \to \infty$. (c) Discuss the validity of the following statement: If $p = 1$ and $q = \infty$, then $f * g(x) \to 0$ as $|x| \to \infty$.

87. Discuss the validity of the following statement: Convolution has a unit in $L^1(\mathbb{R}^n)$.

88. Let f, g be integrable functions on $\mathbb{R}^n$ with nonvanishing integral of the same sign. Prove: (a) $f * g > 0$ on a set of positive measure. (b) If in addition f or g is bounded, $f * g > 0$ in an interval of $\mathbb{R}^n$.

89. Let f be an integrable function on $\mathbb{R}^n$ with $\|f\|_1 = A < 1$ and $f_k = f * \cdots * f$ where the convolution is performed k times. Prove that $\{f_k\}$ converges in $L^1(\mathbb{R}^n)$ and find its limit.

90. Let f, g be compactly supported $C^1(\mathbb{R})$ functions. Prove that $f * g \in C^2(\mathbb{R})$.

91. Let O, O_1 be bounded open subsets of $\mathbb{R}^n$ such that $\overline{O_1} \subset O$. Construct a function $h \in C_0^\infty(\mathbb{R}^n)$ such that $h = 1$ in O_1 and $h = 0$ outside O.

92. For a locally integrable function f on $\mathbb{R}^n$, with $Q(x, h)$ the cube centered at x of sidelength h, let

$$L_f = \left\{ x \in \mathbb{R}^n : f(x) = \lim_{h \to 0+} \frac{1}{h^n} \int_{Q(x,h)} f(y)\, d\lambda_n(y) \right\}$$

be the set where small averages of f converge to f. (a) Give an example of a function f for which $L_f = \emptyset$. (b) Give an example of a function f and $x \in L_f$ such that f is not continuous at x. (c) Given a bounded measurable function φ vanishing for $|x| > 1$ with integral η, let $\varphi_\varepsilon(x) = \varepsilon^{-n}\varphi(x/\varepsilon)$. Prove that $\lim_{\varepsilon \to 0+} f * \varphi_\varepsilon(x) = \eta\, f(x)$ for $x \in L_f$.

93. For $f \in L^p(\mathbb{R})$, $1 \le p \le \infty$, and $h > 0$, put

$$f_h(x) = \frac{1}{h} \int_{x-h/2}^{x+h/2} f(t)\, d\lambda(t)\,.$$

94. Let ψ be a measurable function on $\mathbb{R}^n$ such that for $\delta > 0$, ψ_δ satisfies the following properties: (1) $|\psi_\delta(x)| \le c\min\{\delta^{-n}, \delta|x|^{-n-1}\}$. (2) $\int_{\mathbb{R}^n} \psi_\delta\, d\lambda_n = \eta$. Prove that at each point of continuity x of $f \in L^1(\mathbb{R}^n)$, $\lim_{\delta \to 0+} f * \psi_\delta(x) = \eta f(x)$.

95. Discuss the validity of the following statement: There exists a compactly supported bounded function f on $\mathbb{R}^n$ with nonvanishing integral such that $f * f = f$.

96. Let v be a continuous function on $\mathbb{R}^n$ such that $\int_{\mathbb{R}^n} v\varphi\, d\lambda_n = 0$ for every compactly supported smooth function φ. Prove that $v = 0$.

97. Let $g \in L^p(\mathbb{R}^d)$, $1 \le p \le \infty$, and $f \in L^1(\mathbb{R}^n)$ with $\|f\|_1 < 1$. Prove that the equation $h - f * h = g$ has a unique solution in $L^p(\mathbb{R}^n)$.

98. Suppose that for all compactly supported functions f, g on $\mathbb{R}^n$, $\|f * g\|_r \leq \|f\|_p \|g\|_q$ for some $1 \leq p, q, r < \infty$. Find a relation the indices p, q, r satisfy.

99. Let g be a nonnegative measurable function on $\mathbb{R}^n$ such that $\|f * g\|_p \leq c \|f\|_p$ for all nonnegative $f \in L^p(\mathbb{R}^n)$, $1 \leq p \leq \infty$, and some constant c. Prove that $\|g\|_1 \leq c$.

100. Suppose that $f \in L^p(\mathbb{R}^n)$ and that $\|f * g\|_p \leq c\|g\|_1$ for all $g \in L^1(\mathbb{R}^n)$ and some constant c. Prove that $\|f\|_p \leq c$.

101. Given $f \in L^p(\mathbb{R}^k)$, $1 < p < \infty$, let $\tau_n f(x) = f(x - h_n)$, $n = 1, 2, \ldots$, where $\lim_n |h_n| = \infty$. Prove that $f_n \rightharpoonup 0$ in $L^p(\mathbb{R}^n)$.

102. Let $k \in L^1(\mathbb{R}^n)$ and $T : L^2(\mathbb{R}) \to L^2(\mathbb{R})$ the convolution operator given by $Tf(x) = k * f(x)$. Prove that if T is compact, $k = 0$ a.e.

Chapter 8

Normed Linear Spaces. Functionals

In this chapter we consider normed linear spaces, in particular those which are complete in the metric induced by the norm, or Banach spaces, and the existence and properties of linear functionals on these spaces. So, suppose X is a linear space over the field $\mathbb{R}$ of real scalars or over the field $\mathbb{C}$ of complex scalars; unless pointed out explicitly we restrict ourselves to the real case. A function on X that assumes values in the underlying scalar field is called a *functional*. We say that a nonnegative functional p on X is a *seminorm* provided the following two properties are satisfied: (i) (Triangle inequality) $p(x + y) \leq p(x) + p(y)$, $x, y \in X$. (ii) (Absolute homogeneity) $p(\lambda x) = |\lambda|\, p(x)$, λ scalar, $x \in X$. We say that p is a *norm*, and is denoted $\|\cdot\|$, provided that: (iii) $p(x) = 0$ implies $x = 0$, and, in this case, $(X, \|\cdot\|)$, or plainly X, is called a *normed linear space*. Now, if X is a normed linear space, the nonnegative function d on $X \times X$ given by $d(x, y) = \|x - y\|$, $x, y \in X$, is a metric, called the metric induced by $\|\cdot\|$. When (X, d) is complete we say that X is a *Banach space*.

A useful criterion to decide when a normed linear space is complete is the following. Given $\{x_n\} \subset X$, we say that the series $\sum_n x_n = s$ in X if the sequence $\{s_n\}$ of the partial sums $s_n = x_1 + \cdots + x_n$, $n = 1, 2, \ldots$, converges to s in the norm of X. Along the same lines we say that the series $\sum_n x_n$ is absolutely convergent in X if the numerical series $\sum_n \|x_n\|$ converges. Then a normed linear space X is a Banach space iff every absolutely convergent series converges in X.

Let X be a linear space and M a subspace of X. Given $x, y \in X$, define the equivalence relation $x \sim y$ on M if $x - y \in M$. The equivalence class of

$x \in X$ under this relation is denoted $[x] = x + M$, and the *quotient space* $X/M = \{x + M : x \in X\}$. Then X/M is a linear space, and the *canonical mapping* π of X onto X/M is defined by $\pi(x) = [x]$. When X is a normed linear space, we define $\|[x]\|_{X/M} = \mathrm{dist}(x, M) = \inf_{m \in M} \|x - m\|_X$. Then $\|\cdot\|_{X/M}$ is a well-defined semi-norm on X/M and if M is closed, then $\|\cdot\|_{X/M}$ is a norm on X/M.

We say that a functional L on X is *linear* if, for every $x, y \in X$ and scalar λ, $L(\lambda x + y) = \lambda L(x) + L(y)$. A linear functional L on X is said to be *bounded* if there is a constant c, independent of $x \in X$, such that $|L(x)| \leq c \|x\|$, all $x \in X$; the norm $\|L\|$ of L is defined as the infimum of the constants c for which this inequality holds. Boundedness is equivalent to *continuity*, i.e., if $\lim_n x_n = x$ in X, then $\lim_n L(x_n) = L(x)$, all $x \in X$, or even continuity at a single point.

The Hahn-Banach theorem, or theorems actually, is an indispensable tool in the theory of duality of linear spaces. In the case of arbitrary linear spaces, where no topology is apparent, it assures a plentiful supply of linear functionals, and, in the case of normed linear spaces, under some general domination assumptions, and with the aid of Zorn's lemma or one of its equivalent principles, a supply of bounded linear functionals. More precisely, suppose X is a real linear space and p a sublinear functional on X, i.e., p satisfies $p(x + y) \leq p(x) + p(y)$ and $p(\lambda x) = \lambda p(x)$ for all $x, y \in X$ and $\lambda > 0$. Further, let Y be a subspace of X and ℓ a real linear functional on Y such that $\ell(y) \leq p(y)$ for all $y \in Y$. Then there is a linear functional L on X that extends ℓ, i.e., $L(y) = \ell(y)$, all $y \in Y$, and $L(x) \leq p(x)$, all $x \in X$.

As for the complex version of the Hahn-Banach theorem, it states that if X is a complex linear space, p a *seminorn* on X, i.e., a nonnegative sublinear functional such that $p(\lambda x) = |\lambda| p(x)$ for all $x \in X$ and scalars λ, Y a subspace of X and ℓ a complex linear functional on Y such that $|\ell(y)| \leq p(y)$ for all $y \in Y$, then there is a linear functional L on X that extends ℓ, i.e., $L(y) = \ell(y)$, all $y \in Y$, and $|L(x)| \leq p(x)$, all $x \in X$.

The version that deals with continuous linear functionals is the following: Suppose X is a normed linear space, Y a subspace of X, and ℓ a bounded linear functional on Y. Then there is a norm preserving bounded linear functional L on X that extends ℓ, i.e., $L(y) = \ell(y)$, all $y \in Y$, such that $\|L\| = \|\ell\|$.

Also, there is a geometric version of the Hahn-Banach theorem: If X is a nonempty closed convex subset of a Banach space B and $x_0 \in B \setminus X$, there exists $L \in B^*$ such that $\sup_{x \in C} \Re(L(x)) < \Re(L(x_0))$.

The *conjugate*, or *dual*, space X^* of a normed linear space X is the linear space over the same scalar field as X consisting of all continuous linear functionals L on X. Normed with $\|L\| = \sup_{x \neq 0} |L(x)|/\|x\|$, $(X^*, \|\cdot\|)$ is a Banach space.

The *natural map J_X* of a normed linear space X into its second conjugate space X^{**} (the Banach space of bounded linear functionals on X^*) is defined by $J_X(x)L = L(x)$ for all $L \in X^*$. It is readily seen that $J_X(x)$ is a bounded linear functional on X^* and $\|J_X(x)\|_{X^{**}} = \|x\|$ for all $x \in X$. Thus, J_X establishes a linear isometric embedding from X into X^{**} and, in particular, every normed linear space is a dense subspace of a Banach space, called a completion of X. Now, if J_X is onto X^{**}, X is said to be *reflexive*.

Finally, we say that a nontrivial proper closed subspace M of a Banach space B is *complemented* in B if there exists a closed subspace N of B such that $M \cap N = \{0\}$ and $B = M + N$. In that case we write $B = M \oplus N$.

Problems in this chapter include the consideration of the equivalence of different norms in a linear space; in particular, all norms in a linear space are equivalent iff the space is finite dimensional, Problem 16.

As for Banach spaces, they cannot be expressed as the countable union of closed proper linear subspaces, Problem 19; see also Problem 34. Problem 23 considers the computation of the distance of an element to a proper subspace; different possibilities arise, and the distance is realized if the space is separable and reflexive, Problem 155. The sum of Banach spaces is considered in Problems 37–38, convexity properties in Problems 39–40, and extreme points in Problems 42–43.

The fact that disjoint convex sets in a normed linear space cannot necessarily be separated by a linear functional is covered in Problem 56. Problem 58 establishes that unbounded linear functionals exist. The properties of the kernel of a linear functional are considered in Problems 60–61, and Problem 65 establishes that two functionals that have the same kernel are proportional. The question of the uniqueness of the extension of a bounded linear functional defined on a subspace of a Banach space is addressed in Problems 68–72, and Problem 84 highlights the fact that extensions do not always exist. The question as to whether a linear functional attains its norm is considered in Problem 85; Problem 115 provides an affirmative answer in terms of best approximation and Problem 139 in terms of reflexivity.

Problem 106 establishes that ℓ^1 is not reflexive; in fact, by Problem 109, $c_0^* = \ell^1$. Also, by Problem 114, for no normed linear space X does $X^* = c_0$. Properties of reflexive spaces are discussed in Problems 150–153.

The notion of weak convergence is discussed in Problems 130–138, and quotient spaces in Problems 158–165. Problem 176 establishes that every separable Banach space is isometrically isomorphic to a quotient space of ℓ^1.

The interested reader can further consult, for instance, N. Dunford and J. T. Schwartz, *Linear operators, General theory*, Wiley Classics Library, 1988; K. Yosida, *Functional analysis*, Springer-Verlag, 1980; and J. Conway, *A course in functional analysis*, Springer-Verlag, New York, 1990.

Problems

1. Show that an infinite-dimensional linear space X can be equipped with a norm.

2. Prove that in a normed linear space X, $\lim_n \|x_n - x\| = 0$ implies $\lim_n \|x_n\| = \|x\|$.

3. Given $x_1, \ldots, x_n$ in a normed linear space X, prove that

$$\|x_1 + \cdots + x_n\| \geq \|x_1\| - \left(\|x_2\| + \cdots + \|x_n\| \right).$$

4. Let X be a real or complex normed linear space, $x, y \in X$, and $\lambda \in \mathbb{R}$. Find $\lim_n (\|(n + \lambda)x + y\| - \|nx + y\|)$.

5. Prove that for all x, y in a normed linear space X,

$$\|x\| \leq \max\{\|x - y\|, \|x + y\|\}.$$

Can equality hold for $x \in X$ and a suitable $y \neq 0$?

6. Let X be a normed linear space. Prove that for all nonzero x, y in X, the following inequalities hold:

$$\|x - y\| \geq \frac{1}{2} \max\{\|x\|, \|y\|\} \left\| \frac{x}{\|x\|} - \frac{y}{\|y\|} \right\|,$$

$$\|x - y\| \geq \frac{1}{4} \left(\|x\| + \|y\| \right) \left\| \frac{x}{\|x\|} - \frac{y}{\|y\|} \right\|.$$

Are these inequalities sharp?

7. Let $\{x_n\}$ be a convergent sequence in a normed linear space X. Prove that the sequences $\{v_n\}$ and $\{w_n\}$ given by $v_n = (x_1 + \cdots + x_n)/n$ and $w_n = (x_1 + 2x_2 + \cdots + nx_n)/n^2$, respectively, converge, and find the limits.

8. Let B be a Banach space, $\{x_n\}$ elements in B of norm 1, and $\{\lambda_n\}$ scalars. Discuss the validity of the following statement: $\sum_n \lambda_n x_n$ converges in B iff $\sum_n |\lambda_n| < \infty$.

9. Discuss the validity of the following statements: (a) An absolutely convergent series in a normed linear space X converges. (b) An absolutely convergent series in a Banach space B converges unconditionally, i.e., for every permutation σ of the integers, the series $\sum_n x_{\sigma(n)}$ converges to the same limit.

10. Let $\{x_n\}$ be a sequence in a Banach space B such that given $\varepsilon > 0$, there is a convergent sequence $\{y_n\}$ with $\|y_n - x_n\| < \varepsilon$ for all n. Prove that $\{x_n\}$ converges and give an example that shows that the assertion is false if B is not complete.

11. Let X be a normed linear space. Prove that X is a Banach space iff the unit sphere $S = \{x \in X : \|x\| = 1\}$ equipped with the metric $d(x, y) = \|x - y\|$ is complete.

12. On $\mathbb{R}^2$ let

$$p(x) = \begin{cases} \sqrt{x_1^2 + x_2^2}, & x_1 x_2 \geq 0, \\ \max\{|x_1|, |x_2|\}, & x_1 x_2 < 0. \end{cases}$$

Is $p(x)$ a norm? Is it possible to define a norm $\|\cdot\|$ on $\mathbb{R}^2$ such that the vectors $(1, 0), (0, 1)$ have norm 1 and $\|(1, 1)\| < 1$?

13. Let X be a normed linear space and $\overline{B(x, r)}$ denote the closure of $B(x, r)$, the open ball centered at x of radius r. Prove: (a) $\overline{B(x, r)}$ is the closed ball centered at x of radius r. (b) The diameter of $B(x, r)$ is $2r$. (c) If $B(x, r) \subset B(y, s)$, then $r \leq s$ and $\|x - y\| \leq s - r$. (d) If X is complete, a decreasing sequence of nonempty closed balls has nonempty intersection.

14. Let X be a normed linear space and $\rho : X \to \mathbb{R}$ given by

$$\rho(x) = \frac{\|x\|}{1 + \|x\|}.$$

(a) Is ρ a norm on X? (b) Let $r : X \times X \to \mathbb{R}$ be given by $r(x, y) = \rho(x - y)$. Prove that r is a metric on X. (c) Prove that $\lim_n \|x_n - x\| = 0$ iff $\lim_n r(x_n, x) = 0$. (d) Give an example of a metric in a linear space X that is not associated with a norm as in (b).

15. Let $\{x_1, \ldots, x_n\}$ be linearly independent elements of a normed linear space X. Prove that there is a constant $c > 0$ such that for every choice of scalars $\lambda_1, \ldots, \lambda_n$, $\|\lambda_1 x_1 + \cdots + \lambda_n x_n\| \geq c(|\lambda_1| + \cdots + |\lambda_n|)$.

16. Prove that all norms on a linear space are equivalent iff the space is finite dimensional.

17. Prove that a finite-dimensional subspace of a normed linear space X is closed.

18. Let Y be a subspace of a normed linear space X. Prove: (a) If Y has nonempty interior, $Y = X$. (b) Y^c is empty or dense in X.

19. Let B be a Banach space and $\{X_n\}$ closed linear subspaces of B with $X_n \neq B$ for all n. Prove that $\bigcup_n X_n \neq B$ and give an example that shows that the assumption that B is a Banach space cannot be dropped.

20. Let B be a Banach space. Discuss the validity of the following statements: (a) If X_1, X_2 are closed in B, $X_1 + X_2$ is closed in B. (b) If X_1, X_2 are closed subspaces of B, $X_1 + X_2$ is a closed subspace of B.

21. Let X be an infinite-dimensional normed linear space. Prove that there is a sequence $\{x_n\} \subset X$ such that $\|x_n\| = 1$ for all n, and $\|x_n - x_m\| \geq 1/2$ for all $n \neq m$.

22. Let X be a separable normed linear space and S an uncountable subset of X. Prove that there are a sequence $\{x_n\} \subset S$ and $x \in S$ such that $x_n \to x$ and $x_n \neq x$ for all n.

23. Let B be a Banach space, Y a subspace of B, and $d(x, Y) = \inf_{y \in Y} \|x - y\|$ the distance from $x \in B$ to Y. (a) Prove that if Y is finite dimensional there exists $y_0 \in Y$ with $d(x, Y) = \|x - y_0\|$ and show that y_0 is not necessarily unique. (b) Show that if Y is infinite dimensional, y_0 does not generally exist, even when Y is closed. (c) Suppose that B satisfies the additional property: There is equality $\|x + y\| = \|x\| + \|y\|$ in the triangle inequality iff x and y are linearly dependent. Prove that there is a unique $y \in Y$ such that $d(x, Y) = \|x - y\|$.

24. Let X be a normed linear space and for $K \subset X$ and $x \in X$, let $\mathcal{D}_K(x) = \{y \in X : \|y - x\| = d(x, K)\}$. Prove that if K is convex, $\mathcal{D}_K(x)$ is convex for all $x \in X$.

25. We say that a normed linear space X is *strictly convex* if $\|x\| = \|y\| = 1$ implies $\|x + y\| < 2$. Prove that if K is a convex subset of a strictly convex normed linear space X, $\mathcal{D}_K(x) = \emptyset$ or $\mathcal{D}_K(x)$ contains a single point for every $x \in X$.

26. Let X be a normed linear space and x, y in X such that $\|x + y\| = \|x\| + \|y\|$. Prove that for all $\lambda, \mu \geq 0$, $\|\lambda x + \mu y\| = \lambda\|x\| + \mu\|y\|$.

27. Let X be a normed linear space. Prove that the following statements are equivalent: (a) For all x, y in X, $y \neq 0$, $\|x + y\| = \|x\| + \|y\|$ implies $x = \lambda y$, $\lambda \geq 0$. (b) For all x, y in X, $x \neq y$, $\|x\| = \|y\| = 1$ implies $\|x + y\| < 2$.

28. Discuss the validity of the following statement: A normed linear space X contains linearly independent x, y such that $\|x\| = \|y\| = 1$ and $\|x + y\| = 2$ iff the unit sphere of X contains a line segment.

29. Let $X = \{x \in c_0 : \sum_n 2^{-n} x_n = 0\}$ and $x_0 = (2, 0, \ldots) \in c_0$. Prove that X is a closed subspace of c_0 and compute $d(x_0, X)$. Is the distance attained?

30. Let Y be a proper closed subspace of a normed linear space X. Prove that

$$\sup_{0 \neq x \in X} \frac{d(x, Y)}{\|x\|} = 1 \,.$$

31. Let Y be a proper closed subspace of a normed linear space X. Prove that if $B(x, r) \subset B_Y(0, s) = \{y \in Y : \|y\| < s\}$, then $r \leq s$.

32. Let $\{Y_n\}$ be a sequence of proper closed subsets of a normed linear space X, $r_n \to 0^+$, and $A = \{x \in X : d(x, Y_n) = O(r_n)\}$. Prove that A is of first category in X.

33. Let X be a normed linear space with unit ball $B(0, 1)$. Prove that if for some $r > 1$ the closed ball $\overline{B(0, r)}$ of X can be covered by finitely many translates of $B(0, 1)$, then X is finite dimensional. From this deduce that if a closed ball in X is compact, X is finite dimensional.

34. Let X be a linear space that contains a countable set $\{x_n\}$ such that every $x \in X$ can be written as a finite linear combination of the x_n. Prove that X cannot be equipped with a norm so that $(X, \|\cdot\|)$ is a Banach space.

35. Prove that a Banach space is finite dimensional iff each of its subspaces is closed.

36. We say that $\{B_\alpha\}_{\alpha \in \Lambda}$ is a compatible family of Banach spaces if the following two conditions hold: The B_α are continuously embedded in a Banach space B, and, if a sequence $\{x_n\} \subset B_\alpha \cap B_\beta$ converges to x_α in B_α and to x_β in B_β, then $x_\alpha = x_\beta$. Prove that if $X = \{x \in \bigcap_{\alpha \in \Lambda} B_\alpha : \|x\|_X = \sum_{\alpha \in \Lambda} \|x\|_{B_\alpha} < \infty\}$, then $(X, \|\cdot\|_X)$ is a Banach space.

37. Let B_0, B_1 be Banach spaces continuously embedded in a Banach space B and consider the linear space $B_0 + B_1 = \{x \in B : x = x_0 + x_1, x_0 \in B_0, x_1 \in B_1\}$ and $\|x\|_{B_0 + B_1} = \inf\{\|y\|_{B_0} + \|z\|_{B_1} : x = y + z, y \in B_0, z \in B_1\}$. Prove that $\|\cdot\|_{B_0 + B_1}$ is a norm and $(B_0 + B_1, \|\cdot\|_{B_0 + B_1})$ is a Banach space.

38. Let B_0, B_1 be closed subspaces of a Banach space B. Prove that $B_0 + B_1$ is closed in B iff there exists a constant c such that $\|x\|_{B_0 + B_1} \leq c\|x\|_B$ for all $x \in B_0 + B_1$.

39. Discuss the validity of the following statement: Given a nested decreasing sequence $\{K_n\}$ of nonempty bounded closed convex sets in a Banach space B, $\bigcap_n K_n \neq \emptyset$.

40. Let K be a nonempty convex closed subset of a Banach space B. Discuss the validity of the following statements: (a) K has an element of least norm. (b) An element of least norm in K is unique.

41. Let B be a Banach space and $c_0(B) = \{$sequences $x : x_n \in B$ and $\lim_n \|x_n\| = 0\}$. Prove that endowed with the norm $\|x\|_\infty = \sup_n \|x_n\|$, $(c_0(B), \|\cdot\|_\infty)$ is a Banach space.

42. Let K be a convex subset of a normed linear space X. We say that x is an *extreme point of K* iff $x = \lambda y + (1 - \lambda)z, 0 < \lambda < 1, y, z \in K$, implies $x = y = z$, i.e., x is not in the interior of any line segment joining two distinct points of K. Characterize: (a) $E(B_{\ell^\infty})$, the extreme points in the unit ball of ℓ^∞. (b) $E(B_c)$, the extreme points in the unit ball of the real c. (c) $E(B_{c_0})$, the extreme points in the unit ball of c_0. (d) $E(B_{\ell^1})$, the extreme points in the unit ball of ℓ^1.

43. Characterize: (a) $E(B_{L^\infty(I)})$, the extreme points in the unit ball of real $L^\infty(I)$. (b) $E(B_{C(I)})$, the extreme points in the unit ball of real $C(I)$. (c) $E(B_{L^1(I)})$, the extreme points in the unit ball of real $L^1(I)$.

44. Let $\mathcal{P}_n$ denote the polynomials p of degree at most n normed by $\|p\| = \sum_{k=0}^n |p(k)|$. Is $(\mathcal{P}_n, \|\cdot\|)$ a Banach space?

45. Let $\{p_n\}$ be a sequence of polynomials that converges uniformly to some $f \in C([a,b])$, which is not a polynomial in $[a,b]$. Prove that $\sup_n \text{degree}(p_n) = \infty$.

46. Let w be a strictly positive function on $I = [0, 1]$. Prove: (a) $\|x\|_{\infty,w} = \sup_{t \in I} |x(t)|\, w(t)$ is a norm in $C(I)$. (b) If $\inf_I w = m > 0$ and $\sup_I w = M < \infty$, then $m\,\|x\|_{\infty,w} \le \|x\|_\infty \le M\,\|x\|_{\infty,w}$ and $\|\cdot\|_{\infty,w} \sim \|\cdot\|_\infty$. (c) If $w(t) = t$, $\|\cdot\|_{\infty,w}$ is not equivalent to $\|\cdot\|_\infty$. In fact, $(C(I), \|\cdot\|_{\infty,w})$ is not a Banach space.

47. Recall that $x : I \to \mathbb{R}$ is Lipschitz continuous if for some constant M, $|x(t) - x(s)| \le M\,|t - s|$ for all $t, s \in I$; the infimum over the constants M is called the Lipschitz constant of x. Let $X = \{x : I \to \mathbb{R} : x$ is Lipschitz continuous$\}$. Is X a closed subspace of $C(I)$?

48. Let $X = C(I)$ and $M_\alpha(I) = \{x \in X : x \in C^1(I), |x'(t)| \le \alpha\}$, $\alpha > 0$. Characterize $\overline{M}_\alpha(I)$, the closure of $M_\alpha(I)$ in $C(I)$.

49. For $x \in C(I)$, $0 < \alpha < 1$, put

$$p_\alpha(x) = \sup_{t,s \in I, t \neq s} \frac{|x(t) - x(s)|}{|t - s|^\alpha},$$

and let $C^\alpha(I) = \{x \in C(I) : p_\alpha(x) < \infty\}$. Prove that p_α is a seminorm on $C^\alpha(I)$ and introduce a norm in $C^\alpha(I)$ that turns it into a Banach space.

50. Prove: (a) For $x \in C^\alpha(I)$, $\lim_{\beta \to \alpha^-} \|x\|_\beta = \|x\|_\alpha$. (b) If $x_n \to x$ in $C(I)$ and $\|x_n\|_\alpha \le 1$ for all n, then $x_n \to x$ in $C^\beta(I)$ for $\beta < \alpha$. (c) For $\beta < \alpha$, $S = \{x \in C^\beta(I) : \|x\|_\alpha \le 1\}$ is compact in $C^\beta(I)$.

51. Prove that $C^\alpha(I)$, $0 < \alpha < 1$, is of first category in $C(I)$.

52. For $x \in C(I)$, let $Z(x) = \{t \in I : x(t) = 0\}$ denote the zero set of x. Prove that $\{x \in C(I) : |Z(x)| = 0\}$ is a dense G_δ set in $C(I)$.

53. Let $A = \{e_1, \ldots, e_n, \ldots\}$. Identify $\overline{\mathrm{sp}}(A)$, the closed span of A in ℓ^p for $1 \leq p \leq \infty$.

54. Let $\ell_c^\infty = \{x \in \ell^\infty : x$ is eventually constant, i.e., at most finitely many of the x_n are distinct$\}$. What is the closure of ℓ_c^∞ in ℓ^∞?

55. Consider the unit ball $B_{C(I)}$ in $C(I)$ and $\phi : B_{C(I)} \to \mathbb{R}$ a continuous map. Is ϕ necessarily bounded?

56. Let $A = \{x \in \ell_0^2 : $ if $x_N \neq 0$ and $x_n = 0$ for all $n > N$, then $x_N > 0\}$, and $B = -A$. Prove that A, B are disjoint convex sets, and for any nonzero linear functional L on ℓ^2, $L(A) = L(B) = \mathbb{R}$. Thus A and B cannot be separated by a linear functional.

57. Let X be a real linear space and K a convex subset of X with the property that for any $0 \neq x \in X$, there exists $M(x) > 0$ such that $\{\lambda \in \mathbb{R} : \lambda x \in K\} = (-M(x), M(x))$. Let $x_0 \in X \setminus K$. Prove that there is a linear functional $L : X \to \mathbb{R}$ such that $L(x_0) \geq 1$ and $L(x) < 1$ for all $x \in K$.

58. Let X be an infinite-dimensional normed linear space. Construct an unbounded linear functional on X.

59. Let Y be a subspace of a linear space X and L a linear functional on X that does not map Y onto the scalar field. Prove that L is the 0 functional.

60. Let X be a normed linear space over $\mathbb{C}$ and $L : X \to \mathbb{C}$ a discontinuous linear functional. Prove: (a) $\{L(x) : \|x\| < 1\} = \mathbb{C}$. (b) $K(L)$ is dense in X.

61. Let $0 \neq L$ be a linear functional on a normed linear space X. Prove that the following statements are equivalent: (a) L is continuous. (b) $K(L)$ is a proper closed subspace of X. (c) $K(L)$ is not dense in X.

62. Let L be a linear functional on a normed linear space X. Prove that L is continuous iff $\{x \in X : L(x) = \lambda\}$ is closed for any scalar λ.

63. We say that a subspace Y of a normed linear space X is a *hyperplane* if $Y \neq X$ and there is no proper subspace W of X such that $Y \subsetneq W \subsetneq X$. Let $L \neq 0$ be a linear functional on X. Prove: (a) $K(L)$ is a hyperplane. (b) If L is not continuous, $K(L)$ is dense in X. (c) Y is a hyperplane iff $Y = K(\ell)$ for some linear functional ℓ on X.

64. Let Y be a proper subspace of a normed linear space X. For $x_0 \in X \setminus Y$ let L be the linear functional on $Y + \mathrm{sp}\{x_0\}$ given by $L(y + \lambda x_0) = \lambda$. Prove that L is continuous iff $x_0 \notin \overline{Y}$.

65. Let X be a normed linear space and L, L_1 linear functionals on X. Prove: (a) If $K(L) = K(L_1)$, L and L_1 are proportional. (b) $K(L)$ is connected and $X \setminus K(L)$ is dense in X. (c) If L is continuous, $X \setminus K(L)$ has exactly two connected components.

66. Let X be a normed linear space and L, L_1 nontrivial bounded linear functionals on X such that $|L_1(x)| \leq \varepsilon$ for all $x \in K(L)$ with $\|x\| \leq 1$. Prove that there is a scalar η such that $\|L_1 - \eta L\| \leq \varepsilon$.

67. Let X be a normed linear space and ℓ a mapping of the unit ball $\{\|x\| \leq 1\}$ of X into $\mathbb{R}$ with the property that $\ell(\lambda x + \mu y) = \lambda \ell(x) + \mu \ell(y)$ whenever x, y, and $\lambda x + \mu y$ are in the unit ball. Prove that ℓ may be extended to a bounded linear functional L on X.

68. Let X be a normed linear space. (a) Let Y be a subspace of X and ℓ a continuous linear functional on Y of norm 1. Prove that if X^* is strictly convex, ℓ has a unique extension to a linear functional L on X of norm 1. (b) If X^* is not strictly convex, construct a subspace Y of X and a continuous linear functional on Y of norm 1 admitting two distinct extensions to X of norm 1.

69. Describe an infinite-dimensional subspace X of ℓ^1 such that any bounded linear functional on X has $2^{\aleph_0}$ distinct norm preserving extensions to ℓ^1.

70. Let $Y = \mathbb{R} \times \{0\} \subset \mathbb{R}^2$ and ℓ the linear functional on Y given by $\ell(t, 0) = t$ for $t \in \mathbb{R}$. Discuss the validity of the following statement: ℓ has a unique extension to $(\mathbb{R}^2, \| \cdot \|_p)$, $1 \leq p < \infty$.

71. Let $X_p = (\mathbb{R}^2, \| \cdot \|_p)$, $1 \leq p \leq \infty$. On $Y_p = \{x \in X_p : x_1 - 2x_2 = 0\}$, consider the linear functional ℓ given by $\ell(x_1, x_2) = x_1$. Compute $\|\ell\|$ and determine the norm preserving linear functionals L that extend ℓ to X_p for $p = 1, 2, \infty$.

72. For $A \subset \mathbb{N}$, let 1_A denote the sequence with terms $x_n = \chi_A(n)$, i.e., $x_n = 1$ if $n \in A$ and $x_n = 0$ otherwise. (a) Given $L \in \ell^{\infty*}$, let μ_L denote the set function on $\mathcal{P}(\mathbb{N})$ given by $\mu_L(A) = L(1_A)$. Prove that μ_L is a bounded finitely additive set function on $\mathcal{P}(\mathbb{N})$ such that $\sum_k |\mu_L(A_k)| \leq C$ for all finite partitions $A_1, \ldots, A_n$ of $\mathbb{N}$ and a constant C. (b) Let X denote the subspace of ℓ^∞ defined by $X = \mathrm{sp}\{1_A : A \in \mathcal{P}(\mathbb{N})\}$. Prove: (i) X is dense in ℓ^∞. (ii) If $\mu : \mathcal{P}(\mathbb{N}) \to \mathbb{R}$ is finite, finitely additive, and satisfies the partition property described in (a), μ induces a unique linear functional ℓ on X such that if $x = \sum_n \lambda_n 1_{A_n} \in X$, then $\ell(x) = \sum_n \lambda_n \mu(A_n)$. (iii) ℓ has a unique extension to a linear functional $L : \ell^\infty \to \mathbb{R}$ such that $\mu_L = \mu$.

73. Let $X = M \oplus N$ be a normed linear space where M, N are closed subspaces of X with M finite dimensional. Prove that if $\ell \neq 0$ is a continuous linear functional on M, $L(x) = L(m + n) = \ell(m)$ is a continuous linear functional on X that extends ℓ.

74. Let X be a linear space, M, N subspaces of X, and ℓ_M, ℓ_N linear functionals on M, N, respectively, such that $\ell_M|_{M \cap N} = \ell_N|_{M \cap N}$. Prove that there is a linear functional L on $M + N$ such that $L|_M = \ell_M$ and $L|_N = \ell_N$. Furthermore, if X is a Banach space, M, N, and $M + N$ closed, and ℓ_M, ℓ_N bounded, then L is bounded.

75. Let X be a real normed linear space and $S = \{p : p : X \to \mathbb{R}$ is sublinear, i.e., positively homogenous and subadditive$\}$ partially ordered by pointwise ordering (that is, $p \prec q$ iff $p(x) \leq q(x)$ for all $x \in X$). Prove that the minimal elements of $(S, \prec)$ are precisely the linear functionals on X.

76. Let X be a real linear space and p a sublinear functional on X. Prove that given $x_0 \in X$, there exists a linear functional L on X such that $L(x_0) = p(x_0)$ and $L(x) \leq p(x)$ for all $x \in X$.

77. Let X be a finite-dimensional linear space and $x \neq y \in X$. Prove that there is a linear functional L on X such that $L(x) \neq L(y)$.

78. Let X be a normed linear space and $0 \neq x_0 \in X$. Prove that there exists a bounded linear functional L on X such that $\|L\| = 1/\|x_0\|$ and $L(x_0) = 1$.

79. Let X be a real normed linear space; we say that the norm is smooth at $x \in X$ if for every $y \in X$ the function $\phi(t) = \|x + ty\|$ is differentiable at $t = 0$. For $\|x\| = 1$, a linear functional $L \in X^*$ such that $\|L\| = L(x) = \|x\| = 1$ is called a norming functional at x. Prove that a norming linear functional at x exists and that, if the norm is smooth at x, it is unique.

80. Let Y be a subspace of a normed linear space X and $x \in X$ such that $d(x, Y) = d > 0$. Prove that there exists $L \in X^*$ such that $L(x) = d$, $\|L\| = 1$, and $L(y) = 0$ for $y \in Y$.

81. Let X be a real normed linear space and ℓ, ℓ_1 linear functionals on a subspace Y of X such that $|\ell(y)| + |\ell_1(y)| \leq \|y\|$ for all $y \in Y$. Prove that there exist $L, L_1 \in X^*$ that extend ℓ and ℓ_1 to X, respectively, and verify $|L(x)| + |L_1(x)| \leq \|x\|$, $x \in X$.

82. Let X be a linear space, p, q seminorms on X, and L a linear functional on X such that $|L(x)| \leq p(x) + q(x)$ for all $x \in X$. Prove that there exist linear functionals L_1, L_2 on X such that $L(x) = L_1(x) + L_2(x)$ and $|L_1(x)| \leq p(x)$, $|L_2(x)| \leq q(x)$ for all $x \in X$.

83. Let X be a normed linear space. Discuss the validity of the following statements: (a) For every $x \in X$, there is $0 \neq L \in X^*$ such that $|L(x)| = \|L\| \, \|x\|$. (b) For every $L \in X^*$, there is $0 \neq x \in X$ such that $|L(x)| = \|L\| \, \|x\|$.

84. Let B be a Banach space and X a proper dense subspace of B such that equipped with $\|\cdot\|_X$, $(X, \|\cdot\|_X)$ is complete. Prove that if $\|x\|_B \le C\|x\|_X$ for a constant C and all $x \in X$, there is a continuous linear functional on $(X, \|\cdot\|_X)$ that cannot be extended to a continuous linear functional on $(B, \|\cdot\|_B)$.

85. Let L be the bounded linear functional on $C(I)$ given by $L(x) = \int_I tx(t)\,dt$. (a) Compute $\|L\|$ and find $x \in C(I)$ so that $L(x) = \|L\|$. (b) Consider the subspace $Y = \{x \in C(I) : x(1) = 0\}$ and the restriction $L|_Y = L_1$ of L to Y. Show that $\|L_1\| = \|L\|$ and that $\|L_1\|$ is not attained on Y.

86. Let $\mathcal{P}_n$ denote the space of real polynomials on I of degree $\le n$. Prove: (a) For every $a \in I$, there is a unique $P_{n,a} \in \mathcal{P}_n$ such that $p(a) = \int_I p(t)\,P_{n,a}(t)\,dt$ for all $p \in \mathcal{P}_n$. Furthermore, show that for no finite Borel measure μ on I with $\mu(\{1\}) = 0$ it follows that $p(1) = \int_I p(t)\,d\mu(t)$ for all $p \in \bigcup_n \mathcal{P}_n$. (b) For every $a \in [0,1]$, there is a constant $c_{n,k,a}$ depending on a, k, and n, but independent of $p \in \mathcal{P}_n$, such that for $0 \le k \le n$, $|p^{(k)}(a)| \le c_{n,k,a} \int_0^1 |p(t)|\,dt$, $p \in \mathcal{P}_n$. Is it possible to find a constant c_a, independent of k, n, such that the inequality holds for all p, no matter its degree?

87. Discuss the validity of the following statements: There is a bounded linear functional L on $L^q(I)$, $1 \le q < \infty$, such that $L(x) = x'(0)$: (a) for all $x \in \mathcal{P}_n$, the polynomials of degree $\le n$, (b) for all $x \in \mathcal{P}$, all polynomials, and (c) for all $x \in C^1(I)$.

88. Let L be the functional on $C(I)$ defined by $L(x) = \sum_{k=1}^n \lambda_k\, x(t_k)$ where the λ_k are real numbers and the t_k different points in I. Compute $\|L\|$.

89. Let X be a normed linear space, $M > 0$, and suppose $x \in X$ satisfies $|L(x)| \le M$ for all bounded linear functionals L on X with $\|L\| \le b$. Prove that $\|x\| \le M/b$.

90. Let B be in a real Banach space and $x, y \in B$ with $\|x\| = \|y\| = 1$ such that $\|x + 2y\| = \|x - 2y\| = 3$. Prove that $\|\lambda x + \mu y\| = |\lambda| + |\mu|$ for all reals λ, μ.

91. For $1 \le p < \infty$, $-\infty < s < \infty$, let $h_s^p = \{x : \sum_n ((1+n)^s |x_n|)^p < \infty\}$, $1 \le p < \infty$, and $h_s^\infty = \{x : \sup_n (1+n)^s |x_n| < \infty\}$. Prove: (a) $\|\{(1+n)^s x_n\}\|_{\ell^p} \sim \|\{(1+n^p)^{s/p} x_n\}\|_{\ell^p}$, $1 \le p < \infty$. (b) $\|x\|_{h_s^p} = \|\{(1+n)^s x_n\}\|_{\ell^p}$ is a norm and $(h_s^p, \|\cdot\|_{h_s^p})$ is a Banach space. (c) If $s > r > 0$ and $\|y^n\|_{h_s^p} \le c$ for all n, a subsequence $\{y^{n_k}\}$ of $\{y^n\}$ converges to some $y \in h_r^p$, i.e., closed balls in h_s^p are compact in h_r^p for $s > r > 0$. (d) ℓ^p is dense in h_s^p, all p and s. (e) $h_s^{p*} = h_{-s}^q$ where $1 \le p < \infty$ and q is the conjugate to p.

92. Let X be a normed linear space, $Y \subset X$, and L_0 a linear functional on Y. Prove that the following statements are equivalent: (a) L_0 has a linear extension L to X of norm $\leq \gamma$. (b) For all $m \in \mathbb{N}$ and any $x_1, \ldots, x_m$ in Y and scalars $\lambda_1, \ldots, \lambda_m$, $|\sum_{n=1}^{m} \lambda_n L_0(x_n)| \leq \gamma \|\sum_{n=1}^{m} \lambda_n x_n\|$.

93. Let X be a normed linear space and Y a closed subspace of X. We say that $z \in X$ is orthogonal to Y, and write $z \perp Y$, if $d(z, Y) = \|z\|$. Prove that $z \perp Y$ iff there exists $0 \neq L \in X^*$ such that $Y \subset K(L)$ and $|L(z)| = \|L\| \|z\|$. Also prove that if $Y \neq X$, for any $x_0 \notin Y$, $y_0 \in Y$ is closest to x_0 (i.e., $0 < d(x_0, Y) = \|x_0 - y_0\| \leq \|x_0 - y\|$ for all $y \in Y$) iff $x_0 - y_0 \perp Y$.

94. Let Y be a closed subspace of a normed linear space X. Prove that if every functional $L \in X^*$ that vanishes on Y vanishes on X, then $Y = X$.

95. Let X be a normed linear space and $Y \subset X$. Prove: (a) $x \in X$ is in the closure of $\mathrm{sp}\{Y\}$ iff $L(x) = 0$ for every $L \in X^*$ such that $Y \subset K(L)$. (b) Y is dense in X iff every bounded linear functional L on X that vanishes on Y vanishes identically.

96. Let $L_1, \ldots, L_n$ be bounded linear functionals on an infinite-dimensional normed linear space X. Prove: (a) There exists $0 \neq x \in X$ such that $L_1(x) = \ldots = L_n(x) = 0$. (b) If $L \in X^*$ is such that $L_1(x) = \ldots = L_n(x) = 0$ implies $L(x) = 0$, then L is a linear combination of $L_1, \ldots, L_n$.

97. Let B be a Banach space. Prove that $\{x_1, \ldots, x_n\} \subset B$ are linearly independent iff there exist continuous linear functionals $L_1, \ldots, L_n$ biorthogonal to $x_1, \ldots, x_n$, i.e., for $1 \leq m, k \leq n$,

$$L_m(x_k) = \begin{cases} 1, & m = k, \\ 0, & m \neq k. \end{cases}$$

98. Let B be a Banach space and $\{L_1, \ldots, L_n\} \subset B^*$. Prove: (a) $\{L_1, \ldots, L_n\} \subset B^*$ are linearly independent iff there exist $x_1, \ldots, x_n \in B$ such that

$$L_m(x_k) = \begin{cases} 1, & m = k, \\ 0, & m \neq k. \end{cases}$$

(b) $L \in \mathrm{sp}\{L_1, \ldots, L_n\}$ iff $K(L_1) \cap \ldots \cap K(L_n) \subset K(L)$.

99. Let X be a normed space and $x \in X$ with $\|x\| = 1$. Prove that there is a closed subspace M of X such that $X = M \oplus \mathrm{sp}\{x\}$ and $d(x, M) = 1$.

100. Let $S(0, r)$ be a sphere in a real normed linear space X and $x_0 \in S(0, r)$. Prove there is a hyperplane H_0 that contains x_0 so that the closed ball $\overline{B(0, r)}$ is contained in one of the two semispaces determined by H_0.

101. Let M be a finite-dimensional subspace of a normed linear space X. Prove that M is complemented in X.

102. Let B be an infinite-dimensional Banach space. Construct a strictly decreasing infinite sequence $\{X_n\}$ of infinite-dimensional closed linear subspaces of B.

103. Let X be a normed linear space and B its completion. Prove that $B^* \sim X^*$. Specifically, prove that the mapping $\phi : B^* \to X^*$ given by $\phi(L) = L|_X$ for $L \in B^*$ is a linear isometry.

104. Let Y be a subspace of a normed linear space X. What is the relation of X^* to Y^*?

105. Given Banach spaces $B_1, \ldots, B_N$, consider $\bigoplus_{k=1}^N B_k$, the linear space of N-tuples $x = (x_1, \ldots, x_N)$ with $x_n \in B_n$ for $n = 1, \ldots, N$. This space can be normed by $\|x\|_p = (\sum_{k=1}^N \|x_k\|_{B_k}^p)^{1/p}$ for $1 \le p < \infty$ and $\|x\|_\infty = \max_{1 \le k \le N} \|x_k\|_{B_k}$. Prove: (a) $(\bigoplus_{k=1}^N B_k, \|\cdot\|_p)$ is a Banach space, $1 \le p \le \infty$. (b) The dual of $(\bigoplus_{k=1}^N B_k, \|\cdot\|_p)$ is $(\bigoplus_{k=1}^N B_k^*, \|\cdot\|_q)$ where $1 \le p < \infty$ and p and q are conjugate indices.

106. Prove that $L^1(X)$ is not reflexive.

107. Is the natural map $J_{C(I)} : C(I) \to C(I)^{**}$ a bijection?

108. Prove that c is nowhere dense in ℓ^∞.

109. Prove that $c_0^* = \ell^1$.

110. Discuss the validity of the following statement: There exists a bounded linear functional L on ℓ^∞ such that $\liminf_n x_n \le L(x) \le \limsup_n x_n$ for all $x \in \ell^\infty$.

111. Let L_∞ be the functional on c given by $L_\infty(x) = \lim_n x_n$. (a) Prove that L_∞ is a bounded linear functional and compute its norm. (b) Discuss the validity of the following statement: L_∞ is of the form $L_\infty(x) = \sum_n y_n x_n$ for some $y \in \ell^1$. (c) Prove that there exists L in $\ell^{\infty*}$ such that $L|_c = L_\infty$. (d) Is L in (c) unique? (e) Prove that if L is as in (c) and $x_n \ge 0$ for all n, then $L(x) \ge 0$. (f) Let $S(x) = y$ where $y_n = x_{n+1}$, $n = 1, 2, \ldots$, denote the shift map in ℓ^∞. Prove that there is an extension L of L_∞ with $\|L\| = 1$ that is shift invariant, i.e., $L(S(x)) = L(x)$, $x \in \ell^\infty$.

112. Let $\{\alpha_{n,k}\}$ be a sequence of nonnegative real numbers that satisfies the following two conditions: $\sum_k \alpha_{n,k} = 1$ for all n, and $\lim_n \alpha_{n,k} = 0$ for all k. Define the linear functional L_n on c by $L_n(x) = \sum_k \alpha_{n,k} x_k$, $x \in c$. Finally, let L_∞ be the linear functional on c given by $L_\infty(x) = \lim_n x_n$. Give an example of such a sequence $\{\alpha_{n,k}\}$. Also prove that $\|L_n\| = 1$ for all n, and $\lim_n L_n(x) = L_\infty(x)$ for all $x \in c$.

113. Discuss the validity of the following statements: (a) c and c_0 are topologically isomorphic. (b) c and c_0 are isometrically isomorphic. (c) c^* and c_0^* are isometrically isomorphic.

114. Prove that, for no normed linear space X, $X^* \sim c_0$.

115. Let $0 \neq L$ be a linear functional on a normed linear space X. Prove: (a) $d(x, K(L)) = |L(x)|/\|L\|$, all $x \in X$. (b) L attains its norm at $x \in X$ iff 0 is a best approximation to x from $K(L)$. (c) $z \perp K(L)$ iff $|L(z)| = \|L\|\,\|z\|$. (d) If for a scalar λ, $\Lambda_\lambda = \{x \in X : L(x) = \lambda\}$, then $d(x, \Lambda_\lambda) = |L(x) - \lambda|/\|L\|$ for all $x \in X$.

116. Let X be a normed linear space, $L \in X^*$, and $x_0 \in X$ with $\|x_0\| = 1$ such that $\|x_0 - x\| \geq 1$ for every $x \in K(L)$. Prove that $|L(x_0)| = \|L\|$.

117. Let $X = (C[0,1], \|\cdot\|_1)$ and L the linear functional on X given by $L(x) = x(1/2)$. Prove that L is not bounded.

118. Find the norm of the linear functional L on $C(I)$ given by $L(x) = \int_{[0,1/2]} x(t)\, dt - \int_{[1/2,1]} x(t)\, dt$. Is the norm attained?

119. Let X be a normed linear space, Y a subspace of X, and $\Lambda = \{L \in X^* : \|L\| \leq 1, L = 0 \text{ on } Y\}$. Prove: (a) $d(x, Y) = \sup_{L \in \Lambda} |L(x)|$, all $x \in X$. (b) $\overline{Y} = \bigcap_{L \in \Lambda} K(L)$.

120. Let $X = \{x \in C(I) : x(0) = 0\}$ and $Y = \{x \in X : \int_0^1 x(t)\, dt = 0\}$. Discuss the validity of the following statement: There exists $x \in X$ with $\|x\| = 1$ such that $d(x, Y) \geq 1$.

121. Let L be a bounded linear functional on $C^k(I)$, for some integer $k \geq 1$. Prove there exists a signed Borel measure μ on I and constants $c_0, \ldots, c_{k-1}$ such that $|c_0|, \ldots, |c_{k-1}|, |\mu|(I) \leq c\,\|L\|$, and

$$L(x) = \sum_{n=0}^{k-1} c_n x^{(n)}(0) + \int_I x^{(k)}(t)\, d\mu(t), \quad x \in C^k(I).$$

122. Let X be the space of sequences $X = \{x : \sum_n 2^n |x_n| < \infty\}$. Prove that $\|x\|_X = \sum_n 2^n |x_n|$ is a norm on X and describe X^*.

123. Identify the closure of ℓ^p in c_0.

124. Characterize those sequences $a \neq 0$ such that $A = \{x \in c_0 : \sum_n a_n x_n = 0\}$ is dense in ℓ^p, $1 < p < \infty$.

125. Discuss the validity of the following statement: Given $1 < p < \infty$, there exists a collection of sequences A that is dense in ℓ^r for every $1 < p < r < \infty$ but not in ℓ^p.

126. Let λ be a sequence such that $|\lambda_n| < 1$ for all n and $\lim_n \lambda_n = 0$. Let $\alpha^1 = (1, \lambda_1^1, \lambda_1^2, \ldots)$, and, in general for all $n \geq 1$, let α^n denote the sequence with terms $\alpha_k^n = \lambda_n^{k-1}$, $k = 1, 2, \ldots$. Prove that $A = \mathrm{sp}\{\alpha^1, \alpha^2, \ldots\}$ is dense in ℓ^p, $1 \leq p < \infty$.

127. For $a > 1$, let $x_a \in C([0,1])$ be given by $x_a(t) = 1/(t - a)$. Prove that if $a_n > 1$ for all n and $\lim_n a_n = \infty$, then $X = \mathrm{sp}\{x_{a_n}\}$ is dense in $(C([0,1]), \|\cdot\|_\infty)$.

128. Let $1 < p < \infty$. Prove that $D = \{x$ on $\mathbb{R}^n : x$ is measurable, compactly supported, bounded, and $\int_{\mathbb{R}^n} x(t)\, dt = 0\}$ is dense in $L^p(\mathbb{R}^n)$.

129. Let $1 < p < \infty$. (a) Construct a subspace A that is dense in $L^r(\mathbb{R}^n)$ for all $1 < r < p$, but not in $L^p(\mathbb{R}^n)$. (b) Does there exist A so that A is dense in $L^p(\mathbb{R}^n)$ but not in $L^r(\mathbb{R}^n)$ for any $1 < r < p$?

130. Let $\{x_n\}$ be a sequence in a normed linear space X. Prove that if $x_n \rightharpoonup x$ and $x_n \rightharpoonup y$, then $x = y$.

131. Let X be a normed linear space and suppose that $x_n \rightharpoonup x_0$ in X. Prove that $x_0 \in M = \overline{\mathrm{sp}}\{x_1, \ldots, x_n, \ldots\}$.

132. Discuss the validity of the following statement: If $\{x_n\}$ is a weak Cauchy sequence in a normed linear space X, $\{x_n\}$ converges weakly.

133. Prove that $e_n \rightharpoonup 0$ in ℓ^p, $1 < p \leq \infty$, and in c_0, but not in ℓ^1.

134. Let $v_n = e_1 + \cdots + e_n$. Prove that $\{v_n\}$ does not converge weakly in ℓ^∞.

135. Let $\{x^n\}$ be a sequence that converges weakly to x in ℓ^p, $1 \leq p \leq \infty$, and that in addition satisfies $\lim_n \|x^n\|_p = \|x\|_p$. Discuss the validity of the following statement: $\lim_n \|x^n - x\|_p = 0$.

136. Let X be a normed linear space and Y a dense subset of X^*. Prove that if $\{x_n\} \subset X$ is bounded and $\ell(x_n) \to 0$ for each $\ell \in Y$, then $x_n \rightharpoonup 0$ in X.

137. Let $\{x_n\}$ be a sequence in a normed linear space X. Prove: (a) If $\{x_n\}$ is weakly Cauchy, $\{x_n\}$ is bounded in X. (b) If $x_n \rightharpoonup x$, $\|x\| \leq \liminf_n \|x_n\|$.

138. We say a Banach space B is *uniformly convex* if $\lim_n \|(x_n + x)/2\| = 1$ implies $\lim_n \|x_n - x\| = 0$. Prove that if $x_n \rightharpoonup x$ in B and $\limsup_n \|x_n\| \leq \|x\|$, then $x_n \to x$ in B.

139. Prove that if B is reflexive, every bounded linear functional L on B attains its norm, i.e., $\|L\|_{B^*} = L(x)$ for some $x \in B$ with $\|x\|_B = 1$.

140. Let K be a compact subset of a Banach space B and suppose that $\{x_n\} \subset K$ converges weakly to $x \in K$. Must $\|x_n - x\| \to 0$?

141. Let X be a normed linear space. Discuss the validity of the following statements: (a) A proper subspace Y of X is closed iff Y is weakly closed. (b) The closed unit ball $B(0,1) = \{x \in X : \|x\| \leq 1\}$ is weakly closed. (c) A proper subset Y of X is closed iff Y is weakly closed.

142. Let $(X, \mathcal{M}, \mu)$ be a measure space, $\{u_n\} \subset L^p(X)$ and $\{v_n\} \subset L^q(X)$, such that $u_n \rightharpoonup u$ in $L^p(X)$ and $v_n \to v$ in $L^q(X)$ where $1 < p, q < \infty$ are conjugate indices. Prove that $u_n v_n \rightharpoonup uv$ in $L^1(X)$.

143. Let $X = \{x \in C(I) : x(0) = 0 \text{ and } \int_0^1 x(t)\, dt \geq 1\}$. Prove: (a) X is closed in $C(I)$ and $\|x\|_\infty > 1$ for all $x \in X$. (b) $d(0, X) = 1$. Does there exist $x \in X$ such that $d(0, X) = \|x\|_\infty$? Also: (c) Is X compact?

144. Let $\{x_n\} \subset C(I)$ be given by

$$
x_n(t) = \begin{cases} nt, & t \in [0, 1/n], \\ 2 - nt, & t \in [1/n, 2/n], \\ 0, & \text{otherwise.} \end{cases}
$$

Prove that $x_n \rightharpoonup 0$.

145. Consider $C(I)$ equipped with the sup norm and $\{x_n\} \subset C(I)$. Prove that the following are equivalent: (a) $x_n \rightharpoonup 0$ in $C(I)$. (b) $x_n(t) \to 0$ for all $t \in I$ and $\sup_n \|x_n\|_\infty < \infty$.

146. Consider $C([-1, 1])$ endowed with the sup norm and define the bounded linear functionals L_n, L on $C([-1, 1])$ by $L_n(x) = (n/2) \int_{-1/n}^{1/n} x(t)\, dt$ and $L(x) = x(0)$, $x \in C([-1, 1])$, respectively. Discuss the validity of the following statements: (a) $\|L_n\| = 1$ and $\lim_n L_n(x) = L(x)$ for all $x \in C([-1, 1])$. (b) $\lim_n \|L_n - L\| = 0$.

147. Let Y be a closed subspace of a Banach space X with X^* separable. Prove that Y^* is separable.

148. Let X be a normed linear space and $\{L_n\} \subset X^*$. Prove: (a) If $L_n \rightharpoonup L$ in X^*, $L_n(x) \to L(x)$ for all $x \in X$. (b) If X is reflexive and $L_n(x) \to L(x)$ for all $x \in X$, $L_n \rightharpoonup L$ in X^*.

149. Let B be a Banach space and X a closed subset of B. Prove that $J_B(X)$ is closed in B^{**}.

150. Let B be a Banach space. Prove: (a) If B is reflexive and X is a closed subspace of B, then X is reflexive. (b) If every proper closed subspace of B is reflexive, so is B.

151. Let B be a Banach space. Prove that if B^* contains a proper closed subspace Y that separates points in B, B is not reflexive.

152. Let B be a Banach space. Prove: (a) $M = J_{B^*}(B^*)$ is complemented in B^{***}. Specifically, if $N = J_B(B)^\perp = \{x^{***} \in B^{***} : x^{***}|_{J_B(B)} = 0\}$, then $B^{***} = M \oplus N$. (b) B is reflexive iff B^* is reflexive.

153. Discuss the validity of the following statements: (a) If B^{**} is reflexive, B is reflexive. (b) A Banach space B is reflexive or its second duals are all distinct.

154. Let B be a reflexive Banach space. Prove: (a) A bounded sequence in B has a weakly convergent subsequence. (b) B is weakly complete.

155. Let B be a separable reflexive Banach space and X a subspace of B. Prove that given $x_0 \in B$, there exists $y_0 \in X$ that is closest to x_0.

156. Let $C(X)$ denote the Banach space of real-valued continuous functions on a compact space X equipped with the sup norm and L a linear functional on $C(X)$. Prove that L is bounded and $L(\chi_X) = \|L\|$ iff L is positive, i.e., $L(x) \geq 0$ for all nonnegative real-valued functions $x \in C(X)$.

157. Let X be a subspace of $C(I)$ that contains χ_I and ℓ a positive linear functional on X. Prove that there exists a positive linear functional L on $C(I)$ such that $L|_X = \ell$ and $\|L\| = \|\ell\|_{X^*}$.

158. Let p be a seminorm on a linear space X and $M = \{x \in X : p(x) = 0\}$. Prove that M is a subspace of X and that p is a norm on X/M.

159. Let X be a normed linear space, M a closed subspace of X, and $\pi : X \to X/M$ the canonical map. Prove: (a) $\sup_{x \neq 0} \|\pi(x)\|_{X/M}/\|x\|_X = 1$. (b) For $x \in X$ and $r > 0$, $\pi(B_X(x,r)) = B_{X/M}(\pi(x), r)$. (c) $W \subset X/M$ is open in X/M iff $\pi^{-1}(W)$ is open in X. (d) π is an open mapping, i.e., if U is open in X, then $\pi(U)$ is open in X/M.

160. Let Y be a closed subspace of a normed linear space X. Prove: (a) X is complete iff Y and X/Y are complete. (b) X is separable iff Y and X/Y are separable.

161. Let $X_1, \ldots, X_n$ be subspaces of a normed linear space X. Prove that if $\dim(X/X_k) = 1$, $1 \leq k \leq n$, then $\dim(X/\bigcap_{k=1}^n X_k) \leq n$.

162. Let $B = C(I) = C([0,1])$. Prove that M is closed in B and identify the quotient space B/M where (a) $M = \{x \in C(I) : x(0) = 0\}$, and (b) $A \subset I$ is closed, $M = \{x \in C(I) : x|_A = 0\}$.

163. Let $X = \{x \in C(I) : x$ is constant$\}$. Given $x \in C(I)$, compute the norm $\|\,[x]\,\|$ of x in $C(I)/X$.

164. Let $X = L(I)$ and $Y = \{x \in X : x(t) = 0 \text{ a.e. for } t \in [0, 1/2]\}$. Prove that Y is a closed subspace of X and describe the norm in X/Y.

165. Describe the norm in the nonseparable Banach space ℓ^∞/c_0.

166. Let $M = \{x \in \ell^p : x_{2n} = 0 \text{ for all } n\}$. Prove: (a) M is a closed subspace of ℓ^p. (b) ℓ^p and ℓ^p/M are isometrically isomorphic, $1 \leq p \leq \infty$.

167. Let X be a normed linear space and M a closed subspace of X. The *codimension* $\mathrm{codim}(M)$ of M in X is defined as the quantity $\dim(X/M)$. Prove: (a) $\mathrm{codim}(M) < \infty$ iff there exists a finite-dimensional subspace N of X such that $X = M \oplus N$. (b) If $\mathrm{codim}(M) < \infty$, given $\varepsilon > 0$, there is a finite-dimensional subspace N_1 of X such that each $x \in X$ can be written $x = v_1 + m_1$ with $v_1 \in N_1$, $m_1 \in M$ and $\|v_1\| \leq (1 + \varepsilon)\|x\|$.

168. Let X be a subset of a Banach space B. The *annihilator* $X^\perp$ of X is defined as $X^\perp = \{L \in B^* : L(x) = 0 \text{ for all } x \in X\}$. Prove: (a) $X^\perp$ is a closed subspace of B^*. (b) If X is a closed subspace of B, $X^\perp$ is isometrically isomorphic to $(B/X)^*$. (c) If X is a closed subspace of B and $X^\perp$ is finite dimensional, then X is complemented in B. (d) If X is a closed subspace of B, X^* is isometrically isomorphic to $B^*/X^\perp$.

169. Let B be a reflexive Banach space and X a closed subspace of B. Prove that B/X is reflexive.

170. Let B be a Banach space and X a closed subspace of B. Prove that $(B/X)^*$ is isometrically isomorphic to $X^\perp$.

171. Let B be a Banach space and $X \subset B$ a closed subspace. Prove: (a) X is finite dimensional iff $X^\perp$ has finite codimension in B^*, and in that case $\dim(X) = \operatorname{codim}(X^\perp)$. (b) X has finite codimension in B iff $X^\perp$ is finite dimensional, and in that case $\operatorname{codim}(X) = \dim(X^\perp)$.

172. Let $N \subset B^*$ be a closed subspace. (a) Prove that N is finite dimensional iff $N^\perp$ has finite codimension in B^{**}, and in that case $\dim(N) = \operatorname{codim}(N^\perp)$. (b) Discuss the validity of the following statement: If N has finite codimension, then $N^\perp$ is finite dimensional and $\operatorname{codim}(N) = \dim(N^\perp)$.

173. Let X be a normed linear space and Y a subspace of X. Prove that Y^{**} can be identified with a subspace of X^{**}.

174. Is it possible for a separable and a nonseparable Banach space to have the same dual space?

175. Let B be a Banach space, M, N closed subspaces of B, and $X = \{(x, -x) : x \in M \cap N\} \subset M \times N$. Prove: (a) The mapping $\phi : (M \times N)/X \to M + N$ given by $\phi([m, n]) = m + n$ is well-defined and continuous. (b) $M + N$ is closed in B iff ϕ is an isomorphism of normed spaces when $M + N$ is equipped with the norm in B.

176. Let B be a separable Banach space. Prove that B is isometrically isomorphic to a quotient space of ℓ^1.

Chapter 9

Normed Linear Spaces. Linear Operators

In this chapter we consider linear mappings, or operators, from a normed linear space into another. Let X, Y be linear spaces over the same scalar field. We say that a mapping $T : X \to Y$ is *linear* if $T(x + \lambda y) = T(x) + \lambda T(y)$ for every $x, y \in X$ and scalar λ. When X, Y are normed we say that T is *bounded* if there exists a constant $c > 0$, independent of $x \in X$, such that $\|T(x)\|_Y \leq c \|x\|_X$ for all $x \in X$; $\|T\|$, the *norm* of T, is defined as the infimum of the c for which this inequality holds. For a linear mapping T boundedness is equivalent to *continuity*, i.e., if $\lim_n x_n = x$ in X, then $\lim_n T(x_n) = T(x)$ in Y, all $x \in X$, or even continuity at a single point. Also, $\mathcal{B}(X, Y) = \{T : X \to Y : T \text{ is linear and bounded}\}$ equipped with $\| \cdot \|$ is a normed linear space; for simplicity we denote $\mathcal{B}(X, X) = \mathcal{B}(X)$.

When a linear mapping T is defined in a proper subset of X it is important to specify its *domain* $D(T)$, which is a subspace of X. $R(T)$, the *range* of T, is the image of $D(T)$ by T, i.e., $R(T) = \{y \in Y : T(x) = y \text{ for some } x \in D(T)\}$. The *kernel* $K(T)$ of T is $K(T) = \{x \in D(T) : T(x) = 0\}$. And, by the *graph* $G(T)$ of T we mean the subset of $X \times Y$ given by $G(T) = \{(x, y) \in X \times Y : x \in D(T) \text{ and } y = T(x)\}$.

Given a linear mapping $T : D(T) \subset X \to Y$, we say that a linear operator $S : D(S) \subset X \to Y$ is an *extension* of T if $G(S) \supset G(T)$, i.e., $D(S) \supset D(T)$ and $S(x) = T(x)$ for $x \in D(T)$.

We say that a linear mapping $T : D(T) \subset X \to Y$ is *closed* if $G(T)$ is closed in $X \times Y$. Now, if $G(T)$ is not closed in $X \times Y$, we say that T is *closable* if there is a linear mapping $\overline{T} : D(\overline{T}) \to Y$ such that $G(\overline{T}) = \overline{G(T)}$; $\overline{T}$ is called the closed linear extension, or closure, of T.

A continuous linear mapping $T : X \to Y$ is said to be *compact* if T maps bounded sets in X to precompact sets in Y, i.e., sets with compact closure. Or, if B_X denotes the unit ball in X, $\overline{T(B_X)}$ is compact in Y. Alternatively, given a bounded sequence $\{x_n\}$ in X, $\{T(x_n)\}$ has a convergent subsequence in Y.

The basic principles in the theory of bounded linear operators between Banach spaces include the uniform boundedness principle, open mapping theorem, and closed graph theorem. First, the uniform boundedness principle. Let X and Y be Banach spaces. We say that a family $\mathcal{T} = \{T\} \subset \mathcal{B}(X,Y)$ is pointwise bounded if $\sup_{T \in \mathcal{T}} \|T(x)\|_Y < \infty$ for every $x \in X$. The uniform boundedness principle asserts that, if $\mathcal{T}$ is pointwise bounded, $\mathcal{T}$ is uniformly bounded, i.e., $\sup_{T \in \mathcal{T}} \|T\| < \infty$.

Next, the open mapping theorem. We say that a continuous linear mapping $T : X \to Y$ is open if T maps open sets in X to open sets in Y. The open mapping theorem asserts that, if B, B_1 are Banach spaces and $T : B \to B_1$ is a continuous surjective linear operator, then T is open.

An immediate consequence of the open mapping theorem is the inverse mapping theorem which states that if B, B_1 are Banach spaces and $T : B \to B_1$ is a bijective continuous linear operator, then T is an isomorphism, i.e., $T^{-1} : B_1 \to B$ is a well-defined continuous linear mapping.

Finally, the closed graph theorem asserts that if B, B_1 are Banach spaces and $T : B \to B_1$ is a linear operator with $G(T)$ closed in $B \times B_1$, then T is bounded.

Now, the notion of the adjoint operator of a linear mapping is an extension of the transpose of a matrix. Let $T \in \mathcal{B}(X,Y)$. The *adjoint* T^* of T is defined as the linear mapping $T^* : Y^* \to X^*$ given by $(T^*\ell)(x) = \ell(T(x))$ for all $\ell \in Y^*$ and $x \in X$. Some of the basic properties of the adjoint are: $T^* \in \mathcal{B}(Y^*, X^*)$, $\|T^*\| = \|T\|$, $K(T^*) = R(T)^{\perp}$, and $K(T) = R(T^*)^{\perp}$.

The problems in this chapter are devoted to bounded linear operators, with the sobering thought that unbounded linear operators on infinite-dimensional normed linear spaces abound, Problem 2. Linear mappings defined in Euclidean space, Problem 4, linear mappings from sequences into sequences defined on ℓ^p spaces, Problems 10–12, and linear integral operators defined on continuous functions, Problems 14–15, serve as the prototype for the bounded linear mappings we have in mind. A useful tool in this endeavor is given by Problem 21: Given a continuous linear mapping $T : X \to Y$ that is onto, the mapping $\widetilde{T} : X/K(T) \to Y$ defined by $\widetilde{T}([x]) = T(x)$ is a well-defined linear continuous bijection with $\|\widetilde{T}\| = \|T\|$; see also Problem 166.

Problem 31 poses the question as to whether $C^1(I)$ can be endowed with a norm so that the natural operations there, i.e., differentiation and multiplication by t, are both continuous. Simple criteria for continuity are discussed in Problems 33–35 and Problem 39. The existence of bounded linear extensions of bounded linear mappings is discussed in Problems 42–44 and Problem 47 considers closed linear extensions.

Compact operators are discussed in Problems 53–74; Problem 74 is particularly insightful. Projections are considered in Problems 75–82. The role of category type arguments is highlighted in Problems 87–88, Problem 90, and Problems 108–109. Sequences of operators are covered in Problems 91–92, Problems 94–95, and Problems 100–101. And, equivalent norms in Banach spaces are discussed in Problems 116–122.

The case when a bounded linear operator has a continuous inverse is covered in Problems 131–154 and Problem 173; Problem 139 deals with the existence of a continuous left-inverse or right-inverse. Applications to the theory of equations are given in Problems 155–161. Operators with closed range are considered in Problems 162–164, Problem 168, and Problem 180. And, the properties of Schauder basis are discussed in Problems 183–190.

Problems

1. Let X, Y be linear spaces. Prove that a subspace M of $X \times Y$ is the graph of a linear mapping $T : X \to Y$ iff no element of the form $(0, y)$ with $y \neq 0$ belongs to M.

2. Let X, Y be normed linear spaces, $\dim(X) = \infty$. Construct an unbounded linear operator $T : X \to Y$.

3. Let X, Y be normed linear spaces, $0 \neq x_0 \in X$, and $y_0 \in Y$. Construct a bounded linear operator $T : X \to Y$ such that $T(x_0) = y_0$ and $\|T\| \, \|x_0\|_X = \|y_0\|_Y$.

4. Let $T : \mathbb{R}^n \to \mathbb{R}^m$ be a linear mapping represented by the $m \times n$ matrix (a_{ij}) with respect to the standard bases in $\mathbb{R}^n, \mathbb{R}^m$. Compute the norm of T if: (a) $\mathbb{R}^n$ is equipped with the ℓ^1 norm and $\mathbb{R}^m$ with the ℓ^∞ norm. (b) $\mathbb{R}^n$ is endowed with the ℓ^∞ norm and $\mathbb{R}^m$ with the ℓ^1 norm. (c) $\mathbb{R}^n$ and $\mathbb{R}^m$ are both equipped with the ℓ^1 norm. (d) $\mathbb{R}^n$ and $\mathbb{R}^m$ are both endowed with the ℓ^∞ norm.

5. Let T be a linear mapping from sequences into sequences represented by the infinite scalar matrix (a_{mn}), i.e., $y = T(x)$ iff $y_m = \sum_n a_{mn} x_n$, $m \geq 1$. Prove: (a) $T : c_0 \to \ell^\infty$ with norm η iff $\eta = \sup_{m \geq 1} \sum_n |a_{mn}| < \infty$. (b) $T : c_0 \to c_0$ with norm η iff $\eta = \sup_m \sum_n |a_{mn}| < \infty$ and $\lim_m a_{mn} = 0$ for each n.

6. Let T be the mapping from sequences to sequences given by $T(x) = y$ where $y_n = 2^{-n} \sum_{k=1}^n x_k$. Prove that T is bounded from ℓ^∞ to ℓ^1 and compute its norm.

7. Let T be the linear mapping from sequences into sequences given by $T(x) = y$ where $y_n = x_n/n$, all n, and $1 \leq p \leq \infty$. Discuss for what values of p, T is bounded, onto, and $R(T)$ is dense but not closed. Also, does T^{-1} exist? Is it bounded?

8. On $X = \ell^1$ define $T(x) = y$ by setting $y_n = \lambda_n x_1 + x_n$ where $|\lambda_n| \leq 2^{-n}$. (a) Is T bounded on ℓ^1? What is $\|T\|$? (b) What is $R(T)$? (c) Is T invertible? What is $\sup_{x \neq 0} \|x\| / \|T(x)\|$?

9. Let $X = \{x \in \ell^1 : \sum_n n|x_n| < \infty\}$. Prove: (a) X is dense in ℓ^1. (b) The mapping $T : X \to \ell^1$ given by $T(x) = y$ with $y_n = nx_n$, $n \geq 1$, is closed but not bounded. (c) T^{-1} is surjective but not open. (d) Can X be equipped with a norm so that X becomes a Banach space?

10. Given a sequence λ, let T be the linear mapping from sequences into sequences given by $T(x) = y$ where $y_n = \lambda_n x_n$, $n = 1, 2, \ldots$. Prove: (a) T is bounded from ℓ^p into ℓ^p iff $\lambda \in \ell^\infty$, and in that case $\|T\| = \|\lambda\|_\infty$. Does there exist $0 \neq x \in \ell^p$ such that $\|T(x)\|_p = \|T\| \|x\|_p$? (b) T is 1-1 iff $\lambda_n \neq 0$ for all n. (c) If T is injective, $R(T)$ is dense in ℓ^p. Also, $R(T)$ is not necessarily closed. (d) $R(T) = \ell^p$ and T is a linear homeomorphism iff $\inf_n |\lambda_n| > 0$.

11. Let $1 \leq r < p < \infty$. Given a sequence λ, let T be the linear mapping from sequences into sequences given by $T(x) = y$ where $y_n = \lambda_n x_n$, $n = 1, 2, \ldots$. Prove that T is bounded from ℓ^p into ℓ^r iff $\lambda \in \ell^s$, $1/s = 1/r - 1/p$, and in that case $\|T\| = \|\lambda\|_s$. Can T be a homeomorphism?

12. Let $1 \leq p \leq r \leq \infty$. Given a sequence λ, let T be the linear mapping from sequences into sequences given by $T(x) = y$ where $y_n = \lambda_n x_n$, $n = 1, 2, \ldots$. When is: (a) T densely defined in ℓ^p? $D(T) = \ell^p$? (b) $R(T)$ a dense subspace of ℓ^r? $R(T) = \ell^r$? (c) T bounded? (d) T invertible? In that case compute $\|T^{-1}\|$. (e) T bounded and T^{-1} bounded?

13. Let $k(s,t)$ be a continuous function on $I \times I$ and consider the linear mapping T given by $T(x)(t) = \int_0^1 k(s,t) x(s)\, ds$. Prove that T is continuous from $L^p(I)$ into $C(I)$ for $1 \leq p \leq \infty$, and from $C(I)$ into $C(I)$.

14. Prove that the integral operator T given by

$$Tx(t) = \int_0^1 \frac{x(s)}{|t-s|^{1/3}}\, ds$$

is bounded from $C(I)$ into itself.

15. Let $X = \{x \in C(I) : 0 \le x(t) \le 1 \text{ for all } t \in I\}$ and $k : [0,1] \times [0,1] \to [0,1]$ continuous. (a) Prove that X is closed in $C(I)$. Is X compact? (b) For $x \in X$, let $T(x)(t) = y(t) = \int_0^1 k(x(s), t)\, ds$. Prove that y takes values in $[0,1]$ and is continuous. Also, prove: (c) T is uniformly continuous on X. (d) $T(C)$ is a uniformly equicontinuous subset of $C(I)$. (e) $T(C)$ has a compact closure in C.

16. Let X, Y be normed linear spaces and $T : X \to Y$ a linear mapping. Prove: (a) T is continuous iff $T^{-1}(B_Y(0,1))$ has nonempty interior. (b) T is unbounded iff there exists $\{x_n\} \subset X$ such that $\|x_n\|_X \to 0$ and $\|T(x_n)\|_Y \to \infty$.

17. Let X be a normed linear space with closed unit ball $\overline{B}_X$, B a Banach space, and $T : X \to B$ a continuous injective linear mapping. Prove that if $T(\overline{B}_X)$ is closed in B, X is complete.

18. Prove that $\mathcal{B}(X,Y)$ is a Banach space iff Y is complete.

19. Let X, Y be normed linear spaces with X finite dimensional and $T : X \to Y$ a linear mapping. Prove: (a) $R(T)$ is a closed subspace of Y. (b) T is bounded and assumes its norm.

20. Let X, Y be normed linear spaces and $T : X \to Y$ a linear operator with closed graph and finite-dimensional range $R(T)$. Prove that T is continuous.

21. Let X, Y be normed linear spaces and $T : X \to Y$ a continuous linear mapping. Prove that the mapping $\widetilde{T} : X/K(T) \to Y$ given by $\widetilde{T}([x]) = T(x)$ for $x \in [x]$ is well-defined, linear, and continuous with $\|\widetilde{T}\| = \|T\|$. Moreover, if T is onto, $\widetilde{T}$ is a bijection. Is $\widetilde{T}$ an isomorphism?

22. Let B, B_1 be Banach spaces, $T : B \to B_1$ a surjective bounded linear mapping, and $\{y_n\} \subset B_1$ such that $y_n \to y_0$ in B_1. Prove there exist $\{x_n\} \subset B$ with $x_n \to x_0$ in B and $c > 0$, such that $T(x_n) = y_n$ and $\|x_n - x_0\|_B \le c\,\|y_n - y_0\|_{B_1}$ for all n.

23. Let B, B_1 be Banach spaces and $T : B \to B_1$ a bounded linear operator with $R(T)$ complemented in B_1. Prove that $R(T)$ is closed in B_1.

24. Let B, B_1 be Banach spaces and $T : B \to B_1$ a bounded linear operator. Prove that if $R(T)$ has finite codimension in B_1, $R(T)$ is closed in B_1.

25. Let B be a Banach space, Y a normed linear space, and $T : B \to Y$ a linear mapping with $R(T)$ finite dimensional. Prove: (a) T is continuous iff $K(T)$ is closed. (b) If T is continuous, T is open.

26. Let X be a linear space and $T : X \to X$ a linear mapping. Prove: (a) $K(T) \subset K(T^2) \subset \ldots$, and if $K(T^m) = K(T^{m+1})$ for some $m \geq 1$, then $K(T^n) = K(T^m)$ for all $n > m$. (b) $R(T) \supset R(T^2) \supset \ldots$, and if $R(T^m) = R(T^{m+1})$ for some $m \geq 1$, then $R(T^n) = R(T^m)$ for all $n > m$.

27. Let $T : \ell^p \to \ell^p$, $1 \leq p \leq \infty$, be a bounded linear mapping of norm 1 such that $T(e_n) = e_n$ for all n. Prove that T is the identity in ℓ^p.

28. Let $M : L^q(I) \to L^q(I)$, $1 < q < \infty$, be the linear mapping given by $Mx(t) = tx(t)$ and suppose that $T : L^q(I) \to L^q(I)$ is a bounded linear operator that commutes with M, i.e., $T(Mx) = M(Tx)$, $x \in L^q(I)$. Prove that there is a function y such that $Tx(t) = y(t)x(t)$ for all $x \in L^q(I)$. What can one say about y?

29. Let $(X, \mathcal{M}, \mu)$ be a σ-finite measure space and for $1 \leq p < \infty$ and $y \in L^\infty(X)$, let $T_y : L^p(X) \to L^p(X)$ be given by $T_y(x) = yx$. Prove that if $T : L^p(X) \to L^p(X)$ is a bounded linear mapping that commutes with T_y for all $y \in L^\infty(X)$, then $T = T_{y_T}$ for some $y_T \in L^\infty(X)$ with $\|y_T\|_\infty = \|T\|$.

30. Let X be a normed linear space and $S, T : X \to X$ linear mappings. (a) Prove that if $ST - TS$ commutes with S, $S^n T - T S^n = n S^{n-1}(ST - TS)$, all $n \geq 1$. (b) Do there exist bounded linear operators $S, T : X \to X$ such that $ST - TS = \alpha I$, α a nonzero scalar?

31. Let $Dx(t) = x'(t)$ and $Mx(t) = tx(t)$ denote the derivation and multiplication operators on $C^1(I)$, respectively. Equip $C^1(I)$ with a norm so that D and M are bounded with respect to that norm.

32. Let X, Y be normed linear spaces and $T : X \to Y$ a bounded linear mapping. Prove that $r\|T\| \leq \sup_{y \in B_X(x,r)} \|T(y)\|_Y$ for every $x \in X$ and $r > 0$.

33. Let X, Y be normed linear spaces and $T : X \to Y$ a linear mapping. Prove that T is continuous iff $x_n \to x$ in X implies $T(x_n) \rightharpoonup T(x)$ in Y.

34. Let X, Y be normed linear spaces and $T : X \to Y$ a linear mapping. Discuss the validity of the following statement: T is bounded iff $x_n \rightharpoonup x$ in X implies $T(x_n) \rightharpoonup T(x)$ in Y.

35. Let X, Y be normed linear spaces and $T : X \to Y$ a bounded linear operator. Prove that if $x_n \rightharpoonup x$ in X implies $T(x_n) \to y$ in Y, then $T(x) = y$.

36. Let $T : C(I) \to C(I)$ be a positive linear operator, i.e., $x \geq 0$ implies $T(x) \geq 0$. Prove: (a) T is continuous and $\|T\| = \|T(\chi_I)\|$. (b) If $T(\chi_I) = \chi_I$, and if $\{x_n\} \subset C(I)$ is such that $|x_n(t)| \leq 1$ and $x_n(t) \to x(t)$ for each $t \in I$, then $T(x_n)(t) \to T(x)(t)$ for each $t \in I$.

37. Let B be a Banach space and $T : B \to C(I)$ a linear mapping such that if $\|x_n\| \to 0$ in B, then $T(x_n) \to 0$ pointwise on I. Prove that T is bounded.

38. Given a real normed linear space X, let $X_{\mathbb{C}} = \{x_{\mathbb{C}} = x_1 + ix_2 : x_1, x_2 \in X\}$. Prove: (a) $X_{\mathbb{C}}$ is a complex linear space and $\|x_{\mathbb{C}}\|_{X_{\mathbb{C}},p} = (\|\Re x\|^p + \|\Im x\|^p)^{1/p}$, $1 \le p \le \infty$, with the usual interpretation for $p = \infty$, is a norm in $X_{\mathbb{C}}$. (b) If Y is a complex normed linear space and $T : X \to Y$ is a bounded linear operator, the linear mapping $T_{\mathbb{C}} : X_{\mathbb{C}} \to Y$ given by $T_{\mathbb{C}}(x_{\mathbb{C}}) = T(\Re x_{\mathbb{C}}) + iT(\Im x_{\mathbb{C}})$, $x_{\mathbb{C}} \in X_{\mathbb{C}}$, is bounded and find c_X, C_X such that $c_X \|T\| \le \|T_{\mathbb{C}}\| \le C_X \|T\|$.

39. Let B, B_1 be Banach spaces and $T : B \to B_1$ a linear mapping. Prove that T is continuous under one of the following additional conditions: (a) $L \circ T$ is a bounded linear functional on B for all bounded linear functionals L on B_1. Or: (b) There is a family $\mathcal{F}$ of continuous functionals on B_1 such that $L \circ T$ is continuous for each $L \in \mathcal{F}$ and $\bigcap_{L \in \mathcal{F}} L^{-1}(\{0\}) = \{0\}$.

40. Let X_1, X_2, X_3 be normed linear spaces and $S : X_1 \to X_2$, $T : X_2 \to X_3$ bounded linear mappings. Prove that $TS : X_1 \to X_3$ is a bounded linear operator and $\|TS\| \le \|T\|\,\|S\|$. Can the inequality be strict?

41. Let X be a normed linear space and $T : X \to X$ a bounded linear mapping. Prove that $\lim_n \|T^n\|^{1/n}$ exists and is equal to $\inf_n \|T^n\|^{1/n}$.

42. Let X be a normed linear space, Y a subspace of X, and $S : Y \to \ell^{\infty}$ a bounded linear mapping. Prove that there is a norm preserving linear extension $T : X \to \ell^{\infty}$ of S.

43. Let X, Y be normed linear spaces, $M \subset X$, and $T_0 : M \to Y$ a bounded linear mapping. Prove that T_0 has a bounded extension T to the span of M (in X) iff there exists a constant K such that $\left\| \sum_{n=1}^m \lambda_n T_0(x_n) \right\|_Y \le K \left\| \sum_{n=1}^m \lambda_n x_n \right\|_X$ for arbitrary $x_1, \ldots, x_n$ in M and scalars $\lambda_1, \ldots, \lambda_n$.

44. Let X be a normed linear space with completion B, B_1 a Banach space, and $S : X \to B_1$ a bounded linear mapping. Prove that there is a unique norm preserving linear extension $T : B \to B_1$ of S. Furthermore, if S is an isometry, so is T.

45. Discuss the validity of the following statement: $C^1(I) = \{x \in C(I) : x' \in C(I)\}$ endowed with the norm $\|x\| = \|x\|_{\infty} + \|x'\|_{\infty}$ is a Banach space.

46. Let X be a normed linear space and B a Banach space. Give an example of: (a) A discontinuous linear mapping $T : X \to B$ that has a closed graph. (b) A discontinuous linear mapping $T : B \to X$ that has a closed graph.

47. Let X, Y be normed linear spaces and $T : D(T) \subset X \to Y$ a linear operator. Prove that the following statements are equivalent: (a) T admits a closed linear extension $\overline{T}$. (b) $(0, y) \notin \overline{G(T)}$ for all $0 \neq y \in Y$. (c) If $\{x_n\} \subset D(T), x_n \to 0$ in X, and $T(x_n) \to y$ in Y, then $y = 0$.

48. Let X, Y be normed linear spaces and $T : D(T) \subset X \to Y$ a bounded linear operator. Prove: (a) If $D(T)$ is closed in X, T is closed. (b) If T is closed and Y complete, $D(T)$ is closed in X.

49. Let X, Y be Banach spaces and $T : X \to Y$ a closed linear operator. Prove: (a) $K(T)$ is closed. (b) If $C \subset X$ is compact, $T(C)$ is closed in Y. (c) If $D \subset Y$ is compact, $T^{-1}(D)$ is closed in X.

50. Let X be a normed linear space, $T : D(T) \subset X \to X$ a closed linear mapping, and $T_1 : X \to X$ a bounded linear operator. Prove: (a) $T + T_1$ is closed. (b) If X is a Banach space, $T + T_1 : D(T) \to X$ has a bounded inverse iff $T + T_1$ is a bijection.

51. Let B, B_1 be Banach spaces and $T, T_1 : B \to B_1$ closed linear mappings. Prove that $T + T_1$ is bounded.

52. Let X, Y be normed linear spaces and $T : X \to Y$ a linear operator such that $R(T)$ is closed in Y. Prove that if $\|T(x)\|_Y \geq c\|x\|_X$ for all $x \in D(T)$ and some $c > 0$, T is closed.

53. Let X be a normed linear space with completion B, B_1 a Banach space, and $T : X \to B_1$ a compact linear operator. Prove that the extension $\overline{T}$ of T to B constructed in Problem 44 is compact.

54. Let X, Y be Banach spaces and $T : X \to Y$ a bounded linear operator. Prove: (a) If T is of finite rank, T is compact. (b) If T is compact and $R(T)$ closed, T is of finite rank.

55. Let B be a Banach space and $T_n : B \to B$ compact linear operators such that $T_n(x) \to T(x)$ for all $x \in B$. Is T necessarily compact?

56. Let B be a Banach space and $\{T_n\} \subset \mathcal{B}(B)$ compact operators that converge to T in $\mathcal{B}(B)$. Prove that T is compact.

57. Let k be a measurable function $k : [0, 1] \times [0, 1] \to \mathbb{R}$ such that $\int_I \int_I |k(s, t)|^q \, ds dt < \infty$, $1 < q < \infty$. For $x \in L^p(I)$, $1/p + 1/q = 1$, and $t \in [0, 1]$, let $T(x)(t) = \int_0^1 k(s, t)x(s) \, ds$. Discuss the validity of the following statement: $T : L^p(I) \to L^p(I)$ is compact.

58. Let X be a normed linear space and $T, K : X \to X$ bounded linear mappings with K compact. Prove that KT, TK are compact.

59. Let B be a Banach space, $T_n : B \to B$ bounded linear mappings such that $T_n(x) \to T(x)$ for all $x \in B$, where T is a bounded linear operator on B, and $K : B \to B$ compact. Prove that $\|T_n K - TK\| \to 0$.

60. Given scalar sequences $\{\alpha_n\}$ and $\{\beta_n\}$ such that $|\alpha_{n-1}| \geq |\alpha_n| \to 0$ and $|\beta_{n-1}| \geq |\beta_n| \to 0$, let $T : \ell^2 \to \ell^2$ be given by $T(x) = y$ where $y = (\alpha_1 x_1, \alpha_2 x_2 + \beta_1 x_1, \alpha_3 x_3 + \beta_2 x_2, \ldots)$. Prove that T is compact.

61. Let X be a normed linear space, $x_0 \in X$, and $L \in X^*$. Prove that the mapping $T : X \to X$ given by $T(x) = L(x)\, x_0$ is compact.

62. Let X, Y be normed linear spaces and $T : X \to Y$ a compact linear operator. Prove that if $x_n \rightharpoonup x$ in X, then $T(x_n) \to T(x)$ in Y.

63. Given a sequence λ, let T be the linear mapping from sequences into sequences given by $T(x) = y$, where $y_n = \lambda_n x_n$, $n = 1, 2, \ldots$, and $1 \leq p < \infty$. When is T compact?

64. Let B be an infinite-dimensional Banach space, X a normed linear space, and $T : B \to X$ a linear operator such that $\|T(x)\|_X \geq c\|x\|_B$ for all $x \in B$ and a constant $c > 0$. Prove that T is not compact.

65. Let X be an infinite-dimensional normed linear space and $T : X \to X$ a compact linear operator. Prove that T^{-1} is not bounded.

66. Let $y \in C(I)$ and $T : L^2(I) \to L^2(I)$ be given by $T(x)(t) = y(t)x(t)$. Prove that T is compact iff $y = 0$.

67. Let X be a normed linear space and $T : X \to X$ a compact linear operator. Prove: (a) If $\lambda \neq 0$, $K(\lambda I - T)$ is finite dimensional. (b) If x_K denotes the closest point in $K(\lambda I - T)$ to x, there is a constant c such that $\|x - x_K\| \leq c\,\|(\lambda I - T)(x)\|$ for all $x \in X$.

68. Let X be a normed linear space, $\lambda \neq 0$, and $T : X \to X$ a compact linear operator. Prove that there is an integer $k \geq 1$ such that $K((\lambda I - T)^k) = K((\lambda I - T)^{k+1})$.

69. Let X be a normed linear space, $\lambda \neq 0$, and $T : X \to X$ a compact linear operator. Prove that $R(\lambda I - T) = X$ iff $K(\lambda I - T) = \{0\}$.

70. Let B be a Banach space and $T : B \to B$ a compact linear operator. Prove that for $\lambda \neq 0$, $R(\lambda I - T)$ is closed.

71. Let B be a Banach space, $\lambda \neq 0$, and $T : B \to B$ a compact linear operator. Prove that $K(\lambda I - T)$ is finite dimensional, $R(\lambda I - T)$ has finite codimension, and $\dim(K(\lambda I - T)) = \operatorname{codim}(R(\lambda I - T))$.

72. Let B, B_1 be Banach spaces and $T : B \to B_1$ a bounded linear operator such that $K(T)$ is finite dimensional and $R(T)$ is closed and has finite codimension in B_1. Prove: (a) There are a closed subspace M of B and a finite-dimensional subspace N of B_1 such that $B = M \oplus K(T)$, $B_1 = R(T) \oplus N$, and $T|_M : M \to R(T)$ is an isomorphism. (b) There exist a bounded linear mapping $T_0 : B_1 \to B_0$ and bounded finite rank operators $F_1 : B \to B$ and $F_2 : B_1 \to B_1$ such that $K(T_0) = N$, $R(T_0) = M$,

$T_0 T = I$ on M, $TT_0 = I$ on $R(T)$, and in addition $T_0 T = I - F_1$ on B and $TT_0 = I - F_2$ on B_1.

73. Let X, Y be Banach spaces and $T : X \to Y$ a bounded linear operator with the property that there are bounded linear operators $T_1, T_2 : Y \to X$, and compact linear operators $K_1 : X \to X$, $K_2 : Y \to Y$, such that $T_1 T = I - K_1$ on X and $TT_2 = I - K_2$ on Y. Prove that $K(T)$ is finite dimensional and that $R(T)$ is closed with finite codimension.

74. Let X, Y be Banach spaces and $T : X \to Y$ a bounded linear operator. Prove that the following are equivalent: (a) T is compact. (b) Given $\varepsilon > 0$, there is a finite-dimensional subspace F of Y such that the composition of T with the canonical map $\pi_F : Y \to Y/F$ has norm $\|\pi_F \circ T\| \leq \varepsilon$. (c) Given $\varepsilon > 0$, there is a subspace W of finite codimension in X such that $\|T|_W\| < \varepsilon$.

75. Let X be a linear space. We say that a linear mapping $P : X \to X$ is a *projection* if $P^2 = P$. Prove: (a) If P is a projection, $R(P) = K(I - P)$ and $K(P) = R(I - P)$. (b) If P is a projection, $R(P) \cap K(P) = \{0\}$ and $X = R(P) + K(P)$. (c) If $X = M + N$, where M, N are subspaces of X with $M \cap N = \{0\}$, there is a projection $P : X \to X$ with range $R(P) = M$ and kernel $K(P) = N$.

76. Let B be a Banach space. Prove that the following statements are equivalent: (a) $B = M \oplus N$. (b) There exist bounded projections $P : B \to M$, $Q : B \to N$ such that $P + Q = I$ and $PQ = QP = 0$.

77. Let B be a Banach space and M, N closed subspaces of B such that $M \cap N = \{0\}$. Prove that $M + N$ is closed in B iff $\|m\|_B + \|n\|_B \leq c\|m + n\|_B$, for all $m \in M$ and $n \in N$.

78. Let B be a Banach space and X, Y closed subspaces of B such that $X \cap Y = \{0\}$. Prove that $X + Y$ is closed in B iff $k = \inf\{\|x - y\| : x \in X, y \in Y, \|x\| = \|y\| = 1\} > 0$.

79. Let P be a projection on a normed linear space X. Prove that $\|P(x)\|/\|P\| \leq d(x, K(P)) \leq \|P(x)\|$ for all $x \in X$.

80. Prove that there is no bounded projection $P : c \to c_0$ of norm strictly less than 2 and give an example of one such projection with $\|P\| = 2$.

81. Let X be a Banach space, $P : X \to X$ a projection, and $M = R(P)$, $N = K(P)$. Prove that a bounded linear operator $T : X \to X$ commutes with P iff $T(M) \subset M$ and $T(N) \subset N$.

82. Let X be a normed linear space, X_1 a closed subspace of X, and M a finite-dimensional subspace of X such that $M \cap X_1 = \{0\}$. Prove: (a) $X_2 = X_1 \oplus M$ is a closed subspace of X. (b) The operator $P : X_2 \to X_2$ defined by $P(x) = x$ for $x \in M$ and $P(x) = 0$ for $x \in X_1$ is bounded.

83. Let X be an infinite-dimensional normed linear space and $T : X \to X$ a compact linear operator. Prove that $R(T)$ contains no closed infinite-dimensional subspace.

84. Let B be a Banach space and $T : B \to B$ a bounded linear operator with $\|T\| < 1$. For an integer N consider the Cesàro means $S_N(T) = N^{-1} \sum_{n=0}^{N-1} T^n$, $T^0 = I$, of T. Prove: (a) $\sup_N \|S_N(T)\| \le M = \sup_n \|T^n\| < 1$. (b) $\lim_N \|T S_N(T) - S_N(T)\| = \lim_N \|S_N(T)T - S_N(T)\| = 0$. (c) $X = \{x \in B : \lim_N S_N(T)(x) \text{ exists}\}$ is a closed subspace of B. (d) If $P(x) = \lim_N S_N(T)(x)$, then $TP = PT = P$, and P is the projection onto $K(T - I) \subset X$.

85. Discuss the validity of the following statement: There is a bounded linear operator $T : \ell^\infty \to \ell^\infty$ with $K(T) = c_0$.

86. Prove that the mapping $T : \ell^2 \to \ell^2$ given by $T(x) = \|x\|_{\ell^6} x$ is a continuous bijection but not a homeomorphism.

87. Let B, B_1 be Banach spaces and $T, T_n : B \to B_1$ linear mappings. Prove that $A = \{x \in B : T_n(x) \text{ does not tend to } T(x) \text{ in } B_1\}$ and $C = \{x \in B : \{T_n(x)\} \text{ is not Cauchy in } B_1\}$ are empty or dense in B.

88. Let X be a normed linear space and B a Banach space. Consider $\{T_k\} \subset \mathcal{B}(X, B)$ such that $\sup_k \|T_k\| < \infty$. Let $M = \{x \in X : \lim_k T_k(x) \text{ exists in } B\}$. Prove that M is a closed subspace of X. Conclude that if $\{T_k(x)\}$ converges in B for x in a dense subspace of X, then $\{T_k(x)\}$ converges in B for all $x \in X$.

89. Consider the mappings $\{T_m\}$ from sequences to sequences given by $T_m(x) = y$ where $y_n = x_n$ for $n \le m$, and $y_n = x_m$ for $n > m$. Prove that each $T_m : \ell^1 \to \ell^\infty$ continuously, and that $\|T_m(x) - x\|_\infty \to 0$ as $m \to \infty$ for each $x \in \ell^1$, but $\|T_m - I\| \not\to 0$ as $m \to \infty$.

90. Let $c_{00} = \{x \in c_0 : x_n = 0 \text{ for all } n \text{ sufficiently large}\}$ and consider the functionals L_n on $(c_{00}, \|\cdot\|_\infty)$ given by $L_n(x) = \sum_{k=1}^n x_k$, $n \ge 1$. Prove: (a) L_n is a bounded linear functional and compute $\|L_n\|$. (b) $\{L_n(x)\}$ is bounded and $\lim_n L_n(x)$ exists for each $x \in c_{00}$. Deduce that c_{00} is of first category in itself. (c) $D = \{x \in c_{00} : |L_n(x)| \le 1 \text{ for all } n\}$ is closed, has empty interior, and verifies $c_{00} = \bigcup_n nD$.

91. Let X be a normed linear space and $T_n : X \to X$ bounded linear operators such that $\lim_n T_n(x) = T(x)$ exists for all $x \in X$. Discuss the validity of the following statement: T is a bounded linear operator from X into X.

92. Let B, B_1 be Banach spaces and $T_n, T : B \to B_1$ bounded linear operators. (a) Prove that $\|T_n - T\| \to 0$ iff for every $\varepsilon > 0$ there is an N, depending only on ε, such that for all $n > N$ and all $x \in B$ of norm

1 we have $\|T_n(x) - T(x)\|_{B_1} < \varepsilon$. (b) Discuss the validity of the following statement: $\|T_n - T\| \to 0$ iff $T_n(x) \to T(x)$ in B_1 for each $x \in B$.

93. With T_h given by $T_h(f)(x) = f(x + h)$, define $A_h : C_0(\mathbb{R}) \to C_0(\mathbb{R})$ by

$$A_h = \frac{T_h - I}{h}.$$

Identify the classes of functions in $C_0(\mathbb{R})$ where A_h converges strongly, uniformly.

94. Let B be a Banach space, X a normed linear space, and $T_n : B \to X$ bounded linear operators such that $\lim_n T_n(x) = T(x)$ in X for $x \in B$. Prove : (a) There is a constant $c > 0$ such that $\sup_n \|T_n\| \leq c$. (b) Prove that the conclusion in (a) follows provided that $\ell(T_n(x)) \to \ell(T(x))$ for all $x \in B$, $\ell \in X^*$, instead. (c) $T : B \to X$ is a bounded linear operator and $\|T\| \leq \liminf_n \|T_n\|$.

95. Let B be a Banach space, X a normed linear space, and $T_n : B \to X$ bounded linear operators. Prove that the following statements are equivalent: (a) $\sup_n \|T_n\| < \infty$. (b) If $\|x_n\|_B \to 0$, then $\|T_n(x_n)\|_X \to 0$. (c) If $\sum_n \|x_n\|_B < \infty$, then $\|T_n(x_n)\|_X \to 0$.

96. Let X be a normed linear space and $Y \subset X$. Prove that Y is a bounded subset of X iff $\sup_{y \in Y} |L(y)| < \infty$ for each $L \in X^*$.

97. Let B be a Banach space and $\mathcal{L} \subset B^*$. Prove that $\mathcal{L}$ is bounded iff $\sup_{L \in \mathcal{L}} |L(x)| < \infty$ for all $x \in B$.

98. Let $\mathcal{P}$ be the linear space of polynomials $p(t) = a_0 + a_1 t + \cdots + a_n t^n$ equipped with the norm $\|p\| = \max_k |a_k|$. Discuss the validity of the following statement: $(\mathcal{P}, \|\cdot\|)$ is complete.

99. Let B be a Banach space, X a normed linear space, and $\mathcal{T} = \{T\}$ bounded linear operators $T : B \to X$. Prove that the following statements are equivalent: (a) $\sup_{T \in \mathcal{T}} \|T\| < \infty$. (b) $\sup_{T \in \mathcal{T}} \|T(x)\|_X < \infty$ for all $x \in B$. (c) $\sup_{T \in \mathcal{T}} |L(T(x))| < c_{x,L} < \infty$ for each $x \in B$ and $L \in X^*$, where $c_{x,L}$ is a constant that depends on x and L.

100. Consider the mappings $\{T_n\}$ from sequences to sequences given by $T_n(x) = y$ where $y_k = \lambda_k^n x_k$ for all k. Further suppose $\lim_n \lambda_k^n = \lambda_k$ for all k and $|\lambda_k^n| \leq c$ for all n, k. Let T be given by $T(x) = y$ where $y_k = \lambda_k x_k$ for all k. Let $1 \leq p < \infty$. Discuss the validity of the following statements: (a) $\lim_n T_n(x) = T(x)$ in ℓ^p for every $x \in \ell^p$. (b) $\lim_n \|T_n - T\| \to 0$. (c) $T_n(x) \rightharpoonup T(x)$ in ℓ^p.

101. Let B, B_1 be Banach spaces and $\{T_n\}, T$ bounded linear operators from B to B_1 such that $T_n(x) \rightharpoonup T(x)$ in B_1 for each $x \in B$. Prove that $\{\|T_n\|\}$ is bounded.

102. Let B be a Banach space, $\{x_n\} \subset B$, and $\{L_n\} \subset B^*$. Prove that the following statements are equivalent: (a) If $x_n \to 0$ in B, then $\sum_n |L_n(x_n)|$ converges. (b) $\sum_n L_n$ is absolutely convergent in B^*.

103. Let B be a Banach space and $T : B \to L^1(\mathbb{R}^n)$ a linear mapping. For each $A \in \mathcal{L}(\mathbb{R}^n)$ consider the linear functional L_A on B given by $L_A(x) = \int_A T(x)(t)\,dt$, $x \in B$. Prove that T is bounded iff L_A is bounded for each $A \in \mathcal{L}(\mathbb{R}^n)$.

104. Let $f : I \to \mathbb{R}$ be differentiable. Prove that f is Lipschitz on some open interval $(a, b) \subset [0, 1]$.

105. Let X be a closed subspace of $(C([0, 1]), \|\cdot\|_\infty)$ consisting of differentiable functions on $[0, 1]$ and fix $t_0 \in [0, 1]$. For $x \in X$ and $t \in [0, 1]$, $t \neq t_0$, let

$$\Lambda_t(x) = \frac{x(t) - x(t_0)}{t - t_0}\,.$$

Prove: (a) There exists $M > 0$ such that for all $x \in X$, $\sup_{t \neq t_0} \|\Lambda_t(x)\|_\infty \leq M\|x\|_\infty$. (b) The unit ball of X is equicontinuous in t_0. (c) X is finite dimensional.

106. Let B, B_1 be Banach spaces, X a normed linear space, $T : B_1 \to X$ a 1-1 bounded linear operator, and $S : B \to B_1$ a linear mapping such that $T \circ S : B \to X$ is bounded. Prove that S is bounded.

107. Prove that the closed graph theorem implies the open mapping theorem.

108. Let B be a Banach space, X a normed linear space, and $\mathcal{F} \subset \mathcal{B}(B, X)$. Prove that $G = \left\{ x \in B : \sup_{T \in \mathcal{F}} \|T(x)\|_X < \infty \right\}$ is of first category in B or coincides with B.

109. (Principle of the Condensation of Singularities). Let B be a Banach space, $\{X_m\}$ normed linear spaces, and $T_n^m : B \to X_m$, $n = 1, \ldots$, bounded linear operators. Suppose that for each m there exists $x_m \in B$ such that $\limsup_n \|T_n^m(x_m)\|_{X_m} = \infty$, $m = 1, \ldots$. Prove that $A = \{x \in B : \limsup_n \|T_n^m(x)\|_{X_m} = \infty, \text{ all } m = 1, \ldots\}$ is of second category in B.

110. Let B, B_1 be Banach spaces and $T : B \to B_1$ a bounded linear mapping. Prove: (a) If T is not onto but $R(T)$ is dense in B_1, then $R(T)$ is of first category in B_1, but not nowhere dense in B_1. (b) If $T(B)$ is of the second category in B_1, T is onto.

111. Give an example of a first category set in $L^p(I)$, $1 \leq p < \infty$, which is not nowhere dense.

112. Let B be a Banach space and $T : B \to C(I)$ a linear mapping such that $x_n \to 0$ in B implies $T(x_n) \to 0$ pointwise in I. Prove that T is bounded.

113. Let B be a Banach space and $T : B \to B$ an injective compact linear mapping. Prove that if B is infinite dimensional, T cannot be onto.

114. Let B_1, B_2 be Banach spaces, X a normed linear space, and $T_1 : B_1 \to X$, $T_2 : B_2 \to X$ continuous linear mappings with the property that the equation $T_1(x) = T_2(y)$ has exactly one solution $y \in B_2$ for each $x \in B_1$. Prove that the mapping $T : B_1 \to B_2$ defined by $T(x) = y$ is linear and continuous.

115. Let B, B_1 be Banach spaces and $T : B \to B_1$, $S : B_1^* \to B^*$ linear mappings such that $L(T(x)) = S(L)(x)$ for all $x \in B$, $L \in B_1^*$. Prove that S and T are bounded, with $S = T^*$.

116. Let X be a linear space such that $(X, \| \cdot \|)$ and $(X, \| \cdot \|_1)$ are complete. Prove that the norms are equivalent iff $\|x_n\| \to 0$ implies $\|x_n\|_1 \to 0$.

117. Let $C(I)$ be endowed with a norm $\| \cdot \|$ so that $(C(I), \| \cdot \|)$ is complete and if $\|x_n - x\| \to 0$, a subsequence $\{x_{n_k}\}$ of $\{x_n\}$ satisfies $\lim_{n_k} x_{n_k}(t) = x(t)$ for $t \in I$. Prove that $\| \cdot \| \sim \| \cdot \|_\infty$.

118. Let X be a linear space equipped with two norms $\| \cdot \|_1$ and $\| \cdot \|_2$ such that $(X, \| \cdot \|_1)$ is a Banach space, and assume that the identity mapping $I : (X, \| \cdot \|_1) \to (X, \| \cdot \|_2)$ is continuous. Further assume that, if $B_1(0, r) = \{x \in X : \|x\|_1 < r\}$ and $B_2(0, r) = \{x \in X : \|x\|_2 < r\}$, there exists $r > 0$ such that $B_2(0, 1) \subset B_1(0, r) + B_2(0, 1/2)$. Prove that the two norms are equivalent.

119. Let $(X, \mathcal{M}, \mu)$ be a measure space, $1 \le p, q \le \infty$, and Y a subspace of $L^p(X) \cap L^q(X)$ that is closed in both $L^p(X)$ and $L^q(X)$. Prove that the L^p and L^q norms are equivalent in Y.

120. Let $2 \le p \le \infty$ and X a subspace of $L^p([0, 1])$ that is closed in $L^1([0, 1])$. Prove: (a) If $p < \infty$, $(X, \| \cdot \|_p)$ is isomorphic to a Hilbert space. (b) If $p = \infty$, X is finite dimensional.

121. Let X be a subspace of $(C(I), \| \cdot \|_\infty)$ that is closed relative to the $L^p(I)$ norm for some $1 < p < \infty$. Prove: (a) $\|x\|_p \le \|x\|_\infty$ for each $x \in X$. (b) X is closed in $C(I)$. (c) There exists M such that $\|x\|_\infty \le M \|x\|_p$ for all $x \in X$. (d) For each $t \in [0, 1]$, there is a function $y_t \in L^q(I)$ such that $x(t) = \int_I y_t(s) \, x(s) \, ds$, all $x \in X$. (e) If $x_n \rightharpoonup x$ in $L^p(I)$, $x_n(t) \to x(t)$ for each $t \in I$. (f) If $\{x_n\} \subset X$ and $x_n \rightharpoonup x$ in $L^p(I)$, then $x_n \to x$ in $L^p(I)$. (g) X is finite dimensional.

122. For $x \in \ell^1$, let $\|x\| = 2|\sum_{n=1}^\infty x_n| + \sum_{n=2}^\infty (1 + n^{-1}) |x_n|$. Prove that $\| \cdot \|$ defines a norm in ℓ^1 that is equivalent to $\| \cdot \|_1$.

123. Prove that there exists a positive constant M such that if $p(t) = a_0 + a_1 t + \cdots + a_n t^n$ is a polynomial in $[-1, 1]$ of degree $\leq n$, then $\int_{-1}^{1} |p(t)| \, dt \leq M \max\{|p(t)| : t = 0, 1, \ldots, n\}$.

124. Let X, Y be Banach spaces and $\mathcal{T} = \{T\}$ linear operators $T : X \to Y$ such that $\sup_{T \in \mathcal{T}} \|T(x)\|_Y < \infty$ for every $x \in X$. Prove that $\|x\|_{\mathcal{T}} = \|x\|_X + \sup_{T \in \mathcal{T}} \|T(x)\|_Y$ is a norm on X and that the following statements are equivalent: (a) Each $T \in \mathcal{T}$ is continuous. (b) $\|\cdot\|_{\mathcal{T}} \sim \|\cdot\|_X$. (c) $(X, \|\cdot\|_{\mathcal{T}})$ is complete.

125. Prove that $(C(I), \|\cdot\|_p)$ is not a Banach space, $1 \leq p < \infty$.

126. Prove the following statements: (a) If x is a sequence such that $\sum_n y_n x_n$ converges for each $y \in c_0$, then $x \in \ell^1$. (b) If $\{x^n\} \subset \ell^1$, then $\sum_j y_j x_j^n \to 0$ for every $y \in c_0$ iff $\sup_n \|x^n\|_1 < \infty$ and $\lim_n x_j^n = 0$ for all j.

127. Discuss the validity of the following statement: There is a slowest rate of decay for the terms of an absolutely convergent series; that is, there exists a sequence $a \in \ell^1$ with $a_n > 0$, all n, such that $\sum_n a_n |x_n| < \infty$ iff $x \in \ell^\infty$.

128. Let X, Y be normed linear spaces and $T : X \to Y$ a bounded linear operator. Prove that the following statements are equivalent: (a) $K(T)$ is a finite-dimensional subspace of X and $R(T)$ is closed in Y. (b) Every bounded sequence $\{x_n\}$ in X with $\{T(x_n)\}$ convergent in Y has a convergent subsequence.

129. Let $X = C([0, 1], \|\cdot\|_1)$ and $B : X \times X \to \mathbb{R}$ the bilinear functional given by $B(x, y) = \int_0^1 x(t) y(t) \, dt$. Discuss the validity of the following statements: (a) B is separately continuous. (b) B is jointly continuous.

130. Let X, Y, Z be Banach spaces and $T : X \times Y \to Z$ a separately continuous bilinear mapping, i.e., for each fixed $x \in X$ and $y \in Y$, $T(x, \cdot) : Y \to Z$ and $T(\cdot, y) : X \to Z$ are bounded linear mappings. Prove that T is jointly continuous, i.e., for each $(x_0, y_0) \in X \times Y$, given $\varepsilon > 0$, there exists $\delta > 0$ such that $\max\{\|x_0 - x\|_X, \|y_0 - y\|_Y\} \leq \delta$ implies $\|T(x_0, y_0) - T(x, y)\|_Z \leq \varepsilon$.

131. Let X, Y be normed linear spaces and $T : X \to Y$ a bounded linear operator such that $T^{-1} : R(T) \to X$ is well-defined. Discuss the validity of the following statement: T^{-1} is bounded.

132. Let X, Y be linear spaces with $\dim(X) = \dim(Y) = n$ and $T : X \to Y$ a linear mapping. Prove that T^{-1} is well-defined iff $R(T) = Y$.

133. Let B, B_1 be Banach spaces and $T : B \to R(T) \subset B_1$ a bounded linear operator. Prove that the following statements are equivalent: (a) $T^{-1} : R(T) \to B$ is bounded. (b) There exists a constant $c > 0$ such that $\|T(x)\|_{B_1} \geq c \|x\|_B$ for all $x \in B$. (c) $K(T) = \{0\}$ and $R(T)$ is closed in B_1.

134. Let X be a normed linear space and $T : X \to X$ a bounded linear operator. Prove that if there exists $\{x_n\} \subset D(T)$ such that $\|x_n\| = 1$ and $T(x_n) \to 0$, T does not have a bounded inverse.

135. Let B, B_1 be Banach spaces and $T : D(T) \subset B \to B_1$ a closed linear operator. Prove that if T is invertible, $R(T)$ is closed in B_1.

136. Let X be a linear space and $T, S : X \to X$ linear mappings such that $TS + T + I = 0$ and $ST + T + I = 0$. Prove that T^{-1} is well-defined.

137. Let X be a normed linear space and $T, T_1 : X \to X$ bounded linear operators. Discuss the validity of the following statement: If TT_1 is invertible, T and T_1 are invertible.

138. Let X, Y be normed linear spaces and $T : X \to Y$ a linear mapping. Prove: (a) If T has a unique left inverse S, then $S = T^{-1}$. (b) If T is bounded and onto and S is a bounded left inverse of T, T^{-1} is well-defined and $S = T^{-1}$.

139. Let B, B_1 be Banach spaces and $T : B \to B_1$ a bounded linear operator. Prove: (a) T is right-invertible iff T is onto and $K(T)$ is complemented in B. (b) T is left-invertible iff T is 1-1 and $R(T)$ is complemented in B_1.

140. Let X be a linear space and $S, T : X \to X$ linear mappings. Prove: (a) If $I - TS$ is 1-1, $I - ST$ is 1-1. (b) If $I - TS$ is surjective, $I - ST$ is surjective. (c) $I - TS$ is invertible iff $I - ST$ is invertible.

141. Let X be a normed linear space and $T : X \to X$ a linear mapping with $I - T$ invertible. Prove that T is closed.

142. Let X, Y be normed linear spaces and $T : X \to Y$ a closed linear operator such that $T^{-1} : R(T) \to X$ is well-defined. Prove: (a) T^{-1} is closed. (b) If X, Y are Banach spaces and $R(T) = Y$, then T^{-1} is continuous.

143. Let B be a Banach space, $T : B \to B$ a linear operator such that $\|T(x)\| \geq c\,\|x\|$ for all $x \in B$, and L a bounded linear functional on B. Prove that there exists a bounded linear functional ℓ on B such that $L(x) = \ell(T(x))$ for all $x \in B$.

144. Let X be a linear space and $T : X \to X$ a linear mapping that satisfies the polynomial relation $\sum_{n=0}^{N} c_n T^n = 0$, with $c_0 \neq 0$. Prove that T^{-1} exists.

145. Let $P : B \to B$ be a bounded mapping that satisfies $P^2 = P$, $P \neq 0, I$. Given $\lambda \in \mathbb{C}$ with $|\lambda| < 1$, compute $(\lambda I - P)^{-1}$.

146. Let X be a linear space and $T : X \to X$ a linear operator such that $T \neq 0$ but $T^N = 0$ for some integer $N > 1$. Prove that for a scalar $\lambda \neq 0$, $(T - \lambda I)^{-1} = -\sum_{n=0}^{N-1} \lambda^{-(n+1)} T^n$.

147. Let B be a Banach space, $T : B \to B$ a bounded linear operator with $\|T\| < 1$, and $S_n = \sum_{k=0}^{n} T^k$, $n \geq 1$. Prove that $\{S_n\}$ is a Cauchy sequence in $\mathcal{B}(B)$ and find its limit.

148. Let B be a Banach space and $T \in \mathcal{B}(X)$ with $\|T\| < 1$. Prove that the sequence $\{P_n\}$ given by $P_n = (I + T)(I + T^2)(I + T^4) \cdots (I + T^{2^n})$ has a limit and find it.

149. Let B be a Banach space and $T : B \to B$ a bounded linear operator. Prove that if $\|I - T\| < 1$, T is invertible.

150. Let B be a Banach space and $T : B \to B$ a bounded linear operator. Prove there exists $R > 0$ such that $rI - T$ is invertible for $|r| > R$ and find a bound for $\|(rI - T)^{-1}\|$.

151. Let B be a Banach space and $T : B \to B$ a bounded, invertible, linear operator. Prove that if $S : B \to B$ is a bounded linear mapping satisfying $\|T^{-1}\|^{-1}\|T - S\| < 1$, then S is invertible and $\|S^{-1} - T^{-1}\| \leq \|T^{-1}\|^2\|T - S\|/(1 - \|T^{-1}\|\,\|T - S\|)$.

152. Let B be a Banach space and $T_n, T : B \to B$ bounded linear operators with T invertible. Prove: (a) If $\|T_n - T\| \to 0$, the T_n are invertible for n large enough. (b) If $\|T_n(x) - T(x)\| \to 0$ for all $x \in B$, it does not follow that T_n is invertible for any n.

153. Let B be a Banach space and $T : B \to B$ a bounded, invertible, linear operator. Prove that if $S : B \to B$ is a bounded linear mapping satisfying $\|ST^{-1}\| < 1$, then $T - S$ is invertible and

$$(T - S)^{-1} = T^{-1}\Big(I + \sum_{k=1}^{\infty}(ST^{-1})^k\Big).$$

154. Let B be a Banach space, $T : B \to B$ a bounded invertible linear operator, and $S : B \to B$ a bounded linear operator. Let $\phi(T) = T^{-1}$ denote the inversion mapping on $\mathcal{B}(B)$. Prove: (a) $T + tS$ is invertible for t sufficiently small. (b) $\lim_{t \to 0} \phi(T + tS) = \phi(T)$ in $\mathcal{B}(B)$. (c) $\lim_{t \to 0}(\phi(T + tS) - \phi(T))/t$ exists in B, and find it.

155. Construct a normed linear space X and a bounded linear operator $T : X \to X$ with $\|T\| < 1$ such that the equation $x - T(x) = y$ has no solution for some $y \in X$.

156. Let B be a Banach space and $T : B \to B$ a bounded linear operator with $\|T\| < 1$. Prove that the equation $x - T(x) = y$ has a unique solution $x \in B$ for every $y \in B$ that depends continuously on y.

157. Let T be a linear mapping from sequences into sequences given by $T(e_1) = 0$ and $T(e_k) = 2e_{k-1}$ for $k > 1$. Consider the equation $y =$

$\lambda T(x) + z$, where $z \in \ell^2$ and λ is a scalar. For what values of λ can the equation be solved in ℓ^2?

158. Let X be a normed linear space and $T : X \to X$ a compact linear operator such that the equation $x - T(x) = y$ is solvable for every $y \in X$. Prove: (a) The equation has a solution of minimal norm. (b) If x is a solution of minimal norm, there exists a constant $c > 0$ such that $\|x\| \le c\|y\|$.

159. Let X be a normed linear space and $T : X \to X$ a compact linear operator. Prove that, given $y \in X$, the equation $x - T(x) = y$ has a solution iff $L(y) = 0$ for every linear functional $L \in X^*$ which is a solution of the homogenous equation $(I - T)^*(L) = 0$.

160. Let X be a normed linear space and $T : X \to X$ a compact linear operator. Prove that $L - T^*(L) = \ell$ is solvable for all $\ell \in X^*$ iff $\ell(x) = 0$ for every $x \in X$ such that $x - T(x) = 0$.

161. Let X be a normed linear space and $T : X \to X$ a compact linear operator. Prove that $x - T(x) = y$ is solvable for every $y \in X$ iff the homogenous equation $x - T(x) = 0$ has only the trivial solution $x = 0$. In this case the solution is unique and $I - T$ has a bounded inverse.

162. Let X be a separable Banach space and $T : X \to X$ a bounded linear operator. Prove that if B is the closed unit ball of X, $T(B)$ is closed.

163. Let B be a Banach space and $T : B \to B$ a bounded linear operator which is 1-1 and has closed range. Prove that there exists $\varepsilon > 0$ such that for every linear mapping S with $\|S\| < \varepsilon$, $T + S$ is 1-1 and has closed range.

164. Let $D = \{z \in \mathbb{C} : |z| \le 1\}$ be the closed unit disk in the complex plane and $X = C(D)$ the Banach space of continuous complex-valued functions on D equipped with the sup norm. Consider the operator $A : C(D) \to C(D)$ defined by $Af(z) = z^3 f(z)$. Find $K(A)$ and prove that $R(A)$ is not closed.

165. Let $X = L^2([0,1])$, $k(t,s) = \min\{t,s\}$, $0 \le s, t \le 1$, and $T : X \to X$ given by $Tx(t) = \int_0^1 k(t,s)x(s)\,ds$. (a) Find $R(T)$, compute T^{-1}, and explain why T^{-1} is self-adjoint. (b) Discuss the validity of the following statement: The integral equation $x = y + 2T(x)$ has a unique positive solution x for each positive continuous function $y \in X$.

166. Let X, Y be Banach spaces, $T : X \to Y$ a continuous linear mapping, and $\widetilde{T} : X/K(T) \to R(T)$ the operator defined in Problem 21. Prove that the following are equivalent: (a) $R(T)$ is closed. (b) $\widetilde{T}$ is an isomorphism. (c) There is a constant M such that, for all $x \in X$, there exists $x' \in X$ with $T(x') = T(x)$ and $\|x'\|_X \le M\|T(x)\|_Y$. (d) There is a constant $r > 0$ such that $T(B_X(0,1)) \supset B_{R(T)}(0,r)$. (e) $T(U)$ is open in Y whenever U is an open subset of X (i.e., T is an open mapping).

167. Let X, Y be Banach spaces and $T : X \to Y$ a bounded linear operator. Prove that $R(T)$ is closed in Y iff there is a constant c such that $d(x, K(T)) \leq c\|T(x)\|_Y$, all $x \in D(T)$.

168. Let $T \in \mathcal{B}(X, Y)$. Prove that if T maps bounded closed sets in X onto closed sets in Y, then $R(T)$ is closed.

169. Let B, B_1 be Banach spaces and $T : B \to B_1$ a continuous linear mapping. Prove that if there exists a continuous linear mapping $S : B_1 \to B$ such that $T \circ S = I_{B_1}$, the identity on B_1, then T is open.

170. Let X be a Banach space, Y a normed linear space, and $T : X \to Y$ a bounded linear operator. Prove that if for some $\varepsilon < 1$ and $r > 0$, $T(B_X(0,r))$ contains an ε-net for $B_Y(0,1)$, then $T(B_X(0,R)) \supset B_Y(0,1)$ for some $R > 0$.

171. Prove that the mapping $T : L^1([1, \infty)) \to L^1([1, \infty))$ given by $Tx(t) = t^{-1}x(t)$ is bounded, but not open.

172. Let X, Y be Banach spaces and $\mathcal{U} = \{T \in \mathcal{B}(X, Y) : T^*(Y^*) = X^*\}$. Prove that $\mathcal{U}$ is open to $\mathcal{B}(X, Y)$.

173. Let X, Y be normed linear spaces, $T : X \to Y$ a bounded linear mapping, and $T^* : Y^* \to X^*$ its adjoint. Prove: (a) T^* is surjective iff T is injective and T^{-1} is bounded. (b) If X is reflexive, T is 1-1 iff $\overline{R(T^*)} = X^*$.

174. Let X be a normed linear space and $T : X \to X$ a bounded linear operator that is onto. Prove that there exists $\varepsilon > 0$ such that $T + S$ is onto provided that $\|S\| \leq \varepsilon$.

175. Let X, Y be Banach spaces and $T : X \to Y$ a surjective bounded linear mapping. Prove that $T^* : Y^* \to R(T^*) \subset X^*$ is an isomorphism.

176. Let X be a normed linear space, $Y \subset X$ a subspace of X, and $I : Y \to X$ the inclusion operator. Describe $I^* : X^* \to Y^*$.

177. Let X be a normed linear space and $T : X \to X$ a densely defined linear operator. Prove that T^* is closed.

178. Discuss the validity of the following statements: $T : X \to X$ is an isometry, i.e., a norm-preserving bijection, in a normed linear space X if: (a) $\|T\| = 1$. (b) $\|T\| \leq 1$ and $\|T^{-1}\| \leq 1$.

179. Let X, Y be Banach spaces and $T : X \to Y$ a bounded linear operator. Prove that T is an isometry iff $T^* : Y^* \to X^*$ is an isometry.

180. Let B, B_1 be Banach spaces and $T : B \to B_1$ a bounded linear operator with $R(T)$ closed. Prove that $R(T^*)$ is closed.

181. Let X, Y be Banach spaces and $T, S : X \to Y$ bounded linear operators with $R(T) \subset R(S)$. Prove there exists $M > 0$ such that for all $\ell \in Y^*$, $\|T^*(\ell)\|_{X^*} \leq M \|S^*(\ell)\|_{X^*}$.

182. Let X, Y, Z be Banach spaces and $T : Y \to Z$ a linear, continuous, injective mapping. Prove: (a) Let $S : X \to Y$ be linear. If $TS : X \to Z$ is continuous, S is also continuous. (b) Let $C : X \to Z$ be linear, continuous. If $R(C) \subset R(T)$, there is a continuous linear operator $S : X \to Y$ such that $C + TS = 0 : X \to Z$. (c) Let $S : X \to X$ for $X = L^2(\Omega)$, for a bounded region $\Omega \subset \mathbb{R}^m$. If $S(x)$ is continuous on Ω when $x \in X$ is continuous and bounded, then $S(x)$ is a continuous function for every $x \in X$.

183. Let X be a Banach space. We say that a sequence $\{x_n\} \subset X$ is a *Schauder basis* for X if for each $x \in X$ there exist unique scalars $\lambda_n = \lambda_n(x)$ such that $x = \sum_n \lambda_n x_n$, where the series converges in X. (a) Is every Schauder basis a Hamel basis? Prove: (b) $\{e_n\}$ is a Schauder basis for c_0. (c) If $f_1 = e_1/2$, $f_n = e_n/2 - e_{n-1}$, for $n \geq 2$, then $\{f_n\}$ is linearly independent. Is it a Schauder basis for c_0? (d) If $e = (1, 1, \ldots)$, $e, e_1, \ldots$ is a Schauder basis for c.

184. Let X be a normed linear space with a Schauder basis. Prove that X is separable.

185. Let X be a Banach space, $\{x_n\}$ a Schauder basis for X, and $P_n(x) = \sum_{k=1}^n \lambda_k x_k$ where $x = \sum_k \lambda_k x_k$ is the unique expansion of x. Prove that $\|x\|_1 = \sup_n \|P_n(x)\|$ is a norm on X equivalent to $\| \cdot \|$ and $(X, \| \cdot \|_1)$ is a Banach space.

186. Discuss the validity of the following statement: If X is a Banach space with separable dual and $\{x_n\}$ a Schauder basis for X, then the biorthogonal functionals $\{x_n^*\}$ constructed in Problem 8.97 are a Schauder basis for X^*.

187. Let $\{x_k\}$ be a Schauder basis for a Banach space B and L_n the n-th coordinate functional, i.e., if $x = \sum_k \lambda_k x_k$, then $L_n(x) = \lambda_n$, all n. Prove that L_n is a continuous linear functional and there is a constant M such that $1 \leq \|x_n\| \|L_n\| \leq M$, $n = 1, 2, \ldots$. In particular, if $\inf_n \|x_n\| > 0$, then $\sup_n \|L_n\| < \infty$.

188. Let B, B_1 be Banach spaces and $\{x_n\}, \{y_n\}$ Schauder bases for B and B_1, respectively. Prove that the following statements are equivalent: (a) There exists a continuous linear bijection $T : B \to B_1$ such that $T(x_n) = y_n$ for all n. (b) Given scalars $\{\lambda_n\}$, $\sum_n \lambda_n x_n$ converges in B iff $\sum_n \lambda_n y_n$ converges in B_1.

189. Let $X = C^1(I)$ denote the linear space of differentiable functions $x(t)$ with a continuous derivative $x'(t)$, for $0 < t < 1$ (one-sided at the endpoints??). (a) Prove that $\|x\| = \max_{t \in I} |x(t)| + \max_{t \in I} |x'(t)|$, $x \in X$, defines a norm on X and that $X = (X, \|\cdot\|)$ is a Banach space. (b) Let $\{e_n\}$ be a Schauder basis for $C(I)$. Prove that $\{b_n\}$ given by $b_1(t) = 1$ and $b_n(t) = \int_0^t e_{n-1}(s)\,ds$, $n > 1$, $t \in I$, is a Schauder basis for X.

190. Given an infinite-dimensional Banach space B with a Schauder basis, construct a nonclosable densely defined linear mapping $T : D(T) \subset B \to B$.

191. Let X be a closed subspace of a Banach space B. Prove that the following are equivalent: (a) There exists a bounded projection $P : B \to X$ onto X. (b) For every Banach space Y, every operator $T_0 \in \mathcal{B}(X, Y)$ admits an extension to an operator $T \in \mathcal{B}(B, Y)$.

192. Let B be a Banach space. Prove that there exists a bounded linear mapping $P : B^{***} \to B^{***}$ such that $P^2 = P$, $\|P\| = 1$, $R(P) = B^*$, and $P|_{B^*} = I_{B^*}$.

193. Let B be a Banach space that is not reflexive and $T : B \to B$ a bounded linear operator that is 1-1. Discuss the validity of the following statement: The adjoint T^* of T is 1-1.

194. Let X be a normed linear space, M a closed subspace of X, and $\pi^* : (X/M)^* \to X^*$ the adjoint of the canonical map $\pi : X \to X/M$. Identify $R(\pi^*)$.

195. Let X, Y be Banach spaces and $T : X \to Y$ a bounded linear operator with $R(T)$ closed in Y. Prove: (a) $R(T^*) = K(T)^\perp$. (b) $K(T^*) \sim (Y/R(T))^*$.

196. Let X be a Banach space, Y a normed linear space, and $T : X \to Y$ a bounded linear mapping. Prove that T is open iff $c\|\ell\|_{Y^*} \leq \|T^*(\ell)\|_{X^*}$ for all $\ell \in Y^*$.

197. Let X, Y be Banach spaces, $T : X \to Y$ a bounded linear operator, and $T^* : Y^* \to X^*$ its adjoint. Prove: (a) T is an isomorphic embedding iff $\widetilde{T^*}$ is an isomorphism. (b) $\widetilde{T}$ is an isomorphism iff T^* is an isomorphic embedding.

Chapter 10

Hilbert Spaces

In this chapter we consider linear spaces X over the field $\mathbb{R}$ or $\mathbb{C}$ which is specified as necessary. By an *inner product* on X we mean a scalar-valued function $\langle \cdot, \cdot \rangle$ on $X \times X$ such that: (a) $\langle x, x \rangle \geq 0$ for all $x \in X$ and $\langle x, x \rangle = 0$ iff $x = 0$, (b) $\langle x, y \rangle = \overline{\langle y, x \rangle}$ for all $x, y \in X$, and (c) $\langle \cdot, \cdot \rangle$ is linear in the first variable, i.e., $\langle \lambda x + y, z \rangle = \lambda \langle x, z \rangle + \langle y, z \rangle$ for all $x, y, z \in X$ and scalar λ. Thus, the inner product is skew symmetric, or Hermitian, and satisfies the Cauchy-Schwarz inequality, i.e., $|\langle x, y \rangle| \leq \langle x, x \rangle \langle y, y \rangle$ for all $x, y \in X$.

Now, the expression $\|x\| = \langle x, x \rangle^{1/2}$ is a norm and we say that X is a *Hilbert space* if X, endowed with this norm, is complete. Also, there is a relation between the inner product and the norm known as the "polarization identity". Specifically, $4\langle x, y \rangle = \|x + y\|^2 - \|x - y\|^2$ in a real inner product space, and $4\langle x, y \rangle = \|x + y\|^2 - \|x - y\|^2 - i(\|x + iy\|^2 - \|x - iy\|^2)$ in a complex space.

If $x, y \in X$, an inner product space, then $\|x + y\|^2 + \|x - y\|^2 = 2\|x\|^2 + 2\|y\|^2$, i.e., the "parallelogram law" holds. A normed linear space X is an inner product space iff the parallelogram law holds and, then, the unique inner product that induces the norm is given by the polarization identity.

An important notion in an inner product space is that of orthogonality. We say that $x, y \in X$ are *orthogonal*, and write $x \perp y$, if $\langle x, y \rangle = 0$. Then the Pythagorean theorem holds: If $\{x_n\}_{n=1}^N$ are pairwise orthogonal elements of X, $\|\sum_{n=1}^N x_n\|^2 = \sum_{n=1}^N \|x_n\|^2$.

An interesting question is to decide whether, given a subset M of a Hilbert space H and $x \notin M$, there exists $y \in H$ such that $d(x, H) = \inf\{\|x' - x\| : x' \in H\} = \|x - y\|$, in other words, whether a perpendicular can be dropped from x to M. Simple examples in $\mathbb{R}^2$ when M is an open segment or an arc show that there may exist no $y \in M$ which satisfies this relation, or that there may exist infinitely many such y. In a Hilbert space we have the following result: Let M be a nonempty, complete, convex subset of a Hilbert space H. Then, for every $x \in X$, there exists a unique $y \in M$ such that $d(x, M) = \|x - y\|$.

Then there is the projection theorem. For $x \in X$, let $x^\perp = \{y \in X : \langle x, y \rangle = 0\}$ and, if $M \subset X$, let $M^\perp = \{y \in X : y \perp x \text{ for all } x \in M\}$; $x^\perp$ and $M^\perp$ are closed subspaces of X. Now, if H is a Hilbert space and M a closed subspace of H, every $x \in H$ can be expressed as $x = x_1 + x_2$, where $x_1 \in M$ and $x_2 \in M^\perp$. Furthermore, the representation is unique and $\|x\|^2 = \|x_1\|^2 + \|x_2\|^2$. The mappings $P : X \to M$ and $Q : X \to M^\perp$ given by $P(x) = x_1$ and $Q(x) = x_2$ are linear *projections* and represent the nearest points to x in M and $M^\perp$, respectively. Moreover, $\langle P(x), y \rangle = \langle x, P(y) \rangle$ for all $x, y \in H$. Projections that verify this property are called *orthogonal*.

That Hilbert spaces are reflexive follows from the Fréchet-Riesz representation theorem which states that, if L is a continuous linear functional on a Hilbert space H, there exists a unique $x_L \in H$ with $\|x_L\| = \|L\|$ such that $L(x) = \langle x, x_L \rangle$ for all $x \in H$.

We say that $\{x_\alpha\}_{\alpha \in \Lambda}$ is an *orthogonal system* (OS) in an inner product space X if $x_\alpha \perp x_\beta = 0$ for all $\alpha \neq \beta$ in Λ. An OS is said to be an *orthonormal system* (ONS) if $\|x_\alpha\| = 1$ for all $\alpha \in \Lambda$. Now, if $\{x_\alpha\}$ is an ONS, Bessel's inequality holds, i.e., $\sum_{\alpha \in \Lambda} |\langle x, x_\alpha \rangle|^2 \leq \|x\|^2$ for all $x \in X$. The $\langle x, x_\alpha \rangle$ are called the *Fourier coefficients* of x with respect to $\{x_\alpha\}$ and, in particular, Bessel's inequality implies that for each $x \in X$ all but at most countably many of the Fourier coefficients of x with respect to the ONS $\{x_\alpha\}$ vanish.

We say that an ONS $\{x_\alpha\}$ in a Hilbert space H is *maximal,* or *complete*, and we call it an *orthonormal basis* (ONB), if no nonzero element can be added to it so that the resulting collection of elements is still an ONS in H; by Zorn's lemma a maximal ONS in H always exists. For an ONS $\{x_\alpha\}$ in a Hilbert space H the following properties are equivalent: (i) $\{x_\alpha\}$ is maximal in H. (ii) sp$\{x_\alpha\}$, i.e., the collection of all finite linear combinations of the x_α, is dense in H. (iii) (Plancherel's equality) Equality holds in Bessel's inequality, i.e., $\|x\|^2 = \sum_{\alpha \in \Lambda} |\langle x, x_\alpha \rangle|^2$ for all $x \in H$. (iv) (Parseval's equation) For all $x, y \in H$, $\langle x, y \rangle = \sum_{\alpha \in \Lambda} \langle x, x_\alpha \rangle \overline{\langle y, y_\alpha \rangle}$.

We say that a densely defined linear operator $T : D(T) \subset H \to H$ is *symmetric* if $\langle T(x), y \rangle = \langle x, T(y) \rangle$ for all $x, y \in D(T)$. Recall that, given $T \in \mathcal{B}(H)$, $T^* \in \mathcal{B}(H)$, the adjoint of T, satisfies $\langle x, T(y) \rangle = \langle T^*(x), y \rangle$ for all $x, y \in H$. Furthermore, T^* is unique and $\|T^*\| = \|T\|$. We say that an operator T is *self-adjoint* if $D(T^*) = D(T)$ and $T = T^*$. The condition for a linear operator to be self-adjoint is stronger than to be symmetric: For a symmetric operator T, $D(T^*) \supset D(T)$ and $T^*|_{D(T)}$ coincides with T, i.e., T^* is an extension of T. Now, when T is densely defined, $D(T^*)$ is the maximal subspace of H for which T^* is defined, i.e., $D(T^*) = \{y \in H : \text{there exists } T^*(y) \text{ in } H \text{ such that } \langle T(x), y \rangle = \langle x, T^*(y) \rangle \text{ whenever } x \in D(T)\}$.

Also, we say that a bounded linear operator $T : H \to H$ is of *finite rank* if $R(T)$ is finite dimensional.

A *bilinear*, or *sesquilinear*, mapping $\phi : H \times H \to \mathbb{C}$ is one that is linear in the first variable and conjugate linear in the second variable, i.e., such that $\phi(\lambda x + x', y) = \lambda \phi(x, y) + \phi(x', y)$, $\phi(x, \mu y + y') = \overline{\mu}\phi(x, y) + \phi(x, y')$. As is the case for an inner product, ϕ satisfies the Cauchy-Schwarz inequality $|\phi(x, y)| \leq |\phi(x, x)|^{1/2}|\phi(y, y)|^{1/2}$. Finally, we say that ϕ is *bounded* if $|\phi(x, y)| \leq c\|x\|\|y\|$ for all $x, y \in H$.

The problems in this chapter deal with the closely related concepts of inner product and Hilbert spaces; in fact, by Problem 12 the completion of the former is the latter. Now, closed and bounded subsets of Hilbert spaces have some interesting properties: by Problems 22–23 they do not necessarily have an element of smallest or largest norm. Also, by Problem 21 there is a separable Hilbert space such that the cardinality of an ONB and that of a Hamel basis in the space are different. And, by Problem 33, given a dense subspace M of a separable Hilbert space, there is an ONB consisting of elements in M. Problems 52–53 prove that an ONS sufficiently close to an ONB is itself an ONB.

Problems 35–38 deal with the Riesz-Fréchet representation theorem and its proof. Along similar lines, Problems 38–41 deal with the Hahn-Banach theorem.

Given a subspace M of a Hilbert space, properties of $M^\perp$ are discussed in Problems 60–68, and applications of this concept are given in Problems 70–73. The notion of projection is explored in Problems 75–89; Problem 93 considers when a projection is compact.

Problems 95–107 discuss properties of weak convergence; for instance, given a 2π-periodic function $x \in L^2([0, 2\pi])$, if $x_n(t) = x(nt)$, $n = 1, 2, \ldots$, Problem 104 proves that $\{x_n\}$ converges weakly in $L^2([0, 2\pi])$ and identifies its limit.

Hilbert-Schmidt operators are introduced in Problem 122 and isometries are considered in Problems 124–128. Finite rank operators are discussed in Problems 129–131. Problem 132 establishes that compact operators are the uniform limit of finite rank operators; compact operators are considered in Problems 133–140.

Properties of the adjoint T^* of a linear mapping T are given in Problems 145–148 and Problems 151–153. $D(T^*)$, the domain of T^*, is characterized in Problem 155 and Problem 154 gives an instance when $D(T^*) = \{0\}$. Symmetric operators are covered in Problems 156–161, and bilinear mappings in Problems 172–175.

Problems

1. Let $(X, \|\cdot\|)$ be a normed linear space that satisfies $\|x + y\|^2 + \|x - y\|^2 \leq 2\|x\|^2 + 2\|y\|^2$ for all $x, y \in X$. Prove that X is an inner product space.

2. Let $(X, \|\cdot\|)$ be a normed linear space of dimension greater than 2. Prove that X is an inner product space iff the two-dimensional subspaces of X are inner product spaces.

3. Discuss the validity of the following statement: If X is an inner product space and x, y in X satisfy $\|x + y\|^2 = \|x\|^2 + \|y\|^2$, then x and y are orthogonal.

4. Let X be an inner product space. Prove that the following statements are equivalent: (a) $x \perp y$. (b) $\|x + y\|^2 = \|x\|^2 + \|y\|^2$ and $\|x + iy\|^2 = \|x\|^2 + \|y\|^2$ for all $x, y \in X$. (c) $\|x + y\| = \|x - y\|$ and $\|x + iy\| = \|x - iy\|$ for all $x, y \in X$

5. Let X be an inner product space. Given nonzero x, y in X, prove that the following statements are equivalent: (a) $x \perp y$. (b) For all scalars λ, $\|x + \lambda y\| = \|x - \lambda y\|$. (c) For all scalars λ, $\|x + \lambda y\| \geq \|x\|$.

6. Recall that in a normed linear space X, $x \in X$ is said to be orthogonal to a subspace Y of X if $d(x, Y) = \|x\|$. Let X be an inner product space. Prove that $\langle x, y \rangle = 0$ iff x is orthogonal to the subspace $Y = \mathrm{sp}\{y\}$.

7. Let x, y be nonzero elements of an inner product space X. Prove that $|\langle x, y \rangle| = \|x\| \, \|y\|$ iff $x = \lambda y$ for some scalar λ.

8. Let x, y, z be nonzero elements of an inner product space X. Prove that $\|x - y\| + \|y - z\| = \|x - z\|$ iff $y = \lambda x + (1 - \lambda)z$ for some scalar λ between 0 and 1.

9. Let X be an inner product space and $x_0 \in X$. Find $x \in X$ with $\|x\| = 1$ that is closest to x_0, i.e., so that $\|x - x_0\|^2$ is minimized. What is the distance from x to x_0?

10. Let H be a Hilbert space, B a Banach space, and $T : H \to B$ an isometric linear isomorphism. Prove that B is a Hilbert space.

11. Let λ be a real sequence with $0 < \lambda_k < 1$, all k. On ℓ^2 define the inner product $\langle x, y \rangle_1 = \sum_k \lambda_k x_k \overline{y_k}$. Discuss the validity of the following statement: $X = (\ell^2, \langle \cdot, \cdot \rangle_1)$ is a Hilbert space.

12. Let X be an inner product space and $\overline{X}$ the completion of X. Prove that the inner product on $X \times X$ can be extended to $\overline{X} \times \overline{X}$, and so $\overline{X}$ is a Hilbert space.

13. Discuss the validity of the following statements: (a) Equipped with $\| \cdot \|_\infty$, $C([0,1])$ is a Hilbert space. (b) Equipped with $\| \cdot \|_p$, $\ell^p, p \neq 2$, is a Hilbert space.

14. Let H be a Hilbert space and M a closed subspace of H. Prove that H/M is a Hilbert space.

15. Let B be a Banach space. Discuss the validity of the following statements: (a) If B is isomorphic to a Hilbert space H, then B is a Hilbert space. (b) If there exist a Hilbert space H and a bounded linear operator $T : H \to B$ which is onto, then B is linearly homeomorphic to a Hilbert space. (c) If B is isomorphic to B^*, then B is isomorphic to a Hilbert space.

16. Given x, y in a complex inner product space X, prove that $\langle x, y \rangle = (2\pi)^{-1} \int_0^{2\pi} \|x + e^{it} y\|^2 e^{it} \, dt$.

17. Let X be an inner product space and $x^1, \ldots, x^{100}$ unit vectors in X such that $|\langle x^n, x^m \rangle| = 1/10$ for all $n \neq m$. Find a sharp estimate for $\|x^1 + \cdots + x^{100}\|$.

18. Let $\{x_n\}$ be an unbounded sequence in a Hilbert space H. Prove that $\{\langle x_n, x \rangle\}$ is unbounded for some $x \in H$.

19. Let H be a Hilbert space. (a) Construct a sequence $\{x_n\} \subset H$ that is summable but not absolutely summable. Suppose further that $\sum_n \|x_n\| < \infty$. Prove: (b) $\sum_n x_n$ converges to some $x \in H$ with $\|x\| \leq \sum_n \|x_n\|$. (c) If H is infinite dimensional, the ratio $(\sum_n \|x_n\|)/\|x\|$ may be unbounded. (d) If H is finite dimensional and $\{x_n\}$ is an ONS in H, then $(\sum_n \|x_n\|)/\|x\| \leq \sqrt{\dim(H)}$ and this bound is achieved.

20. Let $\{e_n\}$ be an ONS in a Hilbert space H. Prove that $\sum_n \lambda_n e_n$ converges unconditionally for every $\lambda \in \ell^2$. Is this condition necessary for convergence?

21. Give an example of a separable Hilbert space H such that the cardinality of an ONB and that of a Hamel basis of H are different.

22. Let $\{e_n\}$ be an ONB for a Hilbert space H and $X = \{x \in H : \sum_n (1 + 1/n)^2 |\langle e_n, x \rangle|^2 \leq 1\}$. Prove that X is a closed, bounded, convex subset of H with no element of largest norm.

23. Let $\{e_n\}$ be an ONB for a Hilbert space H and $X = \{x \in H : x = (1 + 1/n)\,e_n, n = 1, 2, \ldots\}$. Prove that X is closed and bounded in H, with no element of smallest norm.

24. Let K be a nonempty, closed, convex subset of a Hilbert space H, and given $x \in H$, let x_K denote the unique element in K such that $d(x, K) = \|x - x_K\|$. Prove: (a) x_K is characterized by the variational inequality: $\Re\langle x - x_K, y - x_K\rangle \leq 0$ for all $y \in K$. (b) Given $x, y \in H$, $\|x_K - y_K\| \leq \|x - y\|$. (c) K has a unique element of smallest norm.

25. Let H be a Hilbert space. Discuss the validity of the following statement: If $\{K_n\}$ is a decreasing sequence of nonempty, bounded, closed convex sets in H, $\bigcap_n K_n \neq \emptyset$.

26. Let H be a real Hilbert space, L a bounded linear functional on H, and K a nonempty closed convex subset of H. Consider the question of minimizing $J(x) = \|x\|^2/2 + L(x)$, $x \in K$. Prove that there is a unique $x_0 \in K$ such that $J(x_0) = \inf_{x \in K} J(x)$.

27. Let K be a closed convex subset of a Hilbert space H and $\{x_n\} \subset K$ such that $x_n \rightharpoonup x$ in H. Prove that $x \in K$.

28. Let H be a separable infinite-dimensional Hilbert space and μ a translation invariant measure such that balls are measurable and μ is finite on finite balls. Prove that μ is identically 0.

29. Let H be an infinite-dimensional separable Hilbert space. Prove that H is isometrically isomorphic to the Hilbert space $H \otimes H$ equipped with the norm $\|(x, y)\|_{H \times H} = (\|x\|_H^2 + \|y\|_H^2)^{1/2}$ for $(x, y) \in H \otimes H$.

30. Let H be a Hilbert space. Prove that H is separable iff every ONB for H is at most countable, and for this it suffices that one such basis is countable.

31. Prove that if a Hilbert space H has a countable ONB, $X \sim \ell^2$.

32. Construct an ONS of polynomials $\{p_n\}$, $n = 0, 1, \ldots$, in real $L^2(I)$ where $p_0(t) = 1$ and $p_n(t)$ is a polynomial of degree n with positive leading coefficient for each $n \geq 1$. Furthermore, prove that the sequence is unique and an ONB for $L^2(I)$.

33. Let M be a dense subspace of a separable Hilbert space H. Prove that H has an ONB consisting of elements in M.

34. Given an ONB $\{e_n\}$ for a Hilbert space H, let $x_n = e_{2n}$, $y_n = \sqrt{1 - 4^{-n}}\,e_{2n} + 2^{-n}e_{2n-1}$, $n \geq 1$, and set $X = \overline{\mathrm{sp}}\{x_1, \ldots, x_n, \ldots\}$ and $Y = \overline{\mathrm{sp}}\{y_1, \ldots, y_n, \ldots\}$. Prove: (a) $\{x_n\}$ and $\{y_n\}$ are ONS's in H, $X \cap Y = \{0\}$, and $\overline{X + Y} = H$. (b) $v = \sum_{n=0}^{\infty} 2^{-n}e_{2n+1} \in H \setminus (X + Y)$ and $X + Y$ is not closed in H.

35. Discuss the validity of the following statement: If L is a continuous linear functional in an inner product space X, there exists $x_L \in X$ such that $L(x) = \langle x, x_L \rangle$ for all $x \in X$.

36. Let H be a separable Hilbert space and L a bounded linear functional on H. Using the fact that H has a countable ONB, give an elementary proof of the Riesz-Fréchet representation theorem. Specifically, prove that there is a unique $x_L \in H$ such that $\|L\| = \|x_L\|$ and $L(x) = \langle x, x_L \rangle$ for all $x \in H$.

37. Give a proof of the Riesz-Fréchet representation theorem using elementary properties of kernels of linear functionals.

38. Let X be an inner product space. Prove that if every continuous linear functional L on X can be represented as $L(x) = \langle x, x_L \rangle$ with $x_L \in X$, X is complete.

39. Let H be a Hilbert space. Prove the following form of the Hahn-Banach theorem: If A, B are disjoint convex subsets of H with A compact and B closed, there exists a hyperplane that separates A and B.

40. Let X be a closed subspace of a Hilbert space H and ℓ a bounded linear functional on X. Without recourse to a Hahn-Banach theorem prove there exists a unique linear functional L on H such that $L|_X = \ell$ and $\|L\| = \|\ell\|$.

41. Let H be a Hilbert space, X a closed subspace of H, and ℓ a bounded linear functional on X. Prove that if $P : H \to X$ denotes the orthogonal projection onto X, then the functional L on H given by $L(x) = \ell(P(x))$ is the unique extension of ℓ to H.

42. Let $\{e_n\}$ be an ONB for a Hilbert space H. Is there a bounded linear functional L on H such that: (a) $L(e_n) = 1/n$ for all n? (b) $L(e_n) = 1/\sqrt{n}$ for all n?

43. Let g be an $L^2(\mathbb{R}^n)$ continuous function such that $g(x) \neq 0$ for a.e. $x \in \mathbb{R}^n$. Discuss the validity of the following statement: $V = \mathrm{sp}\{g(x)e^{ix\cdot\xi} : \xi \in \mathbb{R}^n\}$ is dense in $L^2(\mathbb{R}^n)$.

44. Let $f \in L^2(\mathbb{R}^n)$ be such that $\widehat{f}(\xi) \neq 0$ for a.e. $\xi \in \mathbb{R}^n$, where $\widehat{f}(\xi)$ denotes the Fourier transform of f. Discuss the validity of the following statement: The translates of f span $L^2(\mathbb{R}^n)$.

45. Let $f \in L^1(\mathbb{R}^N) \setminus L^2(\mathbb{R}^N)$. Prove that there is an ONB $\{\varphi_n\}$ for $L^2(\mathbb{R}^N)$ consisting of continuous functions such that $\int_{\mathbb{R}^N} f(x)\varphi_n(x)\,dx = 0$ for all n.

46. Let H be a Hilbert space and $L_1 \neq L_2$ nonzero bounded linear functionals on H such that if $x \in H$ satisfies $|L_1(x)| = \|L_1\|\,\|x\|$, then $L_2(x) = 0$. Prove that if $x \in H$ satisfies $|L_2(x)| = \|L_2\|\,\|x\|$, then $L_1(x) = 0$.

47. Let $\lambda_1, \ldots, \lambda_n$ be nonzero complex numbers and $\{x_k\}$ an ONS in $[0, 2\pi]$. Discuss the validity of the following statement: For some $t \in [0, 1]$, $\left|1 - \sum_{k=1}^n \lambda_k x_k(t)\right| \geq 1$.

48. Let H be a Hilbert space. Prove: (a) If $\|x_n\|, \|y_n\| \leq 1$ and $\langle x_n, y_n \rangle \to 1$, then $\|x_n - y_n\| \to 0$. (b) If $\{x_n\}, x$ are such that $\|x_n\| = \|x\| = 1$ and $|1 - \langle x_n, x \rangle| \leq 1/n$, then $\{x_n\}$ is not an ONB for H.

49. Let $\{e_n\}$ be an ONB for $L^2([0, 1])$. Extend each e_n to $\mathbb{R}$ by making it zero outside $[0, 1]$, and for each integer k define $e_{n,k}(x) = e_n(x - k)$, the translate of e_n by k. Prove that $\{e_{n,k}\}$ is an ONB for $L^2(\mathbb{R})$.

50. Construct an infinite ONS $\{f_n\}$ in a separable Hilbert space H such that Bessel's inequality is strict for some $x \in H$. Also show that although $\sum_{n=1}^\infty \langle x, f_n \rangle f_n$ converges in H, it does not converge necessarily to x.

51. Let $\{e_n\}$ be an ONB for a Hilbert space H. Discuss the validity of the following statements: (a) If $y_n = e_1 + \cdots + e_{n+1}$, $n \geq 1$, $\{y_n\}$ is complete in ℓ^2. (b) If $z_n = e_1 + \cdots + e_n - e_{n+1}$, $n \geq 1$, $\{z_n\}$ is complete in ℓ^2.

52. Let $\{e_n\}$ be an ONB for a Hilbert space H and $\{f_n\} \subset H$ an ONS in H such that $\sum_n \|e_n - f_n\|^2 < 1$. Prove that $\{f_n\}$ is an ONB for H.

53. Let $\{e_n\}$ be an ONB for a Hilbert space H and $\{f_n\}$ an ONS in H such that $\sum_n \|e_n - f_n\|^2 < \infty$. Prove that $\{f_n\}$ is an ONB for H.

54. Let H be a Hilbert space, $\{e_n\}$ an ONB for H, and $f_n = (n^2 + n)^{-1/2}(\sum_{k=1}^n e_k - ne_{n+1})$, $n = 1, 2, \ldots$. Prove that $\{f_n\}$ is an ONB for H.

55. Let $\{e_n\}$ be an ONB for a Hilbert space H and $\{f_n\} \subset H$ such that $\sum_n \|e_n - f_n\|^2 < s^2 < \infty$. Let $T : H \to H$ be the linear mapping defined initially by $T(e_n) = f_n$ and extended to H by linearity. (a) Prove that if $s < 1$, $\|I - T\| \leq s$, and T is bounded and invertible. (b) Give an example of $\{f_n\}$ of norm 1 with dense span in H which does not satisfy $\sum_n \|e_n - f_n\|^2 < \infty$ and such that the corresponding operator T does not have closed range.

56. Let H be a Hilbert space. We say that $\{u_a\}_{a \in A} \subset H$ is stable if there are constants $0 < m < M < \infty$ such that $m \sum_{a \in A} |\lambda_a|^2 \leq \|\sum_{a \in A} \lambda_a u_a\|^2 \leq M \sum_{a \in A} |\lambda_a|^2$ for all $\{\lambda_a\} \in \ell^2(A)$. Prove: (a) Stability implies linear independence. (b) If the normalized vectors $\{u_a\}_{a \in A}$ satisfy $\eta^2 = \sum_{a \neq b} |\langle u_a, u_b \rangle|^2 < 1$, then $\{u_a\}_{a \in A}$ is stable. (c) Let $\{e_n\}$ be an ONS in H and define $u_n = (e_n + e_{n+1})/2$. Is $\{u_n\}$ stable?

57. Prove that the unit ball of an infinite-dimensional Hilbert space contains infinitely many pairwise disjoint translates of a ball of radius $1/4$.

58. Let H be a separable infinite-dimensional Hilbert space. Prove that there are closed linear subspaces $\{X_t\}_{t\in[0,1]}$ of H such that $X_s \subsetneq X_t$ for $0 \le s < t \le 1$.

59. Let M, N be orthogonal closed subspaces of a Hilbert space H. Prove that $M + N$ is closed.

60. Let $X = \ell_0^2$ equipped with the usual ℓ^2 inner product and $M = \{x \in X : \sum_n x_n = 0\}$. Characterize $\overline{M}$ and $M^\perp$. Do the same for $M_1 = \{x \in X : \sum_n x_n/n = 0\}$ and $M_2 = \{x \in X : \sum_n x_n/\sqrt{n} = 0\}$.

61. Let H be a Hilbert space. Prove: (a) If $M \subset H$, $M^\perp = \overline{M}^\perp$. (b) If $X = M$ is a closed subspace of H, $M^{\perp\perp} = M$. (c) If X is a subspace of H, $X^{\perp\perp} = \overline{X}$. (d) If X is a subspace of H, $\overline{X} = H$ iff $X^\perp = \{0\}$.

62. Let M be a subspace of a Hilbert space H. Discuss the validity of the following statement: $H = M \oplus M^\perp$.

63. Let M, N be closed subspaces of a Hilbert space H. Prove: (a) $(M + N)^\perp = M^\perp \cap N^\perp$. (b) $(M \cap N)^\perp = \overline{M^\perp + N^\perp}$.

64. Let H be a Hilbert space and M, N subspaces of H such that $H = M + N$ and $M \perp N$. Prove that $N = M^\perp$.

65. In real $L^2([-1,1])$ consider $M = \mathrm{sp}\{1, t^2, t^4, \ldots\}$. Is M closed? Give an explicit description of the orthogonal projections onto $M^\perp$ and $M^{\perp\perp}$.

66. Let $X = (C([-1,1]), \|\cdot\|_2)$ and consider $M = \{x \in X : x(t) = 0$ for $t \ge 0\}$. Prove that M is a subspace of X, describe $M^\perp$, and discuss the validity of the following statement: If $x \in X$, x can be written uniquely as $x = y + z$ with $y \in M$ and $z \in M^\perp$.

67. Let X be an inner product space, $\{x_1, \ldots, x_n\} \subset X$, and $Y = \mathrm{sp}\{x_1, \ldots, x_n\}$. Identify $Y^\perp$.

68. Suppose that X is a closed subspace of a Hilbert space H. What is the relation between H/X and $X^\perp$?

69. Let H be a Hilbert space and X a linearly independent subset of H. Prove that X is complete iff $\mathrm{sp}\{X\}$ is dense in H.

70. Let M be a closed subspace of a Hilbert space H, $0 \ne x \in H$, and $c = \min\{\|x - y\| : y \in M\}$ and $C = \max\{|\langle x, y\rangle| : y \in M^\perp, \|y\| = 1\}$. Prove that $c = C = \|x_{M^\perp}\|$ where $x = x_M + x_{M^\perp}$ is the unique decomposition of x in $H = M \oplus M^\perp$.

71. In real $L^2(I)$ consider $M = \{x \in L^2(I) : \int_I x(t)\, dt = \int_I tx(t)\, dt = \int_I t^2 x(t)\, dt = 0\}$. Given $y \in L^2(I)$, find the element in M that is closest to y.

72. In real $L^2(I)$ compute $\max \int_I t^3 x(t)\,dt$ and $\min \int_I (t^5 - a - bt)^2\,dt$ subject to the conditions $\int_I t^k x(t)\,dt = 0$, $k = 0, 1, 2$, and $\int_I x(t)^2\,dt = 1$.

73. In real $L^2(I)$ let $M = \{x \in L^2(I) : \int_I x(t)\,dt = 0\}$. Prove that M is a closed subspace of $L^2(I)$, describe $M^\perp$, and find the distance $d(y, M)$ of $y \in L^2(I)$ to M.

74. Let H be a Hilbert space and $0 \neq y \in H$. Prove that

$$d(x, y^\perp) = \frac{|\langle x, y \rangle|}{\|y\|} \quad \text{for all } x \in H.$$

75. In real $L^2([-1, 1])$ let $M = \{x \in L^2([-1, 1]) : x \text{ is even}\}$. Given $y \in L^2([-1, 1])$, find $d(y, M)$.

76. Let H be a Hilbert space, X a closed subspace of H, and suppose that the mapping $P : H \to H$ satisfies the following two properties: $R(P) \subset X$ and $(x - P(x)) \perp X$ for all $x \in H$. Prove: (a) P is linear. (b) $\langle x, P(y) \rangle = \langle P(x), P(y) \rangle = \langle P(x), y \rangle$ for all $x, y \in H$. (c) P is bounded. (d) $P(x)$ is the projection of x onto X, i.e., $\|x - P(x)\| = d(x, X)$ for all $x \in H$. (e) If $\{e_1, \ldots, e_n\}$ is an ONS in H and $X = \text{sp}\{e_1, \ldots, e_n\}$, then $P(x) = \sum_{k=1}^n \langle x, e_k \rangle e_k$.

77. Let H be a Hilbert space and $P : H \to H$ a projection. Prove that P is an orthogonal projection, i.e., $\langle P(x), y \rangle = \langle x, P(y) \rangle$ for all $x, y \in H$, iff $\|P\| = 1$.

78. Let H be a Hilbert space, X a closed subspace of H, and P the orthogonal projection onto X. Prove that $X = \{x \in H : \|P(x)\| = \|x\|\}$.

79. Let H be a Hilbert space, M, N closed subspaces of H, and P_M, P_N the orthogonal projections onto M and N, respectively. Prove that the following statements are equivalent: (a) $P = P_M + P_N$ is the projection onto $M \oplus N$. (b) $P_M P_N = P_N P_M = 0$. (c) $M \perp N$.

80. Let H be a Hilbert space and M, N closed subspaces of H. Prove that $P = P_M P_N$ is a projection iff $P_M P_N = P_N P_M$, i.e., P_M and P_N commute. In that case identify the closed space P projects onto.

81. Let $X = \{x \in \ell^2 : x_1 - x_2 = 0, x_2 + x_3 = 0\}$. Determine the orthogonal projection P onto $X^\perp$.

82. Let M, N be the subspaces of ℓ^2 defined by $M = \{x \in \ell^2 : x_{2n} = 0 \text{ for all } n\}$ and $N = \{y \in \ell^2 : y_{2n-1} = ny_{2n} \text{ for all } n\}$, respectively. Prove that M, N are closed but that $M + N$ is not closed, and thus the sum is not a direct sum. Moreover, prove directly that if P denotes the linear projection $P : M + N \to N$ onto N, then P is not continuous.

83. Let H be a Hilbert space. (a) Let $u \in H$ with $\|u\| = 1$, and $P : H \to H$ be given by $P(x) = \langle x, u \rangle u$. Is P an orthogonal projection onto a subspace of H? (b) Let $u, v \in H$ be linearly independent, and $P : H \to H$ given by $P(x) = \langle x, u \rangle u + \langle x, v \rangle v$ for all $x \in H$. Prove that P is an orthogonal projection iff $\|u\| = \|v\| = 1$ and $u \perp v$.

84. Let H be a Hilbert space and $P : H \to H$ a bounded symmetric linear operator such that P^2 is an orthogonal projection. Is P an orthogonal projection?

85. Let H be a Hilbert space and $P : H \to H$ a bounded self-adjoint operator. Discuss the validity of the following statements: P is an orthogonal projection provided that: (a) $P^3 = P^2$. (b) $P^4 = P^3$. (c) $P^4 = P^2$.

86. Let M, N be closed subspaces of a Hilbert space H. Prove that the following are equivalent: (a) $P_{M^\perp}(N)$ is closed in $M^\perp$. (b) $M + N$ is closed in H. (c) $P_{N^\perp}(M)$ is closed in $N^\perp$.

87. Let $K_1 \subsetneq K_2$ be nonempty closed convex subsets of a Hilbert space H. Prove that $\|P_{K_1}(x) - P_{K_2}(x)\|^2 \leq 2(d(x, K_1)^2 - d(x, K_2)^2)$ for all $x \in H$.

88. Let $\{K_n\}$ be nonempty closed convex subsets of a Hilbert space H. Prove: (a) If $\{K_n\}$ is increasing, then $K = \overline{\bigcup_n K_n}$ is a closed convex subset of H and $\lim_n P_{K_n}(x) = P_K(x)$ for all $x \in H$. (b) If $\{K_n\}$ is decreasing and $K = \bigcap_n K_n$, then $\lim_n P_{K_n}(x) = P_K(x)$ for all $x \in H$ if $K \neq \emptyset$ and $\lim_n d(x, K_n) = \infty$ for all $x \in H$ if $K = \emptyset$.

89. Let $\{X_n\}$ be an increasing sequence of closed subspaces of a Hilbert space H and $X = \overline{\bigcup_n X_n}$. Prove that $\lim_n \|P_{X_n}(x) - P_X(x)\| = 0$ for each $x \in H$, and $K(P_X) = \bigcap_n K(P_n)$.

90. Let H be a separable Hilbert space and $\{X_n\}$ closed subspaces of H such that $X_n \perp X_m$ for $1 \leq n < m$ and $\bigcap_n X_n^\perp = \{0\}$. Prove that every $x \in H$ can be written uniquely as $x = \sum_n x_n$, $x_n \in X_n$, where the sum converges in H.

91. Let H, H_1 be Hilbert spaces and $T : H \to H_1$ a bounded linear operator with $R(T)$ closed in H_1. (a) Prove that $T = T \circ P$ where P is the orthogonal projection onto $K(T)^\perp$. (b) Let $S : K(T)^\perp \to R(T)$ denote the restriction of T to $K(T)^\perp$. Prove that S is bijective and has a bounded inverse.

92. Let M, N be closed subspaces of a Hilbert space H such that $M \cap N^\perp = \{0\}$. Prove that $\dim(M) \leq \dim(N)$.

93. Let H be a Hilbert space, $X \subset H$ a closed subspace of H, and $P = P_X$ the orthogonal projection of H onto X. Prove that P is compact iff X is finite dimensional.

94. Let H be a Hilbert space and $\mathcal{H} = \{H_n\}$ a decomposition of $H = \bigoplus_n H_n$ into nontrivial pairwise orthogonal finite-dimensional subspaces H_n of H. For a sequence λ of positive real numbers define $A_{\lambda, \mathcal{H}} = \{x \in H : x =$

$\sum_n x_n$ with $x_n \in H_n$ and $\|x_n\| \leq \lambda_n$, all n}. Prove: (a) $A_{\lambda,\mathcal{H}}$ is compact iff $\lambda \in \ell^2$. (b) Given a compact $K \subset H$, there exist a decomposition $\mathcal{H}'$ of H and $\mu \in \ell^2$ such that $K \subset A_{\mu,\mathcal{H}'}$.

95. Let $\{e_n\}$ be an ONB for a Hilbert space H. Prove: (a) $e_n \nrightarrow 0$ but $e_n \rightharpoonup 0$ in H. (b) Given $N > 0$, there exists a sequence $\{x^k\}$ of convex combinations of the e_n for $n \geq N$ that converges to 0 in H.

96. Let X be an inner product space and $\{x_n\}, \{y_n\} \subset X$. Discuss the validity of the following statements: $\langle x_n, y_n \rangle \to \langle x, y \rangle$ if: (a) $x_n \to x$ and $y_n \to y$ in X. (b) $x_n \to x$ and $y_n \rightharpoonup y$ in X. (c) $x_n \rightharpoonup x$ and $y_n \rightharpoonup y$ in X.

97. Let H be a Hilbert space and $\{x_n\} \subset H$ such that $x_n \rightharpoonup x$ and $\|x_n\| \to \|x\|$. Prove that $\|x_n - x\| \to 0$.

98. Let H be a Hilbert space and $x \in H$ with $\|x\| = 1$. Prove that if $\{x_n\} \subset H$ is such that $\|x_n\| \leq 1$ for all n and $x_n \rightharpoonup x$, then $x_n \to x$ in H.

99. Let H be a Hilbert space and D dense in H. Prove that $x_n \rightharpoonup x$ in H iff $\|x_n\| \leq c$ and $\langle x_n, y \rangle \to \langle x, y \rangle$ for all $y \in D$.

100. Let $\{e_n\}$ be an ONB for a Hilbert space H, $x_n = \sum_{k=1}^n k^{-1/2} e_k$, and $y_n = n^{-1/2} \sum_{k=1}^n e_k$, $n \geq 1$. Do $\{x_n\}, \{y_n\}$ converge in norm? Weakly?

101. Let $S = \{x_{m,n}\}$ denote the collection of ℓ^2 sequences $x_{m,n}$, $m, n = 1, \ldots, m \neq n$, given by

$$x_{m,n}(k) = \begin{cases} 1, & k = m, \\ m, & k = n, \\ 0, & k \neq m, n. \end{cases}$$

Prove that 0 is in the weak closure of S but no sequence in S converges weakly to 0.

102. Let $f \in L^2(\mathbb{R})$ and $\{\lambda_n\}$ a positive sequence tending to ∞. Prove that each of the following sequences converges to 0 weakly in $L^2(\mathbb{R})$: (a) $\{f(\cdot - \lambda_n)\}$. (b) $\{(1/\sqrt{\lambda_n})f(\cdot/\lambda_n)\}$. (c) $\{f(x)e^{-i\lambda_n x}\}$.

103. Let $\{e_n\}$ be an ONB for a Hilbert space H. Prove that if $\{f_n\} \subset H$ satisfies $\|f_n\| \leq M$ and $f_n \in \{e_1, \ldots, e_n\}^\perp$ for all $n = 1, 2, \ldots$, then $f_n \rightharpoonup 0$ in H.

104. For a 2π-periodic function $x \in L^2([0, 2\pi])$, let $x_n(t) = x(nt)$, $n = 1, 2, \ldots$. Prove that $\{x_n\}$ converges weakly in $L^2([0, 2\pi])$ and find the limit.

105. Let H be a Hilbert space and $\{x_n\} \subset H$ such that $x_n \rightharpoonup x$ in H. Prove that there is a subsequence $\{x_{n_k}\}$ of $\{x_n\}$ such that $N^{-1} \sum_{k=1}^N x_{n_k} \to x$ in H.

106. Let $\{e_n\}$ be an ONB for a Hilbert space H and define $p(x) = \sum_n 2^{-n}|\langle x, e_n \rangle|$, $x \in H$. (a) Prove that $p(x) \leq \|x\|$ for $x \in H$ and that (H, p) is a normed space that is not complete. (b) Prove that for bounded sequences $\{x_k\} \subset H$, $x_k \rightharpoonup x$ in H iff $p(x_k - x) \to 0$. (c) Construct a sequence $\{x_k\} \subset H$ such that $\lim_k p(x_k) = 0$ yet $\{x_k\}$ does not converge weakly to 0 in H. (d) Prove that the unit ball $\{x \in H; \|x\| \leq 1\}$ is compact in (H, p).

107. Let H be an infinite-dimensional Hilbert space. Given $x \in H$, $0 < \|x\| \leq 1$, construct $\{x_n\} \subset H$ such that $\|x_n\| = 1$, all n, and $x_n \rightharpoonup x$ in H.

108. Let H be a Hilbert space and $K \subset H$ be closed, nonempty, and convex. Prove: (a) K is weakly closed. (b) For all $x_0 \in H$ there exists $x \in K$ such that $\|x - x_0\| = \inf\{\|y - x_0\| : y \in K\}$.

109. Let X be an inner product space and $T : X \to X$ a linear mapping. Prove: (a) If the underlying field is $\mathbb{R}$, the polarization identity $4\langle T(x), y \rangle = \langle T(x + y), x + y \rangle - \langle T(x - y), x - y \rangle$ holds for all $x, y \in H$. (b) If the underlying field is $\mathbb{C}$, then $4\langle T(x), y \rangle = \langle T(x+y), x+y \rangle - \langle T(x-y), x-y \rangle + i\langle T(x + iy), (x + iy) \rangle - i\langle T(x - iy), (x - iy) \rangle$ for all $x, y \in H$.

110. Let X be an inner product space and suppose that the linear mapping $T : X \to X$ satisfies $\langle T(x), x \rangle \leq 0$ for all $x \in X$. Prove that $(I - T)$ is injective.

111. Let H be a Hilbert space and $T : H \to H$ a bounded linear operator. Prove that $\|T\| = \sup_{\|x\|=\|y\|=1} |\langle T(x), y \rangle|$.

112. Let $\{e_n\}$ be an ONB for a Hilbert space H. The left shift $L : H \to H$ is the linear mapping given by $L(x) = y$, where if $x = \sum_n x_n e_n$ and $y = \sum_n y_n e_n$, then $y_n = x_{n+1}$ for all $n \geq 1$, and the right shift $R : H \to H$ is the linear mapping given by $R(x) = y$ where $y_1 = 0$ and $y_n = x_{n-1}$ for $n \geq 2$. Prove: (a) L is surjective and $K(L) = \mathrm{sp}\{e_1\} \neq \{0\}$. (b) R is injective and its range $\mathrm{sp}\{e_2, e_3, \ldots\}$ is not dense in H. (c) $R^* = L$ and $R^* R = I$ but $RR^* \neq I$. (d) $\lim_k \|L^k(x)\| = 0$ for each $x \in H$ and $\|L^k\| = 1$ for all k, and so $L^k \not\to 0$. (e) $\lim_k \langle R^k(x), y \rangle = 0$ for all $x, y \in H$ but $R^k \not\to 0$.

113. Let $T : H \to H$ be a bounded linear operator. Prove that for each $x \in H$, $T(x) = \lim_n T_n(x)$ where $\{T_n\}$ are finite rank operators.

114. Let H be a complex Hilbert space and $T, T_n \in \mathcal{B}(H)$ for all n. Prove that $\lim_n T_n(x) = T(x)$ for all $x \in H$ iff $\langle T_n(x), x \rangle \to \langle T(x), x \rangle$ and $\|T_n(x)\| \to \|T(x)\|$ for all $x \in H$.

115. Let X be an inner product space and $T : X \to X$ a bounded linear operator that satisfies $|\langle T(x_0), x_0 \rangle| = \|T\|$ for some $x_0 \in X$ with norm 1. Prove that $T(x_0) = \lambda x_0$ for some scalar λ.

116. Let H be an infinite-dimensional separable Hilbert space. Discuss the validity of the following statement: There is an unbounded linear operator $T : H \to H$ that vanishes on an ONB basis for H.

117. Construct bounded linear operators $\{T_n\}$ from a Hilbert space H into itself such that $\lim_n T_n(x) = 0$ for all $x \in H$ and $\lim_n \|T_n\| = 1$.

118. Let $\{e_n\}$ be an ONB for a Hilbert space H. Does there exist a bounded linear operator $T : H \to H$ such that $\|T(e_n)\| \leq 1$ for all n yet $\|T\|$ is arbitrarily large?

119. Let $\{e_n\}$ be an ONB for a Hilbert space H and $T : H \to H$ such that $\sum_n |\langle e_n, T(e_n)\rangle|^2 < \infty$. Suppose that the restriction of T to some closed subspace X of H is the identity on X. Prove that X is finite dimensional.

120. Let H be a Hilbert space, $\{e_n\}$ an ONB for H, and $\{x_k\}$ such that $\sum_k |\langle x_k, x\rangle|^2 < \infty$ for all $x \in H$. Prove that there exists a unique bounded linear mapping $T : H \to H$ such that $\langle e_n, T(x)\rangle = \langle x_n, x\rangle$ for all $x \in H$.

121. Let H be a real Hilbert space and $u, v \in H$ such that $\langle u, v\rangle \neq 0$. Let $T : H \to H$ be defined by the condition $T(x) = y$ where $y \in H$ is the only point in the line $\ell_x = \{x + tv : t \in \mathbb{R}\}$ which is orthogonal to u. Prove that T is a bounded linear mapping and describe $R(T)$. What if u, v are linearly dependent?

122. Let H be a Hilbert space, $\{e_n\}$ an ONB for H, and $T : H \to H$ a bounded linear operator. Prove: (a) The quantity $\sum_n \|T(e_n)\|^2$, which may be infinite, is independent of the choice of $\{e_n\}$. When the sum is finite we say that T is a Hilbert-Schmidt operator. (b) $\|T\|^2 \leq \sum_n \|T(e_n)\|^2$. (c) Hilbert-Schmidt operators are compact. (d) The composition of Hilbert-Schmidt operators is a Hilbert-Schmidt operator.

123. Let X be an inner product space and $T : X \to X$ a linear mapping. We say that T is a contraction if $\|T\| \leq 1$. Prove that the following statements are equivalent: (a) T is a contraction. (b) $\langle x, T^*T(x)\rangle \leq \langle x, x\rangle$, all $x \in X$. (c) T^* is a contraction.

124. Let H be a real Hilbert space and $\{e_n\}$ an ONB for H. Given $x \in X$, let $T(x) = y$ where $y = \sum_n y_n e_n$ and

$$
y_n = \begin{cases} (\langle x, e_n\rangle - \langle x, e_{n+1}\rangle)/\sqrt{2}, & n \text{ odd}, \\ (\langle x, e_{n-1}\rangle + \langle x, e_n\rangle)/\sqrt{2}, & n \text{ even}. \end{cases}
$$

Prove that $T : H \to H$ is an isometry.

125. Let H be a Hilbert space and $T : H \to H$ a bounded linear operator. Prove that the following are equivalent: (a) T is an isometry. (b) $\langle T(x), T(y)\rangle = \langle x, y\rangle$. (c) $T^*T = I$. (d) If $\{e_\alpha\}$ is an ONB for H, $\{T(e_\alpha)\}$ is an ONS in H.

126. Let H be a complex Hilbert space and $T : H \to H$ a bounded linear operator. We say that T is a partial isometry if T is isometric on the orthogonal complement of $K(T)$. Prove that T is a partial isometry iff T^*T is the projection onto $K(T)^\perp$.

127. Let H be a complex Hilbert space, $T : H \to H$ a linear isometry such that $R(T) \subsetneq H$, $e \perp R(T)$ of norm 1, and $e_n = T^n(e)$ for $n \geq 1$. (a) Evaluate $T^*(e)$ and $T^*(e_n)$ for $n \geq 1$. Prove: (b) e is orthogonal to $\{e_n\}$, $n \geq 1$. (c) $\{e, e_1, \ldots, e_n, \ldots\}$ is an ONS in H.

128. Let $H = L^2([0, 2\pi])$ denote the complex L^2 space with normalized Lebesgue measure, $0 < \alpha < 2\pi$ such that α/π is irrational, and $T : H \to H$ be given by $Tx(t) = x(2\pi + t - \alpha)$ if $0 \leq t < \alpha$ and $Tx(t) = x(t - \alpha)$ if $\alpha \leq t \leq 2\pi$. (a) Prove that $T : H \to H$ is an isometry. (b) Calculate $T(e_n)$ where $e_n(t) = e^{int}$, $n \in \mathbb{Z}$. (c) Let $S_K = K^{-1} \sum_{k=0}^{K-1} T^k$, $K \geq 1$. Verify that $\|S_K\| \leq 1$. Prove that $\lim_K S_K(e_n) = 0$ for all $n \neq 0$ in $\mathbb{Z}$. (d) Prove that for $x \in H$, $\{S_K(x)\}$ converges to a constant function in H and determine it.

129. Let X be a complex inner product space and $T : X \to X$ a bounded linear operator. Prove that T is of rank 1 iff there exist $y, z \in X$ such that $T(x) = \langle x, y \rangle z$, all $x \in X$. Find T^* and compute $\|T\|$.

130. Let H be a Hilbert space and $T : H \to H$ a bounded linear operator. Prove: (a) T is of finite rank iff $T(x) = \sum_{k=1}^{n} \langle x, v_k \rangle w_k$ for all $x \in H$ where $v_k, w_k \in H$. (b) T is of rank n iff T^* is of rank n.

131. Let H be a Hilbert space and $T : H \to H$ a bounded linear operator of finite rank. Prove that T is compact. When does the identity I satisfy this property?

132. Let H be a Hilbert space and $T \in \mathcal{B}(H)$. Prove that T is compact iff T is the uniform limit of finite rank operators.

133. Let $\{e_n\}, \{f_n\}$ be ONSs in a Hilbert space H, $\{\lambda_n\}$ a bounded scalar sequence, and $T : H \to H$ the linear mapping given by $T(x) = \sum_n \lambda_n \langle x, e_n \rangle f_n$. (a) Prove that T is bounded and compute its norm. (b) Find the adjoint T^* of T. (c) Prove that T is compact iff $\lim_n \lambda_n \to 0$.

134. Let H be a Hilbert space and $T : H \to H$ a compact linear mapping. Prove that if $\{e_n\}$ is an ONB for H, then $T(e_n) \to 0$.

135. Let H be a Hilbert space and $T : H \to H$ a compact linear operator. Prove that $R(I - T)$ is closed in H.

136. Let H be a Hilbert space and $T : H \to H$ a bounded linear mapping. Discuss the validity of the following statements: (a) If T^2 is compact, T is compact. (b) If T^*T is compact, T is compact.

137. Let $\{e_n\}$ be an ONB for a Hilbert space H, $T : H \to H$ a bounded linear mapping, and

$$
\mu_n = \sup_{x \perp \{e_1, \ldots, e_n\}, x \neq 0} \frac{\|T(x)\|^2}{\|x\|^2}.
$$

Prove that T is compact iff $\mu_n \to 0$ as $n \to \infty$.

138. Let $\{e_n\}$ be an ONB for a Hilbert space H and $T : H \to H$ the linear mapping given by $T(x) = \sum_n (n+1)^{-1} \langle x, e_{n+1} \rangle e_n$, $x \in H$. Prove that T is compact and find T^*.

139. Let H be a separable Hilbert space and $\mathcal{I} \subset \mathcal{B}(H)$ a nonzero closed two-sided ideal (i.e., for all $S \in \mathcal{I}$ and $T \in \mathcal{B}(H)$, $ST, TS \in \mathcal{I}$). Prove that if $T : H \to H$ is compact, $T \in \mathcal{I}$.

140. On $L^2(I)$ consider the integral operator $T(x)(t) = \int_0^t x(s)\, ds$, $x \in L^2(I)$. (a) Prove that $T : L^2(I) \to L^2(I)$ is compact. (b) Calculate T^*.

141. Let H be a real Hilbert space and $X = \mathrm{sp}\{x_1, \ldots, x_n\}$ where the x_k are linearly independent elements of H. Define the operator $S : X \to X$ by $S(x) = \sum_{k=1}^n \langle x, x_k \rangle x_k$, $x \in X$. Prove: (a) S is self-adjoint and invertible. (b) The orthogonal projection P_X of H onto X is given by $P_X(x) = \sum_{k=1}^n \langle x, S^{-1}(x_k) \rangle x_k$, $x \in H$.

142. Let H be a Hilbert space and $T : H \to H$ a bounded linear mapping. Discuss the validity of the following statements: (a) If $\langle T(x), x \rangle = 0$ for all $x \in H$, then $T = 0$. (b) If $T^2 = 0$, then $T = 0$. (c) If $T^*T = 0$, then $T = 0$. (d) If $S : H \to H$ is a bounded linear mapping and $S^*S + T^*T = 0$, then $S = T = 0$.

143. Let X be an inner product space. Discuss the validity of the following statement: If a linear mapping $T : X \to X$ satisfies $\langle T(x), y \rangle = \langle x, T(y) \rangle$ for all $x, y \in X$, T is bounded.

144. Let H be a Hilbert space and $T : H \to H$ a symmetric mapping. Prove that T is linear and bounded with $\|T\| = \sup_{\|x\|=1} |\langle T(x), x \rangle|$.

145. Let H be a Hilbert space and $T_n : H \to H$ linear mappings such that $T_n(x) \rightharpoonup 0$ and $T_n^*(x) \rightharpoonup 0$ in H for all $x \in H$. Discuss the validity of the following statement: $T_n T_n^*(x) \rightharpoonup 0$ for all $x \in H$.

146. Let H be a Hilbert space and $T : H \to H$ a bounded linear operator. Prove: (a) $T^{**} = T$. (b) $\|T\| = \|T^*\|$ and $\|T^*T\| = \|TT^*\| = \|T\|^2$.

147. Let H be a Hilbert space and $T : H \to H$ compact. Prove that T^* is compact.

148. Let T be a bounded linear operator on a complex Hilbert space H with $\|T\| \leq 1$. Prove: (a) If $T(x) = x$, then $T^*(x) = x$. (b) $K(I - T) = K(I - T^*)$. (c) $H = K(I - T) \oplus \overline{R(I - T)}$.

149. Let H be a complex Hilbert space and $T : H \to H$ a bounded linear operator. Prove that $\|T\|_1 = \sup_{\|x\|=1} |\langle T(x), x \rangle|$ is a norm on $\mathcal{B}(H)$ that satisfies $\|T\|_1 \leq \|T\| \leq 2\|T\|_1$ where 2 is the best possible constant.

150. Let H, H_1 be separable Hilbert spaces and $T : H \to H_1$ a bounded linear operator with the property that there exist a bounded linear operator $B : H_1 \to H$ and compact operators $S : H \to H$ and $S_1 : H_1 \to H_1$ such that $BT = I - S$ and $TB = I_1 - S_1$ where I, I_1 denote the identity operators in H, H_1, respectively. Prove: (a) $K(T)$ is finite dimensional. (b) $R(T)$ is closed in H_1. (c) $R(T)^\perp$ is finite dimensional.

151. Let H be a Hilbert space and $T : H \to H$ a bounded linear operator. Prove: (a) $\overline{R(T)} = K(T^*)^\perp$. (b) $\overline{R(T^*)} = K(T)^\perp$. (c) $R(T)^\perp = K(T^*)$ and $R(T^*)^\perp = K(T)$.

152. Let H be a Hilbert space and $T : H \to H$ a bounded linear operator. Prove: (a) $K(T) = K(T^*T)$. (b) $\overline{R(T^*)} = \overline{R(T^*T)}$.

153. Let H be a Hilbert space and $T : H \to H$ a self-adjoint linear operator. Prove that $K(T) = K(T^n)$ for all $n \geq 1$.

154. Let $\{e_n\}$ be an ONB for $H = L^2(\mathbb{R}^+)$ and $T : D(T) \subset H \to H$ a linear operator given by $Tx(t) = \sum_n x(n)e_n(t)$, $x \in D(T) = C_0(\mathbb{R}^+)$. Prove that $D(T^*) = \{0\}$.

155. Let H be a complex Hilbert space and $T : H \to H$ a densely defined linear operator. Prove that $D(T^*) = \{y \in H : \ell(x) = \langle T(x), y \rangle$ is a bounded linear functional on $D(T)\}$.

156. Let H be a complex Hilbert space and $T : H \to H$ a densely defined linear mapping. Prove that the following are equivalent: (a) T is symmetric. (b) $D(T) \subset D(T^*)$. (c) $\langle T(x), x \rangle \in \mathbb{R}$ for $x \in D(T)$.

157. Let H be a Hilbert space and $T : D(T) \subset H \to H$ a closed symmetric operator. Prove: (a) $R(T + iI)$ and $R(T - iI)$ are closed in H. (b) $K(T^* + iI) = R(T - iI)^\perp$ and $K(T^* - iI) = R(T + iI)^\perp$.

158. Let H be a Hilbert space and $T : D(T) \subset H \to H$ a symmetric operator. Prove that the following are equivalent: (a) T is self-adjoint. (b) T is closed and $K(T^*+iI) = K(T^*-iI) = \{0\}$. (c) $R(T+iI) = R(T-iI) = H$.

159. Let H be a Hilbert space, $T : H \to H$ a linear mapping, and S a symmetric extension of T such that $R(S + iI) = R(T + iI)$. Prove: (a) $S = T$. (b) If T is symmetric, $R(T + iI) = H$, and $R(T - iI) \neq H$, then T has no self-adjoint extension.

160. Let H denote complex ℓ^2, $M = \{x \in \ell_0^2 : \sum_n x_n = 0\}$, and T the mapping from M into sequences given by $T(x) = y$ where

$$y_n = i\left(\sum_{m=0}^{n-1} x_m + \sum_{m=0}^{n} x_m\right), \quad n \geq 0.$$

Prove: (a) $T : \ell^2 \to \ell^2$ is a densely defined symmetric mapping. (b) $R(T+iI)$ is dense in ℓ^2. (c) $e_1 \in D(T^*)$ and $(T^*+iI)(e_1) = 0$. (d) T has no self-adjoint extension to ℓ^2.

161. Let H be a Hilbert space and $S, T : H \to H$ symmetric mappings. Prove that ST is symmetric iff $ST = TS$.

162. Let H be a complex Hilbert space, $S, S_1 : H \to H$ self-adjoint mappings such that $SS_1 = S_1 S$ and $S^2 = S_1^2$, and Q the orthogonal projection onto $K(S + S_1)$. Prove: (a) Every bounded linear operator $T : H \to H$ that commutes with $S + S_1$ commutes with Q. (b) If $S(x) = 0$, then $Q(x) = x$. (c) $S = (I - 2Q)S_1$.

163. Let H be a Hilbert space, $T : D(T) \subset H \to H$ a densely defined continuous linear mapping, and $T^* : D(T^*) \subset H \to H$ the adjoint of T. Prove: (a) $G(T^*)$ is closed in $H \times H$. (b) T^* is densely defined in H iff T is closable. (c) If T is closable, then T^{**} is densely defined and $T^{**} = \overline{T}$ is the closure of T, while $\overline{T}^* = T^*$.

164. Let H be a Hilbert space and $T : H \to H$ a bounded linear operator. Prove that T is invertible iff $m_1 = \inf_{\|x\|=1}\langle T^*T(x), x\rangle$ and $m_2 = \inf_{\|x\|=1}\langle TT^*(x), x\rangle$ are strictly positive.

165. Let T be a bounded linear operator from a real Hilbert space H into itself. Suppose that $|\langle T(x), x\rangle| \geq c\,\|x\|^2$ for all $x \in H$ and a constant $c > 0$. Prove that T is invertible and $\|T^{-1}\| \leq 1/c$.

166. Let H be a Hilbert space. We say that a bounded linear operator $T : H \to H$ is positive if $T = T^*$ and $\langle T(x), x\rangle \geq 0$ for all $x \in H$. Let T be a positive operator. Prove: (a) S^*TS is positive for all $S \in \mathcal{B}(H)$. (b) T^n is positive for all $n \geq 0$. (c) $|\langle T(x), y\rangle|^2 \leq \langle T(x), x\rangle\langle T(y), y\rangle$ for all $x, y \in H$. (d) If $I - T$ is positive, then $\langle T(x), T(x)\rangle \leq \langle T(x), x\rangle$ for all $x \in H$. (e) The following are equivalent: (i) $I - T$ is positive. (ii) $T - T^2$ is positive. (iii) $\|T\| \leq 1$.

167. Let H be a Hilbert space and $T, I - T$ positive mappings on H. Prove: (a) $T^n - T^{n+1}$ is positive for all n. (b) $\{\langle T^n(x), x\rangle\}$ converges for all $x \in H$. (c) $\lim_n T^n(x) = P(x)$ in H for all $x \in H$. (d) If P is as in (c), $P = P^*$ and $TP = P$. (e) Let $x \in H$ such that $T(x) = x$. Prove that $P(x) = x$. Deduce that $P^2 = P$ and that P is the orthogonal projection onto $K(T - I)$.

168. Let T, S be densely defined linear operators on Hilbert space H. Prove that if S is an extension of T, T^* is an extension of S^*.

169. Let H be a Hilbert space and $T : H \to H$ a symmetric linear mapping such that $R(T) = H$. Prove that T is self-adjoint.

170. Let H be a Hilbert space and $T : H \to H$ an injective self-adjoint mapping. Prove that T^{-1} is self-adjoint.

171. Let H be a Hilbert space and $T : H \to H$ a linear mapping. Prove that T is unitary iff T is an inner product preserving surjection.

172. Let X, Y be inner product spaces over the same field and ϕ a separately continuous bilinear mapping from $X \times Y$ into the field, i.e., $\phi(x, \cdot)$ is a continuous linear functional on Y for each fixed $x \in X$ and $\phi(\cdot, y)$ is a continuous linear functional on X for each fixed $y \in Y$, respectively. Discuss the validity of the following statement: ϕ is continuous.

173. Let H be a Hilbert space, $\phi : H \times H \to \mathbb{C}$ a bounded bilinear mapping such that $\phi(x, x) \geq c\|x\|^2$, all $x, y \in H$, and let $T : H \to H$ be given by $\phi(x, y) = \langle T(x), y \rangle$, $x, y \in H$. Prove that T is a bounded linear isomorphism.

174. Let H be a real Hilbert space and $\phi : H \times H \to \mathbb{R}$ a bounded bilinear mapping that is elliptic, i.e., $\phi(x, x) \geq c^2 \|x\|^2$ for all $x \in H$ and some $c > 0$. Prove that, given a bounded linear functional L on H, there is a unique $y_L \in H$ such that $L(x) = \phi(x, y_L)$ for all $x \in H$.

175. Let H be a Hilbert space, $\phi(x, y)$ a positive, bounded, symmetric bilinear form on H, and L a bounded linear functional on H. Consider the following problems: (a) Minimize $\Phi(x) = \phi(x, x)/2 - L(x) + \text{constant}$ over H. (b) Find $x \in H$ satisfying $\phi(x, y) = L(y)$ for all $y \in H$. Prove: (i) $x \in H$ solves (a) iff x solves (b). (ii) There is at most one $x \in H$ that solves (a) and (b). (iii) There is at least one $x \in H$ that solves (a) and (b).

176. Let H be a separable Hilbert space and $T : H \to H$ a linear mapping. Prove that $TT^* = I$ (or $T^*T = I$) and $\langle T(x), T(y) \rangle = \langle x, y \rangle$ for all $x, y \in H$ iff $T(M)$ is a total ONS in H whenever M is a total ONS in H.

177. Let H be a real Hilbert space. Prove that $\text{ext}(\text{ball } H) = \{h \in H : \|h\| = 1\}$.

178. For $c = \{c_n\}$ in ℓ^2, let $K = \{x \in \ell^2 : |x_n| \leq |c_n| \text{ for all } n\}$. (a) Prove that K is compact and convex. (b) Find the extreme points of K. (c) Determine the convex hull of the set E of extreme points and its closure.

Part 2

Solutions

"DON'T PANIC!"
—Douglas Adams,
"The Hitchhiker's Guide to the Galaxy"

Chapter 11

Set Theory and Metric Spaces

Solutions

1. Recall the following relations in a metric space: $X \setminus \overline{A} = \text{int}(X \setminus A)$ and $X \setminus \text{int}(A) = \overline{X \setminus A}$. They imply that $\text{int}(\overline{X \setminus A}) = \text{int}(X \setminus \text{int}(A)) = X \setminus \overline{\text{int}(A)}$, and, consequently, A is an open dense subset of X iff its complement $X \setminus A$ is closed and nowhere dense.

(a) implies (b) Let $\{O_n\}$ be open dense sets in X and put $G = \bigcap_n O_n$. Then $G^c = \bigcup_n (X \setminus O_n)$ where each $X \setminus O_n$ is closed nowhere dense and so G^c is of first category and has empty interior. Now, by the second relation above with $A = G^c$ there, $\overline{G} = X \setminus \text{int}(G^c) = X \setminus \emptyset = X$ and G is dense in X.

(b) implies (c) Let $A = \bigcup_n A_n$ with A_n nowhere dense, all n. Then $\overline{A_n}$ is nowhere dense and $A \subset \bigcup_n \overline{A_n}$; hence $A^c \supset \bigcap_n (X \setminus \overline{A_n}) = G$. Note that since $(X \setminus \overline{A_n})$ is open and dense for all n, $G \subset A^c$ is a dense G_δ subset in X.

(c) implies (a) If A is of first category, $X \setminus A$ contains a dense G_δ set and, consequently, $\overline{(X \setminus A)} = X$. Also, since $\text{int}(A) = X \setminus \overline{(X \setminus A)}$, $\text{int}(A) = \emptyset$.

3. (a) Let $\{O_n\}$ be open dense subsets of O in the induced metric and put $G = \bigcap_n O_n$; we claim that G is dense in O. Let $U = X \setminus \overline{O}$; since O, U are open in X, $U_n = O_n \cup U$ is open in X. We claim that the U_n are dense in X, i.e., given a nonempty open subset V of X, $V \cap U_n \neq \emptyset$. Now, this is clearly true if $V \subset U$. Otherwise $V \cap \overline{O} \neq \emptyset$, and, consequently, $V \cap O \neq \emptyset$. Thus, since $V \cap O$ is open in O, $V \cap O \cap O_n \neq \emptyset$, and, consequently, $V \cap U_n \neq \emptyset$. Therefore $\{U_n\}$ are open dense subsets of X and by Problem 1(b), $H = \bigcap_n U_n = G \cup U$ is dense in X. If V is a nonempty open subset of O, V is an open subset of X and, therefore, $V \cap H \neq \emptyset$. Hence $V \cap G = V \cap H \neq \emptyset$ and G is dense in O.

(b) Let V be a nonempty open subset of X. Then with $F_n' = F_n \cap V$, $V = \bigcup_n F_n'$ where the F_n' are closed in the relative topology of V. Now, by (a) V is a Baire space and, consequently, one of these closed sets, F_m', say, has nonempty interior (in V). Let $O = \text{int}_V(F_m')$. Since V is open, O is open in X and since $O \subset F_m'$ it follows that $O \subset \text{int}(F_m')$. But $O \subset V$ and so $O \subset \text{int}(F_m') \cap V$. Thus $\bigcup_n \text{int}(F_n) \cap V \neq \emptyset$.

4. (a) Let $\{G_n\}$ be dense G_δ subsets of X and consider $F_n = G_n^c$; each F_n is F_σ and nowhere dense. For the sake of argument suppose that $\bigcap_n G_n$ is not dense. Then its complement $\bigcup_n F_n$ contains an open ball and, hence, a smaller nonempty closed ball B, say. Now, since B is closed, B is a complete metric space in its own right and, since $B = \bigcup_n(B \cap F_n)$, B is the countable union of nowhere dense F_σ sets. By the Baire category theorem this cannot happen.

Note that this condition is equivalent to: If $\{C_n\}$ are closed sets, none of which contains a ball, then $\bigcup_n C_n \neq X$. This follows by complementation since $G_n = X \setminus C_n$ is open and dense for each n and, therefore, $\bigcap_n G_n \neq \emptyset$.

(b) For the sake of argument suppose that A and A^c are dense G_δ subsets of X. Then, by (a), $A \cap A^c = \emptyset$ is a dense G_δ subset of X, which is not the case.

(c) For the sake of argument suppose that A is a countable dense G_δ subset of X; then A^c is F_σ. Now, since A is countable, A is F_σ, and since A is dense, A^c is nowhere dense. Thus, contrary to the Baire category theorem, $X = A \cup A^c$ is the countable union of nowhere dense sets.

5. (a) You should not believe everything you read, no such example is possible: By Problem 1(b) the intersection is a dense G_δ set and by Problem 4(c) it cannot be countable.

7. Since the result for $A + B = A - (-B)$ follows from that for $A - B$ we prove the latter. Let $A = G_1 \Delta P_1$, $B = G_2 \Delta P_2$ where G_1, G_2 are nonempty open sets (for otherwise A, B would be of first category) and P_1, P_2 are of first category. First, observe that $G_1 - G_2$, which is a nonempty open set, is contained in $A - B$. Now, if $x \in G_1 - G_2$, $(x + B) \cap A \supset ((x + G_2) \cap G_1) - ((x + P_2) \cup P_1)$ where $(x + G_2) \cap G_1$ is a nonempty open set. Since $(x + P_2) \cup P_1$ is of first category and since a nonempty open set is not of first category, $((x + G_2) \cap G_1) - ((x + P_2) \cup P_1) \neq \emptyset$ and so $(x + B) \cap A \neq \emptyset$. Thus $x \in A - B$ and $G_1 - G_2 \subset A - B$.

9. (a) Since $Y_\alpha = \bigcap_n \bigcup_{q=n}^{\infty} U_\alpha(q)$, Y_α is a G_δ subset of $\mathbb{R}$. Moreover, for $\alpha \geq \beta$ and an integer q, we have $U_\alpha(q) \subset U_\beta(q)$ and so $Y_\alpha \subset Y_\beta$. Thus

$X = \bigcap_{\alpha \in \mathbb{R}^+} Y_\alpha = \bigcap_{n \in \mathbb{N}} Y_{n+1}$ is G_δ in $\mathbb{R}$. And since $\mathbb{Q} \subset X$, X is dense in $\mathbb{R}$.

(b) Let $x \in \mathbb{R}$. Suppose that $x \in X$, let P be a polynomial with real coefficients, and α strictly larger than the degree of P; note that $P(n) < n^\alpha$ for n sufficiently large. Now, since $\{n : d(nx) < n^{-\alpha}\}$ is infinite there exists an integer n such that $P(n) < n^\alpha$ and $d(nx) < n^{-\alpha}$, and, consequently, $d(nx)P(n) < 1$. On the other hand, if $x \notin X$, $x \notin Y_q$ for some integer $q > 0$; since $\{n \in \mathbb{N} : d(nx) < n^{-\alpha}\}$ is a finite set there exists $a > 0$ such that $ad(nx) > 1$ for all n in this set. Then the polynomial $P(t) = t^q + a$ satisfies $P(n)d(nx) > 1$ for all $n \in \mathbb{N}$.

10. We claim that the set of points with at least one of its coordinates rational is of first category in $\mathbb{R}^2$. If $\{r_n\}$ is an enumeration of the rationals in $\mathbb{R}$, let $X_n = \{(r_n, y) : y \in \mathbb{R}\}$ and $Y_n = \{(x, r_n) : x \in \mathbb{R}\}$; each X_n and Y_n is a line in $\mathbb{R}^2$ and is therefore closed and nowhere dense. Now, the complement of the set of points in $\mathbb{R}^2$ with both of their coordinates irrational is $(\bigcup_n X_n) \cup (\bigcup_n Y_n)$ and, therefore, of first category in $\mathbb{R}^2$.

11. (a) Let $P(t) = a_0 t^d + \cdots + a_d$ be an irreducible polynomial of degree $d > 1$ with integer coefficients satisfied by x; since $q^d P(p/q)$ is an integer, if not 0, one has $q^d |P(p/q)| \geq 1$. Now, since $P'(t)$ is bounded near x by M, say, it follows that $q^{-d} \leq |P(x) - P(p/q)| \leq M |x - p/q|$ and, consequently, $|x - p/q| \geq 1/(Mq^d)$.

(b) For an integer n let $A_n = \{x \in \mathbb{R} : \text{there exists a rational } p/q, q > 1,$ such that $|x - p/q| < 1/nq^\tau\}$; clearly A_n is open and since $\mathbb{Q} \subset A_n$, dense in $\mathbb{R}$. Hence, by Problem 1(b), $\mathcal{L} = \bigcap_n A_n$ is a dense G_δ subset of $\mathbb{R}$. Now, $\mathbb{R} \setminus \bigcap_n L_n = \mathcal{D}(\tau)$ is of first category and so is $\mathcal{D} = \bigcup_n \mathcal{D}(n)$. Finally, $\mathcal{L}$ is the complement of $\mathbb{Q}$ in the second category set $\mathbb{R} \setminus \mathcal{D}$ and since $\mathbb{Q}$ is of first category, $\mathcal{L}$ is of second category in $\mathbb{R}$.

12. For the sake of argument suppose that for some $r \in \mathbb{R}$, $r \neq x - y$ for all $x, y \in A^c$. Then $r + A^c \subset A$ and so $r + A^c$ is of first category and by translation so is A^c. Thus $\mathbb{R} = A \cup A^c$ is of first category, which is not the case.

13. Let $d = d_1 \ldots d_n$, $0 \leq d_k \leq 9$, denote a finite pattern of digits; the set $\{d\}$ of all possible finite patterns is then the countable union of finite sets and, hence, countable. Let X_d denote the set of real numbers whose decimal expansion does not contain the pattern d; we claim that X_d is nowhere dense. First, X_d is closed. Indeed, if $x \notin X_d$, the pattern d can be found in x and $x = x_0.x_1 \ldots x_k d_1 \ldots d_n x_{k+n+1} \ldots$, say. Let $\varepsilon = 10^{-(k+n+1)}$ and note that if $|y - x| < \varepsilon$, x and y do not differ in the first $n + k$ digits, y contains the pattern d, and so $y \in X_d^c$, which is therefore open. Next, X_d has empty interior. For the sake of argument suppose that there are

$x \in X_d$ and $\varepsilon > 0$ such that $(x - \varepsilon, x + \varepsilon) \subset X_d$; we may assume that $\varepsilon = 10^{-N}$ for some integer N. Then if the first N digits of y are the same as those of x, the distance between x and y is less than ε. Now consider a number $y = x_0.x_1 \ldots x_N d_1 \ldots d_n$ whose first N digits are those of x followed by the pattern d. Then on the one hand $y \in (x - \varepsilon, x + \varepsilon) \subset X_d$ while at the same time the decimal expansion of y contains the pattern d, which cannot happen. Thus X_d is nowhere dense and X_d^c is open dense. Therefore $\mathcal{A} = \bigcap_d X_d^c$ is the countable intersection of open dense subsets of $\mathbb{R}$ and by Problem 1(b), $\mathcal{A}$ is a dense G_δ set in $\mathbb{R}$.

14. For the sake of argument suppose that each $x \in M_n$ is the limit of a sequence $\{x_k\}$ with $x_k \notin M_n$ for all k. Let $x_0 \in M_1$ and I_0 an open interval containing x_0. Since x_0 is a limit point of $M \setminus M_1$ by the well-ordering of the integers there is a least integer $n_1 > 1$ such that I_0 contains a point $x_1 \in M_{n_1} \setminus M_1$. Let $I_1 \subset \overline{I}_1 \subset I_0$ be an open interval containing x_1 and select a point $x_2 \in M \setminus M_{n_1}$ such that $x_2 \in M_{n_2}$ and $n_2 > n_1$ is the least integer for which this happens; next let I_2 be an open interval containing x_2, $I_2 \subset \overline{I}_2 \subset I_1$ and I_2 contains no point of M_{n_1}. $\overline{I}_2$ contains no point of $M_1 \cup M_2$. Continuing this way one constructs a nested sequence $\{M \cap \overline{I}_k\}$ where each set is closed and bounded and so the intersection contains a point which, by the selection process, is not in M_n for any n. But, since $M = \bigcup_n M_n$, this cannot happen.

17. Let $G/n = \{y \in \mathbb{R} : ny \in G\}$; observe that $D = \limsup_n G/n = \bigcap_m \bigcup_{n=m}^\infty G/n$. Now, since G/n is open for every n, $\bigcup_{n=m}^\infty G/n$ is open for every m; we claim that it is also dense in $(0, \infty)$. For the sake of argument suppose that this is not the case and let $J = (a, b)$ be an open interval such that $J \cap \bigcup_{n=m}^\infty G/n = \emptyset$, i.e, $G \cap \bigcup_{n=m}^\infty (na, nb) = \emptyset$. Let M be the smallest integer such that $M > a/(b - a)$ and note that for $m \geq M$, $ma < (m+1)a < mb$ and, consequently, the above intervals overlap and $\bigcup_{n=m}^\infty (na, nb) = (ma, \infty)$. Therefore $G \cap (ma, \infty) = \emptyset$, which is not the case since G is unbounded. Finally, if $m < M$, $\bigcup_{n=m}^\infty G/n \supset (Ma, \infty)$ and $G \cap \bigcup_{n=m}^\infty (na, nb) \neq \emptyset$. Hence $\bigcup_{n=m}^\infty G/n$ is dense for every m and by Problem 1(b), $\bigcap_m \bigcup_{n=m}^\infty G/n = D$ is dense in $(0, \infty)$.

It is also true that if $\{G_k\}$ is a sequence of open sets unbounded above there is x_0 with the property that nx_0 is in each G_k infinitely often; indeed, it suffices to replace G in the above argument by $G_1, G_1, G_2, G_1, G_2, G_3, G_1, G_2, G_3, G_4, \ldots$.

The result also has the following interesting consequence: If $A \subset \mathbb{N}$ is an infinite set of positive integers, there exists $a > 1$ such that infinitely many $[a^k]$ (here $[x]$ denotes $x -$ fractional part of x) are in A; to obtain this let $G = \bigcup_n (\ln(n), \ln(n+1))$.

18. (a) With $P_0 = [0,1]$ the construction of the P_n proceeds by induction. First, some shorthand notation: let $d_n = a_{n-1} - 2a_n > 0$, all $n \geq 1$. Let $J_{n-1,k}$, $n \geq 1$, $1 \leq k \leq 2^{n-1}$, denote the 2^{n-1} pairwise disjoint closed intervals, each of length a_{n-1}, that comprise P_{n-1}, i.e., $P_{n-1} = \bigcup_{k=1}^{2^{n-1}} J_{n-1,k}$, and let $J_{n,2k-1}$ and $J_{n,2k}$ be the closed intervals of length a_n such that $J_{n,2k-1}$ has the same left endpoint as $J_{n-1,k}$ and $J_{n,2k}$ has the same right endpoint as $J_{n-1,k}$. Note that if $I_{n,k} = J_{n-1,k} \setminus (J_{n,2k-1} \cup J_{n,2k})$, $n > 0$, $1 \leq k \leq 2^{n-1}$, $I_{n,k}$ is the open interval of length d_n having the same midpoint as $J_{n-1,k}$. It is clear that $\{I_{n,k} : n \in \mathbb{N}, 1 \leq k \leq 2^{n-1}\}$ consists of pairwise disjoint open intervals. Also $[0,1] = P_0 \supset P_1 \supset \ldots$, $P_{n-1} \setminus P_n = \bigcup_{k=1}^{2^{n-1}} I_{n,k}$, and $P_0 \setminus P = \bigcup_n (P_{n-1} \setminus P_n)$. In short, P_n is obtained from P_{n-1} by removing the center of $J_{n-1,k}$ for each k, i.e., the open interval $I_{n,k}$, $1 \leq k \leq 2^{n-1}$, and then P is obtained by removing from $[0,1]$ all the open intervals $I_{n,k}$, $n \geq 1, 1 \leq k \leq 2^{n-1}$. Specifically, $P_n = \bigcup_{k=1}^{2^n} J_{n,k}$ and $P = \bigcap_{n=0}^{\infty} P_n$.

For details of this construction consult K. Stromberg, *Introduction to classical real analysis*, Wadsworth International Mathematics Series, 1981.

19. (a) We need to define the sequence $\{a_n\}$ with $2a_n < a_{n-1}$ that corresponds to this construction. Let $a_0 = 1$ and for $n \geq 0$ define a_n by the relation $a_n - 2a_{n+1} = \lambda p^{-(n+1)}$. Note that $1 - 2a_1 = \lambda/p$ or $2a_1 = 1 - \lambda/p$; then $a_1 - 2a_2 = \lambda/p^2$ or $4a_2 = 1 - \lambda/p - (2/p)\lambda/p$ at the next step, and so on. More precisely, by induction it readily follows that

$$2^n a_n = \left(1 - \frac{\lambda}{p}\right) - \left(\frac{2}{p} + \cdots + \left(\frac{2}{p}\right)^{n-1}\right)\frac{\lambda}{p}$$

and, consequently,

$$|P_\lambda| = \lim_n 2^n a_n = \left(1 - \frac{\lambda}{p}\right) - \left(\frac{2}{p-2}\right)\frac{\lambda}{p} = 1 - \frac{\lambda}{p-2}.$$

(b) The construction of the P_n is simplest in this case. Since $a_n = 3^{-n}$ for all n,

$$r_n = \frac{1}{3^{n-1}} - \frac{1}{3^n} = 2\,\frac{1}{3^n}, \quad n = 1, 2, \ldots,$$

and the Cantor discontinuum consists of those $x \in [0,1]$ with expansion $x = \sum_n x_n r_n = 2 \sum_n 3^{-n} x_n$ where $x_n = 0$ or $= 1$.

(e) Note that $C - C = [-1,1]$ iff $C + C = [0,2]$. Indeed, if $C + C = [0,2]$ and $\lambda \in [-1,1]$, then $1 + \lambda \in [0,2]$ and so $1 + \lambda = c_1 + c_2$ for some $c_1, c_2 \in C$. Then $\lambda = c_1 - (1 - c_2) \in C - C$. Since these steps are reversible the statements are equivalent. Now, clearly $C + C \subset [0,2]$. And, given $x \in [0,2]$, $x/2 \in [0,1]$ and by (d), $x/2 = (y+z)/2$ with $y, z \in C$. Thus $x = y + z$ with y, z in C and so $[0,2] \subset C + C$.

(f) First, the rationals. For a positive integer p,

$$x = \frac{2}{3^p} + \frac{2}{3^{2p}} + \cdots = \frac{2}{3^p - 1}, \quad p = 1, 2, \ldots.$$

is in C; for instance, 1, 1/4, 1/13, are in C. Also $0 = 1 - 1$, $3/4 = 1 - 1/4$, $12/13 = 1 - 1/13$ are in C as are $1/39 = 1/(3 \cdot 13)$, and so on.

As for the irrational numbers, let $\{n_k\}$ be given by $n_1 = 1$ and $n_k = n_{k-1} + k$ for $k \geq 2$. Then $x = 2 \sum_k 3^{-n_k} = 2 \left(1/3 + 1/3^3 + 1/3^6 + \cdots \right)$ is in C. Another way to see this is to note that the ternary expansion of x is $.202002000200002\ldots$. Finally, $1/\sqrt{2}$, which satisfies $19/27 < 1/\sqrt{2} < 20/27$, has ternary decimal expansion beginning $.201002\ldots$, and $\pi/4$, whose ternary decimal expansion begins with $.210012\ldots$, are not in C.

(g) The left endpoints of the open intervals removed in the k-th step in the construction of the Cantor discontinuum are those points in C with ternary expansion ending in 1 at the k spot; they can also be thought of as those points with 0 in the k spot followed by a string of 2's. Moreover, since the right endpoints are obtained by adding $1/3^k$ to the left endpoints they are those points in the Cantor discontinuum with ternary decimal expansion with 2 in the k spot and ending in a string of 0's.

20. (a) Let $y = \sum_k a_k/2^k$, $a_k = 0, 1$ for all k, be the dyadic expansion of $y \in I$. Then $x = \sum_k (2a_k)/3^k$ is in C and since

$$f(x) = \frac{1}{2} \sum_k \frac{2a_k}{2^k} = \sum_k \frac{a_k}{2^k} = y,$$

f is onto.

(d) Let $x \in C$ have ternary expansion $x = \sum_n x_n/3^n$; thus $x/3 = \sum_n x_n/3^{n+1}$. Then by the definition of f,

$$f(x/3) = \frac{1}{2} \sum_n \frac{x_n}{2^{n+1}} = \frac{1}{2} \frac{1}{2} \sum_n \frac{x_n}{2^n} = \frac{1}{2} f(x).$$

Next, if $x \in [0, 1] \setminus C$, x and $x/3$ belong to different open intervals and the conclusion follows as before.

(e) Since f is constant in the open components that comprise the complement of the Cantor discontinuum, $f' = 0$ there. Next, let $x \in C$. For each integer n take $x_n = x \pm 2/3^n$ where the sign is chosen so that $x_n \in C$. Then

$$\frac{f(x_n) - f(x)}{x_n - x} = \frac{3^n}{2^{n+1}}$$

and since these quotients increase to ∞ with n, f is not differentiable at any $x \in C$.

21. The statement is true. Let f be the Cantor-Lebesgue function on I extended to be 0 for $x < 0$ and 1 for $x \geq 1$. Let $\{[a_n, b_n]\}$ be an enumeration of the closed subintervals of $[0, 1]$ with rational endpoints $a_n \neq b_n$ and put

$$f_n(x) = f\left(\frac{x - a_n}{b_n - a_n}\right), \quad x \in [0, 1].$$

f_n is a scaling of f on the subinterval $[a_n, b_n]$ and so is nondecreasing and $f_n' = 0$ a.e. Now let $g(x) = \sum_n 2^{-n} f_n(x)$, $x \in [0, 1]$; g is well-defined since $f_n(x) \leq 1$ for all n. Moreover, by Fubini's lemma, $g'(x) = \sum_n f_n'(x) 2^{-n} = 0$ a.e. It only remains to prove that g is strictly increasing. Now, if $0 \leq x < y \leq 1$, pick $[a_n, b_n] \subset [x, y]$, $a_n \neq x, b_n \neq y$. Then

$$\frac{x - a_n}{b_n - a_n} < 0 < 1 < \frac{y - a_n}{b_n - a_n},$$

and so

$$\frac{1}{2^n} f_n(x) = \frac{1}{2^n} f\left(\frac{x - a_n}{b_n - a_n}\right) = 0 < \frac{1}{2^n} = \frac{1}{2^n} f\left(\frac{y - a_n}{b_n - a_n}\right) = \frac{1}{2^n} f_n(y),$$

which implies that $g(x) < g(y)$.

22. Let φ be given by $\varphi(x) = 2 \sum_n x_n / 3^n$. Clearly φ maps X onto C and is 1-1: If $x \neq y$ and if m is the first place where the ternary expansions of x and y differ, then

$$|\varphi(x) - \varphi(y)| \geq 2\left(\frac{1}{3^m} - \sum_{n=m+1}^{\infty} \frac{1}{3^n}\right) = \frac{1}{3^m} > 0.$$

And, since $|\varphi(y) - \varphi(x)| \leq 2 \sum_n |y_n - x_n| / 3^n \leq 2\left(\sum_n (2/3)^n\right) d(y, x)$, φ is continuous.

Next, we verify that $\{x^k\} \subset X$ converges to x in X iff $\lim_k x_n^k = x_n$ for $n = 1, 2, \ldots$. If $d(x^k, x) \to 0$, given $\varepsilon > 0$, there exists N such that $d(x^k, x) < \varepsilon / 2^n$ for $k \geq N$. Then $|x_n^k - x_n| / 2^n \leq d(x^k, x) < \varepsilon / 2^n$, and, consequently, $|x_n^k - x_n| \leq \varepsilon$ for $k \geq N$. Conversely, if $\lim_k x_n^k = x_n$ for each n, given $\varepsilon > 0$, let N be such that $\sum_{n=N}^{\infty} 2^{-n} < \varepsilon / 2$. Next pick k large enough so that $\sum_{n=0}^{N-1} |x_n^k - x_n| / 2^n < \varepsilon / 2$ and observe that $d(x^k, x) \leq \sum_{n=0}^{N-1} |x_n^k - x_n| / 2^n + \sum_{n=N}^{\infty} 1 / 2^n < \varepsilon / 2 + \varepsilon / 2 = \varepsilon$ for those k.

Finally, φ^{-1} is continuous. Suppose that $|\varphi(x^k) - \varphi(x)| \to 0$ as $k \to \infty$ and fix n. Then let $\varepsilon < 3^{-n}$ and pick N so that $|\varphi(x^k) - \varphi(x)| < \varepsilon$ for $k \geq N$. As we saw above, if m is the first index where $x_m^k \neq x_m$, $1/3^m \leq |\varphi(x^k) - \varphi(x)| < \varepsilon < 1/3^n$ for all $k \geq N$ and, therefore, such an index m verifies $m > n$ for that choice of ε, and $x_n^k = x_n$ for all $k \geq N$. In particular, this means that $\lim_k x_n^k = x_n$ and, since n is arbitrary, by the above observation $d(x^k, x) \to 0$ and φ^{-1} is continuous.

23. (a) For the sake of argument suppose that $[0, 1] = \bigcup_n A_n$ where the A_n are pairwise disjoint nonempty closed sets and let $O_n = \text{int}(A_n)$. Then $X = [0, 1] \setminus \bigcup_n O_n = \bigcup_n (A_n \setminus O_n) \neq \emptyset$ is complete with the metric inherited from $[0, 1]$ and, therefore, by the Baire category theorem there exist n_0 and an open interval $J = (a, b)$, say, such that $\emptyset \neq X \cap J \subset A_{n_0}$. We claim that $J \cap O_n = \emptyset$ for $n \neq n_0$. Let $x \in X \cap J$ and suppose that $y \in J \cap O_n$; with no loss of generality we may assume that $x < y$. Then there exists $z \in A_n \setminus O_N$ such that $a < x, z < y < b$. Now, this implies that $z \in X \cap J$, which in turn implies that $z \in A_{n_0}$ but, since A_n is disjoint from A_{n_0}, this cannot happen. Therefore $J \subset X \cup O_{n_0}$, $J = (J \cap X) \cup (J \cap O_{n_0}) \subset A_{n_0}$, and so $J \subset O_{n_0}$, which is not the case since $X \cap J \neq \emptyset$.

(b) By the Baire category theorem one of the sets must contain an interval of positive length and no Cantor set of positive Lebesgue measure does.

24. Since no point $x \in \mathbb{R}$ is a limit point of $P(x)$, to each $x \in \mathbb{R}$ we can associate an interval $J(x)$ with rational endpoints such that $x \in J(x)$ and $J(x) \cap P(x) = \emptyset$. Now, there are countably many intervals with rational endpoints, $J_1, J_2, \ldots$, say, and for every n let $A_n = \{x \in \mathbb{R} : J(x) = J_n\}$. Then $\mathbb{R} = \bigcup_n A_n$ and by the Baire category theorem one of these sets, A_N, say, is of second category. Hence, if x, y are two points in A_N, then $P(x) \cap A_N = P(y) \cap A_N = \emptyset$, x and y are independent, and, consequently, A_N is an independent set.

26. For each $n \in \mathbb{N}$, let $\mathcal{V}_n = \{f \in C(I) : f(x) - f(a) > n(x - a)$ for some $x \in (a, a + 1/n) \cap [0, 1]\}$. Then $G_a = \bigcap_n \mathcal{V}_n$ and by Problem 1(b) it suffices to prove that each $\mathcal{V}_n$ is open and dense in $C(I)$.

First, $\mathcal{V}_n$ is open. Let $f \in \mathcal{V}_n$ and pick $x \in (a, a + 1/n) \cap [0, 1]$ such that $f(x) - f(a) > n(x - a)$. Then find $\varepsilon > 0$ such that $f(x) - f(a) > n(x - a) + 2\varepsilon$ and observe that if $g \in B(f, \varepsilon)$, $g(x) - g(a) \geq f(x) - f(a) - 2\varepsilon > n(x - a)$, which implies that $g \in \mathcal{V}_n$ and $\mathcal{V}_n$ is open.

Next, $\mathcal{V}_n$ is dense. Observe that $\mathcal{PL}(I) = \{g \in C(I) : g$ is piecewise linear$\}$ is dense in $C(I)$ (proof by pictures). Then let $g \in \mathcal{PL}(I)$ and pick $h \in \mathcal{PL}(I)$ such that $\|h\| < \varepsilon$ and $D^+h(a) > n - D^+g(a)$; h is easily constructed graphically, turning rapidly the slopes of the lines that define it, which is always possible since $D^+g(a) < \infty$ for g in $\mathcal{PL}(I)$. Then $D^+(g + h)(a) > n$ and since $g + h \in \mathcal{PL}(I)$, $g + h \in \mathcal{V}_n$. Now, since $\mathcal{PL}(I)$ is dense in $C(I)$ and $g + h \in B(g, \varepsilon)$, this guarantees that $\mathcal{V}_n$ is dense in $C(I)$.

27. For each k let $A_k = \bigcap_{\lambda \in \Lambda} \{x \in X : |f_\lambda(x)| \leq k\}$; since the f_λ are continuous, A_k is the intersection of closed sets and, hence, closed. Now, since each $x \in X$ belongs to A_k provided that $k \geq M(x)$, $X = \bigcup_k A_k$ and by the Baire category theorem A_n has nonempty interior for some n, i.e., a ball $B(x_0, \varepsilon) \subset A_n$ for some $x_0 \in A_n$ and $\varepsilon > 0$. Hence $|f_\lambda(x)| \leq n$ for all $\lambda \in \Lambda$ and $x \in B(x_0, \varepsilon)$ and the conclusion holds for $M = n$.

28. Recall that $f : X \to \mathbb{R}$ is lower semicontinuous if $O_\lambda = \{x \in X : f(x) > \lambda\}$ is open in X for all $\lambda \in \mathbb{R}$. For the sake of argument suppose that f is unbounded on every nonempty open subset on X; we claim that then O_λ is dense in X for all $\lambda \in \mathbb{R}$. Let O be a nonempty open subset of X. Since f is unbounded in O there exists $x \in O$ such that $f(x) > \lambda$, $x \in O \cap O_\lambda$, and O_λ is dense in X. Now, by Problem 1(b), $\bigcap_{n=-\infty}^{\infty} O_n$ is a dense G_δ subset of X, but this is impossible since, as is readily seen, $\bigcap_{n=-\infty}^{\infty} O_n = \emptyset$. Therefore f is bounded on a nonempty open subset O of X.

29. (a) Since $\{f_n\}$ is uniformly bounded to invoke Arzela-Ascoli it suffices to verify that the sequence is equicontinuous. Given $\varepsilon > 0$, let $\delta = \varepsilon/M$. Then, for any $x, y \in \mathbb{R}$ with $|x - y| < \delta$, by the mean value theorem $|f_n(x) - f_n(y)| \leq \varepsilon$ and the verification is complete.

(b) Let $A_{k,\ell} = \{x \in \mathbb{R} : M(x) \leq k, N(x) \leq \ell\}$; since the f_n and f_n' are continuous each $A_{k,\ell}$ is closed for all k, ℓ and $\mathbb{R} = \bigcup_{k,\ell} A_{k,\ell}$. Therefore by the Baire category theorem one of the sets, $A_{k,\ell}$, say, contains an open interval J and, in particular, both $\{f_n\}$ and $\{f_n'\}$ are uniformly bounded on any closed subinterval J_0 of J. Now, as in (a) it readily follows that such a sequence is equicontinuous in J_0 and, consequently, a subsequence $\{f_{n_k}\}$ of $\{f_n\}$ converges uniformly in J_0.

32. By Problem 30 with $f_n = f^{(n)}$ there, f coincides with a piecewise polynomial continuous function in a dense open set $O \subset \mathbb{R}$. Let $F = \{x \in \mathbb{R} :$ there is no neighborhood V_x of x such that $f|_{V_x}$ is a polynomial$\}$; F is the complement of O and, therefore, is closed and nowhere dense.

For the sake of argument suppose that $F \neq \emptyset$. First, note that F has no isolated points because if x_0 is an isolated point of F, f is a polynomial in $(x_0 - \eta, x_0)$ and in $(x_0, x_0 + \eta)$ for some $\eta > 0$ and by continuity also in $(x_0 - \eta, x_0 + \eta)$. Next, let $A_n = \{x \in \mathbb{R} : f^{(n)}(x) = 0\}$; A_n is closed and by assumption $\bigcup_n A_n = \mathbb{R}$. Applying the Baire category theorem to F, $A_n \cap F$ has nonempty interior in F for some n, i.e., there exists a nonempty interval $(x_0 - \delta, x_0 + \delta)$ such that $(x_0 - \delta, x_0 + \delta) \cap F \subset A_n \cap F$. Since $(x_0 - \delta, x_0 + \delta) = ((x_0 - \delta, x_0 + \delta) \cap F) \cup ((x_0 - \delta, x_0 + \delta) \cap O) = A \cup B$, say, the proof will be complete once we verify that f has vanishing derivatives of order $\geq n$ in both A and B for then f coincides with a polynomial in $(x_0 - \delta, x_0 + \delta)$, which is then contained in O, and this gives the desired contradiction.

First, if $x \in A$, $f^{(k)}(x) = 0$ for all $k \geq n$. Indeed, since F has no isolated points there exists a sequence $\{x_k\} \subset A$ with $\lim_k x_k = x$. Thus, since $A \subset A_n$, $f^{(n)}(x_k) = 0$ for all k and, therefore, by Rolle's theorem we can construct a sequence $\{x_k^1\}$ with x_k^1 between x_k and x_{k+1} such that $f^{(n+1)}(x_k^1) = 0$, all k. Since $\lim_k x_k^1 = x$ it follows that $f^{(n+1)}(x) = \lim_k f^{(n+1)}(x_k^1) = 0$. Repeating this argument with $\{x_k^1\}$ in place of $\{x_k\}$ we get that $f^{(n+2)}(x) = 0$, and so on.

Now, suppose that $x \in B$. Then f is a polynomial in some neighborhood of x and since $F \neq \emptyset$ this neighborhood is not the whole interval $(x_0 - \delta, x_0 + \delta)$. Let (a, b) be the maximal subinterval of $(x_0 - \delta, x_0 + \delta)$ containing x on which f is a polynomial. Then a or b, to fix ideas b, say, is in A and so by the above argument $f^{(k)}(b) = 0$ for all $k \geq n$. Now, if f is of degree d in (a, b), $f^{(d)}(x) \neq 0$ for $x \in (a, b)$ and taking limits, $f^{(d)}(a), f^{(b)}(b) \neq 0$. Therefore $d < n$ and, consequently, $f^{(k)}(x) = 0$ for all $k \geq n$ and all x in (a, b). This argument works for all $x \in B$ and combining with the first part we conclude that $f^{(k)}(x) = 0$ for all $k \geq n$ and all $x \in (x_0 - \delta, x_0 + \delta)$. Then, integrating n times and using the fundamental theorem of calculus we deduce that f is a polynomial of degree $\leq n$ in $(x_0 - \delta, x_0 + \delta)$, which is not the case. Therefore $F = \emptyset$.

34. (a) Since χ_O assumes only two values, $x \in D(\chi_O)$ iff every neighborhood of x contains points of O and O^c and so $D(\chi_O)$ is precisely the boundary of O. Thus $D(\chi_O)$ is closed and since as noted above its interior is empty, $D(\chi_O)$ is nowhere dense.

(b) Since by (a) the set of continuity of each χ_{O_n} is a dense open subset of X, by Problem 1(b) their intersection is a dense G_δ set in X and, in particular, not empty.

35. (a) Given an open ball $B(x, r) = \{y \in X : d(x, y) < r\}$, let $w(f; B(x, r))$ denote the oscillation of f in $B(x, r)$, i.e., $w(f; B(x, r)) = \sup\{|f(y) - f(z)| : y, z \in B(x, r)\}$. Now, with $\mathcal{B}_x = \{$open balls $B : x \in B\}$, the oscillation $w(f, x)$ of f at x is defined as $w(f, x) = \inf\{w(f; B) : B \in \mathcal{B}_x\}$; note that f is continuous at x iff $w(f, x) = 0$.

Now, $D(f) = \bigcup_n D_n(f)$ where $D_n(f) = \{x \in X : w(f, x) \geq 1/n\}$. Note that $D_n(f)$ is closed, all n. Indeed, if x is a limit point of $D_n(f)$, then every $B \in \mathcal{B}_x$ contains $y \in D_n(f)$, $x \neq y$, and so $w(f; B) \geq w(f, y) \geq 1/n$. Thus $w(f, x) \geq 1/n$, i.e., $x \in D_n(f)$, and $D_n(f)$ is closed.

(b) Necessity first. By (a) $D(f) = \bigcup_n D_n(f)$ where each $D_n(f)$ is closed. Moreover, since f is continuous on a dense subset of X, any open ball contained in $D(f)$ contains a point of continuity of f and, since $D_n(f) \subset D(f)$, any open ball contained in $D_n(f)$ is also contained in $D(f)$. But $D(f)$ cannot contain any such ball and so $D(f)$ is of first category.

As for sufficiency, by Problem 1(c) if $D(f)$ is of first category in X, its complement contains a dense G_δ subset of X.

36. (a) For the sake of argument suppose such an f exists. Then by Problem 35, $C(f)$ is a G_δ subset of $[0, 1]$, and countable dense at that, but by Problem 4(c) no such set exists.

(b) and (c) The sets in question are countable and dense. Therefore as in (a) it readily follows that f does not exist.

37. The statement is false. Let f be left-continuous everywhere and $\varepsilon_k = 2^{-k}$, $k = 0, 1, \ldots$. Then for any $x_0 \in \mathbb{R}$, let δ_0 be such that $|f(x_0) - f(y)| < \varepsilon_0$ for $y \in J_0 = (x_0 - \delta_0, x_0)$. Next, for each $k \geq 1$ pick $x_k \in J_{k-1}$ and δ'_k such that $|f(x_k) - f(y)| < \varepsilon_k$ for $x_k - \delta'_k < y < x_k$. Now, since J_{k-1} is open, we may pick $\delta_k < \delta'_k$ so that, if $J_k = (x_k - \delta_k, x_k)$, $\overline{J}_k \subset J_{k-1}$; note that $|f(x_k) - f(y)| < \varepsilon_k$ for $y \in J_k$. Then $\{\overline{J}_k\}$ is a nested sequence of nonempty compact sets and by Cantor's nested property $\bigcap_k \overline{J}_k \neq \emptyset$. Let $x \in \bigcap_k \overline{J}_k$ and, given $\varepsilon > 0$, let k be such that $2\varepsilon_k < \varepsilon$. Since $x \in \overline{J}_{k+1} \subset J_k$ we can pick $r_\varepsilon > 0$ such that $(x, x + r_\varepsilon) \subset J_k$ and so, for any $y \in (x, x + r_\varepsilon)$, $|f(x) - f(y)| \leq |f(x) - f(x_k)| + |f(x_k) - f(y)| < 2\varepsilon_k < \varepsilon$ and f is right-continuous at x. Now, since in the above proof J_0 could be an arbitrary open subset of $\mathbb{R}$ it follows that the set of points where f is right-continuous is dense in $\mathbb{R}$.

38. Let $\{r_n\}$ be an enumeration of the rationals and $B = \{x \in A : f'_g(x) < f'_d(x)\}$; note that if $x \in B$ there is a smallest integer k, say, such that $f'_g(x) < r_k < f'_d(x)$. Let $\varphi(x, y) = (f(x) - f(y))/(x - y)$; since $\varphi(x, y) \to f'_g(x)$ as $y \to x^-$ it follows that $\varphi(x, y) < r_k$ for y sufficiently close to x. Thus there is a smallest integer m, say, such that $r_m < x$ and $\varphi(x, y) < r_k$ for all y such that $r_m < y < x$. Similarly, there is a smallest integer n, say, such that $r_n > x$ and $\varphi(x, y) < r_k$ for $x < y < r_n$. Then $f(y) - f(x) > r_k(y - x)$ for $r_m < y < r_n$. We claim that the mapping $x \to (k, m, n)$ is injective. Indeed, if $x \neq x_1$ corresponds to the same triplet, then $f(x_1) - f(x) > r_k(x_1 - x)$ and the opposite inequality must simultaneously hold, which is impossible. Thus B has at most cardinality $\aleph_0^3 = \aleph_0$.

39. Let $\varphi_n(x) = 2^{-n}\chi_{(a_n, \infty)}(x)$ and put $f(x) = \sum_n \varphi_n(x)$.

40. Since Y is separable, given an integer n, there exist open sets $\{O_m^n\} \subset Y$ such that $Y = \bigcup_m O_m^n$ and $\operatorname{diam}(O_m^n) \leq 1/2n$ for all m. Now, $f^{-1}(O_m^n) = \bigcup_k Z_{k,m}^n$, say, where the $Z_{k,m}^n$ are closed in X and since the O_m^n cover Y for each n, $X = \bigcup_{k,m} Z_{k,m}^n$, all n.

Now, since by Problem 3(b) $\bigcup_{k,m} \operatorname{int}(Z_{k,m}^n)$ is dense in X, by Problem 1(b) it suffices to prove that f is continuous on $G = \bigcap_n \bigcup_{m,k} \operatorname{int}(Z_{m,k}^n)$, which is a dense G_δ subset in X. So, let $x \in G$ and note that for every n there exist m, k such that $x \in \operatorname{int}(Z_{m,k}^n)$. Given $\varepsilon > 0$, choose n such that $1/n < \varepsilon$. Then $x \in \operatorname{int}(Z_{m,k}^n)$ for some m, k, n, and, therefore, there exists $\delta > 0$ such that $d(x, y) < \delta$ implies $y \in Z_{m,k}^n$. Hence $f(x), f(y) \in U_m^n$ and, therefore, $d'(f(x), f(y)) < 1/n < \varepsilon$.

Finally, since it is readily seen that the preimage of an open set by a lower semicontinuous function $f : \mathbb{R} \to \mathbb{R}$ is an F_σ set in $\mathbb{R}$, it follows that such an f is continuous on a dense G_δ subset of $\mathbb{R}$.

41. For the sake of argument suppose that f_n are continuous functions such that $\lim_n f_n(x) = \chi_{\mathbb{Q}}(x)$ and let $A_k = \bigcap_{n \geq k} f_n^{-1}([-1/2, 1/2])$; the A_k are the intersection of closed sets and, hence, closed. Moreover, since for $x \in A_k$, $|f_n(x)| < 1/2$ for $n \geq k$, it follows that $\lim_n f_n(x) = \chi_{\mathbb{Q}}(x) = 0$ and A_k consists entirely of irrational numbers. Now, if x is irrational, $x \in A_k$ for sufficiently large k and, therefore, the irrationals in $\mathbb{R}$ can be written as the F_σ set $\bigcup_k A_k$; then the rationals are a G_δ set in $\mathbb{R}$ and by Problem 4(c) this is not the case.

Next, let $f_{m,n}(x) = \cos^{2m}(n!\pi x)$. Note that if x is rational with irreducible expression p/q and $q \leq n$, then $\lim_m \cos^{2m}(n!\pi x) = 1$. In all other cases, i.e., if x is irrational or q does not divide $n!$, $\lim_m \cos^{2m}(n!\pi x) = 0$. Thus with $f_n(x) = \lim_m \cos^{2m}(n!\pi x)$ it follows that $\lim_n f_n(x) = \chi_{\mathbb{Q}}(x)$ and $\chi_{\mathbb{Q}}$ is the limit of functions which are in turn limits of continuous functions.

42. For the sake of argument suppose that d is a metric in $X \times X$ such that convergence in (X, d) is equivalent to pointwise convergence. Now, if $\{f_{m,n}(x)\}$ is as in Problem 41, the pointwise limit $\lim_m f_{m,n}(x) = \varphi_n(x)$, say, exists for all n, and, consequently, $d(f_{m,n}, \varphi_n) \to 0$ as $m \to \infty$. Therefore by the triangle inequality $\lim_n d(\varphi_n, \chi_{\mathbb{Q}}) = 0$ and, using the Cantor diagonal process, it follows that for all m there exists $n(m)$ such that $d(f_{m,n(m)}, \chi_{\mathbb{Q}}) \to 0$ and so, since convergence in the metric is equivalent to pointwise convergence, $\chi_{\mathbb{Q}}$ is the limit of a sequence of continuous functions, which by Problem 41 is not the case.

43. Fix an integer n and for an integer m let $A_m^n = \bigcap_{k=m}^{\infty}\{x \in X : |f_m(x) - f_k(x)| \leq 1/n\}$; by the continuity of the f_k each A_m^n is closed. Now, since the numerical sequence $\{f_m(x)\}$ converges for all $x \in X$ it is Cauchy there and so there exists an integer m (depending on x) such that $|f_m(x) - f_k(x)| \leq 1/n$ for all $k \geq m$, that is to say, $x \in A_m^n$. Therefore $X = \bigcup_m A_m^n$ and if $O_{m,n} = \mathrm{int}(A_m^n)$, by Problem 3(b) $O_n = \bigcup_m O_{m,n}$ is an open dense subset of X.

We claim that f is continuous in $G = \bigcap_n O_n$, which by Problem 1(b) is a dense G_δ subset of X. To see this let $x \in G$ and given $\varepsilon > 0$, pick n such that $1/n < \varepsilon/3$. Now, since $x \in O_n$, $x \in O_{m,n}$ for some m. Moreover, since $O_{m,n} \subset A_m^n$ we have $|f_m(y) - f_k(y)| \leq 1/n$ for $y \in O_{m,n}$ and $k \geq m$. Thus letting $k \to \infty$ it follows that $|f_m(y) - f(y)| \leq 1/n$, an inequality that holds in particular for $y = x$. Next, since f_m is continuous at x there exists a neighborhood V of x, which we may assume is contained in $O_{m,n}$, such that $|f_m(y) - f_m(x)| < \varepsilon/3$ for all $y \in V$. Thus we have proved that for $y \in V$, $|f(y) - f(x)| \leq |f(y) - f_m(y)| + |f_m(y) - f_m(x)| + |f_m(x) - f(x)| < \varepsilon$, which gives the continuity of f at x. So f is at least continuous on the G_δ dense set G, which is of second category.

Note that the result applies to an everywhere differentiable function $f : \mathbb{R} \to \mathbb{R}$. Indeed, consider the sequence

$$f_n(x) = n\left(f\left(x + \frac{1}{n}\right) - f(x) \right), \quad n = 1, 2, \ldots.$$

Then $\{f_n\} \subset C(\mathbb{R})$ and, consequently, $f'(x) = \lim_n f_n(x)$ is continuous on a dense G_δ subset of $\mathbb{R}$.

46. The statement is true. By assumption there is a subsequence $\{f_{n_k}\}$ of $\{f_n\}$ such that $|f_{n_k}(x) - f(x)| \leq 2^{-k}$ for all $x \in I$. Consider now $\{f_{n_{k+1}} - f_{n_k}\}$. By the triangle inequality,

$$|f_{n_{k+1}}(x) - f_{n_k}(x)| \leq |f_{n_{k+1}}(x) - f(x)| + |f(x) - f_{n_k}(x)|$$
$$\leq 2^{-(k+1)} + 2^{-k} = (3/2)\, 2^{-k}$$

and, consequently, by Problem 45, $\sum_k (f_{n_{k+1}}(x) - f_{n_k}(x)) \in \mathcal{B}_1$. Thus $\sum_k (f_{n_{k+1}}(x) - f_{n_k}(x)) = \lim_N \sum_{k=1}^{N}(f_{n_{k+1}}(x) - f_{n_k}(x)) = f(x) - f_{n_1}(x) \in \mathcal{B}_1$ and since $f_{n_1} \in \mathcal{B}_1$, by Problem 45, $f = f - f_{n_1} + f_{n_1} \in \mathcal{B}_1$.

47. Suppose f is lower semicontinuous and for each $r \in \mathbb{Q}$ let $O_r = \{x \in X : f(x) > r\}$; since the O_r are open the sets $F_r = \overline{O_r} \setminus O_r$ are closed and nowhere dense. Thus it suffices to prove that $D(f) \subset \bigcup_{r \in \mathbb{Q}} F_r$. Let $x \in D(f)$. Then there exists $\varepsilon > 0$ such that for every neighborhood U of x and every $r \in (f(x), f(x) + \varepsilon)$, the set $U \cap O_r$ is nonempty. Therefore, if $r \in (f(x), f(x) + \varepsilon)$, $x \in F_r$. Compare with Problem 40.

48. Let $X = \bigcup_n X_n$ where the X_n are closed and nowhere dense and for $x \in X$ let $f(x) = \inf\{n : x \in X_n\}$ and $g = 1 - (1/f)$; g is clearly bounded on X, we claim that g is lower semicontinuous. For this it suffices to prove that f is lower semicontinuous which is obvious since $\{x \in X : f(x) \leq k\} = \bigcup_{n=1}^{k} X_n$ for every $k = 1, \ldots$. Moreover, f is continuous at no point of X. Indeed, if O is a nonempty open subset of X, then no finite number of X_n cover O. Therefore f is not bounded on O. Clearly g is continuous at no point of X.

As an immediate consequence of this result it follows that a metric space (X, d) is of second category iff every semicontinuous function on X is continuous at some point of X.

49. We do the case when f is lower semicontinuous. For each integer n let $f_n : [0, 1] \to \mathbb{R}$ be given by $f_n(x) = \inf_{t \in [0,1]}(f(t) + n|t - x|)$. First, each f_n is continuous. Since a lower semicontinuous function achieves its minimum in $[0, 1]$, f is bounded below and f_n takes finite values. Now, for $x, y \in [0, 1]$, $f_n(x) = \inf_{t \in [0,1]}(f(t) + n|t - x|) \leq f_n(y) + n|y - x|$ and exchanging x and y, also $f_n(y) \leq f_n(x) + n|y - x|$. Therefore $|f_n(x) - f_n(y)| \leq n|x - y|$ and so each f_n is continuous in $[0, 1]$.

Next, $f(x) = \lim_n f_n(x)$ everywhere. Let $x \in [0,1]$; by definition $f_n(x) \le f(t) + n|t - x|$ for all $t \in [0,1]$ and, in particular, setting $t = x$, $f_n(x) \le f(x)$ and so $\limsup_n f_n(x) \le f(x)$. Next, suppose that $f(x) > r$, $r \in \mathbb{R}$. Since f is lower semicontinuous at x there exists $\delta > 0$ such that $f(t) > r$ for all $t \in (x-\delta, x+\delta) \cap [0,1] = J$, say. For these values of t, $\inf_{t \in J}\big(f(t) + n|t-x|\big) \ge r$ and, since $|t - x| \ge \delta$ for $t \in [0,1] \setminus J$, $\inf_{t \in [0,1] \setminus J}\big(f(t) + n|t-x|\big) \ge -M + n\delta$ where $-M$ is a lower bound for f. Thus $f_n(x) \ge r$ for n sufficiently large and so $\liminf_n f_n(x) \ge r$. Hence, since $f(x) > r$ is arbitrary, $\liminf_n f_n(x) \ge f(x)$ and finally, combining these inequalities, $\lim_n f_n(x) = f(x)$.

50. Let $A = \bigcup_n A_n$ where each A_n is closed and nowhere dense. Given $x \in A$, $m_x = \min\{n : x \in A_n\}$ is well-defined, and let $f : X \to \mathbb{R}$ be given by

$$f(x) = \begin{cases} 1/m_x, & x \in A, \\ 0, & x \notin A. \end{cases}$$

We claim that $f \in \mathcal{B}_1$ and $D(f) = A$. First, $D(f) = A$. Let $x \in A$; since A contains no open ball, given $\varepsilon > 0$, there is $y \in B(x,\varepsilon) \setminus A$. Then $|f(y) - f(x)| = |0 - 1/m_x| = 1/m_x$ and f is not continuous at x. Conversely, let $x \in X \setminus A$, $\varepsilon > 0$, and pick N such that $1/N < \varepsilon$. Since $F = \bigcup_{n=1}^{N} A_n$ is closed there exists $\delta > 0$ such that $B(x,\delta) \cap F = \emptyset$. Then $|f(y) - f(x)| < 1/N < \varepsilon$ for all $y \in B(x,\delta)$ and f is continuous at x. Thus $D(f) = A$.

Next, we verify that f is upper semicontinuous. For $r \in \mathbb{R}$ let $B_r = \{x \in X : f(x) \ge r\}$. Then

$$B_r = \begin{cases} X, & r \le 0, \\ \bigcup_{n=1}^{m} A_n, & 1/(m+1) < r \le 1/(m), \\ \emptyset, & r > 1. \end{cases}$$

Thus B_r is closed in all cases, f is upper semicontinuous, and by Problem 49, $f \in \mathcal{B}_1$.

51. Since O can be written as a countable union of pairwise disjoint open intervals and since we can write an open interval $(a,b) = (-\infty, b) \cap (a, \infty)$, it suffices to prove that $f^{-1}((-\infty, q))$ and $f^{-1}((q, \infty))$ are F_σ in X for each rational q. Let $\{f_n\}$ be continuous functions on X such that $\lim_n f_n(x) = f(x)$ for each $x \in X$. Observe that with $\liminf_n\{f_n \le p\} = \bigcup_m \bigcap_{n=m}^{\infty}\{f_n \le p\}$, $f^{-1}((-\infty, q)) = \bigcup_{p \in \mathbb{Q}, \, p<q} \liminf_n\{f_n \le p\}$. Now, the continuity of the f_n implies that the sets $\{f_n \le p\}$ are closed and, therefore, $\liminf_n\{f_n \le p\}$ is an F_σ subset of X, as is $f^{-1}(-\infty, q)$. Similarly, $\{f > q\} = \bigcup_{p \in \mathbb{Q}, \, p>q} \liminf_n\{f_n \ge p\}$ is an F_σ subset of X.

The converse to the statement is also true and is due to Baire. Finally, observe that since $\chi_{\mathbb{Q}}^{-1}((-1/2, 1/2)) = \mathbb{R} \setminus \mathbb{Q}$ is not F_σ, $\chi_{\mathbb{Q}} \notin \mathcal{B}_1$.

53. Given $\varepsilon > 0$, let $B_N = \bigcap_{n=N}^{\infty} \{x \in (0, \infty) : |f(nx)| \leq \varepsilon\}$; since f is continuous, B_N is closed. Now, if $x \in (0, \infty)$, there is N_x such that $|f(nx)| \leq \varepsilon$ for $n \geq N_x$ and so $x \in B_{N_x} \subset \bigcup_N B_N$. Thus $(0, \infty) = \bigcup_N B_N$ and by the Baire category theorem there are N_0 and an interval $J = (a, b)$ such that $|f(nx)| \leq \varepsilon$ for $x \in J$ and all $n \geq N_0$, and, consequently, $|f(x)| \leq \varepsilon$ for all $x \in V = \bigcup_{n \geq N_0} (na, nb)$. Observe that if N is an integer greater than $a/(b-a)$, then for $n > N$ we have $(n+1)a < nb$, and, consequently, $(Na, \infty) \subset V$. Hence, given $\varepsilon > 0$, there exists $A > 0$ such that $x \geq A$ implies $|f(x)| \leq \varepsilon$, i.e., $\lim_{x \to \infty} f(x) = 0$.

54. Such a number x can be written with only a finite number of nonzero decimals, say n (depending on x). Then for some integer m, $x = m/10^n$, x is rational, and the set in question is countable.

55. For the sake of argument suppose that X is countable and let $f : \mathbb{N} \to X$ be a bijection from $\mathbb{N}$ into sequences of 0's and 1's. If $f(n) = (x_1^n, x_2^n, \ldots)$ is the sequence that corresponds to n let y be the sequence with terms $y_n = 1 - x_n^n$, all n. Then y is a sequence of 0's and 1's but since $y_n \neq x_n^n$ for all n, y cannot be any of the sequences in $f(\mathbb{N})$.

56. The statement is true. Since $\mathbb{N}$ and $\mathbb{Q} \cap [0, 1]$ are equivalent we work with the latter set instead. $\mathcal{F}$ is then defined as the set of sequences of rationals in $[0, 1]$ that converge to the irrationals there, one sequence per irrational. Since $\mathrm{card}([0, 1] \setminus \mathbb{Q}) = c$, $\mathcal{F}$ has the right cardinality. Now, if $A, B \in \mathcal{F}$ have infinite intersection, they contain a common subsequence that must converge to the irrational number determined by both A and B. But since these numbers are different, the intersection is finite.

57. Yes. Enumerate $A = \{a_1, a_2, \ldots\}$ and observe that the set of all possible distances $|a_n - a_m|$ is countable. Now, since $\mathbb{R}$ is uncountable there is a real number r that is not equal to any of the distances and so $A \cap (r + A) = \emptyset$.

58. For the sake of argument suppose that $A \cap (-\infty, t)$ or $A \cap (t, \infty)$ is countable for all $t \in \mathbb{R}$; since A is uncountable at most one of these sets can be countable. Let $\{r_n\}$ be an enumeration of the rationals of $\mathbb{R}$ and for every n let A_n be $A \cap (-\infty, r_n)$ if $A \cap (-\infty, r_n)$ is countable and $A \cap (r_n, \infty)$ otherwise. Consider now $A \setminus \bigcup_n A_n$. If $A \setminus \bigcup_n A_n = \emptyset$, A is countable, which is not the case. If not, let $y \in A \setminus \bigcup_n A_n$ and write $A = (\bigcup_{r_n < y} (A \cap (-\infty, r_n))) \cup \{y\} \cup (\bigcup_{r_n > y} (A \cap (r_n, \infty)))$. Now, for all $r_n < y$, $A \cap (-\infty, r_n)$ is countable because $y \notin \bigcup_n A_n$; similarly, for all $r_n > y$, $A \cap (r_n, \infty)$ is countable. Thus $A \setminus \bigcup_n A_n = \{y\}$ and since $A \setminus \bigcup_n A_n$ is countable, A is countable, which is not the case. The second statement follows by applying the first statement twice.

60. Since there are c open dense sets in $\mathbb{R}$ and, consequently, c dense G_δ subsets of $\mathbb{R}$, there are c closed nowhere dense subsets of $\mathbb{R}$. Now, with Ω the first uncountable ordinal let $\{A_\alpha : \alpha < \Omega\}$ be the collection of all closed nowhere dense sets. Inductively choose $x_\alpha \in \mathbb{R}$ as follows: x_1 is any point in A_1 and, having chosen x_β for $\beta < \alpha < \Omega$, let x_α be a point not in $\{x_\beta : \beta < \alpha\} \cup \bigcup_{\beta < \alpha} A_\beta$; this choice is always possible by the Baire category theorem. Then $A = \{x_\alpha : \alpha < \Omega\}$ is the required set.

62. (a) A contains a finite subset A_n of cardinality n for each integer n and $\bigcup_n A_n \subset A$ is countable.

(b) Suppose that $A \cap \mathbb{N} = \emptyset$ and with $B = \{a_1, \ldots, a_n, \ldots\}$ a countable subset of A, let $\varphi : A \cup \mathbb{N} \to A$ be given by

$$\varphi(x) = \begin{cases} x, & x \in A \setminus B, \\ a_{2n-1}, & x = a_n \in B, \\ a_{2n}, & x = n \in \mathbb{N}. \end{cases}$$

Then φ is a bijection and $a + \aleph_0 = \operatorname{card}(A \cup \mathbb{N}) = \operatorname{card}(A) = a$.

(c) Let $\mathcal{F} = \{(X, f) : X \subset A \text{ and } f : X \to \{0, 1\} \times X \text{ is a bijection}\}$; by Problem 61, $X \sim \{0, 1\} \times X$ for $X \subset A$ countable and $\mathcal{F} \neq \emptyset$. Now, $\mathcal{F}$ is partially ordered by set inclusion and extension of functions, i.e., $(X, f) \prec (X', f')$ if $X \subset X'$ and $f'|_X = f$. Moreover, if $\mathcal{C} \subset \mathcal{F}$ is a chain, $(\widetilde{X}, \widetilde{f})$ defined by $\widetilde{X} = \bigcup_{(X, f) \in \mathcal{C}} X$ and $\widetilde{f}(x) = f(x)$ for $x \in X$, $(X, f) \in \mathcal{C}$, is an upper bound of $\mathcal{C}$. Note that $\widetilde{f}$ is well-defined since for $(X, f), (X', f') \in \mathcal{C}$, since $\mathcal{C}$ is a chain we may assume that $(X, f) \prec (X', f')$. So, if $x \in X$, $x \in X'$ and $f'(x) = f(x)$. Thus every chain in $\mathcal{F}$ has an upper bound and, consequently, by Zorn's lemma there is a maximal element (D, g) in $\mathcal{F}$ with $g : D \to \{0, 1\} \times D$ a bijection. Clearly $D \subset A$ has infinite cardinality and we have $A = (A \setminus D) \cup D$. Now, if $A \setminus D$ is finite, by (a) $\operatorname{card}(A) = \operatorname{card}(D)$ and we are done. So, for the sake of argument suppose that $A \setminus D$ is infinite and let B be a countable subset of $A \setminus D$ and $f : B \to \{0, 1\} \times B$ a bijection. Then $B \cap D = \emptyset$, $X = D \cup B \subset A$, and the mapping $h : X \to \{0, 1\} \times X$ given by

$$h(x) = \begin{cases} g(x), & x \in D, \\ f(x), & x \in B, \end{cases}$$

is 1-1 and onto, $(h, X) \in \mathcal{F}$ and, since $h|_D = g$, (h, X) extends (g, D), contrary to its maximality. Therefore this case cannot occur and we have established that $a + a = a$. Finally, since $b + b = b$ and $b \leq a + b \leq b + b = b$, it readily follows that $a + b = b$.

63. (a) Let $\mathcal{A} = \{A_\alpha \subset A : \text{the } A_\alpha \text{ are countable and pairwise disjoint}\}$ and $\mathcal{F} = \{\mathcal{A}\}$; $\mathcal{F}$ consists of all pairwise disjoint collections of countable subsets of A and by Problem 62(a), $\mathcal{F} \neq \emptyset$. Now, $\mathcal{F}$ is partially ordered by set inclusion, i.e., given two collections $\mathcal{A}_1, \mathcal{A}_2$ in $\mathcal{F}$, we say that $\mathcal{A}_1 \prec \mathcal{A}_2$ if every countable subset of A in $\mathcal{A}_1$ is in $\mathcal{A}_2$. And, if $\mathcal{C} = \{P_\alpha\}_{\alpha \in I}$ is a chain in $\mathcal{F}$, the collection of all countable subsets of A which belong to P_α for some $\alpha \in I$ is an upper bound of $\mathcal{C}$. Hence by Zorn's lemma $\mathcal{F}$ has a maximal element $\mathcal{M} = \{M_\alpha\}$, say, comprised of countable subsets of A. Now, if $A \setminus \bigcup_\alpha M_\alpha \neq \emptyset$, $A \setminus \bigcup_\alpha M_\alpha$ is finite or by Problem 62(a) contains a countable set, which is not the case by the maximality of $\mathcal{M}$. Therefore $A = \bigcup_\alpha M_\alpha$ and we have finished.

(b) Let A be a set with at least two elements and consider $\mathcal{F} = \{(X, f) : X \subset A, f : X \to X^c \text{ is injective}\}$; $\mathcal{F}$ is partially ordered by set inclusion and extension of functions and every chain in $\mathcal{F}$ has an upper bound. Therefore by Zorn's lemma $\mathcal{F}$ has a maximal element $(\widetilde{X}, \widetilde{f})$, say. Observe that $\widetilde{X}^c \cap \widetilde{f}(\widetilde{X})^c$ contains at most one element. Indeed, if $x \neq x' \in \widetilde{X}^c \cap \widetilde{f}(\widetilde{X})^c$, then $(\bar{X}, \bar{f})$ defined by

$$\bar{X} = \widetilde{X} \cup \{x\}, \quad \bar{f}\big|_{\widetilde{X}} = \widetilde{f}, \quad \bar{f}(x) = x'$$

belongs to $\mathcal{F}$ and satisfies $(\widetilde{X}, \widetilde{f}) \prec (\bar{X}, \bar{f})$, which is not possible by the maximality of $(\widetilde{X}, \widetilde{f})$. Therefore A is the disjoint union of $\widetilde{X}, \widetilde{f}(\widetilde{X})$ and a set S of at most one element. Since $\widetilde{f}$ is 1-1, $\widetilde{X} \sim \widetilde{f}(\widetilde{X})$ and these sets have the same cardinality. Finally, since A is uncountable, $\widetilde{X}$ and $\widetilde{X}^c = \widetilde{f}(\widetilde{X}) \cup S$ are both uncountable.

64. First, if $A = \{a_1, \ldots, a_n, \ldots\}$ is countable, the mappings $f : A \times \mathbb{N} \to A$ given by $f(a_n, m) = a_{2^n 3^m}$ and $g : A \to A \times \mathbb{N}$ given by $g(a_n) = (a_n, 1)$ are injective and, consequently, by the Cantor-Bernstein-Schröder theorem, there exists a bijection $\phi : A \times \mathbb{N} \to A$. In particular, $\aleph_0 \cdot \aleph_0 = \aleph_0$.

Let $\mathcal{F} = \{(X, f) : X \subset A, f : X \times X \to X \text{ is a bijection}\}$; since for a countable subset X of A, $X \times X \sim X$, $\mathcal{F} \neq \emptyset$. Note that $\mathcal{F}$ is partially ordered by set inclusion and extension of functions and every chain in $\mathcal{F}$ has an upper bound. Therefore, by Zorn's lemma $\mathcal{F}$ has a maximal element (D, f), say. If $\operatorname{card}(D) = \operatorname{card}(A)$ we are done. Now, for the sake of argument suppose that $\operatorname{card}(D) < \operatorname{card}(A)$. Then $A \setminus D \neq \emptyset$ and there is an injection from $A \setminus D$ into D or an injection from D into $A \setminus D$. In the former case there is an injection of $A = (A \setminus D) \cup D$ into $\{0, 1\} \times D$ and, since $D \times D \sim D$, also an injection of A into D and, therefore, by the Cantor-Bernstein-Schröder theorem $A \sim D$, which is not the case.

In the latter case D is equivalent to a subset Y of $A \setminus D$ and put $Z = D \cup Y$; then

$$Z \times Z = (D \times D) \cup (D \times Y) \cup (Y \times D) \cup (Y \times Y).$$

Now, since $D \sim Y$, $(D \times Y) \cup (Y \times D) \cup (Y \times Y) \sim \{0, 1, 2\} \times D$. Also, since $D \times D \sim D$ there is an injection from $\{0, 1, 2\} \times D$ into $D \times D$ and, therefore, an injection from $(D \times Y) \cup (Y \times D) \cup (Y \times Y)$ into $(Y \times Y)$. Hence, by the Cantor-Bernstein-Schröder theorem there is a bijection $f_1 : (D \times Y) \cup (Y \times D) \cup (Y \times Y) \to Y$. Define now $g : Z \to Z \times Z$ by $g|_D = f$ and $g|Y = f_1$. Then $(Z, g) \in \mathcal{F}$ and (D, f) precedes it, but this is not possible since (D, f) is a maximal element of $\mathcal{F}$. In other words, $\mathrm{card}(D) = \mathrm{card}(A)$ and we are done.

The reader will have no difficulty in verifying that given cardinals a, b, $a \cdot b = b \cdot a$, $a \cdot (b \cdot d) = (a \cdot b) \cdot d$ and $a \cdot (b + d) = a \cdot b + a \cdot d$. Also, if $a, b \geq \aleph_0$ and $a \leq b$, then $a \cdot b = b$. The proof follows along similar lines to the ones discussed above; we will not deprive the reader the pleasure of carrying them out.

65. If $\{A_d\}$ are such that $\mathrm{card}(A_d) = a_d$, all $d \leq b$, the sum $\sum_{d \leq b} a_d$ is defined as $\mathrm{card}(\bigcup_{d \leq b}(A_d \times \{d\}))$; manipulating bijections it readily follows that $\sum_{d \leq b} a_d = \mathrm{card}(\bigcup_{d \leq b} B_d)$ where $\{B_d\}_{d \leq b}$ are pairwise disjoint sets with $\mathrm{card}(B_d) = a_d$. It is then clear that $\mathrm{card}(\bigcup_{d \leq b} B_d) \leq b \cdot b = b$.

In particular, with $b = \aleph_0$, if $\{A_n\}$ is any sequence of countable sets, $\mathrm{card}(\bigcup_n A_n) \leq \aleph_0 \cdot \aleph_0 = \aleph_0$. Alternatively, since there exist 1-1 and onto mappings $f_n : A_n \to \mathbb{N}$ for every n, one can define $\phi : \bigcup_n A_n \to \mathbb{N}$ as follows: If $x \in \bigcup_n A_n$, let $x \in A_N$, say, and put $\phi(x) = 2^N 3^{f_N(x)}$; clearly ϕ is 1-1 and onto a subset on $\mathbb{N}$ and so $\bigcup_n A_n$ is at most countable.

66. (a) The statement is false. Let $\mathcal{B}$ be a collection of pairwise disjoint balls in $\mathbb{R}^n$. Since $\mathbb{Q}^n$ is dense in $\mathbb{R}^n$, every ball in $\mathcal{B}$ contains a point in $\mathbb{Q}^n$, and different balls contain different points. Hence, since $\mathbb{Q}^n$ is countable, so is $\mathcal{B}$.

(b) The statement is true. If $S_r = \{x \in \mathbb{R}^n : |x| = r\}$, then $\mathcal{S} = \{S_r : r > 0\}$ is a collection of pairwise disjoint spheres in $\mathbb{R}^n$.

(c) The statement is false. Recall that a figure eight in the plane is a set of the form $D_1 \cup D_2$ where D_1 and D_2 are circles whose bounded disks intersect at exactly one point. Since the plane can be written as a countable union of an increasing sequence of bounded disks it suffices to prove that a bounded disk contains at most countably many pairwise disjoint eights. Then let D be a bounded disk, $\mathcal{C} = \{E_\alpha : E_\alpha \text{ is a figure eight totally contained in } D\}$, and, with $d(E_\alpha)$ denoting the diameter of E_α, $\eta = \{\sup_\alpha d(E_\alpha) : E_\alpha \in \mathcal{C}\}$; since the E_α are totally contained in C, $\eta < \infty$. Let $\mathcal{C}' = \{E_\alpha \in \mathcal{C} : d(E_\alpha) > \eta/2\}$; by the definition of η, $\mathcal{C}' \neq \emptyset$. Observe that no two distinct elements of $\mathcal{C}'$ can be contained in one another (for otherwise the diameter of the larger would exceed η) and that by area considerations $\mathcal{C}'$ contains finitely many E_α. Now let $\mathcal{C}_1 = \mathcal{C} \setminus \mathcal{C}'$; then $\eta_1 = \sup_\alpha\{d(E_\alpha) : E_\alpha \in \mathcal{C}_1\} \leq \eta/2$ and repeating the argument for $\mathcal{C}_1, \eta_1$ in place of $\mathcal{C}, \eta$ we can extract a second pairwise disjoint finite family of E_α with $d(E_\alpha) > \eta_1/2$. The procedure is now clear: having picked $C_1,\ldots,C_{n-1}$ and $\eta_k = \{\sup_\alpha d(C_\alpha) : C_\alpha \in \mathcal{C}_k\}$ with $\eta_1 \geq 2\eta_2 \geq \ldots \geq 2^{n-1}\eta_{n-1}$ and $\eta_k \leq \eta/2^k$ for $1 \leq k \leq n$, pick $\mathcal{C}_n \subset \mathcal{C} \setminus \bigcup_{k=1}^{n-1} \mathcal{C}_k$ and $\eta_n = \sup_\alpha \{d(E_\alpha) : E_\alpha \in \mathcal{C}_n\} \leq \eta_{n-1}/2 \leq \eta/2^n$. Observe that $\mathcal{C}$ can be written as the countable union of the $\mathcal{C}_n$ because, since $\eta_n \to 0$ as $n \to \infty$, if $E_\alpha \in \mathcal{C}$, then $E_\alpha \in \mathcal{C}_n$ the first time that $d(E_\alpha) > \eta_n/2$. Thus $\mathcal{C}$ is the countable union of finite sets and is therefore countable.

67. We consider the rational sequences. Suppose $\{a_n\}$ is such that $a_n = a_N$ for all $n \geq N$. Then there are $\aleph_0^{N-1}$ possible choices for the first $N-1$ terms and $\aleph_0$ choices for a_N, giving a total of $\aleph_0^N = \aleph_0$ possible such sequences. So, since $N \in \mathbb{N}$, there are $\aleph_0^2 = \aleph_0$ such possible sequences.

68. Since continuous functions on I are determined by their values on $\mathbb{Q} \cap [0,1]$, there are $c^{\aleph_0} = 2^{\aleph_0 \cdot \aleph_0} = c$ of them. Now, for the limits of the sequences of continuous functions, there are $c^{\aleph_0} = c$ of them.

69. Let Ω denote the first uncountable ordinal. We first define classes $\mathcal{B}_\alpha$ of cardinality c for each ordinal $\alpha < \Omega$ and then prove that $\mathcal{B}(\mathbb{R}) = \bigcup_{\alpha < \Omega} \mathcal{B}_\alpha$; the cardinality assertion follows readily from this. Let $\mathcal{B}_0$ denote the collection of open intervals in $\mathbb{R}$. In order to define $\mathcal{B}_\lambda$ we need to consider two cases, namely, λ is a countable limit ordinal or $\lambda = \alpha + 1$ is a successor ordinal. In the former case let $\mathcal{B}_\lambda = \bigcup_{\alpha < \lambda} \mathcal{B}_\alpha$ and in the latter let $\mathcal{B}_{\alpha+1}$ be the collection of all subsets of $\mathbb{R}$ that are a countable union of elements in $\mathcal{B}_\alpha$, or a countable intersection of elements in $\mathcal{B}_\alpha$, or differences of two elements in $\mathcal{B}_\alpha$. Clearly the families are increasing in the sense that $\mathcal{B}_\alpha \subset \mathcal{B}_\beta$ for $\alpha < \beta$.

Let $\mathcal{U} = \bigcup_{\alpha < \Omega} \mathcal{B}_\alpha$; we claim that $\mathcal{U} = \mathcal{B}(\mathbb{R})$. The inclusion $\mathcal{U} \subset \mathcal{B}(\mathbb{R})$ is immediate since $\mathcal{B}(\mathbb{R})$ is a σ-algebra and $\mathcal{B}_0 \subset \mathcal{B}(\mathbb{R})$. As for the opposite inclusion, since $\mathcal{B}_0 \subset \mathcal{U}$ it suffices to prove that $\mathcal{U}$ is a σ-algebra. Suppose that $U, V \in \mathcal{U}$. Then for some $\alpha, \beta < \Omega$, $U \in \mathcal{B}_\alpha$ and $V \in \mathcal{B}_\beta$ and so, if $\lambda = \max(\alpha, \beta)$, $U \setminus V \subset \mathcal{B}_{\lambda+1} \subset \mathcal{U}$. Next, suppose that $\{X_n\} \subset \mathcal{U}$ and for each n let $\alpha_n < \Omega$ be such that $X_n \in \mathcal{B}_{\alpha_n}$; note that α_n is countable for each n and that there exists $\gamma < \Omega$ such that $\alpha_n \leq \gamma$ for all n. Then $X_n \in \mathcal{B}_\gamma$ for all n and $\bigcup_n X_n, \bigcap_n X_n \in \mathcal{B}_{\gamma+1} \subset \mathcal{U}$.

Finally, we prove by induction that the cardinality of $\mathcal{B}_\lambda = c$ for all λ. The statement is true for $\mathcal{B}_0$. Also, if the assertion is true for $\mathcal{B}_\alpha$, since countable unions and intersections have cardinality $c^{\aleph_0} = c$, and so do the differences, the statement is also true for $\mathcal{B}_{\alpha+1}$. As for the limiting ordinals, their cardinality is $\aleph_0 \cdot c = c$ and we have finished.

70. Since the complement of a compact set is open there is an injective mapping from $\mathcal{C}$ into $\mathcal{O}$, the open sets in $\mathbb{R}$, and so $\text{card}(\mathcal{C}) \leq \text{card}(\mathcal{O})$. Now, since every open set in $\mathbb{R}$ can be written as at most a countable union of pairwise disjoint open intervals and each such interval contains a distinct rational number, $\operatorname{card}(\mathcal{O}) \leq \operatorname{card}(\mathbb{Q}^{\mathbb{N}}) = \aleph_0^{\aleph_0} \leq c^{\aleph_0} = 2^{\aleph_0 \cdot \aleph_0} = 2^{\aleph_0} = c$. Thus $\operatorname{card}(\mathcal{C}) \leq c$. On the other hand, for each $x > 0$, $[0, x]$ is compact and so $\operatorname{card}(\mathcal{C}) \geq c$. Also, since each $[0, x]$, $x > 0$, is uncountable and has positive Lebesgue measure, the collection of uncountable compact subsets of $\mathbb{R}$ with positive Lebesgue measure has cardinality c.

71. By Problem 70 the class $\mathcal{C}$ of uncountable compact subsets of the line can be indexed by the ordinals less than Ω, the first uncountable ordinal, i.e., $\mathcal{C} = \{K_\alpha : \alpha < \Omega\}$. We may also assume that $\mathbb{R}$ and, consequently, also every K_α, has been well-ordered. Let x_1, y_1 be the first two elements of K_1. Next, if $1 < \alpha < \Omega$ and if x_β and y_β have been chosen for all $\beta < \alpha$, let x_α, y_α be the first two elements of $K_\alpha \setminus \bigcup_{\beta < \alpha} \{x_\beta, y_\beta\}$; such a choice is always possible since the set in question has cardinality c for each α. Now put $B = \{x_\alpha : \alpha < \Omega\}$. Since $y_\alpha \in B^c$, all $\alpha < \Omega$, the intersection property is readily verified.

72. Clearly $a = \operatorname{card}(H) \leq c$. Moreover, since the set of linear combinations of the elements of a finite set with coefficients in $\mathbb{Q}$ is countable but $\mathbb{R}$ is not, a is infinite. For each integer n let $A_n = \{x \in \mathbb{R} : x = \sum_{k=1}^n q_k h_k, q_k$ rational, $h_k \in H\}$; we claim that $\operatorname{card}(A_n) = a$ for all n. Note that listing the rationals $\{r_1, r_2, \ldots\}$ we have $A_1 = \bigcup_n r_n H$, the union being disjoint, and since $\operatorname{card}(r_n H) = \operatorname{card}(H)$ for every $r_n \neq 0$, by Problem 64 it follows that $\operatorname{card}(A_1) = \aleph_0 \cdot \operatorname{card}(H) = \aleph_0 \cdot a = a$. We proceed now by induction; we just saw that the statement is true for $n = 1$. Now, since $A_n \subset A_{n+1} \subset A_n + A_1$, it readily follows that $\operatorname{card}(A_n) \leq \operatorname{card}(A_{n+1}) \leq \operatorname{card}(A_n + A_1) \leq \operatorname{card}(A_n \times A_1) = \operatorname{card}(A_n)$ and if $\operatorname{card}(A_n) = a$, then $\operatorname{card}(A_{n+1}) = a$. Finally, by the definition of Hamel basis, $\mathbb{R} = \bigcup_n A_n$ and $c \leq \operatorname{card}(\bigcup_n A_n) \leq \aleph_0 \cdot a = a$. Therefore $a = c$.

73. By Zorn's lemma $\mathbb{R}$ has a Hamel basis H, say. Let $h_0 \in H$, for $\mathcal{I} \subset H \setminus \{h_0\}$ let $B_\mathcal{I} = \{h_0 + h : h \in \mathcal{I}\}$, and put $H_\mathcal{I} = B_\mathcal{I} \cup (H \setminus \mathcal{I})$. Then $H_\mathcal{I}$ is a Hamel basis of $\mathbb{R}$ and for any two different subsets $\mathcal{I}, \mathcal{I}'$ of $H \setminus \{h_0\}$, $H_\mathcal{I} \neq H_{\mathcal{I}'}$. Since $\operatorname{card}(H) = c$ there are 2^c such subsets and, consequently, 2^c different Hamel bases of $\mathbb{R}$.

74. Since $C + C = [0, 2]$ every $x \in \mathbb{R}$ can be written as $x = n + x_1 + x_2$ where $n \in \mathbb{Z}$ and $x_1, x_2 \in C$. Thus C spans $\mathbb{R}$ as a vector space over $\mathbb{Q}$ and, consequently, by Zorn's lemma C contains a minimal spanning set of $\mathbb{R}$, i.e., a Hamel basis.

75. Let $H = \{h_\alpha\}$ be a Hamel basis of $\mathbb{R}$, h_{α_0} the rational element of H, and $B = \operatorname{sp}\{h_\alpha\}$, $\alpha \neq \alpha_0$. For the sake of argument suppose that for some $J = (a, b)$ and $r \in \mathbb{Q}$, $(a, b) \cap (r + B)$ is of first category and let $E = (a - r, b - r)$; $E \cap B$ is of first category and since $qB = B$ for each rational $q \neq 0$, $q(E \cap B) = (qE) \cap (qB) = (qE) \cap B$ is of first category. Then, with $\{q_n\}$ an enumeration of the rationals, $B = \bigcup_n q_n(E \cap B)$ is of first category and the same is true of the rational translations $A_n = (q_n + B)$, all n, of B. Therefore $\mathbb{R} = \bigcup_n A_n$ is of first category, which is not the case. Thus the sets A_n, all n, will do.

Chapter 12

Measures

Solutions

1. No, let $B = X \setminus A$.

2. Let $X = \{a, b, c\}$, $\mathcal{A}_1 = \{\emptyset, \{a\}, \{b, c\}, X\}$, $\mathcal{A}_2 = \{\emptyset, \{b\}, \{a, c\}, X\}$. Were $\mathcal{A}_1 \cup \mathcal{A}_2$ an algebra it would contain $\{a, b\} = \{a\} \cup \{b\}$, which it does not. On the other hand, if $\{\mathcal{A}_n\}$ is an increasing sequence of algebras, $\mathcal{A} = \bigcup_n \mathcal{A}_n$ is an algebra. First, if $A \in \mathcal{A}$, then $A \in \mathcal{A}_n$ for some n and $A^c \in \mathcal{A}_n \subset \mathcal{A}$. Next, if $A, B \in \mathcal{A}$, then $A \in \mathcal{A}_m$ for some m, $B \in \mathcal{A}_n$ for some n, and $A \cup B \in \mathcal{A}_N$ where $N = \max(m, n)$.

3. First, if $A \in \mathcal{F}$, clearly $A^c \in \mathcal{A}$. Now, if $A = \bigcup_{k=1}^n A_k \in \mathcal{A}$, since $A_k \in \mathcal{F}$, $A_k^c = \bigcup_{\ell=1}^{L_k} B_\ell^k$ with $B_\ell^k \in \mathcal{F}$ for all $1 \le \ell \le L_k$. Therefore $A^c = \bigcap_{k=1}^n A_k^c = \bigcap_{k=1}^n \bigcup_{\ell=1}^{L_k} B_\ell^k = \bigcup_{\ell=1}^{L_k} \bigcap_{k=1}^n B_\ell^k$ and since $\bigcap_{k=1}^n B_\ell^k \in \mathcal{F}$ for all ℓ, $A^c \in \mathcal{A}$. Finally, if $\{A_k\} \subset \mathcal{A}$ is finite, $A = \bigcup_{k=1}^n A_k = \bigcup_{k=1}^n \bigcup_{\ell=1}^{L_k} B_\ell^k$ is a finite union of sets in $\mathcal{F}$ and $A \in \mathcal{A}$. Thus $\mathcal{A}$ is an algebra of subsets of X that contains $\mathcal{F}$ and, therefore, contains $\mathcal{A}(\mathcal{F})$. Moreover, since any algebra of subsets of X that contains $\mathcal{F}$ also contains $\mathcal{A}$, $\mathcal{A}(\mathcal{F})$ contains $\mathcal{A}$ and the two are equal.

4. In fact, $\mathcal{A} = \mathcal{A}(\mathcal{F})$, the algebra generated by $\mathcal{F}$.

In $\mathbb{R}$ note that since $\mathbb{Q} = \bigcup_{x \in \mathbb{Q}} \bigcap_n (x - 1/n, x]$ can be written as the countable union of a countable intersection of sets in $\mathcal{F}$, $\mathbb{Q} \in \mathcal{M}(\mathcal{F})$. However, $\mathbb{Q}$ cannot be written as a finite union of intervals in $\mathcal{F}$ and, therefore, $\mathbb{Q} \notin \mathcal{A}(\mathcal{F})$.

6. (b) Let $\mathcal{M} = \{A \subset X : A$ is countable or $X \setminus A$ is countable$\}$ and $\mathcal{D} = \{\{x, y\} : x, y \in X\}$. First, given $x, y \in X$, $\{x, y\} = \{x\} \cup \{y\} \in \mathcal{M}$ and, consequently, $\mathcal{M}(\mathcal{D}) \subset \mathcal{M}$. Now, given $x \in X$, let $y \neq z \in X$ and observe that $\{x\} = \{x, y\} \cap \{x, z\} \in \mathcal{M}(\mathcal{D})$. Hence the singletons of X belong to $\mathcal{M}(\mathcal{D})$, $\mathcal{M} \subset \mathcal{M}(\mathcal{D})$, and the sets are equal.

Note that the result is not true if $X = \{x, y\}$. Then $\mathcal{M}(\mathcal{D}) = \{\emptyset, \{x, y\}\}$ and $\mathcal{M} = \mathcal{P}(X)$.

7. No. In the setting of Problem 5(b) let $X = \mathbb{R}$ and $A_n = [-n, n]$; we claim that $\bigcup_n \mathcal{S}_{A_n}$ is not a σ-algebra of subsets of X. To see this let $B_m = [0, m]$; then $B_m \subset A_m$ and so $B_m \in \mathcal{S}_{A_m} \subset \bigcup_n \mathcal{S}_{A_n}$. On the other hand, $[0, \infty) = \bigcup_m B_m$ belongs to any σ-algebra that contains $\bigcup_m \mathcal{S}_{A_m}$ but since neither $[0, \infty)$ nor $[0, \infty)^c = (-\infty, 0)$ is contained in A_m for any m, $[0, \infty) \notin \bigcup_m \mathcal{S}_{A_m}$.

Now, a simple condition ensures that the union of finitely many σ-algebras is a σ-algebra: Let $\mathcal{M}_1, \ldots, \mathcal{M}_n$ be σ-algebras of subsets of X such that $\mathcal{M} = \bigcup_{k=1}^{n} \mathcal{M}_k$ is an algebra; then $\mathcal{M}$ is a σ-algebra. First, $\mathcal{M} \neq \emptyset$ and $\mathcal{M}$ is closed under complementation. Next, let $\{A_k\} \subset \mathcal{M}$ and separate the sets into the classes $\mathcal{A}_1 = \{A_k : A_k \in \mathcal{M}_1\}$, $\mathcal{A}_m = \{A_k : A_k \in \mathcal{M}_m \setminus \bigcup_{1 \leq \ell < m} \mathcal{A}_\ell\}$, $1 < m \leq n$. Then $\bigcup_{A_k \in \mathcal{A}_m} A_k \in \mathcal{M}_m$, $1 \leq m \leq n$, and, consequently, $\bigcup_k A_k = \bigcup_{m=1}^{n} \bigcup_{A_k \in \mathcal{A}_m} A_k \in \mathcal{M}$.

8. $\mathcal{S}$ is a σ-algebra of subsets of X iff every point in X is open.

10. First, note that $\mathcal{R}$ is also closed under countable intersections; indeed, if $\{A_n\} \subset \mathcal{R}$, $\bigcap_n A_n = A_1 \setminus \bigcup_n (A_1 \setminus A_n) \in \mathcal{R}$. Next, observe that $\mathcal{R}'$ is closed under countable unions. Let $\{A_n'\} \subset \mathcal{R}'$, $A_n' = A_n^c$ with $A_n \in \mathcal{R}$ for all n. Then $\bigcup_n A_n' = (\bigcap_n A_n)^c$ which, since $\bigcap_n A_n \in \mathcal{R}$, is in $\mathcal{R}'$. Now, if $\{A_n\} \subset \mathcal{S}$ divides the sequence into two parts, one corresponding to those sets in $\mathcal{R}$, which we still call A_n, and the other consisting of those sets in $\mathcal{R}'$, which we rename A_n', then the union can be expressed as $\bigcup_{A_n \in \mathcal{R}} A_n \cup \bigcup_{A_n' \in \mathcal{R}'} A_n'$ where the first union is in $\mathcal{R}$ and the second, as we just proved, is in $\mathcal{R}'$; therefore the union is in $\mathcal{R} \cup \mathcal{R}' = \mathcal{S}$. Finally, by the symmetry in the definition of $\mathcal{S}$, $\mathcal{S}$ is closed under complementation.

12. Let $\{A_n\}$ be a countable family of nonempty pairwise disjoint subsets of X. For each $\mathcal{I} \subset \mathbb{N}$ let $B_{\mathcal{I}} = \bigcup_{n \in \mathcal{I}} A_n$; the $B_{\mathcal{I}}$ are all distinct and there are $2^{\aleph_0} = c$ choices for $\mathcal{I}$.

13. Suppose that $B_x \neq B_y$. Then $B_x \not\subset B_y$ or $B_y \not\subset B_x$; to fix ideas suppose the latter. Then there is $A \in \mathcal{M}$ such that $x \in A$ but $y \notin A$ and, consequently, $y \in X \setminus A \in \mathcal{M}$. Thus, since $x \in A$, $B_x \subset A$ and $B_y \subset X \setminus A$, and so $B_x \cap B_y = \emptyset$ as we wanted to show.

Note that in general $x \neq y$ does not imply $B_x \cap B_y = \emptyset$; consider, for instance, $X = \{0, 1\}$ and $\mathcal{M} = \{\emptyset, X\}$. Also, the B_x need not be measurable. Let X be uncountable and $X_0 \subset X$ such that both X_0 and $X \setminus X_0$ are uncountable. Now, if $\mathcal{C} = \{A \subset X : A \text{ is countable and } A \cap X_0 = \emptyset\}$ and $\mathcal{C}' = \{A \subset X : A^c \in \mathcal{C}\}$, $\mathcal{M} = \mathcal{C} \cup \mathcal{C}'$ is a σ-algebra of subsets of X and for $x \in X_0$, $B_x = X_0$, which is not measurable.

Now, it is always the case that every measurable A is the disjoint union of the B_x with $x \in A$. However, such unions might not be countable and, even if every B_x is measurable, an arbitrary union of B_x is not necessarily measurable. For example, let X be uncountable and $\mathcal{M}$ consist of all countable sets and their complements. Then $B_x = \{x\}$ for all $x \in X$ and every subset $Y = \bigcup_{y \in Y} B_y$ of X is the union of a suitable subcollection of the B_x.

14. No. We claim that $\mathcal{M}$ contains n pairwise disjoint sets for all n and so there are countably many pairwise disjoint sets in $\mathcal{M}$. If $A \in \mathcal{M}$, A, A^c are two disjoint sets in $\mathcal{M}$. Now, if $A_1, \ldots, A_n$ are n pairwise disjoint sets in $\mathcal{M}$, let $A = \bigcup_{k=1}^{n} A_k$; if $X \setminus A \neq \emptyset$, then there are $n+1$ pairwise disjoint sets in $\mathcal{M}$. On the other hand, if $A = X$, let $\mathcal{B} = \{B \in \mathcal{M} : B = \bigcup_{k \in F} A_k$ where F ranges over all nonempty subsets of $\{1, \ldots, n\}\}$; since $\mathcal{B}$ is finite and $\mathcal{M}$ infinite there is a nonempty $C \in \mathcal{M} \setminus \mathcal{B}$. Now, we claim that $A_k \cap C \neq \emptyset$ and $A_k \setminus C \neq \emptyset$ for some k, $1 \leq k \leq n$. For the sake of argument suppose this is not the case and let $\mathcal{I}_1 = \{1 \leq k \leq n : A_k \cap C \neq \emptyset, A_k \setminus C = \emptyset\}$ and $\mathcal{I}_2 = \{1 \leq k \leq n : A_k \cap C = \emptyset, A_k \setminus C \neq \emptyset\}$. It then readily follows that $C \supset \bigcup_{k \in \mathcal{I}_1} A_k$ and that $C^c \supset \bigcup_{k \in \mathcal{I}_2} A_k$, which by complementation gives $C \subset (\bigcup_{k \in \mathcal{I}_2} A_k)^c = \bigcup_{k \in \mathcal{I}_1} A_k$, and, consequently, $C = \bigcup_{k \in \mathcal{I}_1} A_k \in \mathcal{B}$, which is not the case. Therefore for some index k, which we may assume to be n, $A_n \cap C \neq \emptyset$ and $A_n \setminus C \neq \emptyset$. Therefore $A_1, \ldots, A_{n-1}, A_n \cap C, A_n \setminus C$, are $n + 1$ pairwise disjoint sets in $\mathcal{M}$. Hence $\mathcal{M}$ contains countably many pairwise disjoint sets and by Problem 12 is uncountable.

Because of its simplicity we discuss a second approach. For the sake of argument suppose that $\mathcal{M}$ is a countably infinite σ-algebra of subsets of X and for $x \in X$ let B_x be as in Problem 13; since $\mathcal{M}$ is countable B_x is a countable intersection of measurable sets, hence measurable. From Problem 13 we know that if $B_x \neq B_y$, $B_x \cap B_y = \emptyset$. We claim that if $\{B_x : x \in X\}$ is finite, $\mathcal{M}$ is finite. Indeed, note that the definition of B_x implies that if $x \in X$ and $A \in \mathcal{M}$, then $B_x \subset A$ or $B_x \subset X \setminus A$. It thus follows that every element of $\mathcal{M}$ is the union (in fact, pairwise disjoint) of some subcollection of $\{B_x : x \in X\}$. So, $\mathrm{card}(\{B_x : x \in X\}) = n < \infty$, which implies that $\mathrm{card}(\mathcal{M}) \leq 2^n$ and this is not the case. Hence $\{B_x : x \in X\}$ must be infinite and since the B_x are pairwise disjoint, $\mathcal{M}$ is uncountable.

17. Let $\mathcal{S} = \bigcup_{\mathcal{F}} \mathcal{M}(\mathcal{F})$ where $\mathcal{F}$ ranges over all countable families of $\mathcal{E}$; clearly $\mathcal{S} \subset \mathcal{M}(\mathcal{E})$. Conversely, first note that since $\mathcal{E} \subset \bigcup_{\mathcal{F}} \mathcal{F} \subset \mathcal{S}$, $\mathcal{M}(\mathcal{E}) \subset \mathcal{M}(\mathcal{S})$ and, therefore, it suffices to prove that $\mathcal{S}$ is a σ-algebra of subsets of X. Since each $\mathcal{M}(\mathcal{F})$ is closed under complementation, so is $\mathcal{S}$. Let $\{A_k\} \subset \mathcal{S}$; then for each k there is a countable $\mathcal{F}_k \subset \mathcal{E}$ such that $A_k \in \mathcal{M}(\mathcal{F}_k)$. Now, $\mathcal{F} = \bigcup_k \mathcal{F}_k$ is a countable collection of subsets of X and each $A_k \in \mathcal{M}(\mathcal{F})$. Therefore $\bigcup_k A_k \in \mathcal{M}(\mathcal{F}) \subset \mathcal{S}$ and $\mathcal{S}$ is closed under countable unions.

18. (a) Since a σ-algebra of subsets of X that contains $\mathcal{S}$ and V also contains $\mathcal{F}$ it suffices to prove that $\mathcal{F}$ is a σ-algebra of subsets of X; the only condition that requires some thought is that $\mathcal{F}$ is closed under complementation. Note that if $M = (A \cap V) \cup (B \cap V^c)$, $M^c = (A^c \cup V^c) \cap (B^c \cup V) = (A^c \cap B^c) \cup (A^c \cap V) \cup (B^c \cap V^c)$. Moreover, $(A^c \cap B^c) = (A^c \cap B^c) \cap (V \cup V^c) \subset (A^c \cap V) \cup (B^c \cap V^c)$ and so $M^c = (A^c \cap V) \cup (B^c \cap V^c) \in \mathcal{F}$.

(b) Let $\{M_n\}$ be pairwise disjoint sets in $\mathcal{M}(\mathcal{S}, V)$ where $M_n = (A'_n \cap V) \cup (B'_n \cap V^c)$; then $M_n \cap M_k = (A'_n \cap A'_k \cap V) \cup (B'_n \cap B'_k \cap V^c) = \emptyset$ and so $A'_n \cap A'_k \cap V = B'_n \cap B'_k \cap V^c = \emptyset$ for all $n \neq k$.

Now, since $A'_n = (A'_n \setminus (\bigcup_{k \neq n} A'_k)) \cup (A'_n \cap (\bigcup_{k \neq n} A'_k))$ it readily follows that $A'_n \cap V = (A'_n \setminus (\bigcup_{k \neq n} A'_k) \cap V) \cup (A'_n \cap (\bigcup_{k \neq n} A'_k) \cap V)$ where as noted above the second set in the union is empty. Therefore, if $A_n = A'_n \setminus (\bigcup_{k \neq n} A'_k)$, $A_n \cap V = A'_n \cap V$ and, similarly, if $B_n = B'_n \setminus (\bigcup_{k \neq n} B'_k)$, $B_n \cap V = B'_n \cap V$. Finally, $\{A_n\}$ and $\{B_n\}$ consist of pairwise disjoint sets and $M_n = (A_n \cap V) \cup (B_n \cap V^c)$ for all n.

19. We claim that $\lim_n B_n$ exists iff $\lim_n A_n = \emptyset$. Necessity first. Let $B = \lim_n B_n$, $X = B \cup B^c$. First, if $x \in B$, $x \in B_n$ for all $n \geq n_x$ and since $B_{n+1} = B_n \bigtriangleup A_{n+1}$, $x \notin A_{n+1}$ for all $n \geq n_x$ and $x \in \liminf_n A_n^c$. Next, if $x \in B^c$, x belongs to at most finitely many B_n and $x \in B_n^c$ for all $n \geq n_x$. Now, since $B_{n+1} = B_n \bigtriangleup A_{n+1}$, in particular $x \notin A_{n+1} \cap B_n^c$ for all $n \geq n_x$, which, since $x \in B_n^c$, implies that $x \notin A_{n+1}$ for all such n, and so $x \in \liminf_n A_n^c$. Hence $\liminf_n A_n^c = X$ and $\limsup_n A_n = \emptyset$. But then also $\liminf_n A_n = \emptyset$ and $\lim_n A_n = \emptyset$.

Conversely, if $\lim_n A_n = \emptyset$, given $x \in X$, let n_x be such that $x \in A_n^c$ for $n \geq n_x$. Now, there are two possibilities: $x \in B_{n_x - 1}$ or $x \notin B_{n_x - 1}$. In the former case $x \in B_{n_x - 1} \cap A_{n_x}^c$ and, therefore, $x \in B_{n_x} = B_{n_x - 1} \bigtriangleup A_{n_x}$; similarly, $x \in B_n$ for all $n > n_x$ and so $B_{n_x} \subset B_n$ for all $n \geq n_x$. As for the latter case, $x \in B_{n_x - 1}^c \cap A_{n_x}^c$ and, hence, $x \in B_{n_x}^c$; similarly, $x \in B_n^c$ for all $n \geq n_x$ and so $B_{n_x}^c \subset B_n^c$ for all $n > n_x$. Thus $B_n = B_{n_x}$ for all $n \geq n_x$ and $\lim_n B_n$ exists.

21. First, since $\mathcal{N} \neq \emptyset$, $\mathcal{M}_T \neq \emptyset$. Let $A \in \mathcal{M}_T$ and $B \in \mathcal{N}$ such that $A = T^{-1}(B)$. Then $B^c \in \mathcal{N}$ and since $T^{-1}(B^c) = (T^{-1}(B))^c = A^c$, $A^c \in \mathcal{M}_T$. Similarly, let $\{A_n\} \subset \mathcal{M}_T$ and $\{B_n\} \subset \mathcal{N}$ such that $A_n = T^{-1}(B_n)$ for all n. Then $\bigcup_n B_n \in \mathcal{N}$ and $\bigcup_n A_n = \bigcup_n T^{-1}(B_n) = T^{-1}(\bigcup_n B_n) \in \mathcal{M}_T$.

22. We claim that $\mathcal{M}(T^{-1}(\mathcal{C})) = T^{-1}(\mathcal{M}(\mathcal{C}))$. First, Problem 21 gives that $T^{-1}(\mathcal{M}(\mathcal{C}))$ is a σ-algebra of subsets of X. Now, since $\mathcal{C} \subset \mathcal{M}(\mathcal{C})$, $T^{-1}(\mathcal{C}) \subset T^{-1}(\mathcal{M}(\mathcal{C}))$ and, consequently, $\mathcal{M}(T^{-1}(\mathcal{C})) \subset T^{-1}(\mathcal{M}(\mathcal{C}))$. Furthermore, $T^{-1}(\mathcal{C}) \subset \mathcal{M}(T^{-1}(\mathcal{C}))$ and so $T^{-1}(\mathcal{M}(\mathcal{C})) \subset \mathcal{M}(T^{-1}(\mathcal{C}))$ and we have finished.

23. Necessity first. Let $x \in \mathbb{R}$ and for the sake of argument suppose that $f(x) = f(x_1)$ for some $x_1 \neq x$. Then $\{x\} \subsetneq f^{-1}(\{f(x)\})$ and $\{x\} \notin \mathcal{M}_f$, which is not the case. Hence f is injective.

Sufficiency next. Since f is injective, $\{x\} = f^{-1}(\{f(x)\})$ and since the singleton $\{f(x)\}$ is a Borel set in $\mathbb{R}$, $\{x\} \in \mathcal{M}_f$.

24. If f has an interval of constancy the inverse image omits an interval and, consequently, $\mathcal{M}_f \neq \mathcal{B}(\mathbb{R})$. Thus $\mathcal{M}_f = \mathcal{B}(\mathbb{R})$ iff f is injective or, in this case, strictly increasing.

25. f^{-1} is defined on $[0,\infty)$ and is 2-valued, i.e., if $y = f(x)$, then $y = \pm\sqrt{x}$. Given $B \in \mathcal{B}(\mathbb{R})$, let $B_+ = B \cap [0,\infty)$. Then $f^{-1}(B) = f^{-1}(B_+) = -\sqrt{B_+} \cup \sqrt{B_+}$ and $\mathcal{M}_f = \{-\sqrt{B_+} \cup \sqrt{B_+} : B \in \mathcal{B}(\mathbb{R})\}$.

Now, since $g^{-1}((\lambda,\infty)) = (\lambda,\infty)$ is not symmetric about the origin, g is not measurable.

26. First, $\{(x,y) \in \mathbb{R}^2 : x - y = \lambda\}$ is a line with slope 1 shifted by $-\lambda$ in the (x,y) plane. Thus, if $B \in \mathcal{B}(\mathbb{R})$, $F^{-1}(B) = \{(x,y) \in \mathbb{R}^2 : x - y \in B\}$ consists of lines with slope 1 and every possible shift from the set $-B$. In other words, $\mathcal{M}_F$ consists of all 45-degree diagonal stripes in $\mathbb{R}^2$ with base $-B \in \mathcal{B}(\mathbb{R})$.

28. Let $E_1 = A_1$ and for $n > 1$ put $E_n = A_n \setminus \bigcup_{k<n} A_k$; note that $E_n \subset A_n$ and $\bigcup_n E_n = \bigcup_n A_n$. Then, by monotonicity, $\mu(E_n) \leq \mu(A_n)$ and since the E_n are pairwise disjoint, $\mu(\bigcup_n A_n) = \mu(\bigcup_n E_n) = \sum_n \mu(E_n) \leq \sum_n \mu(A_n)$.

30. (a) First, since $X \cap A = A$, $X^c \cap A = \emptyset$, and $\psi(\emptyset) = 0$, we have $\psi(A) = \psi(X \cap A) + \psi(X^c \cap A)$ for all $A \in \mathcal{A}$ and so $X \in \mathcal{C}$. Next, $\mathcal{C}$ is closed under complementation by definition. Finally, to verify that $\mathcal{C}$ is closed under finite unions and, hence, an algebra, it suffices to prove that $\mathcal{C}$ is closed under finite intersections. So, let $C_1, C_2 \in \mathcal{C}$, $C = C_1 \cap C_2$, and $A \in \mathcal{A}$. Since $C_1 \in \mathcal{C}$, $C_2 \cap A \in \mathcal{A}$, and $C_1 \cap (C_2 \cap A) = C \cap A$ we have $\psi(C \cap A) = \psi(C_2 \cap A) - \psi(C_1^c \cap C_2 \cap A)$. Also, since $C_2 \cap C^c = C_2 \cap C_1^c$ and $C_2^c \cap C^c = C_2^c$, and since $\mathcal{A}$ is an algebra, $C^c \cap A \in \mathcal{A}$. Therefore for $C_2 \in \mathcal{C}$, $\psi(C^c \cap A) = \psi(C_2 \cap C_1^c \cap A) + \psi(C_2^c \cap A)$ and, consequently, since $C_2 \in \mathcal{C}$, $\psi(C \cap A) + \psi(C^c \cap A) = \psi(C_2 \cap A) + \psi(C_2^c \cap A) = \psi(A)$. Hence $C \in \mathcal{C}$.

(b) Let $C_1, C_2 \in \mathcal{C}$ be disjoint. Then

$$\psi(C_1 \cap (C_1 \cup C_2)) + \psi(C_1^c \cap (C^1 \cup C_2)) = \psi(C_1 \cup C_2)$$

and since $C_1 \cap (C_1 \cup C_2) = C_1$ and $C_1^c \cap (C_1 \cup C_2) = C_2$ we have $\psi(C_1 \cup C_2) = \psi(C_1) + \psi(C_2)$. Thus ψ is finitely additive on $\mathcal{C}$.

(c) By induction it suffices to prove that if $C_1, C_2 \in \mathcal{C}$ are disjoint, then $\psi((C_1 \cap A) \cup (C_2 \cap A)) = \psi(C_1 \cap A) + \psi(C_2 \cap A)$. First, since $C_1 \in \mathcal{C}$, $\psi(C_1 \cap B) + \psi(C_1^c \cap B) = \psi(B)$. Now, note that if $B = (C_1 \cap A) \cup (C_2 \cap A)$, since $C_1 \cap C_2 = \emptyset$ we have $C_1 \cap B = C_1 \cap A$ and $C_1^c \cap B = C_2 \cap A$. Whence $\psi(C_1 \cap A) + \psi(C_2 \cap A) = \psi(B)$ as required.

31. The condition is necessary by continuity from above. Conversely, given a pairwise disjoint sequence $\{B_k\} \subset \mathcal{M}$, let $A_n = \bigcup_{k=n+1}^{\infty} B_k$; then $\{A_n\}$ is a decreasing sequence with $\bigcap_n A_n = \emptyset$ and so $\lim_n \psi(A_n) = 0$. Then $\psi(\bigcup_k B_k) = \psi(\bigcup_{k=1}^{n} B_k) + \psi(\bigcup_{k=n+1}^{\infty} B_k) = \sum_{k=1}^{n} \psi(B_k) + \psi(A_n)$ for all n and letting $n \to \infty$ it follows that $\psi(\bigcup_k B_k) = \lim_n \sum_{k=1}^{n} \psi(B_k) + 0 = \sum_k \psi(B_k)$.

For instance, if $X = \mathbb{N}$, the set function $\psi(A)$, which $= 0$ if A is finite and $= \infty$ if A is infinite, is additive in $\mathcal{M} = \mathcal{P}(\mathbb{N})$ but since the sequence $A_n = \{1, \dots, n\}$ has limit $\mathbb{N}$ and $\lim_n \psi(A_n) = 0 \neq \infty = \psi(\mathbb{N})$, ψ is not σ-additive.

An equivalent formulation of this result is the following: If $\{B_n\}$ is an increasing sequence of sets in $\mathcal{M}$ with $\bigcup_n B_n = X$, then $\lim_n \psi(B_n) = \psi(X)$. To see this let $A_n = X \setminus B_n$ for all n, and note that $\{A_n\}$ is a decreasing sequence of sets in $\mathcal{M}$ with $\bigcap_n A_n = \emptyset$ and so, by the first part of the argument, $\lim_n \psi(A_n) = 0$.

32. Since μ is a measure it follows that for any decreasing $\{A_n\} \subset \mathcal{M}$ with $\bigcap_n A_n = \emptyset$, $\lim_n \mu(A_n) = 0$. Therefore by assumption $\lim_n \psi(A_n) = 0$ and by Problem 31, ψ is a measure.

33. Necessity is clear by Problem 28. As for sufficiency, given pairwise disjoint $\{A_n\} \subset \mathcal{M}$, by σ-subadditivity $\psi(\bigcup_n A_n) \leq \sum_n \psi(A_n)$. On the other hand, by monotonicity and additivity, $\psi(\bigcup_n A_n) \geq \psi(\bigcup_{n=1}^{k} A_n) = \sum_{n=1}^{k} \psi(A_n)$ for all k, and letting $k \to \infty$, $\psi(\bigcup_n A_n) \geq \sum_n \psi(A_n)$.

34. Since the family of subsets of X which are of first category is closed under countable unions and relative complementation and since the complements of sets of first category are of second category, by Problem 10, $\mathcal{S}$ is a σ-algebra of subsets of X. Now, by the Baire category theorem no subset of X is both of first and second category and μ is well-defined.

Clearly $\mu(\emptyset) = 0$. Let $\{A_n\} \subset \mathcal{S}$ be pairwise disjoint. Then, if each A_n is of first category, $\bigcup_n A_n$ is of first category and $\mu(\bigcup_n A_n) = 0 = \sum_n \mu(A_n)$. On the other hand, if some A_n is of second category, and since no two disjoint subsets of X are of second category there can be only one, we have $\mu(\bigcup_n A_n) = 1 = \sum_n \mu(A_n)$.

35. First, note that if F is closed and B denotes its interior, $F \setminus B$ is nowhere dense. Indeed, since $F \setminus B$ is closed it suffices to verify that its interior is empty and this is clear since any neighborhood of $x \in F \setminus B$, since $x \notin B$, contains a point, and hence a neighborhood, in F^c.

Let $A = G \triangle P \in \mathcal{C}$ with G open and P of first category; then $A^c = G^c \triangle P$ where $G^c = F$ is closed. Now, with B the interior of F, since $B \subset F$, $F = B \triangle (F \setminus B)$ and, consequently, $A^c = G^c \triangle P = B \triangle ((F \setminus B) \triangle P)$ where B is open and $(F \setminus B) \triangle P$ is of first category; thus A^c has the Baire property and $\mathcal{C}$ is closed under complementation. Next, let $A_n = G_n \cup P_n \in \mathcal{C}$ with G_n open and P_n of first category for all n, and $A = \bigcup_n A_n$; note that $G = \bigcup_n G_n$ is open and $P = \bigcup_n P_n$ is of first category. Moreover, $G \setminus P \subset A \subset G \cup P$ and so $G \triangle A \subset P$ is of first category; thus $A = G \triangle (G \triangle A)$ has the Baire property and $\mathcal{C}$ is closed under countable unions. Therefore $\mathcal{C}$ is the σ-algebra generated by the open sets and sets of first category.

36. First, note that for $A \in \mathcal{S}$, (x, y) and (y, x) belong or do not belong to A simultaneously and so $A^c \in \mathcal{S}$; the other properties of σ-algebra are easily verified.

Consider now the half-planes below and above the $\pi/4$-diagonal defined by $L = \{(x, y) \in \mathbb{R}^2 : x \geq y\}$ and $U = \{(x, y) \in \mathbb{R}^2 : y \geq x\}$, respectively. For $A \in \mathcal{B}(\mathbb{R}^2)$ let A^* denote the symmetrization of A, i.e., $A^* = \{(x, y) \in \mathbb{R}^2 : (x, y) \in A$ and/or $(y, x) \in A\}$ (and/or means 'or' not excluding 'and'), and note that $A^* \in \mathcal{S}$. Then $\mu_s(A) = s\mu((A \cap U)^*) + (1 - s)\mu((A \cap L)^*)$, $s \in [0, 1]$, define extensions of μ to $\mathcal{B}(\mathbb{R}^2)$ and these extensions are different for different s.

37. (a) Let $\{A_n\} \subset \mathcal{N}$ be an increasing sequence. Then $\{A_n^c\}$ is a decreasing sequence in $\mathcal{M}_0$ and, therefore, $(\bigcup_n A_n)^c = \bigcap_n A_n^c \in \mathcal{M}_0$; thus $\bigcup_n A_n \in \mathcal{N}$. The proof for a decreasing sequence in $\mathcal{N}$ is analogous and $\mathcal{N}$ is a monotone class.

(b) Let $\{A_n\} \subset \mathcal{N}$ be an increasing sequence. Then $\{A_n \cap C\}$ is an increasing sequence in $\mathcal{M}_0$ for all $C \in \mathcal{C}$ and, therefore, $\bigcup_n (A_n \cap C) = C \cap \bigcup_n A_n \in \mathcal{M}_0$ and $\bigcup_n A_n \in \mathcal{N}$. The proof for a decreasing sequence in $\mathcal{N}$ is analogous and $\mathcal{N}$ is a monotone class.

(c) First, $\mathcal{P}(X)$ is a monotone class that contains $\mathcal{C}$. Next, the intersection of all monotone classes containing $\mathcal{C}$ contains $\mathcal{C}$ and, as is readily seen, is a monotone class; this intersection is $\mathcal{M}_0(\mathcal{C})$, the smallest monotone class of subsets of X that contains $\mathcal{C}$.

(d) Since an algebra is closed under complementation it suffices to prove that $\mathcal{M}_0$ is closed under countable unions. Given $\{A_n\} \subset \mathcal{M}_0$, let $B_n = \bigcup_{k=1}^n A_k$, $n \geq 1$; since $\mathcal{M}_0$ is an algebra $\{B_n\} \subset \mathcal{M}_0$. Thus $\{B_n\}$ is an increasing sequence of sets in $\mathcal{M}_0$ and so $\bigcup_n B_n \in \mathcal{M}_0$. Finally, since $\bigcup_k A_k = \bigcup_n B_n$, $\bigcup_k A_k \in \mathcal{M}_0$.

39. Let $\mathcal{C} = \{[x, \infty) : x \in \mathbb{R}\}$ and λ the Lebesgue measure on $\mathcal{B}(\mathbb{R})$. Then $\mathcal{M}(\mathcal{C}) = \mathcal{B}(\mathbb{R})$ and if $\mu = \lambda$ and $\mu_1 = 2\lambda$, then $\mu = \mu_1$ on $\mathcal{C}$ but $1 = \mu([0, 1]) \neq \mu_1([0, 1]) = 2$.

Now, when $\mathcal{A}$ is an algebra, by Problem 37 it suffices to verify that $\mathcal{C} = \{C \in \mathcal{M}(\mathcal{A}) : \mu(C) = \nu(C)\}$ is a monotone class that contains $\mathcal{A}$. First, let $\{A_k\}$ be an increasing sequence in $\mathcal{C}$. Since μ, ν are measures on $\mathcal{M}(\mathcal{A})$, $\mu(\bigcup_k A_k) = \lim_n \mu(\bigcup_{k=1}^n A_k) = \lim_n \nu(\bigcup_{k=1}^n A_k) = \nu(\bigcup_k A_k)$ and $\bigcup_k A_k \in \mathcal{C}$. Similarly, if $\{A_k\}$ is a decreasing sequence in $\mathcal{C}$, since $\mu(X) = \nu(X) < \infty$, $\mu(\bigcap_k A_k) = \lim_n \mu(\bigcap_{k=1}^n A_k) = \lim_n \mu(A_n) = \lim_n \nu(A_n) = \nu(\bigcap_k A_k)$ and $\bigcap_k A_k \in \mathcal{C}$. Therefore $\mathcal{C}$ is a monotone class containing $\mathcal{A}$ and so it contains $\mathcal{M}_0(\mathcal{A})$ and, consequently, is equal to $\mathcal{M}(\mathcal{A})$. Then $\mu = \nu$.

40. First, note that if $A \in \mathcal{A}$ has $\mu(A) = \nu(A) < \infty$, by Problem 39, μ and ν coincide on $\mathcal{M}(\mathcal{A})_A$. Now, sets in $\mathcal{M}(\mathcal{A})_A$ are of the form $B \cap A$ with $B \in \mathcal{M}(\mathcal{A})$ and so $\mu(B \cap A) = \nu(B \cap A)$ for all $A \in \mathcal{A}$ and $B \in \mathcal{M}(\mathcal{A})$ with $\mu(A) < \infty$. Thus $\mu(B) = \mu(\bigcup_n (A_n \cap B)) = \sum_n \mu(B \cap A_n) = \sum_n \nu(B \cap A_n) = \nu(B)$ for all $B \in \mathcal{M}(\mathcal{A})$ and $\mu = \nu$.

The result may fail if the measures are not σ-finite relative to $\mathcal{A}$. To see this let X be a countable set and $Y \subset X$ such that $\operatorname{card}(Y) = \operatorname{card}(X \setminus Y) = \infty$. Let $\mathcal{A} = \{A \subset X : A \subset Y, A \text{ finite or } A^c \subset Y, A^c \text{ finite}\}$; by an argument similar to that in Problem 5, $\mathcal{A}$ is an algebra of subsets of X. Consider now the measures μ, ν on $\mathcal{M}(\mathcal{A})$ given by $\mu(A) = \operatorname{card}(A)$ and $\nu(A) = \operatorname{card}(A \cap Y)$, $A \in \mathcal{M}(\mathcal{A})$, respectively. Then μ and ν coincide on $\mathcal{A}$. Indeed, if $A \subset Y$, $A \in \mathcal{A}$ is finite and $\mu(A) = \nu(A \cap Y) = \operatorname{card}(A)$. On the other hand, since Y is infinite, if $A^c \subset Y$ is finite, A is infinite and $\mu(A) = \nu(A \cap Y) = \infty$. Finally, since $X \setminus \{x\} \in \mathcal{A}$ for all $x \in Y$, $X \setminus Y = \bigcap_{x \in Y}(X \setminus \{x\}) \in \mathcal{M}(\mathcal{A})$; however, $\mu(X \setminus Y) = \infty$ while $\nu(X \setminus Y) = \operatorname{card}((X \setminus Y) \cap Y) = 0$.

41. (a) Observe that once we prove that $\mu_i(M) + \mu_e(M^c) = \mu(X)$ for $M \subset X$, the original question follows by setting $X = A \cup E$. Now, given $\varepsilon > 0$, let $A \supset M^c$ be such that $\mu(A) \le \mu_e(M^c) + \varepsilon$; then $A^c \subset M$ and $\mu(X) \le \mu_e(M^c) + \mu(A^c) + \varepsilon$. And, since $\mu(A^c) \le \mu_i(M)$, it follows that $\mu(X) \le \mu_e(M^c) + \mu_i(M) + \varepsilon$, which gives one inequality since ε is arbitrary. Conversely, let $A \subset M$, $A \in \mathcal{S}$. Then $M^c \subset A^c$ and so $\mu(X) = \mu(A) + \mu(A^c) \ge \mu(A) + \mu_e(M^c)$. Therefore, taking the sup over these A it follows that $\mu(X) \ge \mu_i(M) + \mu_e(M^c)$, which is the other inequality.

(b) Since the A_k are pairwise disjoint and $\bigcup_{k=1}^K A_k \subset \bigcup_{k=1}^K M_k$ it follows that $\sum_{k=1}^K \mu(A_k) \le \mu(\bigcup_k A_k) \le \mu_i(\bigcup_k M_k)$ for each fixed K; therefore, taking the sup we get $\sum_{k=1}^K \mu_i(M_k) \le \mu_i(\bigcup_k M_k)$ and since K is arbitrary one inequality holds. Conversely, pick $A \in \mathcal{S}$ such that $A \subset \bigcup_k M_k$ and $\mu(A) \ge \mu_i(\bigcup_k M_k) - \varepsilon$; since the A_k are pairwise disjoint and $A \subset \bigcup_k M_k$, $A \cap A_k \subset M_k$ for all k. Thus $\mu_i(\bigcup_k M_k) \le \mu(A) + \varepsilon = \sum_k \mu(M \cap A_k) + \varepsilon$, and since $M \cap A_k \subset M_k$, $\mu_i(\bigcup_k M_k) \le \sum_k \mu_i(M_k) + \varepsilon$, which, since ε is arbitrary, gives the desired inequality.

The proof for μ_e follows along similar lines. Given $\varepsilon > 0$, let $A'_k \supset M_k$, $\mu(A'_k) \leq \mu_e(M_k) + \varepsilon/2^k$. Then $\bigcup_k M_k \subset \bigcup_k A'_k$ and $\mu_e(\bigcup_k M_k) \leq \mu(\bigcup_k A'_k) \leq \sum_k \mu(A'_k) \leq \sum_k \mu_e(M_k)+\varepsilon$. Let $A \supset \bigcup_k M_k$, $\mu(A) \leq \mu_e(\bigcup_k M_k) +\varepsilon$. Then $A \cap \bigcup_k A_k$ contains M_k for all k and since the A_k are pairwise disjoint, $A \cap A_k \supset M_k$. Thus $\sum_k \mu_i(M_k) \leq \sum_k \mu(A \cap A_k) \leq \mu(A) \leq \mu_e(\bigcup_k M_k) + \varepsilon$.

42. Let $\{M_k\}$ be pairwise disjoint sets in $\mathcal{M}(\mathcal{S}, V)$. Then by Problem 18, $M_k = (A_k \cap V) \cup (B_k \cap V^c)$ where each of the sequences $\{A_k\}$ and $\{B_k\}$ consist of pairwise disjoint sets in $\mathcal{S}$. Thus $M_k \cap V = A_k \cap V$ and $\nu_*(\bigcup_k M_k) = \mu_i(\bigcup_k (M_k \cap V)) = \mu_i(\bigcup_k (A_k \cap V))$. Now, since $A_k \cap V \subset A_k$ and the A_k are pairwise disjoint, by Problem 41, $\mu_i(\bigcup_k (A_k \cap V)) = \sum_k \mu_i(A_k \cap V) = \sum_k \mu_i(M_k \cap V) = \sum_k \nu_*(M_k)$ and we have finished. The proof for ν^* is analogous.

43. First, by Problem 42, μ_*, μ^* are measures on $\mathcal{M}(\mathcal{S}, V)$ and by Problem 41, $\mu_*(A) = \mu_i(A \cap V) + \mu_e(A \cap V^c) = \mu(A)$, $A \in \mathcal{S}$; analogously, $\mu^*(A) = \mu(A)$. Thus μ_* and μ^* are extensions of μ. Finally, let $0 < \eta < 1$ be such that $\xi = (1 - \eta)\mu_i(V) + \eta\mu_e(V)$. Then $\nu = (1 - \eta)\mu_* + \eta\mu^*$ is an extension of μ and $\nu(V) = (1 - \eta)\mu_i(V) + \eta\mu_e(V) = \xi$. For further details, including the case of infinite measure, see J. Łoś and E. Marczewski, *Extension of measures*, Fund. Math. **36** (1949), 267–276.

45. First, $\mathcal{N}$ is a σ-algebra of subsets of Y. Indeed, if $B \in \mathcal{N}$, $T^{-1}(B^c) = T^{-1}(B)^c \in \mathcal{M}$ and so $B^c \in \mathcal{N}$. Similarly, if $\{B_n\} \subset \mathcal{N}$, $T^{-1}(\bigcup_n B_n) = \bigcup_n T^{-1}(B_n) \in \mathcal{M}$ and $\bigcup_n B_n \in \mathcal{N}$. In fact, $\mathcal{N}$ is the largest σ-algebra of subsets of Y that makes T measurable.

Also, ν is a measure on $\mathcal{N}$. First, $\nu(\emptyset) = \mu(T^{-1}(\emptyset)) = 0$. Now, if $\{B_n\} \subset \mathcal{N}$ are pairwise disjoint, $\{T^{-1}(B_n)\} \subset \mathcal{M}$ are pairwise disjoint and, consequently, $\nu(\bigcup_n B_n) = \mu(T^{-1}(\bigcup_n B_n)) = \mu(\bigcup_n T^{-1}(B_n)) = \sum_n \mu(T^{-1}(B_n)) = \sum_n \nu(B_n)$.

Now, if μ is a probability measure, $\nu(Y) = \mu(T^{-1}(Y)) = \mu(X) = 1$ and ν is a probability measure. Similarly, if μ is finite or σ-finite, so is ν. Finally, if μ is complete, given $B \in \mathcal{N}$ with $\nu(B) = 0$ and $B_1 \subset B$, since $T^{-1}(B_1) \subset T^{-1}(B)$ and $\mu(T^{-1}(B)) = 0$, then $T^{-1}(B_1) \in \mathcal{M}$ and $\mu(T^{-1}(B_1)) = 0$, which implies that $B_1 \in \mathcal{N}$ and $\nu(B_1) = 0$.

Now, the measure λ induced by $S \circ T$, since $(S \circ T)^{-1}(C) = T^{-1}(S^{-1})(C)$, is defined on the σ-algebra $\{C \subset Z : S^{-1}(C) \in \mathcal{N}\} = \{C \subset Z : (S \circ T)^{-1}(C) \in \mathcal{M}\}$ and is given by $\lambda(C) = \mu(T^{-1}S^{-1}(C))$.

47. T^{-1} is defined on $[0, \infty)$ and is 2-valued, i.e., if $y = T(x)$, then $x = \pm y$. Given $B \in \mathcal{B}(\mathbb{R})$, let $B_+ = B \cap [0, \infty)$. Then $T^{-1}(B) = T^{-1}(B_+) = -B_+ \cup B_+$ and, consequently, $\mathcal{M}_T = \{-B_+ \cup B_+ : B \in \mathcal{B}(\mathbb{R})\}$. Then $\mu(B) = \lambda(T^{-1}(B)) = 2\lambda(B_+)$, $B \in \mathcal{B}(\mathbb{R})$.

48. Suppose first that $\mu(X) < \infty$. Let $\mathcal{F} = \{M \in \mathcal{M} :$ for every $\varepsilon > 0$ there exists $A \in \mathcal{A}$ such that $\mu(M \triangle A) < \varepsilon\}$. Clearly $\mathcal{A} \subset \mathcal{F}$ and, in particular, $X \in \mathcal{F}$. Next, since $M^c \triangle A^c = M \triangle A$, $\mathcal{F}$ is closed under complementation. Finally, let $\{M_n\} \subset \mathcal{F}$ and $M = \bigcup_n M_n$. By continuity from below there exists N such that $\mu(M \setminus \bigcup_{n=1}^{N} M_n) < \varepsilon/2$, pick $A_n \in \mathcal{A}$ such that $\mu(M_n \setminus A_n) \le \varepsilon/2N$, and let $A = \bigcup_{n=1}^{N} A_n \in \mathcal{A}$. Then, since $M \triangle A \subset M \triangle (\bigcup_{n=1}^{N} M_n) \cup ((\bigcup_{n=1}^{N} M_n) \triangle (\bigcup_{n=1}^{N} A_n))$ and $M \triangle (\bigcup_{n=1}^{N} M_n) = M \setminus (\bigcup_{n=1}^{N} M_n)$, we have $\mu(M \triangle A) \le \mu(M \setminus (\bigcup_{n=1}^{N} M_n)) + \sum_{n=1}^{N} \mu(M_n \setminus A_n) \le \varepsilon/2 + N\varepsilon/2N = \varepsilon$. Thus $M \in \mathcal{F}$, which is therefore a σ-algebra of subsets of X. And, since $\mathcal{A} \subset \mathcal{F}$, $\mathcal{M}(\mathcal{A}) = \mathcal{M} = \mathcal{F}$.

Now, in the σ-finite case pick $\{B_n\} \subset \mathcal{A}$ with finite measure such that $X = \bigcup_n B_n$; replacing B_n with $\bigcup_{k=1}^{n} B_k$ for each n if necessary we may assume that the sequence is increasing. Let $M \in \mathcal{M}$ have finite measure. Pick B_N such that $\mu(M \cap B_N) > \mu(M) - \varepsilon/2$ and let λ denote the restriction of μ to B_N. Then λ is a finite measure and so by the above argument there is $A_0 \in \mathcal{A}$ such that $\lambda(M \triangle A_0) = \mu((M \triangle A_0) \cap B_N) < \varepsilon/2$. Note that $A_0 \cap B_N \in \mathcal{A}$ and since

$$M \triangle (A_0 \cap B_N) \subset (M \triangle (M \cap B_N)) \cup ((M \cap B_N) \triangle (A_0 \cap B_N))$$
$$\subset (M \setminus (M \cap B_N)) \cup ((M \triangle A_0) \cap B_N),$$

it follows that $\mu(M \triangle (A_0 \cap B_N)) \le \varepsilon/2 + \varepsilon/2$ and we have finished.

50. The first inequality follows from the σ-subadditivity of μ. As for the second, by Problem 49, $\sum_{k=1}^{n} \mu(A_k) = \mu(\bigcup_{k=1}^{n} A_k) + \sum_{1 \le k < j \le n} \mu(A_k \cap A_j)$. Now, by subadditivity $\mu(\bigcup_{k=1}^{n} A_k) \le \sum_{k=1}^{n} \mu(A_k)$ and since each A_k intersects at most one other A_j with $j \ne k$, $\sum_{1 \le k < j \le n} \mu(A_k \cap A_j) \le \sum_{k=1}^{n} \mu(A_k)$. Therefore $\mu(\bigcup_{k=1}^{n} A_k) \le 2 \sum_{k=1}^{n} \mu(A_k)$ and the conclusion follows letting $n \to \infty$.

51. Let $B_m = \{\sigma : \sigma(m) = m\}$, $1 \le m \le n$. Then $A = (B_1 \cup \ldots \cup B_n)^c$ and by Problem 49, $\mu(A_n) = 1 - \mu(B_1 \cup \cdots \cup B_n) = 1 + \sum_{m=1}^{n} (-1)^m s_m$ where $s_m = \sum_{I \in \mathcal{I}_m^n} \mu(\bigcap_{\ell \in I} B_\ell)$. Now, $\bigcap_{\ell \in I} B_\ell = \{\sigma : \sigma(\ell) = \ell \text{ for } \ell \in I\}$ and so $\bigcap_{\ell \in I} B_\ell$ consists of permutations that are fixed at m elements and permute the remaining $n - m$ elements. Since there are $(n - m)!$ such permutations and each has probability $1/n!$ it follows that $\mu(\bigcap_{\ell \in I} B_\ell) = (n - m)!/n!$ and so $\mu(A_n) = 1 + \sum_{m=1}^{n} (-1)^m \sum_{I \in \mathcal{I}_m^n} (n - m)!/n!$. Now, $\text{card}(\mathcal{I}_m^n) = \binom{n}{m}$ and, consequently,

$$\mu(A_n) = 1 + \sum_{m=1}^{n} (-1)^m \binom{n}{m} \frac{(n - m)!}{n!} = 1 + \sum_{m=1}^{n} (-1)^m \frac{1}{m!} = \sum_{m=0}^{n} (-1)^m \frac{1}{m!}.$$

Whence $\mu(A_n) \to 1/e$ as $n \to \infty$.

54. First, clearly $\mathcal{S}(\mathcal{M}, \mathcal{P}(\mathcal{N})) = \mathcal{S}$ coincides with the σ-algebra of subsets of X discussed in Problem 53. For $A \in \mathcal{S}$ with $A = M \cup B$ where $M \in \mathcal{M}$ and $B \subset N \in \mathcal{N}$, let $\mu_1(A) = \mu(M)$. Note that μ_1 is well-defined; indeed, if also $A = M_1 \cup B_1$, $M_1 \in \mathcal{M}$, $B_1 \subset N_1 \in \mathcal{N}$, then $M \subset M_1 \cup N_1$ and so $\mu(M) \leq \mu(M_1)$. Reversing the roles of M and M_1 also $\mu(M_1) \leq \mu(M)$ and $\mu(M) = \mu(M_1)$.

Next, μ_1 is a measure on $\mathcal{S}$. Clearly $\mu_1(\emptyset) = 0$. Let $\{M_n \cup B_n\} \subset \mathcal{S}$ be pairwise disjoint. Then $B = \bigcup_n B_n \subset N \in \mathcal{N}$ and

$$\mu_1\left(\bigcup_n (M_n \cup B_n)\right) = \mu_1\left(\left(\bigcup_n M_n\right) \cup B\right) = \mu\left(\bigcup_n M_n\right)$$
$$= \sum_n \mu(M_n) = \sum_n \mu_1(M_n \cup B_n).$$

Finally, μ_1 is the only complete measure on $\mathcal{S}$ such that $\mu_1|_{\mathcal{M}} = \mu$. Since for $M \in \mathcal{M}$, $M = M \cup \emptyset$, $\mu_1(M) = \mu(M)$ and μ_1 extends μ. To check that μ_1 is complete let $A = M \cup B \in \mathcal{S}$ with $\mu_1(A) = \mu(M) = 0$, $B \subset N \in \mathcal{N}$, and $A_1 \subset A$. Then $B_1 = (M \cap A_1) \cup (B \cap A_1) \subset M \cup N$ with $\mu(M \cup N) = 0$ and $A_1 = \emptyset \cup B_1 \in \mathcal{S}$ has $\mu_1(A_1) = 0$. Thus μ_1 is complete.

Suppose that μ_2 is another complete extension of μ to $\mathcal{S}$ and note that for $B \subset N \in \mathcal{N}$, $\mu_2(B) \leq \mu_2(N) = \mu(N) = 0$. Thus, if $A = M \cup B \in \mathcal{S}$, $\mu_2(A) \leq \mu_2(M) + \mu_2(B) = \mu(M) = \mu_1(A)$. Moreover, since for $B \in \mathcal{P}(\mathcal{N})$ also $\mu_1(B) = 0$, we can reverse the roles of μ_1 and μ_2 and obtain $\mu_1(A) \leq \mu_2(A)$ as well. In other words, $\mu_1(A) = \mu_2(A)$ for every $A \in \mathcal{S}$ and uniqueness follows.

Finally, μ is σ-finite iff μ_1 is σ-finite. Since $\mathcal{M} \subset \mathcal{S}$ and $\mu_1|_{\mathcal{M}} = \mu$, sufficiency is clear. As for necessity, let $X = \bigcup_n X_n$ where the X_n are pairwise disjoint and $\mu(X_n) < \infty$ for all n. Let $B = X \setminus \bigcup_n X_n$; clearly $B \in \mathcal{S}$ and $\mu_1(B) = 0$ for otherwise B would contain a set $M \in \mathcal{M}$ with $\mu(M) > 0$ disjoint from all the X_n, which is not possible. Then with $X_1' = X_1 \cup B$ and $X_n' = X_n$ for all $n \geq 2$, $\{X_n'\}$ is a pairwise disjoint partition of X with $\mu_1(X_n') < \infty$ for all n.

55. (a) Note that $A = ((A \cup N) \setminus N) \cup (A \cap N)$ where $(A \cup N) \setminus N \in \mathcal{M}$ and, since $A \cap N \subset N$, $A \cap N \in \mathcal{M}$ with $\mu(A \cap N) = 0$.

(b) Since $A \cap B = A \setminus (A \setminus B)$ where $A \in \mathcal{M}$ and $(A \setminus B) \subset A \triangle B$ has measure 0, $A \cap B \in \mathcal{M}$. Now, $B \setminus A \subset A \triangle B$ has measure 0 and so $B = (A \cap B) \cup (B \setminus A)$ is measurable. Moreover, $\mu(B) \leq \mu(A \cap B) + \mu(B \setminus A) \leq \mu(A)$ and since $B \in \mathcal{M}$, exchanging A and B also $\mu(A) \leq \mu(B)$.

56. (a) Let $A = A_1 \cup N$ where $A_1 \in \mathcal{M}$ and $N \subset A_2$, $A_2 \in \mathcal{M}$ with $\mu(A_2) = 0$. Thus $A_1 \subset A \subset A_1 \cup A_2$ and $\mu_1(A) \geq \mu_*(A) \geq \mu(A_1) = \mu(A_1 \cup A_2) \geq \mu^*(A) \geq \mu_1(A)$. Hence $\mu_*(A) = \mu_1(A) = \mu^*(A)$.

(b) Let $\{M_n\}, \{M'_n\} \subset \mathcal{M}$ be such that $M_n \subset A \subset M'_n$ and $\mu_*(A) = \lim_n \mu(M_n) = \lim_n \mu(M'_n) = \mu^*(A)$. Note that $B = \bigcup_n M_n$ and $C = \bigcap_n M'_n$ are in $\mathcal{M}$, $B \subset A \subset C$, and $\mu(M_n) \leq \mu(B) \leq \mu(C) \leq \mu(M'_n)$ for all n. Thus letting $n \to \infty$ it follows that $\mu(B) = \mu(C)$, $\mu(C \setminus B) = 0$, and $A = B \cup (A \setminus B) \in \mathcal{S}$.

59. The statement is true if the sequence is increasing, i.e., for all $k \geq 1$, $\mu_k(A) \leq \mu_{k+1}(A)$, all $A \in \mathcal{M}$, and the statement is false if the sequence is decreasing.

61. Let λ be the set function on $\mathcal{M}$ given by $\lambda(A) = \sup\{\mu(B) - \nu(B) : B \subset A, \nu(B) < \infty\}$, $A \in \mathcal{M}$. Clearly $\lambda(\emptyset) = 0$. We claim that λ is additive. Let $A, B \in \mathcal{M}$, $A \cap B = \emptyset$. Then for $E \subset A \cup B$ with $\nu(E) < \infty$ it follows that $\mu(E) - \nu(E) = \mu(E \cap A) - \nu(E \cap A) + \mu(E \cap B) - \nu(E \cap B) \leq \lambda(A) + \lambda(B)$ and, therefore, taking the sup over such E, $\lambda(A \cup B) \leq \lambda(A) + \lambda(B)$. Next, let $A' \subset A, B' \subset B$ with $\nu(A'), \nu(B') < \infty$ and note that since $A' \cap B' = \emptyset$ and $\nu(A' \cup B') < \infty$, $\mu(A') - \nu(A') + \mu(B') - \nu(B') = \mu(A' \cup B') - \nu(A' \cup B') \leq \lambda(A \cup B)$. Therefore, taking the sup over such A', B', $\lambda(A) + \lambda(B) \leq \lambda(A \cup B)$, and λ is additive.

To prove that λ is a measure we invoke Problem 31. Let $\{A_n\}$ be an increasing sequence with limit $A = \bigcup_n A_n$; we claim that $\lim_n \lambda(A_n) = \lambda(A)$. First, by monotonicity the numerical sequence $\{\lambda(A_n)\}$ is nondecreasing and so it has a limit $\leq \lambda(A)$. Next, let $B \subset A$ with $\nu(B) < \infty$; then $\{B \cap A_n\}$ increases to B and, consequently, $\{\mu(B \cap A_n)\}$ increases to $\mu(B)$, $\{\nu(B \cap A_n)\}$ increases to the finite value $\nu(B)$, and $\{\lambda(B \cap A_n)\}$ increases to a limit $\leq \lambda(B)$. Now, $\lambda(B) = \mu(B) - \nu(B) = \lim_n \mu(B \cap A_n) - \lim_n \nu(B \cap A_n) = \lim_n \lambda(B \cap A_n) \leq \lim_n \lambda(A_n) \leq \lambda(A)$ and, therefore, taking the sup over $B \subset A$, $\lambda(A) \leq \lim_n \lambda(A_n) \leq \lambda(A)$. Thus λ is a measure.

Finally, we must verify that $\mu(A) = \nu(A) + \lambda(A)$ for all $A \in \mathcal{M}$. First, if $\nu(A) = \infty$, $\mu(A) = \infty$ and so $\mu(A) = \nu(A) + \lambda(A)$. Also, if $\nu(A) < \infty$, by the definition of λ, $\mu(A) - \nu(A) \leq \lambda(A)$ and $\mu(A) \leq \lambda(A) + \nu(A)$. As for the opposite inequality note that for $B \subset A$, since $\nu(B) < \infty$, $\mu(B) - \nu(B) + \nu(A) = \mu(B) + \nu(A \setminus B) \leq \mu(B) + \mu(A \setminus B) = \mu(A)$ and so $\lambda(A) + \nu(A) = \sup_{B \subset A}\{\mu(B) - \nu(B) + \nu(A)\} \leq \mu(A)$ and we are done.

To see that λ is not necessarily unique let $\mathcal{M} = \{\emptyset, X\}$ and consider the measures μ, ν on $\mathcal{M}$ given by $\mu(\emptyset) = \nu(\emptyset) = 0$ and $\mu(X) = \nu(X) = \infty$. Then any measure λ on $\mathcal{M}$ given by $\lambda(\emptyset) = 0$ and $\lambda(X) \geq 0$ satisfies $\mu = \nu + \lambda$. Note that the λ constructed above satisfies $\lambda(X) = 0$ and that in this case, as well as in general, is in some sense the smallest measure that works.

On the other hand, suppose that ν is σ-finite and let $\{X_n\}$ be pairwise disjoint sets with $\nu(X_n) < \infty$ such that $X = \bigcup_n X_n$. If λ, λ' are measures so that $\mu = \nu + \lambda = \nu + \lambda'$, then $\mu(A \cap X_n) = \nu(A \cap X_n) + \lambda(A \cap X_n) = \nu(A \cap X_n) + \lambda'(A \cap X_n)$, $A \in \mathcal{M}$, and since $\nu(A \cap X_n) < \infty$, $\lambda(A \cap X_n) =$

$\lambda'(A \cap X_n)$ for all A and n. Hence $\lambda(A) = \sum_n \lambda(A \cap X_n) = \sum_n \lambda'(A \cap X_n) = \lambda'(A)$ and λ is unique.

62. (a) Necessity first. Let $B \in \mathcal{M}$. If $\mu(A \cap B) = 0$ we are done so suppose $\mu(A \cap B) > 0$; then since $A \cap B \subset A$ it follows that $\mu(A \cap B) = \mu(A)$ and, consequently, $\mu(A \setminus B) = \mu(A) - \mu(A \cap B) = 0$.

Sufficiency next. Let $B \subset A$ with $\mu(A) > \mu(B)$. For the sake of argument suppose that $\mu(B) > 0$. Then since $A = B \cup (A \setminus B)$ it follows that $\mu(A \setminus B) = \mu(A) - \mu(B) > 0$, which is not possible since then both $\mu(A \cap B) = \mu(B) > 0$ and $\mu(A \setminus B) > 0$. Therefore $\mu(B) = 0$.

(b) Since μ is nonatomic there is a measurable $B_0 \subset A$ such that $0 < \mu(B_0) < \mu(A) < \infty$. Define now $B_1 \supset B_2 \supset \ldots$ inductively as follows: Having picked B_{n-1}, let $C \in \mathcal{M}$ have $0 < \mu(C) < \mu(B_{n-1})$ and pick B_n to be whichever set, C or $B_{n-1} \setminus C$, with measure at most $\mu(B_{n-1})/2$. Clearly $B_n \subset A$ and $0 < \mu(B_n) \leq \mu(B_0)/2^n$, which may be made arbitrarily small for n sufficiently large.

(c) Given $A \in \mathcal{M}$ with $\mu(A) \leq \eta$, let $\mathcal{F}(A) = \{B \in \mathcal{M} : A \subset B, \mu(B) \leq \eta\}$ and $\psi(A) = \sup\{\mu(B) : B \in \mathcal{F}(A)\}$; note that since $\mu(A) \leq \eta$, $A \in \mathcal{F}(A)$. Also, if $A \subset B$ and $\mu(B) \leq \eta$, then $\psi(A) \geq \psi(B)$. Define now an increasing sequence $A_0 \subset A_1 \subset \ldots$ of measurable sets, each of measure at most η, as follows: Let $A_0 = \emptyset$ and, given A_{n-1}, choose $A_n \in \mathcal{F}(A_{n-1})$ so that $\psi(A_{n-1}) - 1/n \leq \mu(A_n) \leq \psi(A_{n-1})$. Let $A = \bigcup_n A_n$. Note that since the sequence is increasing and $\mu(A_n) \leq \eta$ for all n, $\mu(A) = \lim_n \mu(A_n) \leq \eta$; we claim that $\mu(A) = \eta$. For the sake of argument suppose that $\mu(A) < \eta$. By (a), $X \setminus A$ contains a measurable set C with $0 < \mu(C) < \eta - \mu(A)$. Thus $\mu(A \cup C) < \eta$ and, consequently, $\psi(A_n) - 1/n \leq \mu(A_{n+1}) \leq \mu(A) < \mu(A \cup C) < \eta$. Now, since $A \cup C \in \mathcal{F}(A_n)$ we have $\mu(A \cup C) \leq \psi(A_n)$, which combined with the previous inequality gives $\psi(A_n) - 1/n \leq \mu(A) < \mu(A \cup C) \leq \psi(A_n)$. Hence $0 < \mu(C) = \mu(A \cup C) - \mu(A) < 1/n$, which is impossible for large n. Thus $\mu(A) = \eta$.

This result is true for the Lebesgue measure on $\mathbb{R}^n$ and for nonatomic regular Borel measures μ on $\mathbb{R}^n$ which are inner-regular; see Problem 70.

63. Let $\mu(A) > 0$; then $\nu(A) > 0$ or $\lambda(A) > 0$. Suppose that $\nu(A) > 0$; then there is a ν-atom B such that $B \subset A$. If $\lambda(B) = 0$, B is clearly an atom for μ. On the other hand, if $\lambda(B) > 0$, there exists a λ-atom C such that $C \subset B$; of course $\mu(C) > 0$. If C is a μ-atom we are done. Otherwise, by Problem 62(a) there exists $D \in \mathcal{M}$ such that $\mu(C \cap D) > 0$ or $\mu(C \setminus D) > 0$ and in that case $C \cap D$ or $C \setminus D$ is the required atom for μ.

64. Let $\mathcal{S}$ denote the family of all countable unions of μ-atoms and for $A \in \mathcal{M}$ let $\mu_1(A) = \sup\{\mu(A \cap M) : M \in \mathcal{S}\}$ and $\mu_2(A) = \sup\{\mu(A \cap N) : \mu_1(N) = 0\}$; then $\mu = \mu_1 + \mu_2$. First, μ_1 is purely atomic. Let $A \in \mathcal{M}$.

If $\mu_1(A) > 0$, then $\mu(A \cap M) > 0$ for some $M \in \mathcal{S}$ and since $M = \bigcup_n M_n$ where the M_n are μ-atoms, $\mu(A \cap M_n) > 0$ for some μ-atom M_n. Since $A \cap M_n$ is a μ-atom such that $\mu_1(A \cap M_n) > 0$ and since $\mu_1 \leq \mu$, $A \cap M_n$ is a μ_1-atom contained in A. Next, μ_2 is nonatomic. Suppose that $\mu_2(A) > 0$. Then $\mu(A \cap N) > 0$ for some N such that $\mu_1(N) = 0$. Now, $A \cap N$ is not a μ-atom for otherwise $\mu_1(A \cap N) > 0$. Thus, since $\mu(A \cap N) > 0$ and $A \cap N$ is not a μ-atom there exists $B \in \mathcal{M}$ such that $\mu(A \cap N \cap B) > 0$ and $\mu((A \cap N) \setminus B) > 0$. Now, necessarily $\mu_2(A \cap B) > 0$ and $\mu_2(A \setminus B) > 0$ and, by Problem 62(a), μ_2 has no atoms.

Observe that if μ is atomic and $\mu(A) > 0$, there exist countably many atoms $\{M_n\}$ such that $\mu(A) = \mu(A \cap \bigcup_n M_n)$. Letting

$$A_n = (M_n \setminus (M_1 \cup \ldots \cup M_{n-1})) \cap A$$

and disregarding those A_n, if any, of measure zero, we have the following: If μ is purely atomic and $\mu(A) > 0$, there exist countably many pairwise disjoint atoms $A_n \subset A$ such that $\mu(A) = \mu(\bigcup_n A_n)$.

66. Let $\{r_n\}$ be a sequence that decreases to 0. Since $\bigcap_n A_{r_n} \subset \bigcap_{r>0} A_r$ it follows that $A = \bigcap_{r_n} A_{r_n}$ and so, $A \in \mathcal{M}$. Hence by continuity from above $\mu(A) = \lim_n \mu(A_{r_n})$ and since $\{r_n\}$ is arbitrary, $\lim_{r \to 0} \mu(A_r) = \mu(A)$.

68. The statement is true. Necessity first. For the sake of argument suppose that there exist two indices, which we assume to be $1, 2$, such that $\mu(A_1 \cap A_2) > 0$. Then $\mu(A_1 \cup A_2) < \mu(A_1) + \mu(A_2)$ and so $\mu(A_1 \cup A_2) + \sum_{n=3}^{\infty} \mu(A_n) < \sum_n \mu(A_n) = \mu((A_1 \cup A_2) \cup \bigcup_{n=3}^{\infty} A_n) \leq \mu(A_1 \cup A_2) + \mu(\bigcup_{n=3}^{\infty} A_n)$. Thus canceling the finite quantity $\mu(A_1 \cup A_2)$ we get the strict inequality $\sum_{n=3}^{\infty} \mu(A_n) < \mu(\bigcup_{n=3}^{\infty} A_n)$, which by the σ-subadditivity of μ cannot hold.

Sufficiency next. We begin by constructing sequences $\{B_n\}, \{C_n\}$ as follows. Let $B_1 = \emptyset$ and $B_n = A_n \cap (\bigcup_{m=1}^{n-1} A_m)$ for $n > 1$, and $C_1 = A_1$ and $C_n = A_n \setminus \bigcup_{m=1}^{n-1} A_m$ for $n > 1$; then $A_n = B_n \cup C_n$ for all $n \geq 1$. Moreover, $C_n \cap C_m = \emptyset$ for $n \neq m$ and since $B_n = \bigcup_{m=1}^{n-1}(A_n \cap A_m)$, $\mu(B_n) \leq \sum_{m=1}^{n-1} \mu(A_n \cap A_m) = 0$ for all n and, hence, also $\mu(A_n) = \mu(C_n)$ for all $n \geq 1$. Finally, we claim that $\bigcup_n C_n = \bigcup_n A_n$; it suffices to prove that $\bigcup_n A_n \subset \bigcup_n C_n$, the other inclusion being obvious. Now, if $x \in \bigcup_n A_n$, let m be the smallest index such that $x \in A_m$; then $x \in C_m \subset \bigcup_n C_n$. Hence $\mu(\bigcup_n A_n) = \mu(\bigcup_n C_n) = \sum_n \mu(C_n) = \sum_n \mu(A_n)$.

70. Let $\mathcal{S} = \{B \in \mathcal{B}(\mathbb{R}^n) : \text{inner and outer regularity hold for } B\}$; we claim that $\mathcal{S}$ is a σ-algebra that contains the open sets and so $\mathcal{S} = \mathcal{B}(\mathbb{R}^n)$.

Now, outer regularity holds for open sets. Also, if O is open, $O = \bigcup_k Q_k$ where the Q_k are nonoverlapping closed cubes; let $K_n = \bigcup_{k=1}^{n} Q_k$. Then $\{K_n\}$ is an increasing sequence of compact sets with $O = \bigcup_n K_n$

and, therefore, by continuity from below $\mu(O) = \lim_n \mu(K_n)$. Thus inner regularity also holds for O and $\mathcal{S}$ contains the open sets.

Next, we claim that $\mathcal{S}$ is closed under complementation; clearly it contains $\mathbb{R}^n$ and $\emptyset$. Let A be outer regular and given $\varepsilon > 0$, let $B \supset A$ be an open set such that $\mu(A) \geq \mu(B) - \varepsilon/2$; then $C = B^c \subset A^c$ is closed and $\mu(A^c) = \mu(\mathbb{R}^n) - \mu(A) \leq \mu(\mathbb{R}^n) - \mu(B) + \varepsilon/2 = \mu(C) + \varepsilon/2$. Let $\{K_n\}$ be an increasing sequence of compact sets such that $C = \bigcup_n K_n$. Since μ is finite, $\mu(C) = \lim_n \mu(K_n)$ and so picking n large enough we have $K_n \subset C \subset A^c$ such that $\mu(C) \leq \mu(K_n) + \varepsilon/2$. Hence $\mu(A^c) \leq \mu(K_n) + \varepsilon$ and A^c is inner regular. The fact that if A is inner regular, A^c is outer regular, follows along similar yet simpler lines. Hence $\mathcal{S}$ is closed under complementation.

Finally, let $\{A_n\} \subset \mathcal{S}$ and $A = \bigcup_n A_n$; we claim that $A \in \mathcal{S}$. Given $\varepsilon > 0$, let C_n be a compact set and B_n an open set such that $C_n \subset A_n \subset B_n$ and $\mu(B_n \setminus C_n) \leq \varepsilon/2^{n+1}$. Let $C = \bigcup_n C_n$, $B = \bigcup_n B_n$; then $C \subset A \subset B$ and since $B \setminus C \subset \bigcup_n (B_n \setminus C_n)$, $\mu(B \setminus C) \leq \sum_n \mu(B_n \setminus C_n) \leq \varepsilon/2$ and, consequently, $\mu(B \setminus A) \leq \mu(B \setminus C) \leq \varepsilon/2$ and A is outer regular. Next, since C is not necessarily compact, let $K_n = \bigcup_{m=1}^{n} C_m$; $\{K_n\}$ is an increasing sequence of compact sets and so $\lim_n \mu(K_n) = \mu(C)$; pick N large enough so that $\mu(C \setminus K_N) \leq \varepsilon/2$. Then K_N is compact, $K_N \subset A \subset B$, and $\mu(B \setminus K_N) = \mu(B \setminus C) + \mu(C \setminus K_N) \leq \varepsilon$ and, consequently, $\mu(A \setminus K_N) \leq \mu(B \setminus K_N) \leq \varepsilon$ and A is inner regular. Thus $\mathcal{S}$ is a σ-algebra and $\mathcal{S} = \mathcal{B}(\mathbb{R}^n)$.

In general inner regularity holds for Borel measures that assign a finite measure to compact sets but the same is not true for outer regularity. Consider the Borel measure ν given by $\nu(B) = \operatorname{card}(B \cap \mathbb{Q}^n)$, $B \in \mathcal{B}(\mathbb{R}^n)$; since $\mathbb{R}^n = \bigcup_{q \in \mathbb{Q}^n} (q \cup (\mathbb{Q}^n)^c)$ where $\nu(q \cup (\mathbb{Q}^n)^c) = 1$ for each q, ν is σ-finite. Note that if $B = \{0\}$, then $\nu(B) = 1$. Now, any open set O that contains B must contain an infinite number of rationals and so $\nu(O) = \infty$. Thus $1 = \nu(B) < \inf\{\nu(O) : O \text{ is open and } B \subset O\} = \infty$.

72. For the sake of argument suppose that ψ is not σ-additive. Then by Problem 31 there is a decreasing sequence $\{A_n\}$ of Borel sets with empty intersection such that $\lim_n \psi(A_n) = \inf_n \psi(A_n) = \eta > 0$. Pick now for each n a compact set $K_n \subset A_n$ such that $\psi(A_n) \leq \psi(K_n) + \eta/2^{n+1}$; then $\psi(A_n \setminus \bigcap_{m=1}^{n} K_m) \leq \sum_{m=1}^{n} \psi(A_m \setminus K_m) < \eta/2$. Then, in particular, $\psi(\bigcap_{m=1}^{n} K_m) \neq 0$ and, consequently, $\bigcap_{m=1}^{n} K_m \neq \emptyset$. Finally, $\{\bigcap_{m=1}^{n} K_m\}$ is a decreasing sequence of nonempty compact subsets in the compact space K_1 and, consequently, $\bigcap_m K_m \neq \emptyset$, which is not the case since $\bigcap_m A_m = \emptyset$.

73. (a) For the sake of argument suppose that for some $\eta > 0$ all open cubes of sidelength less than η have finite measure. Now, since K is compact, K can be covered by finitely many open cubes $Q_1, \ldots, Q_N$, say, of sidelength less than η and, therefore, by subadditivity $\mu(K) \leq \sum_{k=1}^{N} \mu(Q_k) < \infty$, which is not the case.

(b) Let K be a compact subset of an open set O. Then there are finitely many open sets $O_1, \ldots, O_N$, say, with $\mu(O_k) = 0$, $1 \leq k \leq N$, such that $K \subset \bigcup_{k=1}^{N} O_k$, and, consequently, $\mu(K) = 0$. Therefore by inner regularity $\mu(O) = \sup\{\mu(K) : K \subset O, K \text{ compact}\} = 0$.

74. No. First, note that given $x \in \mathbb{R}^n$, $\lim_N \mu(\bigcap_{k=N}^{\infty} B(x, 1/k)) = \mu(\{x\}) = 0$ and since μ only assumes the values 0 and 1, $\mu(\bigcap_{k=N}^{\infty} B(x, 1/k)) = 0$ for N large enough. So, since $\mu(\bigcap_{k=N}^{\infty} B(x, 1/k)) = \lim_k \mu(B(x, 1/k))$, $\mu(B(x, 1/k_x)) = 0$ for some k_x sufficiently large. Now, for the sake of argument suppose that μ is a probability measure and pick a closed cube Q in $\mathbb{R}^n$ with positive measure and to each $x \in Q$ assign the open ball $B_x = B(x, 1/k_x)$ with $\mu(B_x) = 0$. Then by Heine-Borel there are finitely many balls $B_{x_1}, \ldots, B_{x_N}$, say, such that $Q \subset \bigcup_{k=1}^{N} B_{x_k}$, and, consequently, $1 = \mu(Q) \leq \sum_{k=1}^{N} \mu(B_{x_k}) = 0$, which is not the case.

75. (a) and (b) Let $\mathcal{F} = \{O \subset \mathbb{R}^n : O \text{ open and } \mu(O) = 0\}$ and $V = \bigcup_{O \in \mathcal{F}} O$; V is an open subset of $\mathbb{R}^n$ and by the inner regularity of μ, $\mu(V) = \sup\{\mu(K) : K \subset V, K \text{ compact}\}$. Now, if $K \subset V$ is compact, K is contained in a finite union of open sets O with $\mu(O) = 0$ and so $\mu(K) = 0$. Therefore $\mu(V) = 0$ and $\mu(\mathbb{R}^n \setminus V) = \eta$.

Let $F = \mathbb{R}^n \setminus V$ and $K \subsetneq F$ a proper compact subset of F; then $W = \mathbb{R}^n \setminus K$ is open and if $\mu(K) = \eta$, $\mu(W) = 0$. Hence $W \subset V$ and $W^c \supset V^c$ or $K \supset F$, contrary to the fact that $K \subsetneq F$. Thus $\mu(K) < \mu(F)$. Also, V is the largest open set with $\mu(V) = 0$.

76. Let μ be the Borel measure given by $\mu(A) = 0$ if $A \cap K = \emptyset$ and $\mu(A) = \infty$ otherwise, $A \in \mathcal{B}(\mathbb{R}^n)$. Now, if $x \notin K$ there is an open neighborhood O_x of x contained in K^c and so $\mu(O_x) = 0$. Otherwise, if $x \in K$ every open neighborhood O_x of x contains x and so $\mu(O_x) = \infty > 0$. Thus $\text{supp}(\mu) = K$.

ν is constructed along similar lines. Let $D = \{d_k\}$ be a countable dense subset of K and for $A \in \mathcal{B}(\mathbb{R}^n)$ let ν be given by $\nu(A) = \sum_{d_k \in A} \nu(\{d_k\})$ where $\nu(\{d_k\}) = 1/2^k$ for all $k = 1, 2, \ldots$. Then $\nu(\mathbb{R}^n) = \nu(K) = 1$ and ν is a probability measure. Now, if $x \in K$ and O_x is a neighborhood of x, since D is dense in K, $O_x \cap D \neq \emptyset$ and so $\nu(O_x) > 0$. Finally, if $x \notin K$ there is an open neighborhood $O_x \subset K^c$ and so $\nu(O_x) = 0$. Thus $\text{supp}(\nu) = K$.

77. $\mu = (\delta_0 + \delta_1)/2$ where δ_0 and δ_1 denote the Dirac measures supported at 0 and 1, respectively.

78. (a) One A_n must be uncountable for otherwise X would be countable. Now, if A_n, A_m are uncountable, $n \neq m$, $X \setminus A_n$ and $X \setminus A_m$ are both countable, and since $X = A_n \cup (X \setminus A_n) \subset (X \setminus A_m) \cup (X \setminus A_n)$, X itself is countable, which it is not.

(b) First, $\mu(\emptyset) = 0$. Next, let $\{A_n\}$ be pairwise disjoint measurable sets and $A = \bigcup_n A_n$. The additivity, and the σ-additivity, of μ are clear since A is always at least countable and so $\mu(A) = \sum_n \mu(A_n) = \infty$. Finally, if $\mu(A) = \infty$, A is an infinite set and given an integer n, A contains a subset A_n of n elements with $0 < \mu(A_n) < \infty$ and μ is semifinite. And, since an uncountable set cannot be written as a countable union of finite sets, μ is not σ-finite.

(c) Only the σ-additivity of ν is not immediate. Let $\{A_n\}$ be pairwise disjoint measurable sets and $A = \bigcup_n A_n$. If A_n is countable for all n, A is countable and $0 = \nu(A) = \sum_n \nu(A_n) = 0$. On the other hand, if A_n is not countable for some n, and by (a) this can only happen for one n, $\nu(A_n) = 1$ and $\nu(A_k) = 0$ for all $k \neq n$. Then $1 = \nu(A) = \sum_k \nu(A_k) = \nu(A_n) = 1$.

Finally, let $A \in \mathcal{M}$ be uncountable with A^c countable and $B \subset A$, $B \in \mathcal{M}$. Then, if B is countable, $\nu(B) = 0$. Otherwise, B^c is countable and $\nu(A \setminus B) = \nu(A \cap B^c) \leq \nu(B^c) = 0$. Thus A is an atom.

79. (a) We claim that $\eta = \sup\{\mu(B) : B \in \mathcal{M}, B \subset A, \mu(B) < \infty\} = \infty$. For the sake of argument suppose that $\eta < \infty$ and pick $\{B_n\} \subset A$ such that $\lim_n \mu(B_n) = \eta$. Let $B = \bigcup_n B_n$ and note that $B \subset A$ and $\mu(B) \geq \eta$ and since the sup is η, $\mu(B) = \eta$. Let $A_1 = A \setminus B$; then $\mu(A_1) = \mu(A) - \mu(B) = \infty$ and since μ is semifinite there is $B_1 \subset A_1$ with $0 < \mu(B_1) < \infty$. Now, since $B \cap B_1 = \emptyset$, it follows that $B \cup B_1 \subset A$ with $\mu(B \cup B_1) > \eta$, which cannot happen. Therefore $\eta = \infty$ and there exists $B \subset A$ with $c < \mu(B) < \infty$.

(b) Let $B = \bigcup_n B_n$ where the B_n correspond to $c = n$ in (a). Then $\mu(B_n) < \infty$ for all n and $\mu(B) = \infty$.

80. (a) That ν is a measure follows by an argument analogous to that in Problem 61 and is therefore omitted. Note that if $\mu(A) < \infty$, $\nu(A) = \mu(A)$. Now, let $\nu(A) = \infty$ and $0 < M < \infty$. Then there exist measurable sets $B_n \subset A$ with $\mu(B_n) < \infty$ and $\mu(B_n) \to \infty$ and, consequently, there exists $B \subset A$ with $M < \mu(B) < \infty$ and since $\nu(B) = \mu(B)$, also $M < \nu(B) < \infty$. Thus ν is semifinite.

(b) As observed in (a) μ and ν coincide on sets of finite measure. If $\mu(A) = \infty$, given $0 < M < \infty$, let $B \subset A$ have $M < \mu(B) < \infty$. Then also $\nu(A) \geq \nu(B) = \mu(B) > M$ and so $\nu(A) = \infty$.

(c) We say that $A \in \mathcal{M}$ is μ-semifinite if $\mu(A) = \infty$ and given $0 < M < \infty$, there exists $B \subset A$ with $M < \mu(B) < \infty$. Define the set function λ on $\mathcal{M}$ by

$$\lambda(A) = \begin{cases} 0, & A \text{ is } \mu\text{-semifinite,} \\ \infty, & \text{otherwise.} \end{cases}$$

Since it is readily seen that $\mu = \nu + \lambda$ it only remains to prove that λ is a measure. Clearly, $\lambda(\emptyset) = 0$. Now, let $\{A_n\} \subset \mathcal{M}$ be pairwise disjoint. If all the A_n are μ-semifinite so is $\bigcup_n A_n$ and, consequently, $\lambda(\bigcup_n A_n) = 0 = \sum_n \lambda(A_n)$. On the other hand, if one A_n is not μ-semifinite, $\bigcup_n A_n$ is not μ-semifinite and $\lambda(\bigcup_n A_n) = \infty = \sum_n \lambda(A_n)$.

81. By Problem 70, μ is inner regular on the Borel sets of finite measure. Now, let $B \in \mathcal{B}(\mathbb{R}^n)$ with $\mu(B) = \infty$; since μ is semifinite, by Problem 79, for every $m \in \mathbb{N}$, there exists $B_m \subset B$, $B_m \in \mathcal{B}(\mathbb{R}^n)$, such that $m < \mu(B_m) < \infty$. And then, by the inner regularity on the Borel sets of finite measure there is a compact $K_m \subset B_m$ such that $m < \mu(K_m) < \infty$.

82. (a) Clearly $\mathcal{M} \subset \mathcal{M}_{loc}$. Next, let $A \in \mathcal{M}_{loc}$ and $M \in \mathcal{M}_f$. Then $A^c \cap M = (A \cap M)^c \cap M \in \mathcal{M}$ and so $A^c \in \mathcal{M}_{loc}$. Finally, let $\{A_n\} \subset \mathcal{M}_{loc}$ and $M \in \mathcal{M}_f$. Then $A_n \cap M \in \mathcal{M}$ for all n and, consequently, $(\bigcup_n A_n) \cap M = \bigcup_n (A_n \cap M) \in \mathcal{M}$. Thus $\mathcal{M}_{loc}$ is a σ-algebra of subsets of X.

(b) By (a) it suffices to prove that $\mathcal{M}_{loc} \subset \mathcal{M}$. Let $X = \bigcup_n X_n$ where $X_n \in \mathcal{M}_f$ for all n. Then for $A \in \mathcal{M}_{loc}$, $A = \bigcup_n (A \cap X_n)$ where $A \cap X_n \in \mathcal{M}$ for all n and, consequently, $A \in \mathcal{M}$.

(c) First, $\nu(\emptyset) = \mu(\emptyset) = 0$. Next, let $\{A_n\} \subset \mathcal{M}_{loc}$ be pairwise disjoint and $A = \bigcup_n A_n$. If $A_n \in \mathcal{M}$ for all n, then $A \in \mathcal{M}$ and $\nu(A) = \mu(A) = \sum_n \mu(A_n) = \sum_n \nu(A_n)$. Now, if $A_n \notin \mathcal{M}$ for some n, then $\nu(A_n) = \infty$ and there are two possibilities: If $A \notin \mathcal{M}$, then $\nu(A) = \infty$ and we have equality and, if $A \in \mathcal{M}$, then necessarily $\mu(A) = \infty$ for, if not, $A_n \cap A = A_n \in \mathcal{M}$, which is not the case. Therefore ν is a measure.

Finally, we prove that $(\mathcal{M}_{loc})_{loc} \subset \mathcal{M}_{loc}$. Let $A \in (\mathcal{M}_{loc})_{loc}$ and $M \in \mathcal{M}_f$; then $\nu(M) = \mu(M) < \infty$ and, consequently, $A \cap M \in \mathcal{M}_{loc}$. Therefore $(A \cap M) \cap M = A \cap M \in \mathcal{M}$ and $A \in \mathcal{M}_{loc}$.

(d) Let $A \in \mathcal{M}_{loc}$ have $\nu(A) = 0$ and $A' \subset A$. Since $\nu(A) = 0$, $A \in \mathcal{M}$ and since μ is complete, $A' \in \mathcal{M}$ and $\mu(A') = 0$. Hence $\nu(A') = \mu(A') = 0$ and the space is complete.

(e) The statement is false. Let $\mathcal{M} = \{\emptyset, X\}$ and μ the measure on $\mathcal{M}$ given by

$$\mu(A) = \begin{cases} 0, & A = \emptyset, \\ \infty, & A = X. \end{cases}$$

Then X is locally μ-null but $\mu(X) \neq 0$.

83. First, $\lambda(\emptyset) = 0$. Now, let $\{A_n\} \subset \mathcal{M}_{loc}$ be pairwise disjoint and $A = \bigcup_n A_n$. If $\lambda(A_n) = \infty$ for some n, by monotonicity $\infty = \lambda(A_n) \leq \lambda(A)$ and we have equality. So assume that $\lambda(A_n) < \infty$ for all n and let $M_n \in \mathcal{M}$ be such that $M_n \subset A_n$ and $\lambda(A_n) \leq \mu(M_n) + \varepsilon/2^n$. Then $\bigcup_n M_n \in \mathcal{M}$ and $\bigcup_n M_n \subset A$ and so $\sum_n \lambda(A_n) \leq \sum_n \mu(M_n) + \varepsilon = \mu(\bigcup_n M_n) + \varepsilon \leq \lambda(A) + \varepsilon$. Hence, since ε is arbitrary, $\sum_n \lambda(A_n) \leq \lambda(A)$.

Next, let $M \subset A$. If $\mu(M) < \infty$, $M \cap A_n \in \mathcal{M}$ for all n and $\mu(M) = \sum_n \mu(M \cap A_n) \leq \sum_n \lambda(A_n)$. And, if $\mu(M) = \infty$, by Problem 79 there exists a sequence $\{M_k\} \subset \mathcal{M}$ that increases to M and $\mu(M_k) < \infty$ for all k. As before $\mu(M_k) \leq \sum_n \lambda(A_n)$ and, consequently, also $\mu(M) \leq \sum_n \lambda(A_n)$. Hence, since $M \subset A$ is arbitrary, $\lambda(A) \leq \sum_n \lambda(A_n)$ and equality holds. Clearly λ extends μ. Finally, that λ is saturated follows as in Problem 82(c).

As for the statement, it is false. Let X_1, X_2 be disjoint uncountable sets, $X = X_1 \cup X_2$, and $\mathcal{M}$ the σ-algebra consisting of the countable subsets of X together with those sets with countable complement. Let μ_0 be the counting measure on $\mathcal{P}(X_1)$ and define μ on $\mathcal{M}$ by $\mu(A) = \mu_0(A \cap X_1)$; clearly μ is a measure on $\mathcal{M}$. Now, if $\mu_0(A) < \infty$ for $A \in \mathcal{M}$, $A \subset X_1$ is finite. Then, for any $Y \subset X$, $A \cap Y$ is $\emptyset$ or a finite subset of X_1, which implies that $A \cap Y \in \mathcal{P}(X_1)$. Thus $\mathcal{M}_{loc} = \mathcal{P}(X)$. Finally, for $x_1 \in X_1$, $\nu(\{x_1\} \cup X_2) = \infty$ and $\lambda(\{x_1\} \cup X_2) = \mu(\{x_1\}) = \mu_0(\{x_1\}) = 1$ and so $\nu \neq \lambda$.

84. It is readily seen that d satisfies the properties of a distance; completeness requires a moment's thought. Let $\{\mu_n\}$ be a Cauchy sequence in $(\mathcal{P}, d)$. Now, since $\{\mu_n(A)\}$ is a numerical Cauchy sequence for all $A \in \mathcal{M}$, $\lim_n \mu_n(A)$ exists, $\mu(A) = \lim_n \mu_n(A)$ is a well-defined set function on $\mathcal{M}$, and $\lim_n d(\mu_n, \mu) = 0$.

We claim that μ is a measure on $\mathcal{M}$. First, $\mu(\emptyset) = 0$. Next, if $A_1, \dots, A_K$ are pairwise disjoint sets in $\mathcal{M}$ we have $\mu_n(\bigcup_{k=1}^{K} A_k) = \sum_{k=1}^{K} \mu_n(A_k)$ for all n and letting $n \to \infty$, $\mu(\bigcup_{k=1}^{K} A_k) = \lim_n \mu_n(\bigcup_{k=1}^{K} A_k) = \lim_n \sum_{k=1}^{K} \mu_n(A_k) = \sum_{k=1}^{K} \mu(A_k)$ and μ is additive. Now, let $\{A_k\} \subset \mathcal{M}$ be a decreasing sequence with $\bigcap_k A_k = \emptyset$. Since $\lim_k \mu_n(A_k) = 0$ for each n, given $\varepsilon > 0$, pick N such that $|\mu(B) - \mu_N(B)| \leq \varepsilon/2$ for all $B \in \mathcal{M}$ and then k_0 such that $\mu_N(A_k) \leq \varepsilon/2$ for all $k \geq k_0$. Then $\mu(A_k) \leq |\mu(A_k) - \mu_N(A_k)| + \mu_N(A_k) \leq \varepsilon$ for all $k \geq k_0$ and $\lim_k \mu(A_k) = 0$. Hence, by Problem 31, μ is a measure.

86. We claim that $\eta = 0$. First, note that for each n, $\liminf_k (A_n \cap A_k^c) = A_n \cap (\liminf_k A_k^c) = A_n \cap (\limsup_k A_k)^c$ and, consequently,

$$\limsup_n \liminf_k (A_n \cap A_k^c) = \limsup_n (A_n \cap (\limsup_k A_k)^c)$$
$$= (\limsup_n A_n) \cap (\limsup_k A_k)^c = \emptyset.$$

Therefore

$$\mu(\limsup_n \liminf_k (A_n \cap A_k^c)) = 0$$

and

$$\limsup_n \mu(\liminf_k (A_n \cap A_k^c)) \leq \mu(\limsup_n \liminf_k (A_n \cap A_k^c)) = 0,$$

provided that $\bigcup_n (\liminf_k A_k^c) \cap A_n \subset \bigcup_n A_n$ has finite measure, which by assumption it does.

87. (a) First, note that for integers n, k,

$$\bigcap_{m=n}^{n+k} A_m \supset A_n \setminus (A_n \setminus A_{n+1}) \setminus \ldots \setminus (A_{n+k-1} \setminus A_{n+k})$$

and, consequently, $\mu(\bigcap_{m=n}^{n+k} A_m) \geq \mu(A_n) - \sum_{m=n}^{n+k-1} \mu(A_m \setminus A_{m+1})$; thus letting $k \to \infty$ we get $\mu(A_n) \leq \mu(\bigcap_{m=n}^{\infty} A_m) + \sum_{m=n}^{\infty} \mu(A_m \setminus A_{m+1})$. Hence, since $\bigcap_{m=n}^{\infty} A_m \subset \liminf_n A_n$ and $\lim_n \sum_{m=n}^{\infty} \mu(A_m \setminus A_{m+1}) = 0$, it follows that $\limsup_n \mu(A_n) \leq \mu(\liminf_n A_n)$. Therefore, by Problem 85(b), $\mu(\limsup_n A_n) \leq \liminf_n \mu(A_n) \leq \mu(\liminf_n A_n)$ and the equality holds.

(b) Similarly, observing that $\bigcup_{m=n}^{n+k} A_m$ is contained in $A_{n+k} \cup (A_n \setminus A_{n+1}) \cup \ldots \cup (A_{n+k-1} \setminus A_{n+k})$, we get that $\mu(\limsup_n A_n) \leq \liminf_n \mu(A_n)$. Hence, by Problem 85(b), $\limsup_n \mu(A_n) \leq \mu(\limsup_n A_n) \leq \liminf_n \mu(A_n)$ and so $\lim_n \mu(A_n)$ exists and is equal to L.

88. Let $B_k = \{x \in X : x \text{ belongs to at least } k \text{ of the } A_n\}$, $k = 1, \ldots, 10$; since $B_k = \bigcup_{1 \leq n_1 < \ldots < n_k \leq 10} (\bigcap_{j=1}^k A_{n_j})$, B_k is measurable. Moreover, since $\sum_{n=1}^{10} \mu(A_n) = \sum_{k=1}^{10} \mu(B_k)$ it follows that $10/3 = \sum_{n=1}^{10} \mu(A_n) \leq \mu(B_1) + \mu(B_2) + \mu(B_3) + 7\mu(B_4)$. But $\mu(B_1) + \mu(B_2) + \mu(B_3) \leq 3$ and so

$$\mu(B_4) \geq 1/21 > 0.$$

The same conclusion does not follow for only 9 sets. Consider Lebesgue measure on $[0,1]$ and $A_1 = [0, 1/3], A_2 = [1/3, 2/3], A_3 = [2/3, 1], A_4 = [0, 1/6] \cup [1/3, 1/2], A_5 = [1/3, 1/2] \cup [5/6, 1], A_6 = [1/6, 1/3] \cup [2/3, 5/6], A_7 = [0, 1/6] \cup [1/2, 2/3], A_8 = [1/6, 1/3] \cup [5/6, 1]$, and $A_9 = [1/3, 1/2] \cup [2/3, 5/6]$. Then $|A_n| = 1/3$ for $1 \leq n \leq 9$ and $|A| = 0$.

89. (a) The relation $\sim$ is clearly reflexive and symmetric. Now, given A, B, C in $\mathcal{M}$, since $A \triangle C \subset (A \triangle B) \cup (B \triangle C)$, $\mu(A \triangle C) \leq \mu(A \triangle B) + \mu(B \triangle C)$ and, consequently, if $A \sim B$ and $B \sim C$, $\mu(A \triangle C) = 0$ and $A \sim C$. Thus transitivity holds and $\sim$ is an equivalence relation in $\mathcal{M}$.

(b) Let $A, B, C \in \mathcal{M}$; as in (a) $\mu(A \triangle B) \leq \mu(A \triangle C) + \mu(C \triangle B)$. Now, if $A \sim C$, $\mu(A \triangle B) \leq \mu(C \triangle B)$ and switching A and C, also $\mu(C \triangle B) \leq \mu(A \triangle B)$. Therefore $\mu(A \triangle B) = \mu(C \triangle B)$ for all measurable B, d is well-defined, and we can simply denote $d([A], [B])$ by $d(A, B)$ where $A \in [A], B \in [B]$.

Also, d satisfies the triangle inequality. Moreover, $d(A, B) \geq 0$ and $d(A, B) = 0$ iff $\mu(A \triangle B) = 0$, i.e., if $A \sim B$, and so $[A] = [B]$. Since clearly $d(A, B) = \mu(A \triangle B) = \mu(B \triangle A) = d(B, A)$, d is a metric. Finally, completeness. Let $\{A_n\}$ be Cauchy in $(\widetilde{\mathcal{M}}, d)$ and pick a subsequence $n_m \to \infty$ such that $d(A_j, A_k) \leq 1/2^m$ for $j, k \geq n_m$. Put $B_m = A_{n_m}$ and $C_m =$

$B_m \triangle B_{m+1}$; then $\sum_m \mu(C_m) < \infty$ and by Borel-Cantelli $\mu(\limsup_m C_m) = 0$. Now, if $x \in (\limsup_m C_m)^c = \liminf_m C_m^c$, $x \in C_m^c$ for $m \geq M$ and since $C_m^c = \{x \in X : \chi_{B_m}(x) = \chi_{B_{m+1}}(x)\}$, $\lim_m \chi_{B_m}$ exists μ-a.e. Since χ_{B_m} only assumes the values 0 and 1, $\lim_m \chi_{B_m} = \chi_A$ where $A = \{x \in X : \lim_m \chi_{B_m}(x) = 1\}$ and $\lim_m B_m = A$.

We claim that $\lim_m d(B_m, A) = 0$. First, observe that for $C \in \mathcal{M}$,

$$\limsup_m C \triangle B_m \subset \limsup_m (C \cap B_m^c) \cup \limsup_m (C^c \cap B_m)$$
$$= (C \cap (\liminf_m B_m)^c) \cup (C^c \cap \limsup_m B_m)$$
$$= C \triangle \lim_m B_m \,.$$

Thus by Problem 85(b) with $C = A$ above,

$$\limsup_m \mu(A \triangle B_m) \leq \mu(A \triangle \limsup_m B_m) = \mu(A \triangle A) = 0 \,,$$

and so $\lim_m d(B_m, A) = 0$. Finally, since a Cauchy sequence in a metric space with a convergent subsequence converges to the same limit as the subsequence, $\lim_n d(A_n, A) = 0$ and the space is complete.

The fact that μ is a finite measure is not essential for the validity of the result. One could also observe that since $\chi_{A \triangle B}(x) = |\chi_A(x) - \chi_B(x)|$, $\mu(A \triangle B) = \int_X |\chi_A - \chi_B| \, d\mu$ and, consequently, $\{A_n\}$ is Cauchy in $(\widetilde{\mathcal{M}}, d)$ iff $\{\chi_{A_n}\}$ is Cauchy in $L^1(X)$. Since $L^1(X)$ is complete, let $f \in L^1(X)$ verify $\lim_n \int_X |\chi_{A_n} - f| \, d\mu = 0$. Let $\{A_{n_k}\}$ be a subsequence of $\{A_n\}$ such that $\lim_{n_k} \chi_{A_{n_k}} = f$ μ-a.e. and note that since $\chi_{A_{n_k}}$ only assumes the values 0 and 1, $f = \chi_A$ where $A = \{x \in X : \lim_{n_k} \chi_{A_{n_k}}(x) = 1\}$. Then $\lim_n d(A_n, A) = \lim_n \int_X |\chi_{A_n} - \chi_A| \, d\mu = 0$ and the space is complete.

As a matter of curiosity note that $d(A, A^c) = \mu(X)$.

90. First, since $A \triangle B = A^c \triangle B^c$, $d(A, B) = d(A^c, B^c)$ and ϕ is an isometry. Since the proof for all mappings is similar we do ϕ_1. Now, since as is readily seen $(A \cup B) \triangle (A_1 \cup B_1) \subset (A \triangle A_1) \cup (B \triangle B_1)$, it follows that $d(A \cup B, A_1 \cup B_1) \leq d(A, A_1) + d(B, B_1)$ and ϕ_1 is continuous.

92. Assume first that $\mu(A_1) < \infty$. Clearly $A_1 = \bigcup_n (A_n \setminus A_{n+1})$ and so $\mu(A_1) = \sum_n \mu(A_n \setminus A_{n+1})$. Now, $\mu(A_1 \setminus A_2) = 1/3$, $\mu(A_2 \setminus A_3) = \mu(A_1 \setminus A_3) - \mu(A_1 \setminus A_2) = 1/2 - 1/3 = (1/3)(1/2)$ and as is readily seen by induction, $\mu(A_n \setminus A_{n+1}) = (1/3)(1/2^{n-1})$, $n \geq 1$. Therefore the sum is $(1/3)(1 + 1/2 + \cdots) = 2/3$.

The result is not true unless μ is a finite measure or $\mu(A_1) < \infty$ as the example $A_1 = [0, \infty)$, $A_2 = [1/3, \infty)$, $A_3 = [1/2, \infty), \ldots$ shows.

93. (a) Let A be a finite set with an even number of elements and $\mathcal{A}$ the collection of all subsets of A that contain an even number of elements. Then $\mathcal{A}$ is a λ-system and a monotone class, but not an algebra or a π-system.

(b) Follows from the fact that for $A \subset B$, $B \setminus A = (A \cup B^c)^c$.

(c) First, suppose (iii) holds. Let $\{B_n\} \subset \mathcal{D}$ be pairwise disjoint and put $A_n = \bigcup_{k=1}^n B_k$; $\{A_n\}$ is an increasing sequence in $\mathcal{D}$ and by (iii) $\bigcup_n A_n = \bigcup_k B_k \in \mathcal{D}$. Conversely, let $\{A_n\} \subset \mathcal{D}$ be an increasing sequence and put $B_1 = A_1$ and $B_n = A_n \setminus A_{n-1}$, $n > 1$; the B_n are pairwise disjoint and by (ii) $\{B_n\} \subset \mathcal{D}$. Then $\bigcup_n B_n = \bigcup_n A_n \in \mathcal{D}$ and (iii) holds.

(d) Clearly the intersection of λ-systems is a λ-system; $\mathcal{D}(\mathcal{F})$ is then the intersection of all the λ-systems containing $\mathcal{F}$.

(e) First, since $X \cap A = A \in \mathcal{D}$, $X \in \mathcal{D}_A$. Next, suppose that $C \subset B \in \mathcal{D}_A$; then $C \cap A \subset B \cap A \in \mathcal{D}$ and $B \cap A \setminus C \cap A = (B \setminus C) \cap A \in \mathcal{D}$ and so $B \setminus C \in \mathcal{D}_A$. Finally, let $\{B_n\}$ be pairwise disjoint sets in $\mathcal{D}_A$; then $\{A \cap B_n\}$ are pairwise disjoint sets in $\mathcal{D}$ and, consequently, $\bigcup_n (A \cap B_n) = A \cap \bigcup_n B_n \in \mathcal{D}$. Thus $\bigcup_n B_n \in \mathcal{D}_A$.

(f) Necessity is obvious. As for sufficiency, first note that if $A, B \in \mathcal{D}$, $A \cup B \in \mathcal{D}$. Now, $A \cap B \in \mathcal{D}$ and since $A \cap B \subset B$, also $B \setminus (A \cap B) \in \mathcal{D}$; similarly, $A \setminus (A \cap B) \in \mathcal{D}$. Then $A \cup B = (B \setminus (A \cap B)) \cup (A \cap B) \cup (A \setminus (A \cap B))$ is the union of three disjoint sets in $\mathcal{D}$ and, consequently, $A \cup B \in \mathcal{D}$. The same argument gives that for any finite $\{A_n\}_{n=1}^N \subset \mathcal{D}$, $\bigcup_{n=1}^N A_n \in \mathcal{D}$. If now $\{A_n\} \subset \mathcal{D}$, let $B_n = (A_1 \cup \ldots \cup A_n) \setminus (A_1 \cup \ldots \cup A_{n-1})$. Then $\{B_n\}$ are pairwise disjoint subsets in $\mathcal{D}$ and so $\bigcup_n B_n = \bigcup_n A_n \in \mathcal{D}$. Finally, if $A \in \mathcal{D}$, since $X \in \mathcal{D}$, $X \setminus A = A^c \in \mathcal{D}$.

94. Clearly $\mathcal{D}(\mathcal{F}) \subset \mathcal{D}$. Now, by Problem 93 to verify that $\mathcal{D}(\mathcal{F})$ is a σ-algebra it suffices to prove that it is closed under intersections. So, let $A, B \in \mathcal{F}$; then $A \cap B \in \mathcal{F} \subset \mathcal{D}(\mathcal{F})$ and $B \in \mathcal{D}(\mathcal{F})_A$. By Problem 93(e) $D(\mathcal{F})_A$ is a λ-system containing $\mathcal{F}$ and, consequently, $\mathcal{D}(\mathcal{F}) \subset \mathcal{D}(\mathcal{F})_A$; hence if $A \in \mathcal{F}$ and $B \in \mathcal{D}(\mathcal{F})$, then $A \cap B \in \mathcal{D}(\mathcal{F})$. So, if $B \in \mathcal{D}(\mathcal{F})$, then $\mathcal{F} \subset \mathcal{D}(\mathcal{F})_B$. But then $\mathcal{D} \subset \mathcal{D}(\mathcal{F})_B$ because $\mathcal{D}(\mathcal{F})_B$ is a λ-system containing $\mathcal{F}$ and $\mathcal{D}(\mathcal{F})$ is the smallest such family. So $\mathcal{D}(\mathcal{F})$ is closed under intersections and we have finished.

95. Assume first that μ, ν are finite measures and $\mu(X) = \nu(X)$. Let $\mathcal{D} = \{A \in \mathcal{M}(\mathcal{F}) : \mu(A) = \nu(A)\}$; we verify that $\mathcal{D}$ is a λ-system and since $\mathcal{F} \subset \mathcal{D}$ implies $\mathcal{M}(\mathcal{F}) \subset \mathcal{D}$, then $\mu = \nu$. First, $\mu(X) = \nu(X)$ and $X \in \mathcal{D}$. Next, if $A, B \in \mathcal{D}$ and $A \subset B$, then $\mu(B \setminus A) = \mu(B) - \mu(A) = \nu(B) - \nu(A) = \nu(B \setminus A)$ and so $B \setminus A \in \mathcal{D}$. Finally, if $\{A_n\} \subset \mathcal{D}$ are pairwise disjoint, then $\mu(A_n) = \nu(A_n)$ for all n and so with $A = \bigcup_n A_n$, $\mu(A) = \lim_N \sum_{n=1}^N \mu(A_n) = \lim_N \sum_{n=1}^N \nu(A_n) = \nu(A)$. The proof for the case when the measures are σ-finite relative to $\mathcal{F}$ follows as in Problem 40.

This condition is necessary. Indeed, consider the counting measure on the integers and the Lebesgue measure which coincide on the π-system in $\mathbb{R}$ consisting of sets of the form $(-\infty, x]$, $x \in \mathbb{R}$, which generate $\mathcal{B}(\mathbb{R})$.

96. Necessity is trivial since $\mathcal{F} \subset \mathcal{M}(\mathcal{F})$ and $\mathcal{F}_1 \subset \mathcal{M}(\mathcal{F}_1)$. Conversely, fix C in $\mathcal{F}_1$ and let λ, ν be the measures on $\mathcal{M}(\mathcal{F})$ given by $\lambda(B) = \mu(B \cap C)$, $\nu(B) = \mu(B)\mu(C)$, $B \in \mathcal{M}(\mathcal{F})$, respectively; it is readily seen that λ and ν are finite measures on $\mathcal{M}(\mathcal{F})$ that agree on $\mathcal{F}$. Then by Problem 40, $\lambda = \nu$ on $\mathcal{M}(\mathcal{F})$, i.e., $\mu(B \cap C) = \mu(B)\mu(C)$ for any fixed C in $\mathcal{F}_1$ and all $B \in \mathcal{M}(\mathcal{F})$. Similarly, fix $B \in \mathcal{M}(\mathcal{F})$ and let λ, ν be the measures on $\mathcal{M}(\mathcal{F}_1)$ given by $\lambda(C) = \mu(B \cap C)$, $\nu(C) = \mu(B)\mu(C)$, $C \in \mathcal{M}(\mathcal{F}_1)$, respectively. By the first part of the argument λ and ν agree on $\mathcal{F}_1$ and, therefore, also on $\mathcal{M}(\mathcal{F}_1)$. Hence they are the same measure. This proves sufficiency.

97. By Problem 45 the set function ν on $\mathcal{M}(\mathcal{F})$ given by $\nu(A) = \mu(\phi^{-1}(A))$, $A \in \mathcal{M}(\mathcal{F})$, is a measure. Since $\phi^{-1}(X) = X$ it follows that $\nu(X) = \mu(X) = 1$ and by assumption that $\mu(F) = \nu(F)$ for all $F \in \mathcal{F}$. Therefore, by Problem 95, $\mu = \nu$.

100. (a) Let $B_n = \bigcup_{m=n}^{\infty} A_m$; clearly $\bigcap_n B_n = \limsup_n A_n$. Now, since $B_n \supset B_{n+1}$ and $\mu(B_1) < \infty$, $\mu(\limsup_n A_n) = \lim_n \mu(B_n)$. Furthermore, by σ-subadditivity, $\mu(B_n) \le \sum_{m=n}^{\infty} \mu(A_m)$, which being the tail of a convergent series goes to 0 as $n \to \infty$. This result is known as the Borel-Cantelli lemma.

(b) Let $B_n = \bigcup_{m=n}^{\infty} A_m$. Note that $B_n \supset B_{n+1}$ and $\mu(B_1) < \infty$ and so $\lim_n \mu(B_n) = \mu(\bigcap_n B_n) = 0$. Furthermore, since $A_n \subset B_n$, $0 \le \mu(A_n) \le \mu(B_n)$ and $\lim_n \mu(A_n) = 0$.

(c) Choose an increasing sequence $\{n_k\}$ such that $\mu(A_{n_k}) \le 2^{-k}$; this is always possible by the assumption. Then $\sum_k \mu(A_{n_k}) < \infty$ and by Borel-Cantelli, $\mu(\limsup_{n_k} A_{n_k}) = 0$.

As for the general case the answer is no. Consider the Lebesgue measure in $[0,1]$. Let $A_1 = [0, 1/2]$, $A_2 = [1/2, 1/2 + 1/3]$, and so on; the addition is mod 1. Now, a divergent series with terms < 1, like $1/n$, wraps around $[0,1]$ infinitely many times. Thus, although $|A_n| = 1/n \to 0$ as $n \to \infty$, each point in $[0,1]$ is in $\limsup_n A_n$.

102. For the sake of argument suppose there exist $\eta > 0$ and a subsequence $n_k \to \infty$ such that $a_{n_k} \ge \eta$ for all n_k, and for each n_k pick an interval $[i_k/n_k, (i_k+1)/n_k]$, $0 \le i_k \le n_k - 1$, such that $\mu([i_k/n_k, (i_k+1)/n_k]) \ge \eta$. Now, since I is compact a subsequence of $\{i_k/n_k\}$ converges to some $x \in I$. Using the same notation for the subsequence, since the intervals $[i_k/n_k, (i_k+1)/n_k]$ shrink to x, by Problem 85(a), $\mu(\{x\}) = \mu(\bigcap_k [i_k/n_k, (i_k+1)/n_k]) \ge \eta$, which is not possible since μ has no atoms.

104. First, assume that $M = X$. Let $B_n = A_n \cap A_{n+1}^c$, $B = \limsup_n B_n$, and $A = \limsup_n A_n$; we claim that $(\liminf_n A_n)^c \subset B \cup A^c$. Indeed, since $(\liminf_n A_n)^c \subset (A \cap (\liminf_n A_n)^c) \cup A^c$ and $(\liminf_n A_n)^c = \limsup_n A_n^c$ it suffices to prove that $C = (\limsup_n A_n) \cap (\limsup_n A_n^c) \subset B$. Let then $x \in C$. If $x \in A_n$, since $x \in \limsup_n A_n^c$ there exists a smallest integer

$m > 0$ such that $x \in A_{n+m}^c$ and, consequently, $x \in A_{n+m-1} \cap A_{n+m}^c$. Now, since $x \in \limsup_n A_n$, repeating this procedure we construct a sequence $n_k \to \infty$ such that $x \in A_{n_k} \cap A_{n_k+1}^c$, and so $x \in B$. Now, taking complements, $A \cap B^c \subset \liminf_n A_n$. Moreover, $\sum_n \mu(B_n) < \infty$ and so by Borel-Cantelli $\mu(B) = \mu(\limsup_n B_n) = 0$ and $\mu(B^c) = \mu(X)$. Therefore $\mu(A) = \mu(A \cap B^c) \le \mu(\liminf_n A_n) = 0$.

In the general case let $\nu(E) = \mu(M \cap E)$ denote the restriction of μ to $\mathcal{M}_M$. Then by the first part of the proof $\mu(M \cap \limsup_n A_n) = \nu(\limsup_n A_n) = 0$. Therefore $\mu(M) + \mu(\limsup_n A_n) = \mu(M \cap \limsup_n A_n) + \mu(M \cup \limsup_n A_n) \le 0 + \mu(X)$ and, consequently, $\mu(\limsup_n A_n) \le \mu(X) - \mu(M)$.

105. Let $B_n = \bigcup_{k \ge n} A_k$; $\{B_n\}$ is a nonincreasing sequence with $\mu(B_1) \le \mu(X) < \infty$ and, consequently, $\mu(\limsup_n A_n) = \lim_n \mu(B_n) \ge \mu(A_n) \ge \eta$. The conclusion is not true if $\mu(X) = \infty$: With μ the Lebesgue measure on $\mathbb{R}$, let $A_n = [n, n+1)$ for all n. Then $\mu(A_n) = 1$ for all n and $\mu(\limsup_n A_n) = 0$.

For the second part of the question consider the binary expansion $x = \sum_n x_n 2^{-n}$, $x_n = 0$ or 1 for all n, of $x \in [0,1]$. Let $A_n = \{x \in [0,1] : x_n = 0\}$; clearly the countable collection of numbers with finite dyadic expansion can be disregarded. Then $|A_n| = 1/2$ for each n and, therefore, $|\bigcap_{k=1}^m A_{n_k}| = 1/2^m$ and $|\bigcap_{n_k} A_{n_k}| = 0$.

106. It suffices to prove that

$$\mu(A_1^c \cap A_{n_1} \cap \ldots \cap A_{n_k}) = \mu(A_1^c)\mu(A_{n_1}) \cdots \mu(A_{n_k}) ,$$

for $2 \le n_1, \ldots, n_k$. First, since $A_1 \cup A_1^c = X$,

$$A_{n_1} \cap \ldots \cap A_{n_k} = \left(A_1^c \cap A_{n_1} \cap \ldots \cap A_{n_k}\right) \cup \left(A_1 \cap A_{n_1} \cap \ldots \cap A_{n_k}\right) .$$

Now, since the union is disjoint, $\mu(A_{n_1} \cap \ldots \cap A_{n_k}) = \mu(A_1^c \cap A_{n_1} \cap \ldots \cap A_{n_k}) + \mu(A_1 \cap A_{n_1} \cap \ldots \cap A_{n_k})$, which since the sets are independent can be rewritten as $\mu(A_{n_1}) \cdots \mu(A_{n_k}) = \mu(A_1^c \cap A_{n_1} \cap \ldots \cap A_{n_k}) + \mu(A_1)\mu(A_{n_1}) \cdots \mu(A_{n_k})$. Thus, rearranging terms,

$$\mu(A_1^c \cap A_{n_1} \cap \ldots \cap A_{n_k}) = (1 - \mu(A_1))\mu(A_{n_1} \cap \ldots \cap A_{n_k})$$
$$= \mu(A_1^c)\mu(A_{n_1}) \cdots \mu(A_{n_k}).$$

108. First, $A^c = \bigcup_m \bigcap_{n=m}^{\infty} A_n^c$. Observe that for fixed m, $\mu(\bigcap_{n=m}^{\infty} A_n^c) = \lim_k \mu(\bigcap_{n=m}^{k} A_n^c) = \lim_k \prod_{n=m}^{k} \mu(A_n^c) = \lim_k \prod_{n=m}^{k} (1 - \mu(A_n))$, which since $1 - x \le e^{-x}$ is bounded by

$$\lim_k \prod_{n=m}^{k} \exp(-\mu(A_n)) = \lim_k \exp\left(-\sum_{n=m}^{k} \mu(A_n)\right) = 0 .$$

Thus $\mu(A) = 1 - 0 = 1$. This result is known as the second Borel-Cantelli lemma and independence is necessary for it to hold. Take $0 < \mu(B) < 1$, $A_n = B$ for all n. Then $\sum_n \mu(A_n) = \infty$ but $A = B$ and $\mu(A) < 1$.

109. Let $E_n = C_{2n} = A_{2n} \cap A_{2n+1}$. Then the E_n are independent and since $\sum_n \mu(E_n) = \infty$, by Problem 108, $\mu(\limsup_n E_n) = 1$. Now, $\limsup_n E_n \subset \limsup_n C_n$ and so $\mu(\limsup_n C_n) = 1$.

110. Sufficiency is clear: Since $\limsup_n A_n \subset \bigcup_n A_n$ it readily follows that $1 = \mu(\limsup_n A_n) \le \mu(\bigcup_n A_n) \le 1$.

As for necessity, first note that $\mu((\bigcup_n A_n)^c) = \mu(\bigcap_n A_n^c) = 0$. Now, by Problem 106 the family $\{A_n^c\}$ is independent and so, $0 = \mu(\bigcap_n A_n^c) = \mu(\bigcap_{n=1}^{k-1} A_n^c)\,\mu(\bigcap_{n=k}^{\infty} A_n^c) = (\prod_{n=1}^{k-1} \mu(A_n^c))\mu(\bigcap_{n=k}^{\infty} A_n^c)$. And, since $\mu(A_n) < 1$ for all n, $\mu(A_n^c) > 0$ for all n, and therefore the first factor above has positive measure, which implies that the second factor is 0 for all $k \ge 1$ and, consequently, $\mu(\liminf_n A_n^c) = \mu(\bigcup_m \bigcap_{n=m}^{\infty} A_n^c) \le \sum_m \mu(\bigcap_{n=m}^{\infty} A_n^c) = 0$. Therefore $\mu(\limsup_n A_n) = 1 - \mu(\liminf_n A_n^c) = 1$.

111. Since T is measure preserving, $\sum_n \mu(T^{-1}(A_n)) = \sum_n \mu(A_n) < \infty$, and so by Borel-Cantelli $\mu(\limsup_n T^{-1}(A_n)) = 0$.

112. (a) implies (b) Suppose that $A \in \mathcal{M}$ with $T^{-1}(A) \subset A$ and with T^0 the identity, let $B = \bigcap_{k=0}^{\infty} T^{-k}(A)$. Then $T^{-1}(B) = B$ and by (a), $\mu(B) = 0$ or 1. Now, by continuity from above, $\mu(A) = \lim_k \mu(T^{-k}(A)) = \mu(B)$ and $\mu(A)$ is 0 or 1.

(b) implies (c) If $A \in \mathcal{M}$ and $T^{-1}(A) \supset A$, then $T^{-1}(A^c) \subset A^c$ and by (b), $\mu(A^c) = 0$ or 1 and, therefore, $\mu(A) = 0$ or 1.

(c) implies (a) If $A \in \mathcal{M}$ and $T^{-1}(A) = A$, then $T^{-1}(A) \supset A$ and by (c) we are done.

113. If $y > x$, $F(x) - F(y) = \sum_{x < n \le y} p_n \ge 0$ and F is nondecreasing. Also, $\lim_{y \to x^+} F(y) - F(x) = \lim_{y \to x^+} \sum_{x < n \le y} p_n$. Note that if y is sufficiently close to x, $(x, y]$ does not contain an integer. Indeed, if x is an integer and $x < y < x + 1$, $(x, y]$ does not contain an integer. And, if x is not an integer and n is the smallest integer $x < n$ as long as $y < n$, $(x, y]$ contains no integer. So $\sum_{x < n \le y} p_n = 0$ for y sufficiently close to x, $\lim_{y \to x^+} F(y) = F(x)$, and F is right-continuous.

Note that we have $n \le x$, and not $n < x$, in the definition to insure the right-continuity of F.

114. Let

$$F(x) = \begin{cases} -\mu\big((x, 0]\big), & x < 0, \\ 0, & x = 0, \\ \mu\big((0, x]\big), & 0 < x. \end{cases}$$

By assumption $F(x)$ is finite for $x \in \mathbb{R}$. Also if $x > 0$ and $\varepsilon_n \to 0^+$, $F(x + \varepsilon_n) = F(x) + \mu\big((x, x + \varepsilon_n]\big)$ and since by continuity from above $\lim_n \mu((x, x + \varepsilon_n]) = \mu(\bigcap_n(x, x + \varepsilon_n]) = \mu(\emptyset) = 0$, F is right-continuous at x. Similarly for the other values of x.

Let μ_F denote the Borel measure associated to F. We claim that $\mu_F = \mu$; by Problem 40 it suffices to verify that μ_F and μ coincide on intervals of the form $(a, b]$. First, if $0 \leq a \leq b$, $\mu_F((a, b]) = F(b) - F(a) = \mu((0, b]) - \mu((0, a]) = \mu((a, b])$. Next, if $a \leq 0 \leq b$, $\mu_F((a, b]) = F(b) - F(a) = \mu((0, b]) - (-\mu((a, 0])) = \mu((a, 0]) + \mu((0, b]) = \mu((a, b])$, and, finally, if $a \leq b \leq 0$, $\mu_F((a, b]) = F(b) - F(a) = -\mu((b, 0]) - (-\mu((a, 0])) = \mu((a, b])$.

115. First, $\psi = \sum_n 2^{-n}\delta_{1/n}$ is a measure and since $\psi(\mathbb{R}) = 1$, ψ is a probability measure. Therefore its distribution function is given by

$$F(x) = \psi\big((-\infty, x]\big) = \begin{cases} 0, & x \leq 0, \\ \sum_{k=n+1}^{\infty} 2^{-k} = 2^{-n}, & 1/(n+1) \leq x < 1/n, \\ 1, & 1 \leq x. \end{cases}$$

116. (a) First, since $\{a\} = \bigcap_n(a - 1/n, a]$,

$$\mu_F(\{a\}) = \lim_n \mu_F((a - 1/n, a]) = \lim_n F(a) - F(a - 1/n) = F(a) - F(a^-).$$

Then,

$$\mu_F((a, b)) = \mu_F((a, b]) - \mu_F(\{b\})$$
$$= F(b) - F(a) - (F(b) - F(b^-)) = F(b^-) - F(a);$$

similarly, $\mu_F([a, b)) = F(b^-) - F(a^-)$ and $\mu_F([a, b]) = F(b) - F(a^-)$.

(b) $\mu(B_1) = F(2) - F(2^-) = 6 - 3 = 3$; $\mu(B_2) = F(3^-) - F(-1/2^-) = 9 - 1/2 = 17/2$; $\mu(B_3) = F(0) - F(-1) + F(2^+) - F(1) = 2 - 0 + 6 - 3 = 5$; and, $\mu(B_4) = F(1/2^-) - F(0^-) + F(2) - F(1) = 9/4 - 1 + 9 - 3 = 29/4$.

118. Not necessarily. Let $F(x) = 0$ if $x < 0$, $F(x) = x$ if $x \in [0, 1]$, and $F(x) = 1$ if $x > 1$. Then for $B = [0, 1]$ we have $\mu_F(B) = 1$ and $\mu_F(\mathbb{R} \setminus B) = \mu((-\infty, 0)) + \mu((1, \infty)) = F(0) - F(-\infty) + F(\infty) - F(1) = 0$ but $\mathbb{R} \setminus B$ is not dense in $\mathbb{R}$.

119. By Problem 64, $\mu_F = \mu_1 + \mu_2$ where μ_1 is nonatomic and μ_2 is atomic and by Problem 40 it suffices to identify μ_1, μ_2 on intervals of the form $(a, b]$. There are three cases, namely, $a \geq 0$, $b \leq 0$, and $a < 0 < b$. In the first case, with $[x] = $ integer part of x, $\mu((a, b]) = F(b) - F(a) = b + [b] - (a + [a]) = (b - a) + [b] - [a]$. Similarly, in the second case, $\mu((a, b]) = F(b) - F(a) = b - a$. Finally, in the third case, $\mu((a, b]) = F(b) - F(a) = b - a + [b]$. Thus in every case $\mu((a, b]) = |(a, b]| + $ number of positive integers in $(a, b]$. Whence $\mu_F = \mu_1 + \mu_2$ where μ_1 is the Lebesgue measure on $\mathcal{B}(\mathbb{R})$ and μ_2 is the counting measure of the positive integers.

120. By the right-continuity of F it follows that $F(x) = 0$ or $F(x) = 1$ for all $x \in \mathbb{R}$. Let $a = \inf\{x \in \mathbb{R} : F(x) = 1\}$. Since $F(x) \to 1$ as $x \to \infty$ we have $a < \infty$; similarly, $a > -\infty$. Then by the right-continuity of F, $F(x) = \chi_{[a,\infty)}(x)$ and, therefore, F is the distribution function of δ_a.

121. (a) First, note that if $t \in [0, F(s)]$, from the definition of $F^{-1}(t)$ it follows that $F^{-1}(t) \leq s$, and, consequently, $[0, F(s)] \subset \{t \in [0, L] : F^{-1}(t) \leq s\}$. Next, if $t > F(s)$, since F is right-continuous and nondecreasing there exists $x > s$ such that $t > F(x)$, and, consequently, $t > F(y)$ for $s \leq y < x$. Hence $F^{-1}(t) \geq x > s$ and $F^{-1}(t) > s$. In other words, $[0, F(s)]^c \subset \{t \in [0, L] : F^{-1}(t) \leq s\}^c$ and taking complements also $\{t \in [0, L] : F^{-1}(t) \leq s\} \subset [0, F(s)]$.

(b) Clearly F^{-1} is Borel measurable on $[0, L]$ and from the definition of μ it follows that $\mu((a, s]) = F(s) - F(a) = F(s) = |[0, F(s)]| = |\{t \in [0, L] : F^{-1}(t) \leq s\}|$.

122. First, as in Problem 121, $[0, F(s)] = \{t \in [0, L] : F^{-1}(t) \leq s\}$. Then $F^{-1}(u)$ can be written as

$$F^{-1}(u) = \begin{cases} x_1, & 0 < u \leq p_1, \\ x_\ell, & \sum_{i=1}^{\ell-1} p_i < u \leq \sum_{i=1}^{\ell} p_i, 2 \leq \ell \leq k. \end{cases}$$

123. First, $[a, b) = \Phi[\Phi^{-1}(a), \Phi^{-1}(b)) = \Phi \circ \Phi^{-1}([a, b))$ and so we have $\mu_{F \circ \Phi}(\Phi^{-1}[a, b)) = F(b) - F(a) = \mu_F([a, b))$. The conclusion now follows from Problem 40. Note that in particular, $\mu_\Phi(\Phi^{-1}(A)) = |A|$ for every Borel $A \subset \mathbb{R}$.

124. First, observe that for no sequence $\{x_n\}$ may it happen that, simultaneously, there exists $\eta > 0$ such that $\varphi(x_n) \geq \eta > 0$ for all n and that the S_{x_n} have bounded overlaps, in the sense that each point belongs to at most N of the S_{x_n}. Indeed, since $\sum_n \varphi(x_n) = \sum_n \mu(S_{x_n}) = \int_{\mathbb{R}^2} \sum_n \chi_{S_{x_n}}(x) \, d\mu(x)$, if the S_{y_n} have bounded overlaps, the expression is bounded by

$$N \int_{\mathbb{R}^2} \chi_{\bigcup_n S_{x_n}}(x) \, d\mu(x) \leq N\mu(\mathbb{R}^2) < \infty$$

and so the sequence has to be finite while, on the other hand, if the sequence is infinite, $\sum_n \varphi(x_n) \geq \sum_n \eta = \infty$ and the S_{x_n} cannot have bounded overlaps.

First, necessity. For the sake of argument suppose that φ is continuous at x and $\varphi(x) = \eta > 0$; then there exists $\delta > 0$ such that $\varphi(y) \geq \eta/2$ for all $y \in B(x, \delta)$. Let $\{x_n\}$ be any points with $|x - x_n| = \delta/2$ for all n. Then the S_{x_n} intersect in exactly 2 points, contrary to the above observation.

Conversely, we claim that if φ is discontinuous at x, $\varphi(x) > 0$. Indeed, there exist $\varepsilon > 0$ and $x_n \in B(x, 1/n)$ with $\varphi(x_n) \geq \varepsilon$, $n = 1, 2, \ldots$. Now, by Problem 85(b), $\mu(\limsup_n S_{x_n}) \geq \varepsilon$. Next, we claim that $\limsup_n S_{x_n} \subset S_x$. Indeed, if $y \in \limsup_n S_{x_n}$, there exists $\{x_{n_k}\}$ such that $y \in \bigcap_k S_{x_{n_k}}$ and each one of them satisfies $|y - x_{n_k}| = 1$; hence

$$|y - x| = |(y - x_{n_k}) + (x_{n_k} - x)| \to 1$$

and so $y \in S_x$. Consequently, $\varphi(x) \geq \mu(\limsup_n S_{x_n}) \geq \varepsilon$ and we have finished.

125. Recall that by one of the characterizations of semicontinuity, a function f on $\mathbb{R}^n$ is lower semicontinuous if $\{f > \lambda\}$ is open for all real λ and upper semicontinuous if $\{f < \lambda\}$ is open for all real λ.

Now, for $n = 1$, let δ be the Dirac measure at 0 and take $O = (0, 1)$. Then

$$\varphi(x) = \delta(x + O) = \begin{cases} 1, & x \in (-1, 0), \\ 0, & x \notin (-1, 0), \end{cases}$$

is not upper semicontinuous and so not continuous.

Nevertheless, we claim that φ is lower semicontinuous. Let $\lambda \in \mathbb{R}$ and suppose that $\varphi(x) > \lambda$. Now, by inner regularity there is a compact $K \subset x + O$ with $\mu(K) > \lambda$. Since $\mathbb{R}^n \setminus (x + O)$ is closed and $K \cap (\mathbb{R}^n \setminus (x + O)) = \emptyset$, there is $\varepsilon > 0$ such that $|y - z| > \varepsilon$ for all $y \in K$ and $z \in \mathbb{R}^n \setminus (x + O)$. Now, suppose that $y \in \mathbb{R}^n$ and $|x - y| < \varepsilon$. Then also $K \subset y + O$ and so $\varphi(y) \geq \mu(K) > \lambda$. Thus $\{\varphi > \lambda\}$ is open and φ is lower semicontinuous.

A similar argument gives that for a closed set C the function $\psi(x) = \mu(x + C)$ is upper semicontinuous.

Note that in particular, if μ is a Borel measure on $\mathbb{R}^n$ that is finite on bounded sets, since the open ball $B(x, r) = x + B(0, r)$, by the above argument it follows that $\mu(B(x, r))$ is lower semicontinuous as a function of x for each $r > 0$.

126. Since $D_k(x) = \lim_{\delta \to 0+} (\sup_{0 < r < \delta} r^{-k} \mu(B(x, r)))$ is the limit of an increasing function of δ it readily follows that $\limsup_{r \to 0+} r^{-k} \mu(B(x, r)) = \inf_{j \geq 1} (\sup_{0 < r < 1/j} r^{-k} \mu(B(x, r)))$ and, therefore, $\sup_{0 < r < 1/j} r^{-k} \mu(B(x, r))$ is a Borel measurable function of x for each $j = 1, 2, \ldots$, and so is the inf of these functions.

Let $f(x) = \lim_{r \to 0} G_r(x)$ where $G_r(x) = \sup_{0 < \rho < r} \rho^{-k} \mu(B(x, \rho))$. By Problem 125, for each ρ the function $\rho^{-k} \mu(B(x, \rho))$ is a lower semicontinuous function of x, and since the sup of a family of lower semicontinuous functions is lower semicontinuous, G_r is lower semicontinuous and Borel measurable. Now, since $f(x) = \lim_n G_{1/n}(x)$, f is the limit of Borel measurable functions and, hence, Borel measurable.

An analogous result with a similar proof holds for lower k-densities defined with $\liminf$ instead of $\limsup$ above.

127. Suppose that μ is a finite measure. By continuity from below $\mu(\mathbb{R}^n) = \lim_{r\to\infty} \mu(Q_r)$ where Q_r is the cube of sidelength r centered at the origin and so, given $\varepsilon > 0$, there exists $r > 0$ such that $\mu(\mathbb{R}^n \setminus Q_r) \le \varepsilon$. Now, consider the translate of Q_r to the cube $Q_r(x_0)$ centered at $x_0 = (0,\ldots,0,r/2)$. Then $Q_{4r}(x_0) \supset Q_r$ and, consequently, by monotonicity and doubling, $\mu(Q_r) \le \mu(Q_{4r}(x_0)) \le c^2\mu(Q_r(x_0)) \le c^2\varepsilon$. Thus $\mu(\mathbb{R}^n) = \mu(Q_r) + \mu(\mathbb{R}^n \setminus Q_r) \le (c^2 + 1)\varepsilon$ and since ε is arbitrary, $\mu(\mathbb{R}^n) = 0$.

Chapter 13

Lebesgue Measure

Solutions

1. Let $\mathbb{Q}^n$ denote the points in $\mathbb{R}^n$ with rational coordinates and for $x, y \in \mathbb{R}^n$ consider the equivalence relation $x \sim y$ iff $x - y \in \mathbb{Q}^n$. A Vitali set $V \subset [0,1)^n$ is one that contains exactly one element from each equivalence class; this construction is made possible by the axiom of choice. Note that $\mathbb{R}^n = V + \mathbb{Q}^n = \bigcup_{v \in V}(v + \mathbb{Q}^n)$ and since $[0,1)^n$ is partitioned into the equivalence classes and each class $v + \mathbb{Q}^n$ is countable, there are c classes.

Let $q_1 = (0, \dots, 0), q_2, \dots$, be an enumeration of $(-1,1)^n \cap \mathbb{Q}^n$ and put $V_k = q_k + V$, $V_1 = V$. Then $V_k \cap V_m = \emptyset$, $k \neq m$, for otherwise two points in V would differ by a nonzero element in $\mathbb{Q}^n$. Also, $[0,1)^n \subset \bigcup_k V_k \subset (-1,2)^n$. Indeed, if $z \in [0,1)^n$, $z - q_k \in V$ for some q_k with $k \geq 1$ and so $z \in \bigcup_k V_k$. And, since $z \in \bigcup_k V_k$ is of the form $z = y + q_k$ with $y \in [0,1)^n$ and $q_k \in (-1,1)^n$, $z \in (-1,2)^n$. Now, by translation invariance, V_k is measurable iff V is measurable and $|V_k| = |V|$ for all k. Thus, if V is Lebesgue measurable, $1 = |[0,1)^n| \leq \sum_k |V_k| = \sum_k |V| \leq 3^n$ where $\sum_k |V|$ is 0 or ∞ and both choices lead to contradiction. Therefore V is Lebesgue nonmeasurable.

Note that if V is a Vitali Lebesgue nonmeasurable subset of $[0,1)^n$, there is a G_δ set G containing V such that $|G| = |V|_e$ and $|G \setminus V|_e > 0$.

2. There are 2^c. Let V denote a Vitali Lebesgue nonmeasurable set in $[0,1)$, $K = \{y + 1 : y \in C\} \subset [1,2]$ where C is the Cantor discontinuum, and $\mathcal{P}(K)$ the collection of subsets of K. Now put $\mathcal{V} = \{V \cup P : P \in \mathcal{P}(K)\}$; each element of $\mathcal{V}$, the disjoint union of a Lebesgue nonmeasurable set V and a measurable subset of K (since $|K| = 0$ all its subsets are measurable), is Lebesgue nonmeasurable. Furthermore, since $\operatorname{card}(K) = c$, $\operatorname{card}(\mathcal{V}) = \operatorname{card}(\mathcal{P}(K)) = 2^c$.

4. First, suppose that $|B|_e < \infty$ and let $\eta = \sup\{|E| : E \subset B, E \in \mathcal{L}(\mathbb{R}^n)\}$; if $\eta = 0$ there is nothing to prove. Otherwise, let $B_k \subset B$ be such that $\{B_k\} \subset \mathcal{L}(\mathbb{R}^n)$ and $|B_k| \to \eta$. Then $|\bigcup_k B_k| \geq \eta$ and since $B_k \subset B$ for all k, $|\bigcup_k B_k| = \eta$. Now, $B_0 = B \setminus \bigcup_k B_k$ cannot contain a measurable subset of positive measure because if it contained one such $A \in \mathcal{L}(\mathbb{R}^n)$, say, then $A \cup \bigcup_k B_k \subset B$ and, contrary to the definition of η, $|A \cup \bigcup_k B_k| = |A| + |\bigcup_k B_k| > \eta$. Thus B_0 only contains sets of measure 0 or Lebesgue nonmeasurable sets. Moreover, since B is Lebesgue nonmeasurable, so is B_0. In case $|B|_e = \infty$ the above argument produces a measurable $A_k \subset B \cap \{|x| < k\}$ such that $(B \cap \{|x| < k\}) \setminus A_k$ contains no measurable set of positive measure. Then $B \setminus (\bigcup_k A_k)$ contains no measurable set of positive measure.

5. Necessity first. For the sake of argument suppose that for each $k \geq 1$ there is $A_k \in \mathcal{L}(\mathbb{R}^n)$, $A_k \subset B$, with $|B \setminus A_k|_e < 1/k$. Let $A = \bigcup_k A_k$. Then $|B \setminus A|_e \leq |B \setminus A_k|_e \leq 1/k$ for all k and, therefore, $|B \setminus A|_e = 0$. Thus $B = (B \setminus A) \cup A$ is the union of a measurable set and a set of measure 0 and so $B \in \mathcal{L}(\mathbb{R}^n)$, which is not the case.

Sufficiency next. For the sake of argument suppose that $B \in \mathcal{L}(\mathbb{R}^n)$. Then, by the inner regularity of the Lebesgue measure, given $\varepsilon > 0$, there is a closed $F \subset B$ such that $|B \setminus F| < \varepsilon$, which is not the case.

7. For the sake of argument suppose that $|A|_e < \infty$ and let O be an open set of finite measure containing A. Then $F = \mathbb{R}^n \setminus O$ is a closed set with $|F| = \infty$ and contains a compact subset K of positive measure. Now, since $F \cap A = \emptyset$, $K \cap A = \emptyset$ and this cannot happen.

A similar argument gives that if a subset A of a bounded interval J in $\mathbb{R}^n$ intersects every compact subset of J of positive measure, then $|A|_e = |J|$.

8. Yes. Let $\mathcal{C}$ denote the family of compact subsets of $[0,1]$ of positive measure; by Problem 1.70, with Ω the first uncountable ordinal, $\mathcal{C} = \{K_\alpha : \alpha < \Omega\}$. Let $\{x_n^0\}$ be a sequence in K_0 and, having picked sequences $\{x_n^\beta\}$ in $K_\beta \setminus \bigcup_{\delta < \beta} \{x_n^\delta\}$ for $\beta < \alpha < \Omega$, let $\{x_n^\alpha\}$ be a sequence in $K_\alpha \setminus \bigcup_{\beta < \alpha} \{x_n^\beta\}$; this choice is always possible since K_α is uncountable and $\bigcup_{\beta < \alpha} \{x_n^\beta\}$, the countable union of countable sets, is countable. Now, let $B_n = \{x_n^\alpha : \alpha < \Omega\}$; the B_n are pairwise disjoint and since each B_n meets every compact subset of $[0,1]$ of positive measure, by Problem 7, $|B_n \cap (0,1)|_e = 1$. Finally, no B_n can be measurable because if it were, by the inner regularity of the Lebesgue measure we would have $|B_n| = \sup\{|K| : K \text{ compact}, K \subset B_n\} = 0$ (since B_n only contains countable compact subsets) but, as noted above, $|B_n|_e = 1$.

10. By the definition of Lebesgue measure, given $\varepsilon > 0$, there are intervals $\{I_k\}$ such that $A \subset \bigcup_k I_k$ and $\sum_k v(I_k) \leq |A| + \varepsilon/2$. Let $\{I_k'\}$ denote open intervals such that $I_k \subset I_k'$ and $|I_k'| = |I_k| + \varepsilon 2^{-(k+1)}$. Now, since A is compact there are finitely many open intervals $I_{n_1}', \ldots, I_{n_N}'$, say, such that $A \subset \bigcup_{k=1}^{N} I'_{n_k}$. Then $\sum_{k=1}^{N} v(I'_{n_k}) \le \sum_k v(I'_k) = \sum_k v(I_k) + \varepsilon/2 \le |A| + \varepsilon$. The conclusion follows since ε is arbitrary.

13. (a) Let $G_A \supset A$, $G_B \supset B$ be G_δ sets with $|G_A| = |A|_e$ and $|G_B| = |B|_e$, respectively. Then, since $|G_A \cup G_B| \le |G_A| + |G_B| = |A|_e + |B|_e = |A \cup B|_e \le |G_A \cup G_B|$ it follows that $|G_A \cup G_B| = |G_A| + |G_B|$. Therefore $|G_A \cap G_B| = 0$ and, since $|A \cap B| \le |G_A \cap G_B|$, also $|A \cap B| = 0$.

(b) Let G_A, G_B be as in (a); since $A \cap G_B \subset G_A \cap G_B$, $A \cap G_B$ is null and, hence, measurable. Now, by a proof by pictures $A = ((A \cup B) \setminus G_B) \cup (A \cap G_B)$ and $A \in \mathcal{L}(\mathbb{R}^n)$. The argument for B is analogous.

14. (a) Since equality holds when $|A| = \infty$ or $|B|_e = \infty$ we may assume that $|A|, |B|_e < \infty$. First, note that $(A \cup B) \cap A = A$ and $(A \cup B) \cap A^c = B \cap A^c$. Therefore, by Carathéodory's characterization $|A \cup B|_e = |A| + |B \cap A^c|_e$. Hence $|A \cup B|_e + |A \cap B|_e = |A| + |A \cap B|_e + |B \cap A^c|_e$ which, by Carathéodory's characterization, equals $|A| + |B|_e$.

(b) First, by Carathéodory's characterization, since $(A \cup B) \cap C = A$ and $(A \cup B) \setminus C = B \setminus C$, $|A \cup B|_e = |A|_e + |B \setminus C|_e$. Now, by Carathéodory again, since $B \cap C = \emptyset$, $|B|_e = |B \setminus C|_e$ and so $|A \cup B|_e = |A|_e + |B|_e$.

15. The statement is not necessarily true if $|A|_e = \infty$.

16. The statement is true. First, necessity follows from the additivity of the Lebesgue measure. Conversely, we prove that $|E|_e = |E \cap A|_e + |E \setminus A|_e$ for every $E \subset \mathbb{R}^n$ and invoke Carathéodory's characterization. First, by the σ-subadditivity of the Lebesgue outer measure, $|E|_e \le |E \cap A|_e + |E \setminus A|_e$, and, in particular, equality holds if $|E|_e = \infty$. On the other hand, if $|E|_e < \infty$ let O be an open set such that $E \subset O$ and $|O| \le |E|_e + \varepsilon$. Now, $O = \bigcup_k Q_k$ is the union of nonoverlapping closed cubes and so by the σ-subadditivity of the outer Lebesgue measure, $|O \cap A|_e \le \sum_k |Q_k \cap A|_e$ and $|O \setminus A|_e \le \sum_k |Q_k \setminus A|_e$. Therefore $|E \cap A|_e + |E \setminus A|_e \le |O \cap A|_e + |O \setminus A|_e \le \sum_k |Q_k \cap A|_e + \sum_k |Q_k \setminus A|_e = \sum_k |Q_k| \le |E|_e + \varepsilon$, which, since ε is arbitrary, implies that $|E \cap A|_e + |E \setminus A|_e \le |E|_e$ and Carathéodory's characterization does the rest. Finally, if $|E|_e = \infty$, let $E_k = E \cap \{|x| \le k\}$; by above $|E \cap A|_e + |E \setminus A|_e \ge |E_k \cap A|_e + |E_k \setminus A|_e = |E_k|_e$ for all k and the conclusion follows letting $k \to \infty$.

17. First, necessity. Given $\varepsilon > 0$, let $\{I_k\}$ be closed intervals such that $A \subset \bigcup_k I_k$ and $|A| \le \sum_k |I_k| \le |A| + \varepsilon/2$. Now, let N be large enough so that $|\bigcup_k I_k \setminus \bigcup_{k=1}^{N} I_k| \le \sum_{k=N+1}^{\infty} |I_k| \le \varepsilon/2$ and observe that $|A \setminus \bigcup_{k=1}^{N} I_k| \le |\bigcup_k I_k \setminus \bigcup_{k=1}^{N} I_k| \le \varepsilon/2$. Moreover, with this choice of N, $|\bigcup_{k=1}^{N} I_k \setminus A| \le |\bigcup_k I_k \setminus A| = |\bigcup_{k=1}^{\infty} I_k| - |A| \le \sum_k |I_k| - |A| < \varepsilon/2$ and combining both estimates we get $|A \triangle \bigcup_{k=1}^{N} I_k| < \varepsilon$.

Next, sufficiency. Given $\varepsilon > 0$, let $\{I_1, \ldots, I_N\}$ be a finite collection of closed intervals such that $|A \triangle \bigcup_{k=1}^{N} I_k|_e < \varepsilon/3$. Now let $\{I_k'\}$ be a collection of open intervals such that $I_k \subset I_k'$ and $|I_k'| \le |I_k| + \varepsilon/3N$, all $k \le N$. Then $I_k' = I_k \cup (I_k' \setminus I_k)$ and, since $\bigcup_{k=1}^{N} I_k' \setminus A \subset (\bigcup_{k=1}^{N} I_k \setminus A) \cup \bigcup_{k=1}^{N} (I_k' \setminus I_k)$, it follows that $|\bigcup_{k=1}^{N} I_k' \setminus A|_e \le |\bigcup_{k=1}^{N} I_k \setminus A|_e + \sum_{k=1}^{N} |I_k' \setminus I_k| < \varepsilon/3 + N\varepsilon/3N = 2\varepsilon/3$. Next, since $|A \setminus \bigcup_{k=1}^{N} I_k'|_e \le |A \setminus \bigcup_{k=1}^{N} I_k|_e < \varepsilon/3$ there is an open set W such that $A \setminus \bigcup_{k=1}^{N} I_k' \subset W$ and $|W| < \varepsilon/3$. Now let $O = W \cup \bigcup_{k=1}^{N} I_k'$; O is open and $A = (A \setminus \bigcup_{k=1}^{N} I_k') \cup (A \cap \bigcup_{k=1}^{N} I_k') \subset W \cup \bigcup_{k=1}^{N} I_k' = O$. Hence $|O \setminus A|_e \le |(\bigcup_{k=1}^{N} I_k') \setminus A|_e + |W \setminus A|_e < 2\varepsilon/3 + |W| < \varepsilon$ and A is measurable.

18. By Problem 17 there exist $\{A_k\}$ where each A_k is a finite union of intervals and $|A \triangle A_k| < 1/k$ for all k. Note that for each $\varepsilon > 0$, $\{|\chi_{A_k} - \chi_A| > \varepsilon\} \subset A_k \triangle A$ and so $|\{|\chi_{A_k} - \chi_A| > \varepsilon\}| \le |A_k \triangle A| \le 1/k$ for all k. Therefore $\{\chi_{A_k}\}$ tends to χ_A in measure and, consequently, there is a subsequence of $\{A_k\}$, which we call simply $\{A_k\}$, such that $\chi_{A_k}(x) \to \chi_A(x)$ for $x \in B^c \subset A$, $|B \cap A| = 0$.

Next, observe that $(\liminf_k A_k) \cap B^c = A \cap B^c$. Indeed, since $\chi_{A_k}(x) \to \chi_A(x)$ for $x \in B^c$, $x \in A \cap B^c$ iff $\chi_{A_k}(x) = 1$ for all k large enough iff $x \in (\liminf_k A_k) \cap B^c$. We claim that this implies that $A \triangle (\liminf_k A_k) \subset B$. Indeed,

$$\left(\liminf_k A_k \right) \cap A^c$$
$$= \left(\left(\liminf_k A_k \right) \cap A^c \cap B \right) \cup \left(\left(\liminf_k A_k \right) \cap A^c \cap B^c \right)$$
$$= \left(\left(\liminf_k A_k \right) \cap A^c \cap B \right) \cup \left(A \cap A^c \cap B^c \right) \subset B,$$

and similarly for $A \cap (\liminf_k A_k)^c$. Thus $|A \triangle (\liminf_k A_k)| \le |B| = 0$.

19. First, since $\{|A_k|_e\}$ is an increasing numerical sequence it has a limit and by monotonicity $\lim_k |A_k|_e \le |\bigcup_k A_k|_e$.

To prove the opposite inequality consider G_δ sets $\{G_k\}$ such that $A_k \subset G_k$ and $|A_k|_e = |G_k|$ for all k. Now, by Problem 2.85(a), $|\liminf_k G_k| \le \liminf_k |G_k|$, and, since $\{A_k\}$ converges to $A = \bigcup_k A_k$, $\liminf_k A_k = A$. Hence $|A|_e = |\liminf_k A_k|_e \le |\liminf_k G_k| \le \liminf_k |G_k| = \liminf_k |A_k|_e = \lim_k |A_k|_e$.

21. (a) The example in Problem 8 works. Also the V_k in Problem 1 are pairwise disjoint and their union is contained in $[0, 2)$, which gives $|\bigcup_k V_k|_e \le 2$, and since $|V_k|_e = |V|_e > 0$ for all k, $\sum_k |V_k|_e = \infty$.

(b) Let $A_k = \bigcup_{m=k}^{\infty} V_m$. Then $A_1 \supset A_2 \supset \ldots$, $|A_1|_e \le 2$, and since $\bigcap_k A_k = \emptyset$, $|\bigcap_k A_k|_e = 0$. Yet $|A_k|_e \ge |V_k|_e = |V|_e > 0$ for all k.

23. First, we claim that $\mu(x + (a, b]) = \mu((a, b])$ for every $x \in \mathbb{R}$. Let $\{d_n\}$ be a sequence in D that increases to x. Then $(a + d_n, b + x] =$

$(a+d_n, b+d_n] \cup (b+d_n, b+x]$ and $\mu((a+d_n, b+x]) = \mu((a, b]) + \mu((b+d_n, b+x])$. Now, since $\{(a+d_n, b+x]\}$ decreases to $x + (a, b]$ and $\{(b+d_n, b+x]\}$ decreases to $\emptyset$, by continuity from below $\mu(x + (a, b]) = \mu((a, b])$ for each $a, b \in \mathbb{R}$ and $x \in \mathbb{R}$.

Next, since $(0, 1] = \bigcup_{k=0}^{n-1}(k/n, (k+1)/n]$ and $(k/n, (k+1)/n] = k/n + (0, 1/n]$ for all such k, $\mu((0, 1]) = \sum_{k=0}^{n-1} \mu((k/n, (k+1)/n]) = n\mu((0, 1/n])$ and so $\mu((0, 1/n]) = (1/n)\mu((0, 1])$. Moreover, since $(0, m/n] = (0, 1/n] \cup \ldots \cup ((m-1)/n, m/n] = (0, 1/m] \cup \ldots \cup ((m-1)/n + (0, 1/n])$, also $\mu((0, m/n]) = m\mu((0, 1/n]) = (m/n)\mu((0, 1])$. Thus, since $(0, 1] = (0, m/n] \cup (m/n, 1]$, $\mu((m/n, 1]) = (1 - m/n)\mu((0, 1])$.

Next, we claim that μ has no atoms. Indeed, since $\{(1-1/n, 1]\}$ decreases to 1, by continuity from above $\mu(\{1\}) = \lim_n \mu((0, 1])/n = 0$. By translation the same is true for every point and by σ-additivity for any countable set.

Next, if $(a, b] = a + (0, b - a]$ is a subinterval of $[0, 1)$ with $b - a$ rational, $\mu((a, b]) = \mu((0, b - a]) = (b - a)\mu((0, 1])$. When $b - a$ is not rational let $\{q_n\}$ be a rational sequence that increases to $b - a$. Then by continuity from below $\mu((0, b-a]) = \lim_n \mu((0, q_n]) = (\lim_n q_n)\mu((0, 1]) = (b-a)\mu((0, 1]) = \mu((a, b])$ and the conclusion holds for this case as well.

We claim that this also holds for $x > 0$. If $x > 1$, let n be the integer such that $x - n \in (0, 1]$. Then

$$(0, x] = \bigcup_{k=1}^{n} \left((k - 1) + (0, 1] \right) \cup \left(n + (0, x - n] \right)$$

where the intervals appearing are pairwise disjoint and so $\mu((0, x]) = cn + c(x - n) = cx$. Finally, suppose that $(x, y]$ is a left-hand open interval in $\mathbb{R}$; then, since $(x, y] = x + (0, y - x]$, $\mu((x, y]) = c(y - x)$ and this is all we need to know about intervals.

Now, since $\mathcal{F} = \{(a, b] : a, b \in \mathbb{R}\}$ is a π-system of subsets of $\mathbb{R}$, by Problem 2.95 the measures coincide on $\mathcal{M}(\mathcal{F}) = \mathcal{B}(\mathbb{R})$.

25. Let K be a compact subset of A; since $\{x_k\}$ is bounded, $\bigcup_k(x_k + K)$ is bounded in $\mathbb{R}^n$. Thus, by Problem 2.85(b),

$$\limsup_k |x_k + K| \le |\limsup_k(x_k + K)|.$$

Now, by the translation invariance of the Lebesgue measure, $|x_k + K| = |K|$ for all k, and so, $\limsup_k |x_k+K| = |K|$. Moreover, since $(x_k+K) \subset (x_k+A)$ it follows that $|\limsup_k(x_k + K)| \le |\limsup_k(x_k + A)|$. Finally, by the regularity of the Lebesgue measure, $|A| = \sup\{|K| : K \subset A, K \text{ compact}\}$ and, therefore, $|A| \le |\limsup_k(x_k + A)|$. This result is from I. Arandelović, *An inequality for the Lebesgue measure*, Univ. Beog. Publ. Elek. Fak. Ser. Math. **15** (2004), 85–86.

26. (a) This is false, as $\mathbb{Q}^n \subset \mathbb{R}^n$ shows. However, by Problem 10 the following is true: $A \subset \mathbb{R}^n$ with bounded closure $\overline{A}$ has $|\overline{A}| = 0$ iff for every $\varepsilon > 0$, there exists a finite pairwise disjoint collection of intervals $\{I_j\}$ in $\mathbb{R}^n$ with $A \subset \bigcup_j I_j$ and $\sum_j |I_j| < \varepsilon$.

(b) This is true. Let $\{q_n\}$ denote an enumeration of the rationals in $\mathbb{R}$ and given $\varepsilon > 0$, let $A = \bigcup_n (q_n - \varepsilon/2^{n+1}, q_n + \varepsilon/2^{n+1})$; A is a dense open set with $|A| \le \varepsilon$. Now, $\partial A = \overline{A} \cap \overline{A^c} = \overline{A^c}$ and since $|\overline{A^c}| = \infty$, $|\partial A| = \infty$.

29. (a) $\mathcal{M}(\mathcal{A}) = \mathcal{P}(\mathbb{Q})$.

(b) The statement is false. Let $X = (a, b] \cap \mathbb{Q}$ and put $\mu([a, b]) = b - a$. Clearly μ extends additively and $\mu(\{q\}) = 0$ for every rational $q \in (a, b]$. However, since $X = \bigcup_{q \in \mathbb{Q} \cap (a,b]} \{q\}$, if μ were a measure on $\mathcal{P}(X)$ it would follow that $\mu(X) = \mu((a, b]) = b - a \ne \sum_{q \in X} \mu(\{q\}) = 0$.

30. The statement is false if A is unbounded. Consider $A = \mathbb{N}$ in $\mathbb{R}$; then $|A| = 0$ and $|O_k| = \infty$ for all $k \ge 2$. And for $n \ge 2$ consider $A = \{0\} \times \mathbb{R}^{n-1}$, which is unbounded and closed in $\mathbb{R}^n$; then $|A| = 0$ and $|O_k| = \infty$ for $k \ge 1$.

On the other hand, $A = \bigcap_k O_k$ if A is bounded. Clearly $A \subset \bigcap_k O_k$. For the sake of argument suppose there is $x \in \bigcap_k O_k \setminus A \subset A^c$; since A^c is open there exists a ball $B(x, r) \subset A^c$ and so $d(x, A) \ge r$. Now pick $k > 1/r$; then, by assumption $x \in O_k$ and $d(x, A) < 1/k < r$, which cannot happen. Therefore $A = \bigcap_k O_k$ and, since A is bounded, $\{O_k\}$ is a decreasing sequence of bounded sets, and by continuity from above $|A| = \lim_k |O_k|$.

31. Since $A \cap B(0, |x|)$ increases to A as $|x| \to \infty$, by continuity from below $\varphi_A(x) \to |A|$ as $|x| \to \infty$. Now, since for $|x| < |y|$ we have $B(0, |y|) \setminus B(0, |x|) = \{z \in \mathbb{R}^n : |x| \le |z| < |y|\}$, it readily follows that $|\varphi_A(y) - \varphi_A(x)| \le |A \cap (B(0, |y|) \setminus B(0, |x|))| \le c \, ||y|^n - |x|^n|$. Hence φ_A is continuous in $\mathbb{R}^n$ and uniformly continuous in $\mathbb{R}$.

32. Let $A \supset F = \bigcup_k F_k$, F_k closed, be an F_σ set such that $|F| = |A|_e$. Now, with $Q_m = [-m, m] \times \cdots \times [-m, m]$, $m \ge 1$, $F_k = \bigcup_m (Q_m \cap F_k)$ is the countable union of compact sets for each k and, therefore, $F = \bigcup_m K_m$ where K_m is compact for each m. Moreover, replacing K_m by $\bigcup_{k=1}^m K_k$ if necessary we may assume that $\{K_m\}$ increases to F and $\{|K_m|\}$ increases to $|F|$. Let ℓ be the smallest integer such that $|K_\ell| > \eta$. Then, with $B(0, r)$ the closed ball of radius r centered at the origin, $\phi(r) = |K_\ell \cap B(0, r)|$ is a continuous function that increases from 0 to $|K_\ell| > \eta$ and, therefore, there exists r such that $\phi(r) = \eta$. Then $K_\ell \cap B(0, r)$ is a compact subset of F with measure η.

33. Let $\varphi(x) = |A \cap (-\infty, x)|$; φ is continuous in $\mathbb{R}$, $\varphi(-\infty) = 0$, and $\varphi(\infty) = |A|$. Therefore, $\varphi(x) = |A|/2$ for some $x \in \mathbb{R}$ and this gives the conclusion.

34. Let $\varphi(r) = |A \cap B(0,r)|$; φ is a continuous function of $r > 0$ that maps onto $[0, \infty)$. Thus there is a positive real number r such that $\varphi(r) = \lambda_1$ and $A_1 = A \cap B(0,r)$ is a Lebesgue measurable subset of A with measure λ_1. Next, having picked pairwise disjoint measurable subsets $A_1, \ldots, A_k$ of A with $|A_j| = \lambda_j$ for $1 \le j \le k$, since $|A| = \sum_{j=1}^{k} |A_j| + |A \cap (\bigcup_{j=1}^{k} A_j)^c|$ and $\sum_{j=1}^{k} |A_j| = \sum_{j=1}^{k} \lambda_j < \infty$, we have $|A \cap (\bigcup_{j=1}^{k} A_j)^c| = \infty$ and, repeating the construction of A_1 with A replaced there by $A \cap (\bigcup_{j=1}^{k} A_j)^c$, we find $A_{k+1} \subset A \cap (\bigcup_{j=1}^{k} A_j)^c \subset A$ such that $|A_{k+1}| = \lambda_{k+1}$. Clearly A_{k+1} is disjoint from $\bigcup_{j=1}^{k} A_j$.

36. (a) The statement is false if $\eta = 1$ because a set of full measure in $[0,1]$ is dense there and, therefore, if closed, it contains the rationals as well. Now, when $0 < \eta < 1$, by Problem 32 there is a compact subset K of $[0,1] \setminus \mathbb{Q}$ of measure $|K| = \eta$.

(b) The proof in $\mathbb{R}^n$ is similar. Given $\varepsilon > 0$, consider the collection $\{I_m\}$ of open intervals centered at points of $\mathbb{Q}^n$ with sidelength $(\varepsilon 2^{-m})^{1/n}$, $|I_m| = \varepsilon 2^{-m}$. Let $O = \bigcup_m I_m$. Clearly O is open and since it contains $\mathbb{Q}^n$, it is dense in $\mathbb{R}^n$; also, $|O| \le \sum_m |I_m| = \varepsilon \sum_m 2^{-m} = \varepsilon$. Observe that we may construct O such that $|O| = \varepsilon$. Indeed, let O_ε be the set just constructed for the value $\varepsilon > 0$ and put $O = (\varepsilon/|O_\varepsilon|)O_\varepsilon$; clearly O is open and by Problem 45, $|O| = \varepsilon$. Finally, O is dense. Given $x \in \mathbb{R}^n$ and $\eta > 0$, pick $z \in O_\varepsilon$ such that $|(|O_\varepsilon|/\varepsilon)x - z| \le \eta(|O_\varepsilon|/\varepsilon)$. Then $|x - (\varepsilon/|O_\varepsilon|)z| \le \eta$ with $(\varepsilon/|O_\varepsilon|)z \in O$.

(c) Let $F_\varepsilon = I \setminus O_\varepsilon$ where O_ε is the dense open set with $|O_\varepsilon| \le \varepsilon$ constructed in (b). Then F_ε is a closed subset of I with measure $|F_\varepsilon| \ge 1 - \varepsilon$ and, since O_ε is dense, F_ε is nowhere dense. Note that if $F = \bigcup_n F_{\varepsilon_n}$ with $\varepsilon_n \to 0$, F is of first category and has measure 1.

(d) The statement is false as the set F in (c) shows.

37. Let $O \supset A$ be an open set with finite measure such that $|O \setminus A| < \varepsilon$. Then $A, (x+A) \subset G = O \cup (x+O)$ and so, $A \cap (x+A) = G \setminus (G \setminus (A \cap (x+A)))$. Now, since $G \setminus (A \cap (x+A)) = (G \setminus A) \cup (G \setminus (x+A))$, $|G \setminus (A \cap (x+A))| \le |G \setminus A| + |G \setminus (x+A)| = |G| - |A| + |G| - |x+A| = 2|G| - 2|A|$, and, consequently, $|A \cap (x+A)| = |G| - |G \setminus (A \cap (x+A))| \ge |G| - (2|G| - 2|A|) = -|G| + 2|A|$. Hence $0 \le |A| - |A \cap (x+A)| \le |G| - |A|$.

Next, G is the disjoint union $G = O \cup ((x+O) \setminus O)$ and so, $|A| - |A \cap (x+A)| \le (|O| - |A|) + |(x+O) \setminus O|$. Thus it suffices to prove that $|(x+O) \setminus O| < \eta$ for finite measure open sets and x sufficiently small. Now, O is the countable union of nonoverlapping closed cubes and so, given $\eta > 0$, there exists N such that $|O \setminus \bigcup_{k=N+1}^{\infty} Q_k| \le \eta$. Then

$$(x + O) \setminus O \subset \left(\left(x + \bigcup_{k=1}^{N} Q_k \right) \setminus O \right) \cup \left(x + \bigcup_{k=N+1}^{\infty} Q_k \right)$$

$$\subset \bigcup_{k=1}^{N} ((x + Q_k) \setminus Q_k) \cup \left(x + \bigcup_{k=N+1}^{\infty} Q_k \right)$$

and, therefore, $|(x + O) \setminus O| \le |\bigcup_{k=1}^{N} ((x + Q_k) \setminus Q_k)| + \eta$. So it suffices to prove that $\lim_{|x| \to 0} |(x + Q) \setminus Q| = 0$ for every bounded closed cube Q, which is a straightforward verification.

38. We may assume that A is bounded. Then, by Problem 37, $|A| \sim |A \cap (A + h)|$ for sufficiently small h and so, repeating this argument $n - 1$ times, $A \cap (h + A) \cap \ldots \cap ((n-1)h + A)$ has positive finite measure $\sim |A|$ for h sufficiently small and, in particular, the intersection is not empty. So, for some $x \in \mathbb{R}$, there are $a_1, \ldots, a_n \in A$ such that $x = a_1, x = a_2 + h, \ldots, x = a_n + (n-1)h$, and, consequently, $x, x - h, \ldots, x - (n-1)h \in A$. The conclusion clearly holds for $a = x - (n-1)h, \ldots, a + (n-1)h = x$, since they all belong to A.

39. Both statements are true and we prove (b). We may assume that A is bounded. First, as in Problem 37, for $h > 0$ small enough $|A| \sim |(A \cap ((h, 0) + A)) \cap ((0, h) + A) \cap ((h, h) + A)|$ and, in particular, the intersection is not empty. If x is in that set there are $a_1, a_2, a_3, a_4 \in A$ such that $x = a_1, x = (h, 0) + a_2, x = (0, h) + a_3, x = (h, h) + a_4$, and so, $x, x - (h, 0), x - (0, h), x - (h, h)$ all belong to A for h sufficiently small and are the vertices of a square of sidelength h.

40. First, since $A = (A \cap (x + A)) \cup (A \cap (x + A)^c)$, by Problem 37, $\lim_{|x| \to 0} |A \cap (x + A)^c| = |A| - \lim_{|x| \to 0} |A \cap (x + A)| = 0$. Also, $x + A = ((x + A) \cap A) \cup ((x + A) \cap A^c)$ and, since $|x + A| = |A|$, again by Problem 36, $\lim_{|x| \to 0} |(x + A) \cap A^c| = \lim_{|x| \to 0} |x + A| - \lim_{|x| \to 0} |(x + A) \cap A| = |A| - |A| = 0$. Thus since $A \triangle (x + A) = (A \cap (x + A)^c) \cup ((x + A) \cap A^c)$, combining the above observations we get $\lim_{|x| \to 0} |A \triangle (x + A)| = 0$.

A direct approach also works. First, since $\{|\chi_{x+A} - \chi_A| > 1/2\} = (x + A) \triangle A$, by Chebychev's inequality and the continuity of integrable functions in the norm we have $|(x + A) \triangle A| \le 2 \int_{\mathbb{R}^n} |\chi_{x+A}(y) - \chi_A(y)| \, dy = 2 \int_{\mathbb{R}^n} |\chi_A(y - x) - \chi_A(y)| \, dy \to 0$ as $|x| \to 0$.

Finally, let $A = \bigcup_n [2n, 2n + 1]$. Then for $0 < \varepsilon < 1$, $A \setminus (\varepsilon + A) = \bigcup_n [2n, 2n + \varepsilon)$ has infinite measure.

41. Let $O \supset A$ be an open set of finite measure. Since $(x + A) \cap A \subset (x + O) \cap O$ it suffices to prove the statement for open sets of finite measure. Now, $O = \bigcup_k Q_k$ is the countable union of nonoverlapping closed cubes and so, given $\varepsilon > 0$, there exists K such that $\sum_{k=K}^{\infty} |Q_k| \le \varepsilon/3$. Write $O = \bigcup_{k=1}^{K-1} Q_k \cup \bigcup_{k=K}^{\infty} Q_k = O_K \cup R_K$, say, where $|R_K| \le \varepsilon/3$. Then

$$(x + O) \cap O \subset ((x + O_K) \cap O_K) \cup ((x + O_K) \cap R_K) \cup ((x + R_K) \cap O_k)$$
$$\cup ((x + R_k) \cap R_K),$$

and, consequently, $|(x+O) \cap O| \le |(x+O_K) \cap O_K| + |R_k| + |x + R_k| + |R_k| \le |(x + O_K) \cap O_K| + \varepsilon$. Thus it suffices to prove the result for finite unions of closed cubes. But then $(x + O_K) \cap O_K = \bigcup_{k,\ell=1}^{K-1} (x + Q_k) \cap Q_\ell$ and we are reduced to proving the result for $A = (x + Q) \cap Q'$ where Q, Q' are closed cubes in $\mathbb{R}^n$. In this case the result is obvious since the intersection is empty for $|x|$ sufficiently large.

42. It suffices to check continuity at 0 since the continuity at $x \in \mathbb{R}^n$ follows from the continuity result for the set $x + B$ at 0. First, $A \cap (x + B) = (A \cap (x + B) \cap B) \cup (A \cap (x + B) \cap B^c)$ and, consequently, $|A \cap (x + B)| = |(A \cap (x + B)) \cap B| + |(A \cap (x + B)) \cap B^c|$. Similarly, $|A \cap B| = |(A \cap B) \cap (x + B)| + |(A \cap B) \cap (x + B)^c|$ and so, $\varphi(x) - \varphi(0) = |A \cap (x + B) \cap B^c| - |(A \cap B) \cap (x + B)^c|$. Therefore $|\varphi(x) - \varphi(0)| \le |A \cap (B \triangle (x + B))|$ which implies, by Problem 40, that $|\varphi(x) - \varphi(0)| \le |(x + B) \triangle B| \to 0$ as $|x| \to 0$.

43. (a) The statement is true. First, observe that T maps compact sets into compact sets. Let $K \subset \mathbb{R}^n$ be compact, $\{y_k\} \subset T(K)$ and $\{x_k\} \subset K$ such that $T(x_k) = y_k$ for all k; since K is compact there are a subsequence $\{x_{k_m}\}$ of $\{x_k\}$ and $x \in K$ such that $x = \lim_{k_m} x_{k_m}$. Then $T(x) = \lim_{k_m} T(x_{k_m})$ and, consequently, $\{T(x_{k_m})\}$ is a subsequence of $\{y_k\}$ that converges to $y = T(x) \in T(K)$ and $T(K)$ is compact. Furthermore, since a closed set in $\mathbb{R}^n$ can be written as a countable union of compact sets, T maps closed sets into F_σ sets and, consequently, F_σ sets into F_σ sets.

Now, let $A \subset \mathbb{R}^n$ be bounded, K a compact set of $\mathbb{R}^n$ that contains A in its interior, and $\{I_k\}$ a covering of A by closed intervals of $\mathbb{R}^n$ contained in K. Since T is locally Lipschitz the image of a set with diameter d contained in K has diameter at most $M_K d$ and, in particular, if $J \subset K$ is a closed interval, $T(J)$ is compact and $|T(J)| \le c_K |J|$ for some constant c_K. Thus $T(A) \subset \bigcup_k T(I_k)$ and $|T(A)|_e \le c_K \sum_k |I_k|$ and, taking the infimum over the coverings of A, $|T(A)|_e \le c_K |A|_e < \infty$. Note that if T is Lipschitz, i.e., $|T(x) - T(y)| \le M|x - y|$ for all x, y in $\mathbb{R}^n$, then $|T(A)|_e \le c|A|_e$ for all $A \subset \mathbb{R}^n$ and some constant c.

(b) The statement is true. Let B_k denote the ball of radius k centered at the origin and $A_k = A \cap B_k$, $k = 1, 2, \ldots$; note that $|A_k|_e \le |A| = 0$ for all k. Then, by (a), $|T(A_k)|_e \le c_k |A_k|_e = 0$ and $|T(A_k)| = 0$ for all k. Finally, since $T(A) \subset \bigcup_k T(A_k)$, $|T(A)|_e \le \sum_k |T(A_k)| = 0$, and $|T(A)| = 0$.

(c) The statement is true.

(d) The statement is false. In $\mathbb{R}$, let $T(x) = x^2$ and $A = \bigcup_k A_k$ with $A_k = [k, k + k^{-3/2}]$; the A_k are pairwise disjoint, $|A_k| = k^{-3/2}$, and $|A| = \sum_k k^{-3/2} < \infty$. Note that, since $T(A_k) = [k^2, k^2 + 2k^{-1/2} + k^{-3})$, the $T(A_k)$ are pairwise disjoint. Then $T(A) = \bigcup_k T(A_k)$, and, consequently, $|T(A)| = \sum_k |T(A_k)| \geq \sum_k k^{-1/2} = \infty$.

44. Since $f'(x) = \sin(x)$ increases in $[0, 1]$ the Lipschitz constant of f is $\sin(1) = .841\ldots$, and by Problem 43(a), $|B| \leq .85\,|A|$.

45. (b) First, necessity. Given $\varepsilon > 0$, put $\eta = \varepsilon/|r|^n$ and let O be an open set such that $A \subset O$ and $|O \setminus A|_e < \eta$. Now, clearly rO is open, $rA \subset rO$, and $r(O \setminus A) = rO \setminus rA$. Hence $|rO \setminus rA|_e = |r(O \setminus A)|_e \leq |r|^n \eta = \varepsilon$ and $rA \in \mathcal{L}(\mathbb{R}^n)$.

Next, sufficiency. If rA is measurable for some $r \neq 0$, by the first part of the argument $(1/r)rA = A$ is also measurable.

46. By Problem 45, $-A = (-1)A \in \mathcal{L}(\mathbb{R}^n)$ and $|-A| = |A|$. Now, if $A = \emptyset$ the statement holds. Otherwise, note that $2|A| = |-A| + |A| = |-A \cap A| + |-A \cup A| \leq |A| + |B|$ and, consequently, $|A| \leq |B|$. Finally, by subadditivity $|B| \leq |-A| + |A| = 2|A|$.

47. With $I_A = A \cap \mathbb{Q}^c$ we have $I_A^c = -I_A = I_{A^c}$, and, consequently, $A \cap I_A = -(A^c \cap I_A^c) = -(A^c \cap I_{A^c})$. Therefore, by Problem 45, $|A| = |A \cap I_A| = |A^c \cap I_{A^c}| = |A^c|$ and so, $|A| = |A^c| = \infty$.

48. Note that $A_1 = (1, \ldots, 1) - A = \{x \in \mathbb{R}^n : x_k = 1 - a_k, 1 \leq k \leq n, a \in A\} \subset I^n$ and so by the translation invariance of the Lebesgue measure and Problem 45, $|A_1| = |A|$. Also, $A_1 \cup B \subset I^n$ and $|A_1 \cap B| = |A_1| + |B| - |A_1 \cup B| > 0$. Hence $A_1 \cap B \neq \emptyset$ and there exist $x, y \in I^n$ with $x \in A, y \in B$ such that $y = (1, \ldots, 1) - x$.

49. (a) Let $Q, Q' \subset \mathbb{R}^n$ be closed cubes of sidelength 1. Then $Q = x + Q'$ for some $x \in \mathbb{R}^n$ and $T(Q) = T(x + Q') = T(x) + T(Q')$. Thus by the translation invariance of the Lebesgue outer measure $|T(Q)|_e = |T(Q')|_e$ and $|T(Q)|_e = \eta$ is independent of the location of the cube Q of sidelength 1 in $\mathbb{R}^n$. Next, let $Q_1, Q_r \subset \mathbb{R}^n$ denote closed cubes of sidelength 1 and r, respectively, and observe that for some $x \in \mathbb{R}^n$, $Q_r = x + r\,Q_1$. Then $T(Q_r) = T(x) + r\,T(Q_1)$ and so, by Problem 45, $|T(Q_r)|_e = |T(x) + r\,T(Q_1)|_e = \eta\, r^n = \eta\,|Q_r|$.

(b) That $T(A)$ is measurable follows from Problem 43. A simple argument gives the result for $A = N$, any part of the boundary of a cube Q, and $A = O$, an open set O in $\mathbb{R}^n$. Finally, $|T(A)| = \inf\{|O| : T(A) \subset O, O$ open$\} = \inf\{|T(V)| : T(A) \subset T(V), V$ open$\} = \eta \inf\{|V| : A \subset V, V$ open$\} = \eta\,|A|$. Alternatively, one may argue as follows. Let μ be the Borel measure on $\mathbb{R}^n$ defined by $\mu(A) = |T(A)|$, $A \in \mathcal{B}(\mathbb{R}^n)$. Then μ is translation invariant and so by the $\mathbb{R}^n$ version of Problem 23, $|T(A)| = \mu(A) = \eta|A|$ for some $\eta > 0$.

It only remains to prove that $\eta = \det(M)$. First, if T is not invertible, i.e., $\det(M) = 0$, $T(\mathbb{R}^n)$ is contained in a hyperplane and so, by Problem 27, $T(A)$ has measure zero for all $A \in \mathcal{L}(\mathbb{R}^n)$ and the conclusion holds with $\eta = 0$.

Next, if T is invertible, by linear algebra M is the product of elementary matrices, i.e., square matrices that implement one of three elementary row operations via left multiplication. These matrices are given below: M_I multiplies a row of a matrix by a nonzero number, M_{II} interchanges rows, and M_{III} adds a row to another. They are:

$$M_I = \begin{bmatrix} c & 0 & \cdots & 0 \\ 0 & 1 & \cdots & 0 \\ \vdots & \vdots & \ddots & \vdots \\ 0 & 0 & \cdots & 1 \end{bmatrix}, \quad M_{II} = \begin{bmatrix} 0 & 1 & \cdots & 0 \\ 1 & 0 & \cdots & 0 \\ \vdots & \vdots & \ddots & \vdots \\ 0 & 0 & \cdots & 1 \end{bmatrix},$$

and

$$M_{III} = \begin{bmatrix} 1 & 1 & \cdots & 0 \\ 0 & 1 & \cdots & 0 \\ & & \ddots & \\ 0 & 0 & \cdots & 1 \end{bmatrix}.$$

In the case of M_I, let $Q = [0,1]^n$ and suppose that $c > 0$. Then $M_I(Q) = [0,c] \times [0,1]^{n-1}$ and $\eta = |M_I(Q)| = c$. Similarly, if $c < 0$, $M_I(Q) = [c,0] \times [0,1]^{n-1}$ and $\eta = |M_I(Q)| = -c$. Hence $\eta = |c|$.

In the case of M_{II}, if $J = \prod_{k=1}^{n}[a_k, b_k)$, $M_{II}(J) = [a_2, b_2) \times [a_1, b_1) \times \prod_{k=3}^{n}[a_k, b_k)$ and so, $|M_{II}(J)| = \eta|J| = |J|$ and $\eta = 1$.

In the case of M_{III} the result follows as for M_{II} once we observe that J is mapped into a parallelogram with the same measure. Alternatively, we may invoke Fubini's theorem. Since $M_{III}(J) = \{(x_1, x_2 + cx_1, x_3, \ldots, x_n) : x_k \in [a_k, b_k), 1 \le k \le n\}$, it follows that

$$|M_{III}(J)| = \prod_{k=3}^{n}(b_k - a_k) \int_{a_1}^{b_1} \left(\int_{a_2+cx_1}^{b_2+cx_2} dx_2 \right) dx_1$$

$$= |c|(b_2 - a_2)(b_1 - a_1) \prod_{k=3}^{n}(b_k - a_k)$$

and, therefore, $|M_{III}(J)| = |\det(M_{III})||J|$.

Finally, by properties of determinants, $\det(M) = \prod_{m=1}^{k} \det(M_m)$. Thus $|T(A)| = |M_1 M_2 \cdots M_k(A)| = |\det(M_1)||M_2 \cdots M_k(A)|$ and, consequently, proceeding this way, $|T(A)| = \prod_{m=1}^{k} |\det(M_m)||A| = |\det(M)||A|$.

50. Rotations are given by orthogonal matrices with determinant ± 1.

51. Rather than using measure theory, one could invoke the Baire category theorem since as is readily seen hyperplanes are closed nowhere dense subsets of $\mathbb{R}^n$.

52. $|\mathcal{E}| = \omega_n |\det(A)|^{-1/2}$ where ω_n denotes the volume of the unit ball in $\mathbb{R}^n$.

54. If $x > 0$, $\mu((0,x]) = x\,\mu((0,1]) = \eta\,|(0,x]|$. Similarly, for $y < 0$, $\mu((y,0]) = |y|\,\mu((0,1]) = \eta\,|(y,0]|$. Thus $\mu((x,y]) = \eta\,|(x,y]|$ for all $x,y \in \mathbb{R}$. The proof now proceeds as in Problem 23.

55. First assume that $0 < |A|_e < \infty$. For the sake of argument suppose that for some $0 < \eta < 1$, $|A \cap Q|_e < \eta|Q|$ for all cubes Q in $\mathbb{R}^n$. Then, given $\varepsilon = (1/\eta - 1)|A|_e > 0$, there is a covering of A by intervals $\{I_k\}$ such that $\sum_k |I_k| \le |A|_e + \varepsilon$. Now, the interior of each I_k is open and, therefore, $I_k = N_k \cup \bigcup_m Q_m^k$ where the Q_m^k are nonoverlapping closed cubes and $|N_k| = 0$ for all k. Hence $A \subset \bigcup_k N_k \cup \bigcup_{k,m} Q_m^k$ and, consequently, $|A|_e \le \sum_{m,k} |A \cap Q_m^k|_e < \eta \sum_{m,k} |Q_m^k| = \eta \sum_k |I_k| \le \eta(|A|_e + \varepsilon) = \eta|A|_e + \eta(1/\eta - 1)|A|_e = |A|_e$, which cannot happen. If $|A|_e = \infty$, let $A_m = A \cap B(0,m)$; then $A_1 \subset A_2 \subset \dots$ and $A = \bigcup_m A_m$. Now, $\lim_m |A_m|_e = |A|_e$ and so $|A_m|_e > 0$ for m sufficiently large. Since $|A_m|_e < \infty$ by the first part of the argument the assertion is true for A_m and, consequently, for any set that contains it, including A.

Finally, since Q can be expressed as a finite union of nonoverlapping subcubes $\{Q_k\}$ obtained by subdividing Q, the conclusion must hold for one of the Q_k as well for, if not, $|A \cap Q|_e \le \sum_k |A \cap Q_k|_e \le \eta \sum_k |Q_k| = \eta|Q|$, which is not the case. Thus Q may be assumed to have arbitrarily small measure.

56. First, let $A \in \mathcal{L}(\mathbb{R}^n)$. Since

$$\frac{|A \cap Q|}{|Q|} + \frac{|A^c \cap Q|}{|Q|} = 1, \quad \text{for all cubes } Q \subset \mathbb{R}^n,$$

it follows that $|A^c \cap Q| \le (1 - \eta)\,|Q|$ for all cubes Q and so by Problem 55, $|A^c| = 0$. Also observe that if $A \subset Q_1$ satisfies $|A \cap Q| \ge b|Q|$ whenever $Q \subset Q_1$, then $|A| = |Q_1|$.

Next, let $n = 1$, $I = [0,1]$, and B the Lebesgue nonmeasurable subset of I with $|B|_e = 1$ constructed in Problem 8; recall that $|B \cap J|_e = |J|$ for all subintervals J of I. Now take $A = (\mathbb{R} \setminus I) \cup B$; we claim that A satisfies $|A \cap J|_e \ge (1/2)|J|$ for all intervals J of $\mathbb{R}$. This is clear if $|J \cap (\mathbb{R} \setminus I)| \ge (1/2)|J|$. Otherwise, $|J \cap I| \ge (1/2)|J|$ and, consequently, $|B \cap J|_e \ge |B \cap (J \cap I)|_e \ge |J \cap I| \ge (1/2)|J|$.

57. By Problem 55, given $0 < \varepsilon < 1$, there is a cube Q centered at x of sidelength $\ell(Q)$, say, such that $|A \cap Q|_e \geq (1 - \varepsilon^n)\,|Q|$. Now let $J = J(x, \varepsilon\ell(I))$ be the cube concentric with Q with sidelength $\ell(J) = \varepsilon\ell(Q)$. Then $(1 - \varepsilon^n)|Q| < |A \cap Q|_e \leq |A \cap J|_e + |A \cap (Q \setminus J)|_e \leq |A \cap J|_e + |Q \setminus J| = |A \cap J|_e + (1 - \varepsilon^n)|Q|$. Thus $|A \cap J|_e > 0$ and, in particular, $A \cap J \neq \emptyset$; let $z \in A \cap J$. Now let $Q' = Q'(z, (1 + \varepsilon)\ell(I))$ be the cube of sidelength $(1+\varepsilon)\ell(Q)$ centered at z. Then, as is readily seen, $Q \subset Q'$ and so $|A \cap Q'|_e \geq |A \cap Q|_e > (1 - \varepsilon^n)|Q| = \psi(\varepsilon)\,|Q'|$ where $\psi(\varepsilon) = (1 - \varepsilon^n)/(1 + \varepsilon)^n$ is continuous and decreases from 1 to 0 as ε increases from 0 to 1 and so, picking ε so that $\psi(\varepsilon) = \eta$, we get $|A \cap Q'|_e > \eta\,|Q'|$.

58. The conclusion follows by Problem 57, or, alternatively, by Problem 38; we illustrate this in $\mathbb{R}$. If $|A| > 0$, A contains a progression $x, x-h, x-2h$, say, and so A contains points $x_1 = x$, $x_2 = x - 2h$, and $x_3 = x - h$ that verify $x_3 = (x_1 + x_2)/2$.

Finally observe that if for $A \subset \mathbb{R}$ with $|A| > 0$ we have that $(x+y)/2 \in A$ whenever $x, y \in A$, A is an interval.

59. Compare with Problem 1.12.

60. The statement is false. If $x = .x_1x_2\ldots$ denotes the dyadic expansion of $x \in [0, 1)$, let $A = \{x \in [0, 1) : x_{2n} = 0 \text{ for all } n \geq 1\}$ and $B = \{x \in [0, 1) : x_{2n-1} = 0 \text{ for all } n \geq 1\}$; we claim that $|A| = |B| = 0$. We do B. First, since half of the points $x \in [0, 1)$ have $x_1 = 0$, $|B| \leq 1/2$, but then, half of the remaining points have $x_3 = 0$ and so, $|B| \leq 1/2^2 = 1/4$. Continuing in this manner $|B| \leq 1/2^n$ for all n and $|B| = 0$. Similarly, $|A| = 0$ and by the translation invariance of the Lebesgue measure, if $D = \bigcup_{n \in \mathbb{Z}}(n + A)$, $|D| = 0$. Therefore we have $|D| = |B| = 0$ and $D + B = \mathbb{R}$.

Of course, a similar result holds for $A = B = C$, the Cantor discontinuum. Also, if $A = C$ and $B = (1/2)C$, then $[0, 1] \subset A + B$.

Now, in $\mathbb{R}^2$ the sets $A = \{0\} \times [0, 1]$ and $B = [0, 1) \times \{0\}$ have measure 0 and $A + B = [0, 1] \times [0, 1]$ has measure 1. On the other hand, $A + B$ is not measurable in general. Indeed, let $V \subset [0, 1]$ be a Lebesgue nonmeasurable set and define $A = V \times \{0\}$ and $B = \{0\} \times [0, 1]$. Then A and B are Lebesgue measurable subsets of $\mathbb{R}^2$ (with $|A| = |B| = 0$) but $A + B$ is Lebesgue nonmeasurable.

61. (a) Observe that for intervals $J = (a, b), J_1 = (c, d)$, $J + J_1 = (a + c, b + d)$. First, suppose that A, B are bounded open sets. Then, since the sum of open intervals is measurable, $A + B$ is measurable. Also, by the translation invariance of the Lebesgue measure, $|(s+t) + (A+B)| = |A+B|$, $|s + A| = |A|$, and $|t + B| = |B|$, and the conclusion is invariant under independent translations of A and B. Thus we can assume that $A \cap B = \emptyset$ and $\sup A = \inf B = 0$. Note that then $A \cup B \subset A + B$; indeed, for $a \in A$

pick $\varepsilon > 0$ such that $(a - \varepsilon, a + \varepsilon) \subset A$ and then, since $\inf B = 0$, $\varepsilon' \in B$ such that $0 < \varepsilon' < \varepsilon$. Then $a = (a - \varepsilon') + \varepsilon' \in A + B$; the proof for $b \in B$ is similar. Hence $|A| + |B| = |A \cup B| \le |A + B|$.

Suppose next that A, B are compact; then $A + B$ is compact. Now, to each point of A associate an open interval centered at the point of radius $1/k$ and denote by A_k the union of these intervals; similarly for B_k. And to each point in $A + B$ associate an open interval of length $2/k$ centered at the point and denote by $(A + B)_k$ the union of those intervals. Now, A_k, B_k, and $(A + B)_k$ are open and bounded and $A_k + B_k \subset (A + B)_k$. Indeed, if $x \in A_k$, let $a \in A$ be such that $|x - a| < 1/k$, and, similarly, if $y \in B_k$ let $b \in B$ such that $|y - b| < 1/k$. Then $|(x + y) - (a + b)| < 2/k$ and $x + y \in (A + B)_k$. Also, $\{A_k\}$ decreases to A and by continuity from above $|A_k| \to |A|$; similarly, $|B_k| \to |B|$ and $|(A + B)_k| \to |A + B|$. Now, by the result for open sets $|A_k| + |B_k| \le |A_k + B_k| \le |(A + B)_k|$ and letting $k \to \infty$, $|A| + |B| \le |A + B|$.

Finally, if A, B are arbitrary measurable of finite measure, by inner regularity, given $\varepsilon > 0$, there exist compact sets $K_A \subset A$ and $K_B \subset B$ such that $|A \setminus K_A| < \varepsilon/2$ and $|B \setminus K_B| < \varepsilon/2$. Then $K_A + K_B \subset A + B$ and $|A + B| \ge |K_A + K_B| \ge |K_A| + |K_B| \ge |A| - \varepsilon/2 + |B| - \varepsilon/2 = |A| + |B| - \varepsilon$, and, therefore, since ε is arbitrary, the conclusion follows. This result is known as the Brunn-Minkowski inequality.

(b) By (a) and Problem 45, $(1 - \eta)|A| + \eta|B| = |(1 - \eta)A| + |\eta B| \le |(1 - \eta)A + \eta B|$. Finally, by the concavity of the function $\ln(x)$, $|A|^{1-\eta}|B|^{\eta} \le (1 - \eta)|A| + \eta|B|$ for $0 < \eta < 1$.

62. Let $A_0 = A \cap [-1, 0]$ and $A_1 = A \cap (0, 1]$; then $|A_0| + |A_1| = |A| > 1$. Now, by the translation invariance of the Lebesgue measure, $A_2 = 1 + A_0 \subset [0, 1]$ is measurable and has measure $|A_2| = |A_0|$. Moreover, since $A_1 \cup A_2 \subset I$ and $|A_1| + |A_2| > 1$, $|A_1 \cap A_2| = |A_1| + |A_2| - |A_1 \cup A_2| > 0$ and so there exists $a \in A_1 \cap A_2$. Then $a \in A_2 = 1 + A_0$, we have $a = 1 + x$ with $x \in A_0 \subset A$, and $1 = a - x \in A - A$.

64. Partition $\mathbb{R}^n$ into cubes with vertices at the points ℓ of the integer lattice $\mathbb{Z}^n$, i.e., $(z_1, \dots, z_n)$, z_i integer for all i. Let $Q_0 = [0, 1)^n$ be the unit cube and note that $\mathbb{R}^n = \bigcup_{\ell \in \mathbb{Z}^n} (\ell + Q_0)$. Then translate the sets $A \cap (\ell + Q_0)$ into the cube Q_0, i.e., consider the sets $A_\ell = \{x - \ell : x \in A \cap Q_\ell\}$. Since $\sum_\ell |A_\ell| = \sum_\ell |A \cap Q_\ell| = |A| > 1$ the A_ℓ cannot be pairwise disjoint and there are $\ell \ne \ell'$ in the integer lattice such that $A_\ell \cap A_{\ell'} \ne \emptyset$. Let $a \in A_\ell \cap A_{\ell'}$. Then $a = x - \ell = x' - \ell'$, where $x, x' \in A$. Thus $0 \ne x - x' = \ell - \ell'$ is in the integer lattice.

Similarly, if $|A| > N$, A contains N points with integer coordinates.

66. By inner regularity there is a compact set $K \subset A$ with $|K| > \eta$. Now, $a = \inf\{x \in \mathbb{R} : x \in K\}$ and $b = \sup\{x \in \mathbb{R} : x \in K\}$ are in K and by the translation invariance of the Lebesgue measure we may assume that $a = 0$. Since $K \cap (b + K) = \{b\}$, $|K \cup (b + K)| = 2|K| > 2\eta$, and, therefore, $K \cup (b + K)$ contains a Lebesgue measurable subset B with $|B| > 2\eta$. Moreover, since K and $K + b$ are contained in $K + K$ it follows that $K \cup (b + K) \subset K + K \subset A + A$ and $B \subset A + A$.

Similarly, since $-K \cap K = \{0\}$, $|-K \cup K| = 2|K| > 2\eta$ and, therefore, $-K \cup K$ contains a Lebesgue measurable subset B with $|B| > 2\eta$. Moreover, since $-K$ and K are contained in $K - K$ it follows that $-K \cup K \subset K - K \subset A - A$ and $B \subset A - A$.

67. We may assume that $|A| < \infty$ and by the regularity of the Lebesgue measure that A is compact; then there is an open set $O \supset A$ such that $|O| < 2|A|$. Note that $d(A, O^c) = \eta > 0$. Let $|x| < \eta$. We claim that A and $x + A$ are contained in O and are not disjoint. For, if they were, they are both measurable with the same measure and, consequently, $2|A| = |A \cup (x + A)| \le |O| < 2|A|$, which cannot happen since $|A| > 0$. Thus there exist $a_1, a_2 \in A$ such that $a_1 = a_2 + x$ and so, $x = a_1 - a_2 \in A - A$.

Alternatively, we may assume that A is bounded and then by Problem 37, $\lim_{|x| \to 0} |A \cap (x + A)| = |A|$. Hence there exists $\delta > 0$ such that $|A \cap (x + A)| \ge |A|/2$ for $|x| < \delta$. But then, for those x, there exist $a_1, a_2 \in A$ such that $a_1 = x + a_2$ and $x \in A - A$.

68. Since the result for the sum $A + B = A - (-B)$ follows from that for the difference we prove the latter. Let $2^n/(2^n + 1) < \eta < 1$. By Problem 57 there exist cubes Q and Q_1 such that $|A \cap Q| > \eta |Q|$ and $|B \cap Q_1| > \eta |Q_1|$, respectively. Suppose that $\ell(Q_1) \le \ell(Q)$ and let k be the nonnegative integer such that $\ell(Q)/2^{k+1} < \ell(Q_1) \le \ell(Q)/2^k$. Now, pick a subinterval of Q of sidelength $\ell(Q)/2^k$ so that the original assumption for Q still holds; as observed in Problem 55 this is possible. Calling this cube Q again for simplicity we have now $\ell(Q_1) \le \ell(Q) < 2\ell(Q_1)$. Now, let x be such that $Q_1' = x + Q_1 \subset Q$ and put $B' = x + B$; by the translation invariance of the Lebesgue measure $|B' \cap Q_1'| = |x + (B \cap Q_1)| = |B \cap Q_1| > \eta |Q_1| = \eta |Q_1'|$.

Next, we want to prove that both A and B' intersect a substantial part of Q and this ensures that their intersection has positive measure; A does by construction. As for B' we have $|B' \cap Q| \ge |B' \cap Q_1'| > \eta |Q_1'| > (\eta/2^n) |Q|$. Thus, by elementary properties of measures, $|A \cap B' \cap Q| = |A \cap Q| + |B' \cap Q| - |(A \cup B') \cap Q| > \eta |Q| + (\eta/2^n) |Q| - |Q| = \left(\eta \left(\frac{2^n + 1}{2^n}\right) - 1\right) |Q| > 0$ and so, by Problem 67, $(A \cap B' \cap Q) - (A \cap B' \cap Q)$ contains an interval about the origin. Now, since $(A \cap B' \cap Q)$ is a subset of both A and B', $A - B'$ contains this difference set and, consequently, $A - B'$ contains an interval about 0. Thus $A - B$ contains an interval about x.

69. Let $\varphi : A \times A \to A - A$ be given by $\varphi(x,y) = x - y$. Now, if $A \in \mathcal{L}(\mathbb{R}^n)$ has $|A| > 0$, $A - A$ contains an interval and $c = \operatorname{card}(A - A) \le \operatorname{card}(A \times A) = \operatorname{card}(A) \le c$.

72. Let V be a Vitali Lebesgue nonmeasurable set in $[0,1)^n$ and for each $q_k \in \mathbb{Q}^n$ let $V_k = q_k + V$; the V_k are Lebesgue nonmeasurable, pairwise disjoint, and by Problem 71 if B is a measurable subset of some V_k, $|B| = 0$. Suppose first that $A \subset [0,1)^n$ has $|A|_e > 0$ and let $A_k = A \cap V_k$. Now, if each A_k is measurable, by Problem 71, $|A_k| = 0$ for all k, and so, $0 = \sum_k |A_k|_e = |\bigcup_k A_k| \ge |A|_e > 0$, which cannot happen. Thus A_k is not Lebesgue measurable for some k and we are done.

Another way to state this result is that if every subset of $A \in \mathcal{L}(\mathbb{R}^n)$ is measurable, then $|A| = 0$.

73. Since $|A_n|_e \ge |V_1|_e$, $|A_n|_e > 0$ for all n. For the sake of argument suppose that A_n is measurable; then $A_n - A_n$ contains a neighborhood of the origin, which is not the case since $A_n - A_n$ only contains the rationals $\pm(r_i - r_j)$, $1 \le i, j \le n$.

As an application of this result we get that Egorov's theorem cannot be improved, i.e., there is a sequence $\{f_k\}$ of functions on $[0,1]$ that does not converge uniformly in any subset of I of positive measure. Indeed, let $f_k = \chi_{\bigcup_{n=k}^{\infty} V_k}$, $k = 1, 2, \ldots$, and pick $0 < \varepsilon < |V|_e$. Fix $x \in I$. Then $x \in V_k$ for only one k and so, $f_n = 0$ for $n > k$, and, consequently, $\lim_n f_n(x) = 0$ for all $x \in I$. Moreover, since $f_k(x) = 1$ for $x \in \bigcup_{n=k}^{\infty} V_n$ and $|\bigcup_{n=k}^{\infty} V_n|_e \ge |V|_e > \varepsilon$, $\{f_k\}$ cannot converge uniformly to 0 outside of a set of measure $< \varepsilon$.

74. (a) The statement is false. Let K be a Cantor set in $[0,1]$ with positive measure and $g(x) = d(x, K)$; g is continuous and vanishes precisely on K. Now let $f(x) = \int_0^x g(t)\, dt$.

(b) The statement is false. Let C be the Cantor discontinuum in $[0,1]$, f the Cantor-Lebesgue function, and $g(x) = (x + f(x))/2$; $g[0,1] \to [0,1]$ is a bijection and has a continuous inverse. Now, $[0,1] \setminus C = \bigcup_n I_n$ where the I_n are pairwise disjoint open intervals with $\sum_n |I_n| = 1$. On each of the I_n we have $2g(x) = x + c$, c a constant, so each I_n gets mapped into an interval of length $|I_n|/2$ and since their images are pairwise disjoint we have $|g([0,1] \setminus C)| = \sum_n |I_n|/2 = 1/2$. Finally, since $g(C) = [0,1] \setminus g([0,1] \setminus C)$, it follows that $2|g(C)| = 2 - 1 = 1 > 0$.

75. Since $O = \bigcup_k (O \cap B(0,k))$ it suffices to consider the case when O is bounded. First, O can be represented as a countable union of nonoverlapping closed cubes, $O = \bigcup_k Q_k$, say. Now, by calculus, if B_k^0 denotes the ball inscribed in Q_k, $|B_k^0|/|Q_k| = \eta < 1$ where $\eta = |B(0,1)|/2^n$ is a dimensional constant. Therefore the balls in $\{B_k^0\}$ are pairwise disjoint and since

$O \setminus \bigcup_k B_k^0 = \bigcup_k (Q_k \setminus B_k^0)$ it follows that $|O \setminus \bigcup_k B_k^0| = \sum_k |Q_k \setminus B_k^0| = \sum_k (|Q_k| - |B_k^0|) = (1 - \eta)|O|$. Next, let $O_1 = O \setminus \bigcup_k \overline{B_k^0}$; O_1 is open and repeating the previous procedure we obtain a sequence of balls $\{B_k^1\}$, disjoint with $\{B_k^0\}$, such that $|O_1 \setminus \bigcup_k B_k^1| = (1 - \eta)|O_1| = (1 - \eta)^2|O|$. Note that $|O \setminus (\bigcup_k B_k^0 \cup \bigcup_k B_k^1)| \leq |O_1 \setminus \bigcup_k B_k^1| = (1 - \eta)^2|O|$. The pattern is now clear and an induction argument produces a family $\{B_k^m\}$, $k, m = 1, 2, \ldots$, of pairwise disjoint open balls such that $|O \setminus \bigcup_{k,m} B_k^m| = 0$.

76. The statement is true. Let $\mathcal{D} = \{D : D$ is a closed disk contained in $Q\}$, $d_1 = \sup_{D \in \mathcal{D}} |D|$, and pick a closed disk $D_1 \in \mathcal{D}$, say, such that $|D_1| > d_1/2$. Now, having picked pairwise disjoint closed disks $D_1, \ldots, D_{n-1}$ contained in Q, we select D_n as follows: Let $\mathcal{D}_n = \{D : D$ is a closed disk contained in $Q \setminus (D_1 \cup \ldots \cup D_{n-1})\}$, $d_n = \sup_{D \in \mathcal{D}_n} |D|$, and pick $D_n \in \mathcal{D}$, say, such that $|D_n| > d_n/2$. We claim that the sequence $\{D_n\}$ has the desired properties; the only property that requires proof is $\sum_n |D_n| = 1$ or, equivalently, $|Q \setminus (\bigcup_n D_n)| = 0$. First, note that since $\sum_n |D_n| \leq |Q|$, $|D_n| \to 0$ as $n \to \infty$ and, since $2|D_n| > d_n$, $d_n \to 0$ as $n \to \infty$. Next, fix N and consider the open set $Q_N = Q \setminus (\bigcup_{n=1}^{N-1} D_n)$. If $x_0 \in Q_N$ there is a closed disk D centered at x_0 contained in Q_N. Because at each step of the selection process we pick the next disk from among the largest available disks and since $d_n \to 0$, $|D|$ is big compared to d_n for n large and, consequently, D meets some disk D_n with $n \geq N$. Let then n_0 be the smallest such index, i.e., $D \cap (\bigcup_{n=1}^{n_0-1} D_n) = \emptyset$ and $D \cap D_{n_0} \neq \emptyset$. Then $2|D_{n_0}| \geq d_{n_0} \geq |D|$ and, consequently, $\sqrt{2}\,\mathrm{radius}(D_{n_0}) \geq \mathrm{radius}(D)$. Now, a simple geometric observation: since $D \cap D_{n_0} \neq \emptyset$, the closed disk concentric with D_{n_0} of radius equal to $\mathrm{radius}(D_{n_0}) + 2\,\mathrm{radius}(D)$ contains D and, consequently, $D_{n_0}^*$, the concentric closed disk with D_{n_0} of radius equal to $(1 + 2\sqrt{2})\,\mathrm{radius}(D_{N_0})$ contains D. Finally, since x_0 is an arbitrary point in $Q \setminus (\bigcup_{n=1}^{N-1} D_n)$ and n_0 may be an arbitrary integer $\geq N$, we conclude that $Q \setminus (\bigcup_{n=1}^{N-1} D_n) \subset \bigcup_{n=N}^{\infty} D_n^*$ and, consequently, $|Q \setminus \bigcup_n D_n| \leq |Q \setminus \bigcup_{n=1}^{N-1} D_n| \leq (1 + 2\sqrt{2})^2 \sum_{n=N}^{\infty} |D_n|$. Thus letting $N \to \infty$ it follows that $|Q \setminus \bigcup_n D_n| = 0$.

78. (a) The construction is easier when $n = 2$ for then we can take a sector A of the unit disk of area πd so that $|A \cap B(0, r)|/|B(0, r)| = d$ for all $r < 1$.

Now, in our case first suppose that $x_0 = 0$, let

$$B_n = \left(-\frac{1}{n}, -\frac{1}{n+1} \right) \cup \left(\frac{1}{n+1}, \frac{1}{n} \right), \quad n \geq 1,$$

and pick measurable sets $A_n \subset B_n$ such that $|A_n| = d|B_n|$ for all n. We claim that $A = \bigcup_n A_n$ has density d at 0. To see this, given $0 < r < 1$, let n be such that $1/(n+1) \leq r < 1/n$ and observe that $\bigcup_{k=n+1}^{\infty} A \cap B_k \subset A \cap [-r, r] \subset$

$\bigcup_{k=n}^{\infty} A \cap B_k$. Therefore $2d/(n+1) = \sum_{k=n+1}^{\infty} |A \cap B_k| \leq |A \cap [-r,r]| \leq \sum_{k=n}^{\infty} |A \cap B_k| = 2d/n$ and, consequently, since $1 - r \leq 1/(n+1)r$ and $1/nr \leq 1 + 2r$, $d(1-r) \leq |A \cap [-r,r]|/(2r) \leq d(1+2r)$ and so, the density of A at the origin is d.

Finally, by the translation invariance of the Lebesgue measure if $x_0 \neq 0$, $x_0 + A$ has the required property.

(b) Let $0 < \alpha, \beta < 1$, $\alpha + \beta = 1$. Then, from right to left, construct nonoverlapping closed intervals I_1 of length $\beta/2$, J_1 of length $\alpha/2$, I_2 of length $\beta/4$, J_2 of length $\alpha/4$, and so on; note that $[0,1] = (\bigcup_k I_k) \cup (\bigcup_k J_k)$. Let $B \subset [-1,1]$ be the symmetric set about the origin such that $B \cap [0,1] = \bigcup_k I_k$; note that for $r < 1$, $|B \cap (-r,r)|/|(-r,r)| = |(\bigcup_k I_k) \cap (0,r)|/|(0,r)|$. Now, let $r_k = \sum_{m=k}^{\infty} \alpha/2^m + \sum_{m=k}^{\infty} \beta/2^m = \alpha/2^{k-1} + \beta/2^{k-1} = 1/2^{k-1}$. Then $|B \cap (0,r_k)| = \sum_{m=k}^{\infty} \beta/2^m = \beta/2^{k-1}$ and so

$$\limsup_{r \to 0} \frac{|B \cap (0,r)|}{|(0,r)|} \geq \lim_{r_k \to 0} \frac{|B \cap (0,r_k)|}{|(0,r_k)|} = \lim_k \frac{\beta/2^{k-1}}{1/2^{k-1}} = \beta.$$

Similarly, with $r_k = \sum_{m=k}^{\infty} \alpha/2^m + \sum_{m=k+1}^{\infty} \beta/2^m = \alpha/2^k + 1/2^k$ now,

$$\liminf_{r \to 0} \frac{|B \cap (0,r)|}{|(0,r)|} \leq \lim_{r_k \to 0} \frac{|B \cap (0,r_k)|}{|(0,r_k)|} = \lim_k \frac{\beta/1/2^k}{1/2^k + \alpha/2^k} = \frac{\beta}{1+\alpha}.$$

80. We may assume that $0 \notin A$. With $B_k = B(0,r/k)$, let $A_k = A \cap B_k$. Then 0 is a point of density of A_k and by Problem 79 we have $-A_k \cap A_k \neq \emptyset$ and $2A_k \cap A_k \neq \emptyset$. We can thus pick $0 \neq y_k \in -A_k \cap A_k$ and $0 \neq z_k \in 2A_k \cap A_k \neq \emptyset$ for all k; note that $|y_k|, |z_k| \leq r/k \to 0$ as $k \to \infty$. For (a) let $x_k = y_k \in A_k \subset A$ for all k and for (b) $x_k = z_k$ so that $x_k = z_k, 2x_k = 2z_k \in A_k \subset A$ for all k.

81. (a) The statement is false. Informally, if $A \subset \mathbb{R}$ consists of pairwise disjoint longer and longer intervals with bigger and bigger gaps between them, the ratio in the definition of $D(A)$ oscillates as $r \to \infty$ and $D(A)$ fails to exist.

(b) Is true and follows readily from the definition of limit.

(c) Is not true. Let $A_n = [n, n+1)$ for all $n \geq 1$, and $A = [1,\infty)$. Then $D(A_n) = 0$ for all n and $D(A) = 1/2$.

82. For $c > 0$, put $P = [-c,c)^n$; it clearly suffices to prove that $|P \setminus \bigcup_{d \in D}(d + A)| = 0$. Now, given $0 < \eta < 1$, by Problem 55 there is a cube Q with $|Q|$ small compared to $|P|$ and $|Q \cap A| > (1 - \eta)|Q|$; note that $|Q \setminus A| < \eta |Q|$. Let $\{y_m + Q\}$ be a collection of finitely many nonoverlapping cubes such that $P = \bigcup_m (y_m + Q)$; note that there is a dimensional constant C such that each of the cubes in the collection is contained in the union $\bigcup_{k=1}^{C}(d_k + Q)$ with $d_k \in D$. Thus combining these observations it follows that there are $d_1, \ldots, d_N \in D$ such that, with a dimensional constant k

independent of Q, $P \subset \bigcup_{1 \leq m \leq N}(d_m + Q)$ and $\sum_{1 \leq m \leq N} |d_m + Q| \leq k\,|P|$. Then $P \setminus \bigcup_{d \in D}(d + A) \subset \bigcup_{1 \leq m \leq N}(d_m + Q) \setminus (d_m + A)$ and by the choice of the intervals and the translation invariance of the Lebesgue measure, $\sum_{1 \leq m \leq N} |(d_m + Q) \setminus (d_m + A)| < \eta \sum_{1 \leq m \leq N} |d_m + Q| \leq \eta\,k\,|P|$. Since η is arbitrary it follows that $|P \setminus \bigcup_{d \in D}(d + A)| = 0$.

83. Since $A = (-1/20, 1/9 + 1/20) \cup (2/9 - 1/20, 1/3 + 1/20) \cup (2/3 - 1/20, 7/9 + 1/20) \cup (8/9 - 1/20, 1 + 1/20)$, $|A| = 38/45$.

84. First, note that P_1 consists of 2 disjoint closed intervals each of length $(1 - \eta_1)2^{-1}$, P_2 consists of 2^2 disjoint closed intervals each of length $((1 - \eta_1)2^{-1} - (1 - \eta_1)2^{-1}\eta_2)2^{-1} = 2^{-2}\prod_{n=1}^{2}(1 - \eta_n)$ and, in general P_k consists of 2^k disjoint closed intervals each of length $2^{-k}\prod_{n=1}^{k}(1 - \eta_n)$, and, consequently, $|P_k| = \prod_{n=1}^{k}(1 - \eta_n)$. Now, since each factor of this product is positive and strictly less than one, $\{|P_k|\}$ is a strictly decreasing sequence that converges to some nonnegative real number and by continuity from above $|P| = \lim_k |P_k|$.

First, suppose that $\sum_n \eta_n$ converges. Then $\lim_n \eta_n = 0$ and by calculus $\ln(1 - \eta_n) \sim -\eta_n$ for n large. Hence $\sum_n \ln(1 - \eta_n)$ converges, and, consequently, $\lim_k |P_k| = \lim_k \prod_{n=1}^{k}(1 - \eta_n)$ exists and is not 0 and $|P| > 0$. On the other hand, if $\sum_n \eta_n$ diverges, there are two cases, $\lim_n \eta_n = 0$ and $\eta_n \not\to 0$. If $\lim_n \eta_n = 0$ as before $\ln(1 - \eta_n) \sim -\eta_n$ for n large, $\sum_n \ln(1 - \eta_n)$ diverges, and $|P| = 0$. On the other hand, if $\eta_n \not\to 0$ there exists a subsequence $\{\eta_{n_k}\}$ of $\{\eta_n\}$ and $\delta > 0$ such that $\eta_{n_k} \geq \delta$ for all n_k. Then $|P_{n_k}| = \prod_{m=1}^{n_k}(1 - \eta_m) \leq (1 - \delta)^k \to 0$ as $n_k \to \infty$ and, therefore, $|P| = 0$.

85. (a) Let $0 \leq k \leq \ell - 1$. Then those $x \in [0,1]$ with $x_n = k$ are of the form

$$
\begin{aligned}
x &= \frac{a_1}{\ell} + \frac{a_2}{\ell^2} + \cdots + \frac{a_{n-1}}{\ell^{n-1}} + \frac{k}{\ell^n} + y \\
&= \frac{a_1\ell^{n-2} + a_2\ell^{n-3} + \cdots + a_{n-1}}{\ell^{n-1}} + \frac{k}{\ell^n} + y
\end{aligned}
$$

where $0 \leq a_j \leq \ell - 1$ for $1 \leq j \leq n - 1$ and $0 < y \leq 1/\ell^n$. Now, since the expression $a_1\ell^{n-2} + a_2\ell^{n-3} + \cdots + a_{n-1}$ assumes all the integer values from 0 to $(\ell - 1)\sum_{j=0}^{n-2}\ell^j = \ell^{n-1} - 1$,

$$
A_n(k) = \bigcup_{j=0}^{\ell^{n-1}-1} \left[\frac{j\ell + k}{\ell^n}, \frac{j\ell + k + 1}{\ell^n}\right), \quad 0 \leq k \leq \ell - 1,
$$

is the pairwise disjoint union of ℓ^{n-1} intervals each of length $1/\ell^n$ and so a Borel set with $|A_n(k)| = 1/\ell$ for all k, n.

(b) A^k is a Borel measurable set of measure 0.

(c) First, note that $B_k = \{x \in [0,1] :$ for every m there is $n \geq m$ such that $x_n = k\}$ and, consequently, $B_k = \limsup_n A_n(k)$ where $A_n(k)$ is defined in (a). Thus by Problem 2.85(b), $|B_k| \geq \limsup_n |A_n(k)| = 1/\ell$.

(d) Let $k = k_1 + k_2$, $0 < k_1, k_2 < k$. For $z \in I$, let $x(z)$ be obtained from z by substituting k_1 for each digit k in the decimal expansion of z, and obtain $y(z)$ by substituting k_2 for each k in the decimal expansion of z and taking 0 for all other digits. Now, there is no problem with 'carry' and, consequently, $z = x(z) + y(z)$ with $x(z), y(z) \in A^k$. Then $A^k + A^k = \{z \in [0,1] : z = x + y$ for some $x \in A^k, y \in A^k\} = [0,1]$.

86. (a) For the sake of argument suppose that x is a limit point of A that is not in A. Then there is a first integer K such that $x_K \neq m, x_K \neq n$ and pick $\varepsilon = \ell^{-(K+1)}$. Then $d(x, A) \geq \ell\varepsilon$. Otherwise, $d(x, A) \geq 2\varepsilon$. In that case $(x - \varepsilon, x + \varepsilon) = \emptyset$, which is not possible.

Also, $A \subset \{x \in I : x_k \neq$ some integer for all $k\}$ and by Problem 85(b), $|A| = 0$. Finally, by Problem 1.55 the collection of sequences consisting of m's and n's is uncountable and so is A.

(b) Informally, the set of $x \in I$ which do not contain m or n in their expansion has measure 0 and to each point in B, by switching the terms in the expansion of x, there corresponds a point where n appears before m. Therefore, modulo a set of measure 0, B contains half of the points in the interval and $|B| = 1/2$. Formally, we prove first that B is measurable; in fact, we claim that B^c is Borel. Let $B_1 = [n/\ell, (n + 1)/\ell)$; $|B_1| = 1/\ell$. Next, consider $B_2 = \bigcup_{k \neq m,n,k=0}^{\ell-1}[kn/\ell, k(n + 1)/\ell)$; B_2 is disjoint from B_1, and has measure $(\ell - 2)/\ell^2$. The process is now clear. Each B_n is the union of pairwise disjoint half-open intervals and B_{n+1} is obtained from B_n by selecting $\ell - 2$ specific subintervals of B_n, each of measure $1/\ell$ the measure of the original interval. Therefore $|B_{n+1}| = ((\ell - 2)/\ell)|B_n|$ and, since $|B_1| = 1/\ell$, $|B_n| = (\ell-2)^{n-1}/\ell^n$. Thus $|B^c| = \sum_n (\ell-2)^{n-1}/\ell^n = 1/2$.

87. $|A| = (\ell - 1)/(2\ell - 1)$.

88. That A is closed and uncountable follows as in Problem 86. Next, divide $[0, 1]$ into 4 intervals of equal length, remove the intervals $(1/4, 1/2)$ and $(3/4, 1)$ and denote the remaining two closed intervals A_1; $A_1 = \{x \in [0, 1] : x_2 = 0\}$. At the second stage divide each of the two remaining closed intervals into 4 intervals of equal length, remove the intervals $(1/4^2, 2/4^2)$, $(3/4^2, 4/4^2)$, $(9/4^2, 10/4^2)$, $(11/4^2, 12/4^2)$, and denote by A_2 the union of 4 remaining closed intervals; $A_2 = \{x \in [0, 1] : x_2 = x_4 = 0\}$. Continuing in this way, at the nth stage the set A_n consists of 2^n closed intervals each of length 4^{-n}. Clearly $A_1 \supset A_2 \supset \ldots$, and $A = \bigcap_n A_n$. Thus, since $|A_n| = 2^n 4^{-n} = 2^{-n}$, $|A| = 0$. Finally, a closed set A of measure 0 is nowhere dense: If J is an open interval, J cannot be a subinterval of A and so, $J \cap A^c$ is nonempty and open, and so it contains a subinterval disjoint from A. Hence A is nowhere dense.

89. A is Lebesgue measurable and $|A| > 0$. To see this let $A_0 = \{x = \sum_k x_k 10^{-k} : x_1 = 0\}$; A_0 is a measurable set with measure $|A_0| = 1/10 = (1 - (9/10)^{2^0})$. Next, $A_1 = \{x = \sum_k x_k 10^{-k} : x_2 = 0 \text{ or } x_3 = 0\}$; A_1 is a measurable set with $|A_1| = (1 - (9/10)^{2^1})$. The process is now clear and we get that $A = \bigcup_{n=0}^{\infty} A_n$ is a measurable set with $|A| = \prod_{n=0}^{\infty}(1 - (9/10)^{2^n})$. Now, the product converges to a nonzero limit iff $\sum_{n=0}^{\infty}(9/10)^{2^n} < \infty$, which is the case since $(9/10)^{2^n} \leq (9/10)^n$ for all n.

91. For $n \geq 1$, let $B_n = \{x = \sum_k x_k 10^{-k} \in I : x_k = 0 \text{ or } x_k = 9$ for all $k \geq n\}$; by Problem 85(b), $|B_n| = 0$. Moreover, since the set of all sequences consisting of 0's and 9's is uncountable, B_n is uncountable for all $n \geq 1$. Thus, if $A_n = \bigcup_{k=-\infty}^{\infty}(k + B_n)$, A_n is an uncountable set of measure 0 for all $n \geq 1$. Let $A = \bigcup_n A_n$; clearly A is an uncountable set of measure 0. Now, given an interval J, let k be such that $|[k, k+1] \cap J| = \eta > 0$ and let N be such that $\eta > 1/10^N$. Then for some integer m, $[m/10^N, (m+1)/10^N] \subset [k, k+1] \cap J$, and, consequently, $A \cap J \supset (k + B_N) \cap ([k, k+1] \cap J)$, which is uncountable, and so is $A \cap J$. Thus A has the desired properties.

93. Fix a cube $Q \subset \mathbb{R}^n$ and let $\varphi(x) = |A \cap (x + Q)|$, $x \in \mathbb{R}^n$. Now, for $d \in D$, $A \cap (d + Q) = (d + A) \cap (d + Q) = d + (A \cap Q)$ and, consequently, $|A \cap (d + Q)| = |A \cap Q|$ for all $d \in D$. Moreover, by Problem 42, $\varphi(x)$ is a continuous function of x which, as we have just seen, is constant in the dense set D. Therefore $\varphi(x)$ assumes the constant value $|A \cap Q|$.

Suppose Q_1 is another cube with the same sidelength as Q, let $x \in \mathbb{R}^n$ be such that $Q_1 = x + Q$, and note that $|A \cap Q_1| = |A \cap (x + Q)| = \varphi(x) = |A \cap Q|$. Thus the expression $|A \cap Q|$ depends on the sidelength but not on the location of Q. Now, let $x, y \in \mathbb{R}^n$ and let $Q(x, \ell), Q(y, \ell)$ be cubes of sidelength ℓ centered at x, y, respectively. Then $|A \cap Q(x, \ell)| = |A \cap Q(y, \ell)|$ and, consequently, the expression

$$\lim_{\ell \to 0} \frac{|A \cap Q(x, \ell)|}{|Q(x, \ell)|}$$

is independent of $x \in \mathbb{R}^n$. Since χ_A is locally integrable, by the Lebesgue differentiation theorem this limit is a.e. equal to 0 or 1. So, if the limit is 0 on a set of positive measure, it is 0 a.e. and $|A| = 0$. On the other hand, if the limit is 1 on a set of positive measure, since $|A^c \cap Q|/|Q| = 1 - (|A \cap Q|/|Q|)$,

$$\lim_{\ell \to 0} \frac{|A^c \cap Q(x, \ell)|}{|Q(x, \ell)|} = 0 \text{ a.e.}$$

and $|A^c| = 0$.

94. The examples include the rationals and the set of reals with infinitely many 7's in their decimal expansion. Now, if A is such a set and t a rational number with only finitely many nonzero digits in its decimal expansion, then $x \in A$ iff $t + x \in A$, i.e., $A = t + A$. The conclusion now follows from Problem 93.

96. Let $p_n \to 0$ be positive periods of A. Then $D = \{m\, p_n : n \in \mathbb{N}, m \in \mathbb{Z}\}$ consists of periods of A and, since $p_n \to 0$, D is dense in $\mathbb{R}$. Therefore $d + A = A$ for all $d \in D$ and, by Problem 93, $|A| = 0$ or $|A^c| = 0$.

97. By assumption $\ln(r) + \ln(A) = \ln(A)$ for all rationals $r > 0$ and so $\ln(A)$ has arbitrary small periods. Then, by Problem 96, $|\ln(A)| = 0$ or $|\ln(A)^c| = 0$. In the former case, since $\exp(\ln(A)) = A$ and e^x is Lipschitz, by Problem 43, $|A| = 0$ and, in the latter case, since $\ln(A)^c = \ln(A^c)$, $|A^c| = |\exp(\ln(A^c))| = 0$.

98. Let Q be a cube in $\mathbb{R}^n$ and $x \in D$. Then

$$A \cap (x + Q) = \big(A \cap (x + A) \cap (x + Q)\big) \cup \big(A \cap (x + A)^c \cap (x + Q)\big)$$

and since $A \cap (x + A)^c \subset A \triangle (x + A)$ it follows that $|(A \cap (x + Q))| = |A \cap (x + A) \cap (x + Q)|$. Similarly, $|(x + A) \cap (x + Q)| = |A \cap (x + A) \cap (x + Q)|$ and, consequently, $|A \cap (x + Q)| = |(x + A) \cap (x + Q)| = |x + (A \cap Q)| = |A \cap Q|$ for all $x \in D$. The proof now proceeds as in Problem 96.

99. Let $T : S^1 \to [0, 2\pi)$ denote the mapping that assigns to $z = (x, y) \in S^1$ the unique $\theta \in [0, 2\pi)$ such that $x = \cos(\theta), y = \sin(\theta)$, and let $\mathcal{M}_T = \{T^{-1}(B) \subset S^1 : B \in \mathcal{B}([0, 2\pi))\}$ denote the σ-algebra of subsets of S^1 induced by T discussed in Problem 2.21.

We define μ on $\mathcal{M}_T$. First, if $A \in \mathcal{M}_T$, $A = T^{-1}(B)$ for $B \in \mathcal{B}([0, 2\pi))$ and since T is a bijection, $T(A) = B \in \mathcal{B}([0, 2\pi))$. The set function μ is then defined by $\mu(A) = (1/2\pi)|T(A)|$, $A \in \mathcal{M}_T$.

It is readily seen that μ is a measure on $\mathcal{M}_T$ and it remains to verify that μ is rotation invariant. Now, the rotation R_α of angle $0 < \alpha < 2\pi$ is given by $R_\alpha(z) = (\cos(\theta + \alpha), \sin(\theta + \alpha))$ for $z = (\cos(\theta), \sin(\theta)) \in S^1$; observe that if $A \in \mathcal{M}_T$ and $B \in \mathcal{B}([0, 2\pi))$ is such that $T(A) = B$, then $T(R_\alpha(A)) = \alpha + B$. Now, $\alpha + B$ can be written as the disjoint union $B_1 \cup B_2$ where $B_1 = (\alpha + B) \cap [0, 2\pi) = (\alpha + B) \cap [\alpha, 2\pi)$ and $B_2 = (\alpha + B) \cap [2\pi, 4\pi)$. By the translation invariance of the Lebesgue measure, $|B_1| = |(\alpha + B) \cap [\alpha, 2\pi)| = |B \cap [0, 2\pi - \alpha)|$ and, since $4\pi - \alpha > 2\pi$, $|B_2| = |(\alpha + B) \cap [2\pi, 4\pi)| = |B \cap [2\pi - \alpha, 2\pi)|$. Therefore, if $A \in \mathcal{M}_T$ and $T(A) = B$, then

$$\mu(R_\alpha(A)) = (1/2\pi)|T(R_\alpha(A))| = (1/2\pi)|\alpha + B|$$
$$= (1/2\pi)(|B \cap [0, 2\pi - \alpha)| + |B \cap [2\pi - \alpha, 2\pi)|) = (1/2\pi)|B| = \mu(A).$$

100. Let $A_n = \{x \in \mathbb{R} : |x - a_n| < \sqrt{\lambda_n}\}$, $n \geq 1$; then $|A_n| = 2\sqrt{\lambda_n}$, $\sum |A_n| < \infty$, and by Borel-Cantelli for x a.e. there exists an integer $n(x)$ such that $|x - a_n| > \sqrt{\lambda_n}$ for $n \geq n(x)$. Therefore for a.e. x and all $n \geq n(x)$, $\lambda_n/|x - a_n| < \sqrt{\lambda_n}$ and the result follows since $\sum_n \sqrt{\lambda_n} < \infty$.

101. Let $H = \{e_\lambda\}_{\lambda \in \Lambda}$ be a Hamel basis of $\mathbb{R}$ over $\mathbb{Q}$ and μ such that e_μ is irrational. We claim that $\mathcal{V} = \mathrm{sp}\{e_\lambda : \lambda \neq \mu\} \notin \mathcal{L}(\mathbb{R})$. For the sake of argument suppose that $\mathcal{V} \in \mathcal{L}(\mathbb{R})$; then, by translation invariance, $qe_\mu + \mathcal{V} \in \mathcal{L}(\mathbb{R})$ and $|\mathcal{V}| = |qe_\mu + \mathcal{V}|$ for all $q \in \mathbb{Q}$. Now, since H is a Hamel basis, it readily follows that $\mathbb{R} = \bigcup_{q \in \mathbb{Q}}(qe_\mu + \mathcal{V})$ where by the uniqueness of the Hamel representation the sets $\{qe_\mu + \mathcal{V}\}$ are pairwise disjoint and so we must have $|\mathcal{V}| > 0$. However, observe that $\mathcal{V} - \mathcal{V}$ cannot contain an interval about the origin. Indeed, if it contained one such interval of length η, say, then for all $q \in \mathbb{Q}$ with $|qe_\mu| < \eta/2$ we would have $qe_\mu = v_1 - v_2$ with $v_1, v_2 \in \mathcal{V}$ and, consequently, $\mathcal{V} \cap (qe_\mu + \mathcal{V}) \neq \emptyset$ for those q's, which, as noted above, cannot happen.

102. $H - H$ cannot contain an interval about the origin.

103. Let $A = \bigcup_{n \in \mathbb{Z}}(nC)$ where C denotes the Cantor discontinuum; A contains a maximal set H, with respect to inclusion, of rationally independent subsets of A. For the sake of argument suppose that A is not a Hamel basis of $\mathbb{R}$ and let $r \in \mathbb{R}$ such that $r \notin \mathrm{sp}\{H\}$, where the span is over $\mathbb{Q}$. Since $A + A = \mathbb{R}$ there exist a_1, a_2 in A such that $a_1 + a_2 = r$. Obviously one of a_1 or a_2, a_1, say, is not in $\mathrm{sp}\{H\}$. Consider $H' = \{a_1\} \cup H \subset A$; H' is also linearly independent over $\mathbb{Q}$ and contains H properly, which cannot happen. Finally, note that A is of first category and $|A| = 0$.

104. Let $\mathcal{F}$ denote the family of compact subsets of $\mathbb{R}$ of positive measure; by Problem 1.70, $\mathcal{F}$ has cardinality c and, therefore, with Ω the first uncountable ordinal, its elements can be listed $\mathcal{F} = \{K_b\}_{b<\Omega}$. We define now the numbers $\{e_i\}_{i<\Omega}$ by transfinite induction as follows. Let e_0 be any nonzero number in K_0. Next, if $d < \Omega$ and if $\{e_b\}_{b<d}$ have been chosen, let B_d be the linear space over $\mathbb{Q}$ spanned by $\{e_b\}_{b<d}$ and pick e_d to be any nonzero number in $K_d \setminus B_d$; this selection is only possible if $K_d \setminus B_d \neq \emptyset$. For the sake of argument suppose that $K_d \setminus B_d = \emptyset$, i.e., $K_d \subset B_d$, for some $d < \Omega$. Now, since $|K_d| > 0$, $K_d - K_d$ contains a neighborhood V of the origin and since B_d is a linear space, $x - y \in B_d$ whenever $x, y \in K_d$ and $V \subset B_d$. Then, since B_d is closed under rational multiplication it follows that $rV \subset B_d$ for every rational r, which is only possible if $B_d = \mathbb{R}$, i.e., if the set $\{e_b\}_{b<d}$ contains a Hamel basis E, say, of $\mathbb{R}$. Since $\mathrm{card}(\{e_b\}_{b<d}) < c$ it follows that $\mathrm{card}(E) < c$, which by Problem 1.72 is not the case. Thus $K_d \setminus B_d \neq \emptyset$ for all $d < \Omega$.

Finally, let $H_0 = \{e_d\}_{d<\Omega}$. H_0 is linearly independent by construction and by Zorn's lemma it can be extended to a Hamel basis H of $\mathbb{R}$. Since

$H_0 \cap K_b \neq \emptyset$ for all $b < \Omega$ also $H \cap K_b \neq \emptyset$ for all $b < \Omega$ or, in other words, $H \cap K \neq \emptyset$ for every $K \in \mathcal{F}$. Thus, by Problem 7, $|H|_e = \infty$ and, therefore, by Problem 102, H cannot be measurable.

105. First, observe that $|f'(x)| \geq 1$ in $(0,1)$. Now, derivatives satisfy the intermediate value property and, therefore, $f'(x)$ cannot vanish or change signs. So, if $f'(x) \geq 1$, $f(x)$ is strictly increasing on $(0,1)$. Thus $f^{-1}([f(x), f(x+h)]) = [x, x+h]$ and, consequently, $|f^{-1}([f(x), f(x+h)])| = f(x+h) - f(x) = h$ for $h > 0$ small enough and $f'(x) = 1$ in $(0,1)$. This means that $f(x) = x + a$ with $a = f(0) \geq 0$ and $1 + a = f(1) \leq 1$; hence $a = 0$ and $f(x) = x$. Similarly, if $f'(x) \leq -1$, $f(x) = 1 - x$. Therefore $f(x) = x$ or $= 1 - x$.

106. Let $O \subset \mathcal{O}$ be open and write $O = \bigcup_k I_k$ where the I_k's are pairwise disjoint open intervals. Then $\varphi^{-1}(O) = \bigcup_k \varphi^{-1}(I_k)$ and, consequently, $|\varphi^{-1}(O)| = \sum_k |\varphi^{-1}(I_k)| = \sum_k |I_k| = |O|$. Now, if N is a null set, given $\varepsilon > 0$, let O be an open set containing N such that $|O| \leq \varepsilon$. Then $\varphi^{-1}(N) \subset \varphi^{-1}(O)$ and, consequently, $|\varphi^{-1}(N)|_e \leq |O| \leq \varepsilon$ and $\varphi^{-1}(N)$ is a measurable set of measure 0. Next, every $A \in \mathcal{L}(\mathbb{R})$ can be written as $A = F \cup N$ where $F \in F_\sigma$ and N null. Thus $\varphi^{-1}(A) = \varphi^{-1}(F) \cup \varphi^{-1}(N)$ and since φ is continuous, $\varphi^{-1}(F)$ is F_σ and since $\varphi^{-1}(N)$ is null, $\varphi^{-1}(A)$ is measurable and it only remains to compute its measure. Let O, O_1 be open sets such that $A \subset O$, $O \setminus A \subset O_1$, and $|O_1| \leq \varepsilon$; note that $|\varphi^{-1}(O \setminus A)| \leq |\varphi^{-1}(O_1)| = |O_1| \leq \varepsilon$. Now, since $\varphi^{-1}(O \setminus A) = \varphi^{-1}(O) \setminus \varphi^{-1}(A)$ it follows that $|\varphi^{-1}(O)| - \varepsilon \leq |\varphi^{-1}(A)| \leq |\varphi^{1-}(O)|$, which, together with $|\varphi^{-1}(O)| = |O|$ and $|O| - \varepsilon \leq |A| \leq |O|$, implies that $|A| - \varepsilon \leq |\varphi^{-1}(A)| \leq |A| + \varepsilon$. Since $\varepsilon > 0$ is arbitrary, $|A| = |\varphi^{-1}(A)|$.

107. The statement is true and by Problem 106 it suffices to prove that $|f^{-1}(J)| = |J|$ for subintervals $J \subset I$.

108. Yes. Let $f(x) = 10^{-N(x)}$ where $N(x)$ is the number of 0's and 9's after the decimal points in the expansion of x; for example, $f(0) = 0 = f(1)$ and $f(1/3) = 1$ since $1/3 = .3333...$ and $N(1/3) = 0$. f is well-defined and assumes values between 0 and 1. The discontinuities of f are precisely the points x where f is finite and this set is of measure 0 and has uncountable intersection with every open interval.

Alternatively, let K_1 be the Cantor discontinuum in $[0,1]$, K_2 the union of similar Cantor sets in each of the intervals in $[0,1] \setminus K_1$, and so on for K_n. Finally, let $K = \bigcup_n K_n$. The function $f(x) = 2^{-n}$ for $x \in K_n$ and 0 for $x \notin K$ satisfies the desired properties.

109. Let $\{I_n\}$ be an enumeration of the open intervals with rational endpoints contained in $[0,1]$ and, for $n = 1, 2, \ldots$, let $\mathcal{P}_n = \{[A] \in \widetilde{\mathcal{L}} : |A \cap I_n| = 0\}$, $\mathcal{Q}_n = \{[A] \in \widetilde{\mathcal{L}} : |A^c \cap I_n| = 0\}$; then $\mathcal{S}^c = (\bigcup_n \mathcal{P}_n) \cup$

$(\bigcup_n \mathcal{Q}_n)$. We claim that both $\mathcal{P}_n$, $\mathcal{Q}_n$ are closed and nowhere dense; since the arguments are similar, we only do $\mathcal{P}_n$. Given $[A] \in \overline{\mathcal{P}_n}$ and $\varepsilon > 0$, let $[A_1] \in \mathcal{P}_n$ be such that $|A \Delta A_1| < \varepsilon$. Then $A \cap I_n \subset ((A \Delta A_1) \cap I_n) \cup (A_1 \cup I_n)$ and so $|A \cap I_n| \leq |A \Delta A_1| < \varepsilon$. Since ε is arbitrary, $|A \cap I_n| = 0$, $[A] \in \mathcal{P}_n$, and $\mathcal{P}_n$ is closed. Next, $\mathcal{P}_n$ has empty interior. Let $[A] \in \mathcal{P}_n$ and $\varepsilon > 0$ arbitrary. Let $J \subset I_n$ be an interval with $|J| < \varepsilon$. Then $d([A], [A \cup J]) < \varepsilon$ and clearly $[A \cup J] \notin \mathcal{P}_n$; hence $\mathcal{P}_n$ has empty interior. Thus $\mathcal{P}_n$, $\mathcal{Q}_n$ are closed and nowhere dense, $\mathcal{S}^c$ is of first category, and since $(\widetilde{\mathcal{L}}, d)$ is a complete metric space $\mathcal{S}$ is of second category.

110. It suffices to prove that $[-L, L] \setminus \bigcup_k A_k$ has Lebesgue measure 0 for each positive integer L.

111. (a) Referring to Problem 1.11, let $\mathcal{D}$ denote the collection of Diophantine numbers; $\mathcal{D}$ is of first category. Now, $\mathcal{D} = \bigcup_\alpha \mathcal{D}(\alpha)$ where $\mathcal{D}(\alpha)$ denotes the Diophantine numbers of exponent α with $\alpha > 2$ and by Problem 110 with $\lambda_q = |q|^{-\alpha}$ there, $\mathcal{D}(\alpha)$ has full measure for $\alpha > 2$ and so does $\mathcal{D}$.

(b) Let $\mathcal{L}$ denote the set of Liouville numbers $\mathcal{L} = \mathbb{R} \setminus (\mathbb{Q} \cup \mathcal{D})$; $\mathcal{L}$ is of second category in $\mathbb{R}$ and since $\mathbb{Q} \cup \mathcal{D}$ has full measure, $\mathcal{L}$ has measure 0. Alternatively, one such set can be constructed as follows. Let $\{O_m\}$ be dense open subsets of $\mathbb{R}^n$ with $|O_m| \leq 1/m$ and put $B = \bigcap_m O_m$; B is a Borel G_δ set and being the intersection of dense sets, by Problem 1.1, is dense in $\mathbb{R}^n$. Moreover, since $|B| \leq |O_m| \leq 1/m$ for all m, $|B| = 0$. Thus B is a set of second category in $\mathbb{R}^n$ with measure 0 and $\mathbb{R}^n \setminus B = \bigcup_m O_m^c$ is the countable union of closed and, since O_m^c does not contain points in $\mathbb{R}^n$ with rational coordinates, also nowhere dense, sets, i.e., $\mathbb{R}^n \setminus B$ is of first category and full measure. Hence $\mathbb{R} = B \cup (\mathbb{R}^n \setminus B)$ is the union of two sets which are small in different senses.

112. The statement is true. Let C denote the Cantor discontinuum; C has measure 0, is nowhere dense, and $C + C = [0, 2]$. Now, also each $nC = \{x \in \mathbb{R} : x = ny, y \in C\}$, $-\infty < n < \infty$, has measure 0 and is nowhere dense. Then $A = \bigcup_n nC$, being the countable union of null sets, has measure 0, and, being the countable union of nowhere dense sets, is of first category. We claim that $A + A = \mathbb{R}$. Indeed, note that $\bigcup_n (nC + nC) \subset A + A = \bigcup_{n,m} (nC + mC)$. Moreover, since $nC + nC = n(C + C) \supset n[0, 2]$, $\bigcup_n (nC + nC) = \mathbb{R}$ and so $A + A = \mathbb{R}$.

113. The statement is false. First, observe that if A, B are arbitrary sets such that $\mathbb{Q} - B \subset B$ and $A \cap B = \emptyset$, then $A + B$ does not contain an open interval. Indeed, it suffices to verify that $(A + B) \cap \mathbb{Q} = \emptyset$. For the sake of argument suppose that there exist $q \in \mathbb{Q}, a \in A$ and $b \in B$ such that $q = a + b$. Then $q - b = a$, which is not possible since $q - b \in B$ and $A \cap B = \emptyset$.

Now, to the construction. Let C be a Cantor discontinuum of positive measure and set $B = (\mathbb{Q} - C) \cup (\mathbb{Q} + C)$; since B is the countable union of translates of C and $-C$, B is of first category. To see that B satisfies $\mathbb{Q} - B \subset B$ let $x = q - b$, $q \in \mathbb{Q}, b \in B$. Since $b \in B$ there exist $q_1 \in \mathbb{Q}$ and $c \in C$ such that $b = q_1 - c$ or $b = q_1 + c$. Thus $x = (q - q_1) + c$ or $x = (q - q_1) - c$ and, in either case, $x \in B$ as we wanted to prove. Now, let $A = B^c$. Since B is of first category, A is of second category. Also, B has positive, and, hence, infinite measure, and A has 0 measure. Finally, $A + B$ contains no interval.

114. (a) The statement is true. First, observe that a nonempty open interval J of $\mathbb{R}$ contains a compact nowhere dense subset of positive measure. A direct proof of this observation without recourse to the Cantor discontinuum goes as follows: Let $J_0 \subset J$ be a closed interval of positive measure, O a dense open subset of $\mathbb{R}$ with $|O| < |J_0|$, and set $K = J_0 \setminus O$; clearly K is a compact nowhere dense subset of J of positive measure. Furthermore, given an integer $n \geq 1$, by subdividing J into n nonempty open intervals (and a finite set of points), it readily follows that J contains n pairwise disjoint compact nowhere dense subsets of positive measure.

Let $\{I_k\}$ denote an enumeration of the open intervals of positive length with rational endpoints in $\mathbb{R}$. We construct n sequences $\{F_m^k\}$, $1 \leq k \leq n$, say, of pairwise disjoint nowhere dense compact sets of positive measure as follows. First, I_1 contains n disjoint compact nowhere dense subsets of positive measure, $F_1^1, \ldots, F_1^n$, say. Next, having chosen $\{F_m^k\}$, $1 \leq k \leq n$, $m < M$, pairwise disjoint compact nowhere dense sets of positive measure, note that $\bigcup_{1 \leq k \leq n, m < M} F_m^k$ is also compact and nowhere dense and, consequently, I_M contains a nonempty open subinterval J_M, say, disjoint from that set. Therefore J_M contains n disjoint closed nowhere dense sets $F_M^1, \ldots, F_M^n$, say, each of positive measure. Now, let $A_k = \bigcup_m F_m^k$, $1 \leq k \leq n$; by construction $A_1, \ldots, A_n$ are pairwise disjoint F_σ subsets of $\mathbb{R}$. Also, if J is a nonempty open interval in $\mathbb{R}$, J contains an open subinterval with rational endpoints I_m, say. But then $|A_k \cap J| \geq |F_m^k \cap I_m| > 0$, $1 \leq k \leq n$.

(b) The statement is true. We construct pairwise disjoint closed nowhere dense sets $\{A_k^n\}$, $1 \leq k \leq n < \infty$, as follows. Let A_1^1 denote a nowhere dense compact subset of I_1 of positive measure. Having constructed the sets $\{A_k^n\}$, $1 \leq k \leq n - 1$, we proceed in the n-th stage as follows. Note that $\bigcup_{k \leq m < n} A_k^m$ is closed and nowhere dense. Thus I_n contains an open subinterval J_n, say, disjoint from $\bigcup_{k \leq m < n} A_k^m$. As noted above there are n pairwise disjoint closed nowhere dense subsets $A_1^n, A_2^n, \ldots, A_n^n$ of J_n, each of positive Lebesgue measure. Now, let $A_k = \bigcup_{n \geq k} A_k^n$, $k = 1, 2, \ldots$; the A_k's are pairwise disjoint F_σ subsets of $\mathbb{R}$. Let J be a nonempty open interval and

k an integer. Then J contains a nonempty open subinterval with rational endpoints I_m, say, that is shorter than each of $I_1, \ldots, I_k$; note that since I_m is shorter than I_i for $i \leq k$, $n > k$. Therefore $|A \cap J| \geq |A_k \cap J| > |A_k^m \cap I_m| > 0$.

115. First, note that by Problem 1.15, D is a countable dense subgroup of the additive group $\mathbb{R}$. Moreover, the countably many sets $\{d + \mathcal{V} : d \in D\}$ are pairwise disjoint and $\mathbb{R} = \bigcup_{d \in D}(d + \mathcal{V})$. For the sake of argument suppose that $\mathcal{V}$ is measurable. Then by the translation invariance of the Lebesgue measure at least one the sets and, hence, also $\mathcal{V}$, has positive Lebesgue measure. Therefore, by Problem 67, $\mathcal{V} - \mathcal{V}$ contains an open interval about the origin and, since D is dense in $\mathbb{R}$, there are $x \neq y \in \mathcal{V}$ such that $x - y \in D$. In other words, x, y are distinct elements of $\mathcal{V}$ that belong to the same equivalence class of D, contrary to the way $\mathcal{V}$ was defined.

Observe that the following is also true: If $G \neq \mathbb{R}$ is a Lebesgue measurable additive subgroup of $\mathbb{R}$, $|G| = 0$. Indeed, were this not the case $G - G$ would contain a neighborhood of the origin and, being a group, G would contain a neighborhood of 0 and, therefore, it would be equal to $\mathbb{R}$.

Chapter 14

Measurable and Integrable Functions

1. Nothing. Let V be a Vitali Lebesgue nonmeasurable subset of I. Then $\chi_V + \chi_{I \setminus V} = \chi_I$ is measurable and $\chi_V + (-\chi_{I \setminus V})$ is not. And $|\chi_V - \chi_{I \setminus V}|$ is measurable yet $|-\chi_V|$ is not. Finally, $\chi_V \chi_{I \setminus V}$ is measurable but $\chi_V (1 + \chi_{I \setminus V})$ is not.

2. It depends on whether the measure is complete. In fact, the following is true: A measure space $(X, \mathcal{M}, \mu)$ is complete iff given f and g on X with f measurable and $g = f$ μ-a.e., g is measurable.

Necessity first. Let $N = \{g \neq f\}$ and note that since $\mu(N) = 0$, $N^c = \{g = f\}$ is measurable. Now, for λ real we have $\{g > \lambda\} = (\{g > \lambda\} \cap N^c) \cup (\{g > \lambda\} \cap N)$ where $\{g > \lambda\} \cap N^c = \{f > \lambda\} \cap N^c \in \mathcal{M}$ and $\{g > \lambda\} \cap N \subset N$ has measure 0; $\{g > \lambda\}$ is then the union of measurable sets and, hence, measurable.

Sufficiency next. Let $A \subset B$ with $\mu(B) = 0$ and consider $f = 0$ and $g = \chi_A + \chi_B$. Then f is measurable, $\{f \neq g\} = B$, and so $f = g$ μ-a.e. Therefore g is measurable, $g^{-1}(\{2\}) = A$ is measurable, and the measure is complete.

3. (a) The measurable functions f are those with $|f|$ even.

4. (a) $f = \chi_{(0,1/2]} + 2\chi_{(1/2,3/4]} + 3\chi_{(3/4,1]}$. (d) $f = \chi_{(0,1/4]} + 2\chi_{(1/4,1/2]} + 3\chi_{(1/2,3/4]} + 4\chi_{(3/4,1]}$.

5. f is measurable iff f is constant on $(-\infty, 0]$ and $(0, \infty)$.

6. If $\{f > 0\} \in \mathcal{M}$, f is measurable.

10. Let $\eta > 0$ be such that $J = (-\eta/2, \eta/2) \subset 2J = (-\eta, \eta) \subset O$. For $r \in \mathbb{R}$ let $A_r = f^{-1}(r + J)$ and note that $\bigcup_{r \in \mathbb{Q}} A_r = f^{-1}(\bigcup_{r \in \mathbb{Q}} r + J) =$

$f^{-1}(\mathbb{R}) = X$. Then $\mu(A_{r_0}) > 0$ for some $r_0 \in \mathbb{Q}$ and with $A = A_{r_0}$ it follows that $f(x), f(y) \in r_0 + J$ for all $x, y \in A$ and, therefore, $f(x) - f(y) \in 2J \subset O$.

11. Let $A_\lambda = \{f \leq \lambda\}$, $\lambda \in \mathbb{R}$, and for the sake of argument suppose that for all real λ not both $\mu(A_\lambda)$ and $\mu(A_\lambda^c) = \mu(\{f > \lambda\})$ are positive. Now, $A_\lambda \subset A_{\lambda'}$ for $\lambda' > \lambda$ and $A = \{\lambda \in \mathbb{R} : \mu(A_\lambda) > 0\}$ is an interval of the form $[a, \infty)$ or (a, ∞) where a is a finite number (if $a = -\infty$, f is the constant function equal to $-\infty$ and if $a = \infty$, f is the constant function equal to ∞). Similarly, let $B_\lambda = A_\lambda^c$, $\lambda \in \mathbb{R}$, and $B = \{\lambda \in \mathbb{R} : \mu(B_\lambda) > 0\}$; B is an interval of the form $(-\infty, b]$ or $(-\infty, b)$ for some finite b. We claim that $a = b$. Indeed, if $a < b$ pick η such that $a < \eta < b$ and observe that $\mu(A_\eta) > 0$ and $\mu(B_\eta) > 0$, which cannot happen. And, if $b < a$ and $b < \eta < a$, then $\mu(A_\eta) = \mu(B_\eta) = 0$, which cannot happen either. Now, $a - 1/n \notin A$ for $n \geq 1$ and so $\mu(A_{a-1/n}) = 0$, which, since $\{A_{a-1/n}\}$ is a nondecreasing sequence, implies $\mu(\{f < a\}) = \lim_n \mu(A_{a-1/n}) = 0$. Similarly, since $\{B_{a+1/n}\}$ is a nondecreasing sequence and $a + 1/n \notin B$, it follows that $\mu(\{f > a\}) = \lim_n \mu(B_{a+1/n}) = 0$. Therefore $\mu(\{f \neq a\}) = \mu(\{f < a\}) + \mu(\{f > a\}) = 0$, which implies that $f = a$ μ-a.e., and this is not the case. Thus $\mu(\{f \leq \lambda\}) > 0$ and $\mu(\{f > \lambda\}) > 0$ for some $\lambda \in \mathbb{R}$.

12. (a) First, since $\mathcal{S}$ is the smallest σ-algebra of subsets of $\mathbb{R}^2$ that contains the open balls centered at the origin, $\mathcal{S} \subset \mathcal{B}(\mathbb{R}^2)$. Now let $\mathcal{C} = \{[0, \eta) : \eta > 0\}$ and recall that $\mathcal{M}(\mathcal{C}) = \mathcal{B}(\mathbb{R}^+)$. Then, since $u^{-1}([0, \eta)) = B(0, \eta)$, it follows that $u^{-1}([0, \eta)) \in \mathcal{S}$ and by Problem 2.22, $u^{-1}(\mathcal{M}(\mathcal{C})) = \mathcal{M}(u^{-1}(\mathcal{C})) = \mathcal{S}$. Therefore u is measurable.

(d) If $f : (\mathbb{R}^2, \mathcal{S}) \to (\mathbb{R}^+, \mathcal{B}(\mathbb{R}^+))$ is measurable, since the singleton $\{f(3, 4)\} \in \mathcal{B}(\mathbb{R}^+)$ by (c) it follows that $f^{-1}(\{f(3, 4)\}) \in \mathcal{S} = \mathcal{C}$. Now, since $(3, 4) \in f^{-1}(\{f(3, 4)\})$ and $u(3, 4) = u(4, 3)$, we have $(4, 3) \in f^{-1}(\{f(3, 4)\})$, which implies that $f(4, 3) = f(3, 4)$. The proof for $(0, 5)$ follows along similar lines.

13. Necessity first. Since $f : (X, \mathcal{S}) \to (\mathbb{R}, \mathcal{B}(\mathbb{R}))$ is measurable, $\{f > \lambda\} \in \mathcal{S}$ for all $\lambda \in \mathbb{R}$. Then by Problems 2.53 and 2.54, $\{f > \lambda\} = F_\lambda \cup N_\lambda$ with $F_\lambda \in \mathcal{M}$ and $\mu_1(N_\lambda) = 0$; moreover, since μ_1 is complete, replacing N_λ by $N_\lambda \setminus F_\lambda$ if necessary, we may assume that $F_\lambda \cap N_\lambda = \emptyset$. Let $N = \bigcup_{\lambda \in \mathbb{Q}} N_\lambda$, observe that $\mu_1(N) = 0$, and define

$$g(x) = \begin{cases} f(x), & x \notin N, \\ -\infty, & x \in N. \end{cases}$$

Clearly $f = g$ μ_1-a.e. Let $\lambda \in \mathbb{R}$ and consider rationals $\{\lambda_n\}$ that decrease to λ. Then $\{g > \lambda\} = \bigcup_n \{g > \lambda_n\}$ where $\{g > \lambda_n\} \in \mathcal{M}$ for all n. Thus $\{g > \lambda\}$ is the countable union of sets in $\mathcal{M}$ and g is measurable.

Conversely, given $\lambda \in \mathbb{R}$, let $N_\lambda = \{f > \lambda\} \setminus \{g > \lambda\}$; by assumption $\mu_1(N_\lambda) = 0$. Then $\{f > \lambda\} = \{g > \lambda\} \cup N_\lambda \in \mathcal{S}$ and f is measurable.

14. With V a Vitali Lebesgue nonmeasurable subset of I, $f(x) = \chi_V(x) - \chi_{V^c}(x)$ will do. Indeed, if $|g(x) - f(x)| < 1$, then $0 < g(x) < 2$ in V and $-2 < g(x) < 0$ in V^c. Thus the sets $\{g > 0\} = V$ and $\{g < 0\} = V^c$ are Lebesgue nonmeasurable.

15. f may be defined in $\mathbb{R}^n$ as follows: Observe that for a nonempty closed set F, $d(x, F)$ is a continuous function that vanishes iff $x \in F$. We may assume $O \neq \mathbb{R}^n$ and $F \neq \emptyset$ (since otherwise we just let $f = 0$ or $f = 1$, respectively), and define

$$f(x) = \frac{d(x, O^c)}{d(x, O^c) + d(x, F)}\,.$$

Since $O^c \cap F = \emptyset$ the denominator does not vanish. Thus f is continuous and $f(x) = 0$ if $x \in O^c$ and $f(x) = 1$ if $x \in F$.

17. First, suppose that $f \geq 0$. Let $\{f_n\}$ be the nondecreasing sequence of nonnegative simple functions assuming rational values described in the introduction. Then $\{f_n\}$ tends to f everywhere: If $f(x)$ is finite, $0 \leq f(x) - f_n(x) \leq 2^{-n}$ for all n and if $f(x) = \infty$, $f_n(x) = n$ for all n. Put $g(x) = \sum_n (f_{n+1}(x) - f_n(x))$; then g is finite a.e. and $g(x) = \lim_N \sum_{n=1}^N (f_{n+1}(x) - f_n(x)) = \lim_N f_{N+1}(x) - f_1(x) = f(x) - f_1(x)$. Thus $f(x) = f_1(x) + \sum_n (f_{n+1}(x) - f_n(x))$, which is as described. Finally, if f is arbitrary, $f = f^+ - f^-$ is the difference of nonnegative functions and has the desired representation.

18. Let $A_1 = \{f \geq 1\}$ and then, assuming that $A_1, \dots, A_{n-1}$ have been defined, let

$$A_n = \left\{ x \in X : f(x) - \sum_{k=1}^{n-1} \frac{1}{k} \chi_{A_k}(x) \geq \frac{1}{n} \right\}.$$

Since $\sum_n n^{-1} = \infty$, $f(x) \in [0, \infty]$ is represented as desired for each $x \in X$.

19. We claim that $\mathcal{D} = \{A \subset X : \chi_A \in \mathcal{V}\}$ is a λ-system in X that contains $\mathcal{F}$. First, since $X \in \mathcal{F}$, $X \in \mathcal{D}$. Next, suppose $A \subset B$, $A, B \in \mathcal{D}$. Then $\chi_A, \chi_B \in \mathcal{V}$ and since $\mathcal{V}$ is a linear space, $\chi_B - \chi_A = \chi_{B \setminus A} \in \mathcal{V}$ and $B \setminus A \in \mathcal{D}$. Finally, let $\{A_n\} \subset \mathcal{D}$ with $A_n \subset A_{n+1}$, $n \geq 1$, and put $A = \bigcup_n A_n$. Then $\chi_{A_n} \in \mathcal{V}$, $\{\chi_{A_n}\}$ increases to χ_A and, therefore, $\chi_A \in \mathcal{V}$ and $A \in \mathcal{D}$. Thus $\mathcal{D}$ is a λ-system in X that contains $\mathcal{F}$ and by the π-λ-theorem, Problem 2.94, $\mathcal{M}(\mathcal{F}) \subset \mathcal{D}$. Therefore $\chi_A \in \mathcal{V}$ for every $A \in \mathcal{M}(\mathcal{F})$.

Next, since $\mathcal{V}$ is a linear space, simple functions $\sum_{n=1}^m \lambda_n \chi_{A_n}$ with λ_n scalar and $A_n \in \mathcal{M}(\mathcal{F})$ are in $\mathcal{V}$. Finally, let f be measurable. Then $f = f^+ - f^-$ where f^+ and f^- are nonnegative and measurable and, hence, monotone limits of simple functions. Now, since simple functions are in $\mathcal{V}$

and since $\mathcal{V}$ is closed under monotone limits, f^+, f^- are in $\mathcal{V}$ and so is their difference, f.

21. Since A remains unchanged if $f(x)$ is replaced by $\sup_{0 \leq y \leq x} f(y)$ we may assume that f is nondecreasing on $[0,1]$. Then $f^{-1}(\{x\}) = [a,b] \subset [0,1]$ with $a < x \leq b$ for each $x \in (0,1] \setminus A$ and so, $(0,1] \setminus A$ can be represented as an at most countable union of disjoint intervals $(a_k, b_k] \subset (0,1]$, say. Hence $[0,1] \setminus A$ is a Borel set and so is A.

22. Consider the family $\{I_{n,k}\}$ of dyadic intervals of $\mathbb{R}$ given by $I_{n,k} = ((k-1)/2^n, k/2^n]$, $k = 0, \pm 1, \pm 2, \ldots$, $n = 1, 2, \ldots$, and, for each n let g_n be the step function $g_n(x) = \sup_{I_{n,k}} f - \inf_{I_{n,k}} f$, $x \in I_{n,k}$. Note that for each $n, m = 1, 2, \ldots$ the set $A_{n,m} = \{x \in \mathbb{R} : g_n(x) < 1/m\}$ is an at most countable union of intervals and so, Borel. Consider now those x which are not endpoints of the $I_{n,k}$, i.e., outside a Borel countable subset of $\mathbb{R}$. For each such x we have $x \in A$ iff $\lim_n g_n(x) = 0$ iff for each $m = 1, 2, \ldots$ there exists an integer k such that $g_n(x) < 1/m$ for all $n \geq k$, i.e., $x \in F = \bigcap_m \bigcup_k \bigcap_{n=k}^{\infty} A_{n,m}$. Then $F = \bigcap_m \liminf_n A_{n,m} \in \mathcal{B}(\mathbb{R})$ and since A differs from F on an at most countable set, $A \in \mathcal{B}(\mathbb{R})$.

24. Let $D(f)$ denote the set of points of discontinuity of f; by assumption $|D(f)| = 0$. Write $\{x \in \mathbb{R}^n : f(x) > \lambda\} = \{x \in \mathbb{R}^n \setminus D(f) : f(x) > \lambda\} \cup \{x \in D(f) : f(x) > \lambda\} = A \cup B$, say. Clearly $|B| = 0$. Now, if $x \in A$, since f is continuous at x there is an open neighborhood O_x of x such that $f(z) > \lambda$ for $z \in O_x$ and so $A \subset \bigcup_{x \in A} O_x = O$ where O is open. Finally, since $A = O \cap (\mathbb{R}^n \setminus D(f))$, A is measurable and so is $\{f > \lambda\} = A \cup B$.

25. Recall that $f(x) = \liminf_{\varepsilon \to 0} \{f(y) : |x - y| < \varepsilon\}$. Given $\lambda \in \mathbb{R}$, it suffices to prove that $A = \{f > \lambda\}$ is Borel; in fact, A is open. To see this let $x \in A$. Then $f(x) > \lambda$ and so $\liminf_{\varepsilon \to 0} \{f(y) : |y - x| < \varepsilon\} > \lambda$. Thus there is $\varepsilon > 0$ such that $\inf\{f(y) : |y - x| < \varepsilon\} > \lambda$. In particular, $f(y) > \lambda$ for all $y \in B(x, \varepsilon)$ and $B(x, \varepsilon) \subset A$.

26. (a) For $n = 1, 2, \ldots$ and $k \in \mathbb{Z}$, let $I_k^n = ((k-1)/n, k/n]$ and consider the functions $f_n(x) = f(k/n)$ in the interval I_k^n for all n, k. Note that $\bigcup_k I_k^n = \mathbb{R}$ for all n, that the f_n are Borel measurable step functions, and, since f is right-continuous, $f(x) = \lim_n f_n(x)$ for all x. Hence f is Borel measurable.

(b) Let $A = \{x \in \mathbb{R} : \limsup_{y \to x} |f(x) - f(y)| = 0\}$ and write $A^c = \bigcup_k B_k$ where $B_k = \{x \in \mathbb{R} : \limsup_{y \to x} |f(x) - f(y)| > 1/k\}$, $k = 1, 2, \ldots$. Fix an integer k; we claim that B_k is at most countable. First, by the right-continuity of f, given $x \in \mathbb{R}$, there exists $\delta = \delta(x) > 0$ such that $|f(x) - f(y)| \leq 1/2k$ for all $y \in I(x) = (x, x + \delta)$. Then for all $y, y' \in I(x)$ it follows that $|f(y) - f(y')| \leq |f(y) - f(x)| + |f(x) - f(y')| \leq 1/k$, and, consequently, $\limsup_{z \to y} |f(z) - f(y)| \leq 1/k$ for all $y \in I(x)$, $y \neq x$. This implies that

$B_k \cap I(x) = \emptyset$. Now let $x, x' \in B_k$ and suppose that $I(x) \cap I(x') \neq \emptyset$. Then $x \in I(x')$ or $x' \in I(X)$ but neither of these alternatives is possible since $B_k \cap I(x) = \emptyset$. Since it is possible to choose a rational point in each of the $I(x)$ for $x \in B_k$, these sets are finite or countable for each k and the same is true for $\bigcup_k B_k$. Hence $|\bigcup_k B_k| = 0$ and f is continuous at each $x \in \mathbb{R} \setminus (\bigcup B_k)$, that is, a.e. in $\mathbb{R}$.

28. Since f is differentiable on $\mathbb{R}$, it is continuous and, hence, Borel measurable, as are $f_n(x) = n(f(x + 1/n) - f(x))$ for all n. Thus $f'(x) = \lim_n f_n(x)$ is Borel measurable on $\mathbb{R}$.

29. The statement is false: With C a Cantor set of positive measure, let $f(x) = \chi_C(x) - \chi_C(x + 1/2)$. However, f agrees a.e. with a function in the class $\mathcal{B}_2$. Indeed, the characteristic functions of closed sets are in $\mathcal{B}_1$, and, therefore, the characteristic functions of F_σ sets are in $\mathcal{B}_2$. Since an arbitrary Lebesgue measurable set is an F_σ set plus a set of measure 0, the statement is true for f the characteristic function of an arbitrary Lebesgue measurable set. It is therefore true for a simple function and the limit of simple functions, i.e., an arbitrary f.

30. Note that $\mathcal{M}$ does not contain all open sets in $\mathbb{R}^n$ because, if it did, it would contain the σ-algebra they generate, $\mathcal{B}(\mathbb{R}^n)$. Furthermore, since an open set in $\mathbb{R}^n$ is the countable union of nonoverlapping closed cubes, $\mathcal{M}$ does not contain a closed cube $Q = [a_1, b_1] \times \cdots \times [a_n, b_n]$, say. Now, given a closed interval $[a, b] \subset \mathbb{R}$, let

$$\varphi_{a,b}(x) = \begin{cases} 0, & x < a - 1, \text{ or } x > b + 1, \\ x - a + 1, & a - 1 \leq x < a, \\ 1, & a \leq x \leq b, \\ -x + b + 1, & b < x \leq b + 1. \end{cases}$$

Then the function $f(x) = \varphi_{a_1,b_1}(x_1) \times \cdots \times \varphi_{a_n,b_n}(x_n)$ is continuous and, since $\{f = 1\} = Q \notin \mathcal{M}$, $f : (\mathbb{R}^n, \mathcal{M}) \to (\mathbb{R}, \mathcal{B}(\mathbb{R}))$ is not measurable.

33. The statement is true. Let $\mathcal{F} = \{A \subset \mathbb{R} : f^{-1}(A) \in \mathcal{L}(\mathbb{R}^n)\}$; we claim that $\mathcal{F}$ is a σ-algebra that contains those sets of the form $[\lambda, \infty)$, which generate the Borel sets, and, therefore, it contains $\mathcal{B}(\mathbb{R})$. Clearly $\emptyset \in \mathcal{F}$. And, if $A \in \mathcal{F}$, since $f^{-1}(A^c) = f^{-1}(A)^c \in \mathcal{L}(\mathbb{R}^n)$, $A^c \in \mathcal{F}$. Finally, if $\{A_k\} \subset \mathcal{F}$, $f^{-1}(\bigcup_k A_k) = \bigcup_k f^{-1}(A_k) \in \mathcal{L}(\mathbb{R}^n)$ and $\bigcup_k A_k \in \mathcal{F}$.

Observe the following consequence of this result. If $g : \mathbb{R} \to \mathbb{R}$ is Borel measurable, then $g \circ f : \mathbb{R}^n \to \mathbb{R}$ is Lebesgue measurable. Indeed, if $B \in \mathcal{B}(\mathbb{R})$, then $g^{-1}(B) \in \mathcal{B}(\mathbb{R})$, and, therefore, $f^{-1}(g^{-1}(B)) \in \mathcal{L}(\mathbb{R}^n)$.

34. Necessity first. For the sake of argument suppose there is a Lebesgue null set N such that $|f^{-1}(N)|_e > 0$. Now, by Problem 3.72 there is a

Lebesgue nonmeasurable set $B \subset f^{-1}(N)$. Let $A = f(B) \subset N$; A is a subset of a set of measure 0 and is therefore measurable, yet $f^{-1}(A) = B \notin \mathcal{L}(\mathbb{R}^n)$.

Sufficiency next. Let $A \in \mathcal{L}(\mathbb{R})$; then $A = F \cup N$ where F is F_σ and $|N| = 0$. Then $f^{-1}(A) = f^{-1}(F) \cup f^{-1}(N)$ where by Problem 33, $f^{-1}(F) \in \mathcal{L}(\mathbb{R}^n)$ and by assumption $|f^{-1}(N)| = 0$. Hence $f^{-1}(A) \in \mathcal{L}(\mathbb{R}^n)$.

35. Since A is contained in a line in $\mathbb{R}^2$ and lines have Lebesgue measure 0, $|A| = 0$, and $A \in \mathcal{L}(\mathbb{R}^2)$. For the sake of argument suppose that $A \in \mathcal{B}(\mathbb{R}^2)$. Then, since $g(x) = (x, x)$ is continuous, by Problem 31, $B = g^{-1}(A) \in \mathcal{B}(\mathbb{R})$, which is not the case.

36. The statement is false. Let f denote the Cantor-Lebesgue function. Then $g(x) = (f(x) + x)/2$ is a strictly increasing function that maps $[0, 1]$ into itself. Now, $g(I) = g(C) \cup g(C^c)$ and it is readily seen that $g(C^c)$ is a Borel subset of I of measure $1/2$. Therefore g maps C into a set of measure $1/2$ and $g(C)$ contains a Lebesgue nonmeasurable set B, say. Let $h = g^{-1}$ and observe that since $h(B) \subset C$, $h(B) \in \mathcal{L}(\mathbb{R})$ with $|h(B)| = 0$. Let $f = \chi_{h(N)}$; then f is Lebesgue measurable, g is continuous, and, since $(f \circ g)^{-1}(\{1\}) = B \notin \mathcal{L}(\mathbb{R})$, $f \circ g$ is not Lebesgue measurable

38. In fact, the following is true: f is measurable iff $\arctan(f)$ is measurable. We have just proved necessity. As for sufficiency, if $\arctan(f)$ is measurable, $\{\arctan(f) > \lambda\}$ is measurable for all $\lambda \in \mathbb{R}$. If $\lambda > \pi/2$, $\{\arctan(f) > \lambda\} = \emptyset$, and if $\lambda < -\pi/2$, $\{\arctan(f) > \lambda\} = X$. Now, for $t \in \mathbb{R}$ there exists $\lambda \in (-\pi/2, \pi/2)$ such that $t = \tan(\lambda)$; then $\{f > t\} = \{\arctan(f) > \lambda\}$, which is measurable.

40. (c) No.

41. Let $L_\lambda = \{f = \lambda\}$; the L_λ are pairwise disjoint and Lebesgue measurable. Therefore, by Problem 2.57, $A = \{\lambda \in \mathbb{R} : |L_\lambda| > 0\}$ is countable. Thus $g(\lambda) = 0$ except for countably many λ and $g = 0$ a.e. This gives the measurability of g and the fact that $\int_\mathbb{R} g(\lambda)\, d\lambda = 0$.

43. Since g is integer-valued it suffices to prove that $A_n = g^{-1}(\{n\})$ is measurable for each integer n. Now, $(x, y) \in A_1$ if $x_1 = y_1$ and A_1 is the union of 3 intervals, $[0, 1/3)^2 \cup [1/3, 2/3)^2 \cup [2/3, 1)^2$, where each interval has sidelength $1/3$ and measure $1/3^2$; thus $|A_1| = 3/3^2 = 1/3$. Similarly, $(x, y) \in A_2$ means that $x_1 \neq y_1$ and $x_2 = y_2$. Now, $x_1 \neq y_1$ means that x and y belong to different thirds of $[0, 1)$ and there are $3 \cdot 2 = 6$ such choices. On each interval we fit three intervals as we did for A_1 and so A_2 is the union of $3(3 \cdot 2)$ intervals each of sidelength $1/3^2$ and size $1/3^4$; thus $|A_2| = 2 \cdot 3^2/3^4 = 2/3^2$. Finally, in general, $(x, y) \in A_n$ if $x_n = y_n$ and $x_k \neq y_k$ for $1 \leq k \leq n - 1$. For each k there are 6 options for pairs of (x_k, y_k) with $x_k \neq y_k$ for a total of 6^{n-1} pairs. Thus A_n can be expressed as the disjoint union of $3 \cdot 6^{n-1}$ squares with sidelength 3^{-n} and consequently

$|A_n| = 3 \cdot 6^{n-1}/3^{2n}$. Furthermore, note that $E = \bigcup_n g^{-1}(\{n\})$ has measure $|E| = \sum_n 3 \cdot 6^{n-1}/3^{2n} = (1/3)\sum_n (2/3)^{n-1} = 1$.

44. (a) Let D denote the dyadic rationals of I, $|D| = 0$, and note that f_n is well-defined for $x \in I \setminus D$, in other words, a.e. Specifically, $f_1(x) = \chi_{(1/2,1)}(x)$, $f_2(x) = \chi_{(1/4,2/4)}(x) + \chi_{(3/4,4/4)}(x)$, and, in general, $f_n(x) = \sum_{j=0}^{2^{n-1}-1} \chi_{(2j+1/2^n, 2j+2/2^n)}(x)$. Clearly the f_n are measurable and integrable.

(b) Let $A = [0,1] \setminus D$. Now, f is the pointwise limit of measurable functions and so measurable, and since $\sum_n x_{\sigma(n)}/2^n \leq 1$, $f \in L^1(A)$. Now, with $\lambda = \sum_n \lambda_n 2^{-n}$ consider $\{x \in A : f(x) < \lambda\} = \{x \in A : \sum_n x_{\sigma(n)} 2^{-n} < \sum_n \lambda_n 2^{-n}\}$ and observe that $A = \bigcup_n A_n$ where $A_n = \{x \in A : x_{\sigma(k)} = \lambda_k$ for $1 \leq k \leq (n-1)$ and $x_{\sigma(n)} < \lambda_n\}$, $n \geq 1$. Now, $\lambda_n = 0, 1$, and if $\lambda_n = 0$, then $A_n = \emptyset$ and $|A_n| = 0$, and if $\lambda_n = 1$, then $A_n = \{x \in A : x_{\sigma(k)} = \lambda_k$ for $1 \leq k \leq (n-1), x_{\sigma(n)} = 0\}$ and $|A_n| = 1/2^n = \lambda_n/2^n$. Hence $|\{f < \lambda\}| = \sum_n \lambda_n/2^n = \lambda$ for all $\lambda \in [0,1] \setminus D$. Finally, if $\lambda \in D$, pick $\{\lambda_n\} \subset [0,1] \setminus D$ such that λ_n increases to λ. Then $\{f < \lambda_n\}$ increases to $\{f < \lambda\}$ and by continuity from below $|\{f < \lambda\}| = \lim_n |\{f < \lambda_n\}| = \lim_n \lambda_n = \lambda$.

This statement can be stated as follows: If μ denotes the Lebesgue measure in $[0,1]$ and $\mu \circ f^{-1}$ the image measure, then $\mu \circ f^{-1} = \mu$.

45. First, note that $f(rx) = rf(x)$ for all rationals r; thus the result follows readily for continuous functions. Furthermore, by linearity the continuity of f at any $x \in \mathbb{R}$ is equivalent to the continuity of f at 0.

(a) By the linearity of f we may assume that $f \in L^1((-\eta, \eta))$ for some $\eta > 0$. Then, since $(x - \eta/2, x + \eta/2) \subset (-\eta, \eta)$ for $x \in (-\eta/2, \eta/2)$, integrating the identity $f(x) + f(y) = f(x+y)$ over $(-\eta/2, \eta/2)$ yields

$$f(x) = -\frac{1}{\eta}\int_{-\eta/2}^{\eta/2} f(y)\, dy + \frac{1}{\eta}\int_{x-\eta/2}^{x+\eta/2} f(y)\, dy, \quad x \in (-\eta/2, \eta/2).$$

Thus f is the integral of an integrable function + a constant in a neighborhood of the origin and, consequently, continuous there.

(b) First, note that f assumes finite values everywhere. Indeed, if f is finite on a set A, then $f(x - y) = f(x) - f(y)$ is finite for all $x, y \in A$ and, consequently, f is finite in $A - A$. Now, if A is a Lebesgue measurable set with $|A| > 0$, $A - A$ contains a neighborhood of the origin $(-\varepsilon, \varepsilon)$, say, and, therefore, since for $0 \neq \eta \in (-\varepsilon, \varepsilon)$ we have $\bigcup_{k \in \mathbb{Z}}(k\eta + (-\varepsilon, \varepsilon)) = \mathbb{R}$, f is finite everywhere.

Let $A_n = \{|f| \leq n\}$ for all n. Since f is finite everywhere we have $\mathbb{R} = \bigcup_n A_n$ and, therefore, $|A_n| > 0$ for some n; then, by the argument above with A_n in place of A there, $A_n - A_n$ contains a neighborhood of the origin $(-\eta, \eta)$, say. Now, since for $\xi \in (-\eta, \eta)$ there are $x, y \in A_n$ such that $\xi = x - y$, we have $|f(\xi)| \leq 2n$, and so f is bounded and, hence, integrable in $(-\eta, \eta)$, and by (a) f is linear.

46. (a) Since $f(0) = 0$ and the continuity of f at any point is equivalent to the continuity at 0, there exist a sequence $x_n \to 0$ and $\eta > 0$ such that $|f(x_n)| \geq \eta$ for all n. Let q_n be rationals such that $x_n/q_n \to 1$; since $x_n \to 0$, $q_n \to 0$. Then $|f(x_n/q_n)| > \eta/q_n$ and the conclusion becomes: there exists a sequence $x_n \to 1$ such that $|f(x_n)| \to \infty$. Now, given $z \in \mathbb{R}$, let $z_n = z - r_n x_n$ where $\{r_n\}$ is a rational sequence such that $r_n f(x_n) \to f(z)$; note that since $f(x_n) \to \infty$, $r_n \to 0$. Then $r_n \to 0$, $z_n \to z$, and $f(z_n) = f(z) - r_n f(x_n) \to f(z) - f(z) = 0$.

(b) First, suppose that $\mathcal{F}$ is linearly independent over $\mathbb{Q}$. Then, if $f(x) = 0$, write $x = \sum_{k=1}^{n} r_k e_{\lambda_k}$ and note that $f(x) = \sum_{k=1}^{n} r_k f(e_{\lambda_k}) = 0$, which implies that $r_k = 0$ for $1 \leq k \leq n$, and, consequently, $x = 0$. Therefore $f^{-1}(\{0\}) = \{0\}$. On the other hand, if $\mathcal{F}$ is linearly dependent over $\mathbb{Q}$ there exist finitely many $e_{\lambda_1}, \dots, e_{\lambda_n}$ and nonzero rationals $r_1, \dots, r_n$ such that $\sum_{k=1}^{n} r_k f(e_{\lambda_k}) = 0$. Then, if $x = \sum_{k=1}^{n} r_k e_{\lambda_k}$ it readily follows that $x \neq 0$ and $f(x) = 0$. Therefore $f(rx) = r f(x) = 0$ for every rational r and since $\{rx : r \text{ is rational}\}$ is dense in $\mathbb{R}$, $f^{-1}(\{0\})$ is dense in $\mathbb{R}$.

(c) Since $f(1) = 1$ and f is not linear, let x_0 be such that $f(x_0) = x_0 + \delta$, $\delta \neq 0$. Given $p \in \mathbb{R}^2$ and $\varepsilon > 0$, choose rationals r, s such that $|p - (r, s)| < \varepsilon/2$. Next, pick a rational $a \neq 0$ such that $|\delta a + (r - s)| < \varepsilon/8$ and a rational b such that $|a(x_0 - b)| < \varepsilon/8$. Observe that, if $x = r + a(x_0 - b)$, then $|x - r| = |a(x_0 - b)| < \varepsilon/8$. Also, $f(x) = r + a(x_0 - b) + a\delta$ and $|f(x) - s| \leq |(r - s) + a\delta| + |a(x_0 - b)| \leq \varepsilon/4$. Thus $z = (x, f(x))$ satisfies $|p - z| < \varepsilon$ and $G(f)$ is dense in $\mathbb{R}^2$.

47. Replacing $f(x)$ by $\pi/2 + \arctan f(x)$ if necessary we may assume that f is nonnegative and bounded. Now, since s/t is irrational, $D = \{x \in \mathbb{R} : x = ns + mt, n, m \in \mathbb{Z}\}$ is a dense subset of $\mathbb{R}$ consisting of periods of f. Let μ be the Borel measure given by $\mu(A) = \int_A f(x)\,dx$, $A \in \mathcal{B}(\mathbb{R})$; as is readily seen μ has the same periods as f and, consequently, by Problem 3.23, μ is a multiple of the Lebesgue measure and $\mu(A) = c|A|$ for all $A \in \mathcal{B}(\mathbb{R})$. Hence $\int_A (f(x) - c)\,dx = 0$ for all measurable A and by Problem 68, $f - c = 0$ a.e. and f is constant a.e. This argument is from R. Cignoli and J. Hounie, *Functions with arbitrarily small periods*, Amer. Math. Monthly **85** (1978), 582–584.

49. Neither. If $f(x) = 1/|x - 1/2|$, f is continuous everywhere except at $x = 1/2$ but for no continuous g on $[0, 1]$ we have $f = g$ a.e.; thus (a) does not imply (b). And, if $f = 1 - \chi_{\mathbb{Q}}$, $g = 1$ is continuous on $[0, 1]$ and $f = g$ a.e. yet f is discontinuous everywhere on $[0, 1]$ and so (b) does not imply (a).

50. The measurable functions are those that are constant except possibly on a countable subset of X. $L^1(X)$ consists of those functions f which are 0 except on a countable subset $\bigcup_j A_j$ of X with $\sum_j |\lambda_j| \mu(A_j) < \infty$, and the sum is the $L^1(X)$ norm of f.

51. The condition is necessary. Indeed, let $A_\lambda = \{|f| > \lambda\}$. Since f is finite μ-a.e., $\chi_{A_\lambda} |f| \to 0$ μ-a.e., and $\chi_{A_\lambda} |f| \le |f|$, and so by the Lebesgue dominated convergence theorem (LDCT) $\lim_{\lambda \to \infty} \int_{\{|f| > \lambda\}} |f| \, d\mu = 0$.

However, the condition is not sufficient as the example $f(x) = x^{-2}$ in $\mathbb{R}$ shows. Nevertheless, it is sufficient if $\mu(X) < \infty$. To see this pick λ large enough so that $\int_{\{|f| > \lambda\}} |f| \, d\mu \le 1$ and note that $\int_X |f| \, d\mu \le \int_{\{|f| \le \lambda\}} |f| \, d\mu + 1 \le \lambda \mu(X) + 1 < \infty$.

52. Note that the B_n are measurable and $\sum_n n \chi_{B_n}(x) \le |f(x)| \le \sum_n (n+1) \chi_{B_n}(x)$. Hence by the monotone convergence theorem (MCT), $\sum_n n \mu(B_n) \le \int_X |f| \, d\mu \le \sum_n (n+1) \mu(B_n)$. Thus $f \in L^1(X)$ implies $\sum_n n \mu(B_n) < \infty$. Conversely, if this last sum is finite, since $\sum_n \mu(B_n) \le \mu(X) < \infty$ it readily follows that $f \in L^1(X)$.

54. First, $\|s_k(f)\|_1 \le \sum_\ell \int_{Q_\ell^k} |Q_\ell^k|^{-1} \int_{Q_\ell^k} |f(y)| \, dy \, dx \le \int_{\mathbb{R}^n} |f(y)| \, dy$. Now, if g is a compactly supported continuous function on $\mathbb{R}^n$, g is uniformly continuous and $s_k(g)$ converges uniformly to g. Moreover, since $|s_k(g)(x)|$ is bounded by $\sup |g|$ uniformly in k and is compactly supported with support independent of k for k large, by the LDCT $s_k(g) \to g$ in $L^1(\mathbb{R}^n)$.

If now $f \in L^1(\mathbb{R}^n)$ is arbitrary, given $\varepsilon > 0$, let g be a compactly supported continuous function on $\mathbb{R}^n$ such that $\|f - g\|_1 \le \varepsilon$. Then

$$\|s_k(f) - f\|_1 \le \|s_k(f) - s_k(g)\|_1 + \|s_k(g) - g\|_1 + \|g - f\|_1.$$

Now, since s_k is linear, $\|s_k(f) - s_k(g)\|_1 \le \|f - g\|_1$ for all k. Thus $\|s_k(f) - f\|_1 \le 2\varepsilon + \|s_k(g) - g\|_1$. Now, since $\|s_k(g) - g\|_1 \to 0$ the conclusion follows. The a.e. convergence follows readily from the Lebesgue differentiation theorem.

55. Since f vanishes at $-1/2$ and $1/2$, letting $f(x) = 0$ for $|x| > 1/2$ we may think of f as a continuous function on $\mathbb{R}$. Now, by the translation invariance of the Lebesgue integral or by making the substitution $u = x - y$, we have $f_k(x) = \int_{x-1/2}^{x+1/2} s_k(y) f(x - y) \, dy$, and since f vanishes for $|x| > 1/2$ it follows that $f_k(x) = \int_{-1}^{1} s_k(y) f(x - y) \, dy$, and, consequently, $|f(x) - f_k(x)| \le \int_{-1}^{1} s_k(y) |f(x) - f(x - y)| \, dy$. Since f is continuous, f is uniformly continuous on any compact interval, in particular $[-2, 2]$, and so given $\varepsilon > 0$, there exists $\delta > 0$ such that $|y| < \delta$ implies $|f(x) - f(x - y)| < \varepsilon$. Therefore $\int_{\{|y| \le \delta\}} s_k(y) |f(x) - f(x - y)| \, dy < \varepsilon \int_{\{|y| < \delta\}} s_k(y) \, dy \le \varepsilon$. Now,

$J = \int_{\{\delta < |y| \le 1\}} s_k(y) |f(x) - f(x - y)| \, dy \le 2M \int_{\{\delta < |y| \le 1\}} s_k(y) \, dy$ where M is a bound for f and picking k large enough, $J \le \varepsilon$.

For instance, $s_k(x) = k\chi_{[0,\,1/k]}$, $k \ge 1$, will do.

56. Since $-|g| \le g \le |g|$ we may assume that g is nonnegative. First, suppose that f is compactly supported and continuous. Now, if $\{t_k\}$ tends to 0, then with $\varphi_k(x) = |f(x + t_k) - f(x)|\, g(x)$ there is a compact K in $\mathbb{R}^n$ such that the φ_k are uniformly bounded, vanish off K, and $\lim_k \varphi_k(x) = 0$ for all x. Then by the LDCT, $\lim_k \int_{\mathbb{R}^n} |f(x + t_k) - f(x)|\, g(x) \, dx = 0$ and since $\{t_k\}$ is arbitrary, $\lim_{t \to 0} \int_{\mathbb{R}^n} |f(x + t) - f(x)|\, g(x) \, dx = 0$. Next, if $f \in L^1(\mathbb{R}^n)$ is arbitrary, let h be a compactly supported continuous function such that $\|f - h\|_1 < \varepsilon/3\|g\|_\infty$, and pick t_0 such that $\int_{\mathbb{R}^n} g(x) |h(x + t) - h(x)| \, dx < \varepsilon/3$ for $|t| < t_0$. Then

$$\int_{\mathbb{R}^n} g(x) |f(x + t) - f(x)| \, dx \le \int_{\mathbb{R}^n} g(x) |f(x + t) - h(x + t)| \, dx$$

$$+ \int_{\mathbb{R}^n} g(x) |h(x + t) - h(x)| \, dx$$

$$+ \int_{\mathbb{R}^n} g(x) |h(x) - f(x)| \, dx$$

$$\le \|g\|_\infty \int_{\mathbb{R}^n} |f(x + t) - h(x + t)| \, dx$$

$$+ \varepsilon/3 + \|g\|_\infty \int_{\mathbb{R}^n} |h(x) - f(x)| \, dx$$

$$\le \varepsilon/3 + \varepsilon/3 + \varepsilon/3.$$

58. (a) That ν is a measure follows as in Problem 2.45; ν is known as the image measure of f and is sometimes denoted ν_f.

(b) Clearly $\varphi \circ f$ is measurable. Next, first suppose that $\varphi = \chi_A$ with $A \in \mathcal{B}(\mathbb{R})$. Then $\int_{\mathbb{R}} \varphi \, d\nu = \nu(A)$ and since $\chi_A \circ f = \chi_{f^{-1}(A)}$, $\int_X \varphi \circ f \, d\mu = \int_X \chi_{f^{-1}(A)} d\mu = \mu(f^{-1}(A))$, the two quantities are equal by definition. By the linearity of the integral and the additivity of μ it follows that the formula is true for a nonnegative simple function φ. Next, let φ be a nonnegative Borel function and $\{\varphi_n\}$ nonnegative simple functions that increase to φ. The result in this case follows by the MCT from $\int_{\mathbb{R}} \varphi_n \, d\nu = \int_X \varphi_n \circ f \, d\mu$ for all n. For functions $\varphi = \varphi^+ - \varphi^-$ of arbitrary sign the result follows from the identities for φ^+ and φ^-. Thus, if for a Borel function φ either integral exists, so does the other and they are equal.

Particular instances of this identity include $\int_X f \, d\mu = \int_{\mathbb{R}} x \, d\nu(x)$ and $\int_X |f| \, d\mu = \int_{\mathbb{R}} |x| \, d\nu(x)$.

59. (c) The proof follows along the lines of Problem 58(b) and we shall be brief. First, let $h = \chi_E$, $E \in B(\mathbb{R}^2)$. Then $\int_X h \, d\mu_\Phi = \mu_\Phi(E) =$

$\mu(\Phi^{-1}(E)) = \int \chi_{\Phi^{-1}(E)} \, d\mu = \int \chi_E(\Phi(x)) \, d\mu(x) = \int_X h \circ \Phi \, d\mu$. By linearity the formula holds for all nonnegative measurable simple functions h. Next, let h be any nonnegative measurable function and pick $\{h_n\}$ simple nonnegative measurable functions that increase to h. By the MCT, $\int h \, d\mu_\Phi = \lim_n \int h_n \, d\mu_\Phi = \lim_n \int_X h_n \circ \Phi \, d\mu = \int_X h \circ \Phi \, d\mu$. Here we used that $h_n(\Phi)$ increases as $n \to \infty$ for all $x \in X$. Finally, for h μ_Φ-integrable write $h = h^+ - h^-$ and note that $\int_{\mathbb{R}^2} h \, d\mu_\Phi = \int_{\mathbb{R}^2} h^+ \, d\mu_\Phi - \int_{\mathbb{R}^2} h^- \, d\mu_\Phi = \int_X h^+ \circ \Phi \, d\mu - \int_X h^- \, d\mu = \int_X h \circ \Phi \, d\mu$.

60. (a) By considering $-\varphi$ if necessary we may assume that φ is increasing. First, suppose that $A = [a, b] \subset J$ is an interval. Then $\varphi^{-1}(A) = [c, d]$ is an interval and since φ is absolutely continuous, $\int_{\varphi^{-1}(A)} \varphi'(x) \, dx = \varphi(d) - \varphi(c) = b - a = |A|$. Next, if A is open, $A = \bigcup_n I_n$ is the pairwise disjoint union of open subintervals of J and since $\varphi^{-1}(\bigcup I_n) = \bigcup \varphi^{-1}(I_n)$, $\int_{\varphi^{-1}(A)} \varphi'(x) \, dx = \int_{\bigcup_n \varphi^{-1}(I_n)} \varphi'(x) \, dx = \sum_n \int_{\varphi^{-1}(I_n)} \varphi'(x) \, dx = \sum_n |I_n| = |A|$. Now, if A is closed, since $J \setminus A$ is open the result holds in this case. Finally, let A be a Borel subset of J and $\varepsilon > 0$. Then there exist a closed set F and an open set O such that $F \subset A \subset O$ and $|O \setminus F| \le \varepsilon$. Note that since φ is continuous and increasing, $\varphi^{-1}(O)$ is open, $\varphi^{-1}(F)$ is closed, and $\varphi^{-1}(F) \subset \varphi^{-1}(A) \subset \varphi^{-1}(O)$. Hence $|A| - \varepsilon \le |F| = \int_{\varphi^{-1}(F)} \varphi'(x) \, dx \le \int_{\varphi^{-1}(A)} \varphi'(x) \, dx \le \int_{\varphi^{-1}(O)} \varphi'(x) \, dx = |O| \le |A| + \varepsilon$ and since ε is arbitrary, $\int_{\varphi^{-1}(A)} \varphi'(x) \, dx = |A|$.

62. Let $g_n(x) = g(x - n)$; then $g_n(x) \ge 0$ for all x, n, $\lim_n g_n(x) = 0$, and $\int_{\mathbb{R}} g_n(x) \, dx = \int_{\mathbb{R}} g(x) \, dx > 0$. For the sake of argument suppose that $h \in L^1(\mathbb{R})$. Then by the LDCT, $\lim_n \int_{\mathbb{R}} g_n(x) \, dx = \int_{\mathbb{R}} \lim_n g_n(x) \, dx = 0$, which is not the case since the limit is $\int_{\mathbb{R}} g(x) \, dx > 0$.

63. Let $A_n = \{f \ge 1/n\}$; $\{A_n\}$ increases to $A = \{f > 0\}$. Now, by Chebychev's inequality $\mu(A_n) < \infty$ for all n, and by the MCT $\lim_n \int_{A_n} f \, d\mu = \int_A f \, d\mu = \int_X f \, d\mu$. Since all the quantities involved are finite, given $\varepsilon > 0$, there exists n such that $\int_{A_n} f \, d\mu > \int_X f \, d\mu - \varepsilon$.

64. The condition, which is known as the absolute continuity of the integral of f, is not sufficient. Indeed, let $\mathcal{M} = \{\emptyset, X\}$ and $\mu(\emptyset) = 0$, $\mu(X) = \infty$. Pick $f = \chi_X$. Then, given $\varepsilon > 0$, since $\emptyset$ is the only set with finite measure, we have $\int_\emptyset f \, d\mu = 0$ yet $f \notin L^1(X)$. More generally, the integral of a constant function f is absolutely continuous yet, in the case of an infinite measure space, f is integrable only when $f = 0$ μ-a.e.

The condition is necessary, and obvious, if f is bounded. For arbitrary f, let $X_n = \{n \le |f| < n + 1\}$, observe that by the MCT $\int_X |f| \, d\mu = \sum_{n=0}^{\infty} \int_{X_n} |f| \, d\mu < \infty$ and pick N such that $\sum_{n=N}^{\infty} \int_{X_n} |f| \, d\mu < \varepsilon/2$. Now, let $0 < \delta < \varepsilon/(2N)$ and $A \in \mathcal{M}$ with $\mu(A) < \delta$. Let $B_N = \bigcup_{k=N}^{\infty} X_n$ and note that since $|f(x)| \leq N$ in B_N^c, $\int_A |f|\, d\mu = \int_{A \cap B_N} |f|\, d\mu + \int_{A \cap B_N^c} |f|\, d\mu \leq \int_{B_N} |f|\, d\mu + N\mu(A) \leq \varepsilon/2 + \varepsilon/2 = \varepsilon$.

65. We show that for any positive c, there exists $\varepsilon > 0$ such that for any sufficiently small $\delta > 0$ there exist a nonnegative integrable function f with $\int_X f\, d\mu = c$ and a measurable set A with $\mu(A) < \delta$ such that $\int_A f\, d\mu > \varepsilon$. We assume that the measure μ is such that given $\delta > 0$, there exists measurable A such that $\mu(A) < \delta$; in particular, μ nonatomic will do. For spaces that don't satisfy this condition there is no such counterexample: If $\delta > 0$ is sufficiently small, then $\mu(A) < \delta$ implies $\mu(A) = 0$ and, consequently, $\int_A f\, d\mu = 0$ for any f and ε. Let c, δ be given and set $\varepsilon = c/2$. Then choose a measurable set A satisfying $0 < \mu(A) < \delta$ and put $f = c\mu(A)^{-1}\chi_A$; f is nonnegative, $\int_X f\, d\mu = c$, and $\int_A f\, d\mu = c > \varepsilon$.

67. For the sake of argument suppose that $\mu(\{f < 0\}) > 0$. Then, if $A_k = \{f < -1/k\}$, $\mu(A_k) > 0$ for some positive integer k and, consequently, $\int_{A_k} f\, d\mu \leq -\mu(A_k)/k < 0$, which cannot happen. Now, if the integral of f over every $A \in \mathcal{M}$ vanishes, by the above argument $f \geq 0$ μ-a.e. and, since the assumption applies to $-f$ as well, also $-f \geq 0$ μ-a.e. Hence $f = 0$ μ-a.e.

Finally, note that the result gives that if f is a nonnegative measurable function on X and $\int_X f\, d\mu = 0$, then $f = 0$ μ-a.e.

68. Let $\mathcal{F} \subset \{A \in \mathcal{M} : \int_A f\, d\mu = 0\}$. From the properties of the integral, $\{A \in \mathcal{M} : \int_A f\, d\mu = 0\}$ is a λ-system and, consequently, by the π-λ theorem, since $\mathcal{M}(\mathcal{F}) = \mathcal{M}$, $\mathcal{M} \subset \{A \in \mathcal{M} : \int_A f\, d\mu = 0\}$ and so by Problem 67, $f = 0$ μ-a.e.

69. The statement is false. Let $\mathcal{M} = \{\emptyset, X\}$, $\mu(\emptyset) = 0$, $\mu(X) = \infty$, and consider $f(x) = 1$ and $g(x) = 2$ on X. Then $\int_\emptyset f\, d\mu = \int_\emptyset g\, d\mu = 0$ and $\int_X f\, d\mu = \int_X g\, d\mu = \infty$, yet $f(x) \neq g(x)$ for all $x \in X$.

On the other hand, the statement is true if f, g are integrable or μ is σ-finite. In the former case, since f, g are finite μ-a.e. we may assume that f and g are finite everywhere. Now, since f and g are measurable, so are $f - g$ and $g - f$ and, consequently, $A = \{f - g > 0\}$ and $B = \{g - f > 0\}$ are measurable. Now, $\int_A (f - g)\, d\mu = 0$ and if $\mu(A) > 0$, since $f - g > 0$ in A it follows that $\int_A (f - g)\, d\mu > 0$, which is not the case. Hence $\mu(A) = 0$. Similarly, $\int_B (g - f)\, d\mu = 0$ and so $\mu(B) = 0$. Now, $A \cup B = \{f \neq g\}$ and, therefore, $\mu(A \cup B) = 0$. Hence $f = g$ μ-a.e.

In the latter case suppose first that $\mu(X) < \infty$. Fix rationals $r > s > 0$, and let $A = \{f > r, g < s\}$. Then $\mu(A) < \infty$ and $r\mu(A) \leq \int_A f\, d\mu = \int_A g\, d\mu \leq s\mu(A)$, which, if $\mu(A) \neq 0$, implies $r \leq s$, which is not the case. Therefore $\mu(A) = 0$. Now, $\{f > g\} = \bigcup_{r,s \in \mathbb{Q}}\{f > r, g < s\}$ and since each set in the right-hand side has measure zero, $\mu(\{f > g\}) = 0$. Exchanging f and g also $A = \{g > f\} = 0$ and so $\mu(\{f \neq g\}) = 0$. Now, if $X = \bigcup_k X_k$

with the X_k pairwise disjoint and $\mu(X_k) < \infty$ for all k, by the argument above $\mu(\{x \in X_k : f(x) \neq g(x)\}) = 0$ for all k and so $\mu(\{f \neq g\}) = 0$.

70. For the sake of argument suppose that $B = \{x \in X : f(x) \notin F\}$ has $\mu(B) > 0$. Since F is closed, $\mathbb{R} \setminus F = \bigcup_n I_n$ where the I_n are pairwise disjoint open intervals of $\mathbb{R}$. Then $B \subset \bigcup_n f^{-1}(I_n)$ and, consequently, $\mu(f^{-1}(I_n)) > 0$ for some n. Now, I_n is a bounded interval or the limit of an increasing sequence $\{J_k\}$ of bounded intervals and in the latter case, by continuity from below, $\mu(f^{-1}(J_k)) > 0$ for some k. Thus $\mu(f^{-1}(J)) > 0$ for some bounded interval $J \subset \mathbb{R} \setminus F$. Now, since μ is semifinite, $f^{-1}(J)$ is of finite measure or there exists $D \subset f^{-1}(J)$ with $0 < \mu(D) < \infty$. Now, if $J = (x - r, x + r)$, note that

$$\left| \frac{1}{\mu(D)} \int_D f \, d\mu - x \right| \leq \frac{1}{\mu(D)} \int_D |f - x| \, d\mu < \frac{1}{\mu(D)} \int_D r \, d\mu = r$$

and, consequently, $(1/\mu(D)) \int_D f \, d\mu \in J$, which since $F \cap J = \emptyset$ cannot happen. Thus $\mu(B) = 0$.

In particular, if for any $A \in \mathcal{M}$ with $0 < \mu(A) < \infty$,

$$\frac{1}{\mu(A)} \int_A f \, d\mu \leq c,$$

the result applies with $F = [c, \infty)$ and, consequently, $f \leq c$ μ-a.e. Note that for the Lebesgue measure it suffices to consider cubes A above for then the conclusion follows from the Lebesgue differentiation theorem.

71. If μ is semifinite, the result follows from Problem 70 with $F = [-c, c]$ there. In the general case, by Chebychev's inequality $\mu(\{|f| > c\}) < \infty$ and if $A = \{f > c\}$ and $B = \{f < -c\}$, $\mu(A), \mu(B) < \infty$. Now, $c\mu(A) < \int_A f \, d\mu \leq c\mu(A)$ and, therefore, $\mu(A) = 0$. Similarly, $c\mu(B) < |\int_B f \, d\mu| \leq c\mu(B)$ and $\mu(B) = 0$.

72. First, the result is false if f is not integrable as the example $f(x) = \sin(x)$, $c = 2\pi$, shows. Now, let $a < b$ be real numbers such that $0 < b - a < c$. Then

$$\int_a^b f(y) \, dy + \int_b^{a+c} f(y) \, dy = \int_a^{a+c} f(y) \, dy = 0$$

and

$$\int_b^{a+c} f(y) \, dy + \int_{a+c}^{b+c} f(y) \, dy = \int_b^{b+c} f(y) \, dy = 0,$$

which combined give $\int_a^b f(y) \, dy = \int_{a+c}^{b+c} f(y) \, dy$.

Next, picking $a = x$ and $b = x + h$ above where $x, x + c$ are Lebesgue points f, and $0 < h < c$, we have

$$\frac{1}{h} \int_x^{x+h} f(y) \, dy = \frac{1}{h} \int_{x+c}^{x+c+h} f(y) \, dy$$

and so letting $h \to 0$, by the Lebesgue differentiation theorem it follows that $f(x) = f(x+c)$ a.e. and f is essentially periodic of period c, and the same is true of $|f|$. But then $\int_0^c |f(y)|\,dy = \int_{nc}^{(n+1)c} |f(y)|\,dy$ for all n, and, consequently, $\int_0^\infty |f(y)|\,dy = \sum_n \int_{nc}^{(n+1)c} |f(y)|\,dy = \infty$ and $f \notin L^1(\mathbb{R})$ unless $f = 0$ a.e.

73. By Problem 51, $\lim_{\lambda \to \infty} \int_{\{|f|>\lambda\}} |f|\,d\mu = 0$ and, therefore, there exists an increasing sequence $\{\lambda_n\}$ with $\lambda_n \to \infty$ such that $\int_{\{|f|>\lambda_n\}} |f|\,d\mu \leq 1/n^3$. Now let $\varphi : \mathbb{R}^+ \to \mathbb{R}^+$ be the continuous piecewise linear function $= 0$ for $t \in [0, \lambda_1]$ and slope n on $(\lambda_n, \lambda_{n+1})$; clearly $\lim_{t \to \infty} \varphi(t)/t = \infty$ and $\varphi(t) \leq 2nt$ for $t \in (\lambda_n, \lambda_{n+1})$. It then follows that $\int_X \varphi(|f|)\,d\mu = \sum_n \int_{\{\lambda_n < |f| \leq \lambda_{n+1}\}} \varphi(|f|)\,d\mu \leq 2\sum_n n \int_{\{|f|>\lambda_n\}} |f|\,d\mu \leq 2\sum_n n^{-2} < \infty$.

75. Equality holds when one function majorizes the other μ-a.e.

77. $\int_I f(x)\,dx = 3$.

78. Since $f(x) = \sum_n n\chi_{[1/10^{n+1}, 1/10^n]}(x)$, f is Borel measurable. Now, by calculus $\sum_n n/r^n = r/(r-1)^2$ for $r > 1$ and so, by the MCT, $\int_I f(x)\,dx = \sum_n n(10^{-n} - 10^{-(n+1)}) = (9/10)\sum_n n10^{-n} = (9/10)(10/9^2) = 1/9$.

79. As in Problem 2.116, the atoms have measure $\mu_F(\{1\}) = F(1) - F(1^-) = 3$, $\mu_F(\{4\}) = 1$, and the intervals have measure $\mu_F((-\infty, 1]) = F(1^-) - F(-\infty) = 0$, $\mu_F((1, 4)) = 0$, and $\mu_F((4, \infty)) = 0$. Hence for a Borel set A in $\mathbb{R}$, $\mu_F(A) = \mu_F(A \cap (-\infty, 1)) + \mu_F(A \cap \{1\}) + \mu_F(A \cap (1, 4)) + \mu_F(A \cap \{4\}) + \mu_F(A \cap (4, \infty)) = 3\chi_A(1) + \chi_A(4)$. Now, the integral $\int_{\mathbb{R}} x^2\,d\mu_F(x)$ is the limit of approximating sums and one readily obtains that it is equal to $3 \cdot 1^2 + 1 \cdot 4^2 = 19$.

80. First, $f(x) = 1$ in $[.7, .8)$, a set of measure $1/10$, $f(x) = 1/2$ in 9 intervals, each of measure $1/100$, $f(x) = 1/3$ in 81 intervals of length $1/1000$, and so on. Hence $\int_I f(x)\,dx = \sum_k k^{-1}(9/10)^{k-1} = (10/9)\sum_k k^{-1}(9/10)^k$. Recall now that $\ln(1-x) = -\sum_k k^{-1}x^k$, $0 < x < 1$. Then $\int_I f(x)\,dx = -(10/9)\ln(1/10) = (10/9)\ln(10)$.

81. Let $I_n = (1/(n+1), 1/n]$. Then, since $f(x) = \sum_n n\chi_{I_n \cap I}(x)$ we have $\int_I f(x)\,dx = \sum_n n \int_{(1/(n+1), 1/n]} dx = \sum_n n\,(n^{-1} - (n+1)^{-1}) = \infty$.

On the other hand, $g(x) = \sum_n n^{-1}\chi_{I_n \cap I}(x)$ and therefore, $\int_I g(x)\,dx = \sum_n n^{-1} \int_I \chi_{I_n \cap I}(x)\,dx = \sum_n n^{-1}(n^{-1} - (n+1)^{-1}) = (\pi^2/6) - 1$.

82. Let $\{f_n\}$ be the sequence given by $f_0(x) = 0$, $f_{n+1}(x) = \sqrt{x + f_n(x)}$, $n = 1, 2, \ldots$; we claim that $\{f_n\}$ is increasing. Clearly $f_1 \geq f_0$ and if $f_n \geq f_{n-1}$, then $f_{n+1}(x) = \sqrt{x + f_n(x)} \geq \sqrt{x + f_{n-1}(x)} = f_n(x)$. Furthermore, the sequence is bounded above by its limit $f(x) = (1 + \sqrt{1 + 4x})/2$. Now, if we denote the integrand by $(1/2 - \varphi(x))^2$, since $\varphi(x)^2 = x + \varphi(x)$ it readily follows that $(1/2 - \varphi(x))^2 = 1/4 + x$ and so by the MCT the integral is equal to 9.

83. $\int_I f(x)\, dx = (\ell - 1)/2(m - 1)$ and $\int_I g(x)\, dx = 0$.

84. The statement is false: The relation is satisfied by the Lebesgue measure and the counting measure on the integers. To see the latter recall that $\sum_{k=1}^n k = n(n+1)/2$ and so $(\sum_{k=1}^n k)^2 = n^2(n+1)^2/4 = z_n$, say. Then $z_{n+1} - z_n = (n+1)^2((n+2)^2 - n^2)/4 = (n+1)^3$. Now, $z_1 = 1$, $z_2 = 1 + 2^3$ and, in general, $z_n = \sum_{k=1}^n k^3$. Hence $\sum_{k=1}^n k^3 = n^2(n+1)^2/4 = (\sum_{k=1}^n k)^2$.

85. First, suppose $f = \chi_A$ is the characteristic function of a measurable set A. Then by Problem 3.49 applied to M^{-1}, $\int_A dx = |\det(M^{-1})|\,|A| = |M^{-1}(A)|$ and since $|\det(M^{-1})| = |\det(M)|^{-1}$ it readily follows that $\int_A dx = |\det(M)|\,|M^{-1}(A)| = |\det(M)| \int_{M^{-1}A} dx$. Next, by the linearity of the integral, if s is a simple function, $\int_{\mathbb{R}} s(x)\, dx = |\det(M)| \int_{\mathbb{R}} s \circ M(x)\, dx$. Suppose now that $f \geq 0$ and let $0 \leq s \leq f$ be a simple function. Then $s \circ M(x) \leq f \circ M(x)$ and, consequently, $\int_A s(x)\, dx = |\det(M)| \int_{\mathbb{R}} s \circ M(x)\, dx \leq |\det(M)| \int_{\mathbb{R}} f \circ M(x)\, dx$. Hence, by the MCT it follows that $\int_A f(x)\, dx \leq |\det(M)| \int_{\mathbb{R}} f \circ M(x)\, dx$. Now, applying this inequality to f replaced by $f \circ M^{-1}$ we get $\int_A f \circ M^{-1}(x)\, dx \leq |\det(M)| \int_{\mathbb{R}} f(x)\, dx$, and since $|\det(M^{-1})| = 1/|\det(M)|$, we finally get $|\det(M^{-1})| \int_A f \circ M^{-1}(x)\, dx \leq \int_{\mathbb{R}} f(x)\, dx$. Equality holds replacing M^{-1} by M. For arbitrary $f = f^+ - f^-$ the conclusion follows by considering f^+, f^- separately.

Finally, if M is not invertible, MA is contained in a lower-dimensional subspace of $\mathbb{R}^n$ for all A, which has measure 0, and the conclusion holds.

88. The integral is equal to $\pi^{n/2} \det(A)^{-1/2}$.

89. Since $f = f^+ - f^-$ it suffices to consider positive f. Also, it suffices to prove that $\int_{\mathbb{R}} f \circ \Phi\, d\mu_{F \circ \Phi} \leq \int_{\mathbb{R}} f\, d\mu_F$ since, replacing Φ by Φ^{-1}, the opposite inequality follows. Finally, by the MCT it suffices to consider the case when f is simple and, by linearity, when $f = \chi_A$ where $A \in \mathcal{B}(\mathbb{R})$. Thus we are reduced to proving that $\int_{\mathbb{R}} \chi_A \circ \Phi\, d\mu_{F \circ \Phi} \leq \int_{\mathbb{R}} \chi_A\, d\mu_F = \mu_F(A)$. Suppose then that A is covered by countably many half-open intervals $\{I_k\}$, say. Then it suffices to prove that $\int_{\mathbb{R}} \chi_A \circ \Phi\, d\mu_{F \circ \Phi} \leq \sum_n \mu_F(I_n)$ for any such covering. By the MCT, $\int_{\mathbb{R}} \chi_A \circ \Phi\, d\mu_{F \circ \Phi} \leq \int_{\mathbb{R}} \sum_n \chi_{I_n} \circ \Phi\, d\mu_{F \circ \Phi} = \sum_n \int_{\mathbb{R}} \chi_{I_n} \circ \Phi\, d\mu_{F \circ \Phi}$ so it suffices to prove that for such an interval $[a, b)$ we have $\int_{\mathbb{R}} \chi_{[a,b)} \circ \Phi\, d\mu_{F \circ \Phi} \leq \mu_F([a, b)) = F(b) - F(a)$. But this follows since $\chi_{[a,b)} \circ \Phi = \chi_{[\Phi^{-1}(a), \Phi^{-1}(b))}$.

90. (a) Let $g(x) = \sum_{n=-\infty}^{\infty} |f(x + n)|$; we claim that g is finite a.e. Since g is periodic of period 1 it suffices to prove that g is finite a.e. in $[0, 1]$. Now, by the MCT, $\int_I \sum_{n=-N}^N |f(x + n)|\, dx = \sum_{n=-N}^N \int_I |f(x + n)|\, dx =$

$\int_{-N}^{N+1} |f(x)|\,dx \to \int_{\mathbb{R}} |f| < \infty$ as $N \to \infty$ and so $\int_I g < \infty$, and, in particular, g is finite a.e. in I.

(b) For each integer n, let $f_n(x) = |f(x+n)|$. Since $|f|$ is nonnegative and measurable, so is each f_n and by the translation invariance of the Lebesgue measure, $\int_{\mathbb{R}} |f(x + n)|\,dx = \int_{\mathbb{R}} |f(x)|\,dx$ for all n. Thus, by the MCT, $\int_{\mathbb{R}} \sum_n f_n(x)\,dx = \sum_n \int_{\mathbb{R}} |f(x + n)|\,dx = \sum_n \int_{\mathbb{R}} |f(x)|\,dx$, which is finite only if $\int_{\mathbb{R}} |f(x)|\,dx = 0$ and, consequently, $f = 0$ a.e.

(c) By the translation invariance of the Lebesgue integral and Problem 86, $\int_{\mathbb{R}} |f(\lambda_n\,(x - x_n))|\,dx = |\lambda_n|^{-1} \int_{\mathbb{R}} |f(x)|\,dx$ for all n. Then, by the MCT, $\int_{\mathbb{R}} \left(\sum_n |f(\lambda_n\,(x - x_n))|\right) dx = \left(\sum_n |\lambda_n|^{-1}\right)\left(\int_{\mathbb{R}} |f(x)|\,dx\right)$.

91. $|A|$ is the Riemann integral of f over $[0, 1]$.

92. The condition is not necessary. Let $f(x) = \sum_n (-1)^n n^{-1} \chi_{(n,n+1)}(x)$ and observe that since $\int_1^{N+1} f(x)\,dx = \sum_{n=1}^N (-1)^n n^{-1}$, the improper integral exists and is equal to $\int_1^\infty f(x)\,dx = \lim_{N\to\infty} \sum_{n=1}^N (-1)^n n^{-1} = \ln(2)$. However, $|f(x)| = \sum_n n^{-1} \chi_{(n,n+1)}(x) \notin L^1(\mathbb{R}^+)$ since by the MCT its integral is infinite.

However, if $f \in L^1(\mathbb{R}^+)$ and the improper Riemann integral of f exists, they are equal. To see this let $f_n = f\chi_{[0,n]}$; then $|f_n| \le |f| \in L^1(\mathbb{R}^+)$ and so, since $\int_0^R f(x)\,dx = \int_{[0,R]} f(x)\,dx$, by the LDCT $\int_{[0,\infty)} f(x)\,dx = \lim_{R\to\infty} \int_0^R f_n(x)\,dx = \lim_{R\to\infty} \int_{[0,R]} f(x)\,dx = \int_{\mathbb{R}^+} f(x)\,dx$.

93. The statement is false. Let $F(x) = x^2 \sin(1/x^2)$, $x \ne 0$, $F(0) = 0$. Then $F'(x) = f(x)$ exists for every $x \in \mathbb{R}$ and is continuous in $(0, 1)$ and, since $\varepsilon^2 \sin(1/\varepsilon^2) \to 0$ as $\varepsilon \to 0$, $\lim_{\varepsilon\to 0} \int_\varepsilon^1 f(x)\,dx = \sin(1)$.

Now,

$$f(x) = \begin{cases} 2x \sin(1/x^2) - (2/x)\cos(1/x^2), & x \ne 0, \\ 0, & x = 0, \end{cases}$$

where $2x \sin(1/x^2)$ is bounded and measurable, hence integrable, over $[0, 1]$. Thus f is integrable iff $(1/x)\cos(1/x^2)$ is integrable. Let $I_k = [1/(k\pi + \pi/6)^{1/2}, 1/(k\pi - \pi/6)^{1/2}]$, $k \ge 1$; the I_k are pairwise disjoint and $|I_k| \sim (\pi k)^{-3/2}$ for k large. Therefore, since $|\cos(1/x^2)| \ge 1/2$ and $1/x \sim (\pi k)^{1/2}$ on I_k, $\int_I |(1/x)\cos(1/x^2)|\,dx \ge (1/2) \sum_k \int_{I_k} (1/x)\,dx \ge \sum_k (\pi k)^{-1} = \infty$. Hence $f \notin L^1([0, 1])$.

However, the statement is true if f is nonnegative. To see this let $0 < \varepsilon_n < b - a$ tend to 0 as $n \to \infty$. Then the functions $f_n = f\chi_{[a+\varepsilon_n,b]}$ are integrable on I and $\int_I f_n(x)\,dx = \int_{a+\varepsilon_n}^b f(x)\,dx$, the integral in the right-hand side in the sense of Riemann. Now, since the f_n increase to f on I, by the MCT $\int_I f(x)\,dx = \lim_n \int_{a+\varepsilon_n}^b f(x)\,dx = \int_a^b f(x)\,dx$.

94. $\int_0^1 x^\eta \ln^n(1/x)\,dx = n!/(\eta+1)^{n+1}$, $n \geq 1$.

95. First, $x^\eta(1-x)^{-1}\ln(x) = \sum_{k=0}^\infty x^{\eta+k}\ln(x) = \sum_{k=0}^\infty f_k(x)$ where $f_k(x) = x^{\eta+k}\ln(x)$ for all $k \geq 0$. Note that $\sum_{k=0}^\infty f_k$ converges to the integrand and by Problem 94 we have

$$\sum_{k=0}^\infty \int_0^1 |f_k(x)|\,dx = \sum_{k=0}^\infty \int_0^1 x^{\eta+k}\ln(1/x)\,dx$$
$$= \sum_{k=0}^\infty 1/(1+\eta+k)^2 < \infty.$$

Then the series converges by the LDCT, the integrand is in $L^1([0,1])$, and

$$\int_0^1 x^\eta(1-x)^{-1}\ln(x)\,dx = \sum_{k=0}^\infty \int_0^1 x^{\eta+k}\ln(x)\,dx = -\sum_{k=0}^\infty (\eta+k)^{-2}\,.$$

96. By Problem 93 and the change of variables $u = \ln(t)$ we get

$$\int_0^x f(t)dt = \lim_{\varepsilon \to 0} \int_\varepsilon^x \frac{1}{t\ln^2(t)}\,dt = \lim_{R \to -\infty} \int_R^{\ln(x)} \frac{1}{u^2}\,du = -\frac{1}{\ln(x)}$$

for $x \in (0, e^{-1})$. Further, f is nonnegative and $\int_\mathbb{R} f(t)dt = \int_0^{1/e} f(t)dt = 1$, and so $f \in L^1(\mathbb{R})$. Now, let $\eta \in (0, 1/e)$. Then the change of variables $u = -\ln(x)$ gives

$$\int_0^\eta F(x)\,dx = -\int_0^\eta \frac{1}{x\ln(x)}\,dx = \int_{-\ln(\eta)}^\infty \frac{1}{u}\,du = \infty\,.$$

From this result it follows that the Hardy-Littlewood maximal function of an integrable function is not necessarily integrable.

97. For each fixed t, $\cos(t)/(t+x)$ is continuous as a function of x and for x in a finite interval $[a, b] \subset \mathbb{R}^+$, is bounded by $1/(t+a) \in L^1([0, \pi/2])$. Hence f is well-defined and continuous on $\mathbb{R}^+$. Now, observe that

$$\int_0^{\pi/2} \frac{1-t^2/2}{x+t}\,dt \leq f(x) \leq \int_0^{\pi/2} \frac{1}{x+t}\,dt\,.$$

Then for $x > 0$ close to 0 we have $\int_0^{\pi/2} 1/(x+t)\,dt = \ln((\pi/2+x)/x) \sim -\ln(2x/\pi)$. Also, $\int_0^{\pi/2} -t^2(x+t)\,dt$ is bounded and so $o(\ln(x))$. It thus follows that $f(x) \sim -\ln(2x/\pi)$ at the origin and, consequently, $\lim_{x\to 0+} f(x) = \infty$. Finally, since $\int_0^{\pi/2} \cos(t)\,dt = 1$ it follows that

$$\frac{1}{x+\pi/2} = \frac{1}{x+\pi/2}\int_0^{\pi/2} \cos(t)\,dt \leq f(x) \leq \frac{1}{x}\int_0^{\pi/2} \cos(t)\,dt = \frac{1}{x}\,,$$

and, therefore, $f(x) \sim 1/x$ at ∞ and $\lim_{x\to\infty} f(x) = 0$.

98. (a) First, since $f_a = f_{1/a}$ for all $a > 0$ it suffices to consider $0 < a \le 1$. Now, f_a is nonnegative, even, and measurable; we are interested in the finiteness of $\int_{\mathbb{R}} f_a(x)dx = 2\int_0^\infty f_a(x)dx$. If $a = 1$, $f_1(x) = 1/2|x| \notin L^1(\mathbb{R})$. Next, when $0 < a < 1$ note that $f_a(x) \le \min\{x^{-a}, x^{-1/a}\}$, $x > 0$, and, consequently, $\int_0^\infty f_a(x)\,dx \le \int_0^1 x^{-a}\,dx + \int_1^\infty x^{-1/a}\,dx = (1+a)/(1-a) < \infty$. Thus $f_a \in L^1(\mathbb{R})$ for every positive $a \ne 1$.

(b) Note that $\int_{\mathbb{R}} |x| f_a(x)\,dx = 2\int_0^\infty x/(x^a + x^{1/a})\,dx$. Now, if $0 < a < 1$, we have that $\int_0^1 x f_a(x)dx \le \int_0^1 f_a(x)\,dx < \infty$, and so, $|x| f_a(x) \in L^1(\mathbb{R})$ iff $I = \int_1^\infty x/(x^a + x^{1/a})\,dx < \infty$. Since $I \sim \int_1^\infty x^{1-1/a}\,dx$, I is finite iff $1 - 1/a < -1$, i.e., $0 < a < 1/2$. Hence $x f_a(x) \in L^1(\mathbb{R})$ iff $a \in (0, 1/2) \cup (2, \infty)$.

(c) Let $h(x, y) = f_a(x)\cos(xy)$, $a \ne 1$. Then for each fixed x, $h(x, y)$ is continuous and $|h(x, y)| \le f_a(x)$ where $f_a \in L^1(\mathbb{R})$ is independent of y. Thus by the LDCT, g_a is continuous on $\mathbb{R}$.

Next, assume that $a > 2$; then it is readily seen that $h_y(x, y)$, the partial derivative of h with respect to y, is $h_y(x, y) = -x f_a(x)\sin(xy)$. Note that this derivative depends continuously of $y \in \mathbb{R}$. On the other hand, $|x f_a(x)\sin(xy)| \le |x| f_a(x)$ and the derivative is bounded by an integrable function since $a > 2$. Thus $g_a \in C^1(\mathbb{R})$ and $g_a'(y) = -\int_0^\infty x f_a(x)\sin(xy)\,dx$ for all $y \in \mathbb{R}$.

99. (a) Note that for $t > 1$ and $x < 0$, $0 \le t^x/(1 + t^2) \le 1/(1 + t^2) \in L^1((1, \infty))$ and $\lim_{x \to -\infty} t^x/(1 + t^2) = 0$. Hence by the LDCT, $\lim_{x \to -\infty} F(x) = 0$. Now, $\lim_{x \to 1^-} t^x/(1 + t^2) = t/(1 + t^2)$ and since $\int_1^\infty t/(1 + t^2) = \infty$, by Fatou's lemma $\lim_{x \to 1^-} F(x) = \infty$.

(b) The statement is proved by recurrence on k.

100. Since cosine is an even function we may assume that $a > 0$. Let

$$f_n(x) = e^{-x^2}\sum_{k=0}^n (-1)^k \frac{a^{2k} x^{2k}}{(2k)!}, \quad x \in \mathbb{R}.$$

The f_n are continuous and $\lim_n f_n(x) = e^{-x^2}\cos(ax)$ for all $x \in \mathbb{R}$. Now,

$$|f_n(x)| \le e^{-x^2}\sum_{k=0}^n \frac{a^{2k}|x|^{2k}}{(2k)!} \le e^{-x^2}\sum_{k=0}^\infty \frac{a^k |x|^k}{(k)!} = e^{-x^2} e^{a|x|}$$

and since $2a|x| \le (a^2 + |x|^2)$, $|f_n(x)| \le e^{-x^2} e^{a|x|} \le e^{a^2/2} e^{-x^2/2} \in L^1(\mathbb{R})$. Therefore by the LDCT,

$$I = \sum_{k=0}^\infty (-1)^k \int_0^\infty \frac{a^{2k} x^{2k}}{(2k)!} e^{-x^2}\,dx = \sum_{k=0}^\infty (-1)^k \frac{a^{2k}}{(2k)!}\int_0^\infty e^{-x^2} x^{2k}\,dx.$$

Now, by a change of variables the inner integrals are equal to $\sqrt{\pi}(2k)!/4^k k!$ for all k and so

$$I = \sum_{k=0}^{\infty}(-1)^k \frac{a^{2k}}{(2k)!}\frac{\sqrt{\pi}(2k)!}{4^k k!} = \sqrt{\pi}\sum_{k=0}^{\infty}(-1)^k\frac{(a^2/4)^k}{k!} = \sqrt{\pi}e^{-a^2/4}.$$

101. Fix a compactly supported smooth function ψ_0 with integral 1. Observe that given a compactly supported function ψ with integral 1, there exists a compactly supported φ such that $\psi = \psi_0 + \varphi'$; indeed, $\varphi(x) = \int_{-\infty}^{x}(\psi(y) - \psi_0(y))\,dy$ is compactly supported, smooth, and has integral 0. Then $0 = \int_{\mathbb{R}} f(x)\varphi'(x)\,dx = \int_{\mathbb{R}} f(x)\psi(x)\,dx - \int_{\mathbb{R}} f(x)\psi_0(x)\,dx$ and $\int_{\mathbb{R}} f(x)\psi(x)\,dx = \int_{\mathbb{R}} f(x)\psi_0(x)\,dx$ is independent of ψ.

104. Let $F(x) = \int_0^x f(t)\,dt$, $G(x) = \int_0^x g(t)\,dt$, $H(x) = \int_0^x f(t)g(t)\,dt$. First, since $0 \le g \le 1$, $0 \le G(x) \le x$, and since f is nonincreasing, $f(x) - f(G(x)) \le 0$. Therefore, since $g \ge 0$,

$$(H - F\circ G)'(x) = g(x)(f(x) - f(G(x))) \le 0,$$

$H - F\circ G$ is decreasing and, in particular, $(H - F\circ G)(x)\big|_0^1 \le 0$. Finally, $\int_0^1 f(x)g(x)\,dx = \int_0^1 f(G(x))g(x)\,dx + \int_0^1 (H - F\circ G)'(x)\,dx \le \int_0^\eta f(x)\,dx$.

105. This is a particular case of Problem 5.77. Moreover, if the functions are of opposite monotonicity, all the inequalities are reversed.

107. The function $f(x) = |p(x)|$ where $p(x)$ is a nonzero polynomial of any degree satisfies $\int_0^\infty |f(x)|\,e^{-ax}\,dx < \infty$, $a > 0$, yet is not integrable.

108. (a) $A = \pi^2/6$.

(b) Since $|f(n^2x)| \le M/(1+n^2x)$ for some constant M the series defining $F(x)$ converges absolutely for $x > 0$. For $t > 0$ let $\nu(t)$ denote the smallest integer greater than or equal to t. Then $f(\nu^2(t)x)$ is a step function equal to $f(n^2x)$ for $n-1 < t \le n$, and $\int_0^N f(\nu^2(t)x)dt = \sum_{n=1}^N f(n^2x)$ for $0 < x \le N$. Moreover, since $\nu(t) > t$, $|f(\nu^2(t)x)| \le M/(1 + \nu^2(t)x) \le M/(1 + t^2x)$ and so $f(\nu^2(t)x)$ is integrable as a function of t for each $x > 0$. Hence, letting $N \to \infty$ it follows that $\int_0^\infty f(\nu^2(t)x)\,dt = \sum_n f(n^2x) = F(x)$.

Now, the change of variables $t = \sqrt{u/x}$ yields

$$\sqrt{x}F(x) = \frac{1}{2}\int_0^\infty u^{-1/2}f(x\nu^2(\sqrt{u/x}))\,du.$$

First, we compute $\lim_{x\to 0^+} x\nu^2(\sqrt{u/x})$. Since $\sqrt{u/x} \le \nu(\sqrt{u/x}) < 1+\sqrt{u/x}$, squaring, $u/x \le \nu^2(\sqrt{u/x}) < 1+2\sqrt{u/x}+u/x$ and, therefore, $u \le x\nu^2(\sqrt{u/x}) < x + 2\sqrt{ux} + u$, which implies that $\lim_{x\to 0^+} x\nu^2(\sqrt{u/x}) = u$. Now, we claim that $u^{-1/2}f(x\nu^2(\sqrt{u/x})) \to u^{-1/2}f(u)$ boundedly as $x \to 0^+$.

Indeed, since $u \leq x\nu^2(\sqrt{u/x})$, $|u^{-1/2}f(x\nu^2(\sqrt{u/x}))| \leq Mu^{-1/2}/(1+u) \in L^1(\mathbb{R}^+)$. Hence by the LDCT, $\lim_{x\to 0^+}\sqrt{x}F(x) = (1/2)\int_0^\infty u^{-1/2}f(u)\,du$.

109. The statement is false. Indeed, for $\mu = \delta$, the Dirac delta at the origin $\int_{B(0,1/n)} f\,d\mu = f(0)$ for all n, which need not be 0; note that if $f(0) = 0$, then $f = 0$ μ-a.e. and the result holds in this case. In general, if $\mu(\{0\}) = 0$, then $f_n = f\chi_{B(0,1/n)} \to 0$ μ-a.e. and since $|f_n| \leq |f| \in L^1(X)$, by the LDCT $\lim_n \int_{B(0,1/n)} f\,d\mu = \int_X \lim_n f_n\,d\mu = 0$.

110. (a) For the sake of argument suppose that for every $x \in I$ there exists $f_x \in C(I)$ such that $f_x(x) \neq 0$ and $\int_X f_x\,d\mu = 0$. Since f_x is continuous $f_x(y) \neq 0$ for y in a neighborhood V_x of x in I. Thus $I \subset \bigcup_{x\in X} V_x$ and so there exist finitely many $x_1, \ldots, x_n$ such that $I \subset \bigcup_{1\leq k\leq n} V_{x_k}$. Now, the function $f(x) = \sum_{k=1}^n f_{x_k}^2(x)$ is strictly positive and by assumption $\int_I f\,d\mu = \int_I \sum_{k=1}^n f_{x_k}^2\,d\mu = \sum_{k=1}^n \int_I f_{x_k}^2\,d\mu = \sum_{k=1}^n (\int_I f_{x_k}\,d\mu)^2 = 0$, which, since $f(x) > 0$ for $x \in I$, can only happen if $\mu(I) = 0$, and this is not the case.

111. Given $k > 0$, there is an open set $O_k' \supset A$ such that $|O_k'| < 2^{-k}$. Let $O_1 = O_1'$ and $O_{k+1} = O_k \cap O_{k+1}'$; $\{O_k\}$ is a decreasing sequence of open sets containing A and $|O_k| < 2^{-k}$, $k \geq 1$. Let $f = \sum_k k\,\chi_{O_k}$; by the MCT, $\int_{\mathbb{R}^n} f(x)\,dx = \sum_k k\,|O_k| < \infty$.

Let $x \in A$ and fix $k \in \mathbb{N}$; since $x \in A \subset O_k$ it follows that $Q(x,\ell) \subset O_k$ for ℓ small enough. Hence $\chi_{Q(x,\ell)} f \geq k\,\chi_{Q(x,\ell)}$, and so,

$$\frac{1}{|Q(x,\ell)|}\int_{Q(x,\ell)} f(y)\,dy \geq \frac{1}{|Q(x,\ell)|}\int_{Q(x,\ell)} k\,dy = k\,.$$

Thus

$$\liminf_{\ell\to 0} \frac{1}{|Q(x,\ell)|}\int_{Q(x,\ell)} f(y)\,dy \geq k\,, \quad \text{all } x \in A\,.$$

Since this holds for every k, the conclusion follows.

113. (a) By the Lebesgue differentiation theorem

$$\lim_{r\to 0^+}\frac{|A \cap B(x,r)|}{|B(x,r)|} = \lim_{r\to 0^+}\frac{1}{|B(x,r)|}\int_{B(x,r)} \chi_A(y)\,dy = \chi_A(x) \quad \text{a.e. } x \in \mathbb{R}^n\,,$$

and, consequently, almost every $x \in A$ is a point of density of A. Moreover, since almost every $x \in A^c$ is a point of density of A^c and

$$\frac{|A \cap B(x,r)| + |A^c \cap B(x,r)|}{|B(x,r)|} = 1$$

it readily follows that $\lim_{r\to 0}|A \cap B(x,r)|/|B(x,r)| = 0$ for a.e. $x \in A^c$ and a.e. $x \in A^c$ is a point of dispersion of A.

(b) Let $f_n(x) = |A \cap B(x,1/n)|/|B(x,1/n)|$; by (a), $f_n \to \chi_A$ for a.e. $x \in A$. Setting $\varepsilon = |A|/2$ in Egorov's theorem it follows that there is

$A_1 \subset A$ such that $f_n \to 1$ uniformly for $x \in A_1^c$ and $|A_1| > |A|/2$. Let $A_0 = A_1^c$; since $\{f_n\}$ converges uniformly in A_0 there is an N such that $|f_n(x) - 1| < 1/2$ for all $x \in A_0$ and $n \geq N$ and so $f_n(x) > 1/2$ for those n. Thus $|A \cap B(x, 1/n)|/|B(x, 1/n)| \geq 1/2$ for all $x \in A_0$ and $n \geq N$.

(c) $|A \setminus \widehat{A}| = 0$.

114. By assumption $|A \cap (a, b)|/|(a, b)| = \eta$ for some constant $0 \leq \eta \leq 1$, all $a, b \in D$. Now, for each $x \in I$ there is a sequence $\{(a_n, b_n)\}$ of intervals with endpoints in D that converges to x and so $\lim_n |A \cap (a_n, b_n)|/|(a_n, b_n)| = \eta$ everywhere. Now, by the Lebesgue differentiation theorem the limit is 0 or 1 a.e., and so, if $\eta = 0$, $|A| = 0$ and if $\eta = 1$, $|A| = 1$.

116. By Problem 3.78(a) there is a measurable $A \subset (0, 1)$ with density $1/4$ at the origin; let $f(x) = \int_{-1}^{x} \chi_A(y)\, dy = |A \cap (-1, x)|$. By a direct computation $f'(0) = \lim_{h \to 0} |A \cap (0, h)|/h = 1/4$.

117. First, for $x, y, z \in \mathbb{R}$, $|x + y - z| \leq |x - z| + |y|$ and, therefore, picking $z \in F$ it follows that $\delta(x + y) \leq \delta(x) + |y|$; in particular, if $x \in F$, $\delta(x + y) \leq |y|$. Let x be a point of Lebesgue density of F and for the sake of argument suppose that $\limsup_{y \to 0} \delta(x + y)/|y| = 2A > 0$ with $A \leq 1/2$. Then, for all $\varepsilon > 0$ there is y such that $|y| < \varepsilon$ and $\delta(x + y)/|y| \geq A$. In particular, $x + y \notin F$ and the ball $B(x + y, A|y|) \subset F^c$. Since x is a point of density of F, for any $0 < \alpha < 1$ we can find $\eta > 0$ such that $|B(x, r) \cap F| \geq \alpha|B(x, r)|$ for all $r < \eta$. Since $B(x + y, A|y|) \subset B(x, 2|y|)$ we have

$$|B(x, 2|y|) \cap F| \leq |B(x, 2|y|)| - |B(x + y, A|y|)| = (1 - (A/2)^d)|B(x, 2|y|)|.$$

Thus, setting $\alpha = 1 - (A/4)^d$ and $|y| < \min\{\varepsilon, \eta\}$, we get

$$\left(1 - (A/4)^d\right)|B(x, 2|y|)| \leq \left(1 - (A/2)^d\right)|B(x, 2|y|)|,$$

which cannot hold if A is positive. So $\delta(x + y)/|y| \to 0$ as $y \to 0$ at the density points x of F, that is, a.e.

118. (a) Yes. The functions $g(x) = \int_0^x f(y)\, dy$ and $h(x) = \int_x^1 f(y)\, dy$ are continuous and satisfy $g(0) = h(1)$ and $g(1) = h(0)$. Thus $g(x) = h(x)$ for some $x \in [0, 1]$.

(b) That f vanishes a.e.

119. By the Lebesgue differentiation theorem almost every point is a Lebesgue point of f; pick $a \leq x < y < b$ as two such points. Then

$$\lim_{h \to 0} \frac{1}{h} \int_{x+h}^{y+h} |f(t + h) - f(t)|\, dt = 0$$

and, consequently,

$$\lim_{h\to 0} \frac{1}{h} \int_{x+h}^{y+h} \left(f(t+h) - f(t) \right) dt = 0.$$

Unraveling the above integral, by the translation invariance of the Lebesgue integral we see that for h small enough the integral is equal to

$$\frac{1}{h} \int_x^y f(t)\, dt - \frac{1}{h} \int_{x+h}^{y+h} f(t)\, dt = \frac{1}{h} \int_x^{x+h} f(t)\, dt - \frac{1}{h} \int_y^{y+h} f(t)\, dt.$$

Finally, since the limit of this expression as $h \to 0$ is 0 and by the Lebesgue differentiation theorem is $f(x) - f(y)$ a.e., we get that $f(x) = f(y)$ a.e. and f is constant.

120. Yes, and in fact A is an interval. Let $\int_0^{1/2} f(x)\, dx = \ell$; if $\ell = L/2$, we are done. If not, $\ell > L/2$ or $\ell < L/2$. Since $\int_X f(x)\, dx = L$ in the former case we have $\int_{1/2}^1 f(x)\, dx < L/2$ and in the latter $\int_{1/2}^1 f(x)\, dx > L/2$. Let $F(t) = \int_t^{t+1/2} f(x)\, dx$; F is continuous (absolutely continuous actually), and considering $F(0)$ and $F(1/2)$, one of these values is $> L/2$ and the other is $< L/2$ and, consequently, $F(t) = L/2$ for some $t \in (0, 1/2)$.

122. The statement is false. Let

$$f_n(x) = \begin{cases} \min(1 - n^2 x, 1 + n^2 x), & -1/n^2 \le x \le 1/n^2, \\ 0, & \text{otherwise.} \end{cases}$$

The graph of f_n is a triangle of width $2/n^2$ and height 1 centered at the origin and so $\int_I f_n(x)\, dx = n^{-2}$. Now let $f(x) = \sum_n f_n(x - n)$, $x \ge 0$; since for any $x \in [0, \infty)$ at most one $f_n(x)$ is nonzero the sum defining f is convergent and continuous. Also, $\int_0^\infty f(x)\, dx = \sum_{n=2}^\infty 2^{-1}(2n^{-2}) < \infty$. However, since $f(n) = 1$ for all n, $f(x) \not\to 0$ as $x \to \infty$.

In fact, a slight modification of the argument shows that given a sequence $x_n \to \infty$, there is a continuous integrable function f on $\mathbb{R}^+$ such that $\lim_n f(x_n) = \infty$.

123. Let $A_n = \{1/n \le |f| \le n\}$. By Chebychev's inequality $\mu(A_n)/n \le \int_X |f|\, d\mu < \infty$ and the first and second conditions hold with $A = A_n$ for any n. Next, let $f_n = f \chi_{A_n^c}$; then $|f_n| \le |f|$ and $f_n \to 0$ μ-a.e. and so, by the LDCT, $\int_X |f_n|\, d\mu \to 0$. Therefore there exists n such that the integrals are less than the given ε; then set $A = A_n$.

124. (a) Let $R > 0$ be large enough so that $\int_{\{|x|>R\}} |f(x)|\, dx < \varepsilon^2$. Then by Chebychev's inequality, $\varepsilon \, |\{|x| > R : |f(x)| > \varepsilon\}| \le \int_{\{|x|>R\}} |f(x)|\, dx < \varepsilon^2$, and so $|f(x)| \le \varepsilon$ except possibly on a set $B \subset \{|x| > R\}$ with measure $|B| < \varepsilon$.

(b) Construct B_k as in (a) for $\varepsilon/2^k$, $k \geq 1$; the conclusion then holds for $B = \bigcup_k B_k$, $|B| \leq \varepsilon$.

125. First, by Problem 86, $\int_{\mathbb{R}} |f(\alpha_n x)|\, dx = |\alpha_n|^{-1} \int_{\mathbb{R}} |f(x)|\, dx$ and so, by the MCT, $\int_{\mathbb{R}} \sum_n \beta_n |f(\alpha_n x)|\, dx = (\int_X |f|\, d\mu) \sum_n \beta_n / |\alpha_n| < \infty$, which implies that $\sum_n \beta_n |f(\alpha_n x)|$ converges a.e. and, consequently, $\lim_n \beta_n |f(\alpha_n x)| = 0$ a.e. In particular, if $\sum_n 1/|\alpha_n| < \infty$, $\lim_n |f(\alpha_n x)| = 0$ a.e.

127. Given $\eta > 0$ and $x_n \to \infty$, let $f_n = |f|\, \chi_{(x_n - \eta, x_n + \eta)}$. Then $f_n(x) \to 0$ for all x and $f_n \leq |f| \in L(\mathbb{R})$ and, consequently, by the LDCT, $\lim_n \int_{\mathbb{R}} f_n(x)\, dx = \lim_n \int_{x_n - \eta}^{x_n + \eta} |f(x)|\, dx = 0$. Now, for the sake of argument suppose that there exist $\varepsilon > 0$ and $x_n \to \infty$ such that $|f(x_n)| > \varepsilon$ for all n. Then by the uniform continuity of f there exists $\eta > 0$ such that $|f(x) - f(y)| \leq \varepsilon/2$ whenever $|x - y| \leq \eta$ and so it follows that $|f(x)| \geq \varepsilon/2$ for $x \in (x_n - \eta, x_n + \eta)$ for all n. Therefore $\int_{x_n - \eta}^{x_n + \eta} |f(x)|\, dx \geq \varepsilon\eta$ for all n, which is not the case since, as we just saw, this quantity goes to 0 for integrable f.

The result applies, for instance, to f nonnegative and integrable on $[0, \infty)$ with a bounded derivative $|f'(x)| < M < \infty$.

128. For the sake of argument suppose there exists $\varepsilon > 0$ such that $\sqrt{n}\, f(n) > \varepsilon$ for all but finitely many n. Now, since $f(x) \geq f(n) - |f(x) - f(n)|$, $\sqrt{n}\, f(x) \geq \varepsilon/2$ for $|x - n| \leq \varepsilon/(2M\sqrt{n})$. Therefore

$$\int_{\mathbb{R}^+} f(x)\, dx \geq \sum_{n=N}^{\infty} \int_{n}^{n + \varepsilon/(2M\sqrt{n})} f(x)\, dx$$

$$\geq \sum_{n=N}^{\infty} \frac{\varepsilon}{2\sqrt{n}} \frac{\varepsilon}{2M\sqrt{n}}$$

$$= \frac{\varepsilon^2}{4M} \sum_{n=N}^{\infty} \frac{1}{n} = \infty,$$

which is not the case.

129. First, we claim that $\int_k^{k+1} \sum_n |f(x + a_n)|\, dx < \infty$ for each finite integer k. To see this let $A_m = \{n \in \mathbb{N} : a_n \in [m, m+1]\}$, $m \geq 1$; by assumption, A_m contains at most N elements. Now, for $a_n \in A_m$,

$$\int_k^{k+1} |f(x + a_n)|\, dx = \int_{k+a_n}^{k+a_n+1} |f(x)|\, dx$$

$$\leq \int_{k+m}^{k+m+2} |f(x)|\, dx,$$

and, consequently, $\sum_{a_n \in A_m} \int_k^{k+1} |f(x+a_n)| \, dx \le N \int_{k+m}^{k+m+2} |f(x)| \, dx$. Next, grouping the a_n into the A_m they belong to,

$$\int_k^{k+1} \sum_n |f(x + a_n)| \, dx = \sum_{n=0}^{\infty} \int_k^{k+1} |f(x + a_n)| \, dx$$

$$= \sum_{m=-\infty}^{\infty} \sum_{a_n \in A_m} \int_k^{k+1} |f(x + a_n)| \, dx$$

$$\le N \sum_{m=-\infty}^{\infty} \int_{k+m}^{k+m+2} |f(x)| \, dx \le 2N \int_{\mathbb{R}} |f(x)| \, dx \,.$$

Then $\sum_n |f(x + a_n)| < \infty$ a.e. on each interval $[k, k+1]$ and, hence, $\lim_n f(x + a_n) = 0$ a.e. on each $[k, k+1]$, and, consequently, a.e. on $\mathbb{R}$.

130. (a) The claim is that $\liminf_{x \to 0+} x \, |f(x)| = 0$. For the sake of argument suppose there exist $\eta, \delta > 0$ such that $x|f(x)| > \eta/2$ whenever $0 < x \le \delta$. Then $\|f\|_1 \ge \int_0^{\delta} |f(x)| \, dx \ge c \int_0^{\delta} x^{-1} \, dx = \infty$, which is not the case.

(b) First, choose $\alpha_k \to 0$ such that $\sum_k \alpha_k \int_I |f_k(x)| \, dx < \infty$; this can be done for any sequence of integrable functions by putting $0 \ne \alpha_k = \beta_k/\|f_k\|_1$ with $\sum_k \beta_k < \infty$. Thus, by (a) applied to $f = \sum_k \alpha_k |f_k|$ there is a decreasing sequence $\{b_n\}$ such that $b_n f(b_n) \to 0$ and so, $\alpha_k \lim_n b_n |f_k(b_n)| = 0$ for all k.

131. The limit is λ.

132. The limit is λ.

133. The limit is 0.

134. No. If $\mu(\{f > 1\}) > 0$, the limit is infinite. Hence $f \le 1$ μ-a.e. and then the limit is ≤ 1.

135. Let $A = \{f < 1\}$, $B = \{f = 1\}$, and $C = \{f > 1\}$. Since $\mu(X) < \infty$, we have $g(f^n) \to g(0)$ boundedly on A and $\lim_n \int_A g(f^n) \, d\mu = g(0)\mu(A)$, and $g(f^n) = g(1)$ on B and $\lim_n \int_B g(f^n) \, d\mu = g(1)\mu(B)$. Finally, $g(f^n)$ increases to $g(\infty)$ in C and by the MCT, $\lim_n \int_C g(f^n) \, d\mu = g(\infty)\mu(C)$. Thus the limit is $g(0)\mu(A) + g(1)\mu(B) + g(\infty)\mu(C)$.

137. Let $A = \{f > 1\}$. Then, by Fatou's lemma, $\int_A \liminf_{n_k} f^{n_k} \, d\mu \le \liminf_{n_k} \int_A f^{n_k} \, d\mu \le c$ and since $\lim_{n_k} f^{n_k}(x) = \infty$ for $x \in A$, this readily implies that $\mu(A) = 0$. Let $B = \{f < 1\}$. Then $c = \int_X f^{n_k} \, d\mu = \int_{\{f=1\}} d\mu + \int_B f^{n_k} \, d\mu$ and, consequently, $\int_B f^{n_k} \, d\mu = \int_B f^{n_\ell} \, d\mu$ for $n_k < n_\ell$. Thus $\int_B f^{n_k}(1 - f^{n_\ell - n_k}) \, d\mu = 0$ and f vanishes μ-a.e. on B or $\mu(B) = 0$. Therefore $f(x) = \chi_M(x)$ μ-a.e. where $M = \{f = 1\}$.

138. If k is even, observe that $\int_X f^k (1-f)^2 \, d\mu = 0$ and so by Problem 67, $f(1-f)$ vanishes μ-a.e. Thus f assumes the values 0 or 1 μ-a.e. and $A = \{f(x) = 1\}$. If k is odd, the conclusion holds provided f is nonnegative.

For two indices the situation is analogous provided that $0 \le f \le 1$ μ-a.e. for then $\int_X f^k(1-f) \, d\mu = 0$ and the integrand is nonnegative and, consequently, vanishes μ-a.e.

139. Let O be an open set of measure $< \varepsilon$ that contains the rationals in I and put $f(x) = \chi_O(x)$. Then $f(x) = 0$ in O^c, a set with measure $\ge 1 - \varepsilon$, and since for every interval $(a,b) \subset I$ the open set $O \cap (a,b) \ne \emptyset$, $\int_a^b f(x) \, dx > 0$.

141. (a) First, note that $\partial B(x,r)$, the boundary of the ball centered at x with radius r, has Lebesgue measure 0. Indeed, since for any $\varepsilon > 0$, $\partial B(x,r) \subset B(x,r) \setminus B(x, r - \varepsilon)$, $|\partial B(x,r)|_e \le |B(x,r) \setminus B(x, r - \varepsilon)| = (r^n - (t - \varepsilon)^n)|B(x,1)|$, which tends to 0 with ε. Now, let $r_k \to r$ be any convergent sequence of nonnegative real numbers. Then clearly $\chi_{B(x,r_k)} \to \chi_{B(x,r)}$ except possibly on the boundary of the ball. Hence $f\chi_{B(x,r_k)} \to f\chi_{B(x,r)}$ a.e. and $|f\chi_{B(x,r_k)}|, |f\chi_{B(x,r)}| \le |f|$ with $f \in L^1(\mathbb{R}^n)$ and, consequently, by the LDCT,

$$g(r_k) = \int_{\mathbb{R}^n} f(z)\chi_{B(x,r_k)}(z) \, dz \to \int_{\mathbb{R}^n} f(z)\chi_{B(x,r)}(z) \, dz = g(r)$$

as $r_k \to r$ and g is continuous.

(b) First, note that for fixed r, if $|x - y| < 2r$, $B(x,r) \cap B(y,r) \ne \emptyset$ and $|B(x,r) \setminus B(y,r)| = |B(y,r) \setminus B(x,r)| = |y - x|$, and so $|B_r(x) \triangle B_r(y)| \le 2|y - x|$. Next, by the absolute continuity of the integral of f, given $\varepsilon > 0$, there is $0 < \delta < 2r$ such that $\int_A |f(z)| \, dz < \varepsilon$ for $|A| < \delta$. In particular, if $x, y \in \mathbb{R}$ with $|x - y| < \delta/2$, we have

$$|g(x) - g(y)| = \left| \int_{B(x,r)} f(z) \, dz - \int_{B(y,r)} f(z) \, dz \right|$$

$$= \left| \int_{B(x,r) \setminus B(y,r)} f(z) \, dz - \int_{B(y,r) \setminus B(x,r)} f(z) \, dz \right|$$

$$\le \int_{(B(x,r) \setminus B(y,r)) \cup (B(y,r) \setminus B(x,r))} |f(z)| \, dz$$

$$= \int_{B(x,r) \triangle B(y,r)} |f(z)| \, dz < \varepsilon$$

since $|B(x,r) \triangle B(y,r)| < 2|x - y| < \delta$. Since this holds for any $x, y \in \mathbb{R}$ with $|x - y| < \delta/2$, g is uniformly continuous on $\mathbb{R}$.

142. Let $r_n \to r$ be any convergent sequence of nonnegative real numbers. Then clearly $\chi_{B_{r_n}} \to \chi_{B_r}$ everywhere except possibly on $\partial B_r = \{|x| =$

$r\}$. Therefore, assuming that ∂B_r is μ-null and that μ is complete, we have $f\chi_{B_{r_n}} \to f\chi_{B_r}$ μ-a.e. and $|f\chi_{B_{r_n}}|, |f\chi_{B_r}| \le |f| \in L(X)$. Thus, by the LDCT, $g(r_n) \to g(r)$.

143. The condition is not necessary if the measure is infinite as the example of the Lebesgue measure on $\mathbb{R}$, $A_n = [n, n+1]$, and f any integrable function shows. However, it is necessary if the measure is finite. Let $\eta > 0$. Then by Chebychev's inequality, $\eta\,\mu(\{x \in A_n : f(x) > \eta\}) \le \int_{\{x\in A_n: f(x)>\eta\}} f\,d\mu \le \int_{A_n} f\,d\mu$, and so $\lim_n \mu(A_n \cap \{f > \eta\}) = 0$ for all $\eta > 0$. Now, since $X = N \cup \bigcup_k \{f > 1/k\}$ with $\mu(N) = 0$, it readily follows from continuity from below that $\lim_k \mu(A_n \cap \{f > 1/k\}) = \mu(A_n)$; let k be large enough so that $\mu(A_n \cap \{f > 1/k\}) \ge \mu(A_n)/2$. Then $\lim_n \mu(A_n) \le 2 \lim_n \mu(A_n \cap \{f > 1/k\}) = 0$.

The condition is sufficient: Since $\lim_n f\chi_{A_n} = 0$ μ-a.e. and $f\chi_{A_n} \le f \in L^1(X)$, by the LDCT $\lim_n \int_{A_n} f\,d\mu = \lim_n \int_X f\chi_{A_n}\,d\mu = 0$.

144. (d) implies (a) Consider the set function $\mu(A) = |\varphi^{-1}(A)|$, $A \in \mathcal{B}(I)$; as in Problem 2.45 μ is a Borel measure on I. Now, μ coincides with the Lebesgue measure on the intervals $[0, x]$ and, therefore, for $0 \le x < y \le 1$, $\mu((x, y]) = \mu([0, y]) - \mu([0, x]) = |[0, y]| - |[0, x]| = |(x, y]|$, i.e., on all intervals. Problem 3.23 implies that $\mu(A) = |A|$ for all Borel A in I.

145. A computation shows that if $I = (a, b)$,

$$\varphi^{-1}(I) = \left(\frac{a - \sqrt{a^2 + 4}}{2}, \frac{b - \sqrt{b^2 + 4}}{2} \right) \cup \left(\frac{a + \sqrt{a^2 + 4}}{2}, \frac{b + \sqrt{a^2 + 4}}{2} \right),$$

and, consequently, $\varphi^{-1}(I)$ is measurable and $|\varphi^{-1}(I)| = (b - a)/2 + (b - a)/2 = |I|$. The conclusion follows now as in Problem 144.

146. (a) implies (b) Given $A \in \mathcal{M}$, let $B = \bigcup_{i=0}^{\infty} T^{-i}(A)$. Then $T^{-1}(B) \subset B$ and, by Problem 2.112(b), $\mu(B) = 0$ or 1. If $\mu(B) = 0$, by monotonicity $\mu(A) = 0$. On the other hand, if $\mu(B) = 1$, since $A \subset B$ and $B \setminus A = \bigcup_{i=1}^{\infty} ((T^{-i}(A)) \setminus A)$, it suffices to prove that $\mu((T^{-i}(A)) \setminus A) = 0$ for all i. We proceed by induction. Since by assumption $\mu(A \triangle T^{-1}(A)) = 0$ the statement is true for $i = 1$. Next, assuming it true for i, note that

$$\mu((T^{-(i+1)}(A)) \setminus A)$$
$$\le \mu(T^{-(i+1)}(A) \setminus (T^{-i}(A) \cup A)) + \mu(((T^{-(i+1)}(A)) \cap T^{-i}(A)) \setminus A)$$
$$\le \mu((T^{-(i+1)}(A)) \setminus (T^{-i}(A))) + \mu((T^{-i}(A)) \setminus A)$$
$$= \mu((T^{-1}(A)) \setminus A) + \mu((T^{-i}(A)) \setminus A) = 0.$$

147. (a) Since ψ is bounded, $\{f\psi(n\cdot)\}$ is uniformly in $L^1(\mathbb{R})$. Now, by the translation invariance of the Lebesgue integral and the assumption on ψ, $\int_{\mathbb{R}} f(x)\psi(nx)\,dx = \int_{\mathbb{R}} f(x + \beta/n)\,\psi(nx + \beta)\,dx = -\int_{\mathbb{R}} f(x + \beta/n)\,\psi(nx)\,dx,$

and, consequently, with $M = \sup |\psi|$,

$$\left| \int_{\mathbb{R}} f(x)\, \psi(nx)\, dx \right| = \frac{1}{2} \left| \int_{\mathbb{R}} f(x)\, \psi(nx)\, dx - \int_{\mathbb{R}} f(x + \beta/n)\psi(nx)\, dx \right|$$
$$\leq \frac{1}{2} \int_{\mathbb{R}} |f(x) - f(x + \beta/n)|\, |\psi(nx)|\, dx$$
$$\leq \frac{M}{2} \int_{\mathbb{R}} |f(x) - f(x + \beta/n)|\, dx\,.$$

The conclusion follows readily from this since integrable functions are continuous in the $L^1(\mathbb{R})$ metric.

(b) First, observe that $\int_0^\beta \varphi(x)\, dx = \int_0^\beta \varphi(x+\beta)\, dx = \int_\beta^{2\beta} \varphi(x)\, dx$; similarly, $\int_{k\beta}^{(k+1)\beta} \varphi(x)\, dx = \int_{(k+1)\beta}^{(k+2)\beta} \varphi(x)\, dx$ for all $k \geq 0$. Let $c = \beta^{-1} \int_0^\beta \varphi(x)\, dx$. Then $\int_0^{2\beta} (\varphi(x) - c)\, dx = \int_0^{2\beta} \varphi(x)\, dx - 2\int_0^\beta \varphi(x)\, dx = 0$ and, similarly, $\int_{2k\beta}^{2(k+1)\beta} (\varphi(x) - c)\, dx = 0$ for all $k \geq 0$. Hence, changing variables, it follows that $\int_{2k\beta/n}^{2(k+1)\beta/n} (\varphi(nx) - c)\, dx = 0$ for all $k, n \geq 0$.

Next, observe that if $h(x) = \sum_{j=1} a_j \chi_{I_j}(x)$ is a step function, then $\lim_n \int_{\mathbb{R}} h(x)(\varphi(nx) - c)\, dx = 0$. Indeed, note that $\int_R h(x)(\varphi(nx) - c)\, dx = \sum_{j=1} a_j \int_{I_j} (\varphi(nx) - c)\, dx$. Now, as observed above the integral on any subinterval of I_j of the form $[2k\beta/n, 2(k+1)\beta/n]$ vanishes, so we are left with an integral over at most two intervals containing the endpoints of I_j, $I_{j,L}, I_{j,R}$, say, each of length not exceeding β/n. Thus, with $M = \sup_{x \in \mathbb{R}} |\varphi(nx) - c|$, $|\int_{I_j} (\varphi(nx) - c)\, dx| \leq 2M\beta/n$ for all j and, consequently, $|\int_{\mathbb{R}} h(x)(\varphi(nx) - c)\, dx| \leq (2M\beta/n) \sum_{j=1} |a_j|$.

Now, given $\varepsilon > 0$, let h be a step function such that $\|f - h\|_1 \leq \varepsilon/2M$. Then, since $\int_{\mathbb{R}} f(x)(\varphi(nx) - c)\, dx = \int_{\mathbb{R}} (f(x) - h(x))\, (\varphi(nx) - c)\, dx + \int_{\mathbb{R}} h(x)(\varphi(nx) - c)\, dx$, we have $| \int_{\mathbb{R}} f(x)(\varphi(nx) - c)\, dx| \leq M \int_{\mathbb{R}} |f(x) - h(x)| + |\int_{\mathbb{R}} h(x)(\varphi(nx) - c)\, dx| = A + B_n$, say. As noted above $A \leq \varepsilon/2$ and $B_n \leq (2M\beta/n) \sum_{j=1} |a_j|$ can be made arbitrarily small provided n is sufficiently large. Therefore $\lim_n \int_{\mathbb{R}} f(x)\varphi(nx)\, dx = c \int_{\mathbb{R}} f(x)\, dx$.

Now, if ψ satisfies (a), $\varphi(x) = |\psi(x)|$ satisfies (b). In particular, $\psi(x) = \sin(x)$ with $\beta = \pi$ will do. It then follows that $\lim_n \int_{\mathbb{R}} f(x) \sin(nx)\, dx = 0$ and $\lim_n \int_{\mathbb{R}} f(x) |\sin(nx)|\, dx = (2/\pi) \int_{\mathbb{R}} f(x)\, dx$. Similar results hold for $\cos(x)$ and $|\cos(x)|$.

148. (a) Follows from Problem 147(b).

(b) We claim that the limit exists and is equal to $c \int_I f(x)\, dx$ where $c = (2\pi)^{-1} \int_0^{2\pi} (2 + \sin(x))^{-1}\, dx$ and this follows from Problem 147(b) with $\beta = 2\pi$ there.

149. Recall that we say that f, g are independent if $\mu(f^{-1}(A) \cap g^{-1}(B)) = \mu(f^{-1}(A))\mu(g^{-1}(B))$ for all Borel A, B in $\mathbb{R}$. Let $A, B \subset \mathbb{R}$ be Borel. Then $\cos^{-1}(A)$ is Borel and $\cos(f)^{-1}(A) = f^{-1}(\cos^{-1}(A))$; similarly, $\sin^{-1}(B)$ is Borel and $\sin(g)^{-1}(B) = g^{-1}(\sin^{-1}(B))$. Then

$$\mu(f^{-1}(\cos^{-1}(A)) \cap g^{-1}(\sin^{-1}(B))) = \mu(f^{-1}(\cos^{-1}(A)))\mu(g^{-1}(\sin^{-1}(B))),$$

and so $\mu(\cos(f)^{-1}(A) \cap \sin(g)^{-1}(B)) = \mu(\cos(f)^{-1}(A))\,\mu(\sin(g)^{-1}(B))$ and $\cos(f)$ and $\sin(g)$ are independent.

150. First, since binary expansions may not be unique, we chose the expansion of x that has the least number of 1's (terminating in 0's). We have to prove that $|f_n^{-1}(A) \cap f_m^{-1}(B)| = |f_n^{-1}(A)|\,|f_m^{-1}(B)|$ for integers $n \neq m$ and Borel subsets $A, B \subset \{0, 1\}$. Since the subsets of $\{0, 1\}$ are $\emptyset, \{0\}, \{1\}, \{0, 1\}$, if one of A or B is $\emptyset$ or $\{0, 1\}$, the condition obviously holds. So we only need to verify the condition when A and B are $\{0\}$ or $\{1\}$; suppose that they are $\{0\}$. With no loss of generality we may assume that $n < m$. Now, since half of the points in $[0, 1]$ have $x_n = 0$ and for the other half $x_n = 1$, clearly $|f_n^{-1}(\{0\})| = 1/2$; similarly, $|f_m^{-1}(\{0\})| = 1/2$. Consider now those points in $f_n^{-1}(\{0\}) \cap f_m^{-1}(\{0\})$, their expansions are like $.x_1 \ldots x_{n-1} 0 x_{n+1} \ldots x_{m-1} 0 x_{m+1} \ldots$, so exactly $1/4$ of the points in $[0, 1]$ have $x_n = 0$ and $x_m = 0$ and so, $|f_n^{-1}(\{0\}) \cap f_m^{-1}(\{0\})| = 1/4$, and this gives the conclusion in this case. A similar argument works for the other subsets of A and B.

151. First suppose that $f(x) = \sum_n a_n \chi_{A_n}(x)$ and $g(x) = \sum_m b_m \chi_{B_m}(x)$ are simple functions. Now, since the a_n's are all different, $A_n = f^{-1}(\{a_n\})$ and, similarly, $B_m = g^{-1}(\{b_m\})$. Then $\int_X \chi_{A_n} \chi_{B_m}\, d\mu = \mu(A_n \cap B_m) = \mu(A_n)\,\mu(B_m)$ and, consequently,

$$\int_X fg\, d\mu = \sum_{n,m} a_n b_m \int_X \chi_{A_n} \chi_{B_m}\, d\mu = \sum_{n,m} a_n b_m \mu(A_n)\,\mu(B_m)$$

$$= \sum_n a_n \mu(A_n) \sum_m b_m \mu(B_m) = \left(\int_X f\, d\mu\right)\left(\int_X g\, d\mu\right).$$

In the general case we first prove that f and g can be approximated by independent simple functions. First, $-M \leq f \leq M$, say, and given $\varepsilon > 0$, let $A_n = f^{-1}([-M + n\varepsilon, -M + (n+1)\varepsilon))$, $a_n = -M + n\varepsilon$, and put $\varphi(x) = \sum_n a_n \chi_{A_n}(x)$; note that since the A_n's are empty whenever $n < 0$ or $n > 2M/\varepsilon$, the sum defining φ has finitely many summands and $|f(x) - \varphi(x)| \leq \varepsilon$, $x \in X$. Similarly, for g let $B_m = g^{-1}([-M' + m\varepsilon, -M' + (m+1)\varepsilon))$, $b_m = -M' + m\varepsilon$, $\psi(x) = \sum_m b_m \chi_{B_m}(x)$, and $|g(x) - \psi(x)| \leq \varepsilon$ for $x \in X$. Now, by independence $\mu(A_n \cap B_m) = \mu(A_n)\mu(B_m)$ for all n, m, and as noted above, $\int_X \varphi\psi\, d\mu = (\int_X \varphi\, d\mu)(\int_X \psi\, d\mu)$.

Finally, since

$$\int_X fg \, d\mu = \int_X (f - \varphi) g \, d\mu + \int_X \varphi(g - \psi) \, d\mu + \int_X \varphi \psi \, d\mu$$

and

$$\int_X \varphi \psi \, d\mu = \left(\int_X \varphi \, d\mu - \int_X f \, d\mu \right) \int_X \psi \, d\mu$$
$$+ \left(\int_X \psi \, d\mu - \int_X g \, d\mu \right) \int_X f \, d\mu + \left(\int_X f \, d\mu \right) \left(\int_X g \, d\mu \right),$$

it readily follows that

$$\int_X fg \, d\mu - \left(\int_X f \, d\mu \right) \left(\int_X g \, d\mu \right) = \int_X (f - \varphi) g \, d\mu + \int_X \varphi(g - \psi) \, d\mu$$
$$+ \left(\int_X \varphi \, d\mu - \int_X f \, d\mu \right) \int_X \psi \, d\mu$$
$$+ \left(\int_X \psi \, d\mu - \int_X g \, d\mu \right) \int_X f \, d\mu$$
$$= I_1 + I_2 + I_3 + I_4,$$

say. First, note that $|I_1| \leq \int_X |f - \varphi| \, |g| \, d\mu \leq \varepsilon \, M' \mu(X)$ and, similarly, $|I_2| \leq \varepsilon \, M \mu(X)$. Next, observe that $|I_3| \leq \int_X |\varphi - f| \, d\mu \int_X |\psi| \, d\mu \leq \varepsilon \, M' \mu(X)^2$ and, similarly, $|I_4| \leq \varepsilon \, M \mu(X)^2$. Finally, $|\int_X fg \, d\mu - \int_X f \, d\mu \int_X g \, d\mu| \leq \varepsilon (M\mu(X) + M'\mu(X) + M\mu(X)^2 + M'\mu(X)^2)$, which can be made arbitrarily small with ε and the conclusion follows.

152. We begin by examining the φ_n more closely; φ_1 is alternatively 1 and -1 on the 10 intervals that comprise $[0, 1]$ and the same is true for any interval of the form $[k, k+1]$ where k is an integer, positive or negative. Thus $|\varphi_1(x)| = 1$ for all x and $\int_{-N}^{N} \varphi_1(x) \, dx = 0$ for all $N \geq 1$. Now, $\varphi_2(x)$ alternates its values between 1 and -1 in each of the intervals of constancy of φ_1 and so $|\varphi_2(x)| = 1$ for all x, $\int_{-N}^{N} \varphi_2(x) \, dx = 0$ and $\int_{-N}^{N} \varphi_1(x)\varphi_2(x) \, dx = 0$ for all $N \geq 1$. A similar situation occurs for all $n \neq m$. In other words, if $n < m$, say, φ_n is constant on the intervals φ_m, changes signs and, consequently, $\int_{-N}^{N} \varphi_n(x)\varphi_m(x) \, dx = 0$ for all $N \geq 1$.

To proceed with the proof let g be a bounded function with support in $[-N, N]$ for some N. Now, since $\|\varphi_n\|_2 = (2N)^{1/2}$ on any $L^2([-N, N])$, $\{(2N)^{-1/2}\varphi_n\}$ is an ON set in $L^2([-N, N])$ and by Bessel's inequality

$$\frac{1}{2N} \sum_n \left| \int_{-N}^{N} g(x)\varphi_n(x) \, dx \right|^2 \leq \int_{-N}^{N} |g(x)|^2 \, dx < \infty.$$

Therefore $\lim_n |\int_{-N}^{N} g(x)\varphi_n(x) \, dx| = 0$.

Now let $f \in L^1(\mathbb{R})$; since compactly supported bounded functions are dense in $L^1(\mathbb{R})$, given $\varepsilon > 0$, there exists one such function g, say, with support contained in $[-N, N]$ for some N and $\|f - g\|_1 < \varepsilon$. It then follows that $\limsup_n |\int_{\mathbb{R}} f(x)\varphi_n(x)\,dx| \le \|f - g\|_1 + \limsup_n |\int_{-N}^{N} g(x)\varphi_n(x)\,dx| < \varepsilon$ and since ε is arbitrary, $\limsup_n |\int_{\mathbb{R}} f(x)\varphi_n(x)\,dx| = 0$.

154. Since $f(x) = \sum_n 2^{-n}\chi_{[r_n,1]}(x)$, by the MCT we have that $\int_I f(x)\,dx = \sum_n 2^{-n}\int_{r_n}^1 dx = \sum_n 2^{-n}(1 - r_n)$, and we can say no more.

155. With $\phi(x) = |x|$ the expression under the integral becomes

$$\frac{\phi(1 + h\,f(x)) - \phi(1)}{h\,f(x)}\, f(x)\,,$$

and tends to $\phi'(1)\,f(x) = f(x)$ at each x where $f(x)$ assumes a finite value, which is μ-a.e. Also, since

$$\frac{\big|\,|1 + a| - |1|\,\big|}{|a|} \le \frac{|a|}{|a|} = 1\,,$$

the integrand is dominated by $|f(x)| \in L^1(X)$ and so by the LDCT the limit can be taken under the integral sign and $L = \int_X f\,d\mu$.

156. (a) For the sake of argument suppose that $n(x) = 1$ for all $x \in X$. Then the A_n are pairwise disjoint and $\infty = \sum_k \mu(A_k) = \mu(\bigcup_k A_k) \le \mu(X) < \infty$, which is not the case.

(b) Observe that $\infty = \sum_k \mu(A_k) = \sum_k \int_X \chi_{A_k}\,d\mu \le \int_X \sum_k \chi_{A_k}\,d\mu \le (\sup_{x \in X} n(x))\mu(X)$.

(c) Let $F = \limsup_n A_n$; by Problem 2.105, $\mu(F) \ge \varepsilon > 0$. Let $x \in F$; then $x \in \bigcup_n A_n$ and so $x \in A_{n_1}$ for some n_1. But then $x \in \bigcup_{n=n_1+1}^{\infty} A_n$ and so $x \in A_{n_2}$ for some $n_2 > n_1$. Continuing this way, x belongs to infinitely many A_n's.

157. Note that if $f(x) = \sum_{m=1}^n \chi_{A_m}(x)$, then $f(x) \ge k$ μ-a.e. Then integrating, $k \le \int_X f\,d\mu = \sum_{m=1}^n \mu(A_m)$ and, therefore, $\mu(A_m) \ge k/n$ for some $1 \le m \le n$.

159. Let $f = \sum_{n=1}^N \chi_{A_n}$; then $\sum_{n=1}^N \mu(A_n) \ge \eta N$. Observe that since $\int_{A^c} f\,d\mu \le aN\mu(A^c)$, it readily follows that $\int_A f\,d\mu = \int_X f\,d\mu - \int_{A^c} f\,d\mu \ge \eta N - aN(1 - \mu(A)) = N(\eta - a) + aN\mu(A)$. Now, since $\int_A f\,d\mu \le N\mu(A)$, we have $N(\eta - a) + aN\mu(A) \le N\mu(A)$, or $(\eta - a) \le (1 - a)\mu(A)$, hence the conclusion.

160. First, by induction, $\int_I f(x)\,x^n\,dx = 0$ for all n. Indeed, with p_0 we get that the integral of f vanishes, with p_1 that the integral of $xf(x)$ is 0, and so on. Now, since continuous functions on I are the uniform limits of polynomials, by the LDCT $\int_I f(x)h(x)\,dx = 0$ for every continuous function h on I. For the sake of argument suppose that $|\{x \in I : f(x) \ne 0\}| > 0$; then, by considering $-f$ if necessary, we may assume that $A = \{x \in I : f(x) > 0\}$ has positive measure. Let $\{h_n\}$ be continuous functions that converge to χ_A in $L^1(I)$; passing to a subsequence if necessary we may assume that $h_n \to \chi_A$ a.e. in I. Now, since χ_A is bounded, by truncating if needed we may assume that $\{h_n\}$ converges to χ_A boundedly and, consequently, by the LDCT, $\int_I f(x)\chi_A(x)\,dx = \lim_n \int_I f(x)h_n(x)\,dx = 0$, which cannot happen unless $f = 0$ a.e. on A.

161. Recall that the Haar functions on $[0,1]$ are $h_0 = 1$,

$$h_{n,k} = \begin{cases} 2^{n/2}, & x \in [\frac{k-1}{2^n}, \frac{k-1/2}{2^n}), \\ -2^{n/2}, & x \in [\frac{k-1/2}{2^n}, \frac{k}{2^n}), \\ 0, & \text{elsewhere.} \end{cases}$$

Now define $H_I(x) = |I|^{-1/2}(\chi_{I_L} - \chi_{I_R})$, I dyadic, with right and left halves I_R and I_L, respectively. This is an ONS.

Write a dyadic interval $J = J_L \cup J_R$, where J_L is the left half of J and J_R is the right half of J. First, observe that if

$$\frac{1}{|J|}\int_J \varphi(x)\,dx = c \quad \text{and} \quad \int_{\mathbb{R}} H_J(x)\varphi(x)\,dx = 0\,,$$

then

$$\frac{1}{|J_L|}\int_{J_L} \varphi(x)\,dx = \frac{1}{|J_R|}\int_{J_R} \varphi(x)\,dx = c\,.$$

Consider H_I where $I = [0,1]$. Then, if $\int_{[0,1]} \varphi(x)\,dx = c$, it readily follows that $2\int_{[0,1/2]} \varphi(x)\,dx = 2\int_{[1/2,1]} \varphi(x)\,dx = c$. Further subdividing the new dyadic intervals that we obtained at each step we have $2^n \int_{[k/2^n, k+1/2^n]} \varphi(x)\,dx = c$ whenever $[k/2^n, k+1/2^n] \subset [0,1]$.

Next, we extend the integral of φ outside $[0,1]$. For this purpose we choose $k = -1, n = 0$, and observe that

$$\int h_{-1,0}(x)\varphi(x)\,dx = \tfrac{1}{2}\int_{[0,1]} \varphi(x)\,dx - \tfrac{1}{2}\int_{[1,2]} \varphi(x)\,dx = 0\,.$$

Therefore $\int_{[1,2]} \varphi(x)\,dx = \int_{[0,1]} \varphi(x)\,dx = c$. Thus, dividing $\int_{[1,2]} \varphi(x)\,dx$ as before we get

$$2^n \int_{[k/2^n, (k+1)/2^n]} \varphi(x)\,dx = c \quad \text{whenever} \quad \left[\frac{k}{2^n}, \frac{k+1}{2^n}\right] \subset [1,2].$$

Continuing to extend and then subdividing the resulting intervals we have

$$2^n \int_{[k/2^n, (k+1)/2^n]} \varphi(x)\,dx = c \quad \text{for} \quad \left[\frac{k}{2^n}, \frac{k+1}{2^n}\right] \subset [0, \infty)\,.$$

Now, for each $x > 0$ we pick dyadic intervals I that contain x, $|I| \to 0$, and apply the Lebesgue differentiation theorem to obtain for almost all x,

$$\varphi(x) = \lim_{|I| \to 0} \frac{1}{|I|} \int_I \varphi(y)\, dy = c\,.$$

Similarly, if $\int_{[-1,0]} \varphi(x)\, dx = d$, $\varphi(x) = d$ for almost all $x \in (-\infty, 0)$. Therefore $\varphi(x) = d\,\chi_{(-\infty,0)}(x) + c\,\chi_{(0,\infty)}(x)$ a.e. in $\mathbb{R}$.

162. First, since $|\Phi(a,x)| \le |a-1| \max(\varphi(x), \varphi(ax))$ for $a \ne 1$, it follows that $\int_{\mathbb{R}^+} |\Phi(a,x)|\, dx \le |a-1| \int_0^\infty \max(\varphi(x), \varphi(ax))\, dx < \infty$ and $\Phi(a,x)$ is integrable. Let $I(a) = \int_0^\infty \Phi(a,x)\, dx$. Now, $\Phi(a,x)$ is differentiable as a function of a with derivative $\Phi_a(a,x) = f'(ax)$ and, since for each $\varepsilon > 0$ we have $|f'(ax)| \le \varphi(\varepsilon x)$ for all x and $a > \varepsilon$ and $\int_0^\infty \varphi(\varepsilon x)\, dx = \varepsilon^{-1} \int_0^\infty \varphi(x)\, dx < \infty$, by the LDCT we can differentiate under the integral sign and $I(a)$ is differentiable on (ε, ∞) with derivative $I'(a) = \int_0^\infty f'(ax)\, dx = a^{-1} \int_0^\infty f'(x)\, dx = a^{-1}(f(\infty) - f(0))$. Finally, since $I(1) = 0$ we conclude that $I(a) = I(a) - I(1) = (f(\infty) - f(0)) \int_1^a (1/y)\, dy = (f(\infty) - f(0)) \ln(a)$. This result is known as Frullani's formula.

163. Necessity first. Given $t > 0$, observe that

$$\frac{1}{f(x)} \int_x^\infty f(s)\, ds \ge \frac{1}{f(x)} \int_x^{x+t} f(s)\, ds \ge t\, \frac{f(x+t)}{f(x)}$$

and, since the left-hand side above goes to 0 as $x \to \infty$, so does the right-hand side.

Sufficiency next. For the sake of argument suppose that for some $t > 0$, there exist $\eta > 0$ and $x_n \to \infty$ such that $f(x_n + t) > \eta f(x_n)$ for all n. Then $\int_{x_n}^{x_n+t} f(s)\, ds \ge t\, f(x_n + t) \ge t\, \eta\, f(x_n)$, and, consequently,

$$\limsup_{x \to \infty} \frac{1}{f(x)} \int_x^\infty f(s)\, ds \ge t\, \eta > 0\,,$$

which is not the case.

165. First, suppose that $A \in \mathcal{M}$ verifies $\lambda(A) \ge \nu(A)$. Then $|\lambda(A) - \nu(A)| = \lambda(A) - \nu(A) = \int_A (f - g)\, d\mu$ and since $\int_X (f - g)\, d\mu = 0$ it follows that

$$\lambda(A) - \nu(A) = (1/2)\left(\int_A (f - g)\, d\mu + \int_{A^c} (g - f)\, d\mu \right)$$

$$\le (1/2)\left(\int_A |f - g|\, d\mu + \int_{A^c} |g - f|\, d\mu \right)$$

$$= (1/2) \int_X |f - g|\, d\mu\,.$$

The case $\lambda(A) \le \nu(A)$ is similar and, consequently,

$$|\lambda(A) - \nu(A)| \le (1/2) \int_X |f - g| \, d\mu \quad \text{for all } A \in \mathcal{M}$$

and so $\|\lambda - \nu\| \le (1/2) \int_X |f - g| \, d\mu$.

As for the reverse inequality, consider the measurable sets $A^+ = \{f \ge g\}$ and $A^- = \{f < g\} = (A^+)^c$. Then

$$\int_X |f - g| \, d\mu = \int_{A^+} (f - g) \, d\mu + \int_{A^-} (g - f) \, d\mu$$
$$= \lambda(A^+) - \nu(A^+) + \nu(A^-) - \lambda(A^-) = 2(\lambda(A^+) - \nu(A^+)).$$

Thus $(1/2) \int_X |f - g| \, d\mu = \lambda(A^+) - \nu(A^+) \le \|\lambda - \nu\|$ and equality holds.

166. First, independently of $x_n \to x^-$ and $y_n \to x^+$, $\lim_{x_n \to x^-} \chi_{(0,x_n]} f = \chi_{(0,x)} f$ and $\lim_{y_n \to x^+} \chi_{(0,y_n]} f = \chi_{(0,x]} f$. Furthermore, both of these sequences are pointwise majorized in absolute value by the integrable function $|f|$ and so by the LDCT,

$$I(x^+) - I(x^-) = \lim_{y_n \to x^+} \int_{\mathbb{R}} \chi_{(0,y_n]} f \, d\mu - \lim_{x_n \to x^-} \int_{\mathbb{R}} \chi_{(0,x_n]} f \, d\mu$$
$$= \int_{\mathbb{R}} \chi_{(0,x]} f \, d\mu - \int_{\mathbb{R}} \chi_{(0,x)} f \, d\mu) = \int_{\mathbb{R}} \chi_{\{x\}} f \, d\mu = f(x)\mu(\{x\}) \,.$$

Therefore, if $\mu(\{x\}) = 0$, then $I(x^+) - I(x^-) = 0$ and $I(x)$ is continuous at x.

168. For each j, let n_j be an increasing sequence such that $\sum_{n=n_j+1}^{\infty} \lambda_n < \Lambda/2^{2j}$. Define $\mu_n = 2^j$ for $n_j < n \le n_{j+1}$. Then $\mu_n \to \infty$ as $n \to \infty$ and

$$\sum_n \lambda_n \, \mu_n = \sum_{j=0}^{\infty} \sum_{n=n_j+1}^{n_{j+1}} \lambda_n \, \mu_n = \sum_{j=0}^{\infty} 2^j \sum_{n=n_j+1}^{n_{j+1}} \lambda_n$$
$$\le \sum_{j=0}^{\infty} 2^j \frac{\Lambda}{2^{2j}} = \Lambda \sum_{j=0}^{\infty} 2^{-j} = 2\Lambda \,.$$

169. We use the conventions that $\ln(0) = -\infty$ and $\int_X g \, d\mu = -\infty$ mean that $\int_X g^+ \, d\mu < \infty$ and $\int_X g^- \, d\mu = \infty$. First, by calculus $\ln(x) \le (x - 1)$ for $x \ge 0$, and so $\int_X \ln(f) \, d\mu \le \int_X (f - 1) \, d\mu < \infty$. Thus $\int_X \ln(f) \, d\mu$ is finite or $-\infty$. In the latter case the inequality holds trivially. In the former case $\{f(x) = 0\}$ has measure 0 and so modifying f on a set of measure 0 we may assume that f is everywhere positive. Finally, by Jensen's inequality $\exp(\int_X \ln(f) \, d\mu) \le \int_X e^{\ln(f)} \, d\mu = \int_X f \, d\mu$ and the conclusion follows by taking ln on both sides.

170. Note that since $\varphi(x)$ is finite a.e.,

$$|G(t) - G(s)| \leq \int_{[0,1]} \big||\varphi(x) - t| - |\varphi(x) - s|\big|\, dx \leq |t - s|$$

and G is uniformly continuous on $\mathbb{R}$.

Next, for $t \in \mathbb{R}$ and $h \neq 0$,

$$\frac{G(t + h) - G(t)}{h} = \int_0^1 \frac{|\varphi(x) - t - h| - |\varphi(x) - t|}{h}\, dx.$$

Let $\{h_n\}$ be a strictly positive sequence decreasing to 0. Then, for $x \in [0, 1]$,

$$\frac{|\varphi(x) - t - h_n| - |\varphi(x) - t|}{h_n} \to \chi_{[\varphi(x),\infty)}(t) - \chi_{(-\infty,\varphi(x))}(t)$$

as $n \to \infty$. Now, since $\big||\varphi(x) - t - h_n| - |\varphi(x) - t|\big| \leq h_n$, by the LDCT, G has a right-hand derivative at t equal to $|\{\varphi \leq t\}| - |\{\varphi > t\}|$; similarly, G has a left-hand derivative at t equal to $|\{\varphi < t\}| - |\{\varphi \geq t\}|$. Therefore G is differentiable at t iff $|\{\varphi = t\}| = 0$.

Chapter 15

L^p **Spaces**

Solutions

1. Yes. Let $X = [0,1]$ and put $f = \chi_V - \chi_{I \setminus V}$ where V is a Vitali Lebesgue nonmeasurable subset of I. Then $f^2 = \chi_I$ is measurable with finite integral but since f is not measurable, $f \notin L^2(X)$.

4. (a) Since for $a > 0$

$$T_a(\lambda) = \begin{cases} \max(-a, \lambda), & \lambda < 0, \\ \min(a, \lambda), & \lambda \geq 0, \end{cases}$$

is continuous, $f_n = T_n \circ f$ is measurable. Now, since $L^p(X)$ functions are finite μ-a.e., the f_n are measurable functions that tend to f μ-a.e. Moreover, $|f_n(x)|^p \leq |f(x)|^p \in L^1(X)$ and the convergence assertion follows from the LDCT.

(b) Since $|f_{n+1}(x)| \geq |f_n(x)|$, $\{|f_n|\}$ is nondecreasing. Therefore, if $f \notin L^p(X)$, $\|f_n\|_p^p \geq \int_{\{|f| \leq n\}} |f|^p \, d\mu \to \infty$ as $n \to \infty$ and the MCT applied to $\{|f|^p \chi_{\{|f| \leq n\}}\}$ gives the conclusion in this case.

5. (a) implies (b) Since $2^{np} \chi_{A_n}(x) \leq |f(x)|^p \chi_{A_n}(x)$, summing over an arbitrary finite $F \subset \mathbb{Z}$ and integrating, $\sum_{n \in F} 2^{np} \mu(A_n) \leq \int_X |f|^p \, d\mu$. Thus taking the sup over F, $I_p = \sum_{n=-\infty}^{\infty} 2^{np} \mu(A_n) \leq \int_X |f|^p \, d\mu$.

6. First, $\|f\|_p^p = p \int_0^1 \lambda^{p-1} \mu(\{|f| > \lambda\}) \, d\lambda + p \int_1^\infty \lambda^{p-1} \mu(\{|f| > \lambda\}) \, d\lambda = I + J$, say. Next, note that $I = p \sum_n \int_{1/n+1}^{1/n} \lambda^{p-1} \mu(\{|f| > \lambda\}) \, d\lambda = \sum_n I_n$, say. To estimate the I_n observe that since $\mu(\{|f| > \lambda\})$ decreases, I_n is bounded below by $(n^{-p} - (n+1)^{-p}) \mu(\{|f| > 1/n\})$ and above by $(n^{-p} - (n+1)^{-p}) \mu(\{|f| > 1/(n+1)\})$. Now, by calculus

$$\left(\frac{1}{n^p} - \frac{1}{(n+1)^p} \right) \sim \frac{1}{n^{p+1}}, \quad n \geq 1,$$

and so $p \int_0^1 \lambda^{p-1} \mu(\{|f| > \lambda\}) \, d\lambda = \sum_n I_n \sim \sum_n n^{-(p+1)} \mu(\{|f| > 1/n\})$ and the conclusion follows for this term. The proof for J is similar.

Note that when $\mu(X) < \infty$, $I \le \mu(X)$ is automatically finite and, therefore, $f \in L^p(X)$ iff $\sum_n n^{p-1} \mu(\{|f| > n\}) < \infty$. Along similar lines, when $0 < p < \infty$ and $\mu(X) < \infty$, $f \in L^p(X)$ iff

$$\sum_n n^{p-1} \ln^p(n) \mu(\{|f| > n \ln(n)\}) < \infty.$$

8. Since $f(x) \sim x^{-\alpha}$ near the origin and $f(x) \sim x^{-\beta}$ at infinity we need $\alpha p < 1$ and $\beta p > 1$, or $\beta^{-1} < p < \alpha^{-1}$.

9. First, (a). By symmetry we may restrict ourselves to the first quadrant. Let $A = \int_0^1 \int_0^1 \int_0^1 (x_1^2 + x_2^4 + x_3^6)^{-p} \, dx_1 \, dx_2 \, dx_3$. The change of variables $x_1 = u, x_2 = v^{1/2}, x_3 = w^{1/3}$ with Jacobian $J = 6^{-1} v^{-1/2} w^{-2/3}$, followed by passing to spherical coordinates transforms the integral into

$$A = \int_0^1 \int_0^1 \int_0^1 \frac{1}{(u^2 + v^2 + w^2)^p} \frac{1}{6 v^{1/2} w^{2/3}} \, dw \, dv \, du$$

$$= \int_0^{\pi/2} \int_0^{\pi/2} \int_0^1 \frac{\rho^2 \sin \phi}{6 \rho^{2p} (\rho \sin \phi \sin \theta)^{1/2} (\rho \cos \phi)^{2/3}} \, d\rho \, d\theta \, d\phi$$

$$= \frac{1}{6} \left(\int_0^{\pi/2} \frac{1}{(\sin \theta)^{1/2}} \, d\theta \right) \left(\int_0^{\pi/2} \frac{(\sin \phi)^{1/2}}{(\cos \phi)^{2/3}} d\phi \right) \left(\int_0^1 \rho^{5/6 - 2p} \, d\rho \right).$$

The integrals involving θ (since $\sin \theta \sim \theta$ near 0) and ϕ (since $\cos \phi \sim 1 - \phi^2/2$ near $\pi/2$) are finite. So the finiteness of A depends on the integral with respect to ρ and this integral is finite iff $5/6 - 2p > -1$, or $0 < p < 11/12$.

A similar reasoning applies to (b) except that now we must consider $\lim_{r \to \infty} \int_1^r \rho^{5/6 - 2p} \, d\rho$, which is finite iff $5/6 - 2p < -1$, or $p > 11/12$.

10. $\int_{\{|x|<1\}} (1 - |x|)^p \, dx = \sum_{k=0}^\infty \int_{\{1-2^{-k}<|x|<1-2^{-(k+1)}\}} (1 - |x|)^p \, dx = \sum_{k=0}^\infty I_k$, say. Now, on each I_k we have $2^{-(k+1)p} \le (1 - |x|)^p \le 2^{-kp}$. Furthermore,

$$|\{1 - 2^{-k} < |x| < 1 - 2^{-(k+1)}\}|$$
$$= c_n ((1 - 2^{-(k+1)})^n - (1 - 2^{-k})^n) \sim c_n n 2^{-k} (1 - 2^{-k})^{n-1} \sim 2^{-k},$$

and, consequently, $I_k \sim 2^{-k} 2^{-kp} \sim 2^{-k(1+p)}$. Therefore $\sum_k I_k < \infty$ iff $1 + p > 0$, or $p > -1$. Note that the range of p is independent of the dimension n.

11. For f apply an argument similar to the proof of Problem 7 with $\varphi(|x|) = |x|^{-\alpha} \ln^\gamma(1/|x|) \chi_{\{|x| \le 1\}}(x)$; since $\sum_k 2^{-k(n-\alpha p)} k^{\gamma p} < \infty$ when $\alpha p < n$, f is in $L^p(\mathbb{R}^n)$ if $p < n/\alpha$. In particular, if $\alpha = 0$, $\ln(x) \in L^p([0,1])$ for $0 < p < \infty$. As for g, consider the function $\varphi(|x|) = |x|^{-\alpha} \ln^{-\gamma}(1/|x|) \chi_{\{|x| \le 1\}}(x)$.

Since $\sum_k 2^{-k(n-\alpha p)}k^{-\gamma p} < \infty$ when $p < n/\alpha$ or $p \leq n/\alpha$ and $p > 1/\gamma$, $g \in L^p(\mathbb{R}^n)$ for those p.

12. For f apply an argument similar to the proof of Problem 7 with $\varphi(|x|) = |x|^{-\beta} \ln^\gamma(|x|)\chi_{\{|x|>1\}}(x)$.

13. In the setting of the Lebesgue measure on $\mathbb{R}^n$ consider the function $f(x) = |x|^{-n/p}(\ln^{-2/p}(1/|x|)\chi_{\{|x|\leq 1\}} + \ln^{-2/p}(|x|)\chi_{\{|x|>1\}})$. Then by Problems 11 and 12, $f \in L^p(\mathbb{R}^n)$ and $f \notin L^q(\mathbb{R}^n)$ for $0 < q \leq \infty$, $q \neq p$.

14. Since $|f|^p \in L^1$ the conclusion follows from Problem 4.73.

15. For $0 < p < q$. Note that f may fail to be in $L^q(X)$ as the function $f(x) = -1 + 1/x^{1/q}$ in $[0,1]$ shows. In this case $|\{f \geq t\}| = |\{x \leq 1/(1+t)^q\}| = 1/(1+t)^q < 1/(1+t^q)$ but $f \notin L^q([0,1])$.

17. Observe that

$$\|f\|_p^p = \int_{\{|f|\leq\lambda\}} |f|^p \, d\mu + \int_{\{|f|>\lambda\}} |f|^p \, d\mu$$
$$\leq \lambda^p \, \mu(\{|f| \leq \lambda\}) + \|f\|_q^p \, \mu(\{|f| > \lambda\})^{1-p/q}.$$

The conclusion follows from the concavity of $t^{1/p}$.

18. Let $A_{k,n} = f^{-1}([k/2^n, (k+1)/2^n))$, $n \geq 1$, and $k = 0, \ldots, 2^n-1$. For $n \geq 1$, let $f_n(x) = \sum_{k=0}^{2^n-1} k^2 4^{-n}\chi_{A_{k,n}}(x)$; $\{f_n\}$ is a sequence of nonnegative measurable functions that increase to f^2 for all $x \in [0,1]$. Furthermore,

$$\int_X f_n \, d\mu = \sum_{k=0}^{2^n-1} \frac{k^2}{8^n} = \frac{2^n(2^n - 1)(2^{n+1} - 1)}{6 \cdot 8^n},$$

and so by the MCT,

$$\int_X f^2 \, d\mu = \lim_n \frac{2^n(2^n - 1)(2^{n+1} - 1)}{6 \cdot 8^n} = \frac{1}{3}.$$

19. (a) Let $A = \{f \geq 0\}$ and $B = \{f < 0\}$. Then for $0 < t < \varepsilon$, $e^{t|f(x)|} = e^{tf(x)}\chi_A(x) + e^{-tf(x)}\chi_B(x)$, and so $\int_X e^{t|f|} \, d\mu \leq \int_A e^{tf} \, d\mu + \int_B e^{-tf} \, d\mu \leq \int_X e^{tf} \, d\mu + \int_X e^{-tf} \, d\mu < \infty$. A similar result holds for $-\varepsilon < t < 0$, and, consequently, $\int_X e^{t|f|} \, d\mu < \infty$ for $|t| < \varepsilon$. In particular, when $t > 0$, $e^{t|f|} < \infty$ μ-a.e. and so $|f| < \infty$ μ-a.e. Therefore the series defining $e^{t|f|}$ is finite μ-a.e. and by the MCT it can be integrated termwise and we have

$$\sum_{k=0}^\infty \frac{1}{k!} t^k \int_X |f|^k \, d\mu = \int_X e^{t|f|} \, d\mu < \infty.$$

Hence $f \in L^k(X)$ for all integers k and by Hölder's inequality $f \in L^p(X)$ for all $0 < p < \infty$. The function $f(x) = \ln(1/|x|)$ in $[0,1]$, $0 < \varepsilon < 1$, shows that f is not necessarily bounded.

(b) Taking the limit under the integral sign,

$$M_f'(t) = \int_X \lim_{h \to 0} \frac{e^{hf(x)} - 1}{h} \, e^{tf(x)} \, d\mu(x) = \int_X f(x) \, e^{tf(x)} \, d\mu(x).$$

The proof for higher order derivatives proceeds by induction, $M_f^k(t) = \int_X f^k \, e^{tf} \, d\mu$, and by the LDCT, $\lim_{t \to 0} M_f^k(t) = \int_X f(x)^k \, d\mu(x)$.

20. (b) Suppose first that f is continuous. Then for all $x \in [0,1]$, $\lim_n \pi_n f(x) = f(x)$ and $|\pi_n f(x)|^p \leq \|f\|_\infty < \infty$, and so by the LDCT, $\lim_n \pi_n(f) = f$ in $L^p(I)$. In the general case, given $\varepsilon > 0$, let g be a continuous function such that $\|f - g\|_p \leq \varepsilon$. Then by (a), $\|f - \pi_n(f)\|^p \leq \|f - g\|_p + \|g - \pi_n(g)\|_p + \|\pi_n(f - g)\|_p \leq 2\varepsilon + \|g - \pi_n(g)\|_p$, and since g is continuous $\|g - \pi_n(g)\|_p < \varepsilon$ for sufficiently large n.

21. First, by Minkowski's inequality and the translation invariance of the Lebesgue measure, $\|g_k\|_p \leq \int_{\mathbb{R}^n} |f_k(y)| \, \|g(\cdot - y)\|_p \, dy \leq c \, \|g\|_p$ for all k. Next, since the f_k have integral η,

$$\eta \, g(x) - g_k(x) = \int_{\mathbb{R}^n} f_k(y) \, (g(x) - g(x - y)) \, dy$$

and by Minkowski's inequality,

$$\|\eta \, g - g_k\|_p \leq \int_{\mathbb{R}^n} |f_k(y)| \, \|g(\cdot) - g(\cdot - y)\|_p \, dy.$$

Now, since $L^p(\mathbb{R}^n)$ functions are continuous in the L^p metric, given $\varepsilon > 0$, let k be large enough so that $\|g(\cdot) - g(\cdot - y)\|_p \leq \varepsilon$ for $|y| < 1/k$. Then for those k, $\|g - g_k\|_p \leq \varepsilon \int_{B(0,1/k)} |f_k(y)| \, dy \leq \varepsilon \, c$.

For example, for $\eta > 0$, in $\mathbb{R}^n$ one may consider $f_k(x) = \eta \, k^n \, \chi_{[0,1/k]^n}(x)$, $k \in \mathbb{N}$. As for $\eta = 0$, $f_k(x) = k^n \left[-\chi_{[-1/k,0]^n}(x) + \chi_{[0,1/k]^n}(x) \right]$, $k \in \mathbb{N}$, will do.

22. The limit is $\|f\|_p$. Suppose first that f is continuous. Then by elementary properties of the Riemann integral, which coincides with the Lebesgue integral since f is continuous, there are $x_{n,k} \in [k/n, (k+1)/n)$ such that

$$n^{p-1} \sum_{k=0}^{n-1} \left(\int_{[k/n,(k+1)/n)} f(y) \, dy \right)^p = \frac{1}{n} \sum_{k=0}^{n-1} f(x_{n,k})^p,$$

and, consequently, the limit in question is

$$\lim_n \left(\frac{1}{n} \sum_{k=0}^{n-1} f(x_{n,k})^p \right)^{1/p} = \left(\int_I f(y)^p \, dy \right)^{1/p}.$$

Next, given $f \in L^p(I)$, let g be continuous with $\|f - g\|_p \le \varepsilon/3$. Then, since $|\,\|a_n\|_{\ell^p} - \|b_n\|_{\ell^p}\,| \le \|a_n - b_n\|_{\ell^p}$, we get

$$\left| \left(n^{p-1} \sum_{k=0}^{n-1} \left(\int_{[k/n,(k+1)/n)} f(y)\,dy \right)^p \right)^{1/p} \right.$$

$$\left. - \left(n^{p-1} \sum_{k=0}^{n-1} \left(\int_{[k/n,(k+1)/n)} g(y)\,dy \right)^p \right)^{1/p} \right|$$

$$\le \left(n^{p-1} \sum_{k=0}^{n-1} \left(\int_{[k/n,(k+1)/n)} [f(y) - g(y)]\,dy \right)^p \right)^{1/p},$$

which by Hölder's inequality is dominated by

$$\left(\sum_{k=0}^{n-1} \int_{[k/n,(k+1)/n)} |f(y) - g(y)|^p\,dy \right)^{1/p} = \left(\int_I |f - g|^p\,dy \right)^{1/p} \le \varepsilon/3.$$

Also, $|\,\|f\|_p - \|g\|_p\,| \le \|f - g\|_p \le \varepsilon/3$. The conclusion follows by combining these estimates.

24. Since $\mathcal{M} = \mathcal{P}(X)$ all functions f on X, which are given by the two values $f(a), f(b)$, are measurable. Now, since $\mu(\{b\}) = \infty$, $L^p(X) = \{f : |f(a)| < \infty, f(b) = 0\}$ for $1 \le p < \infty$ and $\|f\|_p = |f(a)|$. Furthermore, since there are no nontrivial subsets of X of measure 0, $L^\infty(X)$ consists of everywhere bounded functions; that is, $f \in L^\infty(X)$ iff $|f(a)|, |f(b)| < \infty$, and so $L^\infty(X)$ is isomorphic to $\mathbb{R}^2$ endowed with the sup norm.

As for the linear functionals L on $L^p(X)$, $1 \le p < \infty$, they are given by $L(f) = \lambda f(a)$ for some real λ and $\|L\| = |\lambda|$. When $1 < p < \infty$, pick g with $g(a) = \lambda$; then $L(f) = \int_X fg\,d\mu$ and $\|L\| = \|g\|_q$, $1/p + 1/q = 1$.

On the other hand, $L^\infty(X)$ contains properly the dual of $L^1(X)$. Indeed, since $L^1(X)$ is isomorphic to $\mathbb{R}$ its dual is 1-dimensional but $L^\infty(X)$ is isomorphic to $\mathbb{R}^2$ with the sup norm. Actually, the dual of $L^1(X)$ is given by a quotient space of $L^\infty(X)$ under the equivalence relation $f \sim g$ whenever $f(a) = g(a)$.

It may be the case that $\mu(\{a\}) = \mu(\{b\}) = \infty$; then $L^1(X)$ is the trivial space and $L^\infty(X)$ still is $\mathbb{R}^2$. Or $\mu(\{a\}) = 0$ and $\mu(\{b\}) = \infty$. Then, although $f(a)$ is still arbitrary, $\{a\}$ has measure 0 and so the only integrable function is the 0 function and the dual is $\{0\}$. As for $L^\infty(X)$, it is still the same; because of the μ-a.e. equivalence it consists of an arbitrary value $g(a)$ and so the equivalence class is independent of the value of $g(b)$.

25. Not always. In the setting of Problem 24 any $f \in L^p(X)$ verifies $f(b) = 0$, and so if g is given by $g(a) = 1$ and $g(b) = \infty$, then $fg \in$

$L^p(X)$ whenever $f \in L^p(X)$ yet $g \notin L^\infty(X)$. The situation is different if μ is semifinite. For the sake of argument suppose that $g \notin L^\infty(X)$. Then $\{|g| > 2^n\}$ has positive measure for all n and since $\{|g| > 2^n\} = \bigcup_{k=n}^\infty \{2^k < |g| \le 2^{k+1}\}$, there exists a sequence $n_k \to \infty$ such that if $B_k = \{2^{n_k} \le |g| < 2^{n_k+1}\}$, then $\mu(B_k) > 0$. If $\mu(B_k) < \infty$, let $A_k = B_k$, otherwise let $A_k \subset B_k$ have positive finite measure; the A_k are pairwise disjoint. Now put $f(x) = \sum_k 2^{-n_k} \mu(A_k)^{-1/p} \chi_{A_k}(x)$. Then $\|f\|_p^p = \sum_k 2^{-n_k p} < \infty$ and $\int_X |fg|^p \, d\mu = \sum_k 2^{-n_k p} \mu(A_k)^{-1} \int_{A_k} |g|^p \, d\mu \ge \sum_k 1 = \infty$, which is not the case. Hence $g \in L^\infty(X)$.

We claim that $S = \sup\{\|fg\|_p : \|f\|_p = 1\} = \|g\|_\infty$. First, since $\|fg\|_p \le \|g\|_\infty \|f\|_p$, $S \le \|g\|_\infty$. To see that there is equality let $B_n = \{|g| > \|g\|_\infty - 1/n\}$; by the definition of the L^∞ norm, $\mu(B_n) > 0$ for all n. If $\mu(B_n) < \infty$, let $A_n = B_n$; otherwise, there exists $A_n \subset B_n$ with $0 < \mu(A_n) < \infty$. Now put $f_n(x) = \mu(A_n)^{-1/p} \chi_{A_n}(x)$; note that $\|f_n\|_p = 1$ for all n. Then $\|gf_n\|_p^p = \mu(A_n)^{-1} \int_{A_n} |g|^p \, d\mu \ge (\|g\|_\infty - 1/n)^p$ for all n and so $S \ge \|g\|_\infty$.

26. First, necessity. For the sake of argument suppose that X contains a subset A with $\mu(A) = \infty$ such that all its subsets have measure 0 or ∞; clearly $\chi_A \notin L^p(X)$. Now, if $g \in L^q(X)$, $B = \{g \ne 0\}$ is σ-finite and so $\mu(A \cap B) = 0$. Hence $g\chi_A = 0$ μ-a.e. and is therefore integrable. Thus $\chi_A \in L^p(X)$, which is not the case.

Next, sufficiency. We first prove that if a measurable function f on X with $\{f \ne 0\}$ σ-finite is such that its product with every $L^q(X)$ function is integrable, then independent of any condition on μ, $f \in L^p(X)$. Indeed, let $\{f_n\}$ be bounded functions vanishing outside a set of finite measure such that $\{|f_n|^p\}$ increases monotonically to $|f|^p$. The f_n determine bounded linear functionals L_n on $L^q(X)$, say, given by $L_n(g) = \int_X f_n g \, d\mu$, $g \in L^q(X)$, with norm $\|L_n\| = \|f_n\|_p$. Now, for any $g \in L^q(X)$, $|L_n(g)| \le \int_X |fg| \, d\mu < \infty$ because $|f_n| \le |f|$ and fg is integrable by assumption, and so $\{L_n(g)\}$ is bounded. Therefore by the uniform boundedness principle $\{\|L_n\|\}$ is bounded and, consequently, $\int_X |f_n|^p \, d\mu \le M$, say, and since $\{|f_n|^p\}$ increases to $|f|^p$, by the MCT $\int_X |f|^p \, d\mu \le M < \infty$.

Next, we claim that if f is any function whose product with every $L^q(X)$ function is integrable, then $\{f \ne 0\}$ is σ-finite. In fact, let $A_n = \{|f| > 1/n\}$, $n = 1, 2, \ldots$, and for the sake of argument suppose that $\mu(A_n) = \infty$ for some n. Then by Problem 2.79(b), A_n contains a σ-finite subset B such that $\mu(B) = \infty$. Now, $f\chi_B$ satisfies the condition of the above paragraph and so belongs to $L^p(X)$, but this cannot happen since $|f| \ge 1/n$ on B and $\mu(B) = \infty$.

Finally note that not all measures that satisfy the condition are σ-finite: If X is an uncountable set we assign to every finite subset of X a measure equal to the number of points in that set and to all the other sets we assign the measure ∞.

28. The statement is true when $p < \infty$. The situation is different for $p = \infty$. Let $f = \chi_{\mathbb{R}}$ and denote by $B(f, 1)$ the ball in $L^\infty(\mathbb{R})$ centered at f of radius 1. Now, if $g \in B(f, 1)$, then $|1 - g(x)| = |f(x) - g(x)| \leq \|f - g\|_\infty \leq 1$ a.e. and, therefore, $g(x) \geq 0$ a.e. Thus $g \in \mathcal{F}$ and f is an interior point of $\mathcal{F}$.

29. First, necessity. For the sake of argument suppose that

$$\limsup_{|t| \to \infty} |\varphi(t)|^q / |t|^p = \infty$$

and let $|t_k| \to \infty$ be such that $|\varphi(t_k)|^q \geq 2^k |t_k|^p$. Let $\{A_k\}$ be pairwise disjoint measurable subsets of X with $\mu(X)/2^{k+1}|t_k|^p \leq \mu(A_k) \leq \mu(X)/2^k|t_k|^p$ and $f(x) = \sum_k t_k \chi_{A_k}(x)$. Then $|f(x)|^p = \sum_k |t_k|^p \chi_{A_k}(x)$ and, consequently, $\int_X |f|^p \, d\mu \leq \mu(X) \sum_k |t_k|^p/2^k|t_k|^p \leq \mu(X) < \infty$. Now, $|\varphi(f)|^q = |\varphi(t_k)|^q \geq 2^k |t_k|^p$ on A_k. Moreover, since $\mu(A_k) \geq \mu(X)/2^{k+1}|t_k|^p$ it follows that $\|\varphi(f)\|_q^q \geq \sum_k 2^k |t_k|^p/2^{k+1}|t_k|^p = \infty$.

As for sufficiency, since φ is continuous there is a constant $K > 0$ such that $|\varphi(t)|^q \leq K(1 + |t|)^p$ for all $t \in \mathbb{R}$. Thus, if $f \in L^p(X)$, $\varphi(f) \in L^q(X)$.

31. Sufficiency holds. First, note that for $k \geq 0$, $2^{-kr} - 2^{-(k+1)r} = c\, 2^{-kr}$ where $c = (1 - 2^{-r})$. Then $I = c \sum_{k=0}^\infty (2^{-kr} - 2^{-(k+1)r}) \int_{\{|f| \leq 2^k\}} |f|^q \, d\mu = \sum_{k=0}^\infty 2^{-kr} \int_{\{|f| \leq 2^k\}} |f|^q \, d\mu - \sum_k 2^{-kr} \int_{\{|f| \leq 2^{k-1}\}} |f|^q \, d\mu = \int_{\{|f| \leq 1\}} |f|^q \, d\mu + \sum_k 2^{-kr} \int_{\{2^{k-1} < |f| \leq 2^k\}} |f|^q \, d\mu$, which in turn is bounded by $\int_{\{|f| \leq 1\}} |f|^p \, d\mu + \sum_k 2^{-kr} 2^{k(q-p)} \int_{\{2^{k-1} < |f| \leq 2^k\}} |f|^p \, d\mu$, which, since $q - p = r$, is equal to $c \|f\|_p^p$.

To see that necessity does not hold let $f(x) = x^{-1/p}\chi_{[1,\infty)}(x)$. Then, since $f \in L^q(\mathbb{R})$, for $q > p$, $\sum_{k=0}^\infty 2^{-kr} \int_{\{|f| \leq 2^k\}} |f|^q \, d\mu \leq \|f\|_q \sum_{k=0}^\infty 2^{-kr} < \infty$ for all $r > 0$ yet $f \notin L^p(\mathbb{R})$.

32. (a) iff (b) Since $|f|^p \in L^1(X)$ it follows from Problem 4.52.

(a) iff (c) First, since $\alpha > 1$, $\sum_{n > \eta} n^{-\alpha} \sim \eta^{1-\alpha}$, all $\eta > 0$. Now, the series in (c) can be written $\int_X |f|^\gamma (\sum_{|f|^{1/\beta} \leq n} n^{-\alpha}) \, d\mu \sim \int_X |f|^\gamma (|f|^{1/\beta})^{1-\alpha} \, d\mu \sim \int_X |f|^p \, d\mu$ and the equivalence follows.

33. (a) By the translation invariance of the Lebesgue integral τ_y is bounded in $L^p(\mathbb{R}^n)$ with norm 1.

(b) Since $|\int_{\mathbb{R}^n} |\tau_h f(x) - f(x)|^p \varphi(x) \, dx| \leq \|\varphi\|_\infty \int_{\mathbb{R}^n} |\tau_h f(x) - f(x)|^p \, dx$ we may assume that $\varphi = 1$. First, suppose that f is compactly supported and continuous; then $\lim_{|h| \to 0} \sup_{x \in \mathbb{R}^n} |f(x + h) - f(x)| = 0$ a.e. and f is uniformly continuous, which by the LDCT gives the desired result. Now, for arbitrary f, given $\varepsilon > 0$, let g be a continuous function with support in

$B(0, R)$, say, such that $\|f - g\|_p < \varepsilon$ and let $0 < \delta < 1$ be such that, if $|h| < \delta$, $\|g - \tau_h g\|_\infty \leq \varepsilon/(1 + R)^{n/p}$. Moreover, since $\operatorname{supp}(g - \tau_h g) \subset B(0, R + 1)$ it follows that $\int_{\mathbb{R}^n} |g(x) - \tau_h g(x)|^p \, dx \leq \|g - \tau_h g\|_\infty^p |B(0, R + 1)| \leq \varepsilon^p$. Also $\int_{\mathbb{R}^n} |\tau_h f(x) - \tau_h g(x)|^p \, dx \leq \|f - g\|_p^p < \varepsilon^p$. Thus, if $|h| < \delta$, $\|\tau_h f - f\|_p \leq \|\tau_h f - \tau_h g\|_p + \|\tau_h g - g\|_p + \|g - f\|_p \leq 3\varepsilon$.

(c) Observe that if $g \in L^p(\mathbb{R}^n)$ has compact support, $\tau_y g$ and g have disjoint supports for $|y|$ sufficiently large and so by the translation invariance of the Lebesgue integral $\|\tau_y g - g\|_p^p = 2\|g\|_p^p$. Now, given $f \in L^p(\mathbb{R}^n)$ and $\varepsilon > 0$, let g be compactly supported and satisfy $\|f - g\|_p < \varepsilon$. Then $|\,\|\tau_y f - f\|_p - \|\tau_y g - g\|_p\,| \leq \|(\tau_y f - f) - (\tau_y g - g)\|_p \leq \|\tau_y(f - g)\|_p + \|f - g\|_p \leq 2\varepsilon$. Finally, let $|y|$ be large enough so that $\|\tau_y g - g\|_p = 2^{1/p}\|g\|_p$. Then $|\,\|\tau_y f - f\|_p - 2^{1/p}\|f\|_p\,| = |\,\|\tau_y f - f\|_p - 2^{1/p}\|f\|_p - \|\tau_y g - g\|_p + 2^{1/p}\|g\|_p\,| \leq |\,\|\tau_y f - f\|_p - \|\tau_y g - g\|_p\,| + 2^{1/p}|\,\|g\|_p - \|f\|_p\,| \leq 2\varepsilon + 2^{1/p}\varepsilon$, which implies that $\lim_{|y| \to \infty} \|\tau_y f - f\|_p = 2^{1/p}\|f\|_p$.

As for the weak convergence, fix $\psi \in L^q(\mathbb{R}^n)$; we claim that for compactly supported $g \in L^p(\mathbb{R}^n)$, $\lim_{|y| \to \infty} \int_{\mathbb{R}^n} \tau_y g(x)\psi(x) \, dx = 0$. First, given $\varepsilon > 0$, there exists a ball $B = B(0, R)$ such that $\int_{B^c} |\psi(x)|^q \, dx \leq \varepsilon^q$. Suppose $\operatorname{supp}(g) \subset B(0, M)$ and note that if $|x| < M$ and $|y| > M + R$, then $|x - y| \geq |y| - |x| > R$. Therefore

$$\left| \int_{\mathbb{R}^n} \tau_y g(x)\psi(x) \, dx \right| = \left| \int_{\mathbb{R}^n} g(x)\psi(x - y) \, dx \right|$$

$$\leq \|g\|_p \left(\int_{B^c} |\psi(x)|^q \, dx \right)^{1/q} \leq \varepsilon \|g\|_p;$$

since ε is arbitrary it follows that $\lim_{|y| \to \infty} \int_{\mathbb{R}^n} \tau_y g(x)\psi(x) \, dx = 0$. Next, for arbitrary $f \in L^p(\mathbb{R}^n)$, let $\{g_k\}$ be compactly supported functions such that $\|f - g_k\|_p \to 0$. Then

$$\left| \int_{\mathbb{R}^n} \tau_y f(x)\psi(x) \, dx \right| \leq \left| \int_{\mathbb{R}^n} \tau_y(f - g_k)(x)\psi(x) \, dx \right|$$

$$+ \left| \int_{\mathbb{R}^n} \tau_y g_k(x)\psi(x) \, dx \right|,$$

where the second summand goes to 0 with $|y|$ and, by Hölder's inequality and the translation invariance of the Lebesgue integral, the first summand goes to 0 as $k \to \infty$ uniformly in y. Therefore the limit is 0. Now, since the expression converges weakly to 0, the strong limit, if it exists, must be 0 but this is not possible since $\|\tau_y f\|_p = \|f\|_p$.

34. $\mathcal{F}$ is bounded since a sequence $\{f_n\}$ such that $\|f_n\|_p \to \infty$ has no convergent subsequence. For the sake of argument suppose there exist $\varepsilon > 0$, $\{f_n\} \subset \mathcal{F}$, and $|h_n| \to 0$ such that $\|\tau_{h_n} f_n - f_n\|_p \geq \varepsilon$ for all n. Passing to a subsequence if necessary we may assume that $\{f_n\}$ converges to some

$f \in L^p(\mathbb{R}^n)$ in $L^p(\mathbb{R}^n)$. Then

$$\tau_{h_n} f - f = (\tau_{h_n} f - \tau_{h_n} f_n) + (\tau_{h_n} f_n - f_n) + (f_n - f)$$

where

$$\lim_n \|\tau_{h_n} f - \tau_{h_n} f_n\|_p = \lim_n \|f - f_n\|_p = 0.$$

Thus, since $\|\tau_{h_n} f_n - f_n\|_p \geq \varepsilon$, it follows that, contrary to Problem 33(b), $\liminf_n \|\tau_{h_n} f - f\|_p \geq \varepsilon$.

35. First, necessity. Note that with $M = \text{Wk-}L^p(X)$ norm of f, $\mu(\{x \in A : |f(x)| > \lambda\}) \leq \min(M^p \lambda^{-p}, \mu(A))$ and, consequently $\int_A |f|^r \, d\mu \leq r \int_0^\infty \min(M^p \lambda^{-p}, \mu(A)) \, \lambda^{r-1} \, d\lambda \leq \mu(A) \, \eta^r + M^p(r/p - r) \, \eta^{-p+r}$ for $\eta > 0$. Minimizing, or picking η so that the summands are equal, it follows that $\eta \sim \mu(A)^{-1/p}$, which gives the desired estimate.

Next, sufficiency. Given $\lambda > 0$, let $A \subset \{|f| > \lambda\}$ with $\mu(A) < \infty$. By Chebychev's inequality and assumption, $\lambda^r \mu(A) \leq \int_A |f|^r \, d\mu \leq M \, \mu(A)^{1-r/p}$. Thus, clearing out, $\lambda^p \mu(A) \leq M^{p/r}$ for all $A \subset \{|f| > \lambda\}$ with finite measure. Since μ is σ-finite there is a sequence $\{X_n\}$ of sets with finite measure that increases to X and so $\{x \in X_n : |f(x)| > \lambda\}$ increases to $\{x \in X : |f(x)| > \lambda\}$. By the above argument $\lambda^p \mu(\{x \in X_n : |f(x)| > \lambda\}) \leq M^{p/r}$ and by continuity from below, $\lambda^p \mu(\{x \in X : |f(x)| > \lambda\}) \leq M^{p/r}$. Hence $f \in \text{Wk-}L^p(X)$.

The estimate does not imply that $f \in L^p(X)$. Indeed, with $1 = r < p$, let $X = \mathbb{R}$, $f(x) = |x|^{-1/p}$. Then, if $|A| = a$ it is clear that $\int_A f(x) \, dx \leq \int_{[-a/2, a/2]} f(x) \, dx \sim a^{1/q}$ but $f \notin L^p([0,1])$.

36. (a) First, suppose that f is bounded. Then with q the conjugate to p, $\mu(\{|f| > \lambda\})^{1/q} \leq \|f\|_p^{p/q} / \lambda^{p/q}$ and so

$$\|f\|_p^p = p \int_0^\infty \lambda^{p-1} \mu(\{|f| > \lambda\})^{1/p} \mu(\{|f| > \lambda\})^{1/q} \, d\lambda$$

$$\leq p \|f\|_p^{p/q} \int_0^\infty \lambda^{p-1} \lambda^{-p/q} \mu(\{|f| > \lambda\})^{1/p} \, d\lambda.$$

Hence, since p and q are conjugate and $\|f\|_p < \infty$, $\|f\|_p \leq c_p A_p(f)$. For an arbitrary $f \in L^p(X)$, let $f_n = f \chi_{\{|f| \leq n\}}$; f_n is a bounded $L^p(X)$ function and so $\|f_n\|_p \leq c_p A_p(f_n) \leq c_p A_p(f)$. Finally, since $\|f_n\|_p$ increases to $\|f\|_p$, by the MCT $\|f\|_p \leq c_p A_p(f)$.

(b) If $p < r$, $\mu(\{|f| > \lambda\})^{1/p} \leq \|f\|_r^{r/p} / \lambda^{r/p}$ and so $A_p(f)$ is bounded by

$$\mu(X) \int_0^\eta d\lambda + \|f\|_r^{r/p} \int_\eta^\infty \lambda^{-r/p} \, d\lambda = \eta \, \mu(X) + \frac{p}{r-p} \eta^{(r-p)/p} \|f\|_r^{r/p}.$$

The conclusion follows by minimizing with respect to η.

37. $d_1(f, S_1) = 0$.

Next, $d_2(f, S_2)$. S_2 is the unit sphere in $L^2([0,1])$. Let $g \in S_2$; then $\|f - g\|_2^2 = \|f\|_2^2 - 2\int_0^1 f(x)g(x)\,dx + \|g\|_2^2 = 13/12 - 2\int_0^1 f(x)g(x)\,dx \geq 13/12 - 2\|f\|_2 = 13/12 - 1/\sqrt{3}$. Equality is attained at $g = f/\|f\|_2$. Therefore the distance is $\left\|f - (f/\|f\|_2)\right\|_2 = 1 - \|f\|_2 = 1 - 1/\sqrt{12}$.

Finally, if $p = \infty$, let $\|g\|_\infty = r$. Then

$$\int_0^1 (f+g)^2(x)\,dx = \int_0^1 (f^2(x) + 2f(x)g(x) + g^2(x))\,dx$$

$$\leq \int_0^1 f^2(x)\,dx + 2r \int_0^1 |f(x)|\,dx + r^2$$

$$= 1/12 + r/2 + r^2.$$

This inequality becomes equality if $g(x) = -\chi_{[0,1/2]}(x) + \chi_{[1/2,1]}(x)$. Therefore the distance is the positive solution of the equation $1/12 + r/2 + r^2 = 1$, i.e., $r = \sqrt{47/48} - 1/4$.

40. $f(x)^2 = -2\int_x^\infty f'(y)f(y)\,dy = 2\int_{-\infty}^x f'(y)f(y)\,dy$, $x \in \mathbb{R}$. Hence $f(x)^2 \leq \int_{-\infty}^\infty |f'(y)|\,|f(y)|\,dy$ and by the Cauchy-Schwarz inequality, $|f(x)| \leq \|f\|_2\|f'\|_2$.

41. For $f \in \mathcal{F}$ and $-1 \leq x < y \leq 1$, by the Cauchy-Schwarz inequality,

$$|f(y) - f(x)| \leq \int_x^y |f'(t)|\,dt \leq |y - x|^{1/2}\left(\int_x^y |f'(t)|^2\,dt\right)^{1/2}$$

and, therefore, $\eta \leq 1$ for $0 \leq \alpha \leq 1/2$. On the other hand, we claim that $\eta = \infty$ for $1/2 < \alpha \leq 1$. To see this consider the functions $f_\varepsilon(x) = 2c_\varepsilon (x^2 + \varepsilon^2)^{1/4}$, $\varepsilon > 0$, with c_ε chosen appropriately. Then

$$\int_{-1}^1 |f_\varepsilon'(x)|^2 dx = c_\varepsilon^2 \int_{-1}^1 x^2 (x^2 + \varepsilon^2)^{-3/2}\,dx \leq 2c_\varepsilon^2 \int_0^1 (x^2 + \varepsilon^2)^{-1/2}\,dx$$

$$= 2c_\varepsilon^2 \ln\left(x + \sqrt{x^2 + \varepsilon^2}\right)\Big|_0^1 = 2c_\varepsilon^2\left(\ln(1 + \sqrt{1 + \varepsilon^2}) - \ln\varepsilon\right),$$

for small $\varepsilon > 0$ and, therefore, picking $c_\varepsilon = (-\ln\varepsilon)^{-1/2}/2$ the integral is ≤ 1 and $f_\varepsilon \in \mathcal{F}$. However, if $\alpha > 1/2$,

$$\frac{f_\varepsilon(\varepsilon) - f_\varepsilon(0)}{\varepsilon^\alpha} = \frac{1}{4}(-\ln(\varepsilon))^{-1}(2^{1/4} - 1)\varepsilon^{1/2-\alpha} \to \infty \text{ as } \varepsilon \to 0^+.$$

42. Since simple functions are dense in $L^p(\mathbb{R}^n)$ it suffices to approximate χ_A where A is a set of finite measure. Let O be an open set containing A such that $|O \setminus A| < \eta$; then $\|\chi_O - \chi_A\|_p = |O \setminus A|^{1/p} < \eta^{1/p}$. Now, since O is open, $O = N \cup \bigcup_k Q_k$ where the Q_k are pairwise disjoint open cubes and $|N| = 0$. Let $g_m = \sum_{k=1}^m \chi_{Q_k}$. Then by the LDCT (with dominating function χ_O), $\int_{\mathbb{R}^n} |\chi_O(x) - g_m(x)|^p dx \to 0$ as $n \to \infty$. Whence, by the triangle inequality $\|\chi_A - g_m\|_p \leq \|\chi_A - \chi_O\|_p + \|\chi_O - g_m\|_p \leq \eta^{1/p} + \eta$, which can be made arbitrarily small.

43. Since simple functions with rational coefficients are dense in $L^p(X)$ it suffices to approximate the characteristic function χ_A of a set A of finite measure. For this use Problem 2.48.

44. The statement is false if $X = I$, just take $f(x) = g(x) = x^{-1/4}$. On the other hand, the statement is true in ℓ^2 since $\ell^2 \subset \ell^\infty$ and, therefore, $fg \in \ell^2$.

45. Let $f(x) = g(x) = \chi_A(x)$ where $A \subset \mathbb{R}^n$ is measurable. Then the inequality implies $|A|^{1/r} \leq c|A|^{1/p+1/q}$ which, if true for $1 < |A| < \infty$, gives $1/r \leq 1/p + 1/q$, and, if true for $0 < |A| < 1$, gives $1/r \geq 1/p + 1/q$.

47. (a) For the sake of argument suppose that T is not bounded and let $\{f_n\} \subset L^p(X)$, $\|f_n\|_p = 1$, be such that $\|f_n g\|_r \geq 2^n$ for all $n \geq 1$; since $\|f_n\|_p = \| |f_n| \|_p$ and $\|f_n g\|_r = \| |f_n| g \|_r$ we may assume that $f_n \geq 0$. Let $h(x) = \sum_n 2^{-nr} f_n^r$. First, assume that $1 \leq r \leq p$ and observe that by Minkowski's inequality, $\|h\|_{p/r} \leq \sum_n 2^{-nr} \|f_n\|_p^r$ and, consequently, $\sum_n 2^{-nr} f_n^r$ converges μ-a.e. to $h \in L^{p/r}(X)$. Then $f = h^{1/r} \in L^p(X)$, $|fg|^r = \sum_n 2^{-nr} f_n^r |g|^r$, and by the MCT, $\int_X |fg|^r \, d\mu = \sum_n 2^{-nr} \|f_n g\|_r^r \geq \sum_n 2^{-nr} 2^{nr} = \infty$. Thus $T(f) = fg \notin L^r(X)$, which is not the case. Next, if $0 < p < r$, by the concavity of $\varphi(t) = t^{p/r}$, $|h|^{p/r} \leq \sum_n 2^{-np} |f_n|^p$ and, consequently, $\|h\|_{p/r}^{p/r} \leq \sum_n 2^{-np} \|f_n\|_p^p = \sum_n 2^{-np} < \infty$. Then $f = h^{1/r} \in L^p(X)$ and the conclusion follows as before.

(b) Let $\{A_n\} \subset \mathcal{M}$ be an increasing sequence of sets with finite measure such that $X = \bigcup_n A_n$ and put $g_n = \min(|g|, n)\chi_{A_n}$. Then $g_n^{q/p} \in L^p(X)$ and, therefore, $\|gg_n^{q/p}\|_r \leq \|T\|(\int_X g_n^q \, d\mu)^{1/p}$. Hence, since $g_n \leq |g|$ and $q = r + (qr/p)$, $(\int_X g_n^q \, d\mu)^{1/r} \leq (\int_X |g|^r g_n^{qr/p} \, d\mu)^{1/r} \leq \|T\|(\int_X g_n^q \, d\mu)^{1/p}$ and so $(\int_X g_n^q \, d\mu)^{1/q} = (\int_X g_m \, d\mu)^{l/r-l/p} \leq \|T\|$. Then by the MCT $(\int_X g^q \, d\mu)^{l/q} \leq \|T\|$ and $g \in L^q(X)$.

48. Since $|f| \pm f \geq 0$ it follows that $|T(f)| \leq |T(|f|)|$ and so $\|T(f)\|_q \leq \|T(|f|)\|_q$. For the sake of argument suppose that T is unbounded and let $\{g_n\} \subset L^p(\mathbb{R})$ be such that $\|g_n\|_p = 1$ and $\|T(g_n)\|_q \geq n2^n$. Now, the sequence $\{f_n\}$ with $f_n = |g_n|$, all n, satisfies $f_n \geq 0$, $\|f_n\|_p = \|g_n\|_p = 1$, and $n2^n \leq \|T(g_n)\|_q \leq \|T(|g_n|)\|_q = \|T(f_n)\|_q$. Let $f = \sum_n 2^{-n} f_n$; by Minkowski's inequality if $1 \leq p < \infty$ and the concavity of $\varphi(t) = t^p$ if $0 < p < 1$, $f \in L^p(\mathbb{R})$. Moreover, since $f \geq 2^{-n} f_n$ for all n, $T(f) \geq 2^{-n} T(f_n) \geq 0$ for all n and so $\|T(f)\|_q \geq n2^n 2^{-n} = n$ for all n. Thus, although $f \in L^p(\mathbb{R})$, $T(f) \notin L^q(\mathbb{R})$.

49. g is integrable provided $\eta < p \leq \infty$, and h is integrable for any $p > 1$ such that $\eta/2 < p \leq \infty$.

51. First, by Hölder's inequality $\int_X f\chi_{\{f>\lambda\}}\,d\mu \le \|f\|_p\,\mu(\{f>\lambda\})^{1/q}$. Note that since $f\chi_{\{f>\lambda\}} = f$ if $f > \lambda$ and $= 0$ otherwise, we have $f\chi_{\{f>\lambda\}} \ge f - \lambda$, which implies that $\int_X f\chi_{\{f>\lambda\}}\,d\mu \ge \int_X (f-\lambda)\,d\mu$ and, consequently, $\int_X (f-\lambda)\,d\mu \le \mu(\{f>\lambda\})^{1/q}(\int_X f^p\,d\mu)^{1/p}$.

52. Writing the integral of $|g|^p$ in terms of its distribution function we have

$$\int_X |g|^p\,d\mu \le p \int_0^\infty \left(\int_{\{|f|>\lambda^\beta\}} |f|^r\,d\mu \right) \lambda^{p-\alpha-1}\,d\lambda$$

$$= p \int_X |f|^r \left(\int_0^{|f|^{1/\beta}} \lambda^{p-\alpha-1}\,d\lambda \right) d\mu = \frac{p}{p-\alpha} \int_X |f|^{r+(p-\alpha)/\beta}\,d\mu\,.$$

The statement remains true if all the inequalities are reversed.

53. In the case of the L^p spaces the conclusion follows at once from Hölder's inequality. Indeed, $\|f\|_r^r = \int_X |f|^{\eta p+(1-\eta)q}\,d\mu \le \|f\|_p^{\eta p}\,\|f\|_q^{(1-\eta)q}$. In fact, the function $\Phi(r) = \ln(\int_X |f|^r\,d\mu)$ is convex in the interval $[p,q]$. Note that the conclusion also holds for $q = \infty$: If $p < r < \infty$, $\|f\|_r^r \le \|f\|_\infty^{r-p}\|f\|_p^p$ and $\|f\|_r \le \|f\|_\infty^{1-\eta}\|f\|_p^\eta$ with $\eta = p/r$.

The result is also true for the weak L^p spaces. In this case with M_r the weak-$L^r(X)$ norm of f for $r = p, q$, by assumption $\mu(\{|f| > \lambda\}) \le \min(M_p\lambda^{-p}, M_q\lambda^{-q})$ and, consequently,

$$\|f\|_r^r \le rM_p \int_0^\eta \lambda^{r-1}\lambda^{-p}\,dt + rM_q \int_\eta^\infty \lambda^{r-1}\lambda^{-q}\,d\lambda$$

$$= rM_p\frac{\eta^{r-p}}{r-p} + rM_q\frac{\eta^{r-q}}{q-r}\,.$$

Picking η so that the summands are equal it turns out that $M_p\eta^{r-p} = M_q\eta^{r-q}$ and so $\eta = (M_q/M_p)^{1/(q-p)}$, which yields, since $r = \eta p + (1-\eta)q$, $\|f\|_r \le c_{p,q,r}\,M_p^{\eta/r}\,M_q^{(1-\eta)/r}$ with $c_{p,q,r} \to \infty$ as $r \to p$ or $r \to q$.

54. By Hölder's inequality with $1/p + 1/q = 1$ and $p > 1$,

$$\left| \int_{Q(0,2^{\ell+1})\setminus Q(0,2^\ell)} f(x)\,dx \right|$$

$$\le \left(2^{(\ell+1)n} - 2^{\ell n} \right)^{1/q} \left(\int_{Q(0,2^{\ell+1})\setminus Q(0,2^\ell)} |f(x)|^p\,dx \right)^{1/p}\,.$$

Now, the integral above goes to 0 but the factor multiplying it goes to ∞ unless $n = 1$. So, the statement is true if $n = 1$ for $1 \le p < \infty$ and not true for $p = \infty$ as the function $f = 1$ shows.

In $\mathbb{R}^n$ the statement is only true for $p = 1$. To see this let $f(x) = \sum_\ell 2^{-n\ell/(p-\varepsilon)}\chi_{Q(0,2^{\ell+1})\setminus Q(0,2^\ell)}(x)$ where $p - \varepsilon > 1$. Then, as is readily seen,

$\|f\|_p^p \sim \sum_\ell 2^{n\ell(1-p/(p-\varepsilon))} < \infty$ but $\int_{Q(0,2^{\ell+1})\setminus Q(0,2^\ell)} f(x)\,dx \sim 2^{\ell n} 2^{-n\ell/(p-\varepsilon)}$ which tends to infinity as $\ell \to \infty$.

55. Fix α so that $1/q < \alpha < 1/p$ and note that since $|f(x)|^p + x^{-\alpha p} \in L^1(I)$, there exists $\eta > 0$ such that $\int_0^\eta (|f(x)|^p + x^{-\alpha p})\,dx < (\varepsilon/2)^p$. Now put $g(x) = x^{-\alpha}\chi_{[0,\eta)}(x) + f(x)\,\chi_{(\eta,1]}(x)$. Then $\|f - g\|_p^p = \int_0^\eta |f(x) - x^{-\alpha}|^p\,dx \le 2^p \int_0^\eta (|f(x)|^p + x^{-\alpha p})\,dx < \varepsilon^p$ and so $\|f - g\|_p < \varepsilon$. Moreover, since $\alpha q > 1$, $g \notin L^q(I)$ and setting g equal to 0 in a small interval containing the origin and contained in $[0, \eta]$ it follows that $g \in L^q(I)$ and $\|g\|_q > K$. When $f \in C(I)$ an adjustment of the construction gives that g can be chosen to be continuous.

56. (a) Sufficiency is trivial. Next, let h be a nonnegative smooth function with integral 1 supported in $B(0,1)$ and consider $\{h_k\}$ where $h_k(x) = kh(kx)$, $k \ge 1$; h_k is supported in $B(0,1/k)$ and has integral 1. Then by Problem 21, $\lim_k \int_{\mathbb{R}} h_k(y) f(x - y)\,dy = f$ in $L^p(\mathbb{R})$ and, consequently, for a subsequence $\{h_{k_\ell}\}$ of $\{h_k\}$ we have that $\lim_{k_\ell} \int_{\mathbb{R}} h_{k_\ell}(y) f(x - y)\,dy = f(x)$ a.e. Now, by Problem 4.101 if f has weak derivative 0 and ψ is a compactly supported smooth function with integral 1, then $\int_{\mathbb{R}} f(x)\psi(x)\,dx$ is independent of ψ. Hence, with $\psi = h_k$, there is a constant c such that $c = \lim_{k_\ell} \int_{\mathbb{R}} h_{k_\ell}(y) f(x - y)\,dy = f(x)$ a.e.

(b) By (a) $f = c$ a.e. and since $f \in L^p(\mathbb{R})$ with $p < \infty$, $c = 0$.

58. First, necessity. For the sake of argument suppose that $\mu(A) \le M$ for all $A \in \mathcal{M}$ and some constant M. Now, given $f \in L^q(X)$, let $A_n = \{x \in X : |f(x)| > 1/n\}$ for all n. By Chebychev's inequality $\mu(A_n) < \infty$ for all n, and, consequently, $\mu(A_n) \le M$ for all n. Moreover, since $\{A_n\}$ is nondecreasing and $\bigcup_n A_n = \{x \in X : |f(x)| \neq 0\}$ it follows that $\mu(\{f \neq 0\}) = \lim_n \mu(A_n) \le M$ and so, by Hölder's inequality,

$$\int_X |f|^p\,d\mu \le \left(\int_{\bigcup A_n} |f|^q\,d\mu\right)^{p/q} \mu(\{f(x) \neq 0\})^{1-p/q} < \infty$$

contrary to the fact that $L^q(\mu) \not\subset L^p(\mu)$.

Next, sufficiency. First, if X contains sets of arbitrarily large finite measure, by monotonicity $\mu(X) = \infty$. We claim that there exist pairwise disjoint $\{A_n\}$ such that $2^n \le \mu(A_n)$ for all n. To see this pick $A_1 \subset X$ with $2 \le \mu(A_1) < \infty$, and having picked $A_1, \ldots, A_{n-1}$, let $A_n \subset X\setminus\bigcup_{k=1}^{n-1} A_n$ with $2^n < \mu(A_n) < \infty$. The choice is possible since $\mu(X) = \infty$ and $\mu(\bigcup_{k=1}^{n-1} A_n) < \infty$. Finally, let $f(x) = \sum_n \mu(A_n)^{-1/p}\chi_{A_n}(x)$; then $f \in L^p(X) \setminus L^q(X)$ and, consequently, $L^q(X) \not\subset L^p(X)$.

When $q = \infty$ the result is $L^\infty(X) \not\subset L^p(X)$ iff $\mu(X) = \infty$. To prove necessity, for the sake of argument suppose that $\mu(X) < \infty$ and let $f \in L^\infty(X)$. Then, since $|f(x)| \le \|f\|_\infty$ μ-a.e., it follows that $\int_X |f|^p\,d\mu \le$

$\|f\|_\infty^p \mu(X) < \infty$ and so $L^\infty(X) \subset L^p(X)$, which is not the case. As for the sufficiency, if $\mu(X) = \infty$, $\chi_X(x) \in L^\infty(X) \setminus L^p(X)$.

59. By Hölder's inequality $\int_X |f|^p d\mu \le (\int_X |f|^q d\mu)^{p/q} = \int_X |f|^p d\mu$, and so by the case of equality in Hölder's inequality $|f|^p$ and 1 are multiples of each other and, consequently, $|f|$ is constant μ-a.e.

60. The $L^p(X)$ spaces are all equal iff there exist $0 < \varepsilon \le k < \infty$ such that $\varepsilon \le \mu(A) \le k$ for every $\emptyset \ne A \in \mathcal{M}$ and $\mathcal{F} = \{A \subset X : \mu(A) > 0\}$ contains at most finitely many pairwise disjoint sets.

If the spaces are equal, by Problems 57 and 58, X does not contain sets of arbitrarily small or large finite measure and so ε, k exist. For the sake of argument suppose that $\{A_n\}$ are infinitely many pairwise disjoint sets in $\mathcal{F}$; if $p < q$ pick $1/q < \alpha < 1/p$, and let $f = \sum_n n^{-\alpha} \chi_{A_n}$. Then $\int_X |f|^q d\mu = \sum_n n^{-\alpha q} \int_X \chi_{A_n} d\mu \le k \sum_n n^{-\alpha q} < \infty$ and $f \in L^q(X)$; however, $\int_X |f|^p d\mu = \sum_n n^{-\alpha p} \mu(A_n) \ge \varepsilon \sum_n n^{-\alpha p} = \infty$ and $f \notin L^p(X)$.

Conversely, suppose that f is measurable and let $A_n = \{2^n < |f| \le 2^{n+1}\}$, $n = 0, \pm 1, \pm 2, \ldots$. Since the A_n are pairwise disjoint $\mathcal{I} = \{n : \mu(A_n) > 0\}$ is finite and, consequently, $\sum_{n \in \mathcal{I}} 2^{nr} \mu(A_n) \le k \sum_{n \in \mathcal{I}} 2^{nr} < \infty$ for all $r > 0$. Thus by Problem 5(b), $f \in L^r(X)$ and the $L^p(X)$ spaces are all equal.

61. Replacing if necessary $\Psi(p)$ by $\tilde\Psi(p) = \inf\{\Psi(q) : q \ge p\}$, since $\tilde\Psi(p) \le \Psi(p)$, $\lim_p \tilde\Psi(p) = \infty$, and $\tilde\Psi(p)$ is increasing, we may assume that $\Psi(p)$ is increasing. Let $g(x) = \ln(e/x)$; g is unbounded, decreasing, bounded below by 1 on I, and, by Problem 11, $g \in \bigcap_{p<\infty} L^p(I) \setminus L^\infty(I)$. Now let $\varphi : (0, \infty) \to I$ be given by $\int_0^{\varphi(p)} g(x)^p\, dx = 1$; since $g(x)^p > 1$ in I, $0 < \varphi(p) < 1$, φ is well-defined, and, since $g(x)^p < g(x)^q$ for $p < q$, φ is strictly decreasing. Furthermore, by the monotonicity of g, $\lim_{p \to q} \varphi(p)$ exists for all $q \in [0, \infty]$; we claim that $\lim_{p \to 0+} \varphi(p) = 1$ and $\lim_p \varphi(p) = 0$. To see the latter assertion, for the sake of argument assume that $\lim_p \varphi(p) = \eta > 0$. Then, since $\lim_p g(x) = \infty$, for p sufficiently large $\int_0^{\varphi(p)} g(x)^p\, dx \ge \int_0^\eta g(x)^p\, dx > 1$, which is not the case; the former assertion is proved similarly. Thus $\varphi : [0, \infty) \to (0, 1]$; we claim that φ is continuous. We already know that φ is continuous at 0. First, suppose that $p_n \to p^-$; by the monotonicity of φ, $\lim_{p_n \to p} \varphi(p_n) = \eta$ exists. But then $\lim_{p_n \to p} \chi_{[0, \varphi(p_n)]}(x) g(x)^{p_n} = \chi_{[0, \eta]}(x) g(x)^p$, where the convergence is bounded and so, by the LDCT, $\int_0^\eta g(x)^p\, dx = \lim_{p_n \to p} \int_0^{\varphi(p_n)} g(x)^{p_n}\, dx = 1$ and, therefore, by the definition of φ, $\eta = \varphi(p)$. A similar argument gives that if $p = \lim_{p_n \to p^+} p_n$, $\lim_{p_n \to p} \varphi(p_n) = \varphi(p)$ and the continuity follows.

So $\varphi : [0, \infty) \to (0, 1]$ is continuous, decreasing, and onto, and, therefore, $\varphi^{-1} : (0, 1] \to [1, \infty)$ is well-defined, strictly decreasing, continuous, and onto. Note that we may assume that $\Psi(p) \le g(\varphi(p))$ for otherwise we may replace $\Psi(p)$ by $\tilde{\Psi}(p) = \min(\Psi(p), g(\varphi(p)))$, which increases to ∞ with p. Let $0 < \delta < 1$ and define $f(x) = \delta\,\Psi(\varphi^{-1}(x))$, $0 < x \leq 1$. Then $\delta^{-p}\|f\|_p^p = \int_0^{\varphi(p)} |\Psi(\varphi^{-1}(x))|^p\,dx + \int_{\varphi(p)}^1 |\Psi(\varphi^{-1}(x))|^p\,dx = A + B$, say. To estimate A note that $\Psi(p) \leq g(\varphi(p))$ and so $\Psi(\varphi^{-1}(x)) \leq g(x)$ and $A \leq \int_0^{\varphi(p)} g(x)^p\,dx = 1$. As for B, note that for $\varphi(p) \leq x \leq 1$, $\Psi(\varphi^{-1}(x)) \leq \Psi(p)$ and so $B = \int_{\varphi(p)}^1 |\Psi(p)|^p\,dx \leq \Psi(p)^p$. Thus, combining these estimates, $\|f\|_p \leq \delta\,(1 + \Psi(p))^{1/p} \leq \delta\,(1 + \Psi(p))$ and if q is such that $\Psi(q) > \delta/(1-\delta)$, it follows that $\|f\|_p \leq \Psi(p)$ for $p \geq q$.

It remains to verify that $\|f\|_p \to \infty$ as $p \to \infty$. Let t be such that $\varphi^{-1}(t) = \Psi^{-1}(M)$; then $\Psi(\varphi^{-1}(x)) \geq M$ on $[0, t]$ and $\Psi(\varphi^{-1}(x))^p \geq M^p$ on $[0, t]$. Therefore $\|f\|_p \geq t^{1/p} M$ and so $\liminf_{p \to \infty} \|f\|_p \geq M$.

62. Observe that there exist a set A with $\mu(A) > 0$ and constants c, C such that $0 < c \leq |f(x)| \leq C < \infty$ on A; the lower bound follows from the fact that $f \neq 0$ on a set of positive measure and the upper bound from the fact that f, being integrable, is finite μ-a.e. Now, since μ is nonatomic by Problem 2.62(c) A contains subsets of arbitrarily small positive μ measure and so by Problem 57, $L^1(A) \not\subset L^p(A)$, $1 < p < \infty$; in particular, there exists $h \in L^1(A) \setminus L^p(A)$. Let $g = h\chi_A$. Then $\int_{\mathbb{R}^n} |f\,g|\,d\mu \leq C \int_A |h|\,d\mu < \infty$ and $\int_{\mathbb{R}^n} |f|\,|g|^p\,d\mu \geq c \int_A |h|^p\,d\mu = \infty$.

63. (a) The statement is true. Let $\varepsilon > 0$ and $f \in L^\infty(I)$. Put $M = \|f\|_\infty$ and partition $[-M, M]$ with $-M = y_0 < y_1 < \cdots < y_m = M$ so that $y_k - y_{k-1} < \varepsilon$, $k = 1, \ldots, m$. Let $A_k = \{x \in I : y_{k-1} < f(x) \leq y_k\}$ for $k = 1, \ldots, m$, $A_0 = f^{-1}(\{-M\})$, and $g = \sum_{k=0}^m y_k\,\chi_{A_k}$. If $x \in f^{-1}([-M, M]) = \bigcup_{k=0}^m A_k$, pick j such that $x \in A_j$; then $f(x)$ is within ε of y_j and so $|g(x) - f(x)| < \varepsilon$. The set $(\bigcup_k A_k)^c = \{x : |f(x)| > M\}$ has measure 0 and so $\|g - f\|_\infty < \varepsilon$.

(b) The statement is true for p finite, but not for $p = \infty$: The function $\chi_{[0, 1/2]} \in L^\infty(I)$ cannot be approximated uniformly by continuous functions in $[0, 1]$, $\|\chi_{[0,1/2]} - (1/2)\chi_I\|_\infty = 1/2$ and this is the best one can do.

64. (a) If $A(f) = \emptyset$, $\mu(\{|f| > \lambda\}) > 0$ for all λ and $\|f\|_\infty = \infty$.

(b) If $A(f) \neq \emptyset$ it readily follows that $\|f\|_\infty = \inf A(f)$ and, therefore, there are a sequence $\{\lambda_n\} \subset A(f)$ that decreases to $\|f\|_\infty$ and $\{B_n\} \subset \mathcal{M}$ with $\mu(B_n^c) = 0$ such that $|f(x)| \leq \lambda_n$ for $x \in B_n$. Let $B = \bigcap_n B_n$; then $B \in \mathcal{M}$ and $\mu(B^c) \leq \sum_n \mu(B_n^c) = 0$. Now, for $x \in B$ we have $|f(x)| \leq \lambda_n$ for all n and, therefore, letting $n \to \infty$, $|f(x)| \leq \|f\|_\infty$. Since this estimate holds for $x \in B$, that is, μ-a.e., $\|f\|_\infty \in A(f)$.

66. First, let ess inf $f = 0$. Then $\mu(\{f > \lambda\}) > 0$ for every $\lambda > 0$ and, consequently, $\mu(\{1/f < \beta\}) > 0$ for every $0 < \beta < \infty$. Thus ess sup$(1/f) = \infty$ and the conclusion holds in this case.

Next, suppose that ess inf $f > 0$. Then there exists $\lambda > 0$ such that $\mu(\{f < \lambda\}) = 0$. Hence ess inf $f = \sup\{\lambda > 0 : \mu(\{f < \lambda\}) = 0\} = \sup\{1/\beta : \mu(\{f < 1/\beta\}) = 0\} = \sup\{1/\beta : \mu(\{1/f > \beta\}) = 0\} = (\inf\{\beta : \mu(\{1/f > \beta\}) = 0\})^{-1} = 1/\mathrm{ess}\ \sup(1/f)$.

67. (a) implies (b) Recall that by Problem 2.62 if a nonatomic measure assumes a positive finite value, it assumes an infinite number of distinct values. Thus μ is purely atomic and since atoms are pairwise disjoint, there are finitely many of them, $X_1, \ldots, X_n$, say, such that $\mu(X \setminus \bigcup_{k=1}^n X_k) = 0$ and $\mu(X_k \cap X_m) = 0$ for $k \neq m$. Next, let f be a measurable function and A an atom of X. Then $f|_A$ is measurable and, consequently, the limit of a sequence of simple functions $\{f_n\}$, say, defined on A. Since A is an atom, the f_n are constant on A except possibly on a set B_n, say, of measure 0; let $B = \bigcup_n B_n$. Then f is constant on $A \setminus B$ since, if $x, y \in A \setminus B$, $f_n(x) = f_n(y)$ for all n and so $f(x) = \lim_n f_n(x) = \lim_n f_n(y) = f(y)$. Therefore every measurable $f : (X, \mathcal{M}) \to (\mathbb{R}, \mathcal{B}(\mathbb{R}))$ is equal μ-a.e. to a linear combination of the χ_{X_m}, $1 \leq m \leq n$, and $L^\infty(X)$ is finite dimensional.

(b) implies (c) A finite-dimensional linear space is separable.

(c) implies (a) For the sake of argument suppose that $\{X_n\}$ is a countable pairwise disjoint collection of measurable subsets of X with positive measure. For $\lambda = \{\lambda_n\} \in \{0,1\}^{\mathbb{N}}$, let $\ell(\lambda) = \sum_n \lambda_n \chi_{X_n}$; since the X_n are pairwise disjoint $\ell(\lambda)$ is well-defined and $\ell(\lambda) \in L^\infty(X)$. Now, if $\lambda \neq \mu$, we claim that $\|\ell(\lambda) - \ell(\mu)\|_\infty = 1$. Indeed, $\|\ell(\lambda) - \ell(\mu)\|_\infty \leq 1$ and if n is such that $\lambda_n \neq \mu_n$, then, since $\mu(X_n) > 0$, $\|\ell(\lambda) - \ell(\mu)\|_\infty \geq \|\chi_{X_n}\|_\infty = 1$. Finally, observe that the balls $B(\ell(\lambda), 1/2)$ with $\lambda \in \{0,1\}^N$ form an uncountable family of open pairwise disjoint subsets of $L^\infty(X)$, which is not possible since $L^\infty(X)$ is separable.

(a) iff (d) Recall that if (a) holds the measurable functions on X are simple and, therefore, μ-a.e. bounded and (d) holds. Conversely, if (a) doesn't hold there exists a sequence $\{X_n\}$ consisting of pairwise disjoint sets in $\mathcal{M}$ of positive measure. The function $\sum_n n\chi_{X_n}$ is then measurable and unbounded and, consequently, (d) is false.

68. (a) If $A(g)_f = \emptyset$, then $\mu(\{|g| > \lambda\} \cap M) > 0$ for all λ and $M \in \mathcal{M}_f$, and so $\|g\|_{\infty, fin} = \infty$.

(b) If $A(g)_f \neq \emptyset$, it readily follows that $\|g\|_{\infty, f} = \inf A(g)_f$ and, therefore, there exist a sequence $\{\lambda_n\} \subset A(g)_f$ that decreases to $\|g\|_{\infty, loc}$ and $\{B_n\} \subset \mathcal{M}$ such that $\{B_n^c\} \subset \mathcal{N}_f$ and $|g(x)| \leq \lambda_n$ for $x \in B_n$. Let $B = \bigcap_n B_n$; since $\mu(B^c \cap M) \leq \sum_n \mu(B_n^c \cap M) = 0$ for all $M \in \mathcal{M}_f$, $B \in \mathcal{N}_f$. Moreover, for $x \in B$ we have $|g(x)| \leq \lambda_n$ for all n and, therefore, letting $n \to \infty$, $|g(x)| \leq \|g\|_{\infty, loc}$ for $x \in B$, that is, locally μ-a.e., and $\|g\|_{\infty, loc} \in A(g)_f$.

(c) Clearly given $f, g \in L_{loc}^\infty(X)$ and a scalar λ, $\|f + g\|_{\infty,loc} \le \|f\|_{\infty,loc} + \|g\|_{\infty,loc}$ and $\|\lambda f\|_{\infty,loc} = |\lambda| \, \|f\|_{\infty,loc}$; thus $\|\cdot\|_{\infty,loc}$ is a seminorm on $L_{loc}^\infty(X)$ and a norm when identifying functions that coincide except for sets in $\mathcal{N}_f$. To prove completeness suppose that $\sum_k \|f_k\|_{\infty,loc} < \infty$ and let $G(x) = \sum_k |f_k(x)|$. Then by (b), $|f_k(x)| \le \|f_k\|_{\infty,loc}$ for $x \notin B_k$ where $B_k \in \mathcal{N}_f$, and, consequently, $G(x) \le \sum_k \|f_k\|_{\infty,loc} < \infty$ for $x \notin B = \bigcup_k B_k \in \mathcal{N}_f$. Hence $G \in L_{loc}^\infty(X)$ and $\|G\|_{\infty,loc} \le \sum_k \|f_k\|_{\infty,loc}$. Now let $F(x) = \sum_k f_k(x)$; by above the sum converges for $x \notin B$ where $B \in \mathcal{N}_f$, and defines a finite value with $|F(x)| \le G(x)$. Therefore $F \in L_{loc}^\infty(X)$ and since $|F(x) - \sum_{k=1}^n f_k(x)| \le \sum_{k=n+1}^\infty |f_k(x)|$ for $x \notin B \in \mathcal{N}_f$, $\|F - \sum_{k=1}^n f_k\|_{\infty,loc} \le \sum_{k=n+1}^\infty \|f_k\|_{\infty,loc} \to 0$ as $n \to \infty$. Hence $L_{loc}^\infty(X)$ is complete.

(d) For $f \in L^\infty(X)$ we have $|f(x)| \le \|f\|_\infty$ μ-a.e., which implies $|f(x)| \le \|f\|_{loc}^\infty$ locally μ-.a.e., in particular giving the fact that $f \in L_{loc}^\infty(X)$ and the inequality of the norms.

69. The statement is true and the proof follows as in Problem 25.

70. The statement is true and the limit in question is $\gamma\pi$.

73. Since $\|p_m - p_k\|_\infty \le \|p_m - f\|_\infty + \|f - p_k\|_\infty \le 2$ for all m, k large enough and since nonconstant polynomials have infinite $L^\infty(\mathbb{R}^n)$ norm, it readily follows that $p_k(x) - p_m(x) = c_{k,m}$ where $c_{k,m}$ is a constant independent of x, for all m, k large enough. Thus letting $k \to \infty$ it readily follows that $\lim_k p_k(x) - p_m(x) = f(x) - p_m(x) = \lim_k c_{k,m} = c_m$, a constant independent of x. So $f(x) = p_m(x) + c_m$ is a polynomial.

75. If $p = r$ the conclusion follows at once from the LDCT. For $p/r > 1$, $\mu(A_n)^{r/p-1} \int_{A_n} |f|^r \, d\mu \le \mu(A_n)^{r/p-1} (\int_{A_n} |f|^p \, d\mu)^{r/p} \mu(A_n)^{1-r/p}$, by Hölder's inequality, which goes to 0 as $n \to \infty$.

76. The statement is true. Assume first that g is continuous and, hence, bounded in $[0, 1]$; by the LDCT it readily follows that $\lim_{p \to 1} \|g - g_p\|_1 = 0$. But these functions are dense in $L^1([0, 1])$ and so given an arbitrary $f \in L^1([0, 1])$ and $\varepsilon > 0$, pick a continuous function g such that $\|f - g\|_1 < \varepsilon$. Then by the triangle inequality $\|f - f_p\|_1 \le \|f - g\|_1 + \|g - g_p\|_1 + \|g_p - f_p\|_1$, and we are done since, by a change a variables, $\|f_p - g_p\|_1 = \|f - g\|_1 < \varepsilon$.

77. (a) Observe that by Problem 4.27, φ, ψ are Borel measurable and by Problem 4.33, $\varphi(f)\psi(f)$ is Lebesgue measurable. Now, if φ, ψ are monotone in the same sense, $(\psi(s) - \psi(t))(\varphi(s) - \varphi(t)) \ge 0$ for all $s, t \in \mathbb{R}$ and so with $s = f(x)$ and $t = f(y)$ we have $(\psi(f(x)) - \psi(f(y)))(\varphi(f(x)) - \varphi(f(y))) \ge 0$. Whence multiplying out and integrating with respect to x over $\mathbb{R}$ it follows that

$$\int_{\mathbb{R}} \psi(f)\,\varphi(f)\,d\mu - \psi(f(y)) \int_{\mathbb{R}} \varphi(f)\,d\mu - \varphi(f(y)) \int_{\mathbb{R}} \psi(f)\,d\mu + \varphi(f(y))\,\psi(f(y))$$

$$\ge 0 \,,$$

which integrated with respect to y over $\mathbb{R}$ gives

$$2\int_{\mathbb{R}} \psi(f)\,\varphi(f)\,d\mu - 2\Big(\int_{\mathbb{R}} \varphi(f)\,d\mu\Big)\Big(\int_{\mathbb{R}} \psi(f)\,d\mu\Big) \ge 0\,,$$

whence the conclusion.

(b) Follows as (a) with $s = f(x)$ and $t = \int_{\mathbb{R}} f\,d\mu$.

On the other hand, if ψ and φ are monotone in the opposite sense, then $(\psi(s) - \psi(t))(\varphi(s) - \varphi(t)) \le 0$ for $s, t \in \mathbb{R}$, and all the inequalities above are reversed.

78. Note that $\varphi(t) = 1/(1-t)$ is increasing in $(0,1)$, $\psi(t) = 1/(1+t)$ is decreasing in $(0,1)$, and $\varphi(t)\psi(t) = 1/(1-t^2)$. The conclusion follows from Problem 77.

80. The proofs are similar so we only do (a). Since φ is nondecreasing, $\{|f| > t\} = \{\varphi(\lambda\,|f|) > \varphi(\lambda t)\}$ and, consequently,

$$\mu(\{|f| > t\}) = \int_{\{|f|>t\}} d\mu \le \int_{\{|f|>t\}} \frac{\varphi(\lambda\,|f|)}{\varphi(\lambda t)}\,d\mu \quad \text{for all } \lambda, t > 0\,.$$

81. Take $f(t) = x\chi_{[0,\lambda]}(t) + y\chi_{[\lambda,1]}(t)$. Then the left-hand side becomes $\varphi(\lambda x + (1-\lambda)y)$ and the right-hand side becomes $\lambda\varphi(x) + (1-\lambda)\varphi(y)$. Thus φ is convex. Similarly for ψ.

82. Equality clearly holds if f is constant μ-a.e. on X. Conversely, if equality holds for φ it holds for $-\varphi$ and so we may assume that φ is increasing. The case when φ' is increasing or decreasing is handled analogously and we consider the former. Let $\alpha = \int_X f\,d\mu$ and $\psi(t) = \varphi(t) - \varphi(\alpha) - \varphi'(\alpha)(t - \alpha)$; then $\psi(\alpha) = 0$ and $\psi'(t) = \varphi'(t) - \varphi'(\alpha) = 0$ iff $t = \alpha$. We claim that $\psi(t)$ does not change sign and in this case is strictly positive for $t \ne \alpha$. If $t > \alpha$, by the mean value theorem there exists $\alpha < s < t$ such that

$$\frac{\varphi(t) - \varphi(\alpha)}{t - \alpha} = \varphi'(s) > \varphi'(\alpha)$$

and, consequently, $\psi(t) > 0$. Similarly, if $t < \alpha$,

$$\frac{\varphi(t) - \varphi(\alpha)}{t - \alpha} = \varphi'(s) < \varphi'(\alpha)\,,$$

and $\psi(t) > 0$. Let $t = f(x)$ and note that since $\int_X (f - \alpha)\,d\mu = 0$, $\int_X \psi(f)\,d\mu = \int_X \varphi(f)\,d\mu - \varphi(\alpha)$, which is strictly positive if $\mu(\{f \ne \alpha\}) > 0$. However, this is not the case and, consequently, $f = \alpha$ μ-a.e.

83. By calculus $|t^p - s^p| \le p\,|t - s|\,(t^{p-1} + s^{p-1})$, $s, t \ge 0$. Hence putting $s = |g(x)|$, $t = |f(x)|$, and integrating, by Hölder's inequality and

Minkowski's inequality,

$$\int_X \big| |f|^p - |g|^p \big| \, d\mu \leq p \int_X |f - g| \left(|f|^{p-1} + |g|^{p-1} \right) d\mu$$

$$\leq p \, \|f - g\|_p \left(\int_X \left(|f|^{p-1} + |g|^{p-1} \right)^{p/p-1} d\mu \right)^{(p-1)/p}$$

$$\leq p \left(\|f\|_p^{p-1} + \|g\|_p^{p-1} \right) \|f - g\|_p \, ,$$

whence the conclusion.

84. The following proof is due to A. P. Calderón. Since $F(x) = \int_0^1 f(tx) \, dt$, by Minkowski's integral inequality $\|F\|_p \leq \int_0^1 \|f(t\cdot)\|_p \, dt = (\int_0^1 t^{-1/p} \, dt) \|f\|_p = p' \|f\|_p$. To verify that p' is the best possible constant let $f(x) = x^{-\alpha} \chi_I(t)$, $\alpha < 1/p$; then $F(x)$ is strictly positive, $F(x) = (1/(1 - \alpha)) f(x)$ for $0 < x < 1$, and $\|F\|_p \geq (1/(1 - \alpha))\|f\|_p$. Now, by the restriction on α, $1/(1 - \alpha)$ is an arbitrary number $< p'$ and so, given any $c < p'$, there exists $f \in L^p(\mathbb{R}^+)$ such that $\|F\|_p \geq c\|f\|_p$.

Now, since $f(x) \leq \|f\|_\infty$ a.e. on $\mathbb{R}^+$, the estimate is true for $p = \infty$. On the other hand, as Problem 4.96 shows, it does not hold for $p = 1$. Alternatively, given a nonnegative integrable function f, let $a > 0$ be such that $\int_a^\infty f(t) \, dt \leq \|f\|_1/2$. Then for $x > a$, $\|f\|_1 \leq \int_0^x f(t) \, dt + \|f\|_1/2$ and so $F(x) \geq \|f\|_1/2x$. Thus unless f vanishes a.e., F is bounded below by a nonintegrable function and, therefore, $F \notin L^1(\mathbb{R}^+)$.

85. By the arithmetic mean-geometric mean inequality, Problem 107,

$$\left(\prod_{k=1}^n F_k(t) \right)^{1/n} \leq \frac{1}{n} \sum_{k=1}^n F_k(t) = \frac{1}{x} \int_0^x \frac{1}{n} \sum_{k=1}^n f_k(t) \, dt$$

and, therefore, by Hardy's inequality,

$$\int_0^\infty \left(\prod_{k=1}^n F_k(t) \right)^{p/n} dt \leq (p')^p \int_0^\infty \left(\frac{1}{n} \sum_{k=1}^n f_k(t) \right)^p dt \, .$$

This result is from L. Bougoffa, *On Minkowski and Hardy integral inequalities*, J. Inequal. Pure Appl. Math. **7**(2) (2006), Art. 60.

86. The proof of these inequalities follows along the lines as for Hardy's inequalities.

87. By the arithmetic mean-geometric mean inequality,

$$\left(\prod_{k=1}^n F_k(x) \right)^{1/n} \leq \frac{1}{n} \sum_{k=1}^n F_k(x) = \int_0^x \frac{1}{n} \sum_{k=1}^n f_k(t) \, dt \, ,$$

and, therefore, by Problem 86

$$\int_0^\infty x^{-r} \left(\prod_{k=1}^n F_k(x) \right)^{p/n} dx \leq \left(\frac{p}{1 - r} \right)^p \int_0^\infty x^{-r} \left(\frac{1}{n} \sum_{k=1}^n f_k(x) \right)^p dx \, .$$

89. First, for λ, x, y strictly positive, $\lambda K(\lambda x, \lambda y) = \lambda/(\lambda x + \lambda y) = 1/(x + y) = K(x, y)$ and $\lambda K_1(\lambda x, \lambda y) = \lambda/\max(\lambda x, \lambda y) = 1/\max(x, y) = K_1(x, y)$.

Next, with p' the conjugate to p,

$$\gamma_K = \int_0^\infty \frac{1}{1+t} t^{-1/p}\, dt = p' \int_0^\infty \frac{1}{1+x^{p'}}\, dx = \frac{\pi}{\sin(\pi/p')} = \frac{\pi}{\sin(\pi/p)}$$

and

$$\gamma_{K_1} = \int_0^\infty \frac{1}{\max(1,t)} t^{-1/p}\, dt = p + p'.$$

90. Let $t > 0$. Then, if $f(x) > t$ and $g(y) > t$, $h((1-\eta)x + \eta y) \ge f(x)^{1-\eta} g(y)^\eta \ge t$ and, consequently, $(1-\eta)\{f > t\} + \eta\{g > t\} \subset \{h > t\}$. Now, since f, g are lower semicontinuous, $\{f > t\}, \{g > t\}$ are open and $(1-\eta)\{f > t\} + \eta\{g > t\}$ is measurable. Hence by Problems 3.61 and 3.45, $(1-\eta)|\{f > t\}| + \eta|\{g > t\}| \le |(1-\eta)\{f > t\} + \eta\{g > t\}| \le |\{h > t\}|$ and so

$$(1 - \eta) \int_0^\infty \Phi'(t)|\{f > t\}|\, dt + \eta \int_0^\infty \Phi'(t)|\{g > t\}|\, dt$$
$$\le \int_0^\infty \Phi'(t)|\{h > t\}|\, dt.$$

Thus $(1-\eta) \int_{\mathbb{R}} \Phi(f(x))\, dx + \eta \int_{\mathbb{R}} \Phi(g(x))\, dx \le \int_{\mathbb{R}} \Phi(h(x))\, dx$ and the conclusion follows by the concavity of $\ln(t)$. This result is related to the Prékopa-Leindler inequality.

91. Let $d\nu = (f/\|f\|_1)\, d\mu$. Then ν is a probability measure on X and by Jensen's inequality, $(\int_X gf\, d\mu)^p = (\|f\|_1 \int_X g\, d\nu)^p \le \|f\|_1 \int_X g^p\, d\nu = \|f\|_1^{p-1} \int_X g^p f\, d\mu$.

92. The result is true for $1 < p < \infty$ by the case of equality in Hölder's inequality. The result is not true for $p = 1$. Indeed, if $f(x) = 2\chi_{[0,1/2]}(x)$, then for any a with $|a| \le 1$, $g(x) = \chi_{[0,1/2]}(x) + a\chi_{(1/2,1]}(x)$ satisfies $\|g\|_\infty = 1$ and $\int_I f(x)g(x)\, dx = 1 = \|f\|_1$. Nor is it true for $p = \infty$. Let $f(x) = x$, $\|f\|_\infty = 1$, and g any integrable function of norm 1. Then by the LDCT, $\int_{[0,1-1/n]} |g(x)|dx \ge 1/2$ for some integer n. Hence $|\int_I f(x)g(x)\, dx| \le (1 - 1/n) \int_{[0,1-1/n]} |g(x)|\, dx + \int_{[1-1/n,1]} |g(x)|\, dx \le \int_I |g(x)|\, dx - (1/n) \int_{[0,1-1/n]} |g| \le 1 - 1/(2n) < 1$, and no such g exists in this case.

93. $g(x) = c\, x^{1/3}$ for some constant c.

95. (b) If $\int_X f^2 g\, d\mu = 1$, then $1 = |\int_X f^2 g\, d\mu| \le \int_X f^2 |g|\, d\mu \le 1$, and by equality in Hölder's inequality, $|g(x)| = c\,[f(x)^2]^{1/2} = c\,|f(x)|$ μ-a.e. for some nonnegative constant c. Moreover, since $\|f\|_3 = \|g\|_3 = 1$ we have $c = 1$ and so $|f| = |g|$ μ-a.e. Finally, since $\int_X f^2 (|g| - g)\, d\mu =$

$\int_X f^2 |g| \, d\mu - \int_X f^2 g \, d\mu = \int_X |f|^3 \, d\mu - 1 = 0$, and $f^2 (|g| - g) \geq 0$ μ-a.e., it follows that $f^2(|g| - g) = 0$ μ-a.e. and since $|f| = |g|$, $g = |g| = |f|$.

97. Applying Cauchy-Schwarz the integral is bounded by $3^{1/2}\pi^{2/3}$ but the problem is not a misprint. The required estimate follows from Hölder's inequality with conjugate indices 3 and $3/2$.

98. Observe that since $\int_{A \setminus A_\eta} f^p \, d\mu \leq (\eta\alpha)^p \mu(A)$, $\int_{A_\eta} f^p \, d\mu = \int_A f^p \, d\mu - \int_{A \setminus A_\eta} f^p \, d\mu \geq (1 - \eta^p) \alpha^p \mu(A)$. Also, by Hölder's inequality, $\int_{A_\eta} f^p \, d\mu \leq (\int_{A_\eta} f^q \, d\mu)^{p/q} \mu(A_\eta)^{1-p/q} \leq \beta^{p/q} \mu(A_\eta)^{1-p/q} \mu(A)^{p/q}$, which combined with the previous inequality gives $(1 - \eta^p) \alpha^p \mu(A) \leq \beta^{p/q} \mu(A_\eta)^{1-p/q} \mu(A)^{p/q}$. Hence, finally,

$$\left(\frac{\alpha^p (1 - \eta^p)}{\beta^{p/q}} \right)^{q/(q-p)} \mu(A) \leq \mu(A_\eta) \, .$$

101. Since X has finite measure, $L^\infty(X) \subset L^r(X)$ for $0 < r < \infty$, and so we may assume that $r < \infty$. Now, for $f \in L^r(X)$, $\mu(\{|f| = \infty\}) = 0$, and $\max(1, |f(x)|^r)$ is a μ-a.e. integrable majorant for $|f(x)|^p$ for $0 < p < r$. Moreover, since $\lim_{p \to 0^+} |f(x)|^p = \chi_{\{f \neq 0\}}(x)$ μ-a.e., by the LDCT it follows that $\lim_{p \to 0^+} \int_X |f|^p \, d\mu = \int_X \chi_{\{f \neq 0\}} \, d\mu = \mu(\{f \neq 0\})$.

102. Recall that by calculus $\ln(t) \leq t - 1$ for $t \geq 0$ and $\varphi(p) = (t^p - 1)/p$ decreases to $\ln(t)$ for $t > 0$ as $p \to 0^+$. Now, since $|\ln(f)|$ and $|f|^r$ are finite μ-a.e. in X, $0 < f < \infty$ μ-a.e. in X and $(f(x)^p - 1)/p - \ln(f(x)) \geq 0$ decreases pointwise to 0 μ-a.e. in X and has an integrable majorant $(f(x)^r - 1)/r - \ln(f(x))$. Then by Problem 6.102,

$$\lim_{p \to 0^+} \int_X \frac{f^p - 1}{p} \, d\mu = \int_X \ln(f) \, d\mu \, ,$$

and F is right-differentiable at the origin with derivative $\int_X \ln(f) \, d\mu$.

Now, by calculus $t^p |\ln(t)| \leq c(t^s + t^r)$, $s < p < r$, $t > 0$. Thus for each fixed $x \in X$ where $f(x) \neq 0$, $f(x)^p$ has derivative $f(x)^p \ln(f(x))$, which, by the above remark, is integrable. And, since $f^p(x)(f(x)^{r-p} - 1)/r - p$ is integrable, $f^p(x)(f(x)^\varepsilon - 1)/\varepsilon - f^p(x) \ln(f(x)) \geq 0$ decreases pointwise to 0 μ-a.e. in X as $\varepsilon \to 0^+$ and the expression is integrable for $\varepsilon = p - r$. Therefore, by Problem 6.102,

$$\lim_{\varepsilon \to 0^+} \int_X f^p \frac{f^\varepsilon - 1}{\varepsilon} \, d\mu = \int_X f^p \ln(f) \, d\mu \, ,$$

and F has a right-hand derivative at $0 \leq p < r$ equal to $\int_X f^p \ln(f) \, d\mu$.

103. Let $f = \sum_{k=1}^n a_k \chi_{A_k}$; then $\|f\|_p = (\sum_{k=1}^n |a_k|^p \mu(A_k))^{1/p}$ and

$$\ln(\|f\|_p) = \frac{\ln \left(\sum_{k=1}^n |a_k|^p \mu(A_k) \right)}{p} \, .$$

Let $\eta = \sum_{k=1}^{n} \mu(A_k)$. Since $\sum_{k=1}^{n} |a_k|^p \mu(A_k) \to \eta$ as $p \to 0^+$, there are two cases, namely, $\eta < 1$ and $\eta = 1$. In the former case, since $\ln(\eta) < 0$ the limit as $p \to 0^+$ is $-\infty$ and so $\lim_{p \to 0^+} \|f\|_p = e^{-\infty} = 0$. Also, since f vanishes on a set $A = X \setminus \bigcup_{k=1}^{n} A_k$ of positive measure, $\ln(|f|)\,\chi_A = -\infty$, $\exp \int_X \ln(f)\,d\mu = 0$, and we have equality. In the latter case, since $\eta = 1$, evaluating the limit as $p \to 0^+$ gives an indeterminate form and by L'Hôpital,

$$\lim_{p \to 0^+} \ln(\|f\|_p) = \lim_{p \to 0^+} \frac{\sum_{k=1}^{n} \ln(|a_k|)\,|a_k|^p\,\mu(A_k)}{\sum_{k=1}^{n} |a_k|^p\,\mu(A_k)}$$

$$= \frac{\sum_{k=1}^{n} \ln(|a_k|)\,\mu(A_k)}{\sum_{k=1}^{n} \mu(A_k)} = \sum_{k=1}^{n} \ln(|a_k|)\,\mu(A_k) = \int_X \ln(|f|)\,d\mu .$$

Therefore, exponentiating, $\lim_{p \to 0^+} \|f\|_p = \exp(\int_X \ln(|f|)\,d\mu)$.

104. With F as in Problem 102 we have

$$\ln\left(\left(\int_X f^p\,d\mu\right)^{1/p}\right) = \frac{\ln(F(p))}{p} .$$

Thus, since $F(0) = 1$, $\lim_{p \to 0^+} \ln(\|f\|_p) = \lim_{p \to 0^+} \ln(F(p))/p = F'(0)/F(0)$ $= \int_X \ln(f)\,d\mu$, and so the result follows by exponentiating.

105. $\varphi(t) = t\ln(t)$ is convex for $t > 0$.

106. Let $\phi(p) = \int_X f^p\,d\mu$; then $\ln(\phi(p)) = p\ln(\|f\|_p)$. As observed in Problem 102, $f(x)^p$, as a function of p, has derivative $f(x)^p \ln(f(x))$, and since for $p' < p < p''$, $|f^p \ln(f)| \le f^{p'} + f^{p''} \in L^1(X)$, we may differentiate under the integral sign and get $\phi'(p) = \int_E f^p \ln(f)\,d\mu$. A similar argument gives that $\phi'(p)$ is differentiable and $\phi''(p) = \int_E f^p \ln^2(f)\,d\mu$. Now, the second derivative of $\ln(\phi)$ is equal to $(\phi''\phi - \phi'^2)/\phi^2$ and since this quantity is positive, $\ln(\phi)$ is convex.

107. If one of the a_k is 0 there is nothing to prove, so assume that $a_k \ne 0$ for all k and let $x_k = p_k \ln(a_k)$, $1 \le k \le n$; then $\prod_{k=1}^{n} a_k = \exp(\sum_{k=1}^{n} x_k/p_k)$. Now, the measure μ with $\mu(k) = 1/p_k$, $1 \le k \le n$, is a probability measure, and, therefore, since e^t is convex, by Jensen's inequality $\exp(\sum_{k=1}^{n} x_k/p_k) \le \sum_{k=1}^{n} \exp(x_k)/p_k = \sum_{k=1}^{n} a_k^{p_k}/p_k$. Note that the choice $p_k = n$ for all k gives the arithmetic mean-geometric mean inequality.

108. Apply Jensen's inequality for the concave functions $\varphi(t) = \ln(t)$ and $\varphi(t) = t^p$, $0 < p < 1$, respectively, with measure $d\nu = d\mu/\mu(A)$, $\nu(A) = 1$. More precisely,

$$\int_A \ln(f)\,d\mu = \mu(A) \int_A \ln(f)\,d\nu \le \mu(A) \ln\left(\int_A f\,d\nu\right) = \mu(A) \ln(1/\mu(A)) .$$

The other inequality is proved analogously.

110. No.

112. The integral does not exceed 4^η.

114. Let $(X, \mathcal{M}, \mu)$ be a probability measure space, $f \in L^p(X)$, $1 < p < \infty$, and $\lambda \in (0, 1)$. Prove: (a) $(1 - \lambda) \int_X f \, d\mu \leq \int_X f \chi_{\{f \geq \lambda \int_X f \, d\mu\}} \, d\mu$. (b) $\mu(\{f \geq \lambda \int_X f \, d\mu\})^{p-1} \geq (1 - \lambda)^p (\int_X f \, d\mu)^p / \int_X f^p \, d\mu$.

(a) For $\lambda \in (0, 1)$ write $f = f \chi_{\{f < \lambda \int_X f \, d\mu\}} + f \chi_{\{f \geq \lambda \int_X f \, d\mu\}}$. Then $\int_X f \, d\mu \leq \lambda \int_X f \, d\mu + \int_X f \chi_{\{f \geq \lambda \int_X f \, d\mu\}} \, d\mu$ and the conclusion follows.

(b) By Hölder's inequality

$$\left(\int_X f \chi_{\{f \geq \lambda \int_X f \, d\mu\}} \, d\mu \right)^p \leq \mu\left(\left\{ f \geq \lambda \int_X f \, d\mu \right\} \right)^{p-1} \int_X f^p \, d\mu \,,$$

and, therefore, by (a),

$$\mu\left(\left\{ f \geq \lambda \int_X f \, d\mu \right\} \right)^{p-1} \geq \frac{(1 - \lambda)^p \left(\int_X f \, d\mu \right)^p}{\int_X f^p \, d\mu} \,.$$

116. For all $0 < p, q \leq \infty$. First, suppose that $0 < p, q < \infty$, and since the argument is symmetric, that $0 < p \leq q$. Then $[f(x) \, g(x)]^{p/2} \geq 1$ and integrating over X, by the Cauchy-Schwarz inequality,

$$\mu(X) \leq \int_X (f(x) \, g(x))^{p/2} \, d\mu \leq \left(\int_X f(x)^p \, d\mu \right)^{1/2} \left(\int_X g(x)^p \, d\mu \right)^{1/2}$$

and, consequently,

$$\left(\frac{1}{\mu(X)} \int_X f(x)^p \, d\mu \right)^{1/p} \left(\frac{1}{\mu(X)} \int_X g(x)^p \, d\mu \right)^{1/p} \geq 1 \,.$$

Moreover, since by Hölder's inequality

$$\left(\frac{1}{\mu(X)} \int_X g(x)^p \, d\mu \right)^{1/p} \leq \left(\frac{1}{\mu(X)} \int_X g(x)^q \, d\mu \right)^{1/q} \,,$$

combining this estimate with the first inequality gives the conclusion.

The conclusion clearly also holds for either p, q or both equal to ∞.

118. That $\mu(X)$ is finite.

119. (a) Let $r = 1/p$, $s = -q/p$; then $1 < r, s < \infty$ and $1/r + 1/s = 1$. The assumptions imply that $(fg)^p \in L^r(X)$ and $g^{-p} \in L^s(X)$. So, by Hölder's inequality with indices r, s,

$$\int_X f^p \, d\mu = \int_X (fg)^p g^{-p} \, d\mu \leq \left(\int_X (fg)^{p/p} \, d\mu \right)^p \left(\int_X g^{-p(-q/p)} \, d\mu \right)^{-p/q} \,.$$

The conclusion follows by bringing the g term to the left-hand side of the above inequality and raising to the power $1/p$.

(b) For $0 < \|g\|_q < \infty$, the expression for $\|g\|_q$ follows from the case of equality in (a). And, by the case of equality in Holder's inequality, equality holds in (a) iff $a(|f|^p|g|^p)^{1/p} = b(|g|^p)^r$, or $a|fg| = b|g|^q$ μ-a.e. on X for some $a, b > 0$. Since $0 < |g| < \infty$ μ-a.e. on X this is equivalent to $a|f| = b|g|^{q-1}$ or $a^p|f|^p = b^p|g|^{p(q-1)} = b^p|g|^q$ μ-a.e. on X.

120. First, assume $r = \infty$; if $p = \infty$, $M = 1$. Then, suppose $p < \infty$. Let $f \in L^\infty(\mu)$. Then $|f(x)|^p \le \|f\|_\infty^p$ for μ-a.e. $x \in X$ and so $\|f\|_p \le \mu(X)^{1/p}\|f\|_\infty$. Hence $M \le \mu(X)^{1/p}$. Now, if $f = 1$, $\|f\|_p = \mu(X)^{1/p}$ and $\|f\|_\infty = 1$ and, consequently, $M \ge \mu(X)^{1/p}$. Therefore $M = \mu(X)^{1/p}$.

Assume now $r < \infty$ and, since $M = 1$ for $p = r$, also that $p < r$. By Hölder's inequality with indices $s = r/p > 1$ and its conjugate it follows that $\int_X |f|^p \, d\mu \le (\int_X (|f|^p)^{r/p} \, d\mu)^{p/r} \mu(X)^{1-p/r}$ and, consequently, $M \le \mu(X)^{1/r-1/p}$. Now, taking $f = 1$ on X, it follows that $\|f\|_p/\|f\|_r = \mu(X)^{1/p}/\mu(X)^{1/r}$ and so $M = \mu(X)^{1/p-1/r}$.

121. The statement is false: The function $f(x) = x^{-1/p} \in L^r(I)$ with $\|f\|_r = (p/(p-r))^{1/r}$ but $f \notin L^p(I)$. On the other hand, the statement is true if $\sup_{r<p} \|f\|_r < \infty$. First, since by Fatou's lemma $\int_X |f|^p \, d\mu = \int_X \liminf_{r\to p^-} |f|^r \, d\mu \le \liminf_{r\to p^-} \int_X |f|^r \, d\mu \le \sup_{r<p} \int_X |f|^r \, d\mu < \infty$, $f \in L^p(X)$. Next, let $A = \{|f| \le 1\}$, $B = \{|f| > 1\}$. Then $|f|^r$ decreases to $|f|^p$ as r increases to p on A and, therefore, by the MCT, $\int_A |f|^r \, d\mu \to \int_A |f|^p \, d\mu$ as $r \to p^-$. Next, $|f|^r \le |f|^p$ on B and by the LDCT, $\int_B |f|^r \, d\mu \to \int_B |f|^p \, d\mu$ as $r \to p^-$. Finally, since all the quantities involved are finite, $\int_X |f|^r \, d\mu = \int_A |f|^r \, d\mu + \int_B |f|^r \, d\mu \to \int_A |f|^p \, d\mu + \int_B |f|^p \, d\mu = \int_X |f|^p \, d\mu$ as $r \to p^-$. Therefore, as $r \to p^-$,

$$\ln(\|f\|_r) = \frac{1}{r} \ln\left(\int_X |f|^r \, d\mu\right) \to \frac{1}{p} \ln\left(\int_X |f|^p \, d\mu\right) = \ln(\|f\|_p),$$

and the conclusion follows by exponentiating.

122. The statement is false. In $[0,1]$ pick a sequence $\{A_n\}$ of pairwise disjoint measurable subsets of I so that $|A_n| = 2^{-n}$, $n \ge 1$, and put $f(x) = \sum_n n\chi_{A_n}(x)$; f is nonnegative and since the A_n are pairwise disjoint, $|f(x)|^p = \sum_n n^p\chi_{A_n}(x)$. Thus by the MCT, $\int_I |f(x)|^p \, dx = \sum_n n^p 2^{-n} < \infty$, for all $p > 0$. To verify that $f \notin L^\infty(I)$ note that $\{x \in I : f(x) > n+1\} \supset I_n$, and so $|\{x \in I : f(x) > n+1\}| \ge 2^{-n} > 0$ for all large n, and, consequently, $\|f\|_\infty \ge n$ for all large n and $f \notin L^\infty(I)$. Alternatively, by Problem 11 the log function also works.

123. Note that if $0 < \eta < \|f\|_\infty$ and $B = \{|f| > \eta\}$, B has positive measure and since by Chebychev's inequality $\eta\mu(B)^{1/r} \le \|f\|_r$, $\mu(B) < \infty$. Now, since $\eta\mu(B)^{1/p} \le \|f\|_p$ for any $p > 0$ it readily follows that $\eta \le \liminf_{p\to\infty} \|f\|_p$ and since $\eta < \|f\|_\infty$ is arbitrary, $\|f\|_\infty \le \liminf_{p\to\infty} \|f\|_p$. In particular, if $\|f\|_\infty = \infty$ we get that $\|f\|_\infty = \lim_{p\to\infty} \|f\|_p$. So to complete the proof we may also assume that $f \in L^\infty(X)$. Now, for $f \in L^r(X) \cap L^\infty(X)$ and $p > r$, $\|f\|_p^p = \int_X |f|^{p-r}|f|^r \, d\mu \le \|f\|_\infty^{p-r} \int_X |f|^r \, d\mu = \|f\|_\infty^{p-r}\|f\|_r^r < \infty$. Hence $\|f\|_p \le \|f\|_\infty^{1-r/p}\|f\|_r^{r/p}$ and letting $p \to \infty$ it readily follows that $\limsup_{p\to\infty} \|f\|_p \le \|f\|_\infty$.

Finally, if $\|f\|_p = \infty$ for all p it does not follow that $\|f\|_\infty = \infty$. To see this take the Lebesgue measure on $\mathbb{R}$ and $f = 1$. Then $\|f\|_p = \infty$ for all p yet $\|f\|_\infty = 1$.

124. Note that since f is nonzero on a set of positive measure, $0 < \alpha_n \le \|f\|_1 \|f\|_\infty^{n-1} < \infty$ for all n. Now, since $|f|^n = |f|^{(n-1)/2}|f|^{(n+1)/2}$, by Cauchy-Schwarz it readily follows that $\alpha_n \le \alpha_{n-1}^{1/2}\alpha_{n+1}^{1/2}$, and, consequently, $\alpha_n/\alpha_{n-1} \le \alpha_{n+1}/\alpha_n$. Thus the nonnegative sequence in question is nondecreasing and the limit exists. Moreover, $\alpha_{n+1} \le \alpha_n \|f\|_\infty$ and so $\lim_n \alpha_{n+1}/\alpha_n \le \|f\|_\infty$.

We claim that the limit is $\|f\|_\infty$. Assume first that $\mu(X) = 1$. Then by Hölder's inequality with indices $p = 1 + 1/n$ and its conjugate $q = n + 1$, $\alpha_n \|f\|_n = (\int_X |f|^n \, d\mu)^{1+1/n} \le \int_X |f|^{n+1} \, d\mu = \alpha_{n+1}$, and, consequently, $\|f\|_n \le \alpha_{n+1}/\alpha_n$. Thus taking limits, by Problem 123, $\|f\|_\infty \le \lim_n \alpha_{n+1}/\alpha_n$.

Alternatively, there is a proof that does not use Problem 123. Since $\|f\|_\infty > \varepsilon > 0$, let $A \subset X$ with $\mu(A) > 0$ be such that $|f(x)| > \|f\|_\infty - \varepsilon$ for $x \in A$. Then $\alpha_{n+1} \ge \int_A |f|^{n+1} d\mu > (\|f\|_\infty - \varepsilon)^{(n+1)}\mu(A)$ and so $\alpha_{n+1}/\alpha_n \ge (\|f\|_\infty - \varepsilon)(\mu(A)/\mu(X))^{1/(n+1)}$. Now,

$$\lim_n (\|f\|_\infty - \varepsilon)\big(\mu(A)/\mu(X)\big)^{1/(n+1)} = \|f\|_\infty - \varepsilon$$

and so $\lim_n \alpha_{n+1}/\alpha_n \ge \|f\|_\infty - \varepsilon$, and, since $\varepsilon > 0$ is arbitrary, the result follows. This second proof is of interest because the conclusion of Problem 123 follows from this result by general considerations: If $\lim_n \alpha_{n+1}/\alpha_n = L$, then $\lim_n \alpha_n^{1/n} = L$.

Next, in the general case let

$$d\nu = \frac{|f|}{\|f\|_1} \, d\mu, \quad \beta_n = \int_X |f|^n \, d\nu, \quad n = 1, 2, \ldots.$$

Then $\alpha_n = \|f\|_1 \beta_{n-1}$ for all $n > 1$ and by the finite measure case $\alpha_{n+1}/\alpha_n = \beta_n/\beta_{n-1} \to \|f\|_{\infty,\nu}$. Now, let $A_\lambda = \{|f| > \lambda\}$. Note that if $\mu(A_\lambda) = 0$, then $\nu(A_\lambda) = \|f\|_1^{-1} \int_{A_\lambda} |f| \, d\mu = 0$, and, if $\nu(A_\lambda) = 0$, by Chebychev's inequality $\lambda\mu(A_\lambda) \le \nu(A_\lambda) = 0$ and $\mu(A_\lambda) = 0$. Therefore $\|f\|_\infty = \|f\|_{\infty,\nu}$ and $\lim_n \alpha_{n+1}/\alpha_n = \|f\|_\infty$.

125. (a) With the same notation for α_n as in Problem 124 we have that $\alpha_n^{1/n} \to \|f\|_\infty$. Thus, with f there replaced by e^f, it follows that

$$\frac{1}{n} \ln \left(\int_X e^{nf}\, d\mu \right) = \frac{\ln(\alpha_n)}{n} = \ln(\alpha_n^{1/n}) \to \ln\left(\|e^f\|_\infty\right) = \|f\|_\infty .$$

(b) Apply (a) to e^{-h}.

126. Since $\left(\sum_n a_n |x - x_n|^{-1}\right)^\varepsilon \leq \sum_n a_n^\varepsilon |x - x_n|^{-\varepsilon}$, integrating it follows that $\int_I \left(\sum_n a_n^\varepsilon |x - x_n|^{-\varepsilon}\right) dx \leq \sum_n a_n^\varepsilon \int_I |x - x_n|^{-\varepsilon}\, dx < \infty$.

127. Since $e^{t^{1/\eta}} - 1 = \sum_k t^{k/\eta}/k!$, letting $t = \gamma|f(x)|$ and integrating over X,

$$\int_X \left(e^{(\gamma|f(x)|)^{1/\eta}} - 1\right) dx = \sum_k \frac{1}{k!}\, \gamma^{k/\eta} \int_X |f(x)|^{k/\eta}\, d\mu(x)$$

$$\leq \sum_k \left(\frac{\gamma^{1/\eta} c^{1/\eta}}{\eta}\right)^k \frac{k^k}{k!} .$$

Since by Stirling's formula $k! \sim \sqrt{2\pi k}(k/e)^k$, the above sum is dominated by

$$d \sum_k \left(\frac{\gamma^{1/\eta} c^{1/\eta} e}{\eta}\right)^k ,$$

which converges if $\gamma < \eta^\eta/ce^\eta$.

128. We recognize the first quantity as the ℓ^p norm of the numerical sequence $\{\int_X f_n\, d\mu\}$. To estimate it we consider the sup of $\sum_n a_n \int_X f_n\, d\mu$ where $\sum_n a_n^q = 1$, $1/p + 1/q = 1$. Now, $\sum_n a_n \int_X f_n\, d\mu = \int_X (\sum_n a_n f_n)\, d\mu$ $\leq \|\{a_n\}\|_{\ell^q} \int_X \|\{f_n\}\|_{\ell^p}\, d\mu \leq \int_X \|\{f_n\}\|_{\ell^p}\, d\mu$. Thus, taking the sup over sequences $\{a_n\}$ with ℓ^q norm 1, $(\sum_n (\int_X f_n\, d\mu)^p)^{1/p} \leq \int_X (\sum_n f_n^p)^{1/p}\, d\mu$. As for the last inequality, it follows at once from the concavity of $\varphi(t) = t^{1/p}$, i.e., $(\sum_n f_n^p)^{1/p} \leq \sum_n f_n$.

129. (a) By Hölder's inequality, for $t \in I$, $|Tf(t)| \leq t^{1/q}\|f\|_p$, $1/p + 1/q = 1$. Thus $\|T(f)\|_p \leq (\int_0^1 t^{p/q}\, dt)^{1/p}\|f\|_p$ and T is a contraction with $c = (1/p)^{1/p}$.

(b) For $f = \chi_I$, $\|T(f)\|_\infty = \|f\|_\infty$. Now, from $|T(f)(t)| \leq t\|f\|_\infty$ it follows that $T^2(f)(t) \leq (t^2/2)\|f\|_\infty$ and so $\|T^2(f)\|_\infty \leq (1/2)\|f\|_\infty$.

(c) First, $\|T(f)\|_1 \leq \|f\|_1$ and the bound is sharp. Indeed, let $f_n(s) = n\chi_{[0,1/n]}(s)$. Then $T(f_n)(t) = \|f_n\|_1$ for all $t \geq 1/n$ and, consequently, $\|T(f_n)\|_1 \geq (1 - 1/n)\|f_n\|_1$, which, since true for all n, gives that T is not a contraction. Moreover, since $\|T(f)\|_1 \leq \|f\|_1$, $T^2(f)(t) \leq t\|f\|_1$ and so, $\|T^2(f)\|_1 \leq (1/2)\|f\|_1$. Thus $\|T^{2k+1}(f)\|_1 \leq \|T^{2k}(f)\|_1 \leq (1/2)^k\|f\|_1$ and $T^n(f) \to 0$ in $L^1(I)$.

130. True for $1 \leq p < \infty$, not true for $p = \infty$.

132. First assume that $p < q$. Recall that by Problem 131, $\|x\|_q \leq \|x\|_p$. For the sake of argument suppose that $(\ell^p, \|\cdot\|_q)$ is complete; since $(\ell^p, \|\cdot\|_p)$ is complete, by the inverse mapping theorem $\|x\|_p \leq c\|x\|_q$ for some constant c. Now, if x is the sequence whose first n terms are equal to 1 and are equal to 0 otherwise, it follows that $n^{1/p} \leq c\, n^{1/q}$, which does not hold for n large enough. The case $q < p$ follows analogously.

133. Fix n and consider $\sum_{k=1}^n 1/\varphi(k)$. Among all sums with n integers in the denominator, $\sum_{k=1}^n 1/k$ is the largest (smallest denominators) and, therefore, $\sum_{k=1}^n 1/\varphi(k) \leq \sum_{k=1}^n 1/k$. Now, by the Cauchy-Schwarz inequality, $\sum_{k=1}^n 1/k \leq (\sum_{k=1}^n \varphi(k)/k^2)^{1/2}(\sum_{k=1}^n 1/\varphi(k))^{1/2}$ which in turn is bounded by $(\sum_{k=1}^n \varphi(k)/k^2)^{1/2}(\sum_{k=1}^n 1/k)^{1/2}$, and so $\sum_{k=1}^n 1/k \leq \sum_{k=1}^n \varphi(k)/k^2$.

134. A is $\ell^1 \cap \ell^2$.

136. Let $s = p/r$; by assumption $1 < s < 2$ and so its conjugate $s' > 2$. Then $\sum_n |x_n|^r = \sum_n n^{-1/s} n^{1/s} |x_n|^r \leq (\sum_n n|x_n|^{rs})^{1/s}(\sum_n n^{-s'/s})^{1/s'}$, and since $rs = p$ and $s'/s > 1$, we have finished.

137. (a) The statement is true. (b) The statement is false.

138. We claim that A is open in ℓ^2 iff $\inf_n \lambda_n > 0$. First, necessity. Pick $\varepsilon > 0$ such that $\lambda_1 > \varepsilon > 0$. Then x with $x_1 = \lambda_1 - \varepsilon$ and $x_n = 0$, $n \geq 2$, is in A, which is open and, therefore, for some $\delta > 0$ the ball $B(x,\delta) \subset A$. Let $\eta = \min(\varepsilon, \delta/2)$. Then the closed ball $\overline{B(x,\eta)} \subset B(x,\delta) \subset A$. Then y^n with $y_1^n = \lambda_1 - \varepsilon$ and $y_k^n = 0$ for $k \neq n$ and $y_n^n = \eta$ belongs to $\overline{B(x,\eta)} \subset A$, and, therefore, $\lambda_n \geq \eta$ for $n \geq 2$ and so $\inf_n \lambda_n > \eta$.

As for sufficiency, let $\varepsilon > 0$ be such that $\lambda_n > \varepsilon$ for all n. Now, for $x \in A$ pick k such that $\sum_{n=k+1}^\infty x_k^2 < \varepsilon^2/4$. Let $\delta = \min(\varepsilon/2, \lambda_1 - |x_1|, \ldots, \lambda_k - |a_k|)$; since $x \in A$, $\delta > 0$. We claim that $B(x,\delta) \subset A$. Now, if $y \in B(x,\delta)$, $|y_n| \leq |y_n - x_n| + |x_n| < \delta + |x_n| < \lambda_n - |x_n| + |x_n| = \lambda_n$ for all $n \leq k$. Moreover, $|y_n| \leq (\sum_{n=k+1}^\infty |y_n|^2)^{1/2} \leq (\sum_{n=k+1}^\infty |y_n - x_n|^2)^{1/2} + \varepsilon/2 \leq \delta + \varepsilon/2 \leq \varepsilon$ for all $n \geq k+1$; thus $|y_n| < \varepsilon < \lambda_n$ for those n, hence $|y_n| \leq \lambda_n$ for all n and $y \in A$.

139. We do sufficiency. Observe that as in Problems 27(a), C_λ is a closed subset of ℓ^p. To see that C_λ is totally bounded, fix $\varepsilon > 0$, and let N be chosen large enough so that $\sum_{n=N+1}^\infty |\lambda_n|^p < \varepsilon^p/2$. Now, $C_{\lambda,N} = \{y \in \mathbb{R}^N : |y_n| \leq \lambda_n, 1 \leq n \leq N\}$ is a compact subset of $\mathbb{R}^N$ and, hence, totally bounded. Therefore there exist $\{y^1, \ldots, y^M\} \subset \mathbb{R}^N$ with the property that for every $y \in C_{\lambda,N}$, there exists y^m, $1 \leq m \leq M$, such that $\|y - y^m\|_p^p = \sum_{n=1}^{N} |y_n - y_n^m|^p < \varepsilon^p/2$. Now let $x^1, \ldots, x^M \in \ell^p$ be defined as follows:

$$x^m = \begin{cases} y_n^m, & 1 \leq n \leq N, \\ 0, & n > N, \end{cases} \quad \text{for } 1 \leq m \leq M.$$

Now, given $x \in C_{\lambda,N}$, pick x^m, $1 \leq m \leq M$, such that $\sum_{n=1}^{N} |x_n - x_n^m|^p \leq \varepsilon^p/2$, and observe that $\|x - x^m\|_p^p \leq \sum_{n=1}^{N} |x_n - x_n^m|^p + \sum_{n=N+1}^{\infty} |\lambda_n|^p < \varepsilon^p/2 + \varepsilon^p/2 = \varepsilon^p$.

140. The statement is false. First, observe that there is a sequence $\{n_k\}$ such that $\sum_{j=n_k}^{n_{k+1}} \lambda_j^p \geq 1$ for all k. Indeed, since $\lambda \notin \ell^p$ it immediately follows that for any $n_k \in \mathbb{N}$, $\sum_{j=n_k}^{\infty} \lambda_j^p = \infty$ and, therefore, there exists $n_{k+1} \in \mathbb{N}$ such that $\sum_{n_k \leq j < n_{k+1}} \lambda_j^p \geq 1$. Starting with $n_1 = 1$ we can find a sequence $\{n_k\}$ in $\mathbb{N}$ with the required property.

Next, we claim there is a sequence $\{x^k\}$ in M_λ such that $\|x^m - x^n\|_p = 2^{1/p}$ if $m \neq n$. Indeed, with $\{n_k\}$ as above, let

$$\theta_k = \Big(\sum_{n_k \leq j < n_{k+1}} \lambda_j^p \Big)^{-1/p}, \quad k = 1, 2, \ldots.$$

Hence $0 < \theta_k \leq 1$ for all k and x^k defined by

$$x_j^k = \begin{cases} \theta_k \lambda_j, & n_k \leq j < n_{k+1}, \\ 0, & \text{otherwise}, \end{cases}$$

belong to C_λ and have norm 1 and, consequently, are in M_λ. Finally, since $\|x^m - x^n\|_p = 2^{1/p}$ if $m \neq n$, M_λ is not compact.

Note that since C_λ and B are closed, M_λ is closed.

141. First, pick an increasing sequence $\{n_k\}$ such that $b_{n_k} > 2^k$ for all k. Now, let $a_n = 1/2^k$ for $n = n_k$ and $a_n = 0$ otherwise.

Next, pick an increasing sequence $\{n_k\}$, $n_1 = 1$, such that $\sum_{j=n_k}^{n_{k+1}} c_j > k^2$, and define $d_j = 1/k$ when $n_k \leq j < n_{k+1}$.

143. For the sake of argument suppose that $a \notin \ell^q$ and let $n_\ell \to \infty$ be chosen so that $n_1 = 1$, and $w_\ell = \sum_{n_\ell \leq k < n_{\ell+1}} |a_k|^q \geq 2^\ell$ for all $\ell \geq 1$. Note that if x is defined by

$$x_k = \begin{cases} w_\ell^{-1} |a_k|^{q/p}, & n_\ell \leq k < n_{\ell+1}, \\ 0, & \text{otherwise}, \end{cases}$$

then $\|x\|_p^p = \sum_\ell w_\ell^{-p} \sum_{n_\ell \leq k < n_{\ell+1}} |a_k|^q = \sum_\ell w_\ell^{-p+1} \leq \sum_\ell 2^{-(p-1)\ell} < \infty$ and $x \in \ell^p$. However, since $(q/p) + 1 = q$, it follows that $\sum_n x_n a_n = \sum_\ell w_\ell^{-1} \sum_{n_\ell \leq k < n_{\ell+1}} |a_k|^{q/p} a_k = \sum_\ell w_\ell^{-1} w_\ell = \infty$.

145. Let μ be the measure on $X = \{1/n : n \geq 1\}$ given by $\mu(1/n) = b_n$, $n \geq 1$. Then for a function f on X we have $\int_0^1 f\, d\mu = \sum_n f(1/n)\, b_n$, at least if f is nonnegative or, more generally, if the sum in the right-hand side is absolutely convergent. Note that $\mu([0,1]) = \sum_n b_n = 1$.

Now, if $\sum_n (\ln(a_n))\, b_n = -\infty$ there is nothing to prove. Otherwise, since all the terms in the sum are negative it follows that $\sum_n |\ln(a_n)|\, b_n < \infty$ and also trivially $\sum_n a_n b_n < \infty$. So the function f on X given by $f(1/n) = \ln(a_n)$ satisfies $\|f\| = \sum_n |\ln(a_n)|\, b_n < \infty$ and is integrable. Therefore by Jensen's inequality, $\exp(\sum_n (\ln(a_n))\, b_n) \leq \sum_n \exp(\ln(a_n))\, b_n = \sum_n a_n b_n$, and the first assertion follows by taking logarithms on both sides.

Finally, the second inequality follows by exponentiating the first inequality.

146. An example are the sequences $x^n = \{x_k^n\}$ given by

$$x_k^n = \begin{cases} 1/n, & k = 1, \ldots, n^2, \\ 0, & \text{otherwise.} \end{cases}$$

Then $\|x^n\|_1 = \sum_{k=1}^{n^2} (1/n) = n$, $\|x^n\|_2 = (\sum_{k=1}^{n^2} 1/n^2)^{1/2} = 1$, and $\|x^n\|_\infty = 1/n$.

147. A_p is not closed for $1 < p \leq \infty$. On the other hand, A_1 is the kernel of a bounded linear functional on ℓ^1 and so A_1 is closed.

148. Let $F_N = \{x \in \ell_0^p : x_N \neq 0 \text{ and } x_n = 0 \text{ for all } n > N\}$; by definition $\ell_0^p = \bigcup_N F_N$. Thus it suffices to prove that each F_N is nowhere dense in ℓ_0^p. First, the F_N are finite dimensional and, hence, closed in ℓ_0^p. Next, let $x \in F_N$ and $r > 0$. Then $y = (x_1, \ldots, x_N, r/2, 0, \ldots) \in \ell_0^p$, $\|y - x\|_p = r/2$, $y \in B(x, r)$, and $y \notin F_N$, and this is true for every N and every $r > 0$. Thus F_N has no interior points.

149. Let $M_n = \{f \in L^p(I) : \int_I |f(x)|^p\, dx \leq n\}$; by Fatou's lemma M_n is closed in $L^1(I)$. Let $g \in L^1(I) \setminus L^p(I)$. Then for $f \in M_n$ and all $R > 0$, the L^1-ball $B(f, R)$ contains $f + \varepsilon g$ for $\varepsilon < R$ and $B(f, R) \cap M_n^c \neq \emptyset$. Thus M_n has empty interior in $L^1(I)$ and $L^p(I) = \bigcup_n M_n$ is of first category in $L^1(I)$.

150. First note that $(Y, \|\cdot\|_1)$ is a Banach space. For each $n > 1$, let $Y_n = \{f \in Y \cap L^{1+1/n}(X) : \|f\|_{1+1/n} \leq n\}$. Now, if $f \in Y$, let $p > 1$ be such that $f \in L^p(X)$ and n such that $1 + 1/n \leq p$. Then by Hölder's inequality $\|f\|_{1+1/n} \leq (1 + \mu(X))^{1-1/p}\|f\|_p$, and, therefore, $f \in Y_n$ for n sufficiently large. Thus $Y = \bigcup_n Y_n$.

We claim that Y_n is closed in Y for all n. To see this fix n and let $\{f_k\} \subset Y_n$ converge to f in Y. Passing to a subsequence if necessary we may assume that $\lim_k f_k = f$ μ-a.e. and, therefore, by Fatou's lemma,

$\int_X |f|^{1+1/n}\, d\mu \le \liminf_k \int_X |f_k|^{1+1/n}\, d\mu \le n^{1+1/n}$. Thus $f \in Y_n$, which is therefore closed in Y.

Hence, by the Baire category theorem one of the Y_n has nonempty interior in Y. So, there exists $f_0 \in Y_n$ and $r > 0$ such that $B_1(f_0, r) = \{ g \in Y : \|f_0 - g\|_1 < r \} \subset Y_n$. Let $0 \ne f \in Y$; then $f_0 + (\varepsilon/2\|f\|_1)f \in B_1(f_0, r)$ and, therefore, $f_0 + (\varepsilon/2\|f\|_1)f \in Y_n$. Then with $\eta = \varepsilon/2\|f\|_1$, $f = (f_0 + \eta f)/\eta - f_0/\eta$ can be expressed as a difference of functions in $L^{1+1/n}(X)$, and therefore $Y \subset L^{1+1/n}(X)$.

151. Since $|\sin(t)| \le |t|$ for all real t and $|f(x)| < \pi/2$ a.e., $|\sin(f(x))| < 1$ a.e. and, consequently, $|\sin^k(f(x))| \le |\sin(f(x))| \le |f(x)|$ a.e., and $\lim_k \sin^k(f(x)) = 0$ a.e. Thus by the LDCT, $\lim_k \int_{\mathbb{R}^n} \sin^k(f(x))\, dx = 0$.

The conclusion is still true if $f \in L^p(\mathbb{R}^n)$ for eventually $k > p$ and from this point on we have $|\sin^k(f(x))| \le |\sin^p(f(x))| \le |f(x)|^p$ a.e. and the LDCT applies as before.

152. On the other hand, note that $\ell^1 \cap \ell^2$ is closed with respect to the ℓ^1 but not the ℓ^2 norm, and $L^1([0,1]) \cap L^2([0,1])$ is closed with respect to the L^2 but not the L^1 norm.

153. $L^p(X) \cap L^\infty(X)$ is dense in $L^p(X)$ and since it is actually equal to $L^\infty(X)$, also in $L^\infty(X)$. Now, $L^p(X)$ is separable with respect to the L^p norm but $L^\infty(X)$ is not separable with respect to the L^∞ norm.

154. That $\|\cdot\|$ is a norm in $L^p(X) \cap L^r(X)$ follows readily from the fact that $L^p(X)$ and $L^r(X)$ are normed spaces. Next, if $\sum_n \|f_n\| < \infty$, $\sum_n \|f_n\|_p$ and $\sum_n \|f_n\|_r < \infty$, and since $L^p(X)$ and $L^r(X)$ are complete, there exist f and g, say, such that $\sum_n f_n = f$ in $L^p(X)$ and $\sum_n f_n = g$ in $L^r(X)$. But then a subsequence of the partial sums converges to f μ-a.e. and a further subsequence converges to g μ-a.e. Thus, $f = g$ μ-a.e., $f \in L^p(X) \cap L^r(X)$, and the series converges to f in $L^p(X) \cap L^r(X)$, which is therefore complete. Finally, by Problem 53, $\|f\|_q \le \|f\|$ and $L^p(X) \cap L^r(X) \hookrightarrow L^q(X)$.

155. (a) Let $\{K_m\}$ be an increasing sequence of compact subsets of O with $O = \bigcup_m K_m$ and for $f \in L^p_{loc}(O)$ let $q(f) = \sum_m 2^{-m} \min(1, \|f\chi_{K_m}\|_p)$. Note that if $q(f) = 0$, $\|f\chi_{K_m}\|_p = 0$ for all m, $f = 0$ a.e. in K_m, and, since $\bigcup_m K_m = O$, $f = 0$ a.e. in O. Let $d : L^p_{loc}(O) \times L^p_{loc}(O) \to \mathbb{R}^+$ be given by $d(f, g) = q(f - g)$, $f, g \in L^p_{loc}(O)$; it is readily seen that $(L^p_{loc}(O), d)$ is a metric space.

Note that if $\lim_n d(f, f_n) = 0$, $\lim_n \|(f - f_n)\chi_{K_m}\|_p = 0$ for all m and since a compact $K \subset O$ is contained in K_m for all sufficiently large m, $\lim_n \|(f - f_n)\chi_K\|_p = 0$. Conversely, given $\varepsilon > 0$, pick M such that $\sum_{m=M+1}^\infty 2^{-m} \le \varepsilon/2$, and note that $\sum_{m=M+1}^\infty 2^{-m} \min(1, \|(f - f_n)\chi_{K_m}\|_p) \le \varepsilon/2$. Now, since convergence holds for the compact subsets of O, it holds for $K_1, \ldots, K_M$, and we can pick n large enough such that

$\|(f - f_n)\chi_{K_\ell}\|_p \le \varepsilon/2M$ for all $1 \le \ell \le M$ and all k large enough. Hence $d(f, f_n) \le M(\varepsilon/2M) + \varepsilon/2 = \varepsilon$ for n large enough and $\lim_n d(f, f_n) = 0$.

Finally, completeness. If $\{f_n\} \subset (L^p_{loc}(X), d)$ is a Cauchy sequence, then $\{f_n|_{K_m}\}$ is Cauchy in $L^p(K_m)$ for all m, and, therefore, $\{f_n|_{K_m}\}$ converges to ϕ^m, say, in $L^p(K_m)$ for each m. Now, $K_m \subset K_{m'}$ for $m < m'$, and $\lim_n \int_{K_m} |\phi^{m'}(x) - f_n(x)|^p \, dx \le \lim_n \int_{K_{m'}} |\phi^{m'}(x) - f_n(x)|^p \, dx = 0$; thus by the uniqueness of limits in $L^p(K_m)$, $\phi^{m'}|_{K_m} = \phi^m$. Then the function $f : O \to \mathbb{R}$ given by $f|_{K_m} = \phi^m$ is well-defined and since $f\chi_{K_m} \in L^p(K_m)$ for all m, $f \in L^p_{loc}(O)$, and by construction $\lim_n \|(f - f_n)\chi_{K_m}\|_p = 0$ for all m, and so $\lim_n d(f, f_n) = 0$.

(b) Clearly $L^p_c(O) \subset L^p_{loc}(O)$. Now, given $f \in L^p_{loc}(O)$ and $\varepsilon > 0$, let M be such that $\sum_{m>M} 2^{-m} < \varepsilon$. Then $f\chi_{K_M} \in L^p_c(O)$ and $d(f, f\chi_{K_M}) < \varepsilon$.

Note that for $1 \le p < \infty$, one can prove in an analogous fashion that $L^p_{loc}(X)$ is separable, because the $L^p(K_m)$ are.

Chapter 16

Sequences of Functions

1. The statement is not necessarily true. Let $\mathcal{M} = \mathcal{L}([0,1])$, V a Vitali Lebesgue nonmeasurable subset of $[0,1]$, and for $\alpha \in \Lambda$ put $f_\alpha(x) = \chi_V(\alpha)\chi_{\{\alpha\}}(x)$. Since the image of f_α is the single point 0 if $\alpha \notin V$ or the points 0 and 1 if $\alpha \in V$, each f_α is measurable but $\sup_{\alpha \in [0,1]} f_\alpha(x) = \chi_V(x)$ is not. On the other hand, if the f_α are continuous, or lower semicontinuous, and $f = \sup f_\alpha$, then $\{f > \lambda\} = \bigcup_\alpha \{f_\alpha > \lambda\}$ is the union of open sets and, therefore, open, and f is measurable.

2. (a) Let $\mathcal{M} = \mathcal{L}([0,1])$, $\Lambda = [0,1]$, and for $\alpha \in \Lambda$ let $f_\alpha(x) = \chi_{\{\alpha\}}(x)$ and $g_\alpha(x) = 0$. Then $f_\alpha \le g_\beta$ a.e. for any α, β in Λ, $\sup_{\alpha \in \Lambda} f_\alpha(x) = 1$ for any x, and not only $\inf_{\alpha \in \Lambda} g_\alpha(x) = 0$ but also $\sup_{\alpha \in \Lambda} g_\alpha(x) = 0$.

(b) Λ is countable.

3. (a) Let V denote a Vitali Lebesgue nonmeasurable subset of $[0,1]$ and $A_\alpha = -\alpha + \mathbb{Q} = \{x \in \mathbb{R} : x + \alpha \in \mathbb{Q}\}$; each A_α is measurable and $|A_\alpha| = 0$. Let $f_\alpha(x) = \chi_{A_\alpha \setminus V}(x)$; since $|A_\alpha \setminus V| \le |A_\alpha| = 0$, f_α is measurable. Now, if $x \in V$, $f_\alpha(x) = 0$ for all α and $\lim_{\alpha \to \infty} f_\alpha(x) = 0$. On the other hand, if $x \notin V$, setting $\alpha = -x + r$ we get that $f_\alpha(x) = 1$ if r is rational and $f_\alpha(x) = 0$ if r is irrational. Thus, if $x \notin V$, $\liminf_{\alpha \to \infty} f_\alpha(x) = 0 \ne 1 = \limsup_{\alpha \to \infty} f_\alpha(x)$, and if $x \in V$, $\lim_{\alpha \to \infty} f_\alpha(x) = 0$. Then $A = V \notin \mathcal{L}(I)$.

(b) The condition is that the f_α are continuous.

(c) The condition is that Λ is countable.

4. The statement is false. Suppose that $\mathcal{M} \ne \mathcal{P}(X)$, let $B \notin \mathcal{M}$, and put $f_n = \chi_B + (1/n)\chi_{B^c}$. Then $A_n = X \in \mathcal{M}$ for all n, yet $A = B \notin \mathcal{M}$. However, note that $\{f \ge 0\} \in \mathcal{M}$.

5. It depends on whether μ is complete. If not, put $f_n = 0$ for all n, and $f = \chi_B + \chi_A$ where $\mu(B) = 0$ and $A \subset B$, $A \notin \mathcal{M}$. Then $\lim_n f_n = f$

μ-a.e. but f is not measurable. On the other hand, if μ is complete, let $g = \limsup_n f_n$; then g is measurable and since $f = g$ μ-a.e., f is measurable.

6. (a) First, suppose there exist a subsequence $\{f_{n_k}\}$ of $\{f_n\}$ and $\eta > 0$ such that $|f_{n_k}(x) - f(x)| \geq \eta$. It then follows that for each $m \geq 1$, $\sup_{n \geq m} |f_n(x) - f(x)| \geq \eta > 1/k$ whenever $\eta > 1/k$. Thus for some k, $x \in \bigcap_m \{x \in X : \sup_{n \geq m} |f_n(x) - f(x)| \geq 1/k\}$ and, consequently, $x \in B$.

Conversely, if $x \in B$, there is $k \geq 1$ such that $\sup_{n \geq m} |f_n(x) - f(x)| \geq 1/k$ for all $m \geq 1$. Then for any $0 < \eta < 1/k$ it follows that $\sup_{n \geq m} |f_n(x) - f(x)| > \eta$. In other words, for each $m \geq 1$ there exists $n \geq m$ so that $|f_n(x) - f(x)| \geq \eta$ and so $\liminf_n |f_n(x) - f(x)| \geq \eta > 0$.

7. (a) implies (b) and (c) By Egorov's theorem there exists C_k with $\mu(C_k) > 1 - 1/k$ such that $|f_{n_k}| < 1/2^k$ on C_k. Now, let $c_n = 1$ when $n = n_k$ for some k and $n = 0$ otherwise. Then, since $\mu(\bigcup_k C_k) = 1$, $\sum_n c_n f_n(x)$ converges absolutely μ-a.e.

(c) implies (a) For the sake of argument suppose that every subsequence $\{f_{n_k}\}$ of $\{f_n\}$ satisfies $\liminf_{n_k} |f_{n_k}(x)| > 0$ for x in a set B of positive measure. Pick $\{c_{n_k}\}$ such that $\sum_k |c_{n_k}| = \infty$ and observe that $\sum_k |c_{n_k} f_{n_k}(x)| \geq (\liminf_{n_k} |f_{n_k}(x)|) \sum_k |c_{n_k}| = \infty$ on a set B with $\mu(B) > 0$, which is not the case.

9. We may assume that the f_n are measurable and in that case $M(x)$ need not be measurable. On the other hand, if $M(x)$ is measurable, the conclusion follows for arbitrary f_n, measurable or not.

10. The condition is necessary. Since f is finite μ-a.e., given $\varepsilon > 0$, there exists N_ε such that $\mu(\{|f| > N_\varepsilon\}) \leq \varepsilon/3$. By Egorov's theorem there exists $A \subset X$ with $\mu(A) \leq \varepsilon/3$ such that $f_n \to f$ uniformly on $X \setminus A$. Thus there exists N such that, for $n \geq N$, $|f_n| \leq |f| + 1$ on $X \setminus A$. Therefore, on $X \setminus (A \cup \{|f| > N_\varepsilon\})$, $|f_n| \leq |f| + 1 \leq N_\varepsilon + 1$ for all $n \geq N$. Since $|f_n| < \infty$ μ-a.e. for $n = 1, \ldots, N-1$ there exists L_ε such that $\mu(\bigcup_{n=1}^{N-1} \{|f_n| > L_\varepsilon\}) \leq \varepsilon/3$. Let $M_\varepsilon = \max[N_\varepsilon + 1, L_\varepsilon]$ and $B = A \cup \{|f| > N_\varepsilon\} \cup \bigcup_{n=1}^{N-1} \{|f_n| > L_\varepsilon\}$. Then $\mu(B) \leq \varepsilon/3 + \varepsilon/3 + \varepsilon/3 = \varepsilon$, and, on $X \setminus B$, $|f_n(x)| \leq M_\varepsilon$, i.e., $\sup_n |f_n(x)| \leq M_\varepsilon$ for $x \in X \setminus B$. Therefore $\mu(\{\sup |f_n| \leq M_\varepsilon\}) \geq \mu(X \setminus B) \geq 1 - \varepsilon$.

On the other hand, sufficiency is not true in general. For example, consider the Lebesgue measure on $X = [0, 1]$ and the family of dyadic subintervals $\{[k2^{-n}, (k+1)2^{-n}] : 0 \leq k < 2^n, n = 1, 2, \ldots\}$ of $[0, 1]$. This family is countable and can be ordered first by level n and then, at each level n, in an increasing order of k, i.e., going from left to right; for simplicity we denote the resulting family of intervals $\{I_m\}$. Now let $f_m = \chi_{I_m}$; the graphs of the f_m consist of thinner and thinner plateaus of height 1 that sweep across the interval $[0, 1]$. Thus f_m assumes only the values 0 and 1 and $\{f_m(x)\}$ only converges for those $x \in [0, 1]$ which are rationals of the form $k2^{-n}$ for some

$0 \leq k \leq 2^n$, that is to say, in a set of measure 0. Now, the values $M < 1$ are excluded but for any $M \geq 1$ we have $\mu(\{\sup_m |f_m(x)| \leq M\}) = 1 \geq (1 - \varepsilon)$ for any $\varepsilon > 0$.

11. For the sake of argument suppose that a subsequence $\{g_{n_k}\}$ of $\{g_n\}$ converges pointwise on a subset of I of measure 1. Then by Egorov's theorem there is a measurable $A \subset I$ with $|A| > 1/2 + \eta$ where $\{g_{n_k}\}$ converges uniformly and pick k sufficiently large so that $|g_{n_k}(x) - g_{n_{k+1}}(x)| < 1$ for all $x \in A$. Now, $|f_{n_k}(x) - f_{n_{k+1}}(x)| = 1$ on a subset of I of measure $1/2$ and, therefore, by assumption $|g_{n_k}(x) - g_{n_{k+1}}(x)| = 1$ on a subset F of I of measure at least $1/2 - \eta$; thus $|A| + |F| > 1$ and so $|A \cap F| > 0$, which is impossible.

13. Yes.

14. Let $A_n = \{f_n \neq \lambda_n\}$; by Borel-Cantelli $\mu(\limsup_n A_n) = 0$. Therefore $f_n(x) = \lambda_n$ for all $n \geq n_x$, where n_x depends on x for μ-a.e. $x \in X$. Now, since $\sum_n \lambda_n$ converges, $\sum_n f_n(x)$ converges for such x, that is to say, μ-a.e.

15. First, by Chebychev's inequality $\lambda_n \mu(\{|f_n - \int_X f_n \, d\mu| > \lambda_n\}) \leq \int_X |f_n - \int_X f_n| \, d\mu \leq 2\,c$. Hence $\sum_n \mu(\{|f_n - \int_X f_n \, d\mu| > \lambda_n\}) \leq \sum_n 2c/\lambda_n < \infty$ and the conclusion follows from Borel-Cantelli.

16. Necessity first. For the sake of argument suppose $\sum_n \mu(\{f_n > M\}) = \infty$ for all $M > 0$. Then by the second Borel-Cantelli lemma, Problem 2.108, $\mu(\limsup_n A_n) = 1$ for every $M > 0$, i.e., $\limsup_n f_n \geq M$ μ-a.e. for every $M > 0$, which implies that $\limsup_n f_n = \infty$ μ-a.e., which is not the case.

Sufficiency next. Let $A_n = \{f_n > M\}$; by Borel-Cantelli $\mu(\limsup_n A_n) = 0$ and, consequently, $f_n(x) \leq M$ for $n \geq n_x$ μ-a.e. Whence $\sup_n f_n(x) \leq \max_{1 \leq k < n_x} f_k(x) + M < \infty$ μ-a.e.

17. Let $c > 0$. Then $\mu(\{\ln(f_n) > c\ln(n)\}) = \mu(\{f_n > n^c\}) = n^{-5c}$ and by the first and second Borel-Cantelli lemmas,

$$\mu\left(\limsup_n \{\ln(f_n) > c\ln(n)\}\right) = \begin{cases} 0, & \text{if } 5c > 1, \\ 1, & \text{if } 5c \leq 1. \end{cases}$$

Thus $\limsup_n (\ln(f_n(x))/\ln(n)) = 1/5$ μ-a.e.

18. (a) Since $\mu(\{|f_n| > \varepsilon\}) = \mu(\{f_n = 1\}) = p_n$ for all n, $f_n \to 0$ in probability iff $p_n \to 0$.

(b) By the first and second Borel-Cantelli lemmas and (a), for $0 < \varepsilon < 1$,

$$\mu\left(\limsup_n \{|f_n| > \varepsilon\}\right) = \begin{cases} 0, & \text{if } \sum_n p_n < \infty, \\ 1, & \text{otherwise.} \end{cases}$$

Therefore $f_n \to 0$ μ-a.e. iff $\sum_n p_n < \infty$.

20. Since each f_n is finite μ-a.e. and $\mu(X) < \infty$ there are constants b_n such that if $A_n = \{|f_n| < b_n\}$, $\mu(X \setminus A_n) \leq 2^{-n}$ and, consequently, by Borel-Cantelli, $\mu(\limsup_n(X \setminus A_n)) = 0$. The good set is $A = \liminf_n A_n$; then $X \setminus A = \limsup_n(X \setminus A_n)$ and, consequently, $\mu(X \setminus A) = 0$. Now, let $\lambda_n = 1/nb_n$; we claim that $\lim_n \lambda_n f_n(x) = 0$ for $x \in A$. Indeed, if $x \in A$, x belongs to all but finitely many A_n and $x \in A_n$ if $n > N_1 > 0$. Thus $|f_n(x)| < b_n$, which means that $\lambda_n|f_n(x)| < (1/nb_n)b_n = 1/n$ and so if $n > N_2 > N_1 > 0$, $\lim_n \lambda_n|f_n(x)| = 0$ on A, which by the above argument holds μ-a.e.

21. Redefining the f_n on a set of measure 0 if necessary we may assume that the f_n are finite everywhere and $\lim_n f_n(x) = 0$ for every $x \in X$. Let $g_n(x) = \sup_{k \geq n} f_k(x)$, $x \in X$; the g_n are measurable, $g_n(x) \geq f_n(x)$ for all $x \in X$ and n, $g_n(x) \geq g_{n+1}(x)$, and $\lim_n g_n(x) = 0$. So, working with $\{g_n\}$ instead we may assume that the original sequence decreases to 0 everywhere.

We define the λ_n first. Let $n_1 = 1$ and note that since $f_n \to 0$ in measure there is a subsequence $\{f_{n_k}\}$ of $\{f_n\}$ such that $\mu(\{f_{n_k} > 1/k^2\}) \leq 1/2^k$ for $k = 2, 3, \ldots$. The sequence λ is defined in blocks as follows: $\lambda_n = k$ for $n_k \leq n < n_{k+1}$, $k = 1, 2, \ldots$; in other words, the first $n_2 - n_1$ terms of λ are 1, the next $n_3 - n_2$ terms are 2, and so on. Moreover, since $n_k \to \infty$ as $k \to \infty$, $\lim_n \lambda_n = \infty$.

Next, we deal with the convergence of $\{\lambda_n f_n\}$. Let

$$B_m = \bigcup_{n=n_m}^{\infty} \{\lambda_n f_n > 1/m\}, \qquad m = 1, 2, \ldots.$$

Since $\lambda_n = k$ for $n_k \leq n < n_{k+1}$ we have

$$B_m = \bigcup_{k=m}^{\infty} \bigcup_{n=n_k}^{n_{k+1}-1} \{kf_n > 1/m\}$$

and since $\{f_n\}$ is nonincreasing and $k \geq m$ the innermost union above is contained in $\{f_{n_k} > 1/k^2\}$ and so $B_m \subset \bigcup_{k=m}^{\infty}\{f_{n_k} > 1/k^2\}$. Thus $\mu(B_m) \leq \sum_{k=m}^{\infty} \mu(\{f_{n_k} > 1/k^2\}) \leq 2^{-m+1}$, $m = 1, 2, \ldots$. Finally, let $B = \limsup_m B_m$. Since $\sum_m \mu(B_m) < \infty$, by Borel-Cantelli $\mu(B) = 0$. So it only remains to check that $\lim_n \lambda_n f_n(x) = 0$ for $x \in X \setminus B$. Given $\varepsilon > 0$, let m be so large that $1/m \leq \varepsilon$ and $x \notin B_m$; such a choice is possible since x belongs to at most finitely many B_m. Then by the above argument there exists n_m so that $\lambda_n f_n(x) \leq 1/m \leq \varepsilon$ for all $n \geq n_m$, and, consequently, $\lambda_n f_n(x) \to 0$. Note that this result implies Egorov's theorem.

22. The statement is true if $\mu(X) < \infty$. On the other hand, the statement is not true in an infinite measure space: If $f_n = \chi_{[n,\infty)}$ for all n,

$f_n \to 0$ a.e. but $|\{x \in \mathbb{R} : |f_n(x)| > 1/2\}| = \infty$ for all n. Nevertheless, it is true in $\mathbb{R}^N$ if $|f_n(x)| \leq g(x)$ a.e. for all n, where g is a finite a.e. measurable function on $\mathbb{R}^N$ such that $|\{|g| > \varepsilon\}| < \infty$ for each $\varepsilon > 0$. To see this let B_k denote the ball centered at the origin of radius k. Since $|\{|g| > \varepsilon\}| = \lim_k |\{|g| > \varepsilon\} \cap B_k|$, given $\eta > 0$, there exists K such that $|\{|g| > \varepsilon\} \cap B_K^c| < \eta$. Now, $|\{|f_n| > \varepsilon\}| \leq |\{|f_n| > \varepsilon\} \cap B_K| + |\{|g| > \varepsilon\} \cap B_K^c|$ where the second summand can be made arbitrarily small, and by the validity of the statement for spaces of finite measure the first summand goes to 0.

23. (b) The μ-a.e. convergence is a particular case of (a) and if $\mu(X) < \infty$ the convergence in measure follows from Problem 22. Otherwise, given $\varepsilon > 0$, pick m such that $\lambda_n < \varepsilon$ for all $n \geq m$. Then $\mu(\{\sup_{n \geq m} |f_n| > \varepsilon\}) \leq \mu(\bigcup_{n=m}^{\infty}\{|f_n| \geq \lambda_n\}) \leq \sum_{n=m}^{\infty} \mu(\{|f_n| \geq \lambda_n\})$. So, since

$$\lim_m \sum_{n=m}^{\infty} \mu(\{|f_n| \geq \lambda_n\}) = 0,$$

it follows that $\mu(\{\sup_{n \geq m} |f_n| \geq \varepsilon\}) = 0$, which gives convergence in measure.

24. Let $A_n = \{|f_{n+1} - f_n| \geq \lambda_n\}$; by Borel-Cantelli $\mu(\limsup_n A_n) = 0$. So, $|f_{n+1}(x) - f_n(x)| \leq \lambda_n$ for $x \in A$ with $\mu(X \setminus A) = 0$ for all n sufficiently large. Now, since $\sum_n \lambda_n < \infty$, this is readily seen to imply that $\{f_n(x)\}$ is Cauchy for $x \in A$. Indeed, given $x \in A$ and $\varepsilon > 0$, let n_0 be such that $|f_{n+1}(x) - f_n(x)| \leq \lambda_n$ for $n \geq n_0$, let n_1 be such that $\sum_{n=n_1}^{\infty} \lambda_n < \varepsilon$, and $N = \max(n_0, n_1)$. Then, for $m > n > N$, $|f_m(x) - f_n(x)| = |\sum_{k=n}^{m-1} f_{k+1}(x) - f_k(x)| \leq \sum_{k=n}^{\infty} \lambda_k \leq \varepsilon$, $x \in A$. Therefore $\{f_n(x)\}$ is Cauchy for $x \in A$ and there exists f on X such that $\lim_n f_n(x) = f(x)$ for $x \in A$, i.e., μ-a.e. Note that if $\mu(X) < \infty$, convergence in measure follows from Problem 22.

25. By the first Borel-Cantelli lemma, if $\sum_n \mu(\{|f_n - f| > \varepsilon\}) < \infty$, then $\mu(\limsup_n\{|f_n - f| > \varepsilon\}) = 0$ and, hence, $\limsup_n |f_n - f| \leq \varepsilon$ μ-a.e. Applying this to a sequence $\varepsilon_n \to 0$ it follows that $\limsup_n |f_n - f| = 0$ μ-a.e. and the μ-a.e. convergence follows.

26. For each $m, n = 1, 2, \ldots$, let $A_{n,m} = \{|f_n - f| \geq 1/m\}$. Since $f_n \to f$ μ-a.e. it follows that $\inf_{j \geq 1} \mu(\bigcup_{n=j}^{\infty} A_{n,m}) = 0$ for each $m \geq 1$. Let $n_1 = 1$ and for $k > 1$, let n_k be the smallest integer greater than n_{k-1} such that $\mu(\bigcup_{n \geq n_k} A_{n,k}) \leq 2^{-k}$. Given $\varepsilon > 0$, choose an integer $N > \varepsilon^{-1}$. Then $\sum_{k=N}^{\infty} \mu(\{|f_{n_k} - f| \geq \varepsilon\}) \leq \sum_{k=N}^{\infty} \mu(A_{n_k,k}) \leq \sum_{k=N}^{\infty} \mu(\{\bigcup_{n \geq n_k} A_{n,k}\}) < 2$.

27. (a) Replacing f_n by $|f_n - f|$ if necessary we may assume that the f_n are nonnegative and that $f_n \to 0$ in measure. Now, by assumption there exists n_1 such that $\mu(\{f_n > 1\}) \leq 1$ for $n \geq n_1$, and, having chosen

$n_1 < n_2 < \ldots < n_{k-1}$ with the property that $\mu(\{f_{n_j} \geq 1/j\}) \leq 1/j^2$ for $1 \leq j < k$, pick $n_k > n_{k-1}$ so that $\mu(\{f_{n_k} \geq 1/k\}) \leq 1/k^2$. We claim that $\lim_{n_k} f_{n_k} = 0$ μ-a.e. Let $B_k = \{f_{n_k} > 1/k\}$ and $B = \limsup_k B_k$; by construction $\sum_k \mu(B_k) < \infty$, and so by Borel-Cantelli $\mu(B) = 0$. Finally, it remains to verify that $\lim_{n_k} f_{n_k} = 0$ μ-a.e. on $X \setminus B$. Now, if $x \in X \setminus B$, $x \notin B_k$ for all $k \geq k_0$, where k_0 depends on x. In other words, $f_{n_k}(x) < 1/k$, all $k \geq k_0$ and, consequently, $f_{n_k}(x) \to f(x)$ for x outside a set B with $\mu(B) = 0$.

(b) No.

(c) By (a) a subsequence $\{f_{n_k}\}$ of $\{f_n\}$ converges to f μ-a.e. and since $\{f_n\}$ is increasing, the whole sequence converges pointwise μ-a.e. and the MCT does the rest.

28. Pick a subsequence $\{f_{n_k}\}$ of $\{f_n\}$ that converges to f μ-a.e. and let $A = \{x \in X : \lim_{n_k} f_{n_k}(x) = f(x)\}$; then $\mu(A^c) = 0$. Fix $x \in A$. Since $\{f_n\}$ is monotone, $\{f_n(x)\}$ converges to some extended real number $g(x)$, say. But we must have $f_{n_k}(x) \to g(x)$ and so $g(x) = f(x)$. Therefore $f_n \to f$ μ-a.e.

29. For the sake of argument suppose that there exist $\eta > 0$ and $n_k \to \infty$ such that $|f_{n_k}(x) - f(x)| > \eta$. Now, $f_{n_k}(x) - f(x) > \eta$ or $f(x) - f_{n_k}(x) > \eta$ for infinitely many n_k; suppose the former holds. Two more observations: passing to a subsequence if necessary we may assume that $\{f_{n_k}\}$ converges to f μ-a.e., and, since f is continuous at x, we may pick $\delta > 0$ such that $|f(x) - f(y)| \leq \eta/2$ for $|x - y| < \delta$. Finally, write $f_{n_k}(y) - f(y) = f_{n_k}(y) - f_{n_k}(x) + f_{n_k}(x) - f(x) + f(x) - f(y)$ and note that for y in $(x, x + \delta)$, $f_{n_k}(y) - f_{n_k}(x) > 0$. Thus $f_{n_k}(y) - f(y) \geq (f_{n_k}(x) - f(x)) + (f(x) - f(y)) > \eta - \eta/2$ and, consequently, $\{z \in I : |f_{n_k}(z) - f(z)| > \eta/2\} \supset (x, x + \delta)$. Therefore $|\{z \in I : |f_{n_k}(z) - f(z)| > \eta/2\}| \geq \delta > 0$ for all n_k, which cannot happen since $f_n \to f$ in measure.

30. We begin by picking an increasing sequence $n_k \to \infty$ as follows: n_1 is chosen so that $\mu(\{|f_n - f_m| > 1/2\}) \leq 1/2$ for all $n, m \geq n_1$, and, having picked $n_1 < \ldots < n_{k-1}$, let $n_k > n_{k-1}$ be such that $\mu(\{|f_n - f_m| > 1/2^k\}) \leq 1/2^k$ for all $n, m \geq n_k$; this choice is possible since the sequence is Cauchy in measure. Let $B_k = \{|f_{n_k} - f_{n_{k+1}}| > 1/2^k\}$ and $B = \limsup_k B_k$; since $\sum_k \mu(B_k) < \infty$ by Borel-Cantelli $\mu(B) = 0$. Let $x \in X \setminus B$; then x belongs to finitely many k and $\{f_{n_k}\}$ is Cauchy in $X \setminus B$. Let f be the pointwise μ-a.e. limit; by Problem 22, $\{f_{n_k}\}$ converges to f in measure and so the whole sequence converges to f by the general principle that if a subsequence of a Cauchy sequence converges, the whole sequence converges to the same limit.

32. We discuss sufficiency. For the sake of argument suppose that $\{f_n\}$ does not converge to f in measure. Then there exists $\varepsilon > 0$ so that for every integer N, there is $N' > N$ such that $\mu(\{|f_{N'} - f| > \varepsilon\}) \geq \varepsilon$ and, consequently, there is an increasing sequence n_k so that $\mu(\{|f_{n_k} - f| > \varepsilon\}) \geq \varepsilon$. This implies that $\{f_{n_k}\}$ has no subsequence that converges to f μ-a.e., which is not the case.

33. That $\mu(X) < \infty$ is necessary follows from the following example. Consider the Lebesgue measure on $\mathbb{R}$ and let $f_n(x) = g_n(x) = x + 1/n$ for all n, and $f(x) = g(x) = x$; clearly $f_n \to f$ and $g_n \to g$ in measure. On the other hand, $(x + 1/n)^2 - x^2 \geq \varepsilon$ when $x \geq n\varepsilon/2$, and, consequently, $|\{x \in \mathbb{R} : |f_n(x)g_n(x) - f(x)g(x)| \geq \varepsilon\}| = \infty$ for every ε and every n.

And ψ needs to be continuous as the following example shows: Let $\psi(x) = \chi_{\{0\}}(x)$, $f_n(x) = 1/n$, and $f(x) = 0$ for $x \in [0,1]$. Then f_n converges to f in measure but since $\psi(f_n) = 0$ for all n and $\psi(f) = 1$, $\psi(f_n)$ does not converge to $\psi(f)$ in measure.

35. For the sake of argument suppose that such a metric ρ exists. Now, by Problem 2.62(c) for each n there exist pairwise disjoint measurable subsets $\{A_k^n\} \subset X$, $k = 1, \ldots, n$, such that $X = \bigcup_{k=1}^{n} A_k^n$ and $\mu(A_k^n) = \mu(A)/n$ for $k = 1, \ldots, n$. Then the sequence $\{f_m\}$ consisting of the characteristic functions of these sets, ordered first by the level n and then by k, converges to 0 in measure yet tends pointwise to 0 only possibly in a set of measure 0. Then $\rho(f_m, 0) \nrightarrow 0$ and, consequently, there exist $\varepsilon > 0$ and a subsequence $\{f_{m_k}\}$ of $\{f_m\}$ such that $\rho(f_{m_k}, 0) \geq \varepsilon$ for all $k \geq 1$. Now, since $\{f_{m_k}\}$ tends to 0 in measure by Problem 27(a) a further subsequence $\{f_{m_{k_j}}\}$ of $\{f_{m_k}\}$ converges to 0 μ-a.e. This contradicts the inequality $\rho(f_{m_{k_j}}, 0) \geq \varepsilon$ for $j \geq 1$.

36. (a) Since the proofs for d and d_1 follow along similar lines we do d. Clearly $d(f, g) \geq 0$ and if $d(f, g) = 0$, then $|f - g|/(1 + |f - g|) = 0$ μ-a.e., and so $f - g = 0$ μ-a.e. and $f = g$ in $\mathcal{F}$.

Next, $t/(1 + t)$ increases for $t > 0$ and so

$$\frac{|t + s|}{1 + |t + s|} \leq \frac{|t| + |s|}{1 + |t| + |s|} \leq \frac{|t|}{1 + |t|} + \frac{|s|}{1 + |s|}$$

for all $s, t \in \mathbb{R}$. Hence, for $f, g, h \in \mathcal{F}$,

$$\frac{|f - g|}{1 + |f - g|} = \frac{|f - h + h - g|}{1 + |f - h + h - g|} \leq \frac{|f - h|}{1 + |f - h|} + \frac{|h - g|}{1 + |h - g|},$$

and, consequently, integrating, $d(f - g) \leq d(f - h) + d(h - g)$.

(b) The statement is false. For the sake of argument assume that $(C(I), d)$ is complete and consider the sequence

$$f_n(x) = \begin{cases} n^2 x, & 0 \le x \le 1/n, \\ 1/x, & 1/n \le x \le 1. \end{cases}$$

Then, since $d(f_n, f_m) \le d_1(f_n, f_m) \le \int_0^{\vee(1/m, 1/n)} dx = \vee(1/m, 1/n)$, $\{f_n\}$ is Cauchy in $(C(I), d)$, and, consequently, there exists $f \in C(I)$ such that $\lim_n d(f_n, f) = 0$. We claim that $f(x) = 1/x$, $x \in (0, 1]$, and, since $f \notin C(I)$, this cannot happen. Now, if $f(a) \ne 1/a$ for some $a \in (0, 1]$, by continuity there exist $\varepsilon, \delta > 0$ such that $|1/x - f(x)| > \varepsilon$ for $x \in (a - \delta, a]$. Then, if N is large enough so that $1/N < a - \delta$, $f_n(x) = 1/x$ for all $n \ge N$ and $x \in [a - \delta, a)$, and, consequently, $d(f_n, f) \ge \int_{a-\delta}^{a} \varepsilon/(1 + \varepsilon) \, dx$ for these n. Then, since the right-hand side is a positive constant independent of n, f_n cannot converge to f in the metric d.

(c) That convergence in d is equivalent to convergence in d_1 follows readily from the relation $(a \wedge 1)/2 \le a/(1 + a) \le (a \wedge 1)$ valid for $a \ge 0$. Next, since $f_n \to f$ iff $f_n - f \to 0$ for any of the convergences under consideration, we assume that $f = 0$. First suppose that $f_n \to 0$ in measure. Then for $\varepsilon > 0$, $d_1(f_n, 0) = \int_{\{|f_n| \le \varepsilon\}} (|f_n| \wedge 1) \, d\mu + \int_{\{|f_n| > \varepsilon\}} (|f_n| \wedge 1) \, d\mu \le \int_{\{|f_n| \le \varepsilon\}} |f_n| \, d\mu + \int_{\{|f_n| > \varepsilon\}} d\mu \le \varepsilon \mu(X) + \mu(\{|f_n| > \varepsilon\})$. Now, since the last summand goes to zero as $n \to \infty$ it follows that $\limsup_n d_1(f_n, 0) \le \varepsilon \mu(X)$ and, since $\mu(X) < \infty$, $\limsup_n d_1(f_n, 0) = 0$. Thus the limit exists (all integrals are nonnegative) and is 0.

Finally, if $d_1(f_n, 0) \to 0$, then $d(f_n, 0) \to 0$. Now, given $\eta > 0$, let $A_n(\eta) = \{|f_n| \ge \eta\}$. Then since $t/(1 + t)$ increases in $[0, \infty)$,

$$\int_X \frac{|f_n|}{1 + |f_n|} \, d\mu \ge \int_{A_n(\eta)} \frac{|f_n|}{1 + |f_n|} \, d\mu$$

$$\ge \int_{A_n(\eta)} \frac{\eta}{1 + \eta} \, d\mu = \frac{\eta}{1 + \eta} \mu(A_n(\eta)),$$

and, consequently, since $d(f_n, 0) \to 0$, $\lim_n \mu(A_n(\eta)) = 0$ and $f_n \to 0$ in measure.

Observe that by Problem 30, $\mathcal{F}$ endowed with the metrics d, d_1 is a complete metric space and convergence in either metric is equivalent to convergence in measure.

37. (a) implies (b) Let $\lim_n f_n = 0$ in probability and φ be as in (a). For the sake of argument suppose there exist $\eta > 0$ and a subsequence $\{f_{n_k}\}$ of $\{f_n\}$ such that $\int_X \varphi(f_{n_k}) \, d\mu \ge \eta$ for all n_k. Now, $\lim_{n_k} f_{n_k} = 0$ in probability and by Problem 27(a), passing to a subsequence if needed we may assume that $\lim_{n_k} f_{n_k} = 0$ μ-a.e. Hence by (a), $\lim_{n_k} \int_X \varphi(f_{n_k}) \, d\mu = 0$, which is not the case since $\int_X \varphi(f_{n_k}) \, d\mu \geq \eta$. Therefore $\lim_n \int_X \varphi(f_n) \, d\mu = 0$ and by (a), $f_n \to 0$ μ-a.e.

(b) implies (a) If $\lim_n f_n = 0$ μ-a.e., then $\lim_n f_n = 0$ in probability, which by Problem 36 is equivalent to the convergence of $\{f_n\}$ to 0 in φ for $\varphi(x) = |x|/(1 + |x|)$ or $\varphi = \min(1, |x|)$. Conversely, by Chebychev's inequality, if $\{f_n\}$ converges to 0 in φ, then $\lim_n f_n = 0$ in probability and, therefore, by (b), $\lim_n f_n = 0$ μ-a.e.

(b) implies (c) For the sake of argument suppose there is a measurable $A \subset X$ with $\mu(A) > 0$ that contains no atoms. Then, contrary to (b), the sequence $\{f_m\}$ constructed in Problem 35 tends to 0 in probability but not μ-a.e. in A. Thus μ is a purely atomic finite measure with at most countably many atoms.

(c) implies (b) As noted in Problem 5.67, if the X_k are the atoms of X, measurable functions on X are μ-a.e. equal to a linear combination of the functions $\{\chi_{X_k}\}$. Therefore $\lim_n f_n = 0$ in probability implies that if $f_n = \sum_k \lambda_k^n \chi_{X_k}$, $\lim_n \lambda_k^n = 0$ for all k, and, consequently, $\lim_n f_n = 0$ μ-a.e.

38. Pick $n_k \to \infty$ along which the liminf is assumed, i.e., $\lim_{n_k} \int_X f_{n_k} \, d\mu = \liminf_n \int_X f_n \, d\mu$. Since $\lim_{n_k} f_{n_k} = f$ in measure, passing to a subsequence if necessary we may assume that $\lim_{n_k} f_{n_k} = f$ μ-a.e. Then $\liminf_{n_k} f_{n_k} = f$ and by Fatou's lemma $\int_X f \, d\mu \leq \liminf_{n_k} \int f_{n_k} \, d\mu$. Moreover, since this last liminf is actually taken along a convergent sequence, it equals the limit. Specifically, $\int_X f \, d\mu \leq \liminf_n \int_X f_n \, d\mu$.

39. By Problem 27(a) a subsequence $\{f_{n_k}\}$ of $\{f_n\}$ converges to f μ-a.e. and, consequently, $|f| \leq \varphi$ μ-a.e. Now, given $\varepsilon > 0$, $\int_X |f_n - f| \, d\mu \leq \int_{\{|f_n - f| > \varepsilon\varphi\}} |f_n - f| \, d\mu + \int_{\{|f_n - f| \leq \varepsilon\varphi\}} |f_n - f| \, d\mu = I + J$, say. First, $J \leq \varepsilon \int_X \varphi \, d\mu$. Next, given $\eta > 0$, observe that

$$\{|f_n - f| > \varepsilon\varphi\} = (\{|f_n - f| > \varepsilon\varphi\} \cap \{\varphi > \eta\})$$
$$\cup (\{|f_n - f| > \varepsilon\varphi\} \cap \{\varphi \leq \eta\}),$$

and, therefore, since $|f_n - f| \leq 2\varphi$ it follows that

$$I \leq 2 \int_{\{|f_n - f| > \varepsilon\eta\}} \varphi \, d\mu + 2 \int_{\{\varphi \leq \eta\}} \varphi \, d\mu.$$

Now, first pick ε so that J is arbitrarily small and then $\eta > 0$ such that $2\int_{\{\varphi \leq \eta\}} \varphi \, d\mu$ is as small as desired. Finally, since $f_n \to f$ in measure, $\lim_n \mu(\{|f_n - f| > \varepsilon\eta\}) = 0$ and, therefore, by the LDCT the first integral bounding I goes to 0.

As for the example, consider $f_n = n\chi_{[0,1/n]}$ and $f = 0$ in $[0,1]$. Or $f_n = (1/n)\chi_{[0,n]}$, $f = 0$. And it is not necessary to check that φ is not integrable because, if it was, convergence would follow and it does not.

40. First, suppose that $f \notin L^1(X)$. Then by Fatou's lemma, $\infty = \int_X f \, d\mu = \int_X \liminf_n f_n \, d\mu \le \liminf_n \int_X f_n \, d\mu$; thus $\lim_n \int_X f_n \, d\mu = \infty$ and we are done in this case. Now, if $f \in L^1(X)$, by the LDCT $\lim_n \int_X f_n \, d\mu = \int_X f \, d\mu$. Alternatively, the result follows by Fatou's lemma applied to the nonnegative sequence $\{f - f_n\}$.

41. (a) Let $A \subset X$ be measurable. Fatou's lemma then gives $\int_A f \, d\mu \le \liminf_n \int_A f_n \, d\mu$ and $\int_{X \setminus A} f \, d\mu \le \liminf_n \int_{X \setminus A} f_n \, d\mu$. Now, since the f_n and f are integrable and $\lim_n \int_X f_n \, d\mu = \int_X f \, d\mu$, the second inequality can be rewritten as $\int_X f \, d\mu - \int_A f \, d\mu \le \liminf_n (\int_X f_n \, d\mu - \int_A f_n \, d\mu) = \lim_n \int_X f_n \, d\mu - \limsup_n \int_A f_n \, d\mu = \int_X f \, d\mu - \limsup_n \int_A f_n \, d\mu$. Moreover, since $\int_X f \, d\mu < \infty$ this integral can be canceled above and, consequently, $\limsup_n \int_A f_n \, d\mu \le \int_A f \, d\mu$, which combined with $\int_A f \, d\mu \le \liminf_n \int_A f_n \, d\mu$ gives the desired result.

(b) The result still holds.

(c) The conclusion does not hold for functions of variable sign. Consider Lebesgue measure on $[-1, 1]$ and let $f_n(x) = n(\chi_{(0,1/n)}(x) - \chi_{(-1/n,0)}(x))$ and $f(x) = 0$. Then $\lim_n f_n(x) = 0 = f(x)$ for all x, and for every n we have $\int_{[-1,1]} f_n(x) \, dx = 0 = \int_{[-1,1]} f(x) \, dx$. Take $A = [0, 1]$; then $1 = \int_A f_n(x) \, dx \nrightarrow \int_A f(x) \, dx = 0$.

The result may fail if f is not integrable. On $\mathbb{R}^+$ let $f_n(x) = n\chi_{(0,1/n)}(x) + \chi_{[1,n)}(x)$; $f_n(x)$ converges pointwise to $f(x) = \chi_{[1,\infty)}(x)$ on $(0, \infty)$ and we have $\lim_n \int_{\mathbb{R}} f_n(x) \, dx = \int_{\mathbb{R}} f(x) \, dx = \infty$. But for $A = (0, 1)$, $\int_A f_n(x) \, dx = 1$ for all n, and $\int_A f(x) \, dx = 0$.

42. By Fatou's lemma $f \in L^1(X)$. Next, observe that $g + f_n \ge 0, g - f_n \ge 0$ and so, by Problem 38,

$$\int_X g \, d\mu + \int_X f \, d\mu = \int_X (g + f) \, d\mu \le \liminf_n \int_X (g + f_n) \, d\mu$$
$$= \int_X g \, d\mu + \liminf_n \int_X f_n \, d\mu,$$

and, similarly,

$$\int_X g \, d\mu - \int_X f \, d\mu = \int_X (g - f) \, d\mu \le \liminf_n \int_X (g - f_n) \, d\mu$$
$$= \int_X g \, d\mu - \limsup_n \int_X f_n \, d\mu.$$

Whence, combining, $\int_X f \, d\mu \le \liminf_n \int_X f_n \, d\mu \le \limsup_n \int_X f_n \, d\mu \le \int_X f \, d\mu$ and so $\lim_n \int_X f_n \, d\mu = \int_X f \, d\mu$.

43. Pick $n_k \to \infty$ along which the limsup is assumed, i.e.,

$$\limsup_n \int_X |f_n - f| \, d\mu = \lim_{n_k} \int_X |f_{n_k} - f| \, d\mu.$$

Now, by the convergence in measure, passing to a subsequence if necessary we may assume that $\lim_{n_k} f_{n_k} = f$ μ-a.e. and then to a further subsequence if necessary, which we denote again $\{g_{n_k}\}$, so that $g_{n_k} \to g$ μ-a.e. and in $L^1(X)$. Then, since $|f_{n_k}| \leq g_{n_k}$ μ-a.e., $|f| \leq g$ μ-a.e., and $|f_{n_k} - f| \leq g_{n_k} + g \in L^1(X)$, it follows that $g_{n_k} + g \to 2g$ in $L^1(X)$ and by Problem 106 below, $\limsup_n \int_X |f_n - f|\, d\mu = \limsup_{n_k} \int_X |f_{n_k} - f|\, d\mu \leq \int_X \limsup_{n_k} |f_{n_k} - f|\, d\mu = 0$.

44. First, since g is integrable, f_n, f are finite μ-a.e. on X. Now, for $\lambda > 0$, $\{|f_n - f| > \lambda\} \subset \{2g > \lambda\}$ and, consequently, by Chebychev's inequality $\lambda\mu(\{|f_n - f| > \lambda\}) \leq 2\int_X g\, d\mu$. Now let $\lambda_k = 2^{-k}$ and $A_k = \{2g > 2^{-k}\}$, $\mu(A_k) < \infty$. Note that $\{|f_n - f| > 2^{-k}\} \subset A_k$ for all n and $|f_n(x) - f(x)| \leq 2^{-k}$ on $X \setminus A_k$. Now, by assumption $f_n \to f$ pointwise on A_k and, therefore, by Egorov's theorem $f_n \to f$ uniformly on $A_k \setminus B_k$ where $B_k \subset A_k$ and $\mu(B_k) < \varepsilon/2^k$. Therefore there exists N_k such that $|f_n(x) - f(x)| \leq 2^{-k}$ for all $n \geq N_k$ and $x \in A_k \setminus B_k$, and so by the above estimate it follows that $|f_n(x) - f(x)| \leq 2^{-k}$ for $x \in X \setminus A_k$. Let $B = \bigcup_k B_k$; note that $\mu(B) < \varepsilon$. Let $\gamma > 0$ and choose k such that $\gamma \geq 2^{-k}$. Then there exists $N = N_k$ constructed above such that $|f_n(x) - f(x)| < \gamma$ for all $x \in X \setminus B \subset X \setminus B_k$ and all $n \geq N$. Hence $f_n \to f$ uniformly on $X \setminus B$.

45. For the sake of argument suppose such sequences exist. Then $\{f_n\}$ converges boundedly to η and by the LDCT, $\lim_n \int_I f_n(x)\, dx = \eta$, which is not the case since $\int_I f_n(x)\, dx = 0$ for all $n > 1$.

46. Let $I = \int_A \cos(2nx + 2r_n)\, dx$. Since $\cos^2(x) = (1 + \cos(2x))/2$, $\int_A \cos^2(nx + r_n)\, dx = |A|/2 + I/2$. Now, $\cos(2nx + 2r_n) = \cos(2nx)\cos(2r_n) - \sin(2nx)\sin(2r_n)$ and $I = \cos(2r_n)\int_A \cos(2nx)\, dx - \sin(2r_n)\int_A \sin(2nx)\, dx$. By Problem 4.147, or Bessel's inequality applied to the function $\chi_A(x)$, since $|\cos(2r_n)|, |\sin(2r_n)| \leq 1$, $I \to 0$ and, therefore, the limit is $|A|/2$. Moreover, since $\sin^2(nx + r_n) = 1 - \cos^2(nx + r_n)$, $\int_A \sin^2(nx + r_n)\, dx = |A|/2 - \int_A \cos^2(nx + r_n)\, dx$ and $\lim_n \int_A \sin^2(nx + r_n)\, dx = |A|/2$.

Finally, since $\sin^2(nx) = (1 - \cos(2nx))/2$, for $f \in L^1(\mathbb{R})$ it follows that $\lim_n \int_{\mathbb{R}} f(x)\sin^2(nx)\, dx = (\int_{\mathbb{R}} f(x)\, dx)/2$.

47. (a) Suppose there exist a set A of positive finite measure and a subsequence $\{f_{n_k}\}$ of $\{f_n\}$ that converges for x in A and let B be a measurable subset of A. Then by Fatou's lemma $\int_B \lim_{n_k} f_{n_k}\, d\mu \leq \liminf_{n_k} \int_B f_{n_k}\, d\mu = 0$ and since B is arbitrary, $\lim_{n_k} f_{n_k} \leq 0$ μ-a.e. on A. Similarly, since the assumptions hold for $\{-f_n\}$ and $\{f_n\}$ is bounded above, $\lim_{n_k}(-f_{n_k}) \leq 0$ on A, or $\lim_{n_k} f_{n_k} \geq 0$ μ-a.e. on A. Hence $\lim_{n_k} f_{n_k} = 0$ μ-a.e. on A.

(b) Since the Lebesgue measure on $\mathbb{R}$ is σ-finite we may assume that $|A| < \infty$. First, by (a) if $f_{n_k}(x) \to g(x)$ for $x \in A$, $g(x) = 0$ a.e. on A;

we claim that $|A| = 0$. Indeed, since $\sin^2(n_k x) = (1 - \cos(2n_k x))/2$ we have $\int_A \sin^2(n_k x)\, dx = |A|/2 - (\int_A \cos(2n_k x)\, dx)/2$, where the integral in the left-hand side goes to 0 by the LDCT and the integral in the right-hand side goes to 0 as before and, consequently, $|A| = 0$.

48. The statement is false if $\mu(X) = \infty$ as the simple example $\varphi(x) = \sum_n \chi_{[n,n+1)}(x)$ shows. It is however true if $\mu(X) < \infty$. From this result it follows that if $\sum_n |\lambda_n \cos(nx)| < \infty$ on a set of positive measure, then $\sum_n |\lambda_n| < \infty$; the idea being that since $\cos(x)$ is periodic, $|\cos(nx)| \geq c > 0$ on a set A_n with measure bounded below independently of n.

49. In general neither statement implies the other. However, the condition is sufficient under the additional assumption $\int_X |f_n - f|\, d\mu \leq \lambda_n$ with $\sum_n \lambda_n < \infty$. Indeed, if this is the case, by the MCT, $\int_X \sum_n |f_n - f|\, d\mu \leq \sum_n \lambda_n < \infty$ and, consequently, $\sum_n |f_n(x) - f(x)| < \infty$ μ-a.e. Therefore $\lim_n |f_n(x) - f(x)| = 0$ μ-a.e.

50. Suppose that $\mu(\limsup_n A_n) = 0$; then μ-a.e. every point belongs to at most finitely many of the A_n. Thus $\lim_n \chi_{A_n} f = 0$ μ-a.e., $\chi_{A_n} |f| \leq |f| \in L^1(X)$, and the conclusion follows by the LDCT. A similar argument gives that the conclusion holds if $\lim_n \mu(A_n) = 0$. Alternatively, by the absolute continuity of the integral of f, given $\varepsilon > 0$, there exists $\delta > 0$ such that if A is measurable with $\mu(A) < \delta$, then $|\int_A f\, d\mu| < \varepsilon$. Now, since $\lim_n \mu(A_n) = 0$, there is an N such that $\mu(A_n) < \delta$ for $n \geq N$ and so $|\int_{A_n} f\, d\mu| < \varepsilon$ for all $n \geq N$.

51. Since $f_n \to 0$ a.e. and $0 \leq f_n(x)\chi_{[0,N]}(x) \leq \varphi(x)\chi_{[0,N]}(x)$, $1 \leq N < \infty$, and φ is locally integrable in $[0, \infty)$, by the LDCT, $\lim_n \int_0^N f_n(x)\, dx = 0$ for any fixed N. Suppose now that $A_n \cap [N, \infty)$ has positive measure. Then, since $\lambda_n \leq \inf_{x \in A_n \cap [N,\infty)} \varphi(x)$ and φ decreases, this inf is bounded in turn by

$$\inf_{x \in [N, N+|A_n \cap [N,\infty)|]} \varphi(x) = \varphi(N + |A_n \cap [N, \infty)|)$$

$$= \varphi(N + |A_n \cap [N, \infty)|)\frac{N + |A_n \cap [N, \infty)|}{N + |A_n \cap [N, \infty)|}$$

$$\leq \frac{1}{|A_n \cap [N, \infty)|}\varphi(N)N.$$

Therefore, if $A_n \cap [N, \infty)$ has positive measure,

$$\int_N^\infty f_n(x)\, dx \leq \lambda_n |A_n \cap [N, \infty)|$$

$$\leq \frac{1}{|A_n \cap [N, \infty)|}\varphi(N)N |A_n \cap [N, \infty)|$$

$$\leq \varphi(N)N \to 0 \quad \text{as } N \to \infty.$$

The proof can now be finished in one stroke. Given $\varepsilon > 0$, pick N so large that $\int_N^\infty f_n(x)\,dx \le \varepsilon/2$ and then pick n_0 such that $\int_{[0,N]} f_n(x)\,dx \le \varepsilon/2$ for $n \ge n_0$. Then $\int_0^\infty f_n(x)\,dx \le \varepsilon$ for $n \ge n_0$.

54. Clearly $f_n \to 0$ a.e. and $\int_I f_n(x)\,dx = 1/\ln(n) \to 0$. Now, any majorant φ must verify $\varphi(x) \ge \sum_{n=2}^\infty (n/\ln((n)))\,\chi_{(1/(n+1),1/n)}(x)$ and so, by the MCT,

$$\int_I \varphi(x)\,dx \ge \sum_{n=2}^\infty (n/\ln(n)) \int_I \chi_{(1/(n+1),1/n)}(x)\,dx$$
$$= \sum_{n=2}^\infty 1/(n+1)\ln n = \infty\,.$$

Another instance of this result is the following: Let $f \in L^1(I)$ and f_n be given on I by $f_n(x) = f(x + 1/n)$. Then $f_n \to f$ in $L^1(I)$ and if f is sufficiently continuous also a.e., but in general a majorant does not exist. For instance, if $f_n(x) = |x - 1/n|^{-\lambda}$, $x \in I$, $0 < \lambda < 1$, $\{f_n\}$ converges pointwise, in measure, and in $L^1(I)$ on $(0,1)$ but if a majorant φ were to exist, then clearly $\varphi(x) \ge \sum_n \chi_{(1/(n+1),1/n]}(x)f_n(x)$. Now, it readily follows that $\int_{1/(n+1)}^{1/n} f_n(x)\,dx \sim n^{\lambda-1}$ and so $\int_0^1 \varphi(x)\,dx \ge \sum_n n^{\lambda-1} = \infty$.

Finally, if one is content with functions of variable sign, let $f_n(x) = -n\chi_{(-1/n,0)}(x) + n\chi_{(0,1/n)}(x)$, $x \in [-1,1]$, and if one doesn't mind working in an infinite measure space, the sequence $\{g_n\}$ on $\mathbb{R}$ given by $g_n(x) = \chi_{[n,n+1/n]}(x)$ will do.

55. If $a > 0$ the limit is 0, if $a = 0$ the limit is $\pi/2$, and for $a < 0$, the limit is π.

56. Applying L'Hôpital's rule twice it follows that $f(x) = 0$. Thus $\lim_n \int_0^1 f_n(x)\,dx = 1 \ne 0 = \int_0^1 f(x)\,dx$.

57. If $\lambda > 1$ and φ is locally integrable and has polynomial growth at ∞, by the MCT

$$\lim_n \int_0^n \left(1 + \frac{x}{n}\right)^n e^{-\lambda x}\,\varphi(x)\,dx = \int_0^\infty e^{-(\lambda-1)x}\,\varphi(x)\,dx\,.$$

And, if $\lambda = 1$ and $\varphi \in L^1(\mathbb{R}^+)$, by the MCT

$$\lim_n \int_0^n \left(1 + \frac{x}{n}\right)^n e^{-x}\,\varphi(x)\,dx = \int_0^\infty \varphi(x)\,dx\,.$$

58. (a) The limit is equal to $\int_0^1 (1+x)^{-1}\,dx = \ln(2)$.

(b) The statement is false.

59. The limit does not exist.

60. The limit is 0.

62. Necessity first. Note that for $x > 0$,

$$\frac{F(0) - F(x)}{x} = \sum_n \left(\frac{1 - e^{-nx}}{x} \right) a_n = \sum_n A_n(x),$$

say. Moreover, since $\lim_{x \to 0+} A_n(x) = na_n$ and by the mean value theorem $|A_n(x)| \le na_n$ for all n, by the LDCT

$$\lim_{x \to 0+} \frac{F(0) - F(x)}{x} = \sum_n na_n.$$

Sufficiency next. Since $\lim_{x \to 0}(1 - e^{-x})/x = 1$ there exists $\delta > 0$ such that $(1 - e^{-x})/x \ge 1/2$ for $0 < x \le \delta$ and, therefore,

$$\frac{F(0) - F(x)}{x} = \sum_n \left(\frac{1 - e^{-nx}}{x} \right) a_n \ge \frac{1}{2} \sum_{n=1}^{N} na_n, \quad N = [\delta/x].$$

Now, since the right-hand derivative of F exists at the origin, the left-hand side above is bounded by M, say, for $x > 0$ sufficiently small. Therefore, for all such $x > 0$, $\sum_{n=1}^{N} na_n \le 2M$, $N = [\delta/x]$. Finally, since the right-hand side above is independent of x and $N \to \infty$ as $x \to 0$, it follows that $\sum_n na_n < \infty$.

63. $\int_{\mathbb{R}+} f(x)\, dx = \sum_{n=0}^{\infty} (a + nb)^{-2}$.

65. Let $f_n(x) = (1 - \sqrt{\sin(x)})^n \cos(x)$, $x \in [0, \pi/2]$; $f_n(x)$ is nonnegative, measurable, and $\sum_{n=0}^{\infty} f_n(x) = \cos(x)/\sqrt{\sin(x)}$. Then, by the MCT,

$$\sum_{n=0}^{\infty} \int_0^{\pi/2} f_n(x)\, dx = \int_0^{\pi/2} \sum_{n=0}^{\infty} f_n(x)\, dx = \int_0^{\pi/2} \cos(x)/\sqrt{\sin(x)}\, dx = 2.$$

66. Since $x^{-x} = \sum_{n=0}^{\infty} (-x \ln x)^n / n!$, one would like to write

$$\int_0^1 x^{-x}\, dx = \sum_{n=0}^{\infty} \frac{1}{n!} \int_0^1 (-x \ln x)^n\, dx.$$

This is readily justified since $-x \ln x \ge 0$ on $[0, 1]$ and so the MCT applies; of course, one could use the bound $|x \ln x| < 1/e$ and since $\sum_n 1/n! < \infty$ invoke the LDCT. In any case,

$$\int_0^1 (-x \ln x)^n\, dx = \int_0^{\infty} u^n e^{-(n+1)u}\, du$$

$$= \frac{1}{(n+1)^{(n+1)}} \int_0^{\infty} u^n e^{-u}\, du = \frac{n!}{(n+1)^{(n+1)}},$$

and so $\int_0^1 x^{-x}\, dx = \sum_{n=0}^{\infty} 1/(n+1)^{(n+1)} = \sum_n n^{-n}$.

67. Since for continuous functions the Riemann and Lebesgue integrals coincide,

$$\int_a^b \frac{x}{1+x^2}\, dx = \frac{1}{2}\ln\left(\frac{1+a^2}{1+b^2}\right).$$

Thus, if $A_n = [-n, n]$ and $B_n = [-n, 2n]$, the limit along A_n is 0 and along B_n is $\ln(2)$.

68. The limit is $\sum_{j=1}^N a_j\,(n+j)!$.

69. The change of variables $x/k = z$ gives

$$\int_0^{k^2} \frac{2x}{k^2}\left(1 - \frac{x}{k^2}\right)^k dx = \int_0^k 2z\left(1 - \frac{z}{k}\right)^k dz$$

and so, by above, the limit in question is 2.

70. The integrand tends boundedly to $x^{a-1}/(1+x)$ as $r \to \infty$ and is bounded by the limit function, which is integrable on $\mathbb{R}^+$. Thus by the LDCT and a calculation using residues, the limit is

$$\int_0^\infty \frac{x^{a-1}}{1+x}\, dx = \frac{\pi}{\sin(\pi a)}.$$

71. Let $f_n(x)$ denote the integrand above. Then $\lim_n f_n(x) = x^2/(x^4 + x^2 + 1)$ and $|f_n(x)|$ is dominated by the limit function, which is integrable on $\mathbb{R}^+$. Therefore by the LDCT and evaluating the resulting integral by residues,

$$\lim_n \int_0^\infty f_n(x)\, dx = \int_0^\infty \frac{x^2}{x^4 + x^2 + 1}\, dx = \frac{\pi}{2\sqrt{3}}.$$

72. Since $\lim_{\lambda \to \infty}(\sin(|x|/\lambda))/(|x|/\lambda) = 1$ for $x \neq 0$, the integrand above as well as the limit are dominated by $1/(1 + |x|)^{-(n+1)+\alpha}$, which is integrable since $\alpha < 1$. Therefore the limit is $\int_{\mathbb{R}^n} f(x)/(1+|x|)^{n+1}\, dx$.

73. Let $f_n(x)$ denote the integrand above. Since $|\sin(x^{-1}e^x)| \leq 1$ for all x it is readily verified that

$$\lim_n f_n(x) = \begin{cases} 1/(\cos(x) - 2), & 0 < x < 2\pi, \\ 1/2(\cos(x) - 2), & x = 2\pi, \\ 0, & x > 2\pi. \end{cases}$$

To complete the verification of the assumptions of the LDCT we find an integrable function g such that $|f_n(x)| \leq g(x)$ for $n \geq 2$ and $x \in (0, \infty)$. Let $g(x) = (4/3)\chi_{(0,1)}(x) + x^{-2}\chi_{[1,\infty)}(x)$; g is integrable. Now, if $x > 1$ and $n \geq 2$, since $1/(\cos(x) - 2)$ is bounded and

$$1 + x^n + \frac{\sin(x^{-1}e^x)}{4n} \geq x^n \geq x^2,$$

we have $f_n(x) \leq g(x)$. And, if $0 < x \leq 1$,

$$1 + x^n + \frac{\sin(x^{-1}e^x)}{4n} \geq 1 - \frac{|\sin(x^{-1}e^x)|}{4n} \geq 1 - \frac{1}{4n} \geq 1 - \frac{1}{4} = \frac{3}{4}.$$

Hence $f_n(x) \leq g(x)$. Thus, by the LDCT, computing the integral by residues,

$$\lim_n \int_0^\infty f_n(x)\, dx = \int_0^{2\pi} \frac{1}{\cos(x) - 2}\, dx = \frac{-\sqrt{3}}{3}\, \pi\,.$$

74. A monotone function has a right-hand side limit a at 0, $a \leq \infty$. Assume first that f is decreasing. Then $\{g_n\}$ is an increasing sequence that converges to a a.e. and by the MCT, $\lim_n \int_I g_n(x)\, dx = a \int_I dx = a$. On the other hand, if f is increasing, $a < \infty$. Then $\{g_n\}$ is decreasing and converges to a a.e. Furthermore, $g_1 = f$ is integrable and by Problem 102 the limit is a.

75. First, in the notation of Problem 74 with $f(x) = \sin(x)/x$ there, $a = 1$ and $\lim_n \int_0^1 g_n(x)\, dx = 1$. Furthermore, $|\sin(x^n)|/|x|^n \leq |x|^{-2}$ for n large, $\lim_n |\sin(x^n)|/|x|^n = 0$, and by the LDCT $\lim_n \int_1^\infty g_n(x)\, dx = 0$. Thus the limit is 1.

76. The limit exists and is 0.

77. The limit is 0.

78. Since $\sin(x)$ is periodic with period 2π we may assume that $0 \leq r_n \leq 2\pi$. Then

$$\int_0^\infty |e^{-x} \sin^n(x + r_n)|\, dx = e^{r_n} \int_{r_n}^\infty e^{-x} |\sin^n(x)|\, dx$$

$$\leq e^{2\pi} \int_0^\infty e^{-x} |\sin^n(x)|\, dx\,.$$

Now, since $\lim_n |\sin^n(x)| = 0$ a.e., by the LDCT $\lim_n \int_0^\infty e^{-x} |\sin^n(x)|\, dx = 0$ and the same is true for the integral we want to estimate.

79. First, observe that

$$\int_I \frac{(1 - x)^2}{\ln^2(x)}\, \frac{dx}{x} = \int_{-\infty}^0 \frac{(1 - e^u)^2}{u^2}\, du < \infty\,.$$

Next, write

$$\frac{n(1 - x)^2}{(1 + nx) \ln^2(x)} = \frac{n(1 - x)^2}{(1 + nx) \ln^2(x)} \mp \frac{1}{x} \frac{(1 - x)^2}{\ln^2(x)}$$

and note that the integral in question is equal to

$$\int_I \left(\frac{n}{(1+nx)} - \frac{1}{x} \right) \frac{(1-x)^2}{\ln^2(x)} \cos(nx)\,dx + \int_I \frac{1}{x} \frac{(1-x)^2}{\ln^2(x)} \cos(nx)\,dx$$

$$= A_n + B_n\,,$$

say. Since the integrand in B_n is integrable, by Problem 4.147, $\lim_n B_n = 0$. As for A_n, the integrand there is bounded by

$$\frac{1}{(1+nx)} \frac{1}{x} \frac{(1-x)^2}{\ln^2(x)} \le \frac{1}{x} \frac{(1-x)^2}{\ln^2(x)} \in L^1(I)\,,$$

and tends to 0 as $n \to \infty$ for $x \neq 0$, and, consequently, by the LDCT $\lim_n A_n = 0$. Thus the limit is 0.

80. Let $\varphi_n = f_n / \|f_n\|_2$; then $\int_X \varphi_n \varphi_m \, d\mu = 0$ for $n \neq m$ and $\|\varphi_n\|_2 = 1$. Now, since $\|f_n\|_2 \le c$, $\{f_n > \varepsilon\} \subset \{\varphi_n > \varepsilon/c\}$. Then for g bounded, say, by Bessel's inequality $\sum_n |\int_X g \varphi_n \, d\mu|^2 \le \|g\|_2^2$ and with $g = \chi_A$ it follows that $\sum_n |\int_A \varphi_n \, d\mu|^2 \le \mu(A)$ and, consequently, $\lim_n \int_A \varphi_n \, d\mu = 0$. Then if $\varphi_n(x) > \varepsilon/c$ for $x \in A$ with $\mu(A) > 0$, then $\int_A \varphi_n \, d\mu > \varepsilon \, \mu(A)/c$ and since the limit of the integrals is zero this can only happen for finitely many n.

In particular, given $0 < \eta < 1$ and $A \subset \mathbb{R}$ with $|A| > 0$ there can be at most finitely many distinct integers n with the property that $\cos(nx) \ge \eta$ for all $x \in A$.

81. Since $\{y_k\}$ is bounded by Bolzano-Weierstrass it has a subsequence, which we call $\{y_k\}$ again for simplicity, that converges to a limit $y \in \mathbb{R}^n$, say. Then by the continuity of integrable functions in the $L^1(\mathbb{R}^n)$ metric, $\lim_k \int_{\mathbb{R}^n} |f(x+y) - f(x+y_k)| \, dx = 0$. Finally, if a sequence converges in $L^1(\mathbb{R}^n)$, it has an a.e. convergent subsequence, and we have finished.

83. First, since $(f_n^{1/2} - f^{1/2})^2 = f_n + f - 2f_n^{1/2}f^{1/2}$ it follows that $\int_X (f_n^{1/2} - f^{1/2})^2 \, d\mu = 2 - 2\int_X f_n^{1/2}f^{1/2} \, d\mu$. And, since $\lim_n f_n^{1/2}f^{1/2} = f$ μ-a.e., by Fatou's lemma

$$\limsup_n \int_X (f_n^{1/2} - f^{1/2})^2 \, d\mu = 2 - 2\liminf_n \int_X f_n^{1/2}f^{1/2} \, d\mu$$

$$\le 2 - 2\int_X f \, d\mu = 0\,,$$

and, consequently, the limit is 0.

84. (a) Consider the sequence $g_n = 2^p \, |f_n|^p + 2^p \, |f|^p - |f_n - f|^p \ge 0$. Then $\liminf_n g_n = 2 \cdot 2^p \, |f|^p$ and by Fatou's lemma, $2^{p+1} \int_X |f|^p \, d\mu \le 2^p \int_X |f|^p \, d\mu + \liminf_n (2^p \int_X |f_n|^p \, d\mu - \int_X |f_n - f|^p \, d\mu)$. Now, since $2^{p+1} - 2^p = 2^p$ and $f \in L^p(X)$, we have $2^p \int_X |f|^p \, d\mu \le \liminf_n (2^p \int_X |f_n|^p \, d\mu - \int_X |f_n - f|^p \, d\mu)$.

(b) Necessity is evident and will be discussed in the context of Banach spaces. Conversely, if $\lim_n a_n$ exists, $\liminf_n(a_n - b_n) \le \lim_n a_n - \limsup_n b_n$

and so by (a), $2^p \int_X |f|^p \, d\mu \leq 2^p \lim_n \int_X |f_n|^p \, d\mu - \limsup_n \int_X |f_n - f|^p \, d\mu$.
Hence $\lim_n \int_X |f_n - f|^p \, d\mu = 0$.

(c) The statement is false. In $\mathbb{R}^+$ let $f_n(x) = \chi_{(1/n,\infty)}(x)$ and $f(x) = \chi_{\mathbb{R}^+}(x)$. Then $\lim_n f_n(x) = f(x)$ for $x \neq 0$, $1 = \|f\|_\infty = \|f_n\|_\infty$ for all n, yet $\|f - f_n\|_\infty = 1$ for all n.

85. First, note that given $\varepsilon > 0$, there is a constant c_ε such that $\left| |s + t|^p - |t|^p \right| \leq c_\varepsilon |s|^p + \varepsilon |t|^p$ for all real s, t. Indeed, if $0 < p \leq 1$, $|s + t|^p - |t|^p \leq |s|^p$ and the inequality holds for every $\varepsilon > 0$ with $c_\varepsilon = 1$. And, when $1 < p < \infty$, we consider two cases: $|t| \leq |s|/4$ and $|s| \leq 4|t|$. In the former case $\left| |s + t|^p - |t|^p \right| \leq c |s|^p$, c a constant independent of ε, and in the latter case $\left| |s + t|^p - |t|^p \right| \leq p |t + \xi|^{p-1} |s|$ where $|\xi| \leq |s| \leq |t|$, and by Young's inequality, since $q(p - 1) = p$, $\left| |s + t|^p - |t|^p \right| \leq \varepsilon |t|^p + c_\varepsilon |s|^p$.

By virtue of this inequality, since $f_n = f_n - f + f$, with $s = f$ and $t = (f - f_n)$, it follows that $\left| |f_n|^p - |f - f_n|^p - |f|^p \right| \leq |f|^p + |(f_n - f) + f|^p - |f - f_n|^p \leq (1 + c_\varepsilon)|f|^p + \varepsilon |f - f_n|^p$. Now, with $a_+ = \vee(a, 0)$, given $\varepsilon > 0$, let $W_{\varepsilon,n}(x) = \left(||f_n|^p - |f - f_n|^p - |f|^p| - \varepsilon |f - f_n|^p \right)_+$ and observe that since $f_n \to f$ μ-a.e., $W_{\varepsilon,n}(x) \to 0$ μ-a.e. Also, since $W_{\varepsilon,n}(x) \leq (1 + c_\varepsilon)|f(x)|^p \in L^1(X)$, by the LDCT, $\lim_n \int_X W_{\varepsilon,n} \, d\mu = 0$. Finally, since $s \leq (s - t)_+ + t$, it follows that $\left| |f_n|^p - |f - f_n|^p - |f|^p \right| \leq W_{\varepsilon,n}(x) + \varepsilon |f - f_n|^p$, which together with $\|f - f_n\|_p \leq \|f\|_p + c$ implies $\limsup_n \int_X ||f_n|^p - |f - f_n|^p - |f|^p| \, d\mu \leq \varepsilon \limsup_n \int_X |f - f_n|^p \, d\mu \leq \varepsilon(\|f\|_p + c)^p$, and the conclusion follows since ε is arbitrary. This result is from Brezis and Lieb, *A relation between pointwise convergence of functions and convergence of functionals*, Proc. Amer. Math. Soc. **88** (1983), 486–490.

86. (a) The statement is false. To see this let $f_n = \chi_{[0,1/n]}$ for all n and $f = 0$. Given $\varepsilon > 0$, let $A = [0, \varepsilon]$, $|A| = \varepsilon$. Since $f_n = 0$ on A^c for all n with $1/n < \varepsilon$, $\lim_n f_n = 0$ uniformly on A^c and $\lim_n f_n = f$ almost uniformly. However, since $\|f_n\|_\infty = 1$ for all n, $f_n \not\to 0$ in $L^\infty(I)$ as $n \to \infty$.

(b) The statement is true. Let B_n with $\mu(B_n) = 0$ be such that $|f_n(x) - f(x)| \leq \|f_n - f\|_\infty$ for $x \in B_n^c$ and $B = \bigcup_n B_n$, $\mu(B) = 0$. Now, given $\varepsilon > 0$, let N be such that $\|f_n - f\|_\infty \leq \varepsilon$ for $n \geq N$. Then $\sup_{x \in B^c} |f_n(x) - f(x)| \leq \|f_n - f\|_\infty \leq \varepsilon$ for $n \geq N$ and $f_n \to f$ uniformly on B^c.

Sufficiency next. Let $M_n = \sup_{x \in B^c} |f_n(x) - f(x)|$. Then $|f_n(x) - f(x)| \leq M_n$ for all $x \in B^c$ where $\mu(B) = 0$ and $M_n \in \{c : |f_n - f| \leq c \ \mu\text{-a.e.}\}$. Therefore $\|f_n - f\|_\infty = \inf\{c : |f_n - f| \leq c \ \mu\text{-a.e.}\} \leq M_n$.

87. When $p = \infty$ the conclusion follows from Problem 86(b).

88. Since $\left\| \sum_k (f_{k+1} - f_k) \right\|_p \leq \sum_k \|f_{k+1}\|_p + \sum_k \|f_k\|_p < \infty$, it follows that $\sum_k (f_{k+1}(x) - f_k(x))$ is finite μ-a.e. Now, since $f_n(x) = f_1(x) + \sum_{k=1}^{n-1} (f_{k+1}(x) - f_k(x))$, $\lim_n f_n(x) = f_1(x) + \sum_k (f_{k+1}(x) - f_k(x))$ exists and is finite μ-a.e. and so it only remains to prove it is 0. Since $\sum_n \|f_n\|_p < \infty$,

$\lim_n \|f_n\|_p = 0$ and, consequently, a subsequence $\{f_{n_k}\}$ of $\{f_n\}$ converges to 0 μ-a.e. Finally, if a sequence converges, it converges to the same limit as that of any of its convergent subsequences and so $\lim_n f_n = 0$ μ-a.e.

91. By Problem 49 (a) implies (b), so we prove (b). Given $\varepsilon > 0$, let N be such that $\int_X |f_n - f| \, d\mu \leq \varepsilon/2$ for $n \geq N$. Since $f, f_1, \ldots, f_N$ are integrable, their integrals are absolutely continuous and so by picking the smallest δ corresponding to $\varepsilon/2$ for this finite collection of functions it follows that there exists $\delta > 0$ such that $\mu(A) < \delta$ implies $\int_A |f| \, d\mu, \int_A |f_n| \, d\mu \leq \varepsilon/2$, $1 \leq n \leq N$. Also, for $\mu(A) < \delta$ and $n > N$, $\int_A |f_n| \, d\mu \leq \int_A |f_n - f| \, d\mu + \int_A |f| \, d\mu \leq \varepsilon/2 + \varepsilon/2 = \varepsilon$. Since this already holds for $n \leq N$, $\{f_n\}$ is uniformly absolutely continuous.

92. (a) iff (b) First, necessity. For the sake of argument suppose there exist $\eta > 0$, $\{f_n\} \subset \mathcal{F}$, and $\{A_n\}$ with $\mu(A_n) \leq \varepsilon_n$, $\varepsilon_n \to 0$, such that $\eta \leq \int_{A_n} |f_n| \, d\mu$. Let $\delta > 0$ be such that $\sup_{f \in \mathcal{F}} \int_A |f| \, d\mu \leq \eta/2$ whenever $\mu(A) < \delta$. Now pick $0 < \varepsilon_n < \delta$ and note that since $\mu(A_n) \leq \varepsilon_n < \delta$, $\eta \leq \int_{A_n} |f_n| \, d\mu \leq \eta/2$, which cannot happen.

Sufficiency is immediate. Since $\omega(\mathcal{F}, \delta)$ decreases to $\omega(\mathcal{F}) = 0$ as $\delta \to 0$, given $\varepsilon > 0$, $\omega(\mathcal{F}, \delta) \leq \varepsilon$ provided δ is small enough.

(b) iff (c) We actually prove that $\omega(\mathcal{F}) = \widetilde{\omega}(\mathcal{F})$ and, therefore, in particular, $\omega(\mathcal{F}) = 0$ iff $\widetilde{\omega}(\mathcal{F}) = 0$. Note that since $\omega(\mathcal{F}, \varepsilon)$ and $\widetilde{\omega}(\mathcal{F}, R)$ decrease as $\varepsilon \to 0^+$ and $R \to \infty$, respectively, the limits in the definitions of $\omega(\mathcal{F})$ and $\widetilde{\omega}(\mathcal{F})$ coincide with the infima. Since $\mathcal{F}$ is bounded, by Chebychev's inequality there is a constant c such that $\mu(B_{f,R}) \leq c/R$ for all $f \in \mathcal{F}$. Now, given $\varepsilon > 0$, let R be large enough so that $\sup_{f \in \mathcal{F}} \mu(B_{f,R}) \leq \varepsilon$. Then for R sufficiently large, $\widetilde{\omega}(\mathcal{F}) \leq \widetilde{\omega}(\mathcal{F}, R) \leq \omega(\mathcal{F}, \varepsilon)$, and since ε is arbitrary $\widetilde{\omega}(\mathcal{F}) \leq \omega(\mathcal{F})$.

The reverse inequality is also true. Given $\eta > 0$, let $R_\eta > 0$ be such that $\widetilde{\omega}(\mathcal{F}, R) \leq \widetilde{\omega}(\mathcal{F}) + \eta$, all $R > R_\eta$. Then for each $f \in \mathcal{F}$ and $A \in \mathcal{M}$ with $\mu(A) \leq \varepsilon$ it follows that

$$\int_A |f| \, d\mu \leq \int_{B_{f,R}} |f| \, d\mu + \int_{A \cap B_{f,R}^c} |f| \, d\mu \leq \widetilde{\omega}(\mathcal{F}, R) + \varepsilon R \leq \widetilde{\omega}(\mathcal{F}) + \eta + \varepsilon R,$$

and so letting $\varepsilon \to 0$, $\omega(\mathcal{F}) \leq \widetilde{\omega}(\mathcal{F}) + \eta$. Therefore, since η is arbitrary, $\omega(\mathcal{F}) \leq \widetilde{\omega}(\mathcal{F})$.

(c) iff (d) Necessity first. Observe that for each k there exists R_k such that $\sup_{f \in \mathcal{F}} \int_{B_{f,R_k}} |f| \, d\mu \leq 1/2^k$; we may suppose that the sequence $\{R_k\}$ is increasing. For $t > 0$ put $\varphi(t) = t$ if $t < R_1$ and $= t \sum_k \chi_{[R_k, \infty)}(t)$ otherwise; then $\varphi(t)/t = \text{card}\{k : R_k \leq t\} \to \infty$ as $t \to \infty$. Finally, since $\varphi(|f(x)|) = |f(x)| \sum_k \chi_{[R_k, \infty)}|f(x)| = |f(x)| \sum_k \chi_{B_{f,R_k}}(x)$, $\int_X \varphi(|f|) \, d\mu = \sum_k \int_{B_k} |f| \, d\mu \leq \sum_k 2^{-k} < \infty$ and $\sup_{f \in \mathcal{F}} \int_X \varphi(|f|) \, d\mu < \infty$.

Next, sufficiency. Given $\varepsilon > 0$, let t_0 be such that $t/\varphi(t) \leq \varepsilon$ if $t \geq t_0$. Then for $R > t_0$, $\widetilde{\omega}(\mathcal{F}, R) = \sup_{f \in \mathcal{F}} \int_{B_{f,R}} |f| \, d\mu \leq \varepsilon \sup_{f \in \mathcal{F}} \int_X \varphi(|f|) \, d\mu$. Thus, letting $R \to \infty$, $\widetilde{\omega}(\mathcal{F}) \leq \varepsilon \sup_{f \in \mathcal{F}} \int_X \varphi(|f|) \, d\mu$, and since ε is arbitrary $\widetilde{\omega}(\mathcal{F}) = 0$ and (c) follows.

93. The statement is false. On the other hand, an absolutely uniformly continuous family is bounded in $L^1(X)$.

94. First, necessity. (a) is essentially done in Problem 92. As for (b), given $\varepsilon > 0$, let $f_0 = f$ and pick n_0 such that $\|f_n - f_0\|_1 \leq \varepsilon/2$ for $n \geq n_0$. Now, by Problem 4.123 there exist $A_n \in \mathcal{M}$ with $\mu(A_n) < \infty$ and $\int_{A_n^c} |f_n| \, d\mu \leq \varepsilon/2$ for all $n \leq n_0$. Let $A = \bigcup_{n=0}^{n_0} A_n$; then $\mu(A) \leq \sum_{n=0}^{n_0} \mu(A_n) < \infty$ and $A^c \subset A_n^c$ for $0 \leq n \leq n_0$. With this choice of A we have $\int_{A^c} |f_n| \, d\mu \leq \int_{A_n^c} |f_n| \, d\mu \leq \varepsilon/2$ for all $n \leq n_0$ and $\int_{A^c} |f_n| \, d\mu \leq \int_{A_0^c} |f| \, d\mu + \int_X |f_n - f| \, d\mu \leq \varepsilon$ for all $n \geq n_0$. Hence $\mu(A) < \infty$ and $\int_{A^c} |f_n| \, d\mu \leq \varepsilon$ for all n.

Next, sufficiency. Given $\varepsilon > 0$, by (b) there exists $A \in \mathcal{M}$ with $\mu(A) < \infty$ and $\int_{A^c} |f_n| \, d\mu \leq \varepsilon/2$ for all n. Now, by Fatou's lemma applied to $\{|f_n|\chi_{A^c}\}$, $\int_{A^c} |f| \, d\mu \leq \varepsilon/2$, and, therefore, $\int_{A^c} |f_n - f| \, d\mu \leq \varepsilon$ for all n. Thus, since $\mu(A) < \infty$, we have reduced the problem to the finite measure case. Now, by (a) there exists $\delta > 0$ such that $\mu(B) \leq \delta$ implies $\int_B |f_n| \, d\mu \leq \varepsilon$ for all n. With this choice of δ, by Egorov's theorem there exists a set B with $\mu(B) < \delta$ such that $\{f_n\}$ converges to f uniformly on $A \setminus B$. Therefore by Problem 119, $\int_{A \setminus B} |f_n - f| \, d\mu \to 0$ as $n \to \infty$ and so it remains to estimate the integral over B. First, note that by Fatou's lemma $\int_B |f| \, d\mu \leq \liminf_n \int_B |f_n| \, d\mu \leq \varepsilon$. Therefore $\int_B |f_n - f| \, d\mu \leq \int_B |f_n| \, d\mu + \int_B |f| \, d\mu \leq 2\varepsilon$ is arbitrarily small. Thus combining the estimates over A^c, $A \setminus B$, and B, we have $\lim_n \int_X |f_n - f| \, d\mu = 0$. Also, $\int_X |f| \, d\mu \leq \int_X |f - f_n| \, d\mu + \int_X |f_n| \, d\mu < \infty$ and $\|f\|_1 \leq \liminf_n \|f_n\|_1$.

Note that if $\mu(X) < \infty$, (b) holds automatically. If the measure is infinite, (b) is necessary as the example $f_n = \chi_{[n,n+1)}$ shows: $\{f_n\}$ is bounded in $L^1(\mathbb{R})$, uniformly absolutely continuous, and converges to $f = 0$ a.e. but $\lim_n \int_{\mathbb{R}} |f_n(x) - f(x)| \, dx = 1$.

95. We discuss sufficiency. Let $\{f_{n_k}\}$ be a subsequence of $\{f_n\}$ that converges to f μ-a.e. Then by Fatou's lemma $\int_X |f| \, d\mu \leq \liminf_k \int_X |f_{n_k}| \, d\mu < \infty$, where the last inequality follows from Problem 93 since uniformly continuous families in $L^1(X)$ are bounded in $L^1(X)$. Thus f is integrable. Next, since $\{|f_n|, |f|\}$ is uniformly absolutely continuous, given $\varepsilon > 0$, let δ be the value corresponding to $\varepsilon/4$ for this family. Now, since f_n converges to f in measure, pick N such that $\mu(\{|f_n - f| > \varepsilon/2\}) < \delta$ for $n > N$. Then with $B_n = \{|f_n - f| > \varepsilon/2\}$ and $n \geq N$, $\int_X |f_n - f| \, d\mu \leq \int_{B_n} (|f_n| + |f|) \, d\mu + \int_{B_n^c} |f_n - f| \, d\mu \leq 2\varepsilon/4 + \varepsilon/2 = \varepsilon$.

96. If $\mathcal{F}$ is uniformly absolutely continuous we pick $\{f_k\}$ for the subsequence and $A_k = \emptyset$ for all k. Otherwise, by Problem 92(b), $\omega(\mathcal{F}) > 0$ and there exist a subsequence of $\{f_n\}$, which we denote $\{f_n\}$ for simplicity, and measurable sets $\{B_n\}$ such that $\mu(B_n) < 2^{-n}$ and $\int_{B_n} |f_n|\,d\mu > \omega(\mathcal{F}) - 2^{-n}$ for all n. Let $n_1 = 1$. Since $\lim_k \mu(\bigcup_{n=k}^{\infty} B_n) = 0$ and $f_{n_1} \in L^1(X)$, it follows that $\int_{\bigcup_{n=k}^{\infty} B_n} |f_{n_1}|\,d\mu \to 0$ as $k \to \infty$ and so, there exists $n_2 > n_1$ such that $\int_{B_{n_1} \setminus \bigcup_{n=n_2}^{\infty} B_n} |f_{n_1}|\,d\mu > \omega(\mathcal{F}) - 2^{-n_1}$. Let $A_1 = B_{n_1} \setminus \bigcup_{n=n_2}^{\infty} B_n$. Applying the same argument to f_{n_2}, pick $n_3 > n_2$ so that $\int_{B_{n_2} \setminus \bigcup_{n=n_3}^{\infty} B_n} |f_{n_2}|\,d\mu > \omega(\mathcal{F}) - 2^{-n_2}$ and let $A_2 = B_{n_2} \setminus \bigcup_{n=n_3}^{\infty} B_n$. In this way we obtain a subsequence $\{f_{n_k}\}$ of $\{f_n\}$ and pairwise disjoint measurable sets $\{A_k\}$ such that $\int_{A_k} |f_{n_k}|\,d\mu > \omega(\mathcal{F}) - 2^{-n_k}$ for all k. We claim that $\{\chi_{A_k^c} f_{n_k}\}$ is uniformly absolutely continuous. For the sake of argument suppose there exist $\eta > 0$, a subsequence of $\{f_{n_k}\}$, which we denote $\{f_{n_k}\}$ for simplicity, and measurable sets $\{B_k\}$ with $\lim_k \mu(B_k) = 0$ such that $\int_{B_k} \chi_{A_k^c} |f_{n_k}|\,d\mu > \eta$ for all k. Then $\int_{B_k \cup A_k} |f_{n_k}|\,d\mu = \int_{A_k} |f_{n_k}|\,d\mu + \int_{B_k} \chi_{A_k^c} |f_{n_k}|\,d\mu > \omega(\mathcal{F}) - 2^{-n_k} + \eta$ for all k, which, since $\omega(\mathcal{F}) - 2^{-n_k} + \eta > \omega(\mathcal{F})$ for k large enough, contradicts the definition of $\omega(\mathcal{F})$. This result is known as the Rosenthal splitting or Biting lemma.

97. (a) implies (b) Let $\{f_{n_k}\}$ be a subsequence of $\{f_n\}$ that converges to f μ-a.e. Then by Fatou's lemma $\int_X |f|^p\,d\mu \le \liminf_k \int_X |f_{n_k}|^p\,d\mu < \infty$, where the last inequality follows from Problem 93 since uniformly continuous families in $L^p(X)$ are bounded in $L^p(X)$. Thus $f \in L^p(X)$. Now, since $f \in L^p(X)$ and $\{|f_n|^p\}$ is uniformly absolutely continuous, it readily follows that $\{|f_n - f|^p\}$ is uniformly absolutely continuous. Since $f_n \to f$ in probability iff $f_n - f \to 0$ in probability we can assume that $f = 0$ μ-a.e. and, consequently, we need to prove that $\int_X |f_n|^p\,d\mu \to 0$. Fix $\varepsilon > 0$ and observe that

$$\int_X |f_n|^p\,d\mu = \int_X |f_n|^p \chi_{\{|f_n|^p \le \varepsilon/2\}}\,d\mu + \int_X |f_n|^p \chi_{\{|f_n|^p > \varepsilon/2\}}\,d\mu$$

$$\le \varepsilon/2 + \int_X |f_n|^p \chi_{\{|f_n|^p > \varepsilon/2\}}\,d\mu\,.$$

Let $\rho > 0$ be such that $\sup_n \int_X |f_n|^p \chi_A\,d\mu < \varepsilon/2$ whenever $\mu(A) \le \rho$. Convergence in probability now implies that there exists n_0 such that for $n \ge n_0$ we have $\mu(\{|f_n|^p > \varepsilon/2\}) \le \rho$. It then follows that $\int_X |f_n|^p\,d\mu \le \varepsilon$ for $n \ge n_0$.

(b) implies (c) It is trivial and holds in the general context of Banach spaces.

(c) implies (a) For $M \geq 1$ define the function $\psi_M : [0, \infty) \to [0, \infty)$ by

$$
M(x) = \begin{cases} x, & 0 \leq x \leq M - 1, \\ 0, & M \leq x < \infty, \end{cases}
$$

interpolated linearly in $x \in (M - 1, M)$. Observe that for a given $\varepsilon > 0$, the LDCT ensures the existence of a constant $M > 0$ (which we fix throughout) such that $\int_X |f|^p \, d\mu - \int_X \psi_M(|f|^p) \, d\mu < \varepsilon/2$. Convergence in probability together with continuity of M imply that $\psi_M(f_n) \to \psi_M(f)$ in probability for all M. Hence, from the boundedness of ψ_M and the LDCT, it follows that $\int_X \psi_M(|f_n|^p) \, d\mu \to \int_X \psi_M(|f|^p) \, d\mu$. Now, by (c) and the above remark, there exists n_0 such that $\int_X |f_n|^p \, d\mu - \int_X |f|^p \, d\mu < \varepsilon/4$ and $\int_X \psi_M(|f|^p) \, d\mu - \int_X \psi_M(|f_n|^p) \, d\mu < \varepsilon/4$ for $n \geq n_0$. Therefore for $n \geq n_0$, $\int_X |f_n|^p \chi_{\{|f_n|^p > M\}} \, d\mu \leq \int_X |f_n|^p \, d\mu - \int_X \psi_M(|f_n|^p) \, d\mu \leq \varepsilon/2 + \int_X |f|^p \, d\mu - \int_X \psi_M |f|^p \, d\mu \leq \varepsilon$. Finally, to get the uniform absolute continuity of the entire sequence, we choose an even larger value of M to get $\int_X |f_n|^p \chi_{\{|f_n|^p > M\}} \, d\mu \leq \varepsilon$ for the remaining $n < n_0$.

98. First, note that since $\mathcal{F}$ is bounded, by Problem 38, $g \in L^1(X)$.

(a) implies (b) By Problem 96 there are a subsequence $\{f_{n_k}\}$ of $\{f_n\}$ and pairwise disjoint $\{A_k\} \subset \mathcal{M}$ with $\lim_k \mu(A_k) = 0$ such that $\{\chi_{A_k^c} f_{n_k}\}$ is uniformly absolutely continuous and $\lim_k \int_{A_k} f_{n_k} \, d\mu = \omega(\mathcal{F})$. Furthermore, since $\{|\chi_{A_k^c} f_{n_k} - g| > \eta\} \subset A_k \cup \{|f_{n_k} - g| > \eta\}$ for all $\eta > 0$, $\lim_k \chi_{A_k^c} f_{n_k} = g$ in measure and, consequently, by Problem 97, $\lim_k \chi_{A_k^c} f_{n_k} = g$ in $L^1(X)$. Hence, since $\int_X f_{n_k} \, d\mu = \int_X \chi_{A_k} f_{n_k} \, d\mu + \int_X \chi_{A_k^c} f_{n_k} \, d\mu$ and by assumption $\lim_{n_k} \int_X f_{n_k} \, d\mu = \lim_n \int_X f_n \, d\mu$ exists, $\lim_n \int_X f_n \, d\mu = \lim_{n_k} \int_X f_{n_k} \, d\mu = \omega(\mathcal{F}) + \int_X g \, d\mu$.

In general, let $\mathcal{F}' = \{f_{n_k}\}$ be a subsequence of $\{f_n\}$. Since $f_{n_k} \to g$ in measure, by the above argument with $\mathcal{F}'$ in place of $\mathcal{F}$, it follows that $\lim_{n_k} \int_X f_{n_k} \, d\mu = \omega(\mathcal{F}') + \int_X g \, d\mu$. Finally, since $g \in L^1(X)$ and $\lim_n \int_X f_n \, d\mu = \lim_{n_k} \int_X f_{n_k} \, d\mu$, it follows that $\omega(\mathcal{F}) = \omega(\mathcal{F}')$.

(b) implies (c) Let $\mathcal{F}' = \{f_{n_k}\}$ and $\mathcal{F}'' = \{f_{n_\ell}\}$ be subsequences of $\{f_n\}$ along which $\{\int_X f_n \, d\mu\}$ assumes its $\liminf$ and $\limsup$, respectively. Then by (b), $\omega(\mathcal{F}') = \omega(\mathcal{F}'')$, and $\liminf_n \int_X f_n \, d\mu = \lim_{n_k} \int_X f_{n_k} \, d\mu$ and $\limsup_n \int_X f_n \, d\mu = \lim_{n_\ell} \int_X f_{n_\ell} \, d\mu$.

(c) implies (a) Let $\mathcal{F}' = \{f_{n_k}\}$ and $\mathcal{F}'' = \{f_{n_\ell}\}$ be subsequences of $\{f_n\}$ that satisfy (c). Then, as in (a) implies (b), we get that $\lim_{n_k} \int_X f_{n_k} \, d\mu = \omega(\mathcal{F}') + \int_X g \, d\mu$ and $\lim_{n_\ell} \int_X f_{n_\ell} \, d\mu = \omega(\mathcal{F}'') + \int_X g \, d\mu$. Therefore, since $\omega(\mathcal{F}') = \omega(\mathcal{F}'')$ it follows that $\liminf_n \int_X f_n \, d\mu = \lim_{n_k} \int_X f_{n_k} \, d\mu = \omega(\mathcal{F}') + \int_X g \, d\mu = \omega(\mathcal{F}'') + \int_X g \, d\mu = \lim_{n_\ell} \int_X f_{n_\ell} \, d\mu = \limsup_n \int_X f_n \, d\mu$ and (a) holds. This result is from H.-A. Klei, *Convergences faible, en mesure et au sens de Cesaro dans* $L^1(\mathbb{R})$, C. R. Acad. Sci. Paris, Ser. I **315** (1992), 9–12.

99. (a) As in Problem 97, $g \in L^1(X)$. Let $\mathcal{F}' = \{|f_{n_k} - g|\}$ be a subsequence of $\{|f_n - g|\}$ along which $\{\int_X |f_n - g|\, d\mu\}$ assumes its $\limsup$. Then since $|f_n - g| \to 0$ in measure, along the lines of Problem 98,

$$\limsup_n \int_X |f_n - g|\, d\mu = \lim_{n_k} \int_X |f_{n_k} - g|\, d\mu = \omega(\mathcal{F}') \le \omega(\mathcal{F}).$$

(b) Since $g \in L^1(X)$, if $\mathcal{F}' = \{|f_n - g|\}$ it follows that $\omega(\mathcal{F}') = \omega(\mathcal{F})$. Therefore since $|f_n - g| \to 0$ in measure, by Problem 98(b),

$$\lim_n \int_X |f_n - g|\, d\mu = \omega(\mathcal{F}') = \omega(\mathcal{F}).$$

(c) The condition is clearly necessary. As for sufficiency, first, since $\{|\,|g| - |f_n|\,| > \eta\} \subset \{|g - f_n| > \eta\}$, $|f_n| \to |g|$ in measure and by Problem 38, $\int_X |g|\, d\mu \le \liminf_n \int_X |f_n|\, d\mu$ and, consequently, $\int_X |g|\, d\mu = \lim_n \int_X |f_n|\, d\mu$. The conclusion follows now from Problem 97.

100. First, f is integrable by Fatou's lemma. Next, note that

$$\int_X |f_n - f|\, d\mu = \int_X (|f_n - f| - |f_n| + |f|)\, d\mu + \int_X |f_n|\, d\mu - \int_X |f|\, d\mu$$

$$= A_n + \int_X |f_n|\, d\mu - \int_X |f|\, d\mu$$

where by Problem 84, $\lim_n A_n = 0$. Thus the limit in the left-hand side of the equality exists iff the limit in the right-hand side exists and we have $\lim_n \int_X |f_n - f|\, d\mu = \lim_n \int_X |f_n|\, d\mu - \int_X |f|\, d\mu$. By Problem 99(b) (which applies since in a finite measure space μ-a.e. convergence implies convergence in measure), the limit in the left-hand side of the equality is $\omega(\mathcal{F})$.

102. (a) The conclusion is not necessarily true unless $f_n \in L^1(X)$ for some n.

104. Since $\{\varphi_n\}$ is an increasing sequence, $\varphi = \sup_n \varphi_n$ is integrable by the MCT. The conclusion follows now from the LDCT. In particular, note that since $\lim_n \int_0^1 nx^n/(1+x)\, dx = 1/2$, $\varphi(x) = \sup_n nx^n/(1+x) \notin L^1([0,1])$.

105. The statement is false in an infinite measure space.

106. Let $g = \sup_n f_n$ and note that if $h_n = g - f_n$, then $h_n \ge 0$ μ-a.e. and since $\liminf_n h_n = g - \limsup_n f_n$ and $\liminf_n \int_X h_n\, d\mu = \int_X g\, d\mu - \limsup_n \int_X f_n\, d\mu$, by Fatou's lemma $\int_X g\, d\mu - \int_X \limsup_n f_n\, d\mu \le \int_X g\, d\mu - \limsup_n \int_X f_n\, d\mu$, and the result follows by canceling the finite quantity $\int_X g\, d\mu$.

Let $\{A_n\} \subset \mathcal{M}$ and $f_n = \chi_{A_n}$; then $\limsup_n \mu(A_n) \le \mu(\limsup_n A_n)$ and equality does not necessarily hold: If $A_n = [0, 1/2)$ for n odd and $=$

$[1/2, 1]$ for n even, then $1/2 = \limsup_n \int_0^1 \chi_{A_n}(x)\,dx < \int_0^1 \limsup_n \chi_{A_n}(x)\,dx = 1$.

Finally, to see failure consider Lebesgue measure in $[0,1]$ and $f_n(x) = 2^{n+1}\chi_{[2^{-(n+1)},2^{-n}]}(x)$. Then $\sup_n f_n(x) = \sum_n 2^{n+1}\chi_{[2^{-(n+1)},2^{-n}]}(x)$ is not integrable, $\int_0^1 f_n(x)\,dx = 1$ for all n, and since $\lim_n f_n(x) = 0$ a.e. on $[0,1]$, $\int_0^1 \limsup_n f_n(x)\,dx = 0$.

107. (c) First, necessity. Given $\varepsilon > 0$, by Problem 4.123 there exist finite measure sets A_n such that $\int_{A_n^c} |f_n|^p\,d\mu < \varepsilon$ for all n. Let $A = A_0 \cup \ldots \cup A_N$ where A_0 corresponds to f and N is large enough such that $\|f_n - f\|_p \le \varepsilon$ for all $n > N$. Then $\mu(A) < \infty$ and $\int_{A^c} |f_n|^p\,d\mu < \varepsilon$ for all $n \le N$, and if $n > N$, $(\int_{A^c} |f_n|^p\,d\mu)^{1/p} \le (\int_{A^c} |f_n - f|^p\,d\mu)^{1/p} + (\int_{A^c} |f|^p\,d\mu)^{1/p} \le \varepsilon/2 + \varepsilon/2 = \varepsilon$.

Next, sufficiency. Given $\varepsilon > 0$, with A the set above, put $A_{mn} = \{|f_m - f_n| \ge (\varepsilon/\mu(A))^{1/p}\}$, and let δ be as in given in (b). By (a) there exists N such that $|A_{mn}| < \delta$ for all $m, n \ge N$. Hence

$$\|f_m - f_n\|_p^p \le \int_{A_{mn}} |f_m - f_n|^p\,d\mu + \int_{A \setminus A_{mn}} |f_m - f_n|^p\,d\mu$$

$$+ \int_{A^c} |f_m - f_n|^p\,d\mu$$

$$\le \int_{A_{mn}} 2^p(|f_m|^p + |f_n|^p)\,d\mu + \int_{A \setminus A_{mn}} \varepsilon/\mu(A)\,d\mu$$

$$+ \int_{E^c} 2^p(|f_m|^p + |f_n|^p)\,d\mu$$

$$\le 2^{p+1}\varepsilon + \varepsilon + 2^{p+1}\varepsilon,$$

and $\{f_n\}$ is Cauchy in $L^p(\mathbb{R}^n)$.

108. Passing to a subsequence if necessary we may assume that $\lim_n f_n = f$ μ-a.e. Now, since $\{f_n\}$ is Cauchy in $L^p(X)$ there is a subsequence $\{f_{n_k}\}$ of $\{f_n\}$ such that $\|f_{n_{k+1}} - f_{n_k}\|_p \le 2^{-k}$ for all k. Observe that by Minkowski's integral inequality when $1 < p < \infty$ and the concavity of t^p when $0 < p < 1$, $g = |f_{n_1}| + \sum_k |f_{n_{k+1}} - f_{n_k}| \in L^p(X)$ and, consequently, the series $f_{n_1} + \sum_k (f_{n_{k+1}} - f_{n_k})$ converges absolutely μ-a.e. in X with limit $f_{n_1} + \lim_N \sum_{k=1}^N (f_{n_{k+1}} - f_{n_k}) = \lim_N f_{n_{N+1}} = f$ μ-a.e. Then $|f_{n_k} - f| = |\sum_{m=k-1}^\infty (f_{n_{m+1}} - f_{n_m})| \le g$ and $|f_{n_k}| \le h = |f| + g \in L^p(X)$ μ-a.e. for all n_k.

109. (a) and (b) follow readily by known properties of Cèsaro means.

110. All one can say is that $|f_n| \to 1$.

112. Since a subsequence $\{f_{n_k}\}$ of $\{f_n\}$ tends to f μ-a.e., $f \ge \eta$ μ-a.e. Now, if c is the Lipschitz constant for φ, $|\varphi(f(x)) - \varphi(\eta)| \le c|f(x) - \eta|$ and, consequently, $|\varphi(f(x))| \le c|f(x) - \eta| + \varphi(\eta) = \psi(x) \in L^1(X)$; thus

$\varphi(f) \in L^1(X)$ and, similarly, the $\varphi(f_n)$ are integrable for all n. Now, since $f_n(x), f(x) \geq \eta$, $|\varphi(f(x)) - \varphi(f_n(x))| \leq c|f(x) - f_n(x)|$ and, consequently, integrating, $\int_X |\varphi(f) - \varphi(f_n)|\, d\mu \leq c \int_X |f - f_n|\, d\mu \to 0$ as $n \to \infty$.

114. Note that $s = p/r > 1$ and $s' = p/(p-r)$ are conjugate Hölder indices. Now, since $f_n g_n - fg = (f_n - f)g_n - f(g - g_n)$ it follows that

$$\|f_n g_n - fg\|_r^r \leq c \int_X |f_n - f|^r |g_n|^r\, d\mu + c \int_X |f|^r |g - g_n|^r\, d\mu$$
$$\leq c \|\, |f_n - f|^r \|_s \|\, |g_n|^r \|_{s'} + c \|\, |f|^r \|_s \|\, |g - g_n|^r \|_{s'}.$$

Then, since $rs = p$ and $rs' = q$, and $\|g_n\|_q, \|f\|_p \leq c$, it readily follows that

$$\|f_n g_n - fg\|_r^r \leq c(\|f_n - f\|_p^r \|g_n\|_q^r + \|f\|_p^r \|g - g_n\|_q^r)$$
$$\leq c(\|f_n - f\|_p^r + \|g - g_n\|_q^r) \to 0 \quad \text{as } n \to \infty.$$

Particular cases include $r = 1$, $q = p'$, and $r = p/2$, $q = p$. The result is also true for the limiting case $r = p$, $q = \infty$.

115. By calculus for $s, t > 0$ and $r > 1$, $|s^r - t^r| \leq r|s - t|(s + t)^{r-1}$.

117. The statement is true for $1 < p < \infty$ but not true for $p = 1$.

119. If $\mu(X) = \infty$, f is not necessarily in $L^p(X)$.

120. When $p = \infty$, let A be such that $\lim_n f_n(x) = f(x)$ for $x \in A$ and $\mu(A^c) = 0$. Now, for all n, let $A_n \in \mathcal{M}$ be such that $\mu(A_n^c) = 0$ and $|f_n(x)| \leq \|f_n\|_\infty$ for $x \in A_n$, and put $B = A \cap (\bigcap_n A_n)$; $B \in \mathcal{M}$ and $\mu(B^c) = 0$. Then for $x \in B$, $|f(x)| = \lim_n |f_n(x)| = \liminf_n |f_n(x)| \leq \liminf_n \|f_n\|_\infty$, and, consequently, $\|f\|_\infty \leq \liminf_n \|f_n\|_\infty$.

Clearly there may be strict inequality: For $p < \infty$, let $f_n = n^{1/p}\chi_{(0,1/n)}$ and $f = 0$ a.e., and for $p = \infty$, let $f_n = \chi_{(0,1/n)}$ and $f = 0$ a.e.

122. Let n_k be such that $\lim_{n_k} \int_X f_{n_k}\, d\mu = \limsup_n \int_X f_n\, d\mu$ where the possibility that the limsup is infinite is allowed, and choose a further subsequence, which we denote again $\{n_k\}$, so that $g_{n_k} \to g$ μ-a.e. and in $L^1(X)$, and observe that $\liminf_{n_k}(g_{n_k} - f_{n_k}) = \lim_{n_k} g_{n_k} - \limsup_{n_k} f_{n_k} \geq g - \limsup_n f_n$. Now, since $g_{n_k} - f_{n_k} \geq 0$ by Fatou's lemma it follows that $\int_X \liminf_{n_k}(g_{n_k} - f_{n_k})\, d\mu \leq \liminf_{n_k} \int_X (g_{n_k} - f_{n_k})\, d\mu$, where by the previous remark the left-hand side majorizes $\int_X g\, d\mu - \int_X \limsup_n f_n\, d\mu$ and by assumption the right-hand side is equal to $\int_X g\, d\mu - \limsup_n \int_X f_n\, d\mu$. Therefore $\int_X g\, d\mu - \int_X \limsup_n f_n\, d\mu \leq \int_X g\, d\mu - \limsup_n \int_X f_n\, d\mu$ and the result follows by canceling the finite quantity $\int_X g\, d\mu$.

123. That $\|g_n\|_\infty \leq M$ for all n.

124. (b) No.

127. Nothing. On $\mathbb{R}$ let $I_n = [n, n+1]$, $f_n(x) = \chi_{I_n}(x)/|x - n|^{1/2}$; then $\lim_n f_n(x) = 0$ everywhere. Next, with $0 < \beta_n < 1/2$ for all n, $\sum_n \beta_n <$

∞, let $g(x) = \sum_n \beta_n \chi_{I_n}(x)/|x - n|^{1/2 - \gamma_n}$ where $0 < \gamma_n < 1/2$ are to be chosen. Now, $\int_{\mathbb{R}} g(x)\,dx = \sum_n \beta_n \int_{I_n} 1/|x - n|^{1/2 - \gamma_n}\,dx \leq 2 \sum_n \beta_n < \infty$ and, since $f_n(x)g(x) = \beta_n \chi_{I_n}(x)/|x - n|^{1 - \gamma_n}$ it readily follows that $A_n = \int_{\mathbb{R}} f_n(x)g(x)\,dx = \beta_n/\gamma_n < \infty$ for all n. Therefore, if $\gamma_n = \beta_n^{1/2}$, $A_n = \beta_n^{1/2} \to 0$; if $\gamma_n = \beta_n$, $A_n = 1$ for all n; and if $\gamma_n = \beta_n^2$, $A_n = 1/\beta_n \to \infty$.

129. (a) Let $A = \{0 < f < \infty\}$ and $B = \{x \in X : \lim_n f_n(x) = f(x)\}$, $\mu(B^c) = 0$. We claim that $\chi_{A_n} \to \chi_A$ μ-a.e. First, if $\chi_A(x) = 0$, $f(x) = 0$, $x \notin A_n$ for all n, and $\chi_{A_n}(x) = 0$ for all n. Next, if $x \in A \cap B$, then $0 < f(x)$, $f_n(x) < 2f(x)$ for all n large, and $\chi_{A_n}(x) = 1$ for those n. Thus $\chi_{A_n}(x) \to \chi_A(x)$ for $x \in A \cap B$, i.e., μ-a.e. Then $\chi_{A_n} f_n \to \chi_A f$ μ-a.e., where the convergence is bounded since μ-a.e., $\chi_{A_n} f_n, f \leq 2f \in L^1(X)$. The conclusion follows from the LDCT.

(b) Since $\int_X f_n\,d\mu \to \int_X f\,d\mu$, by (a), $\int_{A_n^c} f_n\,d\mu \to 0$. Now, $f_n(x) \geq 2f(x)$ for $x \in A_n^c$, and, therefore, $|f_n(x) - f(x)| \leq f_n(x)$ in A_n^c and $\int_{A_n^c} |f_n - f|\,d\mu \leq \int_{A_n^c} f_n\,d\mu \to 0$. Furthermore, $\{\chi_{A_n}|f_n - f|\}$ tends to 0 μ-a.e. boundedly and by the LDCT, $\int_{A_n} |f_n - f|\,d\mu \to 0$. The result follows by combining the estimates.

130. Put $\phi_n(x) = 2n\chi_{(-1/n, 1/n)}(x)$ and let

$$f_n(x) = 3^{-1}(\phi_n(1/5 - x) + \phi_n(2/5 - x) + \phi_n(3/5 - x)).$$

The f_n satisfy the desired conditions.

131. (a) The statement is true.

(b) The statement is false. Let $f_n(x) = \sin(2\pi nx)$; by the Riemann-Lebesgue lemma $\lim_n \int_I f_n(x)f(x)\,dx = 0$ for $f \in L^1([0,1])$. However, as we saw in Problem 4.147, $\int_0^1 |\sin(2\pi nx)|\,dx \not\to 0$ as $n \to \infty$.

132. The statement is true. We do the case of a single function f first. For the sake of argument suppose that $\eta = \liminf_{x \to 0} x\,|f(x)| > 0$; then there exists $\delta > 0$ such that $x|f(x)| > \eta/2$ whenever $0 < x \leq \delta$. But then $\int_0^1 |f(x)|\,dx \geq \int_0^\delta |f(x)|\,dx \geq c\int_0^\delta 1/x\,dx = \infty$, which is not the case. Therefore $\liminf_{x \to 0} x\,|f(x)| = 0$ and $\{a_n\}$ exists. Next, choose $\alpha_k > 0$ for all k such that $\sum_k \alpha_k \int_0^1 |f_k(x)|\,dx < \infty$; $\alpha_k = \beta_k/\|f_k\|_1$ with $\sum_k \beta_k < \infty$ will do. We can then apply the previous result to the integrable function $f(x) = \sum_k \alpha_k |f_k(x)|$ and obtain a sequence $\{a_n\}$ decreasing to 0 such that $a_n f(a_n) \to 0$. Moreover, since for all k, $a_n \alpha_k |f_k(a_n)| \leq a_n f(a_n)$, $\lim_n a_n \alpha_k |f_k(a_n)| = 0$ and since $\alpha_k \neq 0$, $\lim_n a_n |f_k(a_n)| = 0$ for all k.

A similar argument gives that if $f \in L(\mathbb{R})$, there exists a sequence $\{a_n\}$ that increases to ∞ such that $|a_n|\,|f(a_n)| \to 0$ as $a_n \to \infty$.

133. Since $\varphi(0) = 0$ and $\varphi'(0) = 1$ we get for $x \neq 0$,

$$\eta\varphi(x/\eta) = x\,\frac{\varphi(x/\eta) - \varphi(0)}{x/\eta} \to x, \quad \text{as } \eta \to \infty.$$

Moreover, since $\varphi(0) = 0$, the relation is also valid at $x = 0$. Also, if $|\varphi'(x)| \leq M$, say, as above we get that $\eta\varphi(x/\eta) \leq M\,x$; this relation clearly holds in case $\varphi(x)/x$ is bounded. Therefore, if $x > 0$, $n\varphi((x/n)^\alpha) = n^{1-\alpha}n^\alpha\varphi((x/n)^\alpha)$ and so

$$\lim_n n\varphi((x/n)^\alpha) = \begin{cases} \infty, & 0 < \alpha < 1, \\ x, & \alpha = 1, \\ 0, & 1 < \alpha < \infty. \end{cases}$$

First, suppose that $0 < \alpha < 1$. Since $\int_X f\,d\mu > 0$ there exists A with $\mu(A) > 0$ such that $f(x) > 0$ for $x \in A$. Then $\lim_n n\varphi((f(x)/n)^\alpha) = \infty$, $x \in A$, and by Fatou's lemma $\lim_n \int_A n\varphi((f/n)^\alpha)\,d\mu = \infty$.

Next, if $1 \leq \alpha < \infty$, to apply the LDCT we estimate $n\varphi((f/n)^\alpha) \leq c\,f \in L^1(X)$. We consider the two cases, namely, $f(x) \leq n$ and $f(x) > n$. In the former case $(f(x)/n)^\alpha \leq f(x)/n$, and since φ is nondecreasing and $\varphi(x)/x \leq M$, we have $n\varphi((f(x)/n)^\alpha) \leq n\varphi((f(x)/n))(f(x)/f(x)) = n\varphi((f(x)/n))/(f(x)/n)f(x) \leq c\,f(x)$. In the latter case, since $\varphi(x) \leq c$ and $f(x) > n$ we have $n\varphi((f(x)/n)^\alpha) \leq c(n/f(x))f(x) \leq c\,f(x)$. So the assumptions of the LDCT hold and so does the conclusion.

As for functions that satisfy the assumptions, they include $\arctan(x)$ and $x/(1+x)$. The conclusion also holds for $\ln(1+x)$, which satisfies all the required properties except for boundedness; this fact is only needed for $1 \leq \alpha < \infty$. Then by elementary calculus, $1 + t^\alpha \leq (1+t)^\alpha$, $t > 0$, and so $n\ln(1 + (f(x)/n)^\alpha) \leq n\ln((1 + f(x)/n)^\alpha) = \alpha\,n\,\ln(1 + f(x)/n) \leq \alpha\,f(x)$ whenever $f(x) < \infty$, which since f is integrable holds μ-a.e. So we can apply the LDCT with majorant αf to get the limit for $\alpha \geq 1$.

135. Let c be such that $\int_{\{|y|\leq M\}} dy = cM^n$ for all $M > 0$. Then for k sufficiently large, $3/4 \leq \int_{\{|y|\leq 1/M\}} \varphi_k(y)\,dy \leq \int_{\{|y|\leq 1/M,\,\varphi_k(y)\leq M^n/4c\}} \varphi_k(y)\,dy + \int_{\{\varphi_k > M^n/4c\}} \varphi_k(y)\,dy \leq 1/4 + \int_{\{\varphi_k > M^n/4c\}} \varphi_k(y)\,dy$. We then have $1/2 \leq \int_{\{\varphi_k > M^n/4c\}} \varphi_k(y)\,dy$, and so $\int_{\mathbb{R}^n} \varphi_k^p(y)\,dy \geq \int_{\{\varphi_k > M^n/4c\}} \varphi_k(y)^{p-1}\varphi_k(y)\,dy \geq (M^n/4c)^{p-1} \int_{\{\varphi_n > M\}} \varphi_k(y)\,dy \geq M^{n(p-1)}/(2(4c)^{p-1})$.

136. By Hölder's inequality it suffices to prove that $\lim_n \|f_n - 1/2\|_2 = 0$.

137. Let $I_{k,m} = (2^{-m}(k-1), 2^{-m}k)$ for $m = 1, 2, \dots$ and $k = 1, \dots, 2^m$ denote the dyadic subintervals of I. Then

$$x_m = \begin{cases} 0, & x \in I_{k,m}, k \text{ odd}, \\ 1, & x \in I_{k,m}, k \text{ even}. \end{cases}$$

Now, for integers $n > m$, each $I_{k,m}$ is represented as the union of 2^{n-m} subintervals $I_{j,n}$ plus a finite number of their endpoints. On the interval $I_{k,m}$ the function $f_n(x) = x_n(x) - 1/2$ alternates between $-1/2$ and $1/2$ and has zero integral while $f_m(x)$ is constant, specifically, 0 or 1. Hence for $n > m$, $\int_0^1 f_m(x)f_n(x)\,dx = \sum_{k=1}^{2^m} \int_{I_{k,m}} f_m(x)f_n(x)\,dx = 0$. By symmetry the functions $\{f_k\}$ are orthogonal in $L^2([0,1])$. Further, note that $S_n(x) - 1/2 = n^{-1}\sum_{m=1}^n f_m(x)$. Then $\int_0^1 |S_n(x) - 1/2|^2\,dx = \int_0^1 (n^{-1}\sum_{m=1}^n f_m(x))^2\,dx = n^{-2}\int_0^1 \sum_{k,m=1}^n f_k(x)f_m(x)\,dx = n^{-2}\int_0^1 \sum_{m=1}^n f_m^2(x)\,dx = n^{-2}(n/2) \to 0$ as $n \to \infty$.

140. Let $f_n(x) = n^2\chi_{[0,1/n]}(x)$, $n \geq 1$. Then $f_n/(1 + f_n) \leq 1$ and $\lim_n(f_n/(1 + f_n)) = 0$ a.e. in I; hence the conclusion follows by LDCT. The f_n can be easily modified to be continuous on $[0,1]$.

141. (a) Clearly (X, ρ) is a metric space. Let $x^n = e_1 + \cdots + e_n$ denote the sequence with terms $= 1$ for $k \leq n$ and $= 0$ otherwise; since $\rho(x^n, x^m) = 2^{-1}\sum_{k=n+1}^m 2^{-k} \to 0$ as $n, m \to \infty$, $\{x^n\}$ is Cauchy in (X, ρ). Were $\{x^n\}$ to converge, the limit would be the sequence x with $x_k = 1$ for all k, but $x \notin X$.

(b) Let Y denote the collection of all sequences; we claim that (Y, ρ) is a complete metric space and that X is dense in Y. The density is readily seen since for $y \in Y$, $\sum_{k=n+1}^\infty 2^{-k}|y_k|/(1 + |y_k|) \leq \sum_{k=n+1}^\infty 2^{-k} = 2^{-n}$, and, consequently, y is approximated by the truncates $\{y^n\}$ of y given by $y_k^n = y_k$ for $k \leq n$ and $y_k^n = 0$ otherwise.

As for completeness, let $\{x^n\}$ be a Cauchy sequence in (Y, ρ). First, we claim that $\{x_k^n\}$ is a Cauchy sequence in $\mathbb{R}$ for each $k \geq 1$. Indeed, given $0 < \varepsilon < 1$, fix $k \geq 1$, let ε' be such that $2^k\varepsilon' < \varepsilon/2$, and pick N such that $\rho(x^p, x^q) \leq \varepsilon'$ for $p, q \geq N$. Then, in particular, $2^{-k}|x_k^p - x_k^q|/(1 + |x_k^p - x_k^q|) \leq \rho(x^p, x^q) \leq \varepsilon'$, and so $|x_k^p - x_k^q|/(1 + |x_k^p - x_k^q|) \leq 2^k\varepsilon' \leq \varepsilon/2$ for $p, q \geq N$. Therefore, since $\varepsilon < 1$ it follows that $|x_k^p - x_k^q| \leq \varepsilon$ for $p, q \geq N$, and $\{x_k^p\}$ is Cauchy in $\mathbb{R}$ and, hence, convergent. Let x_k^∞ denote the limit for each k and $x^\infty = \{x_k^\infty\}$. Finally, given $\varepsilon > 0$, pick N so that $\sum_{k=N+1}^\infty 2^{-k} \leq \varepsilon/2$ and note that

$$\rho(x^n, x^\infty) \leq \sum_{k=1}^N 2^{-k}\frac{|x_k^n - x_k^\infty|}{1 + |x_k^n - x_k^\infty|} + \frac{\varepsilon}{2},$$

which, since $\lim_n |x_k^n - x_k^\infty| = 0$ for all $k \leq N$, implies that $\lim_n \rho(x^n, x^\infty) = 0$. Therefore (Y, ρ) is a complete space.

(c) $(X, \|\cdot\|_\infty)$ is not complete. For example, consider the sequence $x^n = (1, 1/2, \ldots, 1/n, 0, \ldots)$; $\{x^n\}$ is a Cauchy sequence that does not converge in X, and its limit in Y is $x^\infty = (1, 1/2, \ldots)$. And, the closure of X in the topology defined by $\|\cdot\|_\infty$ in Y is c_0.

144. Let $f_n = g_n + h_n$ where we set $g_n = f_n \chi_{\{f_n \leq M\}}(x)$ and $h_n(x) = f_n \chi_{\{f_n > M\}}(x)$; it suffices to prove that $\sum_n g_n(x), \sum_n h_n(x) < \infty$ μ-a.e. First, by the MCT, $\int_X \sum_n g_n \, d\mu = \sum_n \int_X g_n \, d\mu \leq \sum_n \alpha_n < \infty$, and, consequently, the integrand, $\sum_n g_n$, is finite μ-a.e. Next, let $A_n = \{f_n > M\}$ and $A = \limsup_n A_n$; by Borel-Cantelli $\mu(A) = 0$. In other words, μ-a.e. every $x \in X$ belongs to finitely many A_n, and, consequently, $\sum_n h_n(x) \leq$ [finite number] $M < \infty$ μ-a.e.

145. The statement is false.

146. First, if $\alpha < N/p$, changing variables, $\|f_n\|_p = n^{\alpha - N/p}\|f\|_p \to 0$ as $n \to \infty$. Next, let $\alpha = N/p$. Suppose first that f vanishes for $|x| > K$, say, and note that for $\varphi \in L^q(\mathbb{R}^N)$, by Hölder's inequality, $|\int_{\mathbb{R}^N} f_n(x)\varphi(x) \, dx| \leq \|f_n\|_p (\int_{\{|x| \leq K/n\}} |\varphi(x)|^q \, dx)^{1/q} \to 0$ as $n \to \infty$. For arbitrary f, given $\varepsilon > 0$ and $\varphi \in L^q(\mathbb{R}^N)$, let g be a function vanishing outside a compact set such that $\|f - g\|_p \leq \varepsilon/2\|\varphi\|_q$. Then with $g_n(x) = n^{N/p}g(nx)$ it readily follows that $\int_{\mathbb{R}^N} f_n(x)\varphi(x) \, dx = \int_{\mathbb{R}^N} (f_n(x) - g_n(x))\, \varphi(x) \, dx + \int_{\mathbb{R}^N} g_n(x)\varphi(x) \, dx = A_n + B_n$, say. Now, by Hölder's inequality $|A_n| \leq \|f_n - g_n\|_p \|\varphi\|_q = \|f - g\|_p \|\varphi\|_q \leq \varepsilon/2$ and since by the above $\lim_n B_n = 0$, it follows that $|\int_{\mathbb{R}^N} f_n(x)\varphi(x) \, dx| < \varepsilon$ for n large enough, and we have weak convergence to 0 in this case.

Finally, if $\alpha > N/p$ and q is the conjugate to p, pick $0 < \beta < N/q$ such that $\alpha - N/p > N/q - \beta$. Then for $f = \chi_B$ the characteristic function of the unit ball in $\mathbb{R}^N$ and $\varphi(x) = |x|^{-\beta}\chi_B(x) \in L^q(\mathbb{R}^N)$ note that $\int_{\mathbb{R}^N} f_n(x)\varphi(x) \, dx = n^\alpha \int_{B(0,1/n)} |x|^{-\beta} \, dx = c_{\beta,N}\, n^{\alpha+\beta-N}$, which is unbounded as $n \to \infty$ since $\alpha + \beta - N = (\alpha - N/p) - (N/q - \beta) > 0$. Therefore $\{f_n\}$ cannot converge weakly.

148. Since $|\int_A f_n \, d\mu| \leq \mu(A)^{1/r'}\|f_n\|_r \to 0$ as $n \to \infty$, the conclusion follows from Problem 147.

149. Note that $X = A_n \cup B_n \cup N_n$ where $A_n = \{f_n = 0\}$, $B_n = \{|f_n| \geq n\}$, and $\mu(N_n) = 0$ for all n. By Chebychev's inequality, $n^p \mu(B_n) \leq \int_{B_n} |f_n|^p \, d\mu \leq c^p$ and, in particular, $\lim_n \mu(B_n) = 0$. Now, for $g \in L^q(X)$, by Problem 4.143, $\lim_n \int_{B_n} |g|^q \, d\mu = 0$ and so by Hölder's inequality $|\int_X f_n g \, d\mu| \leq c(\int_{B_n} |g|^q \, d\mu)^{1/q} \to 0$ as $n \to \infty$.

150. Given $\eta > 0$, let $B_n = \{f_n > \eta\}$; since $\{f_n\}$ is decreasing, $\{B_n\}$ decreases. Moreover, $\mu(B_1)\eta^p \leq \int_{B_1} f_1^p \, d\mu < \infty$ and $\mu(B_1) < \infty$, and so with $B = \bigcap_n B_n$, $\lim_n \mu(B_n) = \mu(B)$. Now, let $g = \chi_B$; $g \in L^q(X)$ where q is the conjugate to p, and so by assumption $\lim_n \int_X f_n g \, d\mu = 0$. Since $\{f_n\}$ is monotone, $\lim_n f_n(x) = f(x)$ exists for all $x \in X$, and, by definition $f(x) \geq \eta$ on B. Thus by the MCT, $0 = \lim_n \int_X f_n g \, d\mu = \int_B f \, d\mu \geq \eta \mu(B)$, which gives $\lim_n \mu(B_n) = \mu(B) = 0$ and $f_n \to 0$ in measure.

That $\{f_n\}$ is decreasing is necessary as the example $f_n = \chi_{[n,n+1)}$ on $\mathbb{R}$ shows.

151. Since weakly convergent sequences are norm bounded by Fatou's lemma, $\|g\|_p \leq \liminf_n \|f_n\|_p < \infty$ and $g \in L^p(X)$. To prove that $f = g$ μ-a.e. it suffices to verify that $\int_X (f - g)\psi \, d\mu = 0$ for all ψ in a dense class of functions of $L^q(X)$ where q is the conjugate to p. Now, by weak convergence $\int_X (f-g)\psi \, d\mu = \lim_n \int_X (f_n-g)\psi \, d\mu$. Essentially as in Problem 147 this limit is 0 if X has finite measure. Next, given $\varepsilon > 0$, pick δ so that $\int_A |\psi|^q \, d\mu \leq (\varepsilon/2c)^q$, for all A with $\mu(A) < \delta$. Now invoke Egorov with $\eta = \delta$. In other words, f_n converges uniformly to g in a set G of measure at least $\mu(X) - \eta$. Then $\int_X (f_n - g)\psi \, d\mu = \int_G (f_n - g)\psi \, d\mu + \int_{X\setminus G} (f_n - g)\psi \, d\mu$. Since ψ is bounded, the first integral goes to 0 with n by the LDCT. As for the second term, by Hölder's inequality it is controlled by

$$(\|f_n\|_p + \|g\|_p)\left(\int_{X\setminus G} |\psi|^q \, d\mu \right)^{1/q} \leq 2c\varepsilon/2c = \varepsilon.$$

So, $\int_X (f - g)\psi \, d\mu = 0$.

Finally, since $L^q(X)$ functions vanish off a σ-finite set, the σ-finiteness assumption is not really needed.

152. For the sake of argument suppose that $A = \{|f| > \lambda\}$ has positive measure. Let $g = \operatorname{sgn}(f)\chi_A$ (note that $g \in L^q(X)$ where $1/p + 1/q = 1$), $A_n = \int_{\{|f_n|<\lambda\}} g f_n \, d\mu$, and $B_n = \int_{\{|f_n|\geq\lambda\}} g f_n \, d\mu$. Note that $\limsup_n |A_n| \leq \lambda\mu(A)$. Now, since weakly convergent sequences are norm bounded, $\|f_n\|_p \leq c$ for all n, and, consequently, by Hölder's inequality $|B_n| \leq c\,\mu(\{|f_n| \geq \lambda\})^{1/q}$ which tends to 0 as $n \to \infty$. Hence, since $\int_A |f| \, d\mu = \lim_n (A_n + B_n)$, $\lambda\mu(A) < \int_A |f| \, d\mu \leq \lambda\mu(A)$, which is not the case. Therefore $\mu(A) = 0$ and $\|f\|_\infty \leq \lambda$.

153. The statement is false. Let $X = [0,1]$, $\varphi(t)$ the periodic function of period 1 given on $[0,1]$ by $\varphi(t) = \chi_{(0,1/2)}(t) - \chi_{(1/2,1]}(t)$, and for $n \geq 1$, let $f_n(t) = \varphi(2^n t)$, $t \in I$. We claim that $\lim_n \int_0^1 f_n(t)\chi_A(t) \, dt = 0$ for every measurable subset A of $[0,1]$; the verification of this claim can be carried out in steps, first for dyadic intervals, then intervals, open sets, sets of measure 0, and, finally, all measurable sets. Note that $|f_n(t)| = 1$ a.e. for all n and t, and so $\|f_n\|_p \leq c$ for all n. Thus by Problem 147, $f_n \rightharpoonup 0$ in L^p but because $|f_n(t)| = 1$ a.e. for all n and t, no subsequence $\{f_{n_k}\}$ of $\{f_n\}$ converges to $f = 0$ a.e.

154. (a) By Fatou's lemma $f \in L^p(X)$. Assume first that $\mu(X) < \infty$. Let $E_N = \bigcap_{n=N}^\infty \{|f_n - f| \leq 1\}$ and for $g \in L^q(X)$, with q the conjugate to p, let $g_N = g\chi_{E_N}$, $N \geq 1$; note that since $\lim_n f_n = f$ μ-a.e., $\lim_N g_N = g$ μ-a.e., and since $|g_N| \leq |g|$ μ-a.e. for all N, by the LDCT, $\lim_N g_N = g$

in $L^q(X)$. Write $\int_X f_n g \, d\mu - \int_X fg \, d\mu = \int_X f_n(g - g_N) \, d\mu + \int_X f_n g_N \, d\mu - \int_X fg_N \, d\mu + \int_X f(g_N - g) \, d\mu = A_{n,N} + B_{n,N} - C_N + D_N$, say, and let $\varepsilon > 0$ be given. Then by Hölder's inequality $|A_{n,N}| \le \|f_n\|_p \|g_N - g\|_q \le c\|g_N - g\|_q \le \varepsilon$ for all n, provided N is sufficiently large; similarly, $|D_N| \le \|f\|_p \|g_N - g\|_q \le \varepsilon$ provided N is large enough. Finally, since $\{(f_n - f)g_N\}$ are integrable functions majorized by the integrable function g_N and $\lim_n (f_n - f)g_N = 0$ μ-a.e. for each N, $B_{n,N} - C_N \to 0$ as $n \to \infty$ for each N. Thus picking first N and then n it readily follows that $|\int_X f_n g \, d\mu - \int_X fg \, d\mu| \le 3\varepsilon$ for n sufficiently large and $f_n \rightharpoonup f$ in $L^p(X)$.

When X has infinite measure, since $|g|^q \in L^1(X)$, by Problem 4.123, there exists A with $\mu(A) < \infty$ such that $\int_{X \setminus A} |g|^q \, d\mu \le \varepsilon^q$. Then

$$\int_X (f_n - f)g \, d\mu = \int_A (f_n - f)g \, d\mu + \int_{X \setminus A} (f_n - f)g \, d\mu = I + J,$$

say. By Hölder's inequality $J \le 2c\|g\chi_{X \setminus A}\|_q \le 2c\varepsilon$. As for I, we use the finite measure space result.

Note that it may be the case that $f_n \not\rightharpoonup f$ in $L^p(X)$ as the example $f_n = n^{1/p}\chi_{(0,1/n)}$, $f = 0$, in $L^p(I)$ shows.

And, when $p = 1$, the functions $f_n(x) = n\chi_{[0,1/n]}(x)$, $f = 0$, satisfy $\|f_n\|_1 = 1$ for all n, $f_n \to f$ a.e., but since $\int_{[0,1]} f_n(x) \, dx = 1$ for all n, $\{f_n\}$ does not converge weakly to 0 in $L^1(I)$.

(b) The statement is true.

(c) The statement is not true if the measure is infinite.

156. (a) Recall that by Problem 2.89, $\widetilde{\mathcal{M}}$ endowed with the metric $d(A, B) = \mu(A \triangle B)$ is a complete metric space. For $\varepsilon > 0$ and $N = 1, 2, \ldots,$ let $F_N = \{A \in \widetilde{\mathcal{M}} : |\int_A (f_m - f_n) \, d\mu| \le \varepsilon \text{ for all } n, m \ge N\}$. Since $\{\int_A f_n \, d\mu\}$ is Cauchy for each $A \in \widetilde{\mathcal{M}}$, it follows that $\widetilde{\mathcal{M}} = \bigcup_N F_N$. Now, each F_N is closed and $\widetilde{\mathcal{M}}$ is complete and so by the Baire category theorem some F_{N_0} has a nonempty interior, i.e., there exist $A_0 \in F_{N_0}$ and $r > 0$ such that $\mu(A \triangle A_0) < r$ implies $A \in F_{N_0}$.

Now let $B \in \mathcal{M}$ with $\mu(B) < r$; then $B = (A_0 \cup B) \setminus (A_0 \setminus B)$ and $\mu((A_0 \cup B) \triangle A_0), \mu((A_0 \setminus B) \triangle A_0) < r$. Then for $n, m \ge N_0$ we have $|\int_B (f_n - f_m) \, d\mu| = |\int_{A_0 \cup B} (f_n - f_m) \, d\mu - \int_{A_0 \setminus B} (f_n - f_m) \, d\mu| \le 2\varepsilon$. Applying the same argument to the sets $B \cap \{f_n - f_m \ge 0\}$ and $B \cap \{f_n - f_m \le 0\}$ yields $\int_B |f_n - f_m| \, d\mu \le \varepsilon$ whenever $\mu(B) < r$ and $m, n \ge N_0$. Now, the family $F = \{f_1, \ldots, f_{N_0}\}$ is absolutely uniformly continuous and so there exists $0 < s \le r$ such that $\int_B |f_k| \, d\mu \le \varepsilon$ whenever $1 \le k \le N_0$ and $\mu(B) < s$. Finally, if $\mu(B) < s$ and $k > N_0$, $\int_B |f_k| \, d\mu \le \int_B |f_k - f_{N_0}| \, d\mu + \int_B |f_{N_0}| \, d\mu \le 4\varepsilon + \varepsilon$. Thus $\{f_n\}$ is uniformly absolutely continuous.

(b) Let $\psi(A) = \lim_n \int_A f_n \, d\mu$, $A \in \mathcal{M}$; ψ is an additive set function on $\mathcal{M}$. We will use Problem 2.32. Let $\{A_k\} \subset \mathcal{M}$ be such that $\mu(A_k) \to 0$. Since $\{f_n\}$ is uniformly absolutely continuous, given $\varepsilon > 0$, there exists $\delta > 0$ such that $|\int_A f_n \, d\mu| \le \varepsilon$ provided $\mu(A) \le \delta$, and, in particular, $\psi(A) \le \varepsilon$ whenever $\mu(A) \le \delta$. Thus if N is sufficiently large so that $\mu(A_k) \le \delta$ for all $k \ge N$, $\psi(A_k) \le \varepsilon$ for those A_k and $\lim_k \psi(A_k) = 0$. Then by Problem 2.32, ψ is a measure on $\mathcal{M}$, which as pointed out in the argument above, is absolutely continuous with respect to μ. Thus by the Radon-Nykodym theorem there exists $f \in L^1(X)$ such that $\psi(A) = \int_A f \, d\mu$ for all $A \in \mathcal{M}$. Now, since by Problem 5.63(a) simple functions are dense in $L^\infty(X)$, it readily follows that $\lim_n \int_X f_n g \, d\mu = \int_X f g \, d\mu$ for all $g \in L^\infty(X)$ and $f_n \rightharpoonup f$ in $L^1(X)$.

157. (a) Let $\{\chi_{A_n}\} \subset \mathcal{F}$ and $\chi_{A_n} \to g$ in $L^1([0,1])$. Then g is measurable and since a subsequence $\{\chi_{A_{n_k}}\}$ of $\{\chi_{A_n}\}$ converges to g a.e., g takes on the values 0 and 1 a.e. and is, therefore, the characteristic function of a measurable set. Thus $\mathcal{F}$ is closed.

Next, if $\{\chi_{A_m}\} \subset \mathcal{F}_n$ and $\chi_{A_m} \to g$ in $L^1([0,1])$, then $g = \chi_A$ for some $A \in \mathcal{L}([0,1])$. Now fix $k \ge n$. Passing to a subsequence if necessary we may assume that $\chi_{A_m} \to \chi_A$ μ-a.e. Now, $(f_k - f)\chi_{A_m}$ converges to $(f_k - f)\chi_A$ boundedly and so by the LDCT, $\int_I (f_k - f)\chi_{A_m} \, dx \to \int_I (f_k - f)\chi_A \, dx$ and, therefore, $|\int_I (f_k - f)\chi_A \, dx| \le \varepsilon$. Thus $g = \chi_A \in \mathcal{F}_n$ and $\mathcal{F}_n$ is closed.

Moreover, since $f_k \rightharpoonup f$ in $L^1(I)$ and $\chi_A \in L^\infty(I)$, given $\varepsilon > 0$, $|\int_I (f_k - f)\chi_A \, dx| \le \varepsilon$ for all k large enough, i.e., $k \ge N$.

(b) Since $\mathcal{F}$ is closed in $L^1(I)$, it is complete and $\mathcal{F} = \bigcup_n \mathcal{F}_n$. Therefore by the Baire category theorem one of the sets, $\mathcal{F}_n$, say, has nonempty interior and there exist $E \subset [0,1]$ with $\chi_E \in \mathcal{F}_n$ and $\delta > 0$ such that $B(\chi_E, \delta) = \{\chi_A : \|\chi_E - \chi_A\|_1 < \delta\} \subset \mathcal{F}_n$ or, in other words, if $\|\chi_A\|_1 = |A| \le \delta$, then $\chi_E + \chi_A \in \mathcal{F}_n$. Therefore $|\int_F (f_k - f) \, dx| \le \varepsilon$ for all $k \ge n$ and $\chi_F \in B(\chi_E, \delta)$. In particular, this applies to $F = E \cup A$ and $F = E \setminus A$ and since $A = (E \cap A) \setminus (E \setminus A)$, it readily follows that if $|A| \le \delta$, $|\int_A (f_k - f) \, dx| \le 2\varepsilon$ for all $k \ge n$. Now, note that this gives the uniform absolute continuity of $\{f_k - f\}$ since A_k^+ and A_k^- defined as the subsets of A where $f_k - f \ge 0$ (resp. $f_k - f < 0$) have measure less than or equal to $|A|$. Finally, since $f \in L^1(I)$, this readily implies the uniform absolute continuity of $\{f_k\}$.

158. For $\varphi \in L^1(I)$ we compute $I_n = \int_I f_n(x)\varphi(x) \, dx$; we do the case n even, the case n odd being similar. Then

$$I_n = \sum_{k=0}^{n/2-1} \int_{2k/n}^{2k+1/n} (\varphi(x) - \varphi(x + 1/n)) \, dx,$$

and, consequently, by the continuity of $L^1(I)$ functions in the metric, $|I_n| \leq \int_0^{1-1/n} |\varphi(x) - \varphi(x + 1/n)|\,dx \to 0$ as $n \to \infty$. Then, since $L^q(I) \subset L^1(I)$ for $q \geq 1$, $f_n \rightharpoonup 0$ in $L^p(I)$ for all $1 \leq p < \infty$.

159. Fix $0 < t < 1$ and let $\varphi(x)$ denote the one-periodic extension of $\chi_{[0,t)}(x) - t\chi_{[0,1]}(x)$ to $\mathbb{R}$. Then $\varphi(nx) = \sum_{k=0}^{n-1} \chi_{[k/n,(k+t)/n)}(x) - t\chi_I(x)$ on I and since $\int_I \varphi(x)\,dx = 0$, by Problem 4.147, $\lim_n \int_0^1 \varphi(nx)f(x)\,dx = 0$ for $f \in L^p([0,1])$, $1 \leq p \leq \infty$. Thus $\varphi(nx) \rightharpoonup 0$ in $L^p(I)$, $1 \leq p < \infty$, or, equivalently, $\sum_{k=0}^{n-1} \chi_{[k/n,(k+t)/n)}(x) \rightharpoonup t\chi_I(x)$ in $L^p(I)$ for $1 \leq p < \infty$. Moreover, since $\chi_{[1-t,1]}(x) = (1-t)\chi_I(x) - \chi_{[0,t]}(x) + t\chi_I(x)$, if $\psi(x)$ denotes the 1-periodic extension of $\chi_{[1-t,1]}(x)$ to $\mathbb{R}$, then $\psi(nx) \rightharpoonup (1-t)\chi_I(x)$ in $L^p(I)$ for $1 \leq p < \infty$.

160. (a) By Problem 5.29, if $u \in L^p(X)$, then $T(u) \in L^q(X)$. Let $\{u_n\}, u \in L^p(X)$ be such that $u_n \to u$ in $L^p(X)$. Then, by Problem 108 there exist a subsequence $\{u_{n_k}\}$ of $\{u_n\}$ and $h \in L^p(X)$ such that $u_{n_k} \to u$ μ-a.e. and $|u_{n_k}| \leq h$. Since $|\varphi(u_{n_k}) - \varphi(u)|^q \to 0$ μ-a.e. and $|\varphi(u_{n_k}) - \varphi(u)|^q \leq C(h^p + |u|^p + 1) \in L^1(X)$, by the LDCT, $\int_X |\varphi(u_{n_k}) - \varphi(u)|^q d\mu \to 0$ as $k \to \infty$, and so $T(u_{n_k}) \to T(u)$ in $L^q(X)$. Similarly, from each subsequence of $\{T(u_n)\}$ we may extract a subsequence that converges to $T(u)$ in $L^q(X)$, and, consequently, $T(u_n) \to T(u)$ in $L^q(X)$.

(b) For a fixed $t \in [0,1]$ and $\alpha, \beta \in \mathbb{R}$, let $\xi(x)$ denote the 1-periodic extension of the function $\alpha\chi_{[0,t]}(x) + \beta\chi_{[1-t,1]}(x)$ to $\mathbb{R}$, and set $u_n(x) = \xi(nx)$ for all n. By Problem 159, $u_n \rightharpoonup (\alpha t + \beta(1-t))\chi_I$ in $L^p(I)$. Moreover, since $T(\xi(nx)) = \varphi(\alpha)\chi_{[0,t]}(nx) + \varphi(\beta)\chi_{[1-t,1]}(nx)$, by Problem 159, $T(u_n) \rightharpoonup (\varphi(\alpha)t + \varphi(\beta)(1-t))\chi_I$ in $L^q(I)$.

Thus, by assumption, $\varphi(\alpha t + \beta(1-t)) = \varphi(\alpha)t + \varphi(\beta)(1-t)$ for all α, β and for all $t \in [0,1]$. If $T(0) = 0$, letting $\beta = 0$ it follows that $\varphi(\alpha t) = t\varphi(\alpha)$ for all $t \in [0,1]$. Furthermore, picking $t = 1/\lambda$ if $\lambda > 1$ and $\lambda\alpha$ in place of α, we get that $\varphi(\lambda\alpha) = \lambda\varphi(\alpha)$ for $\lambda \in \mathbb{R}^+$. In particular, if $t = 1/2$, $\varphi(\alpha + \beta)/2 = \varphi((\alpha + \beta)/2) = (\varphi(\alpha) + \varphi(\beta))/2$, $\varphi(\alpha + \beta) = \varphi(\alpha) + \varphi(\beta)$, and, therefore, T is linear. On the other hand, if $T(0) \neq 0$, the argument gives that T is affine.

161. When $p = 2$ the result follows from Problem 10.97. Next observe that for $p > 2$, by calculus considerations there exists $c > 0$ such that $c|t|^p + pt + 1 \leq |1 + t|^p$ for all real t. Thus, $\eta(t) = h(t)/|t|^p$ is positive and well-defined for all $t \neq 0$, and since $\eta(t) \to 1$ as $|t| \to \infty$ and $\lim_{t \to 0} \eta(t) = \lim_{t \to 0} |t + 1|^{p-2}/|t|^{p-2} = \infty$, there exists $c > 0$ such that $\eta(t) \geq c$. Now, replacing t by $(t/s) - 1$ it readily follows that for all $s, t \in \mathbb{R}$, $|t|^p \geq c|t - s|^p + |s|^p + p(t - s)|s|^{p-1}\mathrm{sgn}(s)$, and, therefore, with $s = f$ and $t = f_n$ we have $|f_n|^p \geq c|f_n - f|^p + |f|^p + p(f_n - f)|f|^{p-1}\mathrm{sgn}(f)$. Since $|f|^{p-1}\mathrm{sgn}(f) \in L^q(X)$, where q is the conjugate to p, the conclusion follows by integrating the above estimate and letting $n \to \infty$.

The result is also true for $1 < p < 2$, and the proof follows along the same lines, although it is more complicated.

Chapter 17

Product Measures

1. By Problem 2.3 it suffices to verify that $\mathcal{R}$ is a π-system that contains $X \times Y$ and that the complement of every set in $\mathcal{R}$ can be expressed as a finite union of elements in $\mathcal{R}$.

2. The statement is false if $A = \emptyset$ because then $A \times B = \emptyset \in \mathcal{M} \otimes \mathcal{N}$ for all $B \subset Y$; similarly if $B = \emptyset$. On the other hand, the statement is true if $E = A \times B \neq \emptyset$. In that case pick $a \in A, b \in B$ and note that $E_a = B \in \mathcal{N}$ and $E^b = A \in \mathcal{M}$; sufficiency is obvious.

3. (a) The statement is false: $\mathcal{B}(\mathbb{R}^m) \times \mathcal{B}(\mathbb{R}^n)$ is not a σ-algebra of subsets of $\mathbb{R}^{m+n}$. Indeed, if 0_k denotes the origin in $\mathbb{R}^k$, $0_{m+n} = 0_m \times 0_n \in \mathcal{B}(\mathbb{R}^n) \times \mathcal{B}(\mathbb{R}^m)$. For the sake of argument suppose that $\{0_{m+n}\}^c = A \times B$ with $A \in \mathcal{B}(\mathbb{R}^m)$ and $B \in \mathcal{B}(\mathbb{R}^n)$. Since the projection of $\{0_{m+n}\}^c \subset \mathbb{R}^m \times \mathbb{R}^n$ into $\mathbb{R}^m$ is $\mathbb{R}^m$ and likewise the projection into $\mathbb{R}^n$ is $\mathbb{R}^n$, it readily follows that $A = \mathbb{R}^m$ and $B = \mathbb{R}^n$ but this is not the case since $0_{m+n} \notin \{0_{m+n}\}^c$.

A proof by pictures also works. Let $A = (0,1), B = (2,3); A, B \in \mathcal{B}(\mathbb{R})$. Were $\mathcal{B}(\mathbb{R}) \times \mathcal{B}(\mathbb{R})$ an algebra, $(A \cup B) \times (A \cup B) \setminus A \times A$ would be an element of $\mathcal{B}(\mathbb{R}) \times \mathcal{B}(\mathbb{R})$ but clearly this set cannot be expressed as a product of sets in $\mathcal{B}(\mathbb{R})$.

(b) The statement is true.

4. The statement is false. Let $A \subset \mathbb{R}, A \notin \mathcal{B}(\mathbb{R})$, and put $C = A \times \{0\}$; since $C^0 \notin \mathcal{B}(\mathbb{R})$, $C \notin \mathcal{B}(\mathbb{R}^2)$. Let B be a rotation of C with angle θ, $0 < \theta < \pi/2$. Since balls are rotation invariant so are open sets and, hence, $\mathcal{B}(\mathbb{R}^2)$ is rotation invariant and $B \notin \mathcal{B}(\mathbb{R}^2)$. Now, B_x consists of a single point or is $\emptyset$ for all $x \in \mathbb{R}$ and so $B_x \in \mathcal{B}(\mathbb{R})$ for all $x \in \mathbb{R}$. Similarly, $B^y \in \mathcal{B}(\mathbb{R})$ for all $y \in \mathbb{R}$.

5. Let $\mathcal{P}_k = \left\{ \prod_{\ell=1}^{k} [a_\ell, b_\ell) : a_\ell < b_\ell, \right\}$. If $C \in \mathcal{P}_{m+n}$,

$$C = \left(\prod_{\ell=1}^{m} [a_\ell, b_\ell) \right) \times \left(\prod_{\ell=m+1}^{n} [a_\ell, b_\ell) \right) \in \mathcal{L}(\mathbb{R}^m) \times \mathcal{L}(\mathbb{R}^n)$$

and, consequently,

$$\mathcal{B}(\mathbb{R}^{m+n}) = \mathcal{M}(\mathcal{P}_{m+n}) \subset \mathcal{M}\big(\mathcal{L}(\mathbb{R}^m) \times \mathcal{L}(\mathbb{R}^n)\big) = \mathcal{L}(\mathbb{R}^m) \otimes \mathcal{L}(\mathbb{R}^n).$$

To show that the inclusion is proper consider $A \neq \emptyset$ with $\lambda_m(A) = 0$, $B \in \mathcal{L}(\mathbb{R}^n) \setminus \mathcal{B}(\mathbb{R}^n)$, and note that $A \times B \in (\mathcal{L}(\mathbb{R}^m) \times \mathcal{L}(\mathbb{R}^n)) \setminus \mathcal{B}(\mathbb{R}^{m+n})$.

Next, let $A \times B \in \mathcal{L}(\mathbb{R}^m) \times \mathcal{L}(\mathbb{R}^n)$ and write $A \times B = (A \times \mathbb{R}^n) \cap (\mathbb{R}^m \times B)$. Now, $A = F_A \cup N_A$ where $F_A \in \mathcal{B}(\mathbb{R}^m)$ is F_σ and $\lambda_m(N_A) = 0$. Then

$$A \times \mathbb{R}^n = (F_A \times \mathbb{R}^n) \cup (N_A \times \mathbb{R}^n)$$

where $F_A \times \mathbb{R}^n \in \mathcal{B}(\mathbb{R}^{m+n})$ and $\lambda_{m+n}(N_A \times \mathbb{R}^n) = 0$ and so $A \times \mathbb{R}^n \in \mathcal{L}(\mathbb{R}^{m+n})$; similarly, $\mathbb{R}^m \times B \in \mathcal{L}(\mathbb{R}^{m+n})$ and, consequently, $A \times B \in \mathcal{L}(\mathbb{R}^{m+n})$. Therefore $\mathcal{L}(\mathbb{R}^m) \otimes \mathcal{L}(\mathbb{R}^n) = \mathcal{M}\big(\mathcal{L}(\mathbb{R}^m) \times \mathcal{L}(\mathbb{R}^n)\big) \subset \mathcal{L}(\mathbb{R}^{m+n})$. To see that the inclusion is proper let $B \notin \mathcal{L}(\mathbb{R}^m)$ and note that since a y-section of $B \times \{0_n\} \notin \mathcal{L}(\mathbb{R}^m)$, $B \times \{0_n\} \notin \mathcal{L}(\mathbb{R}^m) \otimes \mathcal{L}(\mathbb{R}^n)$.

6. Let $\overline{\mathcal{M}}$ denote the completion of a σ-algebra $\mathcal{M}$. By Problem 5, $\mathcal{L}(\mathbb{R}^{m+n}) = \overline{\mathcal{B}(\mathbb{R}^{m+n})} \subset \overline{\mathcal{L}(\mathbb{R}^m) \otimes \mathcal{L}(\mathbb{R}^n)} \subset \overline{\mathcal{L}(\mathbb{R}^{m+n})} = \mathcal{L}(\mathbb{R}^{m+n})$.

Note that since $\mathcal{B}(\mathbb{R}^{m+n}) = \mathcal{B}(\mathbb{R}^m) \otimes \mathcal{B}(\mathbb{R}^n)$, the completions of $\mathcal{B}(\mathbb{R}^m) \otimes \mathcal{B}(\mathbb{R}^n)$ and $\mathcal{L}(\mathbb{R}^m) \otimes \mathcal{L}(\mathbb{R}^n)$ are the same.

9. (a) In fact, both product σ-algebras are equal to $\mathcal{M}$, the σ-algebra generated by the parallelepipeds $\{A_1 \times A_2 \times A_3 : A_k \in \mathcal{M}_k, k = 1, 2, 3\}$.

(b) Recall that $\mu_1 \otimes \mu_2$ is σ-finite. Next, we define $\mu_1 \otimes \mu_2 \otimes \mu_3$ on $\mathcal{M}_1 \otimes \mathcal{M}_2 \otimes \mathcal{M}_3$ as the product measure $(\mu_1 \otimes \mu_2) \otimes \mu_3$. This measure assigns the value $(\mu_1 \otimes \mu_2)(A_1 \times A_2)\mu_3(A_3) = \mu_1(A_1)\mu_2(A_2)\mu_3(A_3)$ to $A_1 \times A_2 \times A_3$ and is the only σ-finite measure that does so. Note that $\mu_1 \otimes (\mu_2 \otimes \mu_3)$ agrees with $(\mu_1 \otimes \mu_2) \otimes \mu_3$ on the parallelepipeds $A_1 \times A_2 \times A_3$ and, therefore, they agree on the σ-algebra they generate, i.e., $\mathcal{M}_1 \otimes \mathcal{M}_2 \otimes \mathcal{M}_3$.

11. (a) $\mathcal{P}(\mathbb{N}) \otimes \mathcal{P}(\mathbb{N}) = \mathcal{P}(\mathbb{N}^2)$.

(b) $\mu \otimes \mu$ is the counting measure on $\mathbb{N}^2$.

12. Since $\mu \otimes \nu(E) = \int_X \nu(E_x) \, d\mu(x) = 0$ it readily follows that $\int_Y \mu(E^y) \, d\nu(y) = \mu \otimes \nu(E) = 0$, and, consequently, $\mu(E^y) = 0$ for ν-a.e. $y \in Y$.

14. (a) Since λ_{m+n} is the completion of $\lambda_m \otimes \lambda_n$ there exists $E \in \mathcal{L}(\mathbb{R}^{m+n})$ such that $A \subset E$ and $\lambda_{m+n}(E) = 0$ and, therefore, by Tonelli's theorem $0 = \int_{\mathbb{R}^{m+n}} f \, d\lambda_{m+n} = \int_{\mathbb{R}^m} \int_{\mathbb{R}^n} f_x \, d\lambda_n \, d\lambda_m = \int_{\mathbb{R}^n} \int_{\mathbb{R}^m} f^y \, d\lambda_m \, d\lambda_n$. Therefore $\lambda_n(E_x) = \lambda_n(E^y) = 0$ for λ_n-a.e. $x, y \in \mathbb{R}^n$, respectively, and, since $A_x \subset E_x$ and $A_y \subset E_y$ and λ_n is complete, $\lambda_n(A_x) = \lambda_n(A^y) = 0$ for λ_n-a.e. $x, y \in \mathbb{R}^n$, respectively.

(b) First, by Tonelli's theorem,

$$\lambda_{m+n}(E) = \int_{\mathbb{R}^m} \lambda_n(E_x) \, d\lambda_m(x) = \int_{\mathbb{R}^n} \lambda_m(E^y) \, d\lambda_n(y) \, .$$

Now, since for almost every $x \in \mathbb{R}^m$, $\lambda_n(E_x) = 0$, we have $\int_{\mathbb{R}^m} \lambda_n(E_x) \, d\lambda_m(x) = 0$, and, consequently, $\lambda_{m+n}(E) = 0$ and $\int_{\mathbb{R}^n} \lambda_m(E^y) \, d\lambda_n(y) = 0$, which implies that for almost every $y \in \mathbb{R}^n$, $\lambda_m(E^y) = 0$.

15. Let $T : \mathbb{R}^2 \to \mathbb{R}^2$ be given by $T(x, y) = (x, y - x)$; T is an invertible linear transformation given by a matrix with determinant 1. Moreover, since $A \in \mathcal{B}(\mathbb{R}^2)$ and T^{-1} is continuous, $B = T(A) \in \mathcal{B}(\mathbb{R}^2)$ and by Problem 3.49, $\lambda_2(B) = \lambda_2(A)$. Now, since B^y is finite for any $y \in \mathbb{R}$, by Problem 3(b) and Tonelli's theorem it follows that $\lambda_2(B) = \int_{\mathbb{R}} \int_{\mathbb{R}} \chi_B(x, y) \, d\lambda(x) d\lambda(y) = 0$ and so $\lambda_2(A) = 0$.

17. First, $\eta^2 = \mu \otimes \nu(E) = \int_X \nu(E_x) \, d\mu(x) \geq \int_A \nu(E_x) \, d\mu(x) \geq \eta \, \mu(A)$ and so $\mu(A) \leq \eta$. To see that this bound is sharp consider the Lebesgue measure on I, $E = [0, 1/2] \times [0, 1/2]$, and $\eta = 1/2$. Now, for each $x \in I$, A is $[0, 1/2]$ or $\emptyset$ according to whether $0 \leq x \leq 1/2$ or $x > 1/2$, respectively. Therefore $A = [0, 1/2]$ and $\lambda(A) = 1/2 = \eta$.

20. We claim that if $p(x, y)$ is a nonidentically 0 polynomial of degree n in two real variables, the closed set $E = p^{-1}(\{0\})$ has Lebesgue measure 0. To see this observe that for each fixed x the section $p_x(y) = p_0(x) + p_1(x)y + \cdots + p_k(x)y^k$ is a polynomial in y of degree $\leq n$ with coefficients polynomials in x of degree $\leq n$ and, consequently, it has at most finitely many roots or is identically 0. In the first case E_x is finite and has Lebesgue measure 0. But since p is not identically 0 there are only finitely many values of x such that p_x is identically zero (namely, the simultaneous roots of all the $p_i(x)$, which are not identically 0). Thus $\lambda(E_x) = 0$ a.e. and $\lambda \otimes \lambda(E) = \int_{\mathbb{R}} \lambda(E_x) \, d\lambda(x) = 0$.

21. (a) $A \in \mathcal{B}(\mathbb{R}^+) \otimes \mathcal{P}(\mathbb{N})$ iff $A = \bigcup_{n \in \mathbb{N}} B_n \times \{n\}$ where $B_n \in \mathcal{B}(\mathbb{R}^+)$ for all n.

(b) Let ν denote the counting measure on $(\mathbb{N}, \mathcal{P}(\mathbb{N}))$. Let $A = \bigcup_{n \in \mathbb{N}} A_n \times \{n\} \in \mathcal{B}(\mathbb{R}^+) \otimes \mathcal{P}(\mathbb{N})$. If π exists,

$$\pi(A) = \sum_{n \in \mathbb{N}} \pi(A_n \times \{n\}) = \sum_{n \in \mathbb{N}} \int_{[0, \infty)} \chi_{A_n}(t) \, e^{-t} \frac{t^n}{n!} \, d\mu(t) \, ,$$

which gives the uniqueness of π. Moreover, this formula suggests the definition of the desired measure.

(c) If $\mu = \delta_0$, the delta at the origin, then $\pi(A \times \{n\}) = \delta(A)\delta(\{n\})$, the product of measures. On the other hand, if $\mu = \delta + \delta_2$, then $\pi(\{0\} \times \{n\})/$

$\pi(\{0,2\} \times \{n\})$ depends nontrivially on n (just check $n = 1$ and $n = 0$) and π is not a product of measures.

22. Let $0 \le a < \infty$. Since $\chi_{A_a}(x,y) = \chi_A(x)\,\chi_{[0,a]}(y)$, A is measurable and by Tonelli's theorem $\mu \otimes \lambda(A_a) = a\,\mu(A)$. Now, letting $a \to \infty$, $\mu \otimes \lambda(A_\infty) = \infty$ if $\mu(A) > 0$ and $= 0$ otherwise.

23. Let $\phi : (X \times \mathbb{R}, \mathcal{M} \otimes \mathcal{B}(\mathbb{R})) \to (\mathbb{R}^2, \mathcal{B}(\mathbb{R}^2))$ be given by $\phi(x,y) = (f(x), y)$; since for $A, B \in \mathcal{B}(\mathbb{R})$, $\phi^{-1}(A \times B) = f^{-1}(A) \cap B$, ϕ is Borel measurable. Now, since $\psi : (\mathbb{R}^2, \mathcal{B}(\mathbb{R}^2)) \to \mathbb{R}$ given by $\psi(x,y) = x - y$ is continuous, $\Phi : (X \times \mathbb{R}, \mathcal{M} \otimes \mathcal{B}(\mathbb{R})) \to (\mathbb{R}, \mathcal{B}(\mathbb{R}))$ given by $\Phi(x,y) = \psi \circ \phi(x,y) = f(x) - y$ is measurable and $\Gamma(f) = \Phi^{-1}(\{0\}) \in \mathcal{M} \otimes \mathcal{B}(\mathbb{R})$ is measurable.

Next, by Tonelli's theorem, $\mu \otimes \lambda(\Gamma(f)) = \int_X \int_{\mathbb{R}^+} \chi_{\Gamma(f)}(x,y)\, d\lambda(y) d\mu(x)$. Now, for each fixed $x \in X$, $(\chi_{\Gamma(f)})_x(y) = \{f(x)\}$ if $y = f(x)$ and $= \emptyset$ otherwise, and so $\int_{\mathbb{R}^+} (\chi_{\Gamma(f)})_x(y)\, d\lambda(y) = 0$. Hence $\mu \otimes \lambda(\Gamma(f)) = \int_X 0\, d\mu(x) = 0$.

24. The statement is true.

25. First, $R(A, f) = E(f) \cup F(f)$ where $E(f) = \{(x,y) \in X \times [0, \infty] : 0 \le y \le f(x) < \infty\}$ and $F(f) = \{(x,y) \in X \times [0, \infty] : 0 \le y < \infty, f(x) = \infty\}$. Then with the notation of Problem 23, $E(f) = \Phi^{-1}([0, \infty))$ and $F(f) = f^{-1}(\{\infty\}) \times [0, \infty]$ are $\mathcal{M} \otimes \mathcal{B}$ measurable and, therefore, $R(A, f)$ is measurable and by Tonelli's theorem $\mu \otimes \lambda(R(f)) = \int_X \int_{\mathbb{R}} \chi_{R(f)}(x,y)\, d\lambda(y)\, d\mu(x)$. Now, for fixed $x \in X$, $\chi_{R(f)}(x,y) = 1$ iff $\chi_{[0,f(x)]}(y) = 1$, and so

$$\mu \otimes \lambda(R(f)) = \int_X \int_{\mathbb{R}} \chi_{[0,f(x)]}(y)\, d\lambda(y)\, d\mu(x)$$
$$= \int_X \lambda([0, f(x)))\, d\mu(x) = \int_X f(x)\, d\mu(x).$$

If the statement $y \le f(x)$ is replaced by $y < f(x)$, the result still holds since $E = (\psi \circ \phi)^{-1}([0, \infty))$ is measurable and $\lambda([0, f(x)]) = \lambda([0, f(x)))$.

26. (a) $\lambda \otimes \delta(E) = 2$.

(b) $\int_{\mathbb{R}} f\, d(\lambda \otimes \delta) = \pi$.

28. First, since $((X \times X) \setminus \Delta)_x = X \setminus \{x\}$, by Tonelli's theorem $\mu \otimes \nu((X \times X) \setminus \Delta) = \int_X \nu(X \setminus \{x\})\, d\mu(x) = 0$, and so $\nu(X \setminus \{x\}) = 0$ μ-a.e. Pick one such x and observe that since ν is nontrivial, $\nu(\{x\}) > 0$. Now, if there exists $x_1 \ne x$ such that $\nu(X \setminus \{x_1\}) = 0$, then $\nu(X \setminus \{x\}) \ge \nu(\{x_1\}) > 0$, which is not the case. Thus $x = a$ is unique and $\nu = \alpha \delta_a$ with $\alpha = \nu(\{a\}) > 0$. Finally, again by Tonelli's theorem $\mu \otimes \nu((X \times X) \setminus \Delta) = \alpha \int_X \mu(X \setminus \{y\})\, d\delta_a(y) = \alpha\mu(X \setminus \{a\}) = 0$, and as before $\mu = \beta \delta_a$ with $\beta = \mu(\{a\})$.

29. The statement is not necessarily true: The counting measure on $\mathbb{R}^n$ is translation-invariant but not a multiple of λ_n; of course the counting measure is not σ-finite. We claim that if μ is a σ-finite translation-invariant measure on $\mathcal{L}(\mathbb{R}^n)$, then $\mu = c\,\lambda_n$ for some constant c. To see this let $A \in \mathcal{L}(\mathbb{R}^n)$ be a set with $\mu(A) < \infty$, $Q = [0,1]^n$ the unit cube, and observe that since $\int_{\mathbb{R}^n} \chi_Q(y)\,d\lambda(y) = 1$, by the translation-invariance of μ and Tonelli's theorem we have $\mu(A) = \int_{\mathbb{R}^n} \chi_A(x)\,d\mu(x) = \int_{\mathbb{R}^n} \chi_Q(y) \int_{\mathbb{R}^n} \chi_A(x)\,d\mu(x)\,d\lambda(y) = \int_{\mathbb{R}^n} \int_{\mathbb{R}^n} \chi_Q(y)\,\chi_{y+A}(x)\,d\mu(x)\,d\lambda(y) = \int_{\mathbb{R}^n} \int_{\mathbb{R}^n} \chi_Q(y)\,\chi_A(x - y)\,d\lambda(y)\,d\mu(x) = \int_{\mathbb{R}^n} \int_{\mathbb{R}^n} \chi_Q(x - y)\,\chi_A(y)\,d\lambda(y)\,d\mu(x) = \int_{\mathbb{R}^n} \int_{\mathbb{R}^n} \chi_{y+Q}(x)\,\chi_A(y)\,d\lambda(y)\,d\mu(x)$. Finally, by Tonelli's theorem and the translation invariance of μ, $\mu(A) = \int_{\mathbb{R}^n} \chi_A(y) \int_{\mathbb{R}^n} \chi_{y+Q}(x)\,d\mu(x)\,d\lambda(y) = \mu(Q)\,\lambda_n(A)$.

30. (a) The statement is false. Take $N = 1$ and $f = \chi_B$ with B as in Problem 4.

(b) The statement is true.

31. Pick a countable dense subset $D = \{d_k\}$ of $\mathbb{R}^n$ and for each $m \geq 1$ consider the sets $A_m^1 = B(d_1, 1/m)$ and $A_m^\ell = B(d_\ell, 1/m) \cap \bigcap_{j=1}^{\ell-1} B^c(d_j, 1/m)$, $\ell > 1$; note that $A_m^\ell \in \mathcal{B}(\mathbb{R}^n)$ for all ℓ and $A_m^\ell = \{x \in \mathbb{R}^n : d_\ell$ is the first $d_k \in D$ such that $|x - d_k| < 1/m\}$. Now, let $f_m : [a,b] \times \mathbb{R}^n \to \mathbb{R}$ be defined as follows: Given $x \in \mathbb{R}^n$, pick ℓ so that $x \in A_m^\ell$ and put $f_m(t,x) = f(t, d_\ell)$; since f_t is continuous, $\lim_m f_m(t,x) = f(t,x)$ for every $(t,x) \in [a,b] \times \mathbb{R}^n$.

Next, we claim that the f_m are $\mathcal{L}([a,b]) \otimes \mathcal{B}(\mathbb{R}^n)$ measurable for all m. To this end pick an open set $U \subset \mathbb{R}$. Then $f_m^{-1}(U) = \{(t,x) \in [a,b] \times \mathbb{R}^n : f_m(t,x) \in U\} = \bigcup_\ell \{(t,x) \in [a,b] \times \mathbb{R}^n : x \in A_m^\ell$ and $f_m(t,x) \in U\} = \bigcup_\ell \left(\{t \in [a,b] : f(t, d_\ell) \in U\} \times A_m^\ell\right)$. Now, by assumption $\{t \in [a,b] : f(t, d_\ell) \in U\} = (f^{d_\ell})^{-1}(U) \in \mathcal{L}([a,b])$ for all ℓ, and, therefore, $E_\ell = \{t \in [a,b] : f(t, d_\ell) \in U\} \times A_m^\ell \in \mathcal{L}([a,b]) \times \mathcal{B}(\mathbb{R}^n)$ for all ℓ, and $f_m^{-1}(U) = \bigcup_\ell E_\ell \in \mathcal{L}([a,b]) \otimes \mathcal{B}(\mathbb{R}^n)$. Finally, since the limit of measurable functions is measurable, f is measurable.

32. Since λ_{m+n} is the completion of $\lambda_m \otimes \lambda_n$ there exists $E \in \mathcal{L}(\mathbb{R}^{m+n})$ such that $A \subset E$ and $\lambda_{m+n}(E) = 0$. Then by Tonelli's theorem $0 = \int_{\mathbb{R}^{m+n}} \chi_E\,d\lambda_{m+n} = \int_{\mathbb{R}^m} \int_{\mathbb{R}^n} (\chi_E)_x\,d\lambda_n\,d\lambda_m = \int_{\mathbb{R}^n} \int_{\mathbb{R}^m} \chi_E^y\,d\lambda_m\,d\lambda_n$, and, consequently, $\int_{\mathbb{R}^n} (\chi_E)_x\,d\lambda_n = \lambda_n(E_x) = 0$ for λ_n-a.e. $y \in \mathbb{R}^n$, and, similarly, $\lambda_m(E^y) = 0$ for λ_m-a.e. $x \in \mathbb{R}^m$. Hence since $A_x \subset E_x$ and λ_n is complete it follows that $\lambda_n(A_x) = 0$ for λ_n-a.e. $y \in \mathbb{R}^n$ and, similarly, since $A_y \subset E_y$, $\lambda_m(A^y) = 0$ for λ_m-a.e. $y \in \mathbb{R}^n$.

35. Since f is integrable, $f_x \in L^1(Q_n)$ for λ_m-a.e. $x \in Q_m$. Now, for those sections f_x the conclusion holds for λ_n-a.e. $y \in Q_n$ by the Lebesgue differentiation theorem. Let $E = \{(x,y) \in Q_n \times Q_m :$ the conclusion does not hold$\}$. Then $\lambda_n(E_x) = 0$ and by Problem 12, $\lambda_m \otimes \lambda_n(E) = 0$, and since λ_{m+n} is the completion of $\lambda_m \otimes \lambda_n$, it follows that $\lambda_{m+n}(E) = 0$.

38. The statement is false.

39. To verify that the statement is true consider

$$f(x,y) = \begin{cases} (n+1), & x \in (1/(n+1), 1/n], \ y \in \mathbb{Q} \cap [0,1], \\ 0, & x = 0, \text{ or } y \notin \mathbb{Q} \cap [0,1]. \end{cases}$$

40. For every fixed $y \in [0,1]$, $\{x \in [0,1] : f(x,y) \neq 0\} = \{x \in [0,1] : x \prec y\}$ is an initial segment of this well-ordered set and, consequently, at most countable and of Lebesgue measure 0. Thus $\int_0^1 \int_0^1 f(x,y)\, d\lambda(x)\, d\lambda(y) = 0$. Similarly, for each $x \in [0,1]$ the set $\{y \in [0,1] : f(x,y) \neq 1\} = \{y \in [0,1] : y \prec x\}$ is at most countable and $\int_0^1 \int_0^1 f(x,y)\, d\lambda(y)\, d\lambda(x) = 1$. Hence the iterated integrals are not equal and S is not measurable.

41. (a) Δ is closed and hence Borel measurable.

(b) For a fixed $x \in [0,1]$, $\chi_\Delta(x,y) = 1$ for $y = x$ and $= 0$ elsewhere, and, consequently, since μ is the counting measure we have $\int_I \chi_\Delta(x,y)\, d\mu(y) = 1$ for all $x \in [0,1]$, and, therefore, $\int_I \int_I \chi_\Delta(x,y)\, d\mu(y)\, d\lambda(x) = 1$. On the other hand, for a fixed $y \in [0,1]$, $\chi_\Delta(x,y) = 1$ for $x = y$ and $= 0$ elsewhere, and this function is a.e. 0 with respect to the Lebesgue measure. Then $\int_I \chi_\Delta(x,y)\, d\lambda(x) = 0$ and the other iterated integral is 0.

(c) Since $\chi_\Delta(x,y)$ is nonnegative, bounded, and measurable, if it had a finite integral, by Tonelli's theorem the iterated integrals would be finite and all three equal. Since this is not the case the double integral is ∞.

For $E \in \mathcal{B} \otimes \mathcal{B}$ let $\nu_1(E) = \lambda \otimes \mu(C)$, the product measure, $\nu_2 = \int \left(\int \chi_E(x,y)\, d\mu \right) d\mu$, and $\nu_3 = \int \left(\int \chi_E(x,y)\, d\mu \right) d\mu$. That ν_2 and ν_3 are measures follows from the MCT and that they are all different follows from (b).

Finally, the reader should consider why Fubini's theorem does not apply.

43. The statement is not true. Let

$$f(x,y) = \sum_{n=0}^{\infty} a_n \chi_{[n,n+1) \times [n,n+1)}(x,y) - a_n \chi_{[n,n+1) \times [n+1,n+2)}(x,y),$$

where $\lim_n a_n = s > 0$.

44. (a) Let $B = \{x \in \mathbb{R}^{n+1} : x_1^2 + \cdots + x_n^2 + t^2 \leq 1\}$. Then $B^t = \{(x_1, \ldots, x_n) \in \mathbb{R}^n : x_1^2 + \cdots + x_n^2 \leq 1 - t^2\}$, $\lambda_n(B^t) = v_{n-1}(\sqrt{1-t^2})$, and by Tonelli's theorem v_{n+1} is equal to

$$\int_{\mathbb{R}^{n+1}} \chi_B\, d\lambda_{n+1} = \int_{-1}^{1} \int_{\mathbb{R}^n} \chi_{B^t}(x)\, d\lambda_n(x)\, d\lambda(t) = \int_{-1}^{1} v_n(\sqrt{1-t^2})\, d\lambda(t).$$

Now, by Problem 4.85, $v_n(\sqrt{1-t^2}) = (1-t^2)^{n/2} v_n(1)$ and so $v_{n+1} = v_n \int_{-1}^{1} f(t)^n\, d\lambda(t)$ with $f(t) = \chi_{[-1,1]}(t)(1-t^2)^{1/2}$; since $f(t) < 1$ for $t \neq 0$, $\lim_n f(t)^n = 0$ a.e.

(b) Note that

$$v_n = v_{n-1} \int_{-1}^{1} f(t)^{n-1} \, d\lambda(t) = \ldots = \prod_{k=0}^{n-1} \int_{-1}^{1} f(t)^k \, d\lambda(t) \,,$$

and, consequently, $A^n v_n = \prod_{k=0}^{n-1} (A \int_{-1}^{1} f(t)^k \, d\lambda(t))$. Now, by (a) it follows that $A \int_{-1}^{1} f^k(t) \, d\lambda(t) < s < 1$ for k sufficiently large and, therefore, the product diverges to 0.

45. The statement is true.

46. Picking $a = x, b = -x$, and integrating over $[0,1]$ it readily follows that $J = \int_0^1 \int_0^1 |f(t+x) - f(t-x)| \, d\lambda(t) \, d\lambda(x) \leq c$, and, consequently, by Tonelli's theorem $F(t,x) = |f(t+x) - f(t-x)| \in L^1([0,1] \times [0,1])$. Let M be the matrix

$$M = \begin{pmatrix} 1 & 1 \\ -1 & 1 \end{pmatrix}.$$

Then the linear transformation with matrix M maps $[0,1] \times [0,1]$ into the diamond-shaped region D in (ξ, η) space with vertices $(0,0)$, $(1,1)$, $(1,-1)$, and $(2,0)$, and so $\int_D |f(\xi) - f(\eta)| \, d\lambda(\xi) \otimes \lambda(\eta) = 2J \leq 2c$. Thus $|f(\xi) - f(\eta)| \in L^1(D)$ and by Fubini's theorem there exists $\xi_0 \in (1/2, 3/2)$ such that $|f(\xi_0) - f(\eta)| \in L^1(D_{\xi_0})$ where $\lambda(D_{\xi_0}) > 1$. Hence, since $|f(\eta)| \leq |f(\eta) - f(\xi_0)| + |f(\xi_0)|$, $f \in L^1(D_{\xi_0})$. And, since $f(\eta)$ is periodic of period 1 and $\lambda(D_{\xi_0}) > 1$ it follows that $f \in L^1([0,1])$.

47. First, by Tonelli's theorem, $\int_0^\infty \int_0^\infty \chi_{[t/b, t/a]}(x) \, |g(t)| \, x^{-1} \, d\lambda(x) \, d\lambda(t) = \ln(b/a) \int_0^\infty |g(t)| \, d\lambda(t) < \infty$. Thus, since $f(bx) - f(ax) = \int_{ax}^{bx} g(t) \, d\lambda(t)$, by Fubini's theorem

$$\int_0^\infty \frac{f(bx) - f(ax)}{x} \, d\lambda(x) = \int_0^\infty \frac{1}{x} \int_0^\infty \chi_{[ax,bx]}(t) g(t) \, d\lambda(t) \, d\lambda(x)$$

and as before $J = \ln(b/a) \int_0^\infty g(t) \, d\lambda(t)$.

This expression is known as Frullani's formula; see also Problem 4.162. In the particular case when f is continuously differentiable with compact support, $J = \ln(b/a) f(0)$.

50. $\int_{\mathbb{R}} f \, d\lambda = \pm 2$.

51. (a) By Tonelli's theorem the three integrals, whether finite or infinite, are equal.

(b) By Tonelli's theorem it suffices to verify that an iterated integral of f is finite. Now, if $a \neq 1$ we have

$$\int_I f(x,y) \, d\lambda(x) = -\frac{(1-xy)^{1-a}}{(1-a)y} \Big|_0^1 = \frac{1 - (1-y)^{1-a}}{(1-a)y} \,.$$

This function is undetermined as $y \to 0$, we calculate its asymptotic behaviour with the aid of L'Hôpital:

$$\lim_{y \to 0} \frac{1 - (1-y)^{1-a}}{(1-a)y} = \lim_{y \to 0} \frac{(1-a)(1-y)^{-a}}{(1-a)} = \lim_{y \to 0}(1-y)^{-a},$$

and, therefore, the integral converges for $a < 1$. Thus the integral is finite for $a \in (0,1)$. Next, we examine the case $a = 1$. Then

$$\int_I f(x,y)\, d\lambda(x) = -\frac{\log(1-xy)}{y}\Big|_0^1 = -\frac{\log(1-y)}{y}.$$

Now, this function converges to 1 as $y \to 0$ and tends logarithmically to ∞ as $y \to 1$; thus the asymptotic behavior as $y \to 1$ is integrable and the integral converges. Hence the integral converges for $a \in (0,1]$.

52. $\int_{\mathbb{R}^+ \times \mathbb{R}^+} f\, d(\lambda \otimes \lambda) = (\alpha - 1)^{-1}(\alpha - 2)^{-1}$.

53. $\int_C f\, d\lambda_3 = \sum_{n=0}^{\infty}(n+1)^{-3} < \infty$.

54. (a) Let $A_1 = \{(x,y,z) \in J : \max(y^b, z^c) \le x^a\}$, $A_2 = \{(x,y,z) \in J : \max(x^a, z^c) \le y^b\}$, and $A_3 = \{(x,y,z) \in J : \max(x^a, y^b) \le z^c\}$; it is readily seen that the A_k are pairwise disjoint and $J = A_1 \cup A_2 \cup A_3$. We consider the integral of f extended over A_1, the other integrals being treated similarly. Observe that if $(x,y,z) \in A_1$, $x^a \le x^a + y^b + z^c \le 3x^a$ and, consequently, it suffices to consider the integral of x^{-a} over A_1. Now, by Problem 9 and Tonelli's theorem, $\int_J \chi_{A_1}(x,y,z)x^{-a}\, d\lambda_3(x,y,z) = \int_{(0,1)} x^{-a} \int_{(0,1)\times(0,1)} \chi_{(0,x^{a/b})}(y)\, \chi_{(0,x^{a/c})}(z)\, d\lambda_2(y,z)\, d\lambda(x)$ and this integral is equal to $\int_{(0,1)} x^{-a}\, x^{a(1/b+1/c)}\, d\lambda(x)$, which is finite iff $a(1/b + 1/c - 1) > -1$, i.e., for $1/a + 1/b + 1/c > 1$.

Note that if $a = 1$, the integral converges for any $b, c > 0$. This is because $\int_0^1 (x + y^b + z^c)^{-1}\, d\lambda(x) = \ln(1 + y^b + z^c) - \ln(y^b + z^c)$ has an integrable singularity in $(0,1) \times (0,1)$.

(b) A similar argument gives that $f \in L^1(\mathbb{R}^3 \setminus J)$ iff $1/a + 1/b + 1/c < 1$.

56. The statement can be stated as follows. Let μ_1, μ_2 be finite Borel measures on $\mathbb{R}$ and let $\Phi : R \times R \to R$ be given by $\Phi(x,y) = x + y$. Then the measure $\nu(A) = \mu_1 \otimes \mu_2(\Phi^{-1}(A))$ on the Borel sets satisfies $\int_{\mathbb{R}} f\, d\nu = \int_{\mathbb{R}^2} f(x+y)\, d\mu_1(x)\, d\mu_2(y)$ for all $f \in L^1(\nu)$.

59. $J = c \int_{\mathbb{R}} d\mu(y)$.

60. Let $A \in \mathcal{L}(\mathbb{R})$ be bounded and consider the integral

$$J = \int_{\mathbb{R}} \int_A \frac{|\ln(|x-t|)|}{|x-t|^{1/2}}\, d\lambda(x)\, d\mu(t).$$

Since $|\ln(|x|)|/|x-t|^{1/2}$ is locally integrable on $\mathbb{R}$ the innermost integral above is finite and, therefore, $J < \infty$. Therefore by Fubini's theorem, $f \in$

$L^1(A)$ and so f is finite λ-a.e. there. Hence, since A is arbitrary, f is finite λ-a.e. in $\mathbb{R}$.

61. (b) Suppose first that $x \in F^c$. Then the ball $B = B(x, \delta(x)/2) \subset F^c$ and for all $y \in B$, $\delta_F(y) \geq |x - y|$. Hence

$$M(x) \geq \int_B \delta_F(y)|x - y|^{-(n+1)} \, d\lambda(y) \geq \int_B |x - y|^{-n} \, d\lambda(y) = \infty.$$

As for $x \in F$, it suffices to prove that $\int_F M(x) \, d\lambda(x) < \infty$. First, since $\delta_F(y) = 0$ for $y \in F$, by Tonelli's theorem

$$\int_F M(x) \, d\lambda(x) = \int_{F^c} \delta_F(y) \int_F |x - y|^{-(n+1)} \, d\lambda(x) \, d\lambda(y).$$

Now, since $F \subset \{x \in \mathbb{R}^n : |x - y| \geq \delta_F(y)\}$ the innermost integral is bounded by $\int_{\{|x-y|>\delta_F(y)\}} |x - y|^{-(n+1)} \, d\lambda(x) = c\delta_F(y)^{-1}$, and, therefore, $\int_F M(x) \, d\lambda(x) \leq c\lambda(F^c)$.

63. $I = 0$.

64. For the sake of argument suppose that $f_{xy}(a, b) - f_{yx}(a, b) > 0$. Then by continuity this expression is > 0 on an interval $J = [a, a + h] \times [b, b + h]$, say. Therefore $I = \int_J (f_{xy} - f_{yx}) \, d(\lambda \otimes \lambda) = \int_J f_{xy} \, d(\lambda \otimes \lambda) - \int_J f_{yx} \, d(\lambda \otimes \lambda) = I_1 - I_2 > 0$. Now, by Fubini's theorem, $I_1 = \int_b^{b+h} \int_a^{a+h} f_{xy} \, d\lambda(x) \, d\lambda(y) = \int_b^{b+h} (f_y(a + h, y) - f_y(a, y)) \, d\lambda(y) = f(a + h, b + h) - f(a, b + h) - f(a + h, b) + f(a, b)$. A similar argument gives that $I_2 = I_1$ and so $I = I_1 - I_2 = 0$, which is not the case.

65. $\int_I \int_I \sin(\pi t \chi_A(x)) \, d\lambda(x) \, d\lambda(t) = \int_I \int_I \sin(\pi t \chi_A(x)) \, d\lambda(t) \, d\lambda(x) = 2\lambda(A)/\pi$.

66. $J = \pi^2/2$, $K = \pi^2/4$, and $L = K/2$.

68. Note that for fixed $N > 0$, by Problem 6.64,

$$A(N) = \int_0^N \frac{\sin(x)}{x} \, d\lambda(x) = \int_0^N \int_0^\infty \sin(x) e^{-xt} \, d\lambda(t) \, d\lambda(x),$$

and, consequently, since Fubini's theorem applies, it follows that $A(N) = \int_0^\infty \int_0^N \sin(x) e^{-xt} \, d\lambda(x) \, d\lambda(t)$.

Now, by calculus (or differentiating with respect to N), the innermost integral in $A(N)$ is equal to

$$\int_0^N \sin(x) e^{-xt} \, d\lambda(x) = \frac{1 - \cos(N)e^{-Nt} - \sin(N)te^{-Nt}}{(1 + t^2)},$$

and, therefore, $A(N)$ is equal to

$$\int_0^N \frac{1}{1+t^2}\, d\lambda(t) - \cos(N) \int_0^\infty \frac{e^{-Nt}}{1+t^2}\, d\lambda(t) - \sin(N) \int_0^\infty t\, \frac{e^{-Nt}}{1+t^2}\, d\lambda(t)$$
$$= a_N - c_N - d_N\,,$$

say. Since $\lim_N a_N = \pi/2$, and by the LDCT, $\lim_N c_N = \lim_N d_N = 0$, we get

$$\int_0^\infty \frac{\sin(x)}{x}\, d\lambda(x) = \lim_N \int_0^N \frac{\sin(x)}{x}\, d\lambda(x) = \frac{\pi}{2}\,.$$

The motivated reader will note that a similar argument gives that

$$\int_0^\infty \left(\frac{\sin(x)}{x}\right)^2 d\lambda(x) = \pi\,.$$

69. The integral is equal to $2^{-1}\ln(2)$.

70. Let $f \in L^p(X)$. First, observe that since by Problem 5.23, $\{f \neq 0\}$ is σ-finite, we may assume that μ is σ-finite. As in Problem 25, let $\mu \otimes \lambda$ denote the product measure on $X \times \mathbb{R}^+$ and $R(f) = \{(x,t) \in X \times \mathbb{R}^+ : 0 \leq t < |f(x)|\}$, and define $h(x,t) = p\,t^{p-1}\chi_{R(f)}(x,t)$. Then, since for fixed $x \in X$, $\chi_{R(f)}(x,t) = 1$ iff $\chi_{[0,f(x)]}(t) = 1$ it follows that $\int_X \int_0^\infty h(x,t)\, d\lambda(t)\, d\mu(x) = \int_X \int_0^{|f(x)|} p\,t^{p-1}\, d\lambda(t)\, d\mu(x) = \int_X |f|^p\, d\mu$. Hence, by Tonelli's theorem we have $\int_X |f|^p\, d\mu = \int_0^\infty \int_X h(x,t)\, d\mu(x)\, d\lambda(t) = \int_0^\infty p\,t^{p-1}\mu(\{|f| > t\})\, d\lambda(t)$.

In general, if $\varphi : \mathbb{R}^+ \to \mathbb{R}^+$ is an increasing differentiable function, then $\int_X \varphi \circ f\, d\mu = \int_{\varphi^{-1}(0)}^{\varphi^{-1}(\infty)} \varphi'(t)\, \mu(\{f > t\})\, d\lambda(t)$.

71. With $f = f^+ - f^-$, $\int_X f\, d\mu = \int_X f^+\, d\mu - \int_X f^-\, d\mu$. Now, since $f^+, f^- \geq 0$, by Problem 70, we have $\int_X f\, d\mu = \int_0^\infty \mu(\{f^+ > t\})\, d\lambda(t) - \int_0^\infty \mu(\{f^- > t\})\, d\lambda(t) = \int_0^\infty \mu(\{f > t\})\, d\lambda(t) - \int_0^\infty \mu(\{-f > t\})\, d\lambda(t)$ and the conclusion follows since $\int_0^\infty \mu(\{-f > t\})\, d\lambda(t) = \int_{-\infty}^0 \mu(\{f < t\})\, d\lambda(t)$.

72. By Problem 4.57, $\nu(A) = \int_A g(x)\, d\mu(x)$ is a measure on $\mathcal{M}$ and by Problem 70, $\int_X fg\, d\mu = \int_0^\infty \nu(\{f > t\})\, d\lambda(t) = \int_0^\infty \varphi(t)\, d\lambda(t)$.

73. Note that $f(x) = \int_0^\infty \chi_{F_t}(x)\, d\lambda(t)$ and $g(x) = \int_0^\infty \chi_{G_s}(x)\, d\lambda(s)$. Then by Tonelli's theorem

$$\int_X fg\, d\mu = \int_X \left(\int_0^\infty \chi_{F_t}(x)\, d\lambda(t)\right)\left(\int_0^\infty \chi_{G_s}(x)\, d\lambda(s)\right) d\mu(x)$$
$$= \int_0^\infty \int_0^\infty \int_X \chi_{F_t}(x)\chi_{G_s}(x)\, d\mu(x)\, d\lambda(t)\, d\lambda(s)\,.$$

Now, since $\chi_{F_t}(x)\chi_{G_s}(x) = \chi_{G_s \cap F_t}$, the innermost integral is equal to $\mu(G_s \cap F_t)$ and the conclusion follows.

Moreover, since $F_s \subset F_t$ or $F_t \subset F_s$, $\mu(F_s \cap F_t) = \min(\mu(F_s), \mu(F_t))$, and $\int_X f^2 \, d\mu = \int_0^\infty \int_0^\infty \min(\mu(F_s), \mu(F_t)) \, d\lambda(s) \, d\lambda(t)$.

78. $J = \alpha \int_{\mathbb{R}^+} f(t) \, t^{\alpha_1 + \alpha_2 + \alpha_3 - 1} \, d\lambda(t)$.

79. Let $f \in L^p(Y)$ and $g \in L^q(X)$ with $\|g\|_q \leq 1$. Then $|k(x, y)| \, |f(y)|^p$ is measurable and if $I = (\int_{X \times Y} |k(x, y)| \, |f(y)|^p \, d\mu(x) \otimes \nu(y))^{1/p}$, by Tonelli's theorem

$$I = \left(\int_X \int_Y |k(x, y)| |f(y)|^p \, d\mu(x) \, d\nu(y) \right)^{1/p} \leq A^{1/p} \left(\int_X |f(y)|^p \, d\nu(y) \right)^{1/p}.$$

Similarly, if $J = (\int_{X \times Y} |k(x, y)| \, |g(x)|^q \, d\mu(x) \otimes \nu(y))^{1/q}$,

$$J = \left(\int_Y \int_X |k(x, y)| |g(x)|^q \, d\nu(y) \, d\mu(x) \right)^{1/q} \leq B^{1/q} \left(\int_Y |g(x)|^q \, d\mu(x) \right)^{1/q}.$$

Now, $|k(x, y)| \, |f(y)| \, |g(x)|$ is measurable and by Hölder's inequality it follows that $\int_{X \times Y} |k(x, y)| |f(y)| |g(x)| \, d\mu(x) \otimes \nu(y) \leq I \cdot J \leq A^{1/p} B^{1/q} \|f\|_p < \infty$. Therefore $k(x, y) f(y) g(x) \in L^1(X \times Y)$ and, consequently, by Fubini's theorem $\int_Y k(x, y) f(y) g(x) \, d\nu(y)$ exists for μ-a.e. $x \in X$ and is measurable and integrable, and since g is arbitrary, $Tf(x) = \int_Y k(x, y) f(y) \, d\nu(y)$ is measurable and finite μ-a.e. in X. Hence by the converse of Hölder's inequality the above estimate gives that $Tf \in L^p(X)$ and satisfies $\|Tf\|_p \leq A^{1/p} B^{1/q} \|f\|_p$. The result is known as Schur's lemma.

80. First, by the Cauchy-Schwartz inequality $|Tf(x)|$ does not exceed

$$\left(\int_{\mathbb{R}^n} |k(x, y)| \, |f(y)|^2 w(y)^{-1} \, d\lambda_n(y) \right)^{1/2} \left(\int_{\mathbb{R}^n} |k(x, y)| w(y) \, d\lambda_n(y) \right)^{1/2}$$

$$\leq A^{1/2} w(x)^{1/2} \left(\int_{\mathbb{R}^n} |k(x, y)| \, |f(y)|^2 w(y)^{-1} \, d\lambda_n(y) \right)^{1/2}.$$

Therefore, squaring and integrating, by Tonelli's theorem

$$\|Tf\|_2^2 \leq A \int_{\mathbb{R}^n} w(x) \int_{\mathbb{R}^n} |k(x, y)| \, |f(y)|^2 w(y)^{-1} \, d\lambda_n(y) \, d\lambda_n(x)$$

$$= A \int_{\mathbb{R}^n} |f(y)|^2 w(y)^{-1} \int_{\mathbb{R}^n} |k(x, y)| \, w(x) \, d\lambda_n(x) \, d\lambda_n(y)$$

$$= A^2 \int_{\mathbb{R}^n} |f(y)|^2 w(y)^{-1} w(y) \, d\lambda_n(y).$$

81. Necessity first. If $A, B \in \mathcal{B}(\mathbb{R})$, by independence $\mu_\Phi(A \times B) = \mu(\Phi^{-1}(A \times B)) = \mu(f^{-1}(A) \cap g^{-1}(B)) = \mu_f(A) \mu_g(B) = \mu_f \otimes \mu_g(A \times B)$. Hence μ_Φ coincides with $\mu_f \otimes \mu_g$ on the measurable rectangles of $X \times X$ and by the uniqueness of product measures they are equal.

Sufficiency next. If $A, B \in \mathcal{B}(\mathbb{R})$, $\mu_\Phi(A \times B) = \mu(f^{-1}(A) \cap g^{-1}(B))$. Thus $\mu_f \otimes \mu_g(A \times B) = \mu(f^{-1}(A) \cap g^{-1}(B))$, and f, g are independent.

82. Let $h(s,t) = st$ and $\Phi(x) = (f(x), g(x))$; h is continuous, Φ is measurable, $h \circ \Phi$ is measurable, and $h \circ \Phi(f,g) = fg$. Then by Problem 4.58 and Problem 81, $J = \int_X fg \, d\mu = \int_{\mathbb{R}^2} h(s,t) \, d\mu_\Phi(s,t) = \int_{\mathbb{R}^2} st \, d\mu_f \otimes \mu_g(s,t)$.

Now, $|h|$ is continuous, hence $\mathcal{D}$-measurable, and by Tonelli's theorem $\int_{\mathbb{R}^2} |h(s,t)| \, d\mu_\Phi(s,t) = \int_{\mathbb{R}^2} |st| \, d\mu_f \otimes \mu_g(s,t) = (\int_{\mathbb{R}} |s| \, d\mu_f(s))(\int_{\mathbb{R}} |t| \, d\mu_g(t)) = \|f\|_1 \|g\|_1 < \infty$. Thus Fubini's theorem applies and, consequently, $J = \int_{\mathbb{R}} \int_{\mathbb{R}} st \, d\mu_g(t) \, d\mu_f(s) = (\int_{\mathbb{R}} s \, d\mu_f(s))(\int_{\mathbb{R}} t \, d\mu_g(t)) = (\int_X f \, d\mu)(\int_X g \, d\mu)$.

83. Sufficiency is true in general and necessity is false (take $g = -f$) unless the functions are independent. If this is the case, let $F(x) = (f(x), g(x))$, $x \in X$. Then by Problem 4.59(c) and Problem 81, $J = \int_X |f + g|^p \, d\mu = \int_{\mathbb{R}^2} |s + t|^p \, d\mu_F(s,t) = \int_{\mathbb{R}^2} |s + t|^p d\mu_f \otimes d\mu_g(s,t)$ and by Tonelli's theorem and Problem 4.58(b),

$$J = \int_{\mathbb{R}} \int_{\mathbb{R}} |s + t|^p d\mu_f(s) \, d\mu_g(t) = \int_{\mathbb{R}} \int_X |f(x) + t|^p \, d\mu(x) \, d\mu_g(t) < \infty.$$

Thus $\int_X |f(x) + t|^p \, d\mu(x) < \infty$ for μ_g-a.e. t and fixing one such t,

$$\int_X |f(x)|^p \, d\mu(x) = \int_X |f(x) + t - t|^p \, d\mu(x)$$
$$\leq 2^p \left(\int_X |f(x) + t|^p \, d\mu(x) + |t|^p \mu(X) \right) < \infty.$$

Finally, $g = (g + f) - f$ is in $L^p(X)$.

84. As in Problem 83, $\int_X |f + g|^p \, d\mu = \int_{\mathbb{R}} \int_X |s + g(y)|^p \, d\mu(y) \, d\mu_f(s)$. Now, for $s \in \mathbb{R}$, by Jensen's inequality we have $\int_X |g(y) + s|^p \, d\mu(y) \geq |\int_X (g(y) + s) \, d\mu(y)|^p = |\int_X g(y) \, d\mu(y) + s|^p = |s|^p$. Hence integrating, $\int_{\mathbb{R}} \int_X |s + g(y)|^p \, d\mu(y) \, d\mu_f(s) \geq \int_{\mathbb{R}} |s|^p \, d\mu_f(s) = \int_X |f|^p \, d\mu$. The conclusion follows by combining these estimates.

85.

$$f_a * f_b(x) = \left(\frac{2\pi}{a + b} \right)^{n/2} f_{ab/(a+b)}(x).$$

86. (c) The result is not true in general but is true if $g(x) \to 0$ as $|x| \to \infty$.

87. The statement is false. For the sake of argument suppose that f is an integrable function such that $f * g = g$ for all $g \in L^1(\mathbb{R}^n)$. Then if g is bounded, by Problem 86, $f * g$ is continuous and, therefore, $g = f * g$ is continuous. Just pick $g = \chi_A$ to be the characteristic function of a set A of finite measure to see that this is not always the case.

88. (a) Since by Fubini's theorem $\int_{\mathbb{R}^n} (f * g) \, d\lambda_n = (\int_{\mathbb{R}^n} f \, d\lambda_n)(\int_{\mathbb{R}^n} g \, d\lambda_n) > 0$, $f * g > 0$ in a set of positive measure. Similarly, if f, g have integrals of different sign, $f * g < 0$ in a set of positive measure.

(b) If one of the functions is bounded the convolution is continuous and, therefore, if positive at a point, positive in an interval. In particular note that if $f = \chi_A$, $g = \chi_B$ where A, B are Lebesgue measurable sets of positive measure, then $\chi_A * \chi_B$ is a continuous not identically 0 function with support included in $A + B$. Now, if $x \in A + B$ is such that $\chi_A * \chi_B(x) \neq 0$, by continuity there exists an open set U containing x such that $\chi_A * \chi_B(y) > 0$ for all $y \in U$, and so $U \subset A + B$.

89. $\lim_k f_k = 0$ in $L^1(\mathbb{R}^n)$.

91. First, since $d(\overline{O_1}, O^c) > 0$, there exists an open set O_2 such that $\overline{O_1} \subset O_2 \subset \overline{O_2} \subset O$; pick $\varepsilon < \min(d(O_1, \partial(O_2)), d(O_2, \partial(O)))$. Next, let $\varphi(x)$ be a $C_0^\infty(\mathbb{R}^n)$ function supported in the unit ball $B(0,1)$ of $\mathbb{R}^n$ with integral 1; for instance, $\varphi(x) = ce^{-1/(1-|x|^2)}\chi_{B_1(0)}(x)$ where c is chosen so that φ has integral 1 will do. Then $\varphi_\varepsilon(x) = \varepsilon^{-n}\varphi(x/\varepsilon)$ is $C_0^\infty(\mathbb{R}^n)$, has integral 1, and is supported in $\overline{B(0,\varepsilon)}$.

Now let $h(x) = \chi_{O_2} * \varphi_\varepsilon(x)$. Since O_2 is bounded and $\varphi \in C_0^\infty(\mathbb{R}^n)$, $h \in C_0^\infty(\mathbb{R}^n)$. Furthermore, if $x \in O_1$ and $t \in B(0,\varepsilon)$, $\chi_{O_2}(x - t) = 1$, and so $h(x) = \int_{B(0,\varepsilon)} \chi_{O_2}(x - t)\varphi_\varepsilon(t)\,d\lambda_n(t) = \int_{B(0,\varepsilon)} \varphi_\varepsilon(t)\,d\lambda_n(t) = 1$. And, if $x \notin O$ and $t \in B(0,\varepsilon)$, then $\chi_{O_2}(x - t) = 0$, and, consequently, $h(x) = \int_{B(0,\varepsilon)} \chi_{O_2}(x - t)\varphi_\varepsilon(t)\,d\lambda_n(t) = 0$.

92. (a) Since by the Lebesgue differentiation theorem for λ_n-a.e. $x \in \mathbb{R}^n$, $f(x) = \lim_{h \to 0^+} h^{-n} \int_{Q(x,h)} f(y)\,d\lambda_n(y)$, such an f does not exist.

(b) Let $f(x) = \chi_{(-1,0)}(x) - \chi_{(0,1)}(x)$. Then $0 = h^{-1} \int_{-h/2}^{h/2} f(x)\,d\lambda(x) \to f(0)$ as $h \to 0$, yet f is not continuous at 0.

(c) Assume that $|\varphi| \leq M$ in $\mathbb{R}^n$ and let $x \in L_f$. Then

$$|f * \varphi_\varepsilon(x) - \eta f(x)| \leq \int_{\mathbb{R}^n} |f(x - y) - f(x)|\,|\varphi_\varepsilon(y)|\,d\lambda_n(y)$$

$$\leq M\varepsilon^{-n} \int_{B(0,\varepsilon)} |f(x - y) - f(x)|\,d\lambda_n(y)$$

where $\lambda_n(B(0,\varepsilon)) = c_n\varepsilon^n$. Hence

$$\limsup_{\varepsilon \to 0} |f * \varphi_\varepsilon(x) - f(x)|$$

$$\leq \limsup_{\varepsilon \to 0} c_n M |B(0,\varepsilon)|^{-n} \int_{B(0,\varepsilon)} |f(x - y) - f(x)|\,d\lambda_n(y) = 0$$

and we have finished.

94. Many classical kernels satisfy the restrictions we have imposed. Three important examples for the case $n = 1$ include the Poisson kernel, the Fejér kernel, $(1/\pi)(\sin x/x)^2$, and the Gauss-Weierstrass kernel, $1/\sqrt{\pi y}e^{-x^2/y}$.

95. For the sake of argument suppose such an f exists. Then by Fubini's theorem $\int_{\mathbb{R}^n} f * f \, d\lambda_n = (\int_{\mathbb{R}^n} f \, d\lambda_n)^2 = \int_{\mathbb{R}^n} f \, d\lambda_n$, and, therefore, $\int_{\mathbb{R}^n} f \, d\lambda_n = 1$. Now, let $\varepsilon > 0$ and g a compactly supported continuously differentiable function. Then, since

$$\varepsilon^{-n} f(x/\varepsilon) = \varepsilon^{-n} \int_{\mathbb{R}^n} f(x/\varepsilon - y) f(y) \, d\lambda_n(y),$$

multiplying through by $g(z - x)$ and integrating it follows that

$$f_\varepsilon * g(z) = \int_{\mathbb{R}^n} \varepsilon^{-n} \int_{\mathbb{R}^n} f(x/\varepsilon - y) f(y) \, d\lambda_n(y) \, g(z - x) \, d\lambda_n(x).$$

Now, switching the order of integration, which is legitimate since both functions are compactly supported and bounded,

$$f_\varepsilon * g(z) = \int_{\mathbb{R}^n} f(y) \, \varepsilon^{-n} \int_{\mathbb{R}^n} f(x/\varepsilon - y) g(z - x) \, d\lambda_n(x) \, d\lambda_n(y).$$

The change of variables $x - \varepsilon y = w$ gives that $f_\varepsilon * g(z)$ is equal to

$$\int_{\mathbb{R}^n} f(y) \varepsilon^{-n} \int_{\mathbb{R}^n} f(w/\varepsilon) g(z - w + \varepsilon y) \, d\lambda_n(w) \, d\lambda_n(y)$$

$$= \int_{\mathbb{R}^n} f(y) \, \varepsilon^{-n} \int_{\mathbb{R}^n} f(w/\varepsilon) g(z - w) \, d\lambda_n(w) \, d\lambda_n(y)$$

$$+ \int_{\mathbb{R}^n} f(y) \, \varepsilon^{-n} \int_{\mathbb{R}^n} f(w/\varepsilon)(g(z - w + \varepsilon y) - g(z - w)) \, d\lambda_n(w) \, d\lambda_n(y)$$

$$= A_\varepsilon + B_\varepsilon,$$

say.

Since f has integral 1, by Problem 94 and Fubini's theorem $\lim_{\varepsilon \to 0} A_\varepsilon = g(z) \, \lambda_n$-a.e. Moreover, since $|g(z - w + \varepsilon y) - g(z - w)| \le c\varepsilon |y|$ for all $y, z, w \in \mathbb{R}^n$, it readily follows that

$$|B_\varepsilon| \le c \int_{\mathbb{R}^n} \varepsilon^{-n} |f(w/\varepsilon)| \int_{\mathbb{R}^n} |f(y)| \, \varepsilon |y| \, d\lambda_n(y) \, d\lambda_n(w) = c\varepsilon.$$

Therefore $\lim_{\varepsilon \to 0} B_\varepsilon = 0$, and, hence, $f * g = g \, \lambda_n$-a.e. for such g.

Next, for an arbitrary $h \in L^1(\mathbb{R}^n)$, pick a continuously differentiable function g vanishing off a compact set such that $\|h - g\|_1 \le \varepsilon$. Then $f * h - h = f * (h - g) + f * g - h = f * (h - g) + g - h$, and, therefore, $\|f * h - h\|_1 \le \|f * (h - g)\|_1 + \|h - g\|_1 < c\varepsilon$. Thus $\|f * h - h\|_1 = 0$ and $f * h = h \, \lambda_n$-a.e. By Problem 87 this is impossible.

96. Let ψ be a compactly supported smooth function and for $x \in \mathbb{R}^n$ and $t > 0$, let $\varphi(\cdot) = t^{-n} \psi(x - \cdot/t)$. Then by assumption $v * \psi_t(x) = 0$ for all x, t, and so by Problem 94, $v = 0$.

A strengthening of Problem 94 gives the same conclusion for a locally integrable function v on $\mathbb{R}^n$.

97. Let $T : L^p(\mathbb{R}^n) \to L^p(\mathbb{R}^n)$ be the mapping given by $T(h) = f * h$; then $\|T(h)\|_p \leq \|f\|_1\|h\|_p$ and $\|T\| \leq \|f\|_1 < 1$. By Problem 9.156 the equation $h - T(h) = g$ has a unique solution $h \in L^p(\mathbb{R}^n)$ that depends continuously on g for every $g \in L^p(\mathbb{R}^n)$.

100. If $p < \infty$ pick $g = \varphi_k$ where $\{\varphi_k\}$ is an approximate identity. Then $\|f * \varphi_k\|_p \leq c\|\varphi_k\|_1 = c$, and letting $k \to \infty$ it follows that $\|f\|_p \leq c$. On the other hand, if $p = \infty$, let $g = |Q(x,r)|^{-1}\chi_{Q(x,r)}$, where $Q(x,r)$ is an arbitrary cube centered at x of sidelength r, $\|g\|_1 = 1$. Then

$$\left| \int_{\mathbb{R}^n} fg \, d\lambda_n \right| = \frac{1}{|Q(0,r)|} \left| \int_{Q(0,r)} f(x-y) \, d\lambda_n(y) \right| \leq c,$$

and consequently by the Lebesgue differentiation theorem $\|f\|_\infty \leq c$.

101. Let $g \in L^q(\mathbb{R})$ where q is the conjugate to p. Then with $h(x) = f(-x)$ it follows that $\int_{\mathbb{R}^k} \tau_n f(x)g(x) \, d\lambda_n(x) = h * g(h_n)$ and by Problem 86, $\lim_n h * g(h_n) = 0$.

102. Let $J = [a,b]$ be a fixed interval, $\delta = b - a$, $f = \chi_J$, and $f_n(x) = f(x + n\delta)$, $n = 0, \pm 1, \ldots$. Each of these functions has $\|f_n\|_2 = \sqrt{\delta}$ and $\int_R f_m f_n = \delta\delta_{m,n}$, and so they form a bounded sequence. Since by Problem 101, $f_n \rightharpoonup 0$, by Problem 9.62 it follows that $T(f_n) \to 0$ in L^2.

Now, since $Tf_n = T\tau_{n\delta}f = \tau_{n\delta}Tf$, we get

$$\|Tf\|_2 = \|\tau_{n\delta}Tf\|_2 = \|Tf_n\|_2 \to 0 \quad \text{as } n \to \infty,$$

and, consequently, $\|Tf\|_2 = 0$ and $\int_a^b k(y-z) \, d\lambda(y) = 0$ for all a, b and a.e. $z \in \mathbb{R}$. Hence $k = 0$ a.e.

Chapter 18

Normed Linear Spaces. Functionals

1. Let $\{e_\lambda\}_{\lambda\in\Lambda}$ be an infinite Hamel basis for X; then each $x \in X$ has a unique representation as a finite sum $x = \sum_{\lambda\in\Lambda} x_\lambda e_\lambda$. Now, for any collection of positive real numbers $\{\alpha_\lambda\}_{\lambda\in\Lambda}$ the expression $\|x\|_\alpha = \sum_{\lambda\in\Lambda} \alpha_\lambda |x_\lambda|$ is a norm on X and the norms induced by a bounded and an unbounded $\{\alpha_\lambda\}$ are not equivalent. Note that when $\alpha_\lambda = 1$ for all λ, $\|e_\lambda - e_\mu\| = 2$ for $\lambda \neq \mu$ and $(X, \|\cdot\|)$ is not separable.

4. The limit is $\lambda \|x\|$.

5. Since $2x = (x - y) + (x + y)$ the triangle inequality gives the sharper estimate $2\|x\| \leq \|x - y\| + \|x + y\|$. As for the case of equality, in $X = L^\infty([0, 1])$ let $x = \chi_{[0,1/2)}$ and $y = \chi_{[1/2,1]}$.

6. Since $\max\{\|x\|, \|y\|\} \geq (\|x\| + \|y\|)/2$ the second inequality follows from the first. Now, assume that $\max\{\|x\|, \|y\|\} = \|x\|$ and set $\lambda = \|x\|/\|y\|$. Then

$$\|x - y\| = \| [x - \lambda y] - (1 - \lambda) y\| \geq |\, \|x - \lambda y\| - \|(1 - \lambda)y\| \,|$$

$$\geq \|x - \lambda y\| - \|(1 - \lambda)y\| = \|x\| \left\| \frac{x}{\|x\|} - \frac{y}{\|y\|} \right\| - |(1 - \lambda)| \|y\|$$

$$= \|x\| \left\| \frac{x}{\|x\|} - \frac{y}{\|y\|} \right\| - |\,\|x\| - \|y\|\,| \geq \|x\| \left\| \frac{x}{\|x\|} - \frac{y}{\|y\|} \right\| - \|x - y\|,$$

and, consequently,

$$2\|x - y\| \geq \|x\| \left\| \frac{x}{\|x\|} - \frac{y}{\|y\|} \right\|.$$

To see that $1/2$ is the best possible constant suppose that the inequality holds for a constant M, and for $x \in X$ with $\|x\| = 1$ put $y = -x/n$, $n \geq 2$.

Then $\|y\| = 1/n$, $\max\{\|x\|, \|y\|\} = 1$, and $\|x - y\| = (1 + 1/n)$. Thus

$$\max\{\|x\|, \|y\|\} \left\| \frac{x}{\|x\|} - \frac{y}{\|y\|} \right\| = 2,$$

and so, for all n, $1 + 1/n \geq 2M$, which implies that $2M \leq 1$ and $M \leq 1/2$. The same example shows that the constant in the second inequality is best possible.

7. Since the proofs are similar we only consider $\{w_n\}$. Observe that we may assume that $\lim_n x_n = 0$. Indeed, if the limit is x, say,

$$w_n = \frac{1}{n^2} \left((x_1 - x) + 2(x_2 - x) + \cdots + n(x_n - x) \right) + \frac{n(n+1)}{2} \frac{1}{n^2} x,$$

and so the resulting limit differs from the one obtained for sequences tending to 0 by $x/2$. Now, we claim that if $x_n \to 0$, $w_n \to 0$. Given $\varepsilon > 0$, let N be such that $\|x_n\| < \varepsilon$ for $n \geq N$. Then for any such n,

$$\|w_n\| \leq \frac{1}{n^2} \left(\|x_1\| + 2\|x_2\| + \cdots + N\|x_N\| \right)$$
$$+ \frac{1}{n^2} \left((n - N) + (n - N + 1) + \cdots + n \right) \varepsilon$$

where the second term is less than $\varepsilon/2$ and the first goes to zero with n. Thus $\lim_n w_n = (\lim_n x_n)/2$. A similar argument gives that $\lim_n v_n = \lim_n x_n$.

8. The condition is sufficient but not necessary: Take $x_n = x$ for all n and let $\sum_n \lambda_n$ be a series that is conditionally, but not absolutely, convergent; for example, $\lambda_n = (-1)^n/n$ for $n \geq 1$ will do.

9. (a) The statement is false. Obviously X cannot be complete but other than that any space will do. If X is not complete there is a Cauchy sequence $\{x_n\} \subset X$ that does not converge in X. Pick an increasing sequence $\{n_k\}$ such that $\|x_n - x_m\| \leq 2^{-k}$ for all $m, n \geq n_k$, and consider the series $\sum_k (x_{n_k+1} - x_{n_k})$; since $\sum_k \|x_{n_k+1} - x_{n_k}\| \leq \sum_k 2^{-k} < \infty$ it converges absolutely. However, it does not converge in X for if it did the partial sums $\sum_{k=1}^{j} x_{n_k+1} - x_{n_k} = x_{n_{j+1}} - x_{n_1}$ would converge, $\{x_{n_{j+1}}\}$ would converge, and, therefore, the whole sequence would converge, which it does not.

(b) The statement is true. Since B is complete, $y_n = \sum_{k=1}^{n} x_k$ converges to $x \in B$. Let $\sigma : \mathbb{N} \to \mathbb{N}$ be a bijection; we claim that $y_n^\sigma = \sum_{k=1}^{n} x_{\sigma(k)}$ converges to x. Let n be large enough so that $\|x - y_n\| < \varepsilon$ and pick N large enough so that all the terms $x_1, \ldots, x_n$ appear in the sum that defines y_N^σ. Then, for $m \geq N$, $\|x - y_m^\sigma\| \leq 2 \sum_{k=N+1}^{\infty} \|x_k\| < \varepsilon$. Thus $\lim_n y_n^\sigma = x$ and the convergence is unconditional.

11. Necessity is obvious: S is closed and so complete. Conversely, suppose that S is complete and let $\{x_n\}$ be a Cauchy sequence in X. Since $|\|x_n\| - \|x_m\|| \leq \|x_n - x_m\|$, the numerical sequence $\{\|x_n\|\}$ is Cauchy in

$\mathbb{R}$; let η be its limit. If $\eta = 0$, $x_n \to 0$. If $\eta > 0$, there is a constant $c > 0$ such that $1/c \leq \|x_n\| \leq c$. Normalize the Cauchy sequence and note that

$$\frac{x_n}{\|x_n\|} - \frac{x_m}{\|x_m\|} = x_n\left(\frac{1}{\|x_n\|} - \frac{1}{\|x_m\|}\right) + \frac{1}{\|x_m\|}\left(x_n - x_m\right)$$

is the sum of two Cauchy sequences and, hence, Cauchy in S. If x is the limit, $x_n = \|x_n\|\,(x_n/\|x_n\|) \to \eta\, x$ in X and X is complete.

12. Recall that a bounded open set O in $\mathbb{R}^2$ that is convex, symmetric $(O = (-1)O)$, and contains the origin is the unit ball of a norm $\|\cdot\|$ on $\mathbb{R}^2$; indeed, $\|(x_1, x_2)\| = \inf\{\lambda > 0 : (x_1, x_2) \in \lambda O\}$ defines a norm in $\mathbb{R}^2$ and, in fact, generates all norms there. Therefore p is a norm in $\mathbb{R}^2$. As for the second question, let $\eta = \|(1,1)\| < 1$. If the norm exists, then $\|(1/\eta, 1/\eta)\| = 1$, and observe that the convex hull of $(1,0)$, $(1/\eta, 1/\eta)$, $(0,1)$, $(-1,0)$, $(-1/\eta, -1/\eta)$, $(-1,0)$, and $(1,0)$ is then the unit ball of a norm $\|\cdot\|$ in $\mathbb{R}^2$ that satisfies the required properties.

13. (d) Let $\{B(a_k, r_k)\}$ be a decreasing sequence of nonempty closed balls. By (a), $r_1 \geq r_2 \geq \ldots$, and the radii form a decreasing sequence which is bounded below and, hence, has a limit; in particular, $\{r_k\}$ is a Cauchy sequence. Now, by (c), $\|a_n - a_m\| \leq |r_n - r_m| \to 0$ as $n, m \to \infty$, and, consequently, the sequence $\{a_n\}$ consisting of the centers of the balls is Cauchy in the Banach space X, and so it converges to a point $a \in X$, say. We claim that $a \in B(a_n, r_n)$ for all n. Indeed, for each n, $a = \lim_{m > n, m \to \infty} a_m \in B(a_n, r_n)$ and, since the balls are nested and closed, a is in each B_n and, consequently, in their intersection.

14. (a) Since ρ does not satisfy the homogeneity condition $\rho(\lambda x) = \lambda\rho(x)$, $\lambda > 0$, it is not a norm.

(c) Since $t/(1 + t)$ decreases to 0 as $t \to 0^+$, $r(x_n, x) \to 0$ whenever $\|x_n - x\| \to 0$. Conversely, since for $t \leq 1$, $t \leq 2t/(1 + t)$, $r(x_n, x) \to 0$ implies $\|x_n - x\| \to 0$.

(d) Since $\rho(x) \leq 1$ any unbounded metric on X, for instance, $d(x, y) = \|x - y\|$, will do.

15. If $|\lambda_1| + \cdots + |\lambda_n| = 0$ the inequality holds for any $c > 0$, so it suffices to prove that there exists $c > 0$ such that $\|\mu_1 x_1 + \cdots + \mu_n x_n\| \geq c$ whenever $|\mu_1| + \cdots + |\mu_n| = 1$. For the sake of argument suppose this is not the case. Then for each integer k there exist $\mu_k^1, \ldots, \mu_k^n$ with $|\mu_k^1| + \cdots + |\mu_k^n| = 1$ yet $\|\mu_k^1 x_1 + \cdots + \mu_k^n x_n\| < 1/k$. Now, since the $\{\mu_k^i\}$ are contained in a compact subset of the scalar field, by repeated application of Bolzano-Weierstrass (first for $i = 1$, then for $i = 2$, and so on) there is a subsequence $\{k_m\}$, say, such that $\{\mu_{k_m}^i\}$ converges for $1 \leq i \leq n$. Let $\lim_{k_m} \mu_{k_m}^i = \alpha_i$, $1 \leq i \leq n$; note that since the generating μ sum in modulus to 1 not all the α_i can be zero. Now, by Problem 2, $\|\mu_{k_m}^1 x_1 + \cdots + \mu_{k_m}^n x_n\| \to \|\alpha_1 x_1 + \cdots + \alpha_n x_n\|$ and by construction $\|\mu^1_{k_m} x_1 + \cdots + \mu^n_{k_m} x_n\| \to 0$. Thus $\|\alpha_1 x_1 + \cdots + \alpha_n x_n\| = 0$, which implies that $\alpha_1 x_1 + \cdots + \alpha_n x_n = 0$, and so by the linear independence all the α_i are 0, which is not the case.

16. When X is an infinite-dimensional linear space we have constructed nonequivalent norms in Problem 1. On the other hand, if X is finite dimensional let $X = \mathrm{sp}\{x_1, \ldots, x_n\}$ and suppose that X is endowed with norms $\|\cdot\|$ and $\|\cdot\|_1$, say; we prove that one norm dominates the other and since they are arbitrary, we are done. Let $c = \max_{1 \leq k \leq n} \|x_k\|_1$. Then for $x = \sum_{k=1}^n \lambda_k x_k$ we have $\|x\|_1 \leq c \sum_{k=1}^n |\lambda_k|$, and by Problem 15 there is a constant c_1 such that this expression is dominated by $c\,c_1 \|\sum_{k=1}^n \lambda_k x_k\| = c\,c_1 \|x\|$.

18. (b) Assume $Y^c \neq \emptyset$ and let $x_0 \in Y^c$. For $x \in Y$, $x + \lambda x_0$ is in Y^c for all scalars λ, for, if $x + \lambda x_0 \in Y$, since x is in the subspace Y, $\lambda x_0 \in Y$, which in turn implies that $x_0 \in Y$, and this is not the case. Now, as $\lambda \to 0$, $x + \lambda x_0 \in Y^c$ converges to x, and so $x \in \overline{Y^c}$. Thus $Y \subset \overline{Y^c}$ and $X = Y \cup Y^c \subset \overline{Y^c}$, which gives $X = \overline{Y^c}$.

19. By Problem 18(a) X_n contains no ball and by Problem 1.3(b), $\bigcup_n X_n \neq B$. As for the example, in $(C(I), \|\cdot\|_\infty)$ let $X = \mathrm{sp}\{1, t, \ldots, t^n, \ldots\}$ and $X_n = \mathrm{sp}\{1, \ldots, t^n\}$. Each X_n is finite dimensional, hence closed, and each $x \in X$ is in some X_n, i.e., $X = \bigcup_n X_n$.

20. (a) The statement is false. For instance, in ℓ^p, $1 \leq p < \infty$, let $X_1 = \{e_1, \ldots, e_n, \ldots\}$ and $X_2 = \{-e_2 + 2^{-1} e_1, \ldots, -e_n + n^{-1} e_1, \ldots\}$. X_1, X_2 are closed since they have no limit points and 0, the only limit point of $X_1 + X_2$, does not belong to $X_1 + X_2$.

(b) Although as is readily seen $X_1 + X_2$ is a subspace of B, it may happen that $X_1 + X_2$ is dense but not closed in B. In c_0, let $X_1 = \{x \in c_0 : x_{2n} = 0$ for all $n\}$ and $X_2 = \{x \in c_0 : x_{2n} = x_{2n-1}/n$ for all $n\}$. We claim that $X_1 + X_2$ is dense in, but not equal to, c_0. To see this we first characterize the sequences in $X_1 + X_2$, namely, $x \in c_0$ is in $X_1 + X_2$ iff $n x_{2n} \to 0$. Indeed, if $x \in c_0$ and $n x_{2n} \to 0$, let y^1 be given by $y^1_{2n} = 0$, $y^1_{2n-1} = x_{2n-1} - n x_{2n}$ for all n, and y^2 the sequence with terms $y^2_{2n} = x_{2n}$ and $y^2_{2n-1} = n x_{2n}$ for all n. Then $y^1 \in X_1, y^2 \in X_2$ and $x = y^1 + y^2$. On the other hand, if $x \in X_1 + X_2$, then $x \in c_0$ and the condition is clearly satisfied. From this it readily follows that $X_1 + X_2$ is a proper subset of c_0 so it remains to prove that it is dense. Let $x \in c_0$ and fix $\varepsilon > 0$. Then for some N, $|x_n| < \varepsilon$ for $n \geq N$. Defining the sequence y with terms $y_n = x_n$ for $n < N$ and $y_n = 0$ for $n \geq N$, we have $\|x - y\|_\infty < \varepsilon$. From the above it is clear that $y \in X_1 + X_2$, and so $X_1 + X_2$ is dense. Since the sum is still a subspace we have shown that an infinite-dimensional Banach space may contain a dense subspace that is not closed. As we shall see later the kernel of a discontinuous linear functional on B shares this property.

On the other hand, if X_2, say, is finite dimensional, $X_1 + X_2$ is closed. It suffices to prove that $X_1 + \mathrm{sp}\{x\}$, $x \in X_2 \setminus X_1$, is closed. By Hahn-Banach there is a bounded linear functional L on B with $L(x) = 1$ that vanishes on X_1. Suppose that $\{y_n\}$ is a sequence in $X_1 + \mathrm{sp}\{x\}$ that converges to y; we claim that $y \in X_1 + \mathrm{sp}\{x\}$. Now, $y_n = x_n + \lambda_n x$ for some $x_n \in X_1$ and scalars λ_n and so $\lambda_n = L(y_n) \to L(y)$. Therefore, since X_1 is closed, $x_n = y_n - \lambda_n y \to y - L(y)x \in X_1$. Thus $y = (y - L(y)x) + L(y)x \in X_1 + \mathrm{sp}\{x\}$. Alternatively, since X is complete and X_1 is closed in X, X/X_1 is complete and the projection $\pi : X \to X/X_1$ is bounded. Since X_2 is finite dimensional, $\pi(X_2)$ is finite dimensional, hence closed in X/X_1. Then $X_1 + X_2 = \pi^{-1}(\pi(X_2))$ is the inverse image of a closed set under a continuous map, and so is closed.

21. The assertion that for some infinite-dimensional normed space X there is a bounded sequence with no convergent subsequence has an explicit answer: Consider $\{1, t, t^2, \ldots\}$ in the unit ball of $C([0, 1])$.

23. (a) Let $Y = \mathrm{sp}\{x_1, \ldots, x_n\}$. For a scalar n-tuple $\lambda = (\lambda_1, \ldots, \lambda_n)$, let $\phi(\lambda) = \lambda_1 x_1 + \cdots + \lambda_n x_n$. Then $d(x, Y) = \inf_\lambda \|x - \phi(\lambda)\|$; we claim that there exists λ_0 such that $d(x, Y) = \|x - \phi(\lambda_0)\|$. Now, we may assume that the x_k are linearly independent and that ϕ is a continuous B-valued function, linear in λ. Linearity is clear and continuity follows since $\|\phi(\lambda) - \phi(\mu)\| \leq n(\max_{1 \leq k \leq n} |\lambda_k - \mu_k|)$. Let η denote the infimum in question; clearly $\eta \leq \|x\|$. To compute η we may restrict ourselves to $\mathcal{A} = \{\lambda : \|x - \phi(\lambda)\| \leq \|x\|\}$ which, by continuity, is compact. Indeed, $\mathcal{A}$ is closed and by Problem 15, $|\lambda_1| + \cdots + |\lambda_n| \leq c\|\phi(\lambda)\| \leq c(\|\phi(\lambda) - x\| + \|x\|) \leq 2c\|x\|$ for $\lambda \in \mathcal{A}$. Now let $\{\lambda_k\} \subset \mathcal{A}$ be such that $\lim_k \|x - \phi(\lambda_k)\| = \eta$. By Bolzano-Weierstrass there is a subsequence, which we call again $\{\lambda_k\}$ to avoid unnecessary waste of time and energy writing two subscripts in what follows (which, by the way, may be less of a waste of time and energy than writing the preceding remark), so that $\lambda_k \to \lambda \in \mathcal{A}$. Thus by the continuity of ϕ, $\eta \leq \|x - \phi(\nu)\| \leq \|x - \phi(\lambda_k)\| + \|\phi(\lambda_k) - \phi(\lambda)\| \to \eta$ as $k \to \infty$ and the infimum is attained, i.e., it is a minimum.

Now, y_0 is not necessarily unique: Let $B = \mathbb{R}^N$ equipped with the ℓ^∞ norm, $x = (1, 0, \ldots, 0)$, and $Y = \{y \in B : y_1 = 0\}$. Then $d(x, Y) = 1$ and the infimum is attained at the vectors $(0, 1, 0, \ldots, 0), \ldots, (0, \ldots, 0, 1)$.

(b) Let Y be the subspace of $B = C([-1, 1])$ defined by $Y = \{y \in B : \int_{-1}^0 y(t)\, dt = \int_0^1 y(t)\, dt = 0\}$; we compute $d(x, Y)$ where $x \in B$ satisfies $\int_{-1}^0 x(t)\, dt = -1$ and $\int_0^1 x(t)\, dt = 1$. Since $\int_{-1}^0 (x(t) - y(t))\, dt = -1$ it readily follows that $\inf_{t \in [-1, 0]} (x(t) - y(t)) \leq -1$, and by continuity equality holds only if $y(t) = x(t) + 1$ for all $t \in [-1, 0]$. Similarly, since $\int_0^1 (x(t) - y(t))\, dt = 1$, $\sup_{t \in [0, 1]} (x(t) - y(t)) \geq 1$ with equality holding only if $y(t) = x(t) - 1$ for all

$t \in [0,1]$. Thus $\operatorname{dist}(x, Y) \geq 1$ and there is no $y \in Y$ such that $\|x - y\| = 1$ because if such y existed we would have both $y(0) = x(0) + 1$ and $y(0) = x(0) - 1$. However we can define $y(t)$ as $x(t) + 1$ for $t \in [-1, 0]$ and $x(t) - 1$ for $t \in (0, 1]$ and then change $y(t)$ in a small neighborhood of 0 to make it continuous and approximating this way we get that $\inf_{z \in Y} \|x - z\| = 1$.

(c) As for uniqueness, suppose $\|x - \phi(\lambda)\| = \|x - \phi(\mu)\| = \eta$, say. If $\eta = 0$, $x - \phi(\lambda) = x - \phi(\mu)$, and, consequently, $\phi(\lambda) = \phi(\mu)$, which, by the linear independence, implies $\lambda = \mu$. Otherwise, by the definition of η, $2\eta = \|x - \phi(\lambda)\| + \|x - \phi(\mu)\| \geq \|2x - [\phi(\lambda) + \phi(\mu)]\| = 2\|x - \phi((\lambda + \mu)/2)\| \geq 2\eta$, we have equality in the triangle inequality above, and, consequently, by assumption, $x - \phi(\lambda) = \alpha[x - \phi(\mu)]$ for some scalar α. If $\alpha = 1$, $\lambda = \mu$ as before. If, on the other hand, $\alpha \neq 1$, it readily follows that $x = \phi(\lambda - \alpha\mu)/(1 - \alpha)$ and $\eta = 0$, which is not the case.

25. If $y, z \in \mathcal{D}_K(x)$, by Problem 24, $(y + z)/2 \in \mathcal{D}_K(x)$, and, consequently, $\|(x - y)/d(x, K)\| = \|(x - z)/d(x, K)\| = 1$. Thus $(x - y)/d(x, K) + (x - z)/d(x, K) = (2/d(x, K))(x - (y + z)/2)$ has norm 2. This not being the case implies that no two such points exist.

26. By the triangle inequality $\|\lambda x + \mu y\| \leq \lambda\|x\| + \mu\|y\|$. Conversely, suppose that $\lambda \geq \mu \geq 0$. Then $\|\lambda x + \mu y\| = \|\lambda(x + y) - (\lambda - \mu)y\| \geq \lambda\|x + y\| - (\lambda - \mu)\|y\| = \lambda(\|x\| + \|y\|) - (\lambda - \mu)\|y\| = \lambda\|x\| + \mu\|y\|$ and equality holds.

27. (b) implies (a) Suppose $\|x + y\| = \|x\| + \|y\|$ with $y \neq 0$. If $x = 0$ we have $x = 0 \cdot y$ and we are done. So assume that $x \neq 0$. Consider $x/\|x\|$, $y/\|y\|$, and for the sake of argument suppose that $x/\|x\| \neq y/\|y\|$. Then by assumption $x/\|x\| + y/\|y\| < 2$. On the other hand, since

$$\frac{x}{\|x\|} + \frac{y}{\|y\|} = \frac{1}{\|x\|}(x + y) - \left(\frac{1}{\|x\|} - \frac{1}{\|y\|}\right)y,$$

from the triangle inequality it readily follows that

$$\left\|\frac{x}{\|x\|} + \frac{y}{\|y\|}\right\| \geq \left\|\frac{1}{\|x\|}(x + y)\right\| - \left\|\left(\frac{1}{\|x\|} - \frac{1}{\|y\|}\right)y\right\|$$

$$= \frac{1}{\|x\|}\|x + y\| - \|y\|\left(\frac{1}{\|x\|} - \frac{1}{\|y\|}\right)$$

$$= \frac{1}{\|x\|}(\|x\| + \|y\|) - \|y\|\left(\frac{1}{\|x\|} - \frac{1}{\|y\|}\right) = 2,$$

which is not the case. Thus $x/\|x\| = y/\|y\|$, which implies that $x = \lambda y$ with $\lambda = \|x\|/\|y\| \geq 0$.

28. The statement is true.

29. $d(x_0, X) = 1$ and for no $x \in X$ we have $\|x_0 - x\|_\infty = d$. Indeed, if $x \in X$, assuming as we may that $x_1 \geq 0$, we have $x_1 \geq 1$ and $|x_n| \leq 1$ for all $n \geq 2$. But then as we saw above $|\sum_{k=2}^{N} 2^{-k} x_k| < 1/2$ for all N, and, therefore, the distance cannot be attained for any $x \in X$.

Note that for $x' = (1, -1, -1, \ldots)$ we have $\|x_0 - x'\|_\infty = 1$, and the minimum is attained at this sequence $x' \in c$, but $\|x_0 - x\|_\infty > 1$ for all $x \in X$.

31. It is a matter of rescaling. Note that $d(ax, Y) = |a|\, d(x, Y)$. Hence

$$\sup_{x \in B(0,r)} d(x, Y) = \sup_{x \in B(0,1)} d(rx, Y) = r \sup_{x \in B(0,1)} d(x, Y)$$

and by the continuity of $d(x, Y)$,

$$\sup_{\|x\|<1} d(x, Y) = \sup_{\|x\|\leq 1} d(x, Y) = \sup_{\|x\|=1} d(x, Y)$$
$$= \sup_{x \neq 0} d(x/\|x\|, Y) = \sup_{x \neq 0} d(x, Y)/\|x\| .$$

Thus by Problem 30, $\sup_{x \in B(0,r)} d(x, Y) = r$ when Y is a linear subspace that is not dense; in other words, if $B(0, r) \subset B_Y(0, s)$, and more generally by translation, if $B(x, r) \subset B_Y(0, s)$, then $r \leq s$. Indeed,

$$B(0, r) \subset (B(x, r) + B(-x, r))/2 \subset (B_Y(0, s) - B_Y(0, s))/2 \subset B_Y(0, s),$$

and, consequently, $r \leq s$.

32. Observe that $d(x, Y_n) = O(r_n)$ iff $\limsup_n d(x, Y_n)/r_n < \infty$, i.e., iff there exist integers m, M such that $d(x, Y_n) \leq M\, r_n$ for all $n \geq m$. Let $A_m = \bigcap_{n=m}^{\infty} \overline{B_{Y_n}(0, r_n)}$, m integer. Then observe that the last condition is equivalent to $d(x, Y_n) = O(r_n)$ iff $x \in M(\bigcap_{n=m}^{\infty} \overline{B_{Y_n}(0, r_n)}) = M\, A_m$, say. Thus, since each A_m is closed, $A = \bigcup_{m,M} M\, A_m$ is the countable union of closed sets. For the sake of argument suppose that some A_m has nonempty interior and let $B(x, r) \subset A_m$. Then for all $n \geq m$, $B(x, r) \subset \overline{B_{Y_n}(0, r_n)}$ and by Problem 31, $r \leq r_n$ and since $r_n \to 0$, $r = 0$.

33. Assume that $\overline{B(0, r)}$ can be covered by the n translates $x_1 + B(0, 1), \ldots, x_n + B(0, 1)$, say, of $B(0, 1)$, and let $Y = \mathrm{sp}\{x_1, \ldots, x_n\}$; Y is a closed subspace of X. For the sake of argument suppose that Y is contained properly in X and let θ, r_1 be such that $1 < \theta < r_1 < r$. Then by Problem 30 there exists $x \in X$ with $\|x\| = 1$ and $\|x - y\| \geq \theta$ for all y in Y. Now, $\|r_1 x\| = r_1 < r$, i.e., $r_1 x \in B(0, r)$, and $\|r_1 x - r_1 y\| \geq r_1 \theta > 1$ for all $y \in Y$. In particular, taking $y = r_1^{-1} x_j \in Y$ it follows that $\|r_1 x - x_j\| > 1$ for each $j = 1, \ldots, n$. Hence the sets $x_1 + B(0, 1), \ldots, x_n + B(0, 1)$ do not cover $\overline{B(0, r)}$, which is not the case. Therefore X is spanned by n vectors and X is finite dimensional.

Now, if a closed ball in X is compact, since by translation and rescaling all closed balls in a normed linear space are homeomorphic, $\overline{B(0,r)}$ is compact and can be covered by finitely many translates of the unit ball $B(0,1)$. Therefore the first part of the argument applies and X is finite dimensional.

34. For the sake of argument suppose that X endowed with the norm $\|\cdot\|$ is complete. Let $X_n = \operatorname{sp}\{x_1, \ldots, x_n\}$, $n \geq 1$; X_n is a finite-dimensional and, hence, closed subspace of X for all n. Now, each X_n has empty interior. Indeed, let $x \in X \setminus X_n$ (if no such x exists, X already is finite dimensional). Then for $y \in X_n$ and $\varepsilon > 0$, $y + \varepsilon x \notin X_n$, and so the interior of X_n is empty. Now, by assumption $X = \bigcup_n X_n$ is a countable union of nowhere dense sets and so of first category; this contradicts the Baire category theorem unless X is finite dimensional. In particular, if a Banach space B contains a countable spanning set, B is finite dimensional.

36. To verify that X is complete, let $\sum_n \|x_n\|_X < \infty$, or more precisely, $\sum_n \sum_{\alpha \in \Lambda} \|x_n\|_{B_\alpha} < \infty$. Now, since $\|x_n\|_{B_\alpha} \leq \|x_n\|_X$ for all α, $\sum_n \|x_n\|_{B_\alpha} < \infty$ for each α and since each B_α is complete, $\sum_{n=1}^N x_n$ converges to x_α, say, in B_α. By the compatibility condition $x_\alpha = x_\beta$ for all $\alpha, \beta \in \Lambda$ and calling the common value x, $\sum_{n=1}^N x_n \to x$ in B_α for all α. Now, $\|x - \sum_{n=1}^N x_n\|_{B_\alpha} = \|\sum_{n=N+1}^\infty x_n\|_{B_\alpha} \leq \sum_{n=N+1}^\infty \|x_n\|_{B_\alpha}$, and, consequently, summing over α and invoking Tonelli's theorem, $\|x - \sum_{n=1}^N x_n\|_X \leq \sum_{n=N+1}^\infty \|x_n\|_X \to 0$ as $N \to \infty$. Thus $\lim_N \sum_{n=1}^N x_n = x$ in X and X is complete.

37. The homogeneity and triangle properties of $\|\cdot\|_{B_0+B_1}$ are readily verified. Let $\|x\|_{B_0+B_1} = 0$ and pick sequences $\{y_n\}$ in B_0 and $\{z_n\}$ in B_1 such that $x = y_n + z_n$ with $\|y_n\|_{B_0} \to 0$ and $\|z_n\|_{B_1} \to 0$. Then $y_n + z_n \to 0$ in B and since $x = y_n + z_n$, $y_n + z_n \to x$ in B; thus $x = 0$. Finally, completeness. Let $\sum_n \|x_n\|_{B_0+B_1} < \infty$, and let $x_n = y_n + z_n$ with $y_n \in B_0$, $z_n \in B_1$, and $\|y_n\|_{B_0}, \|z_n\|_{B_1} \leq \|x_n\|_{B_0+B_1} + 2^{-n}$. Then $\sum_n \|y_n\|_{B_0}, \sum_n \|z_n\|_{B_1} < \infty$ and since B_0 and B_1 are complete there exist $y \in B_0$ and $z \in B_1$ such that $\|y - \sum_{n=1}^N y_n\|_{B_0}, \|z - \sum_{n=1}^N z_n\|_{B_1} \to 0$. Then $y + z \in B_0 + B_1$, $(y+z) - \sum_{n=1}^N x_n = (y - \sum_{n=1}^N y_n) + (z - \sum_{n=1}^N z_n)$, and so

$$\left\| (y+z) - \sum_{n=1}^N x_n \right\|_{B_0+B_1} \leq \left\| y - \sum_{n=1}^N y_n \right\|_{B_0} + \left\| z - \sum_{n=1}^N z_n \right\|_{B_1} \to 0$$

as $N \to \infty$.

38. Necessity first. Note that $B_0 + B_1$ is complete with both the norm inherited from B and the norm introduced in Problem 37. Let $x \in B_0 + B_1$ and $x = x_0 + x_1$ with $x_0 \in B_0$ and $x_1 \in B_1$. Since B_0 and B_1 are continuously embedded in B, $\|x\|_B \leq \|x_0\|_B + \|x_1\|_B \leq c(\|x_0\|_{B_0} + \|x_1\|_{B_1})$ and, taking the inf over all possible decompositions of x in $B_0 + B_1$ it follows that

$\|x\|_B \le c\|x\|_{B_0+B_1}$. Now, since $B_0 + B_1$ is complete with respect to both norms by the inverse mapping theorem there exists a constant c such that $\|x\|_{B_0+B_1} \le c\|x\|_B$, which is what we wanted to prove.

Conversely, let $\{x_n\}$ be a sequence in $B_0 + B_1$ that converges to $x \in B$. Then $\{x_n\}$ is Cauchy in B, and, consequently, $\{x_n\}$ is Cauchy in $B_0 + B_1$, which by Problem **37** is complete. Let $y = \lim_n x_n$ in $B_0 + B_1$. Now, $\|y - x\|_{B_0+B_1} \le \|y - x_n\|_{B_0+B_1} + c\|x_n - x\|_B$ where the first term goes to 0 since $\|y - x_n\|_{B_0+B_1} \to 0$ and the second term goes to 0 since $\|x_n - x\|_B \to 0$; thus $\|y - x\|_{B_0+B_1} = 0$ and $x = y \in B_0 + B_1$.

39. The statement is false: In $C(I)$ let $K_n = \{x \in C(I) : x(0) = 1, 0 \le x(t) \le 1 \text{ if } t \in [0, 1/n] \text{ and } x(t) = 0 \text{ if } t \in [1/n, 1]\}$.

40. (a) The statement is false. In $C(I)$ let $B = \{x \in C([0,1]) : x(1) = 0\}$ and $K = \{x \in B : \int_0^1 tx(t)\,dt = 1\}$; B is a closed subspace of $C(I)$, and so a Banach space, and K is closed and convex. And, since for all $x \in K$ we have $\|x\|_\infty > \inf_{y \in K} \|y\|_\infty$, K has no element of least norm.

(b) The statement is false. In $(\mathbb{R}^n, \|\cdot\|_\infty)$ let A be the closed unit ball centered at $2e_n = (0, \ldots, 2)$. Note that if $x_n = 1$ and $|x_k| \le 1$ for each k, then $\|x\| = 1$ and $\|x - 2e_n\| = 1$; thus $x \in A$ and the distance from x to 0 is 1. Now, if $y \in A$, then $\|y - 2e_n\| \le 1$, and so $2 = \|-2e_n\| = \|y - 2e_n - y\| \le 1 + \|y\|$; that is, $\|y\| \ge 1$. Therefore the minimum distance 1 is attained at each $x \in \mathbb{R}^n$ with $x_n = 1$ and $|x_k| \le 1$ for each $1 \le k \le n$.

42. (a) We claim that $x \in E(B_{\ell^\infty})$ iff $|x_n| = 1$ for all n. For the sake of argument suppose that $x \in E(B_{\ell^\infty})$ and $|x_N| < 1 - \varepsilon < 1$ for some N. Then

$$x = \frac{1}{2}(x + \varepsilon\, e_N) + \frac{1}{2}(x - \varepsilon\, e_N)$$

can be written as a convex combination of distinct elements of B_{ℓ^∞}, which is not the case. Conversely, note that if a scalar λ with $|\lambda| = 1$ is a convex combination of distinct α and β, then $|\alpha| > 1$ or $|\beta| > 1$. Thus, if $x = \lambda y + (1 - \lambda)z$ with $|x_n| = 1$ for all n and $0 < \lambda < 1$, then $x_N \ne z_N$ for some N, and $|y_N| > 1$ or $|z_N| > 1$.

(b) If $x \in E(B_c)$, then $x \in E(B_{\ell^\infty})$ and we have that both $\lim_n x_n$ exists and $|x_n| = 1$ for all n. Therefore, if x is an extreme point, then $x_n = \pm 1$ for all n, and there exists N such that either $x_n = 1$ or $x_n = -1$ for all $n \ge N$; these are the extreme points in the unit ball of real c.

(c) $E(B_{c_0}) = \emptyset$.

(d) $E(B_{\ell^1})$ consists of those $x \in \ell^1$ with $\|x\|_1 = 1$ such that all the terms except one, x_n, say, vanish, and $|x_n| = 1$.

43. (a) $E(B_{L^\infty(I)}) = \{x \in B_{L^\infty(I)} : |x(t)| = 1 \text{ a.e.}\}$. First, if $x \in B_{L^\infty(I)}$ and there exist $\varepsilon > 0$ and $B \subset I$ with $|B| > 0$ such that $|x(t)| < 1 - \varepsilon$ on B, with

$$y(t) = \begin{cases} x(t), & t \notin B, \\ x(t) + \varepsilon, & t \in B, \end{cases} \qquad z(t) = \begin{cases} x(t), & t \notin B, \\ x(t) - \varepsilon, & t \in B, \end{cases}$$

it readily follows that $y, z \in B_{L^\infty(I)}$, $x \neq y, z$, and $x = (y + z)/2$, which is not the case.

Next, let $|x(t)| = 1$ a.e. and suppose that $x = \lambda y + (1 - \lambda)z$ for some y, z in $L^\infty(I)$ and $0 \leq \lambda \leq 1$. Note that we may assume that $|y(t)| = 1$ a.e. Indeed, if this is not the case, $|\{|y| > 1\}| > 0$ or $|\{|y| < 1\}| > 0$; in the former case $\|y\|_\infty > 1$ and $y \notin B_{L^\infty(I)}$ and in the latter $\|z\|_\infty > 1$ and $z \notin B_{L^\infty(I)}$; similarly, $|z(t)| = 1$ a.e. Now let t be such that $x(t) = 1$. Since $|y(t)|, |z(t)| \leq 1$ and $1 = \lambda y(t) + (1 - \lambda)z(t)$ it follows that $y(t) = z(t) = 1$. Analogously, if t is such that $x(t) = -1$, then $y(t) = z(t) = -1$. Therefore $x = y = z$ a.e. and $x \in E(B_{L^\infty(I)})$. A similar, yet more involved, proof gives the conclusion for complex-valued functions.

(b) $E(B_{C(I)}) = \{-\chi_I, \chi_I\}$.

(c) We claim that $E(B_{L^1(I)}) = \emptyset$. If $x = 0$ a.e. write $x = (y + z)/2$ with $y(t) = 1, z(t) = -1$ a.e. Otherwise, if $0 \neq x \in L^1(I)$, let $F(t) = \int_0^t |x(s)| \, ds$; F is a continuous function of t with $F(0) = 0$ and $F(1) = \|x\|_1$, and, therefore, there exists $\lambda \in (0, 1)$ such that $F(\lambda) = \|x\|_1/2$. If $\|x\|_1 \leq 1$ let

$$y(t) = \begin{cases} 2x(t), & 0 \leq t \leq \lambda, \\ 0, & \lambda < t \leq 1, \end{cases} \qquad z(t) = \begin{cases} 0, & 0 \leq t \leq \lambda, \\ 2x(t), & \lambda < t \leq 1. \end{cases}$$

Then, since $\|x\|_1 \neq 0$, $x \neq y, z$, $\|x\|_1 = \|y\|_1 = \|z\|_1$, and $x = (y + z)/2$.

44. To check that the given expression is a norm it suffices to verify that $\|p\| = 0$ implies $p = 0$. If $\|p\| = 0$, $p(0) = \ldots = p(n) = 0$, p has $n + 1$ roots, and, therefore, vanishes identically. Now, B is a finite-dimensional normed linear space and so by Problem 17 a Banach space.

46. (a) and (b) are clear. As for (c), if $\|x\|_{\infty,w} = 0$, $x(t)w(t) = 0$ for all $t \in I$, which implies that $x(t) = 0$ for $t > 0$, and, by continuity, $x(0) = 0$. Thus $x = 0$; the other properties of the norm are readily verified. Finally, the following is a Cauchy sequence with respect to $\| \cdot \|_{\infty,w}$ that does not converge in $C(I)$: Let

$$x_n(t) = \begin{cases} n^{1/2}, & t \in [0, 1/n], \\ t^{-1/2}, & t \in (1/n, 1]. \end{cases}$$

Fix N and let $n > m \geq N$. It then readily follows that $x_n(t) - x_m(t) = (n^{1/2} - m^{1/2})\chi_{[0,1/n]}(t) + (t^{-1/2} - m^{1/2})\chi_{(1/n,1/m]}(t)$, and, consequently, $\|x_n - x_m\|_{\infty,w} = \sup_{t \in I} t \, |x_n(t) - x_m(t)|$ does not exceed

$$\sup_{t \in I} [\, t \, n^{1/2}\chi_{[0,1/n]}(t) + t^{1/2}\chi_{(1/n,1/m]}(t)\,] \leq n^{-1/2} + m^{-1/2} \leq 2N^{-1/2},$$

which can be made arbitrarily small for N sufficiently large. Thus $\{x_n\}$ is a Cauchy sequence with respect to the metric induced by $\|\cdot\|_{\infty,w}$ which does not converge to any x uniformly because if it did it would follow that $x(0) = \lim_n x_n(0) = \infty$.

47. X is a subspace of $C(I)$ but is not closed.

48. With the notation of Problem 47, $\overline{M}_\alpha(I) = L_\alpha(I)$, the space of Lipschitz functions with Lipschitz constant $\le \alpha$.

49. It is readily seen that $p_\alpha(x)$ is nonnegative, homogeneous with $|\lambda|$, and subadditive, and thus a seminorm. On the other hand, $p_\alpha(x) = 0$ for any constant function x. We claim that $\|x\|_\alpha = |x(0)| + p_\alpha(x)$ is a norm on $C^\alpha(I)$. Since $\|x\|_\alpha = 0$ implies that $|x(0)| = p_\alpha(x) = 0$, if $\|x\|_\alpha = 0$, then x is the constant $x(0) = 0$, and $x = 0$; the other conditions of norm are readily verified. To prove that $C^\alpha(I)$ is complete, let $\{x_n\} \subset C^\alpha(I)$ be such that $\sum_n \|x_n\|_\alpha < \infty$. Since $\sum_n |x_n(0)| < \infty$, $x(0) = \sum_n x_n(0)$ is well-defined, and since $\sum_n |x_n(t)| \le \sum_n |x_n(0)| + |t|^\alpha \sum_n p_\alpha(x_n) < \infty$, $x(t) = \sum_n x_n(t)$ is well-defined for every $t \in I$. We claim that $\|x - \sum_{n=1}^N x_n\|_\alpha \to 0$. Clearly $|x(0) - \sum_{n=1}^N x_n(0)| \le \sum_{n=N+1}^\infty |x_n(0)| \to 0$. Also,

$$\left| \left(x(t) - x(s) \right) - \left(\sum_{n=1}^N x_n(t) - \sum_{n=1}^N x_n(s) \right) \right|$$
$$\le \sum_{n=N+1}^\infty |x_n(t) - x_n(s)| \le |t - s|^\alpha \sum_{n=N+1}^\infty p_\alpha(x_n),$$

and, consequently, $p_\alpha(x - \sum_{n=1}^N x_n) \le \sum_{n=N+1}^\infty p_\alpha(x_n) \to 0$ as $N \to \infty$. Thus $\sum_n x_n$ converges in $C^\alpha(I)$, and $C^\alpha(I)$ is complete.

50. (b) First, note that $x \in C^\alpha(I)$ and $\|x\|_\alpha \le 1$; indeed, for $t \ne s$,

$$\frac{|x(t) - x(s)|}{|t - s|^\alpha} = \lim_n \frac{|x_n(t) - x_n(s)|}{|t - s|^\alpha} \le 1.$$

Next, let $\varepsilon > 0$. Then there exists N such that $\|x_n - x\|_\infty < \varepsilon$ for $n > N$. Now, if $|t - s| < \varepsilon$, since $\left| (x(t) - x_n(t)) - (x(s) - x_n(s)) \right| \le |t - s|^\alpha p_\alpha(x - x_n)$, it readily follows that $p_\beta(x - x_n) \le \varepsilon^{\alpha-\beta} p_\alpha(x - x_n) \le \varepsilon^{\alpha-\beta}(\|x_n\|_\alpha + \|x\|_\alpha) \le 2\varepsilon^{\alpha-\beta}$. And, when $|t - s| \ge \varepsilon$,

$$\frac{|(x(t) - x_n(t)) - (x(s) - x_n(s))|}{|t - s|^\beta} \le 2\varepsilon^{-\beta}\|x - x_n\|_\infty \le 2\varepsilon^{1-\beta}.$$

Thus combining these estimates, $p_\beta(x - x_n) \le 2\varepsilon^{\alpha-\beta} + 2\varepsilon^{1-\beta}$, and, consequently, $\|x - x_n\|_\beta = |x(0) - x_n(0)| + p_\beta(x - x_n) \le \varepsilon + 2\varepsilon^{\alpha-\beta} + 2\varepsilon^{1-\beta}$.

(c) Let $\{x_n\}$ be a sequence in S with $\|x_n\|_\beta \le 1$. Given $\varepsilon > 0$, let $\delta = \varepsilon^{1/\alpha}$ and note that since by the definition of S, $\|x_n\|_\alpha \le 1$ for all n, $|x_n(t) - x_n(s)| \le \|x_n\|_\alpha |t - s|^\alpha \le \varepsilon$ for $|t - s| < \delta$. Thus $\{x_n\}$ is uniformly bounded and uniformly equicontinuous and, consequently, by the Arzela-Ascoli theorem $\{x_n\}$ has a uniformly convergent subsequence $\{x_{n_k}\}$, say. By (b) $\{x_{n_k}\}$ converges in $C^\beta(I)$.

51. In fact, $C^\alpha(I)$ is an F_σ set with empty interior. Let $F_n = \{x \in C^\alpha(I) : \|x\|_\alpha \leq n\}$; then $C^\alpha(I) = \bigcup_n F_n$. As is readily seen each F_n is closed under uniform convergence. For the sake of argument suppose that $C^\alpha(I)$ contains an open ball. Then it contains a ball centered at 0 but this is not possible since $\{t^{\alpha/2}/n\}$ is a sequence in $C(I) \setminus C^\alpha(I)$ that converges to 0.

52. Let $O_n = \{x \in C(I) : |Z(x)| < 1/n\}$; we claim that O_n is an open dense subset of $C(I)$. It is clearly dense since it contains all polynomials, which have a finite zero set. Now, given $x \in O_n$, let $G \supset Z(x)$ be an open set such that $|G| < 1/n$; observe that since $I \setminus G$ is compact and x does not vanish in $I \setminus G$, x has a positive minimum there and $\eta = \inf_{t \in I \setminus G} |x(t)| > 0$. We claim that if $\|x - y\|_\infty < \eta$, $y \in O_n$. Indeed, let $t \in Z(y)$. Then, since $\|x - y\|_\infty < \eta$, $|x(t)|_\infty = |x(t) - y(t)|_\infty < \eta$ and so $t \notin I \setminus G$, i.e., $t \in G$ and $|Z(y)| < 1/n$. This means that $y \in O_n$, which is therefore open. Finally, by Problem 1.4(a), $\bigcap_n O_n = \{x \in C(I) : |Z(x)| = 0\}$ is a dense G_δ subset of $C(I)$.

53. The closed span is ℓ^p for $1 \leq p < \infty$ and c_0 for $p = \infty$.

54. The closure is c.

55. Only if ϕ is linear. By Problem 21 there exists $\{x_n\} \subset B_{C(I)}$ such that $\|x_n - x_m\|_\infty \geq 1/2$ for $n \neq m$. Fix a continuous function f on $\mathbb{R}^+$ such that $f(0) = 1$ and $f(t) = 0$ for $t > 1/6$. Now, for any $x \in B_{C(I)}$, consider the functional ϕ_x on $B_{C(I)}$ given by $\phi_x(y) = f(\|y - x\|_\infty)$; note that since $\|y - x\|_\infty$ is continuous in y and the composition of continuous functions is continuous, ϕ_x is a continuous function on $B_{C(I)}$. Let ϕ be the functional on $B_{C(I)}$ given by $\phi = \sum_n n\phi_{x_n}$. This series converges and its sum is continuous because by construction for any $x \in B_{C(I)}$ there is an open neighborhood U_x of x such that all except for finitely many terms in the series vanish on U_x. On the other hand, $\phi(x_n) = n$, and ϕ is unbounded.

56. Given a linear functional $0 \neq L$ on ℓ^2, $L(e_k) \neq 0$ for some integer k. Now, for every $\lambda \in \mathbb{R}$, $\lambda e_k + e_{k+1} \in A$ and so $L(\lambda e_k + e_{k+1}) = \lambda L(e_k) + L(e_{k+1}) \in L(A)$. Moreover, since A is convex, $L(A)$ is convex, i.e., an interval, and since $L(e_k) \neq 0$ the endpoints of the interval go to $\pm\infty$ with $|\lambda| \to \infty$, and, consequently, $L(A) = \mathbb{R}$. Similarly for B.

In general, convex sets can be separated when one, A, say, has an absorbing point, i.e., there exists $x_0 \in A$ such that for any $x \in X$ we can find a scalar λ so that $x \in x_0 + \lambda(A - x_0)$.

57. Note that $p(x) = 1/M(x)$, $p(0) = 0$, is a norm. First, p is nonnegative and positively homogenous. As for the triangle inequality, it is readily seen if one of x, y or $x + y$ is 0, so we assume that $x, y, x + y \neq 0$. Then, since $\lambda x \in K$ for $0 < \lambda < 1/p(x)$ and $\mu y \in K$ for $0 < \mu < 1/p(y)$, for all such λ, μ, by the convexity of K,

$$\frac{\mu}{\lambda + \mu} \lambda x + \frac{\lambda}{\lambda + \mu} \mu y = \frac{\lambda \mu}{\lambda + \mu} (x + y) \in K,$$

which implies that $\lambda \mu / (\lambda + \mu) \leq M(x + y)$, and, consequently, $p(x + y) \leq 1/\lambda + 1/\mu$, and since $1/\lambda$ and $1/\mu$ can be picked arbitrarily close to $p(x)$ and $p(y)$, respectively, the triangle inequality holds.

Now, since $x_0 \notin K$ it readily follows that $M(x_0) \leq 1$, and so $p(x_0) \geq 1$. Thus by Hahn-Banach there exists a linear functional L on X with $\|L\| = 1$ such that $L(x_0) = p(x_0) \geq 1$. In particular, since $M(x) > 1$, and so $p(x) < 1$, it readily follows that $|L(x)| \leq \|L\| p(x) < 1$ for $x \in K$.

58. Let $H = \{x_n\} \cup \{x_\lambda\}$ be a necessarily infinite Hamel basis for X and define $L(x_n) = n\|x_n\|$ and $L(x_\lambda) = 0$ for $x_\lambda \in H \setminus \{x_n\}$. Now, given $x \in X$, x can be written uniquely as $x = \alpha_1 x_{\lambda_1} + \cdots + \alpha_n x_{\lambda_n}$ with $x_{\lambda_1}, \ldots, x_{\lambda_n} \subset H$ and scalars $\alpha_1, \ldots, \alpha_n$. Then it is clear that $L(x) = \alpha_1 L(x_{\lambda_1}) + \cdots + \alpha_n L(x_{\lambda_n})$ defines a linear functional on X such that $|L(x_n)| = n\|x_n\|$. Thus $\|L\| \geq n$ for all n and L is unbounded.

59. For the sake of argument suppose that $L(y) \neq 0$ for some $y \in Y$ and given a scalar λ, let $x_\lambda = (\lambda/L(y))y$. Then $x_\lambda \in Y$, $L(x_\lambda) = L(\lambda/L(y)y) = \lambda$, and L maps Y onto the scalar field, which is not the case. Therefore $L(y) = 0$ for all $y \in Y$.

60. (a) First, $L(0) = 0$. Next, since L is discontinuous, given a scalar $z \neq 0$, there exists $x_z \in X$ with $\|x_z\| = 1$ such that $|L(x_z)| > |z|$; note that, in particular, $L(x_z) \neq 0$. Now, let $x = (z/L(x_z)) x_z$. Then $L(x) = z$ and $\|x\| = (|z|/|L(x_z)|)\|x_z\| < 1$. The argument actually gives that if $L(B(0, r))$ is unbounded for some $r > 0$, then $L(B(0, r)) = \mathbb{C}$.

(b) Let $\{x_n\} \subset X \setminus K(L)$ be such that $\|x_n\| = 1$ and $|L(x_n)| > n$. Then for $y \in X$, $\{y - (L(y)/L(x_n))x_n\} \subset K(L)$ and converges to y, whence the density of $K(L)$.

61. (a) implies (b) $K(L)$ is a closed subspace of X and since L is nontrivial, is proper.

(b) implies (c) Since $\overline{K(L)} = K(L)$ is a proper subspace of X, it is not dense.

(c) implies (a) This has been done in Problem 60(b).

62. Necessity is clear. Conversely, Problem 61 gives the result if $\lambda = 0$. So suppose that $L^{-1}(\{\lambda\})$ is closed for some $\lambda \neq 0$. For the sake of argument suppose that L is not bounded; then there is a sequence $\{x_n\}$ in X with $\|x_n\| = 1$ and $|L(x_n)| > n$. Let $y_n = \lambda x_n/L(x_n)$; then $\|y_n\| \to 0$ while $L(y_n) = \lambda$ for all n. Now, since $L^{-1}(\{\lambda\})$ is closed, it follows that $L(0) = \lambda$, which is not the case since L is linear.

63. (a) $Y = K(L)$ is a subspace of X and since $L \neq 0$, $Y \neq X$. Let W be a subspace of X that contains Y properly and $w \in W \setminus Y$. Then $L(w) \neq 0$ and so $L(x - (L(x)/L(w))w) = 0$ for all $x \in X$. Thus $x - (L(x)/L(w))w \in Y \subset W$. So $x = x - (L(x)/L(w))w + (L(x)/L(w))w \in W$ and $W = X$ as required.

(b) Since $0 \neq L$ is discontinuous, $K(L)$ is not closed; let $y_0 \in \overline{K(L)} \setminus K(L)$. In particular, $L(y_0) \neq 0$ and for any $y \in X$, $L(y - (L(y)/L(y_0))y_0) = 0$ and, consequently, $X = K(L) + \mathrm{sp}\{y_0\}$. In fact, since $K(L) \cap \mathrm{sp}\{y_0\} = \{0\}$, X is the direct sum of these subspaces. Moreover, since $\overline{K(L)}$ is a subspace of X, we have $X = K(L) + \mathrm{sp}\{y_0\} \subset \overline{K(L)}$ and $K(L)$ is dense in X.

(c) Sufficiency follows from (a). Next, if Y is a hyperplane, let $\psi : X/Y \to F$ be a linear bijection into the scalar field F, and put $\ell = \psi \circ \pi$, where $\pi : X \to X/Y$ is the canonical map; we claim that ℓ is a linear functional on X with $K(\ell) = Y$. Indeed, if $x \in Y$, then $\pi(x) = [0]$, so $\ell(x) = \psi([0]) = 0$, and $x \in K(\ell)$. Finally, if $x \in K(\ell)$, then $\psi(\pi(x)) = \ell(x) = 0 = \psi([0])$, and, consequently, since ψ is a bijection, $\pi(x) = [0]$, $x \in Y$, and $K(\ell) \subset Y$.

Note that, by (b), if Y is a hyperplane in a normed linear space X, then Y is closed or dense in X.

65. (a) For $x \notin K(L)$, as noted in Problem 63(b), $X = K(L) \oplus \mathrm{sp}\{x\}$ and so, given $y \in X$, we have $y = y_K + \lambda x$ where $y_K \in K(L)$ and λ is a scalar. Now let ℓ be the linear functional on X given by $\ell = L_1(x)L - L(x)L_1$. Then $\ell(x) = L_1(x)L(x) - L(x)L_1(x) = 0$ and since $K(L) = K(L_1)$, $\ell(y_K) = L_1(x)L(y_K) - L(x)L_1(y_K) = 0$. Thus ℓ is identically 0, $L_1 = (L_1(x)/L(x))L$, and the functionals are proportional.

Note that if $L_1 \neq 0$ it suffices to assume that $K(L) \subset K(L_1)$. Indeed, since $K(L)$ is a hyperplane and $K(L_1) \neq X$, by Problem 63, $K(L_1) = K(L)$.

(b) A hyperplane is convex and, hence, connected. Alternatively, it is clear that for $x, y \in K(L)$, the segment $\gamma = \{z \in X : z = tx + (1 - t)y, 0 \leq t \leq 1\}$ is contained in $K(L)$, which is therefore path connected, and hence connected.

Next, let $y \in X$. If $y \in X \setminus K(L)$ there is nothing to prove. Otherwise, let $y = y_K + \lambda x$ with $y_K \in K(L)$, put $y_n = y_K + (\lambda + 1/n)x$, and observe that $L(y_n) = (\lambda + 1/n)L(x) \neq 0$ for n sufficiently large and so $y_n \in X \setminus K(L)$ for those n. Finally, $\lim_n \|y - y_n\| = \lim_n \|x\|/n = 0$.

(c) When L is continuous, $X \setminus K(L) = \{x \in X : L(x) < 0\} \cup \{x \in X : L(x) > 0\}$ is the union of two disjoint open sets, which are also connected since they are path connected as before.

66. If $0 \neq x \in K(L)$, $x/\|x\| \in K(L)$ and has norm 1, and so $|L_1(x/\|x\|)| \leq \varepsilon$; thus $|L_1(x)| \leq \varepsilon\|x\|$ for all $x \in K(L)$. By Hahn-Banach there is an extension ℓ, say, of $L_1|_{K(L)}$ to X with norm $\leq \varepsilon$. Now, since L_1 and ℓ coincide in $K(L)$, $K(L_1 - \ell) \supset K(L)$, and by Problem 65(a), $L_1 - \ell = \eta L$ for some scalar η. Thus $\|L_1 - \eta L\| = \|\ell\| \leq \varepsilon$.

Note that since $|\,\|L_1\| - \|\eta L\|\,| \leq \|L_1 - \eta L\| \leq \varepsilon$, ε and η are related by the inequalities $-\|L_1\| - \varepsilon \leq |\eta|\|L\| \leq -\|L_1\| + \varepsilon$.

67. Let $L(x) = \|x\|\ell(x/\|x\|)$.

68. (b) Let $L_1, L_2 \in X^*$ with norm 1 be such that $(L_1 + L_2)/2 \in S$, the unit sphere of X^*. Put $Y = K(L_2 - L_1)$ and let $\ell = L_1|_Y$ denote the restriction of L_1 to Y; we claim that ℓ is a continuous linear functional on Y of norm 1 that admits two distinct extensions to X of norm 1. Let $x \in Y$. Then $L_2(x) = L_2(x) - L_1(x) + L_1(x) = L_1(x)$ and clearly L_1 and L_2 extend ℓ to X; it only remains to prove that $\|\ell\| = 1$. To see this we first construct a sequence $\{x_n\} \subset Y$ such that $\|x_n\| \to 1$ and $\ell(x_n) \to 1$. Since $(L_1 + L_2)/2 \in S$ there exists $\{y_n\} \subset X$ of norm 1 such that $(L_1(y_n) + L_2(y_n))/2 \to 1$, in other words, $L_1(y_n) + L_2(y_n) \to 2$; since L_1 and L_2 have norm 1 each of the terms converges to 1. Now, since $L_1 \neq L_2$ there exists $u \in X$ such that $L_1(u) - L_2(u) = 1$. Let $x_n = y_n + (L_2(y_n) - L_1(y_n))u$. It then readily follows that $L_1(x_n) = L_2(x_n)$, i.e., $x_n \in Y$. Also, since $|L_2(y_n) - L_1(y_n)| \to 0$, $\lim_n \|x_n\| = \lim_n \|y_n\| = 1$, and since $\ell(x_n) = (L_2(y_n) - L_1(y_n))\ell(u) + \ell(y_n)$, $\lim_n \ell(x_n) = \lim_n \ell(y_n) = \lim_n L_1(y_n) = 1$.

69. Let $X = \{x \in \ell^1 : x_{2n} = 0, n = 1, 2, \ldots\}$.

70. No matter what ℓ^p norm is considered in $\mathbb{R}^2$, $\|\ell\| = 1$. Now, for $p > 1$, with q the conjugate to p, an extension L of ℓ of norm 1 is given by $L(x_1, x_2) = ax_1 + bx_2$, where $|a|^q + |b|^q = 1$. Moreover, since $L|_Y = \ell$, $L(x_1, 0) = ax_1 = x_1$, and $a = 1$, which together with $|a|^q + |b|^q = 1$ implies $b = 0$, and the extension is unique.

On the other hand, when $p = 1$ an extension L of ℓ is given by $L(x_1, x_2) = ax_1 + bx_2$ with $\max(|a|, |b|) = 1$, which together with $a = 1$ gives any b with $|b| \leq 1$, and so there are infinitely many extensions to $(\mathbb{R}^2, \|\cdot\|_1)$.

72. (a) If $A, B \subset \mathbb{N}$ are disjoint, then $1_{A \cup B} = 1_A + 1_B$, and, consequently, $\mu_L(A \cup B) = L(1_{A \cup B}) = L(1_A) + L(1_B) = \mu_L(A) + \mu_L(B)$. Moreover, for every $A \in \mathcal{P}(\mathbb{N})$, $|\mu_L(A)| \leq \|L\|\,\|1_A\|_\infty \leq \|L\| < \infty$ and μ_L is finitely additive and bounded. Finally, if $A_1, \ldots, A_n$ is a finite partition of $\mathbb{N}$, pick $\lambda_k = \pm 1$ so that $I = \sum_k |\mu(A_k)| = \sum_k \lambda_k \mu(A_k)$. Then $I = \sum_k \lambda_k L(1_{A_k}) = L(\sum_k \lambda_k 1_{A_k}) \leq \|L\|$ and the conclusion holds with $C \leq \|L\|$.

(b) (i) Let $x \in \ell^\infty$ and given $\varepsilon > 0$ and $N \in \mathbb{Z}$, put $A_N = \{k \in \mathbb{Z} : \varepsilon N \le x_k < \varepsilon(N+1)\}$; note that the A_N are pairwise disjoint and that $B = \{N \in \mathbb{Z} : A_N \ne \emptyset\}$ is a finite set. Therefore the sequence $y = \sum_{N \in \mathbb{Z}} \varepsilon N \, 1_{A_N}$ is well-defined, belongs to X, and $0 \le x_k - y_k < \varepsilon$ for all $k \in \mathbb{N}$, i.e., $\|x - y\|_\infty \le \varepsilon$.

(ii) Let ℓ be the linear functional on X defined as follows: If $x = \sum_n \lambda_n 1_{A_n} \in X$ with $\{A_n\}$ pairwise disjoint, then $\ell(x) = \sum_n \lambda_n \mu(A_n)$; we claim that ℓ is well-defined. Note that if $x = \lambda_1 1_{A_1} + \lambda_2 1_{A_2}$ where A_1, A_2 are arbitrary, then $x = \lambda_1 1_{A_1 \setminus A_2} + \lambda_2 1_{A_2 \setminus A_1} + (\lambda_1 + \lambda_2) 1_{A_1 \cap A_2}$, and so

$$\ell(x) = \lambda_1 \mu(A_1 \setminus A_2) + \lambda_2 \mu(A_2 \setminus A_1) + (\lambda_1 + \lambda_2)\mu(A_1 \cap A_2)$$
$$= \lambda_1(\mu(A_1 \setminus A_2) + \mu(A_1 \cap A_2)) + \lambda_2(\mu(A_2 \setminus A_1) + \mu(A_1 \cap A_2))$$
$$= \lambda_1 \mu(A_1) + \lambda_2(A_2).$$

From this it readily follows that if $x = \sum_i a_i 1_{A_i} = \sum_j b_j 1_{B_j}$, then $\sum_i a_i \mu(A_i) = \sum_j b_j \mu(B_j)$, and, therefore, ℓ is well-defined on X. Also, ℓ is linear by construction.

(iii) Since ℓ is linear, to prove that it admits a continuous linear extension to $\ell^\infty = \overline{X}$ it suffices to prove that it is continuous. Let $x = \sum_k a_k 1_{A_k} \in X$. As we saw above we may assume that the A_k are pairwise disjoint and then $\|x\|_\infty = \sup_k |a_k|$. Therefore $|\ell(x)| \le \|x\|_\infty \sum_k |\mu(A_k)| \le C\|x\|_\infty$.

74. Let $L(m+n) = \ell_M(m) + \ell_N(n)$, $m + n \in M + N$; we claim that L is well-defined. Indeed, if $m + n = m_1 + n_1$, say, then $m - m_1 = n_1 - n \in M \cap N$, and, consequently,

$$L(m + n) = \ell_M(m) + \ell_N(n)$$
$$= \ell_M(m_1 + (m - m_1)) + \ell_N(n_1 + (n - n_1))$$
$$= \ell_M(m_1) + \ell_N(n_1),$$

since $\ell_M(m - m_1) + \ell_N(n - n_1) = \ell_M(m - m_1) + \ell_N(-(m - m_1)) = 0$. L is linear, $L|_M(m) = \ell_M(m)$, and $L|_N(n) = \ell_N(n)$.

Now, in the normed case, if ℓ_M, ℓ_N are continuous, it follows that $|L(m + n)| \le \|\ell_M\|\,\|m\| + \|\ell_N\|\,\|n\| \le c(\|m\| + \|n\|)$ and since $M + N$ is closed, by Problem 38, $|L(m + n)| \le c\|m + n\|$.

75. First, suppose p is a linear functional on X and let q be a sublinear functional such that $q \prec p$. By sublinearity $q(0) = 0 \le q(x) + q(-x)$, and so $q(x) \ge -q(-x)$. Then $q(x) \le p(x) = -p(-x) \le -q(-x) \le q(x)$ for all $x \in X$, $p = q$, and p is minimal.

Next observe that if p is a sublinear functional on X, then $q : X \to \mathbb{R}$ given by $q(x) = \inf\{p(x + \lambda z) - \lambda p(z) : \lambda \ge 0\}$, where z in X is fixed, is a sublinear functional on X that precedes p. First, q is positively homogenous. Indeed, since $q(0) = \inf\{p(\lambda z) - \lambda p(z) : t \ge 0\} = 0$, we have $q(0x) = 0q(x)$.

And, if $\mu > 0$,

$$q(\mu x) = \inf\{p(\mu x + \lambda z) - \lambda p(z) : \lambda \geq 0\}$$
$$= \inf\{\mu\big(p(x + (\lambda/\mu)z) - (\lambda/\mu)p(z)\big) : \lambda \geq 0\} = \mu q(x).$$

Next, subadditivity. Let $x, y \in X$ and, given $\varepsilon > 0$, pick $\lambda, \mu \geq 0$ such that $p(x + \lambda z) - \lambda p(z) \leq q(x) + \varepsilon/2$ and $p(y + \mu z) - \mu p(z) \leq q(y) + \varepsilon/2$. Then, adding these inequalities gives $q(x) + q(y) \geq p(x + y + (\lambda + \mu)z) - (\lambda + \mu)p(z) - \varepsilon \geq q(x + y) - \varepsilon$, and q is sublinear. Finally, setting $\lambda = 0$ in the definition of q it readily follows that $q(x) \leq p(x)$ for all $x \in X$ and $q \prec p$.

Suppose now that p is minimal. Then with q as above, we have $q = p$ for an arbitrary fixed $z \in X$. In particular, when $\lambda = 1$ and $z = -x$ we get $p(x) = q(x) \leq p(x - x) - p(-x) = -p(-x)$. And, since as observed above for sublinear functionals $-p(-x) \leq p(x)$, we have $p(x) = -p(-x)$ for all $x \in X$. To see that p is homogenous let $\lambda < 0$. Then $p(\lambda x) = (-\lambda)p(-x) = \lambda p(x)$ for all $x \in X$. Finally, additivity follows from $p(x) \leq p(x + y) + p(-y) = p(x + y) - p(y)$ since sublinearity guarantees the reverse inequality. Hence p is a linear functional on X.

77. If x, y are linearly dependent, i.e., $x = \lambda y$, $\lambda \neq 1$, let $\{x, x_2, \dots, x_n\}$ be a basis of X and for $z = \mu x + \sum_{k=2}^{n} \mu_k x_k \in X$ let $L(z) = \mu$. Then L is a linear functional on X and $L(x) = 1 \neq \lambda = L(y)$. On the other hand, if x, y are linearly independent, let $\{x, y, x_3, \dots, x_n\}$ be a basis of X and for $z = \mu x + \lambda y + \sum_{k=3}^{n} \mu_k x_k$ in X let $L(z) = \lambda$. Then $L(x) = 0 \neq 1 = L(y)$.

79. First, that a norming linear functional L at x exists follows by a rescaling of the functional constructed in Problem 78. Now, if L is a norming functional at x, for $y \in X$ and $t > 0$, we have

$$\frac{\|x + ty\| - 1}{t} = \frac{\|x + ty\| - L(x)}{t} \geq \frac{L(x + ty) - L(x)}{t} = L(y),$$

and, replacing t with $-t$ above the inequality reverses and therefore for all $y \in X$ and for all t we have

$$\frac{\|x + ty\| - 1}{t} \geq L(y) \geq \frac{\|x - ty\| - 1}{-t}.$$

Therefore, if the norm is smooth at x, letting $t \to 0$ it follows that $L(y) = \phi'(0)$ and L is unique.

81. Note that $u = \ell + \ell_1$ is a linear functional on Y such that $|u(y)| \leq |\ell(y)| + |\ell_1(y)| \leq \|y\|$ for all $y \in Y$; similarly, $u' = \ell - \ell_1$ is a linear functional on Y such that $|u'(y)| \leq \|y\|$ for all $y \in Y$. Then by Hahn-Banach there are linear functionals U, U' on X that extend u and u', respectively, and satisfy $|U(x)|, |U'(x)| \leq \|x\|$ for all $x \in X$. Now let $L = (U + U')/2$ and $L_1 = (U - U')/2$; L, L_1 are linear functionals on X of norm ≤ 1. Also, $L(y) = \ell(y)$

and $L_1(y) = \ell_1(y)$ for $y \in Y$, and so $L, L_1 \in X^*$ extend ℓ and ℓ_1, respectively. Finally, $|L(x)| + |L_1(x)| = (|U(x) + U'(x)| + |U(x) - U'(x)|)/2$, and, since $(|a+b| + |a-b|)/2 = \max(|a|, |b|)$, $|L(x)| + |L_1(x)| \leq \max(|U(x)|, |U'(x)|) \leq \|x\|$, $x \in X$.

82. Note that $p(x) + q(y)$, $x, y \in X$, is a seminorm on the linear space $X \times X$. Now, the diagonal Δ of $X \times X$ is a linear subspace of $X \times X$ and $\ell(x, x) = L(x)$ is a linear functional on Δ that satisfies $|\ell(x, x)| \leq p(x) + q(x)$ for all $(x, x) \in \Delta$. Then by Hahn-Banach there is a linear functional ℓ^* on $X \times X$ that extends ℓ and satisfies $|\ell^*(x, y)| \leq p(x) + q(y)$ for all $(x, y) \in X \times X$. Let $L_1(x) = \ell^*(x, 0)$ and $L_2(y) = \ell^*(0, y)$. Then $|L_1(x)| \leq p(x)$, $|L_2(x)| \leq q(x)$, and $L(x) = \ell(x, x) = \ell^*(x, x) = \ell^*(x, 0) + \ell^*(0, x) = L_1(x) + L_2(x)$ for all $x \in X$.

83. (a) The statement is true.

(b) The statement is false.

84. For the sake of argument suppose that every continuous linear functional on $(X, \| \cdot \|_X)$ can be extended to a bounded linear functional on $(B, \| \cdot \|_B)$. Let $R : B^* \to X^*$ denote the restriction mapping given by $R(L) = L|_X$, $L \in B^*$; by assumption R is onto. Furthermore, if $L \in B^*$ and $R(L) = L|_X = 0$, since X is dense in B, $L = 0$ and R is injective. Finally, if $x \in X$ and $L \in B^*$, then $|L|_X(x)| = |L(x)| \leq \|L\|_{B^*}\|x\|_B \leq C\|L\|_{B^*}\|x\|_X$, and so $\|R\| \leq C$. Thus R is a bounded bijection and by the inverse mapping theorem $R^{-1} : X^* \to B^*$ is bounded, and, consequently, there exist c, C such that $c\|L\|_{B^*} \leq \|L|_X\|_{X^*} \leq C\|L\|_{B^*}$ for all $L \in B^*$.

Now, let $x \in X$. By Problem 83(a) there is $L \in X^*$ with $\|L\|_{X^*} = 1$ such that $|L(x)| = \|x\|_X$, and, by assumption, L can be extended to a continuous linear functional ℓ, say, on $(B, \| \cdot \|_B)$. Then $\|x\|_X = |L(x)| = |\ell(x)| \leq \|\ell\|_{B^*}\|x\|_B$ and since $\|x\|_B \leq C\|x\|_X$ it follows that $\| \cdot \|_B \sim \| \cdot \|_X$ on X. Moreover, since $(X, \| \cdot \|_X)$ is complete, so is $(X, \| \cdot \|_B)$. Thus X is closed in B, and being dense, $X = B$, which is not the case.

85. (a) It is readily seen that $\|L\| = 1/2$ and that if $x = \chi_I$, $L(x) = 1/2$.

86. (a) Given $a \in I$, let $\Phi : \mathcal{P}_n \to \mathbb{R}^{n+1}$ be the mapping given by $\Phi(p) = (x_0, \ldots, x_n)$ where $x_k = \int_I p(t)\,(t-a)^k\,dt$, $k = 0, 1, \ldots, n$. We claim that $K(\Phi) = \{0\}$, the zero polynomial. Indeed, let $p(t) = \sum_{k=1}^{n} c_k\,(t-a)^k \in \mathcal{P}_n$, and note that if $\Phi(p) = (0, \ldots, 0)$, then $\Phi(p) = \int_I p(t)\,(t-a)^k\,dt = 0$ for all k, $0 \leq k \leq n$, and, consequently, $\int_I p(t)^2\,dt = 0$, which implies that $p = 0$. Therefore by dimensional considerations Φ maps $\mathcal{P}_n$ onto $\mathbb{R}^{n+1}$ and there exists $P_{n,a} \in \mathcal{P}_n$ such that $\int_I P_{n,a}(t)\,dt = 1$, $\int_I (t-a)^k\,P_{n,a}(t)\,dt = 0$, $0 < k \leq n$. Now, if p is a polynomial of degree $\leq n$, write $p(t) = \sum_{k=1}^{n} d_k(t-a)^k$ and note that $\int_I p(t)\,P_{n,a}(t)\,dt = p(a)$.

As for uniqueness, if two polynomials, $P_{n,a}, Q_{n,a}$, say, satisfy the property, then $\int_I p(t)\,(P_{n,a}(t) - Q_{n,a}(t))\,dt = 0$ for all $p \in \mathcal{P}_n$, and so $\int_I (P_{n,a}(t) - Q_{n,a}(t))^2\,dt = 0$ and $P_{n,a} = Q_{n,a}$.

Finally, for the sake of argument suppose that μ is a finite Borel measure on I with $\mu(\{1\}) = 0$ such that $p(1) = \int_I p(t)\,d\mu(t)$ for all $p \in \bigcup_n \mathcal{P}_n$. Then with $p(t) = t^k$, by the LDCT it follows that $1 = \lim_k \int_I t^k\,d\mu(t) = 0$, which is not the case.

(b) It suffices to prove the result for $k = 0$, deduce from this the result for $k = 1$, and then iterate. Let $P_{n,a} \in \mathcal{P}_n$ be the polynomial in (a) and $c_{n,0,a} = \sup_{t \in I} |P_{n,a}(t)|$; by the representation formula $|p(a)| \leq c_{n,0,a} \int_I |p(t)|\,dt$. Since for $p \in \mathcal{P}_n$ also $p' \in \mathcal{P}_n$, by (a), $p'(a) = \int_I p'(t)\,P_{n,a}(t)\,dt$, and so integrating by parts, $p'(a) = P_{n,a}(1)\,p(1) - P_{n,a}(0)\,p(0) - \int_I p(t)\,P'_{n,a}(t)\,dt$ and the conclusion holds with $c_{n,1,a} = (c_{n,0,0} + c_{n,0,1})c_{n,0,a} + \sup_I |P'_{n,a}(t)|$.

Finally, for no universal constant c_a the inequality holds for all polynomials. Indeed, let $a = 1$ and $p_k(t) = t^k$. Then $\|p_k\|_1 = 1/(k+1)$ and the inequality $1 = |p_k(1)| \leq M/(k+1)$ cannot hold for all k for a fixed finite constant M.

87. (a) The statement is true.

(b) and (c) are false. For the sake of argument suppose that $|L(x)| \leq c\|x\|_q$ for $x \in \mathcal{P}$ and consider the sequence $\{x_n\} \subset L^q(I) \cap C^1(I)$ given by $x_n(t) = (1-t)^n$, $n \geq 1$. Then $L(x_n) = x'_n(0) = -n$ and $\|x_n\|_q = (nq+1)^{-1/q} \leq 1$ for all n. Thus $n \leq c$ for all n, which is not the case.

88. The norm is $\sum_{k=1}^n |\lambda_k|$.

91. (a) For $s \in \mathbb{R}$, the function $f(t) = (1+t)^s(1+t^p)^{-s/p}$, $t \in [0, \infty)$, is nonvanishing, continuous, and tends to 1 as $t \to \infty$, and, consequently, there exist constants $c, C > 0$ such that $c < f(t) < C$ for all $t > 0$.

(c) Follows from a diagonal argument. Note that $|(1+k)^s y_k^n| \leq c$ for all n and k, and so the sequence $\{y_k^1, \ldots, y_k^n, \ldots\}$ is bounded for each $k = 1, 2, \ldots$. In particular, since $\{y_1^1, \ldots, y_1^n, \ldots\}$ is bounded there exists a subsequence $\{n_1\}$ such that $\{y_1^{n_1}\}$ converges to y_1, say. Having picked successively refined subsequences $\{n_1\}, \ldots, \{n_k\}$ with the property that $\{y_j^{n_k}\}$ converges for $1 \leq j \leq k$, since as noted above $\{y_{k+1}^{n_k}\}$ is bounded, there is a subsequence $\{n_{k+1}\}$ of $\{n_k\}$ such that $\{y_{k+1}^{n_{k+1}}\}$ converges to a scalar y_{k+1}, say, and, since $\{n_{k+1}\}$ is a refinement of $\{n_k\}$ it follows that $\{y_j^{n_{k+1}}\}$ converges for $1 \leq j \leq k+1$. Once the iterative process is completed we claim that the subsequence $\{y^{n_k}\}$ of $\{y^n\}$ converges in h_r^p and for this it suffices to prove that it is Cauchy in h_r^p, i.e., given $\varepsilon > 0$, there exists N such that $\|y^{n_k} - y^{n_\ell}\|_{h_r^p} \leq \varepsilon$ for $n_k, n_\ell > N$. Let K be such that $1/(1+K)^{p(s-r)} < (\varepsilon/c)^p$. Then we can find N such that if $n_m, n_\ell > N$, $|y_k^{n_m} - y_k^{n_\ell}|^p \leq \varepsilon/(1+K)^r K$ for $1 \leq k \leq K$, and observe that

$$\|y^{n_m} - y^{n_\ell}\|_{h_s^p}^p = \sum_{k=1}^{K}(1+k)^{pr}|y_k^{n_m} - y_k^{n_\ell}|^p + \sum_{k>K}(1+k)^{pr}|y_k^{n_m} - y_k^{n_\ell}|^p$$

$$\leq \sum_{k=1}^{K}\varepsilon K^{-1} + \sum_{k>K}(1+k)^{ps}(1+K)^{-p(s-r)}|y_k^{n_m} - y_k^{n_\ell}|^p$$

$$\leq \varepsilon + \varepsilon c^{-p}\|y^{n_m} - y^{n_\ell}\|_{h_s^p}^p \leq \varepsilon + 2\varepsilon = 3\varepsilon.$$

Therefore the sequence is Cauchy in h_r^p and, hence, convergent.

(d) Finite linear combinations of the e_n are in h_s^p and so if $x \in h_s^p$ and $\varepsilon > 0$, we can find a linear combination y of the e_n such that $\|(1+n)^s x_n - y_n\|_{\ell^p} < \varepsilon$. Then setting $z_n = y_n/(1+n)^s$, $z \in h_s^p$ and $\|z - x\|_{h_s^p} < \varepsilon$.

(e) First, assume that $L_y(x) = \sum_n y_n x_n$ and $\|y\|_{h_{-s}^q} \leq 1$; L_y is well-defined as the sum converges by Hölder's inequality. Therefore $|L_y(x)| = \left|\sum_n (1+n)^{-s}y_n(1+n)^s x_n\right| \leq \|y\|_{h_{-s}^q}\|x\|_{h_s^p}$.

Next, observe that every linear functional L on h_s^p is of the form L_y for $y \in h_{-s}^q$, i.e., $L(x) = L_y(x)$ for all $x \in h_s^p$ and $\|L\|_{h_s^{p*}} = \|y\|_{h_{-s}^q}$. Consider the sequence $\{x^N\} \subset h_s^p$ given by $x_n^N = (1+n)^{-sq}|y_n|^{q-1}\mathrm{sgn}(y_n)$, $n \leq N$, and $x_n^N = 0$ otherwise. Then $L(x^N) = \sum_{n=1}^{N}(1+n)^{-sq}|y_n|^q$ and $\|x^N\|_{h_s^p} = (\sum_{n=1}^{N}(1+n)^{-sq}|y_n|^q)^{1/p}$ and so,

$$|L(x^N)| = \sum_{n=1}^{N}(1+n)^{-sq}|y_n|^q \leq \|L\|_{h_s^{p*}}\left(\sum_{n=1}^{N}(1+n)^{-sq}|y_n|^q\right)^{1/p}.$$

Hence $(\sum_{n=1}^{N}(1+n)^{-sq}|y_n|^q)^{1/q} \leq \|L\|_{h_s^{p*}}$ and taking the sup over N, $\|y\|_{h_{-s}^q} \leq \|L\|_{h_s^{p*}}$. Finally, as noted above, if $L = L_y$, $\|L_y\|_{h_s^{p*}} = \|y\|_{h_{-s}^q}$.

92. Suppose (a) holds and let L be the bounded linear functional that extends L_0. Then given $x_1, \ldots, x_n \in Y$ and scalars $\lambda_1, \ldots, \lambda_n$, $\sum_{n=1}^{m}\lambda_n L_0(x_n) = L(\sum_{n=1}^{m}\lambda_n x_n)$ and, therefore, $|\sum_{n=1}^{m}\lambda_n L_0(x_n)| \leq \gamma\|\sum_{n=1}^{m}\lambda_n x_n\|$.

Conversely, let $W = \mathrm{sp}\{x_1, \ldots, x_m\}$ and define the functional L_W on W by setting $L_W(x) = \sum_{n=1}^{m}\lambda_n L_0(x_n)$ for $x = \sum_{n=1}^{m}\lambda_n x_n$. Note that L_W is well-defined: If $\sum_{n=1}^{m}\lambda_n x_n = 0$, by assumption $|\sum_{n=1}^{m}\lambda_n L_0(x_n)| \leq \gamma\|\sum_{n=1}^{m}\lambda_n x_n\| = 0$, and so $L_W(\sum_{n=1}^{m}\lambda_n x_n) = 0$. Moreover, from the definition of L_W it readily follows that $\|L_W\| \leq \gamma$. Then by Hahn-Banach there exists $L \in X^*$ such that $L|_W = L_W$ and $\|L\| = \|L_W\| \leq \gamma$.

93. By Problem 80, given $x_0 \notin Y$, there exists $L \in X^*$ such that $\|L\| = 1$, $Y \subset K(L)$, and $|L(x_0)| = d(x_0, Y)$. Observe that if $\|L\| = 1$ and $Y \subset K(L)$, then $|L(x_0)| \leq d(x_0, Y)$, and so $|L(x_0)| = d(x_0, Y)$ is the largest possible value. Also, since $z \perp Y$ iff $d(z, Y) = \|z\|$, it readily follows that

$z \perp Y$ iff there exists $0 \neq L \in X^*$ such that $Y \subset K(L)$ and $|L(z)| = \|L\| \, \|z\|$, and that $d(x_0, Y) = \|x_0 - y_0\|$ iff $x_0 - y_0 \perp Y$.

95. A particular case of interest is $Y = \{x_n\}$, a sequence in X. Then $x \in X$ is the limit of finite linear combinations of the x_n iff $L(x) = 0$ for all bounded linear functionals L on X such that $L(x_n) = 0$ for all n.

96. (a) For the sake of the argument suppose that $L_k(x) = 0$ for $k = 1, \ldots, n$ implies $x = 0$. Assuming as we may that the L_k are linearly independent, we claim that $\dim(X) \leq n$. If $\dim(X) = m > n$, let $x_1, \ldots, x_m$ be linearly independent elements of X and consider the system of linear equations $\sum_{k=1}^{m} a_k L_i(x_k) = 0$, $i = 1, \ldots, n$, with unknown scalars $a_1, \ldots, a_m$. Since $n < m$ the system has a nontrivial solution $\lambda_1, \ldots, \lambda_m$, say, and if $x = \sum_{k=1}^{m} \lambda_k x_k$, $x \neq 0$, by construction $L_k(x) = 0$ for $k = 1, \ldots, n$, which by assumption implies that $x = 0$, which is not the case.

(b) Let F denote the scalar field. Assume that the L_k are linearly independent and let $T : X \to F^{n+1}$ be given by $T(x) = (L(x), L_1(x), \ldots, L_n(x))$; T is a bounded linear mapping and since $(1, 0, \ldots, 0) \notin R(T)$, $R(T)$ is a proper subspace of F^{n+1}. Therefore there exists a nonzero linear functional on F^{n+1} that vanishes on $R(T)$. In other words, there exist scalars $\lambda, \lambda_1, \ldots, \lambda_n$, not all 0, such that $\lambda L(x) + \sum_{k=1}^{n} \lambda_k L_k(x) = 0$ for all $x \in X$. If $\lambda = 0$, this gives a nontrivial relation in the L_k, contrary to their linear independence. Thus $\lambda \neq 0$ and $L(x) = -\lambda^{-1} \sum_{k=1}^{n} \lambda_k L_k(x)$.

98. (b) Necessity is clear. If $x \in K(L_1) \cap \cdots \cap K(L_n)$, then $L(x) = \lambda_1 L_1(x) + \cdots + \lambda_n L_n(x) = 0$ and $K(L_1) \cap \cdots \cap K(L_n) \subset K(L)$.

As for sufficiency, observe that if $x_1, \ldots, x_n$ are as in (a), any $x \in X$ has a unique representation $x = \lambda_1 x_1 + \cdots + \lambda_n x_n + y$ where $\lambda_k \in F$ and $y \in K(L_1) \cap \cdots \cap K(L_n)$. Indeed, let $y = x - L_1(x)x_1 - \cdots - L_n(x)x_n$. Then $L_m(y) = L_m(x) - L_1(x)L_m(x_1) - \cdots - L_n(x)L_m(x_n) = L_m(x) - L_m(x)L_m(x_m) = 0$, $m = 1, \ldots, n$, and $y \in K(L_1) \cap \ldots \cap K(L_n)$. Hence $x = \lambda_1 x_1 + \cdots + \lambda_n x_n + y$ where $\lambda_m = L_m(x)$. On the other hand, if $x = \lambda_1 x_1 + \cdots + \lambda_n x_n + y$, then $\lambda_m = L_m(x)$, $1 \leq m \leq n$, and the representation is unique. Finally, in view of this representation $L(x) = L_1(x)L(x_1) + \cdots + L_n(x)L(x_n) + L(y)$ where, since $y \in K(L_1) \cap \ldots \cap K(L_n) \subset K(L)$, $L(y) = 0$. Thus $L = L(x_1)L_1 + \cdots + L(x_n)L_n$ and $L \in \mathrm{sp}\{L_1, \ldots, L_n\}$.

100. Let L be the linear functional constructed in Problem 99, $M = K(L)$, and define $H_0 = x_0 + M$, i.e., $H_0 = \{x \in B : L(x) = r\}$.

101. Let $M = \mathrm{sp}\{x_1, \ldots, x_n\}$, $L_k \in X^*$, $1 \leq k \leq n$, as in Problem 97, and $N = \bigcap_{k=1}^{n} K(L_k)$; N is the intersection of closed sets and, hence, closed. Let $x \in M \cap N$. Since $x \in M$ there are scalars $\lambda_1, \ldots, \lambda_n$ such that $x = \lambda_1 x_1 + \cdots + \lambda_n x_n$, and since $x \in N$, applying L_k to this expression it follows that $\lambda_k = L_k(x) = 0$ for each k. Thus $x = 0$ and $M \cap N = \{0\}$.

Finally, if $x \in X$, let $y = \sum_{k=1}^{n} L_k(x)x_k \in M$; then $L_k(x - y) = 0$ for all k, and, consequently, $x - y \in N$. Therefore $X = M \oplus N$.

102. Let L_1 be a nontrivial bounded linear functional on B; such a functional exists since by Hahn-Banach $B^* \neq \emptyset$. Let $X_1 = K(L_1)$; by Problem 63(a), X_1 is a proper closed subspace of codimension 1 in B and so it is infinite dimensional and can take the place of B in the previous argument. Next, let L_2 be a nontrivial continuous linear functional on X_1 and $X_2 = K(L_2)$; X_2 is a proper closed subspace of X_1 of codimension 1 in X_1. Keep going. This gives the desired sequence.

103. Let $L \in B^*$; then $\|L\| = \sup_{x \in B, \|x\|=1} |L(x)| \geq \sup_{x \in X, \|x\|=1} |L(x)| = \|L|_X\|$. Let $x \in B$, $\|x\| = 1$, and pick a sequence $\{x_n\} \subset X$ that converges to x in B. Then $|L(x)| \leq |L(x - x_n)| + |L(x_n)| \leq |L(x - x_n)| + \|L|X\| \, \|x_n\|$, and since L is continuous and $\|x_n - x\| \to 0$ implies $\|x_n\| \to \|x\|$, taking the sup we get $\|L\| \leq \|L|_X\|$. Thus the two norms are the same and $B^* \sim X^*$ by means of the restriction map.

104. It depends on whether Y is endowed with the norm induced by X or not. In the former case, given $\ell \in Y^*$, by the Hahn-Banach theorem there exists $L \in X^*$ with $L|_Y = \ell$ and $\|L\|_{X^*} = \|\ell\|_{Y^*}$, and, therefore, we may think of Y^* as consisting of restrictions of elements in X^*, and so smaller than X^*. In the latter case, let X be equipped with norm $\| \cdot \|_X$ and Y with a different norm $\| \cdot \|_Y$ so that the topology induced by Y is stronger than the subspace topology X induces in Y. Also assume that Y is dense in X with the space topology. Then the mapping $\psi : X^* \to Y^*$ given by $\psi(L) = L|_Y$ is a well-defined bounded linear injection and we may then think of Y^* as bigger than X^*. A typical instance of this is h_s^2, $s > 0$, viewed as a subspace of ℓ^2 but with its stronger topology. Then $h_s^{2*} = h_{-s}^2$, and so $h_s^2 \subset \ell^2 = \ell^{2*} \subset h_{-s}^2$.

105. (a) To prove completeness, let $\{x^n\} \subset \bigoplus_{k=1}^{N} B_k$ be absolutely convergent, i.e., $\sum_n \|x^n\|_p < \infty$. If $x^n = (x_1^n, \ldots, x_N^n)$ it readily follows that $\sum_n \|x_k^n\|_{B_k} \leq \sum_n \|x^n\|_p < \infty$, and, consequently, since the B_k are Banach spaces, $\sum_n x_k^n$ converges to X_k in B_k, say, for $1 \leq k \leq N$. Let $x = (X_1, \ldots, X_N)$. Then $\|x - \sum_{n=1}^{M} x^n\|_p = (\sum_{k=1}^{N} \|X_k - \sum_{n=1}^{M} x_k^n\|_{B_k}^p)^{1/p} \to 0$ as $M \to \infty$ and $\sum_n x^n$ converges to x in $\bigoplus_{k=1}^{N} B_k$.

(b) Given $L_k \in B_k^*$, $k = 1, \ldots, N$, let $L(x) = \sum_{k=1}^{N} L_k(x_k)$; we claim that L is a linear functional on $\bigoplus_{k=1}^{N} B_k$ that is continuous with respect to $\| \cdot \|_p$ for each $1 \leq p < \infty$, and $\|L\| = (\sum_{k=1}^{N} \|L_k\|_{B_k^*}^q)^{1/q}$ where q is conjugate to p. Indeed, by Hölder's inequality, $|L(x)| = |\sum_{k=1}^{N} L_k(x_k)| \leq \sum_{k=1}^{N} \|L_k\|_{B_k^*} \|x_k\|_{B_k} \leq (\sum_{k=1}^{N} \|L_k\|_{B_k^*}^q)^{1/q} (\sum_{k=1}^{N} \|x_k\|_{B_k}^p)^{1/p}$, and so $\|L\| \leq (\sum_{k=1}^{N} \|L_k\|_{B_k^*}^q)^{1/q}$ with the obvious interpretation for $q = \infty$.

For the opposite inequality recall that for $1 < p < \infty$, if s_k, t_k are positive and $s_k = t_k^{q/p}$, $1 \le k \le N$, there is equality in Hölder's inequality. Let $\varepsilon > 0$. Then there are $x_k \in B_k$ such that $\|x_k\|_{B_k} = \|L_k\|_{B_k^*}^{q/p}$ and $L(x_k) \ge \|L_k\|_{B_k^*}\|x_k\|_{B_k} - \varepsilon$, $1 \le k \le n$, and, consequently,

$$|L(x)| = \sum_{k=1}^{N} L_k(x_k) \ge \sum_{k=1}^{N} \|L_k\|_{B_k^*}\|x_k\|_{B_k} - N\varepsilon$$

$$= \Big(\sum_{k=1}^{N} \|L_k\|_{B_k^*}^{q}\Big)^{1/q}\Big(\sum_{k=1}^{N} \|x_k\|_{B_k}^{p}\Big)^{1/p} - N\varepsilon,$$

which since ε is arbitrary gives $\|L\| \ge (\sum_{k=1}^{N} \|L_k\|_{B_k^*}^{q})^{1/q}$. If $p = 1$ clearly $\|L\| \ge \max_{1 \le k \le n} \|L_k\|_{B_k^*}$.

Given a continuous linear functional L on $\bigoplus_{k=1}^{N} B_k$ equipped with $\|\cdot\|_p$ for a fixed $1 \le p < \infty$, let L_k be the linear functional on B_k given by $L_k(x_k) = L((0,\ldots,0,x_k,0,\ldots,0))$, $k = 1,\ldots,N$. Note that $L_k \in B_k^*$ and $\|L_k\|_{B_k^*} \le \|L\|$, $1 \le k \le n$. Then $L(x) = \sum_{k=1}^{N} L_k(x_k)$ and by the above argument $\|L\| = (\sum_{k=1}^{N} \|L_k\|_{B_k^*}^{q})^{1/q}$, which gives the desired characterization.

An interesting application of this result is the following: If $(X, \mathcal{M}, \mu)$ is a measure space and X is the disjoint union of $X_1, \ldots, X_N$, then $L^p(X) = (\bigoplus_{k=1}^{N} L^p(X_k), \|\cdot\|_p)$, $1 \le p \le \infty$.

106. Considering sequences first, since ℓ^1 can be identified via the natural map with a subspace of $\ell^{\infty*}$ it suffices to show that this subspace is proper; Problem 111(b) is relevant here. Or by Problem 139 it suffices to verify that some bounded linear functional on ℓ^1 does not attain its norm; Problem 83(b) is relevant here. Similarly, for $L^1(I)$; since with the usual integral pairing $L^1(I)^* = L^\infty(I)$, it suffices to exhibit a functional L on $L^\infty(I)$ that is not of the form $L(x) = \int_I x(t)y(t)\,dt$ with $y \in L^1(I)$ for all $x \in L^\infty(I)$. Since $\chi_{[0,1/2]} \in L^\infty(I) \setminus \overline{C(I)}$, by Hahn-Banach there is a bounded linear functional ℓ on $L^\infty(I)$ such that $\ell|_{C(I)} = 0$ and $\ell(\chi_{[0,1/2]}) = 1$. For the sake of argument suppose that $\ell(x) = \int_I x(t)y(t)\,dt$ for $y \in L^1(I)$ and all $x \in L^\infty(I)$. Let $\{x_n\} \subset C(I)$ be given by $x_n(t) = 1$ if $t \in [0, 1/2 - 1/n]$, $x_n(t) = 0$ if $t \in [1/2 + 1/n, 1]$, and piecewise linear in between. Then $\chi_{[0,1/2]}y = \lim_n x_n y$ a.e. in I, and, therefore, by the LDCT, $1 = \ell(\chi_{[0,1/2]}) = \int_I \chi_{[0,1/2]}(t)y(t)\,dt = \lim_n \int_I x_n(t)y(t)\,dt = \lim_n \ell(x_n) = 0$, which is not the case. Finally, since $L^\infty(I) = L^1(I)^*$, $L^\infty(I)$ is not reflexive either.

107. No.

108. We invoke the fact that a subset A of a metric space is nowhere dense if A^c is open and dense. So, let $c = A$; A^c is open. Indeed, given

$x \in A^c$, there exist a subsequence $\{x_{n_k}\}$ of $\{x_n\}$ and $\eta > 0$ such that $|x_{n_k} - x_{n_m}| > \eta$ for all k, m. We claim that if $\|x - y\|_\infty < \eta/3$, $y \in A^c$. For the sake of argument suppose that one such $y \in A$ and pick M large enough so that $|y_{n_k} - y_{n_m}| < \eta/3$ for all $k, m \geq M$. Then $|x_{n_k} - x_{n_m}| \leq |x_{n_k} - y_{n_k}| + |y_{n_k} - y_{n_m}| + |y_{n_m} - x_{n_m}| < \eta$, which is not the case. Also, A^c is dense in ℓ^∞; let $x \in \ell^\infty$ and $\varepsilon > 0$ be given. If $x \in A^c$, put $y = x$. Otherwise, define y with terms $y_n = x_n$, n odd, and $y_n = x_n + (\varepsilon/2)$, n even. Then $\|x - y\|_\infty \leq \varepsilon/2$, and, clearly, $y \in A^c$.

109. First, $\ell^1 \subset c_0^*$. Indeed, since $\sum_n x_n y_n$ converges for $x \in c_0$ and $y \in \ell^1$, $L_y(x) = \sum_n x_n y_n$ defines a linear functional on c_0, which, since $|L_y(x)| \leq \|y\|_1 \|x\|_\infty$, is bounded with $\|L_y\| \leq \|y\|_1$. Actually $\|L_y\| = \|y\|_1$; let $x \in c_0$ be given by $x_n = y_n/|y_n|$ if $n \leq N$ and $y_n \neq 0$, and $x_n = 0$ otherwise. Then $|L_y(x)| = \sum_{n=1}^N |y_n|$, and since $\|x\|_\infty = 1$ and N is arbitrary, $\|L_y\| \geq \|y\|_1$.

Conversely, given $L \in c_0^*$, let $y_n = L(e_n)$, $n = 1, 2, \ldots$; we claim that $y \in \ell^1$ and $L(x) = L_y(x)$ for all $x \in c_0$. The latter assertion is obvious. As for the former, fix N and as before define $x \in c_0$ with $\|x\|_\infty = 1$ and $L(x) = \sum_{n=1}^N |y_n|$. Since L is bounded, $\sum_{n=1}^N |y_n| = |L(x)| \leq \|L\|_{c_0^*}$ for all N, $y \in \ell^1$, and, since $L = L_y$, as before $\|L\| = \|y\|_1$.

110. Let p be the functional on ℓ^∞ given by $p(x) = \limsup_n x_n$; p is clearly sublinear. Let ℓ be the linear functional on $\{0\}$ given by the zero functional; then $\ell(0) = 0 = \limsup_n 0_n$ and by Hahn-Banach ℓ can be extended to a linear functional L on ℓ^∞ such that $L(x) \leq p(x) = \limsup_n x_n$ for all $x \in \ell^\infty$. Applying this to $-x$ we get $-L(x) = L(-x) \leq \limsup_n(-x_n) = -\liminf_n x_n$, and, therefore, $\liminf_n x_n \leq L(x) \leq \limsup_n x_n$. Finally, since $\limsup_n x_n \leq \sup_n x_n \leq \sup_n |x_n|$ and also $\liminf_n x_n \geq \inf_n x_n \geq -\sup_n |x_n|$, we have $|L(x)| \leq \|x\|_\infty$ and L is bounded with norm ≤ 1.

111. (a) Clearly L_∞ is a bounded linear functional with norm ≤ 1. Moreover, since the sequence in c with $x_n = 1$ for all n satisfies $\|x\| = 1$ and $L_\infty(x) = 1$, $\|L_\infty\| = 1$. Note that L_∞ vanishes in c_0.

(b) The statement is false.

(c) We first define a positively homogenous, subadditive functional $p : \ell^\infty \to \mathbb{R}$ such that $L_\infty(x) \leq p(x)$ for $x \in c$. Of course $p(x) = \|x\|_\infty$ or $p(x) = \limsup_n x_n$ will do but, since an attractive feature of $p(x)$ is being as small as possible, let

$$p(x) = \limsup_n \frac{1}{n} \sum_{k=1}^n x_k, \quad x \in \ell^\infty .$$

It is readily seen that p has the desired properties and if for $x \in \ell^\infty$ we let $y_n = n^{-1} \sum_{k=1}^n x_k$, then $\liminf_n x_n \leq \liminf_n y_n \leq \limsup_n y_n \leq$

$\limsup_n x_n$. This implies in particular that $L_\infty(x) = p(x)$ for $x \in c$ and by Hahn-Banach L_∞ has an extension L to ℓ^∞ such that $L|_c = L_\infty$ and $L(x) \le p(x)$ for $x \in \ell^\infty$. Moreover, applying this observation to $-x$, since $-L(x) = L(-x) \le p(-x) = -\liminf_n n^{-1} \sum_{k=1}^n x_k$, it readily follows that

$$-\|x\|_\infty \le \liminf_n \frac{1}{n} \sum_{k=1}^n x_k$$

$$\le L(x) \le \limsup_n \frac{1}{n} \sum_{k=1}^n x_k \le \|x\|_\infty$$

and, consequently, $|L(x)| \le \|x\|_\infty$.

A similar argument applies in the following setting: Let $X = \{x \in \ell^\infty : \lim_n (-1)^n x_n \text{ exists}\}$. Then there exists a bounded linear functional L on ℓ^∞ such that $L(x) = \lim_n (-1)^n x_n$ for $x \in X$.

(d) The extension is not unique.

(e) In particular, if $x_n \ge 0$ for all n, then $\liminf_n n^{-1} \sum_{k=1}^n x_k \ge 0$ and by (c), $L(x) \ge 0$.

(f) Let $X = R(S - I) = \{y \in \ell^\infty : \text{there exists } x \text{ in } \ell^\infty \text{ such that } y_n = x_{n+1} - x_n \text{ for all } n\}$; X is a subspace of ℓ^∞. We claim that if $e = (1, 1, \ldots)$, $d(e, X) = 1$. First, $e \notin X$; this is clear since, if $(S - I)(x) = e$, then $x_n = x_1 + n$ for all n and $x \notin \ell^\infty$. Next, since X is a subspace, $d(e, X) \le d(e, 0) = \|e\| = 1$. Suppose that actually $d(e, X) < 1$. Then there exists $y \in \ell^\infty$ such that $\|e - y\|_\infty = \eta < 1$ and so $|1 - y_n| < \eta$ for all n. Therefore there exists $x \in \ell^\infty$ such that $|1 - (x_{n+1} - x_n)| \le \eta$ for all n, and, consequently, $x_{n+1} \ge x_n + (1 - \eta)$, and $x_{n+1} > n(1 - \eta) + x_1$, contrary to the fact that $x \in \ell^\infty$. Finally, since $L_\infty(e) = 1$ and $\|L_\infty\| = 1$, by Hahn-Banach L_∞ can be extended to a functional L on ℓ^∞ so that $L(e) = 1$, $\|L\| = 1$, and $L(y) = 0$ for all $y \in X$. But for any $x \in \ell^\infty$, $y = S(x) - x \in X$, and so $L(y) = L(S(x)) - L(x) = 0$.

112. An example of such a sequence is $\alpha_{n,k} = n^{-1}(1 - 1/n)^{k-1}$, $k \ge 1$, $n \ge 2$.

113. (a) The statement is true.

(b) The statement is false. An isometric isomorphism maps the unit ball of one space, B_{c_0}, say, bijectively onto the unit ball of the other, B_c in this case, and so it maps extreme points into extreme points. Recall that $E(B_{c_0}) = \emptyset$ whereas $(1, 1, \ldots) \in E(B_c)$.

(c) The statement is true. Given $x \in c$, let $x_\infty = \lim_n x_n$, and for $x \in c$, a scalar λ, and $y \in \ell^1$, let $L_{\lambda,y}(x) = \lambda x_\infty + \sum_n y_n x_n$; each $L_{\lambda,y}$ defines a linear functional on c and these functionals are different for different pairs (λ, y). We claim that $L_{\lambda,y}$ is bounded and $\|L_{\lambda,y}\| = |\lambda| + \|y\|_1$; $\|L_{\lambda,y}\| \le |\lambda| + \|y\|_1$

being obvious. As for the other inequality, let x^N be the sequence in c given by

$$x_k^N = \begin{cases} \operatorname{sgn}(y_k), & 1 \leq k \leq N, \\ \operatorname{sgn}(\lambda), & k > N. \end{cases}$$

Then $\|x^N\|_\infty = 1$, $x_\infty^N = \operatorname{sgn}(\lambda)$, and, consequently, $L_{\lambda,y}(x^N) = |\lambda| + \sum_{k=1}^N |y_k| + \sum_{k=N+1}^\infty a_k \operatorname{sgn}(\lambda)$. Therefore, since $y \in \ell^1$, $\lim_N L_{\lambda,y}(x^N) = |\lambda| + \|y\|_1$.

Furthermore, we claim that a bounded linear functional L on c is of the form $L_{\lambda,y}$ for some scalar λ and $y \in \ell^1$. To prove this we verify that the sequence y with terms $y_n = L(e_n)$ for all n is in ℓ^1 and $\|y\|_1 \leq \|L\|$. If x^N is the sequence in c with terms $x_n^N = \operatorname{sgn}(L(e_n))$ for $n \leq N$ and $x_n^N = 0$ otherwise, then $\|x^N\|_\infty = 1$ and $|L(x^N)| = |L(\sum_{n=1}^N x_n e_n)| = \sum_{n=1}^N |L(e_n)| \leq \|L\|$. Since this is true for all N, $\sum_n |L(e_n)| \leq \|L\|$. Also, if $e = (1, 1, \ldots)$, $|L(e)| \leq \|L\|$.

Finally, for $x \in c$ write $x = x_\infty e + \sum_n (x_n - x_\infty) e_n$ where the series converges in ℓ^∞ and, hence, in c, and so $L(x) = (L(e) - \sum_n L(e_n)) x_\infty + \sum_n L(e_n) x_n$. Thus with $\lambda = L(e) - \sum_n L(e_n)$ and $y \in \ell^1$ with $y_n = L(e_n)$ for all n, $L = L_{\lambda,y}$ and $\|L\| = |\lambda| + \|y\|_{\ell^1}$. Thus the norm in c^* is an ℓ^1 norm and c^* is isometrically isomorphic to ℓ^1, which is the dual of c_0.

114. For the sake of argument suppose X is such a space; since by Problem 103 the dual of a normed linear space and its completion are the same we can assume that X is a Banach space; also via the natural map X can be identified with a closed subspace of $X^{**} = c_0^* = \ell^1$. Fix n and consider $e_n \in c_0$. Since $\|e_n\| = \sup_{x \in X, \|x\|_X = 1} |e_n(x)| = 1$ there exist $\{v^k\} \subset X$ with $\|v^k\| = 1$ for all k, such that $e_n(v^k) \to 1$ as $k \to \infty$. Viewing the v^k as sequences in ℓ^1 this means that $\|v^k\|_1 = 1$ and $e_n(v^k) = v_n^k \to 1$ as $k \to \infty$ and together these relations imply that $v_m^k \to 0$ as $k \to \infty$ for $m \neq n$. Thus $v^k \rightharpoonup e_n$ in ℓ^1, and, consequently, by Schur's lemma $v^k \to e_n$ in ℓ^1, which, since X is closed in ℓ^1, implies that $e_n \in X$. But finite linear combinations of the e_n are dense in ℓ^1 and so $X = \ell^1$. However, $\ell^{1*} = \ell^\infty$, which is strictly larger than c_0.

115. (b) Since $0 \in K(L)$, $d(x, K(L)) \leq \|x\|$ for all $x \in X$. By (a), a bounded linear functional L attains its norm at x, i.e., $|L(x)| = \|L\| \|x\|$ iff $d(x, K(L)) = \|x\|$, which is precisely the case when 0 is a best approximation to x from $K(L)$.

(c) By definition $z \perp K(L)$ iff $\|z\| = d(z, K(L))$, and by (a) this is the case iff $|L(z)| = \|L\| \|z\|$.

(d) With x such that $L(x) \neq 0$, put $y = \lambda x / L(x) \in \Lambda_\lambda$; in particular, $\Lambda_\lambda \neq \emptyset$. Then $\Lambda_\lambda = y + K(L)$, and so by (a), $d(x, \Lambda_\lambda) = d(x - y, K(L)) = |L(x - y)| / \|L\| = |L(x) - \lambda| / \|L\|$.

118. The norm is 1 and is not attained.

119. (a) Since for any $L \in \Lambda$ and $y \in Y$, $|L(x)| = |L(x - y)| \le \|x - y\|$, taking the sup over $L \in \Lambda$ and then the inf over $y \in Y$ gives $\sup_{L \in \Lambda} |L(x)| \le d(x, Y)$. As for the opposite inequality, if $d(x, Y) > 0$, by Hahn-Banach there is a bounded linear functional L on X with norm $\|L\| = 1/d(x, Y)$ so that $L(x) = 1$ and L vanishes on Y. Therefore $d(x, Y) = |L(x)|/\|L\|$ and since $L/\|L\| \in \Lambda$, $d(x, Y) \le \sup_{Y \in \Lambda} |L(x)|$.

(b) If $Y = X$ there is nothing to prove. Otherwise, if $L \in \Lambda$, $\overline{Y} \subset \overline{K(L)} = K(L)$, and so $\overline{Y} \subset \bigcap_{L \in \Lambda} K(L)$. Conversely, suppose that $x \notin \overline{Y}$. Then by Problem 80 there exists $L \in X^*$ such that $\|L\| \le 1$, $L(x) = d(x, \overline{Y}) > 0$, and $L(y) = 0$ for $y \in \overline{Y}$. In particular, $x \notin K(L)$ and so $x \notin \bigcap_{L \in \Lambda} K(L)$. Hence $\bigcap_{L \in \Lambda} K(L) \subset \overline{Y}$.

120. The statement is false.

122. $X^* = \{y : |y_n| \le c2^n$ for some constant c for all $n\}$ with $\|y\|_{X^*} = \sup_n 2^{-n} |y_n|$.

123. ℓ^p is dense in c_0.

124. A is dense in ℓ^p if $\{a_n\} \notin \ell^q$, where q is the conjugate to p.

125. The statement is true. By Problem 5.131(b), with q the conjugate to p, there is a sequence $a \in \ell^q \setminus \bigcup_{s<q} \ell^s$. Let $A = \{x \in c_0 : \sum_n a_n x_n = 0\}$; by Problem 124, A is not dense in ℓ^p. If now $p < r < \infty$ and s denotes the conjugate to r, $1 < s < q$, and since $a \notin \ell^s$, by Problem 124, A is dense in ℓ^r.

127. Let L be a continuous linear functional on $C([0,1])$ such that $L(x_{a_n}) = 0$ for all n. Now, $\lim_m \sum_{k=0}^m t^k/a_n^k = -x_{a_n}(t)/a_n$ uniformly in $[0,1]$, and so by the continuity of L it follows that $\sum_{k=0}^\infty L(t^k)/a_n^k = -L(x_{a_n})/a_n = 0$ for all n. Moreover, since $|L(t^k)| \le \|L\|$ for all k, the function $g(z) = \sum_{k=0}^\infty L(t^k)z^k$ is analytic in the open unit disk $|z| < 1$ and vanishes for $z = 1/a_n$ for all n. Since 0 is an accumulation point of $\{1/a_n\}$, by elementary properties of analytic functions $g(z)$ vanishes identically. Thus $L(t^k) = 0$ for all $k \ge 0$ and since polynomials are dense in $C([0,1])$, L is the zero functional. Therefore by Problem 95(b), X is dense in $C([0,1])$.

128. Let L be a bounded linear functional on $L^p(\mathbb{R}^n)$ that vanishes on D and $y \in L^q(\mathbb{R}^n)$ such that $L(x) = \int_{\mathbb{R}^n} x(t)y(t)\,dt$ for all $x \in L^p(\mathbb{R}^n)$, $1/p + 1/q = 1$. Now, if I_1, I_2 are bounded cubes in $\mathbb{R}^n$, then $x(t) = |I_1|^{-1}\chi_{I_1}(t) - |I_2|^{-1}\chi_{I_2}(t)$ is in D, and, consequently, $\int_{\mathbb{R}^n} x(t)y(t)\,dt = |I_1|^{-1}\int_{I_1} y(t)\,dt - |I_2|^{-1}\int_{I_2} y(t)\,dt = 0$. In other words, the averages of y are constant and by the Lebesgue differentiation theorem y is constant a.e. But since $y \in L^q(\mathbb{R}^n)$

the constant is 0, $y = 0$ a.e., and L vanishes for all $x \in L^p(\mathbb{R}^n)$. Hence by Problem 95(b), D is dense in $L^p(\mathbb{R}^n)$.

129. (a) Let $Q \subset \mathbb{R}^n$ denote the cube centered at the origin of side-length 1 and q the conjugate exponent to p. By Problem 5.11 the function $y(t) = |t|^{-n/q}(\ln(1/|t|))^{-2/q}\chi_Q(t) \in L^q(\mathbb{R}^n) \setminus \bigcup_{q<s} L^s(\mathbb{R}^n)$. Let $A = \{x$ on $\mathbb{R}^n : x$ is measurable, bounded, and $\int_{\mathbb{R}^n} y(t)x(t)\,dt = 0\}$. Now, for $1 < r < p$, let L be a bounded linear functional on $L^r(\mathbb{R}^n)$ that vanishes on A and $z \in L^s(\mathbb{R}^n)$, where s is the conjugate to r, such that $L(x) = \int_{\mathbb{R}^n} z(t)x(t)\,dt$ for all $x \in L^r(\mathbb{R}^n)$. Since $\chi_J \in A$ for all cubes $J \subset Q^c$, $\int_J z(t)\,dt = 0$ for such cubes and by the Lebesgue differentiation theorem $z = 0$ a.e. in Q^c. Next, $x(t) = (\int_{J_1} y(u)\,du)^{-1}\chi_{J_1}(t) - (\int_{J_2} y(u)\,du)^{-1}\chi_{J_2}(t) \in A$ for all cubes $J_1, J_2 \subset Q$, and, therefore, $L(x) = (\int_{J_1} y(u)\,du)^{-1}\int_{J_1} z(t)\,dt - (\int_{J_2} y(u)\,du)^{-1}\int_{J_2} z(t)\,dt = 0$ for those cubes. Consequently, the expression $(\int_J y(u)\,du)^{-1}\int_J z(t)\,dt$ is independent of the cubes $J \subset Q$ and $\int_J z(t)\,dt = c\int_J y(u)\,du$ for all those cubes J and some constant c. Hence by the Lebesgue differentiation theorem $z(t) = cy(t)$ a.e. in Q and since $y \notin L^s(Q)$, $c = 0$, and then $z(t) = 0$ in Q. Thus $z = 0$ a.e. and L is the zero functional. Consequently, by Problem 95(b), A is dense in $L^r(\mathbb{R}^n)$.

(b) Note that $y(t) = |t|^{-n/q}\chi_{Q^c}(t) \in \bigcap_{q<s} L^s(\mathbb{R}^n) \setminus L^q(\mathbb{R}^n)$ where Q is as in (a). Let $A = \{x$ on $\mathbb{R}^n : x$ is measurable, compactly supported, bounded, and $\int_{\mathbb{R}^n} y(t)x(t)\,dt = 0\}$. Now, for $1 < r < p$, let s be the conjugate to r, and L the linear functional on $L^r(\mathbb{R}^n)$ given by $L(x) = \int_{\mathbb{R}^n} y(t)x(t)\,dt$, $x \in L^r(\mathbb{R}^n)$. L is bounded and, therefore, since $A \subset K(L)$, A is not dense in $L^r(\mathbb{R}^n)$. On the other hand, as in (a) it readily follows that if L is a bounded linear functional on $L^p(\mathbb{R}^n)$ that vanishes on A and $L(x) = \int_{\mathbb{R}^n} z(t)x(t)\,dt$ for $x \in L^p(\mathbb{R}^n)$, then $z(t) = 0$ in Q and $z(t) = cy(t)$ in Q^c, and, therefore, $z = 0$ a.e. Thus L is the 0 functional and by Problem 95(b), A is dense in $L^p(\mathbb{R}^n)$.

131. For the sake of argument suppose that $x_0 \notin M$. Then $d = d(x_0, \overline{M}) > 0$ and by Problem 80 there exists a bounded linear functional L on X such that $\|L\| = 1$, $L(y) = 0$ for all $y \in \overline{M}$, and $L(x_0) = d$. Now, since $\{x_n\} \subset M$, $L(x_n) = 0$ for all n, and since $x_n \rightharpoonup x_0$ we have $L(x_n) \to L(x_0)$. Therefore $L(x_0) = 0$, which is not the case since $L(x_0) = d \neq 0$.

132. The statement is false.

134. For the sake of argument suppose that $v_n \rightharpoonup v$ in ℓ^∞; we claim that $v = (1, 1, \ldots)$. Let $L_k \in \ell^{\infty*}$ be given by $L_k(x) = x_k$ for all k. Then $\lim_n L_k(v_n) = L_k(v)$. Now, $L_k(v_n) = 1$ for all $n \geq k$, and, consequently, $\lim_n L_k(v_n) = 1$ for all k and $v = (1, 1, \ldots)$. However, for the functional L_∞ defined in Problem 111, $L_\infty(v_n) = 0$ for each n whereas $L_\infty(v) = 1$, which is not possible. Thus $\lim_n L_\infty(v_n) \neq L_\infty(v)$ and $\{v_n\}$ does not converge weakly to v in ℓ^∞.

135. The statement is true for $1 \leq p < \infty$. The statement is not true for $p = \infty$. Let $x^n = e_1 - e_n$ and $x = e_1$. Then $\|x^n\|_\infty = \|x\|_\infty = 1$ and by Problem 133, $x^n \rightharpoonup x$ but $\|x^n - x\|_\infty = \|e_n\|_\infty = 1$ for all n. And since $c_0^* = \ell^1$, it is not true for c_0 either.

136. Let $L \in X^*$; with $M = \sup_n \|x_n\|$, given $\varepsilon > 0$, there exists $\ell \in Y$ such that $\|L - \ell\| \leq \varepsilon/2M$. Therefore $|(L - \ell)(x_n)| \leq \|L - \ell\| \|x_n\| \leq (\varepsilon/2M)M = \varepsilon/2$ for all n. Pick N such that $|\ell(x_n)| \leq \varepsilon/2$ for all $n \geq N$; then $|L(x_n)| \leq |L(x_n) - \ell(x_n)| + |\ell(x_n)| \leq \varepsilon/2 + \varepsilon/2 = \varepsilon$ for all $n \geq N$.

137. (a) Let $L \in X^*$. Then $\{L(x_n)\}$ is a scalar Cauchy sequence and, hence, bounded, i.e., $|L(x_n)| \leq c_L < \infty$ for all n and some constant c_L. With J_X the natural map note that $|J_X(x)(L)| = |L(x)| \leq c_L$. Since X^* is a Banach space, by the uniform boundedness principle $\|J_X(x_n)\| \leq c < \infty$ for all n, and since $\|x_n\| = \|J_X(x_n)\|$, $\{\|x_n\|_X\}$ is bounded.

(b) By Hahn-Banach there exists $L \in X^*$ with $\|L\| = 1$ such that $|L(x)| = \|x\|$. Since $x_n \rightharpoonup x$, $L(x_n) \to L(x)$, and so $|L(x_n)| \to |L(x)|$; also, $|L(x_n)| \leq \|L\| \|x_n\| = \|x_n\|$. Hence $\|x\| = |L(x)| = \lim_n |L(x_n)| = \liminf_n |L(x_n)| \leq \liminf_n \|x_n\|$.

Since $\{\sin(nt)\}$ converges weakly to 0 in $L^2(0, 2\pi)$ it is possible to have strict inequality in the above equality. In particular, $x_n \rightharpoonup x$ does not imply $\|x_n\| \to \|x\|$.

139. By Problem 80, given $L \in B^*$, there exists $x^{**} \in B^{**}$ with $\|x^{**}\|_{B^{**}} = 1$ such that $x^{**}(L) = \|L\|_{B^*}$. Now, since B is reflexive let $x \in B$ be such that $x^{**} = J_B(x)$. Then $\|x\|_B = \|x^{**}\|_{B^{**}} = 1$ and $L(x) = J_B(x)(L) = x^{**}(L) = \|L\|_{B^*}$.

140. Yes.

141. (a) The statement is true. First, necessity. For the sake of argument suppose that $\{x_n\} \subset Y$ is such that $x_n \rightharpoonup x$ but $x \notin Y$. Since Y is closed, by Hahn-Banach there is a bounded linear functional L on X with $L(x) = 1$ that vanishes on Y. Therefore, since $L(x_n) = 0$ for all n, $L(x) \neq \lim_n L(x_n)$, which is not possible since $x_n \rightharpoonup x$. Conversely, let $\{x_n\} \subset Y$ and $\lim_n x_n = x \in X$. Then for every bounded linear functional L, $|L(x_n) - L(x)| = |L(x_n - x)| \leq \|L\| \|x - x_n\| \to 0$. Hence $x_n \rightharpoonup x$, which since Y is weakly closed gives that $x \in Y$.

(b) The statement is true. We prove that $B(0, 1)^c$ is weakly open. To see this let $x \in B(0, 1)^c$ and set $\varepsilon = \|x\| - 1 > 0$. By Problem 80 there exists $L \in X^*$ such that $\|L\| = 1$ and $L(x) = \|x\| = 1 + \varepsilon$. Then $O = \{y \in X : |L(y - x)| < \varepsilon\}$ is an open neighborhood of x in the weak topology, and we claim that $O \subset B^c(0, 1)$. Indeed, for any $y \in B(0, 1)$, we have

$|L(y)| \leq \|L\| \, \|y\| \leq 1$, and hence $|L(y - x)| \geq L(x) - |L(y)| \geq \varepsilon$, so $y \notin O$. Thus $B(0, 1)$ is weakly closed.

(c) The statement is false.

143. (a) That X is closed in $C(I)$ follows readily from the fact that convergence in $C(I)$ is uniform convergence. Suppose that $x \in C(I)$ vanishes at 0 and $\|x\|_\infty \leq 1$; by continuity there exists $\eta \in (0, 1]$ such that $|x(t)| \leq 1/2$ on $[0, \eta]$. Therefore $\int_0^1 |x(t)| \, dt = \int_0^\eta |x(t)| \, dt + \int_\eta^1 |x(t)| \, dt \leq \eta/2 + (1 - \eta) = 1 - \eta/2 < 1$, and, consequently, $x \notin X$.

(c) Let $x_n(t) = nt, n \geq 2$; then $x_n \in X$ and $\|x_n\|_\infty = n$. Hence X is not bounded and, therefore, not compact.

147. Let $y^* \in Y^*$; by Hahn-Banach y^* can be extended to $x^* \in X^*$, say, with no increase in norm. Now, X^* is separable and so given $\varepsilon > 0$, there is x_ε^* in a countable dense subset of X^* with $\|x^* - x_\varepsilon^*\|_{X^*} \leq \varepsilon$. Note that the restriction $x_\varepsilon^*|_Y$ of x_ε^* to Y satisfies $\|y^* - x_\varepsilon^*|_Y\|_{Y^*} \leq \|x^* - x_\varepsilon^*\|_{X^*} \leq \varepsilon$.

148. (a) Given $x \in X$, since $J_X(x) \in X^{**}$ and $L_n \rightharpoonup L$, $\lim_n J_X(x)(L_n) = J_X(x)(L)$. Moreover, since $J_X(x)(L_n) = L_n(x)$ for all n and $J_X(x)(L) = L(x)$, $\lim_n L_n(x) = L(x)$.

(b) Given $x^{**} \in X^{**}$, let $x \in X$ be such that $x^{**} = J_X(x)$. Then, since $x^{**}(L_n) = J_X(x)(L_n) = L_n(x)$ and $x^{**}(L) = J_X(x)(L) = L(x)$, $\lim_n x^{**}(L_n) = \lim_n L_n(x) = L(x) = x^{**}(L)$, and, consequently, $L_n \rightharpoonup L$ in X^*.

149. Let $\{x_n\} \subset X$ and $J_B(x_n) \to y \in B^{**}$. Then $\{J_B(x_n)\}$ is a Cauchy sequence in B^{**} and since J_B is an isometric isomorphism, $\{x_n\}$ is Cauchy in X. Since X is complete, $\lim_n x_n = x \in X$ exists and by the continuity of J_B, $\lim_n J_B(x_n) = J_B(x)$. By the uniqueness of limits $y = J_B(x) \in J_B(X)$, which is therefore closed in B^{**}.

150. (a) Let $x^{**} \in X^{**}$ and Λ the functional on B^* given by $\Lambda(L) = x^{**}(L|_X)$, $L \in B^*$; Λ is linear and continuous and since B is reflexive, $\Lambda = J_B(x)$ for some $x \in B$. Thus $x^{**}(L|_X) = \Lambda(L) = L(x)$ for all $L \in B^*$ and some $x \in B$; we claim that $x \in X$. For the sake of argument suppose that $x \notin X$; by Hahn-Banach there is a bounded linear functional L on B with $L|_X = 0$ and $L(x) \neq 0$ but this is impossible since then $0 \neq L(x) = x^{**}(L|_X) = x^{**}(0) = 0$. Therefore we have $x^{**}(\ell) = \ell(x)$, $\ell = L|_X, L \in B^*$. But by Hahn-Banach every bounded linear functional on X can be written as $L|_X$ with $L \in B^*$. Thus $x^{**}(\ell) = \ell(x)$ for all $\ell \in X^*$ where $x \in X$. This proves that $x^{**} = J_X(x)$ with $x \in X$ and X is reflexive.

(b) Recall that by Problem 63, $B = K(L) \oplus \mathrm{sp}\{x_0\}$ where L is a nontrivial bounded linear functional on B and $x_0 \notin K(L)$. Now, since $K(L)$ is a proper closed subspace of B, it is reflexive, and since reflexivity is preserved under isomorphisms, so is $\mathrm{sp}\{x_0\} \sim \mathbb{R}$. Finally, since the direct sum of reflexive spaces is reflexive, B is reflexive.

151. For the sake of argument suppose that B is reflexive. Since $B^* \backslash Y \neq \emptyset$, by Problem 80 there is a bounded linear functional $0 \neq x^{**}$ on B^* that vanishes on Y. Furthermore, since B is reflexive $x^{**} = J_B(x)$ for some $0 \neq x \in B$. In particular, this means that $x^{**}(L) = L(x)$ for all $L \in B^*$, and, consequently, $L(x) = 0$ for all $L \in Y$. Therefore, contrary to the properties of Y, x and 0 cannot be separated by a functional in Y.

152. First, $M = J_{B^*}(B^*)$ and $N = J_B(B)^{\perp}$ are closed subspaces of B^{***}. Suppose that $b^{***} \in M \cap N$. Then, since $b^{***} \in M$, $b^{***} = J_{B^*}(b^*)$ for some $b^* \in B^*$. Now, by the definition of J_{B^*}, $J_{B^*}(b^*)(x^{**}) = x^{**}(b^*)$ for all $x \in X^{**}$, and so, in particular, $J_{B^*}(b^*)(J_B(x)) = J_B(x)(b^*)$ for all $x \in X$. Also, since $b^{***} \in N$, $0 = b^{***}(J_B(x)) = J_{B^*}(b^*)(J_B(x)) = J_B(x)(b^*)$, which by the definition of the natural map is equal to $b^*(x)$. Therefore $b^*(x) = 0$ for all $x \in X$, which implies that $b^* = 0$, and so $b^{***} = J_{B^*}(0) = 0$ and $M \cap N = \{0\}$.

Next, given $x^{***} \in B^{***}$, let x^* be the functional on B given by $x^*(x) = x^{***}(J_B(x))$; clearly x^* is linear and bounded and thus in B^*. Write $x^{***} = J_{B^*}(x^*) + (x^{***} - J_{B^*}(x^*))$ where $J_{B^*}(x^*) \in J_{B^*}(B^*) = M$. We claim that $x^{***} - J_{B^*}(x^*) \in N$. To see this note that as above, $x^*(x) = J_{B^*}(x^*)(J_B(x)) = x^{***}(J_B(x))$, and, consequently, $(x^{***} - J_{B^*}(x^*))J_B(x) = 0$ for all $x \in B$. Hence $(x^{***} - J_{B^*}(x^*))|_{J_B(B)} = 0$ and $x^{***} - J_{B^*}(x^*) \in N$.

(b) B is reflexive iff $J_B(B)^{\perp} = \{0\}$. Therefore, since $B^{***} = J_{B^*}(B^*) \oplus J_B(B)^{\perp}$, B is reflexive iff B^* is reflexive.

153. Both statements are true.

154. (a) Given a bounded sequence $\{x_n\} \subset B$, let $X = \overline{\mathrm{sp}}\{x_n\}$; by considering linear combinations of the x_n with rational coefficients it follows that X is separable. And by Problem 150, X is reflexive. Now, since $X^{**} = X$ is separable, so is X^*; let $\{L_k\}$ be a dense subset of X^*. Since $\{x_n\}$ is bounded the scalar sequence $\{L_1(x_n)\}$ is bounded and, hence, a subsequence $\{x_{n_1}\}$ of $\{x_n\}$ has the property that $\lim_{n_1} L_1(x_{n_1}) = \eta_1$, say, exists. Continuing this way we obtain sequences $\{x_{n_k}\} \subset \{x_{n_{k-1}}\} \subset \cdots \subset \{x_n\}$ such that $\lim_{n_k} L_k(x_{n_k}) = \eta_k$, say, exists for all k. Now, for the diagonal sequence $\{x_{k_k}\}$ we have $L_n(x_{k_k}) \to \eta_n$ as $k \to \infty$ for any n. Since $\{x_{k_k}\}$ is bounded and $\{L_n\}$ is dense in X^* it readily follows that for any $L \in X^*$, $x_{k_k}(L) = L(x_{k_k}) \to x^{**}(L)$ where $x^{**} \in X^{**} = X$. Thus we conclude that there exists $x^{**} \in X$ that satisfies $L(x_{k_k}) \to L(x^{**})$ for all $L \in X^*$. Finally, for an arbitrary $L \in B^*$, let $L_0 = L|_X$ and note that $L_0 \in X^*$ and $L(x_{k_k}) = L_0(x_{k_k}) \to L_0(x^{**}) = L(x^{**})$. Since X is separable, $\{J(x_n)\}$ has a convergent subsequence $\{J(x_{n_k})\}$, say, which converges to $J(x_0)$, $x_0 \in X$.

But since X is reflexive, $x_{n_k} \rightharpoonup x_0$ in X and the conclusion follows since $B^* \subset X^*$.

Also, note that the closed unit ball of $C(I)$ is closed and bounded but $x_n(t) = t^n$ is a sequence of functions of norm 1 that does not admit a convergent subsequence; thus it is not compact.

155. If $x_0 \in X$ let $y_0 = x_0$. Otherwise $d(x_0, X) > 0$ and let $\{y_n\} \subset X$ be such that $\|x_0 - y_n\| \to d(x_0, X)$ and $\|x_0 - y_n\| \leq 1 + d(x_0, X)$ for all n. Thus $\|y_n\| = \|y_n - x_0 + x_0\| \leq \|y_n - x_0\| + \|x_0\| \leq 1 + d(x_0, X) + \|x_0\|$ and $\{y_n\}$ is bounded. Now, by Problem 154(a) a subsequence of $\{y_n\}$, which for simplicity we denote $\{y_n\}$ again, converges weakly to some $y_0 \in B$. Note that $y_0 \in X$ for otherwise $d(y_0, X) > 0$ and then there exists an $L \in B^*$ such that $\|L\| = 1$, $X \subset K(L)$, and $|L(y_0)| = d(y_0, X)$. But then $0 = L(y_n) \to L(y_0) \neq 0$, which is a contradiction. Finally, for any $L \in B^*$ with $\|L\| = 1$ we have $|L(x_0 - y_0)| = \lim_n |L(x_0 - y_n)| \leq \lim_n \|x_0 - y_n\| = d(x_0, X)$. Since L is arbitrary we get $\|x_0 - y_0\| \leq d(x_0, Y) \leq \|x_0 - y_0\|$, which gives $\|x_0 - y_0\| = d(x_0, X)$.

158. Let $x, y \in M$ and λ a scalar. Then $p(\lambda x + y) \leq |\lambda| p(x) + p(y) = 0$, and, therefore, $\lambda x + y \in M$, and M is a subspace. First, we verify that $v : X/M \to \mathbb{R}$ given by $v([x]) = p(x)$ is well-defined. Note that if $[x_1] = [x_2]$, then $x_1 - x_2 \in M$ and $p(x_1 - x_2) = 0$, and so by the triangle inequality $p(x_1) \leq p(x_2) + p(x_1 - x_2) = p(x_2)$; similarly, exchanging x_1 and x_2, $p(x_2) \leq p(x_1)$. Thus $p(x_1) = p(x_2)$, which implies that $v([x_1]) = v([x_2])$ and v is well-defined. Next, v is a norm. Clearly $v \geq 0$ and if $v([x]) = 0$, $p(x) = 0$, and $x \in M$. Therefore, since $0 = x + (-x)$, $0 \in x + M$, i.e., $[x]$ is the zero element of X/M. Now, let λ be a scalar and $[x] \in X/M$. Then $\lambda[x] = [\lambda x]$, and, consequently, $v(\lambda[x]) = p(\lambda x) = |\lambda| p(x) = |\lambda| v([x])$. Finally, since $v([x]+[y]) = v([x+y]) = p(x+y) \leq p(x)+p(y) = v([x])+v([y])$, v satisfies the triangle inequality.

159. (b) By translation it suffices to consider the case $x = 0$. Then, if $[y] \in \pi(B_X(0, r))$, let $[y] = [z]$ for $z \in B_X(0, r)$. Hence $\|[y]\|_{X/M} = \|[z]\|_{X/M} \leq \|z\|_X < r$ and $[y] \in B_{X/M}([0], r)$.

Next, if $[y] \in B_{X/M}([0], r)$ pick $m \in M$ such that $\|y - m\|_X < r$. Then $y - m \in B_X(0, r)$, and $[y] = [y - m] \in \pi(B_X(0, r))$.

(d) Let $U \subset X$ be open; note that $\pi^{-1}(\pi(U)) = U + M = \bigcup_{m \in M}(m+U)$ where each $m + U$, being the translate of the open set U, is itself open and, therefore, $\pi^{-1}(\pi(U))$ is open. By (c), $\pi(U)$ is open in X/M.

160. (a) First, necessity. Clearly Y is complete. As for X/Y, let $\sum_{n=1}^{\infty} \|[x_n]\|_{X/Y} < \infty$, and let $y_n \in Y$ be such that $\|x_n + y_n\|_X \leq 2 \|[x_n]\|_{X/Y}$ for each n. Then $\sum_{n=1}^{\infty} \|x_n + y_n\|_X < \infty$, $\{x_n + y_n\}$ is absolutely convergent in X, and since X is complete, there exists $x \in X$ such that

$x - \sum_{n=1}^{N}(x_n + y_n) = (x - \sum_{n=1}^{N} x_n) - \sum_{n=1}^{N} y_n \to 0$ in X. Therefore, since $\sum_{n=1}^{N} y_n \in Y$, it follows that

$$\left\| [x] - \left[\sum_{n=1}^{N} x_n\right] \right\|_{X/Y} \leq \left\| \left(x - \sum_{n=1}^{N} x_n\right) - \sum_{n=1}^{N} y_n \right\|_X \to 0,$$

and, consequently, $[\sum_{n=1}^{N} x_n] \to [x]$ in X/Y. Hence X/Y is complete .

Sufficiency next. Let $\sum_n \|x_n\|_X < \infty$. Then $\sum_n \| [x_n] \|_{X/Y} < \infty$ and since X/Y is complete there exists $[x] \in X/Y$ such that

$$\lim_N \left\| [x] - \sum_{n=1}^{N} [x_n] \right\|_{X/Y} = 0 .$$

Now, since $[x] - \sum_{n=1}^{N} [x_n] = [x - \sum_{n=1}^{N} x_n]$, for each N there is $y_N \in Y$ such that $\| y_N - (x - \sum_{n=1}^{N} x_n)\|_X \leq 2 \| [x] - \sum_{n=1}^{N}[x_n] \|_{X/Y}$. We claim that $\{y_N\}$ is Cauchy in X, and, hence, convergent in Y. Indeed, let $N < M$, and observe that $y_M - y_N = y_M \mp (x - \sum_{n=1}^{M} x_n) - y_N \pm (x - \sum_{n=1}^{N} x_n) = y_M - (x - \sum_{n=1}^{M} x_n) - y_N + (x - \sum_{n=1}^{N} x_n) - \sum_{n=N-1}^{M} x_n$, and, consequently,

$$\|y_M - y_N\|_X \leq 2 \left\| [x] - \sum_{n=1}^{M} [x_n] \right\|_{X/Y} + 2 \left\| [x] - \sum_{n=1}^{N} [x_n] \right\|_{X/Y} + \sum_{n=N+1}^{M} \|x_n\|_X .$$

Since $\sum_n \|x_n\|_X < \infty$ the expression in the right-hand side above goes to 0 as $N, M \to \infty$ and $\{y_N\} \subset Y$ is Cauchy and, hence, convergent to $y \in Y$, say. It is now clear that $\lim_N \|(x-y) - \sum_{n=1}^{N} x_n\|_X = 0$ and we have finished.

161. The proof proceeds by induction. The statement is true for $n = 1$. Assume then that $\dim(X/\bigcap_{k=1}^{n-1} X_k) \leq n - 1$ and consider X_n. There are two cases. First, if $X_n \supset \bigcap_{k=1}^{n-1} X_k$, then $\bigcap_{k=1}^{n-1} X_k = \bigcap_{k=1}^{n} X_k$ and $\dim(X \setminus \bigcap_{k=1}^{n} X_k) \leq n - 1 < n$. Next, suppose that $X_n \not\supseteq \bigcap_{k=1}^{n-1} X_k$ and let $x_0 \in \bigcap_{k=1}^{n-1} X_k \setminus X_n$. Since $\dim(X/\bigcap_{k=1}^{n-1} X_k) \leq n-1$ there exist $x_1, \ldots, x_{n-1}$ in X such that for any $y \in X$, $y = \sum_{k=1}^{n-1} a_k x_k + z'$, $z' \in \bigcap_{k=1}^{n-1} X_k$. Since $\dim(X/X_n) = 1$ and $x_0 \notin X_n$, we have $z' = a_0 x_0 + z$, $z \in X_n$. But $x_0 \in \bigcap_{i=1}^{n-1} X_i$, so we get $z = z' - a_0 x_0 \in \bigcap_{i=1}^{n-1} X_i$, and thus for any $y \in X$, $y = \sum_{i=1}^{n-1} a_i x_i + a_0 x_0 + z$, $z \in \bigcap_{i=1}^{n} X_i$. It then follows that $\dim(X/\bigcap_{i=1}^{n} X_i) \leq n$.

162. (a) $B/M = \mathbb{C}$.

164. Let $\{x_n\} \subset Y$ converge to $x \in X$; since a subsequence $\{x_{n_k}\}$ of $\{x_n\}$ converges to x a.e., $x(t) = 0$ a.e. in $[0, 1/2]$ and $x \in Y$, which is therefore closed. Now, $x = x \chi_{[0,1/2]} + x \chi_{[1/2,1]}$ and since $x \chi_{[1/2,1]} \in Y$, $\inf_{y \in Y} \|x + y\|_X = \|x \chi_{[0,1/2]}\|_X$, and $\| [x] \|_{X/Y} = \int_0^{1/2} |x(t)| \, dt$.

165. Since $x \in \ell^\infty$ can be written as $x = y + z$ where $y_m = x_m$ for $m \leq n$ and $y_m = 0$ otherwise, and since $y \in c_0$, $\| [x] \|_{\ell^\infty/c_0} = \inf_n \sup_{m \geq n} |x_m| = \limsup_n |x_n|$.

Now, since ℓ^∞ is not separable and c_0 has a Schauder basis and is therefore separable, by Problem 160(b), ℓ^∞/c_0 is not separable,

167. (a) Necessity first. Let $\operatorname{codim}(M) = n$ and suppose that $X/M = \operatorname{sp}\{[x_1], \ldots, [x_n]\}$. We claim that the x_k, $1 \leq k \leq n$, are linearly independent. This is clear since $\sum_{k=1}^n \lambda_k x_k = 0$ implies $\sum_{k=1}^n \lambda_k [x_k] = [0]$, which since the $[x_k]$ are linearly independent gives $\lambda_k = 0$, $1 \leq k \leq n$. Let $N = \operatorname{sp}\{x_1, \ldots, x_n\}$. Then for $x \in X$, $[x] = \sum_{k=1}^n \lambda_k [x_k] = [\sum_{k=1}^n \lambda_k x_k]$, and so $x - \sum_{k=1}^n \lambda_k x_k \in M$. Thus $x = m + \sum_{k=1}^n \lambda_k x_k \in M + N$. Finally, clearly $M \cap N = \{0\}$.

Sufficiency next. Let $N = \operatorname{sp}\{x_1, \ldots, x_n\}$ and let $\phi : N \to X/M$ be given by $\phi(x_k) = [x_k]$, i.e., ϕ is the restriction of the canonical mapping to N. Now, since $M \cap N = \{0\}$, ϕ is injective, and since $X = M + N$, ϕ is surjective. Hence by the inverse mapping theorem, ϕ has a continuous inverse, ϕ is a bijection, and, consequently, $\dim(X/M) = \operatorname{codim}(M) = n$.

(b) We may assume that $\varepsilon < 1$, and in that case $(1 - \varepsilon/2)^{-1} < 1 + \varepsilon$. Let $N = \operatorname{sp}\{x_1, \ldots, x_n\}$ be as in (a), $[x'_k] = [x_k]/\| [x_k] \|_{X/M}$, $1 \leq k \leq n$, and observe that since X/M is finite dimensional its closed unit ball $B_{X/M}(0, 1)$ is compact and so, given $\varepsilon/2$, we can complete $\{B_{X/M}([x'_k], \varepsilon/2)\}$, $1 \leq k \leq n$, to an $\varepsilon/2$-net $\{B_{X/M}([y_1], \varepsilon/2), \ldots, B_{X/M}([y_m], \varepsilon/2)\}$ of $B_{X/M}(0, 1)$. Let $N_1 = \operatorname{sp}\{y_1, \ldots, y_m\}$; then $X = M + N_1$, with M and N_1 not necessarily disjoint. Let ϕ denote as above the restriction of the canonical mapping to N_1. Then by Problem 9.170 with $R = (1 - \varepsilon/2)^{-1}$ there,

$$\phi(B_{N_1}(0, (1 - \varepsilon/2)^{-1})) \supset B_{X/M}(0, 1)$$

and since $(1 - \varepsilon/2)^{-1} < 1 + \varepsilon$, given $x \in X$, there exists $v \in B_{N_1}(0, 1 + \varepsilon)$ such that $[x/\|x\|] = [v]$. Hence for some $m \in M$, $x/\|x\| - v = m$, or $x = \|x\|v + \|x\|m = v_1 + m_1$ where $\|v_1\| = \|x\| \, \|v\| \leq (1 + \varepsilon)\|x\|$.

168. (b) Define $\phi : (B/X)^* \to X^\perp$ as follows: Given $\Lambda \in (B/X)^*$, let λ be the functional on B given by $\lambda(x) = \Lambda([x])$, $x \in B$; $\phi(\Lambda) = \lambda$ is a surjective linear isometry.

(c) By (b), $X^\perp$ is isometrically isomorphic to $(B/X)^*$, which is therefore finite dimensional. Hence B/X is finite dimensional and the conclusion follows from Problem 167(a).

(d) For $L \in B^*$, let $[L]$ denote its image in $B^*/X^\perp$ and let $\phi : B^*/X^\perp \to X^*$ be given by $\phi([L]) = L|_X$; we claim that ϕ is a well-defined isometric isomorphism. First, if $[L] = [L_1]$, $L - L_1 \in X^\perp$ and so $(L - L_1)(x) = 0$ for all $x \in X$, i.e., $L|_X = L_1|_X$; thus ϕ is well-defined and clearly linear. Now, if $\phi([L]) = 0$, then $L|_X = 0$, $L \in X^\perp$, $[L] = [0]$ and ϕ is injective. And, if

$\ell \in X^*$, by Hahn-Banach there is a norm preserving linear extension L of ℓ to B. Thus $\phi([L]) = L|_X = \ell$ and ϕ is surjective.

Next, the isometry part. If $L \in B^*$, given $\varepsilon > 0$, pick $L_1 \in X^\perp$ such that $\|L - L_1\|_{B^*} \leq \|[L]\|_{B^*/X^\perp} + \varepsilon$. Then for $x \in X$, $|L(x)| = |(L - L_1)(x)| \leq \|L - L_1\|_{B^*}\|x\|_X$, and so $\|\phi([L])\|_{X^*} \leq \|L - L_1\|_{B^*} \leq \|[L]\|_{B^*/X^\perp} + \varepsilon$, which, since ε is arbitrary, gives $\|\phi([L])\|_{X^*} \leq \|[L]\|_{B^*/X^\perp}$. On the other hand, if $\ell \in X^*$, by Hahn-Banach there is a norm preserving extension $\ell_1 \in B^*$ of ℓ. If $\ell = \phi([L])$, $L \in B^*$, it follows that $L|_X = \ell_1|_X$ and so $L - \ell_1 \in X^\perp$. Thus $\|[L]\|_{B^*/X^\perp} \leq \|L - (L - \ell_1)\|_{B^*} = \|\ell_1\|_{B^*} = \|\ell\|_{X^*} = \|\phi([L])\|_{X^*}$, which combined with the opposite inequality gives $\|[L]\|_{B^*/X^\perp} = \|\phi([L])\|_{X^*}$ and we have finished.

169. Since by Problem 168(a), $X^\perp$ is a closed subspace of B^*, and by Problem 152(b), B^* is reflexive, by Problem 150, $X^\perp$ is reflexive. Now, by Problem 168(b), $(B/X)^* \sim X^\perp$ is reflexive, and by Problem 152(b) so is B/X.

170. With $\pi : B \to B/X$ the quotient map, define $\phi : (B/X)^* \to X^\perp$ as follows: For $\lambda \in (B/X)^*$, let $\phi(\lambda) = \lambda \circ \pi$, i.e., $\phi(\lambda)(x) = \lambda([x])$ for all $x \in B$. Then $\phi(\lambda) \in B^*$ and since $\pi(x) = [0]$ for all $x \in X$ it follows that $\phi(\lambda) \in X^\perp$. For $\lambda \in (B/X)^*$ we have

$$\|\phi(\lambda)\| = \sup_{x \in B_B(0,1)} |\lambda(\pi(x))| = \sup_{[x] \in B_{B/X}(0,1)} |\lambda([x])| = \|\lambda\|$$

where the second inequality follows on the one hand from the fact that $\|\pi(x)\| \leq \|x\|$ and on the other hand from the fact that for any $[x] \in B_{B/X}(0,1)$ there is a sequence $\{y_n\} \subset X$ such that $\lim_n \|x + y_n\| \leq 1$. Thus ϕ is an isometric embedding.

If $L \in X^\perp \subset B^*$ we define $\lambda \in (B/X)^*$ by $\lambda([x]) = L(x)$. First note that, since $L(x + y_1) = L(x + y_2)$ for all $y_1, y_2 \in X$, λ is well-defined. Moreover, since L is linear, λ is linear and $|\lambda([x])| = |L(x)|$ for all $x \in B$, and, therefore, $\|\lambda\| = \|L\|$. Finally, since $(\pi(\lambda), x) = (\lambda, \pi(x)) = L(x)$ for all $x \in B$, it follows that $\phi(\lambda) = L$ and ϕ is surjective.

171. (a) First necessity. By linear algebra X is finite dimensional iff X^* is finite dimensional and in that case $\dim(X) = \dim(X^*)$. Now, by Problem 168, X^* is finite dimensional iff $X^\perp$ has finite codimension and then $\dim(X^*) = \dim(B^*/X^\perp) = \operatorname{codim}(X^\perp)$. All the steps are reversible and sufficiency holds as well.

(b) By Problem 170, $X^\perp$ is finite dimensional iff $(B/X)^*$ is finite dimensional, which, in turn, is equivalent to (B/X) being finite dimensional, i.e., X being of finite codimension in B. As above, $\dim(X^\perp) = \operatorname{codim}(X)$.

173. Y^{**} can be identified with the subspace $Y^{\perp\perp}$ of X^{**}. More precisely, with J_Y denoting the restriction of J_X to Y, there is an isometric isomorphism $U : Y^{**} \to Y^{\perp\perp}$ with the property that $U \circ J_Y = J_X|_Y$. Indeed, let $S : Y \to X$ be the natural embedding; then $S^{**} : Y^{**} \to Y^{\perp\perp} \subset X^{**}$. Moreover $S^{**} \circ J_Y = J_X \circ S = J_X|_Y$, as is shown by the following ("algebraic") argument: for $z \in Y$ and $L \in X^*$, $(S^{**} \circ J_Y(z))(L) = (J_Y(z))(S^*(L)) = (S^*(L))(z) = L(S(z)) = (J_X \circ S(z))(L)$. Moreover, by standard arguments, $\|S^{**}\| = \|S\| = 1$. It remains to show that S^{**} is an isometry onto $Y^{\perp\perp}$, i.e. that $\|S^{**}\theta\| \geq \|\theta\|$ for all $\theta \in Y^{**}$.

If $\theta \in Y^{**}$ and $\|\theta\| = 1$, then, for all $r < 1$, there exists $h \in Y^*$ with $\|h\| = 1$ and $\theta(h) > r$. By Hahn-Banach we can extend h to $f \in X^*$ satisfying $\|f\| = \|h\|$. Expressing the notion of "extension" in other language, we have $f \circ S = h$, or $S^* f = h$. We thus have $\|S^{**}\theta\| \geq (S^{**}\theta)(f) = \theta(S^* f) = \theta(h) \geq r$. Thus $\|S^{**}\theta\| = 1$ and S^{**} is isometric.

Now suppose that $\xi \in Y^{\perp\perp} \subset X^{**}$. We want to find $\theta \in Y^{**}$ with $S^{**}(\theta) = \xi$. We notice that if $f_1, f_2 \in X^*$ are such that $f_1|_Y = f_2|_Y$, then $f_1 - f_2 \in Y^\perp$ and so $\xi(f_1) = \xi(f_2)$ because $\xi \in Y^{\perp\perp}$. For $h \in Y^*$ we can unambiguously define $\theta(h) = \xi(f)$, where f is any element of X^* extending h. It follows from this definition that $\theta(f|_Z) = \xi(f)$, i.e., $S^{**}(\theta) = \xi$.

174. The duals of $C(I)$ and of $C(I) \oplus c_0$ are isometrically isomorphic.

176. Fix a countable dense subset $\{x_n\}$ of the closed unit ball of B and let $T : \ell^1 \to B$ be the linear mapping given by $T(\lambda) = \sum_n \lambda_n x_n$, $\lambda \in \ell^1$; since $T(e_n) = x_n$ and $\|T(\lambda)\| \leq \sum_n |\lambda_n| \|x_n\| \leq \|\lambda\|_{\ell^1}$, T is bounded with norm 1. We claim that T is onto. First observe that, given $x \in B$ with $\|x\| \leq 1$, there is a subsequence $\{x_{n_k}\}$ of $\{x_n\}$ such that $\|x - \sum_{j=0}^{k} 2^{-j} x_{n_j}\| \leq 2^{-(k+1)}$ for all $k \geq 0$. Indeed, when $k = 0$, by the density assumption there is x_{n_0} such that $\|x - x_{n_0}\| \leq 2^{-1}$. Next, having chosen $x_{n_0}, \ldots, x_{n_{k-1}}$ such that $\|x - \sum_{j=0}^{k-1} 2^{-j} x_{n_j}\| \leq 2^{-k}$, let $y = x - \sum_{j=0}^{k-1} 2^{-j} x_{n_j}$. Then $\|2^k y\| \leq 1$ and again by density there exists $x_{n_k} \neq x_{n_j}$ for $0 \leq j \leq k - 1$, such that $\|2^k y - x_{n_k}\| \leq 2^{-1}$ or, equivalently, $\|x - \sum_{j=0}^{k} 2^{-j} x_{n_j}\| \leq 2^{-(k+1)}$. Now, $x = \sum_{j=0}^{\infty} 2^{-j} x_{n_j}$, and, therefore, if λ is the sequence defined by $\lambda_n = 2^{-j}$ if $n = n_j$ and $\lambda_n = 0$ otherwise, then $\lambda \in \ell^1$ and $T(\lambda) = \sum_{j=0}^{\infty} 2^{-j} x_{n_j} = x$. Next, if $x \neq 0$ and $\|x\| \neq 1$, let $x' = x/\|x\|$, $\|x'\| = 1$, and let $\lambda' \in \ell^1$ be such that $T(\lambda') = x'$. Then by the linearity of T, if $\lambda = \|x\|\lambda' \in \ell^1$, $T(\lambda) = \|x\|T(\lambda') = \|x\|x' = x$. Finally, if $x = 0$, then $T(0) = 0$. Hence, T is surjective.

Now, let $\widetilde{T} : \ell^1/K(T) \to B$ be the mapping defined in Problem 9.21; $\widetilde{T}$ is a continuous bijection of norm 1 and so, by the open mapping theorem, $\widetilde{T}$ is an isomorphism.

It remains to prove that $\widetilde{T}$ is an isometry. First note that if $\|x\| = 1$ the above argument with 2 replaced by $r > 1$ gives that there exists a subsequence $\{x^r_{n_j}\}$ such that $\|x - \sum_{j=0}^{k} r^{-j} x^r_{n_j}\| \leq r^{-(k+1)}$, and so, if λ^r is the sequence defined by $\lambda^r_n = r^{-j}$ if $n = n_j$ and $\lambda^r_n = 0$ otherwise, then $\lambda^r \in \ell^1$ and $T(\lambda^r) = x$. Then, with λ, λ^r as defined above, $\lambda - \lambda^r \in K(T)$ and, consequently, $\| [\lambda] \|_{\ell^1/K(T)} \leq \|\lambda - (\lambda - \lambda^r)\|_{\ell^1} = \|\lambda^r\|_{\ell^1} \leq \sum_{j=0}^{\infty} r^{-(j+1)} = r/(r-1)$. Therefore, letting $r \to \infty$, $\| [\lambda] \|_{\ell^1/K(T)} \leq 1$.

Finally, given $\lambda \in \ell^1$, let $\lambda' = \lambda/\|T(\lambda)\|$ and $x' = T(\lambda')$. Since $\|x'\| = 1$, applying the above to λ gives $\| [\lambda] \|_{\ell^1/M} \leq \|T(\lambda)\| = \|\widetilde{T}([\lambda])\|$. And since $\|\widetilde{T}([\lambda])\| \leq \|[\lambda]\|$, we have equality.

Chapter 19

Normed Linear Spaces. Linear Operators

Solutions

2. Let $\{x_n\}$ be a countable subset of a (necessarily infinite) Hamel basis H for X, fix $0 \neq y \in Y$, and let $T : H \to Y$ be given by $T(x_n) = n\|x_n\|_X\, y$ and $T(x_\alpha) = 0$ for $x_\alpha \in H \setminus \{x_n\}$. Given $x \in X$, $x = \lambda_1 x_{\alpha_1} + \cdots + \lambda_n x_{\alpha_n}$ is a finite linear combination of elements in H and $T(x) = \lambda_1 T(x_{\alpha_1}) + \cdots + \lambda_n T(x_{\alpha_n})$ defines an unbounded linear mapping from X into Y. Indeed, given $c > 0$, let $n > c/\|y\|_Y$. Then $\|T(x_n)\|_Y = n\|x_n\|_X\,\|y\|_Y > c\|x_n\|_X$ and, therefore, $\|T\| > c$.

4. (a) Let $\eta = \max_{1 \leq i \leq m, 1 \leq j \leq n} |a_{ij}|$. Then it readily follows that $\|T(x)\|_\infty \leq \max_{1 \leq i \leq m} \sum_{j=1}^{n} |a_{ij}|\,|x_j| \leq \eta\,\|x\|_1$ and $\|T\| \leq \eta$. If $\eta = 0$ all the entries of T are 0 and $T = 0$. Otherwise, if $\eta > 0$, pick $1 \leq I \leq m, 1 \leq J \leq n$, such that $\eta = |a_{IJ}|$, and $x \in \mathbb{R}^n$ with $x_j = \operatorname{sgn}(a_{IJ})$ if $j = J$ and $x_j = 0$ otherwise. Then $\|x\|_1 = 1$ and $\|T(x)\|_\infty = \|T\| = \eta$.

(b) Let $\eta = \sum_{i=1}^{m} \sum_{j=1}^{n} |a_{ij}|$. Then for $x \in \mathbb{R}^n$ it follows that $\|T(x)\|_1 = \sum_{i=1}^{m} |\sum_{j=1}^{n} a_{ij} x_j| \leq \eta\,\|x\|_\infty$, and so $\|T\| \leq \eta$. That equality is possible if all the entries of A are nonnegative is seen by picking $x = (1, \ldots, 1)$, $\|x\|_{\ell^\infty} = 1$.

(c) and (d) The reader should consider why the answers for (c) and (d) are the same.

5. (b) Necessity first. Since $c_0 \subset \ell^\infty$, the statement concerning $\|T\|$ follows from (a). Also, if $T(x) \in c_0$ whenever $x \in c_0$, $\lim_m |\sum_n a_{mn} x_n| = 0$, and letting $x = e_n \in c_0$ it follows that $\lim_m |a_{mn}| = 0$ for all n.

As for sufficiency, by (a) it remains to prove that $R(T) \subset c_0$. Let $x \in c_0$. Given $\varepsilon > 0$, let N be such that $|x_n| < \varepsilon/(2\eta + 1)$ for $n \geq N$.

Since $\lim_m a_{mn} = 0$, there exists M such that $|a_{mn}| < \varepsilon/(2N)$ for each $m > M$ and $n \le N$. Then for $m > M$, $\sum_n |a_{mn}| |x_n| = \sum_{n=1}^{N} |a_{mn}| |x_n| + \sum_{n=N+1}^{\infty} |a_{mn}| |x_n| < \varepsilon(2N)^{-1} \sum_{n=1}^{N} 1 + \varepsilon(2\eta + 1)^{-1} \sum_{n=N+1}^{\infty} |a_{mn}| \le \varepsilon/2 + \varepsilon\eta(2\eta + 1)^{-1} < \varepsilon$. Thus, given $\varepsilon > 0$, there exists an integer M such that $\sum_n |a_{mn}| |x_n| < \varepsilon$ for $m > M$.

6. A direct proof works. Alternatively, we may use the infinite-dimensional version of Problem 4(b). Let

$$A = \begin{bmatrix} 1/2 & 0 & 0 & \cdots & \cdots & \cdots \\ 1/2^2 & 1/2^2 & 0 & \cdots & \cdots & \cdots \\ \vdots & \vdots & \ddots & \ddots & & \\ 1/2^n & \cdots & \cdots & 1/2^n & 0 & \cdots \\ \vdots & \vdots & \vdots & \vdots & \vdots & \vdots \end{bmatrix}.$$

Then $T(x) = Ax$ and the norm of T is obtained by summing the entries of A. Thus summing first over the columns, $\|T\| = \sum_{k=0}^{\infty} 2^{-k} = 2$.

7. Note that when $p = \infty$, T is bounded with $\|T\| = 1$ and injective, but since $y = (1, 1, \dots) \in l^\infty \setminus R(T)$, not surjective. Also T^{-1}, which is defined on $R(T) \subset l^\infty$, since $\|T^{-1}(e_n)\|_\infty = n$, is not bounded.

Now, when $1 \le p < \infty$, $y = (1, 1/2, 1/3, \dots) \in \overline{R(T)} \setminus R(T)$. To see this consider the ℓ^p sequences x^n with terms $x_k^n = 1$ for $1 \le k \le n$ and $x_k^n = 0$ otherwise, $n = 1, 2, \dots$. Then $y^n = T(x^n) \to y$ in ℓ^p, and so $y \in \overline{R(T)}$. On the other hand, if $y = T(x)$, then $x = (1, 1, \dots)$, which is not in ℓ^p.

8. (c) T is invertible: If $y \in X$ and

$$x_n = \begin{cases} (1/(1 + \lambda_1))y_1, & n = 1, \\ (-\lambda_n/(1 + \lambda_1))y_1 + y_n, & n \ge 2, \end{cases}$$

then $x \in X$ and $T(x) = y$; thus $R(T) = X$. And, $\sup_{x \ne 0} \|x\|/\|T(x)\| = \max\{|1 + \lambda_1|^{-1}(1 + \sum_{n=2}^{\infty} |\lambda_n|), 1\}$.

10. (a) The norm of T is achieved iff the ℓ_∞ norm of λ is achieved, i.e., there exists a nonzero $x \in \ell^p$ such that $\|T(x)\|_p = \|T\| \|x\|_p$ iff there exists a positive integer k such that $|\lambda_k| = \|\lambda\|_\infty$.

11. By Problem 5.46, $\|T(x)\|_r \le \|\lambda\|_s \|x\|_p$, and T is bounded with $\|T\| \le \|\lambda\|_s$. If T is bounded, since $T(x) \in \ell^r$ whenever $x \in \ell^p$, by Problem 5.47(b), $\lambda \in \ell^s$, $\|\lambda\|_s \le \|T\|$ holds, and, consequently, $\|T\| = \|\lambda\|_s$.

Now, T is 1-1 iff $\lambda_n \ne 0$ for all n. Next, $R(T)$ is dense in ℓ^r. Indeed, given $y \in \ell^r$, consider the truncations $y^n = (y_1, \dots, y_n, 0, \dots)$ of y. Then $\|y - y^n\|_r \to 0$ and if $x^n = (y_1/\lambda_1, \dots, y_n/\lambda_n, 0, 0, \dots)$, $x^n \in \ell^p$, $T(x^n) = y^n$, and $\|T(x^n) - y\|_r \to 0$. Hence $R(T)$ is dense in ℓ^r and if $R(T)$ is closed, then T is onto ℓ^r and T is a linear homeomorphism. Then T^{-1} is given by $T^{-1}(x) = y$ where $y_n = x_n/\lambda_n$, and if $T^{-1} : \ell^r \to \ell^p$ is bounded, since $\ell^p \hookrightarrow \ell^r$, T^{-1} maps ℓ^r continuously into ℓ^r and by Problem 10(a), $\|T^{-1}\| = 1/\inf_n |\lambda_n| < \infty$, which is impossible since $\lambda \in \ell^s$.

12. (a) Since T is defined on ℓ_0^p, it is densely defined if $p < \infty$. Note then that $D(T) = \ell^p$ iff $\lambda \in \ell^\infty$. Sufficiency first. If $\lambda \in \ell^\infty$, then $\sum_n |\lambda_n x_n|^r \leq \|\lambda\|_\infty^r \|x\|_p^r < \infty$, and $D(T) = \ell^p$. On the other hand, if $\lambda \notin \ell^\infty$, let $\{\lambda_{n_k}\}$ be a subsequence of λ such that $|\lambda_{n_k}| \geq 3^k$ for all k and observe that for $x = \sum_k 2^{-k} e_{n_k}$, $\|x\|_p^p = \sum_k 2^{-kp} < \infty$, yet $\|T(x)\|_r^r \geq \sum_k (3/2)^{kr} = \infty$ if $r < \infty$, and $\|T(x)\|_\infty \geq \sup_k (3/2)^k = \infty$ if $r = \infty$. Hence $x \notin D(T)$ and so $D(T) \neq \ell^p$.

Lastly, if $p = r = \infty$, $D(T)$ is dense in ℓ^p iff $D(T) = \ell^p$, which happens iff $\lambda \in \ell^\infty$ (done in (b)) below.

(b) If $\lambda_n = 0$ for some n, $\|e_n - y\|_r \geq 1$ for all $y \in R(T)$ and $R(T)$ is not dense in ℓ^r. On the other hand, if $\lambda_n \neq 0$ for all n, $R(T)$ contains the subspace of all finitely nonzero sequences, which is dense in ℓ^r if $r < \infty$. Hence if $1 \leq p \leq r < \infty$ and $\lambda_n \neq 0$ for all n, $R(T)$ is dense in ℓ^r.

Now, if $p < r = \infty$ and $y = T(x)$, then $\|x\|_p^p = \sum_{n=1}^\infty |\lambda_n|^{-p} |y_n|^p \leq \|y\|_\infty^p \sum_{n=1}^\infty |\lambda_n|^{-p}$. Hence, if $\sum_n |\lambda_n|^{-p} < \infty$, then $R(T) = \ell^\infty$. On the other hand, if $\sum_n |\lambda_n|^{-p} = \infty$ and $\sum_n |\lambda_n|^{-p} |y_n|^P < \infty$, then $\liminf_n |y_n| = 0$, which implies that if $e = (1, 1, \ldots)$, $\|e - y\|_\infty \geq 1$, and so $R(T)$ is not dense in ℓ^∞.

Finally, if $p = r = \infty$, as above it follows that $R(T)$ is dense in ℓ^∞ iff $\inf_n |\lambda_n| > 0$, in which case $R(T) = \ell^\infty$.

Now, if T is onto we must have $\lambda_n \neq 0$ for all n. If $r < \infty$, let $t = r/p$ and $s = r/(r - p)$ be conjugate exponents. Given $y \in \ell^r$, let x be given by $x_n = \lambda_n^{-1} y_n$ for all n. Then by Hölder's inequality $\sum_n |x_n|^p = \sum_n |\lambda_n|^{-p} |y_n|^p \leq (\sum_n |\lambda_n|^{-ps})^{1/s} (\sum_n |y_n|^r)^{p/r}$, and, therefore, a sufficient condition for T to be onto is that $\sum_n |\lambda_n|^{-ps} < \infty$. This condition is also necessary. For the sake of argument suppose that $\sum_n |\lambda_n|^{-ps} = \infty$ and let $n_\ell \to \infty$ be chosen so that $n_1 = 1$ and $w_\ell = (\sum_{n_\ell \leq k < n_{\ell+1}} |\lambda_k|^{-ps})^{1/r} \geq 2^{ps\ell/r}$ for all $\ell \geq 1$. Clearly, if y^ℓ is given by

$$y_k^\ell = \begin{cases} |\lambda_k|^{-sp/r}/w_\ell, & n_\ell \leq k < n_{\ell+1}, \\ 0, & \text{otherwise}, \end{cases}$$

$\|y^\ell\|_r = 1$ for all ℓ. Put $y = \sum_\ell 2^{-\ell} y^\ell$ and note that $y_k = 2^{-\ell} y_k^\ell$ for $n_\ell \leq k < n_{\ell+1}$ for all ℓ, and $\|y\|_r \leq \sum_\ell 2^{-\ell} = 1$. Observe that if x has terms $x_n = \lambda_n^{-1} y_n$ for all $n \geq 1$, then $T(x) = y$. Moreover, since

$$x_k = \begin{cases} |\lambda_k|^{-sp/r}/2^\ell w_\ell \lambda_k, & n_\ell \leq k < n_{\ell+1}, \\ 0, & \text{otherwise}, \end{cases}$$

and $p(1 + sp/r) = pr/(r - p) = sp$, a simple computation gives that $\|x\|_p^p = \sum_{\ell=1}^\infty 2^{-\ell p} w_\ell^{-p} \sum_{n_\ell \le k < n_{\ell+1}} |\lambda_k|^{-ps} = \sum_{\ell=1}^\infty 2^{-p\ell} w_\ell^{r-p}$, which since $w_\ell^{r-p} \ge 2^{p\ell}$, implies $\|x\|_p = \infty$. This is not the case since T is onto.

Next, if $p < r = \infty$, we claim that $R(T) = \ell^\infty$ iff $\sum_n |\lambda_n|^{-p} < \infty$. If $y = T(x)$, then $\|x\|_p^p = \sum_n |\lambda_n|^{-p}|y_n|^p \le (\sum_n |\lambda_n|^{-p})\|y\|_\infty^p$. Hence $R(T) = \ell^\infty$ and the condition is sufficient. On the other hand, if $\sum_n |\lambda_n|^{-p} = \infty$ and $\sum_n |\lambda_n|^{-p}|y_n|^p < \infty$, then $\liminf_n |y_n| = 0$, which implies that with $e = (1, 1, \ldots)$, $\|e - y\|_\infty \ge 1$, and $R(T)$ is not dense in ℓ^∞.

And, if $p = r = \infty$, as pointed out above $R(T)$ is dense in ℓ^∞ iff $\inf_n |\lambda_n| > 0$, and then $R(T) = \ell^\infty$.

Next, if $R(T) = \ell^r$, by (b) it follows that $\lambda_n \ne 0$ for all n. Let $y \in \ell^r$ and $x \in \ell^p$ be such that $y = T(x)$. If $p = r$ it follows as in (a) that the condition is $\inf_n |\lambda_n| > 0$. Otherwise, if $p < r < \infty$, the answer is given in Problem 5.47, namely, $\sum_{n=1}^\infty |\lambda_n|^{-ps} < \infty$.

Finally if $p < r = \infty$, by Problem 5.47, $R(T) = \ell^\infty$ iff $\sum_j |\lambda_j|^{-p} < \infty$.

(c) From $\|T\| \ge \|T(e_n)\| = |\lambda_n|$, for all n, it follows that $\|T\| \ge \|\lambda\|_\infty$. Also, as in (a), $\|T(x)\|_r \le \|\lambda\|_\infty\|x\|_p$, and so $\|T\| \le \|\lambda\|_\infty$. Hence $\|T\| = \|\lambda\|_\infty$ for all $1 \le p \le r < \infty$.

(d) Since $K(T) = \{x \in D(T) : x_n = 0 \text{ whenever } \lambda_n \ne 0\}$, T is invertible for all p, r, such that $\lambda_n \ne 0$ for all n. If $p = r$, by (c), $\|T^{-1}\| = \sup_n |\lambda_n|^{-1} = 1/\inf |\lambda_n|$. If $p < r$ we put $t = r/p, s = t/(t - 1) = r/(r - p)$. Assume first that $\sum |\lambda_j|^{-ps} < \infty$. For any $y \in \ell^r$, by Hölder's inequality it follows that $\|T^{-1}(y)\|_p^p = \sum_n |\lambda_n|^{-p}|y_n|^p \le (\sum_n |\lambda_n|^{-ps})^{1/s}(\sum_n |y_n|^{pt})^{1/t} = (\sum_n |\lambda_n|^{-ps})^{1/s}\|y\|_r^p$. Hence $\|T^{-1}\|^{ps} \le \sum_n |\lambda_n|^{-ps}$. Next, let $y = \{\eta_n\}$ where $\eta_n = |\lambda_n|^{-s/t}$ for all n. Then $\|y\|^r = \sum_n |\eta_n|^r = \sum_n |\lambda_n|^{-ps} < \infty$, and $\|T^{-1}(y)\|_p^p = \sum_n |\lambda_n|^{-p}|\eta_n|^p = \sum_n |\lambda_n|^{-p}|\lambda_n|^{-ps/t} = \sum_n |\lambda_n|^{-p(1+s/t)} = \sum_n |\lambda_n|^{-ps} = \|y\|_r^r$, which implies that $\|T^{-1}\|/\|y\|_r^r = \|y\|_r^{r/p-1}$, and, consequently, $\|T^{-1}\|^{ps} \ge \|y\|_r^r = \sum_n |\lambda_n|^{-ps}$. Thus $\|T^{-1}\| = (\sum_n |\lambda_n|^{-ps})^{1/ps}$.

Assume now that $\sum_n |\lambda_n|^{-ps} = \infty$. For the y_k with norm 1 we found in (b) we have $\|T^{-1}(y_k)\|^p = w_k^{-p} \sum_{n_k \le j < n_{k+1}} |\lambda_j|^{-p}|\lambda_j|^{-ps/t} = w_k^{-p} \sum |\lambda_j|^{-ps} = w_k^{r-p} \to \infty$. Thus the statement is true for all p and r with $p < r$.

(e) If $p = r$, from (c), it follows that T and T^{-1} are bounded iff, for some $c > 1$, $c^{-1} \le |\lambda_n| \le c$ for all n. In case $p < r$ it is impossible for both T and T^{-1} to be defined and bounded. From (c) and (d) we see that T is bounded if all $\lambda_n \ne 0$ and $|\lambda_n| \le c$ for some c, and T^{-1} is bounded if all $\lambda_n \ne 0$ and $\sum_n |\lambda_n j|^{-ps} < \infty$ ($s = 1$ if $r = \infty$), respectively. Now, if $|\lambda_n| \le c$, then $\sum_n |\lambda_n|^{-ps} = \infty$, and if $\sum_n |\lambda_n|^{-ps} < \infty$, then $|\lambda_n| \to \infty$.

17. Let $\{x_n\} \subset X$ be such that $M = \sum_n \|x_n\|_X < \infty$. Then, since T is continuous, $\sum_n \|T(x_n)\|_B \leq M\|T\| < \infty$, and since B is complete there exists $y \in B$, say, such that $\lim_N \sum_{n=1}^N T(x_n) = y$ in B. Now, the sequence of partial sums $\{\sum_{n=1}^N T(x_n)\}$ is contained in $T(M\overline{B}_X)$, which is closed since $T(\overline{B}_X)$ is closed, and, therefore, $y \in T(M\overline{B}_X)$. Hence there exists $x \in M\overline{B}_X \subset X$ such that $y = T(x)$; we claim that $\lim_N \sum_{n=1}^N x_n = x$ in X. To see this, given $\varepsilon > 0$, let N, k be large enough so that $\|\sum_{n=N}^k x_n\|_X < \varepsilon$. Then $\sum_{n=N}^k x_n \in \varepsilon\overline{B}_X$ and $T(\sum_{n=1}^N x_n) - T(\sum_{n=1}^k x_n) \to T(x) - T(\sum_{n=1}^k x_n)$ as $N \to \infty$; thus, as before $T(x) - T(\sum_{n=1}^k x_n) \in T(\varepsilon\overline{B}_X)$ and there is $z_k \in \varepsilon\overline{B}_X$ such that $T(x - \sum_{n=1}^k x_n) = T(z_k)$ for all large k. Therefore, since T is injective, $x - \sum_{n=1}^k x_n = z_k \in \varepsilon\overline{B}_X$ and $\|x - \sum_{n=1}^k x_n\|_X = \|z_k\|_X \leq \varepsilon$, k large enough. In other words, $\lim_k \sum_{n=1}^k x_n = x$ in X.

19. (b) Since all norms on X are equivalent, $\|x\|_X \sim \|x\|_1 = \sum_{k=1}^n |\lambda_k|$ where $x = \sum_{k=1}^n \lambda_k x_k \in \mathrm{sp}\{x_1, \ldots, x_n\}$. Then

$$\|T(x)\|_Y \leq \max_{1 \leq k \leq n} \|T(x_k)\|_Y \|x\|_1 \leq c\|x\|_X$$

and T is bounded. Moreover, the expression $\|x\|_X + \|T(x)\|_Y$ is a norm on X which is equivalent to $\|x\|_X$. Now let $\{x_n\} \subset X$ so that $\|T(x_n)\|_Y / \|x_n\|_X \to \|T\|$ and observe that since $S = \{x \in X : \|x\|_X = 1\}$ is compact, a subsequence $\{x_{n_k}/\|x_{n_k}\|_X\}$ of $\{x_n/\|x_n\|_X\}$ converges to $x \in S$, say. Then $\|T(x)\|_Y = \lim_{n_k} \|T(x_{n_k})\|_Y / \|x_{n_k}\|_X = \|T\|$ and T assumes its norm.

20. Note that the closed graph theorem does not apply since the spaces involved are not necessarily complete.

21. Since $K(T)$ is a closed subspace of X, $X/K(T)$, equipped with the norm defined in Problem 8.159, is a normed linear space. Now, if $[x] = [y]$, then $x - y \in K(T)$, and so $T(x - y) = 0$, and, therefore, $\widetilde{T}([x]) = \widetilde{T}([y])$ and $\widetilde{T}$ is well-defined. Also, since $\widetilde{T}([x] + \lambda[y]) = \widetilde{T}([x + \lambda y]) = T(x + \lambda y) = T(x) + \lambda T(y) = \widetilde{T}([x]) + \lambda\widetilde{T}([y])$, $\widetilde{T}$ is linear.

Next, note that if $y \in [x]$, $\|\widetilde{T}([x])\| = \|T(y)\| \leq \|T\| \|y\|$, and since y is arbitrary, $\|\widetilde{T}([x])\| \leq \|T\| \|[x]\|$ and $\|\widetilde{T}\| \leq \|T\|$. Conversely, let $x \in X$ with $\|x\| \leq 1$. Then $\|[x]\| \leq 1$ and $\|T(x)\| = \|\widetilde{T}([x])\| \leq \|\widetilde{T}\|$ and taking the sup over $\|x\| \leq 1$ it follows that $\|T\| \leq \|\widetilde{T}\|$.

Finally, since $\widetilde{T}([x]) = \widetilde{T}([y])$ implies $x - y \in K(T)$, it follows that $[x] = [y]$ and $\widetilde{T}$ is injective. Also, if T is onto, $\widetilde{T}$ is clearly onto.

Now, not every linear bijection is an isomorphism. To see this let X be a linear space endowed with the norms $\|\cdot\|$ and $\|\cdot\|_1$ such that $\|\cdot\|$ dominates but is not equivalent to $\|\cdot\|_1$. Then the identity $I : (X, \|\cdot\|) \to (X, \|\cdot\|_1)$ is a continuous surjection with $K(I) = \{0\}$ but the inverse of the induced bijection $\widetilde{I}$, which is essentially I because $K(I)$ is trivial, is not continuous.

22. As in Problem 21, if $\pi : B \to B/K(T)$ denotes the canonical projection of B into $B/K(T)$ and $\widetilde{T}$ the unique continuous mapping such that $T = \widetilde{T} \circ \pi$, $\widetilde{T}$ is an isomorphism with bounded inverse $S :$ $B_1 \to B/K(T)$, say. So, if $y_n \to y_0$ in B_1, $S(y_n) = [z_n] \to S(y_0) = [z_0]$ for some $z_n, z_0 \in B$. By the continuity of S, $\| [z_n] - [z_0] \|_{B/K(T)} = \| [z_n - z_0] \|_{B/K(T)} \leq C \|y_n - y_0\|_{B_1}$ for all n, and so there exists $\{a_n\} \subset K(T)$ such that $\|z_n - z_0 - a_n\|_B \leq 2\,C \|y_n - y_0\|_{B_1}$. Let $x_n = z_n - a_n$, $x_0 = z_0$. Then $x_n \to x_0$ in B, $T(x_n) = T(z_n) = y_n$, $T(x_0) = T(z_0) = y_0$, and $\|x_n - x_0\|_B \leq c \|y_n - y_0\|_{B_1}$.

23. Let Y be a closed subspace of B_1 such that $B_1 = R(T) \oplus Y$. With $\widetilde{T} : B/K(T) \to B_1$ the operator introduced in Problem 21, consider the mapping $S : B/K(T) \oplus Y \to B_1$ between Banach spaces given by $S([x], y) = \widetilde{T}([x]) + y$. Then S is a bounded linear isomorphism and hence, by the open mapping theorem, open. Thus $R(T) = S(B/K(T) \oplus \{0\})$ is closed.

Observe that it is not always the case that if a subspace X of a Banach space B is complemented in B, i.e., there exists a closed subspace M of B such that $B = X \oplus M$, then X is closed. To see this let L be an unbounded linear functional on B and put $X = K(L)$. Then as in Problem 8.63 there is a one-dimensional, and hence closed, subspace M of B such that $B = X \oplus M$, but, since L is unbounded, X cannot be closed.

25. (a) and (b) If T is continuous, $K(T) = T^{-1}(\{0\})$ is closed. Conversely, if $K(T)$ is closed, consider the Banach space $B/K(T)$ and the mapping $\widetilde{T}$ introduced in Problem 21; $\widetilde{T}$ is a linear bijection between finite-dimensional spaces and $T = \widetilde{T} \circ \pi$. By Problem 19, $\widetilde{T}$ is continuous and since π is continuous, so is T. Finally, since $\widetilde{T}$ is a continuous isomorphism, $\widetilde{T}$ is open, and so is T.

27. Since $\mathrm{sp}\{e_1, \ldots, e_n, \ldots\}$ is dense in ℓ^p, $1 \leq p < \infty$, the conclusion follows readily in that case. Now, if $p = \infty$, we first claim that $T(e) = e$, where $e = (1, 1, \ldots)$. Let $T(e) = (a_1, \ldots, a_n, \ldots)$, say. Then, since $\|T(e)\|_\infty \leq 1$, $|a_n| \leq 1$, and, in particular, $a_n \leq 1$ for all n. Next, since $(1, \ldots, 1, -1, 1, \ldots) = e - 2e_n$ and $T(e_n) = e_n$, $\|T(e) - 2e_n\|_\infty \leq 1$, and, in particular, $|a_n - 2| \leq 1$, i.e., $-1 \leq a_n - 2 \leq 1$, and so $1 \leq a_n$ for each n. Therefore $a_n = 1$ for all n, and $T(e) = e$. Next, observe that if $x \in \ell^\infty$ and $x_n \geq 0$ for all n, then the terms of $T(x)$ are nonnegative. To see this let $y = 1 - x/\|x\|_\infty$ and note that $\|y\|_\infty = \sup_n (1 - x_n/\|x\|_\infty) = 1 - \inf_n (x_n/\|x\|_\infty)$. Then $\|T(y)\|_\infty = \|T(e) - T(x/\|x\|)\|_\infty \leq 1 - \inf_n (x_n/\|x\|)$ for all n. Hence $(T(e) - T(x/\|x\|_\infty))_n = 1 - T(x/\|x\|_\infty)_n \leq 1 - \inf_n (x_n/\|x\|)$, and, consequently, $0 \leq \inf_n (x_n/\|x\|_\infty) \leq T(x)_n/\|x\|_\infty$, which implies that $T(x)_n \geq 0$ for all n. Moreover, this implies that $T(x) \geq x$ termwise. Indeed, if $T(x)_n < x_n$ for some n, then since $x - x_n e_n \geq 0$ termwise, $T(x - x_n e_n)_n =$

$T(x)_n - x_n \geq 0$, which is not the case. Thus $T(x) \geq x$. Suppose now that for some $x \in \ell^\infty$ with $0 \leq x_n \leq 1$ we have $T(x)_k > x_k$ for some k, and let $y = (1 - x_1, \ldots, 1 - x_n, \ldots)$. Then $y_n \geq 0$ for all n, and, consequently, $T(y)_n \geq y_n$ for all n. However, since $T(y)_k = 1 - T(x)_k < 1 - x_k = y_k$, this doesn't hold for the index k. Therefore $T(x) = x$. Finally, for an arbitrary sequence $x \in \ell^\infty$, let $x = x_1 - x_2$ where all the terms of x_1 and x_2 are ≥ 0. Then $\|x_1\|_\infty, \|x_2\|_\infty \leq \|x\|_\infty$, and

$$T(x) = \|x\|_\infty \left(T\left(\frac{x_1}{\|x\|_\infty} \right) - T\left(\frac{x_2}{\|x\|_\infty} \right) \right) = \|x\|_\infty \left(\frac{x_1}{\|x\|_\infty} - \frac{x_2}{\|x\|_\infty} \right) = x\,.$$

28. By Problem 5.25, $y \in L^\infty(I)$ and $\|y\|_\infty = \|T\|$.

30. (a) We proceed by induction. The case $n = 1$ is trivial. Suppose now the relation holds for n and note that $S^{n+1}T - TS^{n+1} = S^n ST - TSS^n = S^n(ST - TS) + S^n TS + (ST - TS)S^n - STS^n$. Now, since $ST - TS$ commutes with S, $(ST - TS)S^n = S^n(ST - TS)$. Also, by the induction assumption, $S^n TS - STS^n = S(S^{n-1}T - TS^{n-1})S = S((n-1)S^{n-2}(ST - TS))S = (n-1)S^n(ST - TS)$. Whence combining both expressions we have $S^{n+1}T - TS^{n+1} = 2S^n(ST - TS) + (n-1)S^n(ST - TS) = (n+1)S^n(ST - TS)$, which is what we set out to prove.

(b) First, dividing by α and incorporating α into T we may assume that $\alpha = 1$. For the sake of argument suppose that $ST - TS = I$. By (a) we have $S^n T - TS^n = nS^{n-1}$, and, consequently, it follows that $n\|S^{n-1}\| \leq \|S\|\,\|S^{n-1}\|\,\|T\| + \|T\|\,\|S\|\,\|S^{n-1}\| = 2\|S\|\,\|T\|\,\|S^{n-1}\|$. Thus $\|S^{n-1}\| = 0$ for some large n, or $n \leq 2\|S\|\,\|T\|$ for all n, which is impossible. Working back, by the induction condition $S^n = 0$ implies $S^{n-1} = 0$, and so finally, $S = 0$, which is not the case.

31. Since $DM - MD = I$, by Problem 30 no such norm exists.

34. The statement is true.

38. (a) To verify that $X_{\mathbb{C}}$ is a complex linear space offers no difficulty. The only property that offers any difficulty in proving that $\|\cdot\|_{X_{\mathbb{C}},p}$ is a norm is the triangle inequality, and it follows at once from the fact that $\|\cdot\|$ is a norm and the triangle inequality for the ℓ^p spaces.

(b) Clearly $c_X = 1$. Next note that

$$\|T_{\mathbb{C}}(x_{\mathbb{C}})\|_Y = \|T(\Re x_{\mathbb{C}}) + iT(\Im x_{\mathbb{C}})\|_Y$$
$$\leq \|T(\Re x_{\mathbb{C}})\|_Y + \|T(\Im x_{\mathbb{C}})\|_Y \leq \|T\| \left(\|\Re x_{\mathbb{C}}\|_X + \|\Im x_{\mathbb{C}}\|_X \right),$$

and so

$$C_X = \sup_{x_{\mathbb{C}} \in X_{\mathbb{C}}} \frac{\|\Re x_{\mathbb{C}}\|_X + \|\Im x_{\mathbb{C}}\|_X}{\|x_{\mathbb{C}}\|_{X_{\mathbb{C}},p}}\,.$$

39. (b) Let $x_n \to x$ in B and $T(x_n) \to y$ in B_1. Note that on the one hand, $L(T(x_n)) \to L(y)$, and on the other, $L(T(x_n)) \to L(T(x))$. Therefore by the uniqueness of limits, $L(T(x)) = L(y)$. Thus $T(x) - y \in K(L)$ for all $L \in \mathcal{F}$, and, therefore, $T(x) - y \in \bigcap_{L \in \mathcal{F}} L^{-1}(\{0\}) = \{0\}$ and $T(x) = y$. The continuity of T follows by the closed graph theorem.

40. The inequality can be strict. For example, consider $S, T : \mathbb{R}^2 \to \mathbb{R}^2$, endowed with an ℓ^p norm, say, the matrices of norm 1 defined below and the resulting matrix for ST,

$$S = \begin{bmatrix} 1 & 0 \\ 0 & 0 \end{bmatrix}, \qquad T = \begin{bmatrix} 0 & 0 \\ 0 & 1 \end{bmatrix}, \qquad ST = \begin{bmatrix} 0 & 0 \\ 0 & 0 \end{bmatrix}.$$

41. Fix a positive integer m and for $n \in \mathbb{N}$, write $n = mq_n + r_n$ with $0 \le r_n < m$. Then $\|T^n\| \le \|T^m\|^{q_n} \|T\|^{r_n}$, and since $q_n/n \to 1/m$ and $r_n/n \to 0$ as $n \to \infty$, it follows that $\limsup_n \|T^n\|^{1/n} \le \|T^m\|^{1/m} \le \|T\| < \infty$. This readily gives $\limsup_n \|T^n\|^{1/n} \le \inf_m \|T^m\|^{1/m} \le \liminf_n \|T^n\|^{1/n}$.

42. Observe that S is of the form $S(x) = y$ where $y_n = \ell_n(x)$ is a bounded linear functional on Y with norm $\le \|S\|$ for all n. Then, by Hahn-Banach, there are norm preserving linear functionals L_n that extend ℓ_n to X for all n. The mapping $T : X \to \ell^\infty$ defined by $T(x) = y$ with $y_n = L_n(x)$ for all n gives the desired extension of S to X.

44. Given $x \in B$, let $\{x_n\} \subset X$ converge to $x \in B$. In particular, $\{x_n\}$ is Cauchy in X and by continuity $\|S(x_n) - S(x_m)\|_{B_1} \le \|S\| \, \|x_n - x_m\|_X$. Therefore $\{S(x_n)\}$ is Cauchy in B_1 and we put $T(x) = \lim_n S(x_n)$. And, if S is an isometry, $\|T(x)\|_{B_1} = \lim_n \|S(x_n)\|_{B_1} = \lim_n \|x_n\|_X = \|x\|_B$ and T is norm preserving.

45. The statement is false. Let $T : C^1(I) \to C(I)$ be given by $T(x) = x + x'$; then T is linear and closed but not bounded, and, therefore, by the closed graph theorem, $(C^1(I), \|\cdot\|)$ is not a Banach space.

46. (a) As proved in Problem 45, the differential operator $T : C^1(I) \to C(I)$ is closed but not bounded; in this case the domain $X = C^1(I)$ is not complete. An example along similar lines has been discussed in Problem 9, with $X = \{x \in \ell^1 : \sum_n n|x_n| < \infty\}$ and $T : X \to \ell^1$ given by $T(x) = y$, where $y_n = nx_n$, $x \in X$.

(b) Let B be an infinite-dimensional Banach space, H a (necessarily infinite) Hamel basis for B such that $\|h\| = 1$ for all $h \in H$, and $\|\cdot\|_1$ the norm on B given by $\|x\|_1 = \sum_n |a_n|$ where $x = \sum_n a_n h_n$, $h_n \in H, 1 \le n \le N$. Let X denote the linear space B endowed with the metric $\|\cdot\|_1$, and $T : B \to X$ the identity map; T is 1-1, onto, and as is readily verified, closed. However T is not bounded, for otherwise since B is complete it would follow that X is complete, which is not the case.

47. (a) implies (b) This is clear since $(0, y) \notin G(\overline{T}) = \overline{G(T)}$ for all $y \neq 0$.

(b) implies (c) Since $(x_n, T(x_n)) \in G(T)$ for all n, $(0, y) \in \overline{G(T)}$, and, therefore, by assumption $y = 0$.

(c) implies (a) Let $D = \{x \in X : \text{there exists } y \in Y \text{ such that } (x, y) \in \overline{G(T)}\}$ and define $\overline{T} : D \to Y$ by $\overline{T}(x) = y$. Now, if $y_1 \in Y$ is such that $(x, y_1) \in \overline{G(T)}$, since $\overline{G(T)}$ is a linear subspace, $(x, y) - (x, y_1) = (0, y - y_1) \in \overline{G(T)}$, and so there exist $(x_n, T(x_n)) \in G(T)$ such that $(x_n, T(x_n)) \to (0, y - y_1)$, and, consequently, by assumption, $x_n \to 0$ and $T(x_n) \to y - y_1$. Therefore $y - y_1 = 0$, and $\overline{T}$ is well-defined. Next, $\overline{T}$ is linear and closed by definition. To see that $\overline{T}$ extends T we must verify that $D(T) \subset D$ and $\overline{T}(x) = T(x)$ if $x \in D(T)$. Now, $x \in D(T)$ implies $(x, T(x)) \in G(T) \subset \overline{G(T)}$, and so $x \in D$. Then by the definition of D and $\overline{T}$ it follows that $\overline{T}(x) = T(x)$ for $x \in D(T)$.

Note that $\overline{T}$ is the minimal extension of T, i.e., if T_1 is a closed extension of T, then (i) $D(\overline{T}) \subset D(T_1)$, and (ii) $T_1(x) = \overline{T}(x)$ for all $x \in D(\overline{T})$. To see this observe that since T_1 is closed, $G(T) \subset G(T_1) = \overline{G(T_1)}$, and, consequently, $\overline{G(T)} = G(\overline{T}) \subset G(T_1)$, which gives (i). And, for all $x \in D(\overline{T})$, $(x, y) \in G(\overline{T})$ implies $(x, y) \in G(T_1)$. Thus $y = \overline{T}(x)$ implies $y = T_1(x)$, which gives (ii).

49. (a) Since $G(T)$ and $X \times \{0\}$ are closed in $X \times Y$, $G(T) \cap (X \times \{0\}) = K(T) \times \{0\}$ is closed in $X \times Y$. Let $\phi : X \to X \times Y$ denote the linear mapping given by $\phi(x) = (x, 0)$; ϕ is 1-1, bounded, and onto $X \times \{0\}$, and so by the inverse mapping theorem ϕ^{-1} is continuous. Hence $K(T) = \phi^{-1}(K(T) \times \{0\})$ is the inverse image of a closed set and, consequently, closed.

50. (b) If X is a Banach space, by (a), $T + T_1$ is continuous, and if $(T + T_1)^{-1}$ is bounded, $T + T_1$ is a bijection. Conversely, since $G((T + T_1)^{-1}) = \{(v, w) \in X \times X : (w, v) \in G(T + T_1)\}$ is simply $G(T + T_1)$ with the coordinates switched, $(T_1 + T)^{-1}$ has a closed graph, and hence is continuous by the closed graph theorem. Alternatively, one may check that $G(T^{-1}) = \{(T(x), x) : x \in D(T)\} \subset Y \times X$ is closed, and, since the mapping $\phi : X \times Y \to Y \times X$, defined by $\phi(x, y) = (y, x)$, is an isometry and $G(T)$ is closed, it follows that $G(T^{-1})$ is closed.

51. Let $S : B \times B_1 \to B \times B_1$ be the linear mapping given by $S(x, y) = (x, y + T_1(x))$. Note that S is closed: If $(x_n, y_n) \to (x, y)$ and $S(x_n, y_n) = (x_n, y_n + T_1(x_n)) \to (w, z)$, then $w = x$, and, since T_1 is closed, $z = y + \lim_n T_1(x_n) = y + T_1(x)$. Therefore by the closed graph theorem S is continuous. Now, in this case, if $x_n \to x$, then $(x_n, y) \to (x, y)$, and $S(x_n, y) = (x_n, y + T_1(x_n)) \to S(x, y) = (x, y + T_1(x))$, and, therefore, T_1 is continuous. The conclusion now follows by Problem 50(a).

53. Let $\{\overline{x}_n\}$ be a bounded sequence in $\overline{X}$; since X is dense in B there are $\{x_n\} \subset X$ such that $\overline{x}_n - x_n \to 0$. Clearly $\{x_n\}$ is bounded and, consequently, $\{T(x_n)\}$ has a convergent subsequence $\{T(x_{n_k})\}$, say, that converges to $y \in B_1$. Now, since $\|\overline{x}_{n_k} - x_{n_k}\| \to 0$, then $\overline{T}(\overline{x}_{n_k}) - T(x_{n_k}) = \overline{T}(\overline{x}_{n_k}) - \overline{T}(x_{n_k}) \to 0$, and, consequently, $\overline{T}(\overline{x}_{n_k}) \to y$. Thus $\overline{T}$ is compact.

54. (a) Let $B \subset X$ be bounded. Then $T(B) \subset T(X)$ is a bounded set in a finite-dimensional Banach space, and so has compact closure.

(b) Note that $R(T)$ is a Banach space. Now, if B is the open unit ball in X, by the open mapping theorem the image $T(B)$ is open in $R(T)$. Thus there is a closed ball of nonzero radius contained in $T(B)$ and since T is compact, this ball is compact. But by Problem 8.33, a ball is compact iff the space, $R(T)$ in this case, is finite dimensional.

55. No.

56. Given $\varepsilon > 0$, let n be such that $\|T_n - T\| < \varepsilon$. Now, with B_X a ball in X, since T_n is compact, $T_n(B_X)$ is precompact. Hence $T_n(B_X) \subset \bigcup_{n=1}^N B(y_n, \varepsilon)$, say. But for all $x \in B_X$, $\|T_n(x) - T(x)\| < \varepsilon$, and, therefore, $T(B_X) \subset \bigcup_{n=1}^N B(y_n, 2\varepsilon)$. Since this is true for all $\varepsilon > 0$, $T(B_X)$ is precompact.

57. The statement is true.

59. By Problem 94(a), $\|T_n\| \leq c$ for all n and some constant $c > 0$. For the sake of argument suppose that there exist $\eta > 0$ and a subsequence $\{T_{n_k} K\}$ of $\{T_n K\}$ such that $\|T_{n_k} K - TK\| > \eta$. Let $\{x_{n_k}\} \subset B$ with $\|x_{n_k}\| = 1$ such that $\|T_{n_k} K(x_{n_k}) - TK(x_{n_k})\|_B > \eta$. Since K is compact, passing to a subsequence if necessary we may assume that $\{K(x_{n_k})\}$ converges to $y \in B$, say. Then $T_{n_k} K(x_{n_k}) - TK(x_{n_k}) = T_{n_k} K(x_{n_k}) - T_{n_k}(y) + T_{n_k}(y) - T(y) + T(y) - TK(x_{n_k})$, where $\|T_{n_k} K(x_{n_k}) - T_{n_k}(y)\|_B \leq c\|K(x_{n_k}) - y\|_B$ is arbitrarily small for n_k large, and the same is true for $\|T(y) - TK(x_{n_k})\|_B$ since T is bounded. Also, for n_k sufficiently large, $\|T_{n_k}(y) - T(y)\|_B$ is arbitrarily small. Thus combining these estimates, $\|T_{n_k} K(x_{n_k}) - TK(x_{n_k})\|_B$ is arbitrarily small, which is not the case since it exceeds $\eta > 0$.

62. Since $\{\|x_n\|\}$ is bounded a subsequence $\{T(x_{n_k})\}$ of $\{T(x_n)\}$ converges to some $y \in Y$. Now, since T is continuous, by Problem 34, $T(x_{n_k}) \to T(x)$ in Y, and, therefore, $y = T(x)$. The same argument applies to any subsequence $\{x_{n_k}\}$ of $\{x_n\}$, i.e., $\{T(x_{n_k})\}$ has a subsequence that converges to $T(x)$, and, consequently, $\lim_n T(x_n) = T(x)$.

63. T is compact iff $\lim_n \lambda_n = 0$.

64. For the sake of argument suppose that T is compact. Now, by Problem 8.21 there is a bounded sequence $\{x_n\} \subset B$ with no convergent subsequence. Let $\{T(x_{n_k})\}$ be a convergent subsequence of $\{T(x_n)\}$ and observe that since $\|T(x_{n_k}) - T(x_{n_\ell})\|_X \geq c\|x_{n_k} - x_{n_\ell}\|_B$, $\{x_{n_k}\}$ is a Cauchy subsequence of $\{x_n\}$, and, hence, convergent in B, which is not the case.

65. For the sake of argument suppose that T^{-1} is bounded. Then by Problem 58, $TT^{-1} = I$ is compact, but the identity is not compact unless X is finite dimensional.

67. (a) Let $\{x_n\} \subset K(\lambda I - T)$ be a bounded sequence; passing to a subsequence if necessary we may assume that $\{T(x_n)\}$ converges, and since $\lambda x_n = T(x_n)$, that $\{x_n\}$ converges to some $x \in X$. By the continuity of T, $T(x) = \lim_n T(x_n)$, $(\lambda I - T)(x) = \lim_n (\lambda I - T)(x_n) = 0$, and $x \in K(\lambda I - T)$. Thus the unit ball of $K(\lambda I - T)$ is compact and, consequently, by Problem 8.33, $K(\lambda I - T)$ is finite dimensional.

(b) For the sake of argument suppose there is a sequence $\{x_n\}$ such that $\|(I - T)(x_n)\| < \|x_n - x_{n_K}\|/n$; in particular, $x_n - x_{n_K} \neq 0$, and if $y_n = (x_n - x_{n_K})/\|x_n - x_{n_K}\|$, $\|(I - T)(y_n)\| < 1/n$, and so $(I - T)(y_n) \to 0$. Now, by the compactness assumption there is a subsequence $\{y_{n_k}\}$ of $\{y_n\}$ such that $\{T(y_{n_k})\}$ converges in Y, and so $y_{n_k} = (y_{n_k} - T(y_{n_k})) + T(y_{n_k})$ converges to some y with $\|y\| = 1$. Then by the continuity of T, $y - T(y) = \lim_k (y_{n_k} - T(y_{n_k})) = 0$ and $y \in K(I - T)$. Thus we have shown that $((x_{n_k} - x_{n_K})/\|x_{n_k} - x_{n_K}\|) - y \to 0$, or, in other words,

$$\frac{(x_{n_k} - x_{n_K}) + \|x_{n_k} - x_{n_K}\|\, y}{\|x_{n_k} - x_{n_K}\|} \to 0\,,$$

and, in particular, $\|(x_{n_k} - x_{n_{k_K}}) + \|x_{n_k} - x_{n_{k_K}}\|\, y\| < \|x_{n_k} - x_{n_{k_K}}\|$ for sufficiently large k, and this cannot happen by the definition of $\{x_{n_{k_K}}\}$, since $x_{n_{k_K}} + \|x_{n_k} - x_{n_{k_K}}\|\, y \in K(I - T)$ is closer to x_{n_k} than $x_{n_{k_K}}$.

69. This result shows that for $\lambda \neq 0$, $(\lambda I - T)$ is injective iff it is surjective, in which case, when X, Y are Banach spaces, $(\lambda I - T)^{-1}$ is bounded.

71. Since $x \in K(\lambda I - T)$ implies that $x = \lambda^{-1}T(x)$, the identity acts as a compact operator on $K(\lambda I - T)$, which is therefore finite dimensional. And, since $R(\lambda I - T)$ is closed and $T^* : B^* \to B^*$ is compact, from $K(\lambda I - T^*) \sim B/R(\lambda I - T)$ it follows that $\operatorname{codim}(R(\lambda I - T)) = \dim(K(\lambda I - T^*)) < \infty$.

Now let $K(\lambda I - T) = \operatorname{sp}\{x_1, \ldots, x_m\}$, $m \geq 1$, and $B/R(\lambda I - T) = \operatorname{sp}\{[y_1], \ldots, [y_n]\}$, $n \geq 1$, where all the vectors involved are linearly independent. We claim that $m = n$.

First, suppose that $m \leq n$. By Problem 8.101, $K(\lambda I - T)$ is complemented in B. Let $B = K(\lambda I - T) \oplus X$, where X is a closed subspace of B, and let F be the finite-rank operator which is 0 on X and $F(x_k) = y_k$, $1 \leq k \leq m$. The operator $T' = T + F$ is compact. Now, if $(\lambda I - T')(x) = 0$, then $(\lambda I - T)(x) = -F(x) \in R(\lambda I - T) \cap \operatorname{sp}\{y_1, \ldots, y_n\} = \{0\}$, and so $x \in K(\lambda I - T)$, $x = \sum_{k=1}^m \alpha_k x_k$, and $F(x) = \sum_{k=1}^m \alpha_k y_k = 0$. Then by the linear independence of the y_k, all the $\alpha_k = 0$ and $x = 0$. Thus $\lambda I - T'$ is injective and by Problem 69, $\lambda I - T'$ is onto, and, consequently, $m = n$.

Along the same lines, if $m \geq n$, let $F(x_k) = y_k$ for $1 \leq k \leq n$ and $F(x_k) = y_n$ for $i \geq n$. With $T' = T + F$, it readily follows that $\lambda I - T'$ is surjective, so by Problem 69, injective, and, therefore, $m = n$.

This result is the Fredholm alternative for compact operators T and $\lambda \neq 0$: $\lambda I - T$ is bijective or has nontrivial kernel and nontrivial cokernel of the same dimension.

72. (a) By Problem 8.101, $K(T)$ is complemented in B; let M be a closed subspace of B with $B = M \oplus K(T)$. Now, $T|_M : M \to R(T)$, the restriction of T to M, is 1-1 on M, bounded, and $R(T|_M) = R(T)$. To see the last assertion note that clearly $R(T|_M) \subset R(T)$. And, if $y \in R(T)$, let $x \in X$ be such that $T(x) = y$. Then $x = x_0 + x_1$ where $x_0 \in M$ and $x_1 \in K(T)$, and so $T(x_0) = T(x) = y \in R(T|_M)$. Hence, since $R(T)$ is closed, by the inverse mapping theorem $T|_M^{-1} : R(T) \to M$ is bounded, and therefore $T|_M$ is an isomorphism.

(b) Since $R(T)$ has finite codimension in B_1, by Problem 8.167(a), $R(T)$ is complemented in B_1; let N be a finite-dimensional subspace of B_1 such that $B_1 = R(T) \oplus N$. Next, let $P : B_1 \to B_1$ be the bounded linear operator defined in Problem 82 such that $P(y) = y$ if $y \in N$ and $P(y) = 0$ if $y \in R(T)$; clearly $(I - P) : B_1 \to R(T)$ is bounded and vanishes on N. Lastly, let $T_0 = T|_M^{-1} \circ (I - P) : B_1 \to B$; note that T_0 vanishes on N, coincides with $T|_M^{-1}$ on $R(T)$, and, therefore, is bounded. All the properties listed above are readily verified.

74. (a) iff (b) Necessity first. Since T is compact, given $\varepsilon > 0$, there is a finite ε-net of balls for $T(B_X(0,1))$, i.e., $T(B_X(0,1)) \subset \bigcup_{k=1}^n B_Y(y_k, \varepsilon)$. Let $F = \text{sp}\{y_1, \ldots, y_n\}$; F is a finite-dimensional subspace of $R(T)$ with dimension $\leq n$ and $d(y, F) \leq \varepsilon$ for all $y \in R(T)$. It then readily follows that $\|\pi_F \circ T\| < \varepsilon$.

Conversely, note that $T(B_X(0,1))$ is a bounded subset of Y that lies within an ε-neighborhhood of a bounded subset of F. And, since F is finite dimensional, bounded subsets of F are totally bounded, $T(B_X(0,1))$ is contained in the union of finitely many balls of radius 2ε in Y, and T is compact.

(b) iff (c) Necessity first. Since $T^* : Y^* \to X^*$ is compact, given $\varepsilon > 0$, by (a) there is a finite-dimensional subspace Z of X^* such that $\|\pi_Z \circ T^*\| < \varepsilon$. Let $W = Z_\perp = \{x \in X : \ell(x) = 0 \text{ for all } \ell \in Z\}$; W is a closed linear subspace of X. By linear algebra, W has finite codimension in X equal to the dimension of Z, and there is a natural isomorphism between $(X/W)^*$ and Z. In particular, $W^\perp = Z$ in this case, since $Z \subset W^\perp$ automatically, and Z and $W^\perp \sim (X/W)^*$ have the same finite dimension. Let $A : W \to X$ be the inclusion mapping; then $(T \circ A)^* = A^* \circ T^*$, where by Problem 176, A^* corresponds to the restriction mapping from X^* onto $W^* \sim X^*/W^\perp = X^*/Z$. By construction $\|A^* \circ T^*\| \le \varepsilon$, which implies that $\|T \circ A\| \le \varepsilon$. In other words, $\|T|_W\| \le \varepsilon$.

Sufficiency next. Let F be the subspace $T(V)$ of Y. Since by Problem 8.167(b) with $\varepsilon = 1$ there, each $x \in X$ can be written as $x = v + w$ with $v \in V, w \in W$ and $\|w\|_X \le 3\|x\|_X$, it follows that $\|\pi_F(T(x))\|_{Y/F} \le \|T(w)\|_Y \le 3\varepsilon\|x\|_X$. Thus (b), and hence (a), holds.

76. (a) implies (b) Given $x = m + n \in B$, let $P : B \to M$, $Q : B \to N$, be given by $P(x) = m$, $Q(x) = n$, respectively. By Problem 75, P, Q are projections and $P + Q = I$. Also, $PQ(x) = PQ(m + n) = P(n) = 0$ for all $x \in X$, and, similarly, $QP = 0$. Next, continuity. Let $T : M \times N \to B$ be given by $T(m, n) = m + n$; T is clearly linear, 1-1, and onto. Moreover, since $\|T(m,n)\|_B = \|m + n\|_B \le \|m\|_B + \|n\|_B = \|(m,n)\|_{M \times N}$, T is bounded. Then by the inverse mapping theorem T^{-1} is bounded and, consequently, $\|P(x)\|_B + \|Q(x)\|_B = \|m\|_B + \|n\|_B \le c\|m + n\|_B = c\|x\|_B$. Alternatively, the boundedness of P follows from the closed graph theorem. Suppose that $x_n \to 0$ and $P(x_n) \to y$. Since $R(P) = M$ is closed, $y \in R(P)$. Also, $P(x_n - P(x_n)) = P(x_n) - P^2(x_n) = 0$ and $x_n - P(x_n) \to -y$. Therefore $y \in K(P)$. Since $K(P) \cap R(P) = \{0\}$, it follows that $y = 0$. Thus P is closed and by the closed graph theorem, bounded.

(b) implies (a) By Problem 75 it only remains to prove that M, N are closed. But this is clear since P, Q are continuous and $N = K(P)$, $M = K(Q)$.

Another way of stating this problem is that a closed subspace M of a Banach space B is complemented in B iff M is the range of a continuous projection on B. By Problem 8.101 this remark applies to a finite-dimensional subspace M of an infinite-dimensional Banach space B. In fact, if M is a finite-dimensional subspace of B and N is any closed subspace of B with $M \cap N = \{0\}$, then $M + N$ is a direct sum in B. For, if $M + N$ is not a direct sum, the projection P onto M is discontinuous and there exist sequences $\{x_n\} \subset M$ and $\{y_n\} \subset N$ such that $\|x_n + y_n\| \to 0$ and $\|P(x_n + y_n)\| = \|x_n\| \to \infty$; considering $\{x_n/\|x_n\|\}, \{y_n/\|x_n\|\}$ if necessary we may assume that $\|x_n + y_n\| \to 0$ and $\|x_n\| = 1$ for all n. Since M is finite dimensional, there is a convergent subsequence $\{x_{n_k}\}$. If the limit of this subsequence is x we have $x \in M$ and $\|x\| = 1$. But since $\|x_n + y_n\| \to 0$, x is also the limit of the sequence $\{y_{n_k}\}$ and hence in N. Thus $0 \ne x \in M \cap N$, which is not the case.

Finally, note that if $B = M \oplus N$, M is isomorphic to B/N. Indeed, the projection $P : B \to M$ is continuous, has kernel N, and is onto. Then the mapping $\widetilde{P} : B/N \to M$ defined in Problem 21 is an isomorphism.

77. Necessity first. Since $M + N$ is a closed subspace of a Banach space, it is a Banach space. Then by Problem 76 the projection $P_M : M \oplus N \to M$ given by $P_M(m + n) = m$ is bounded and, consequently, $\|m\|_B = \|P_M(m + n)\|_B \leq \|P_M\| \, \|m + n\|_B$ for all $m \in M, n \in N$. Similarly for $Q_N(m + n) = n$.

Sufficiency next. Let $\{z_n\} \subset M + N$ be such that $z_n \to z$ for some $z \in B$. Now, $z_n = x_n + y_n$, $x_n \in M$, $y_n \in N$, and, by assumption, $\|x_n - x_m\| \leq c\|z_n - z_m\|$. Hence $\{x_n\}$ is Cauchy in M, and so convergent to $x \in M$, say. Moreover, since N is closed, $y_n = z_n - x_n \to z - x \in N$, and $z = x + (z - x) \in M + N$ as required.

78. Necessity first. If $x \in X$, $y \in Y$, both of norm 1, by Problem 77 there exists c such that $c\|x + (-y)\| = c\|x - y\| \geq 2$, and so $k \geq 2/c > 0$.

Sufficiency next. For the sake of argument suppose that $X + Y$ is not closed. Then by Problem 77, given $n \geq 1$, there exist $x_n \in X$, $y_n \in Y$, such that $\|x_n\| + \|y_n\| > 2n\|x_n + y_n\|$, and, consequently, there exists a subsequence $\{x_{n_k}\}$ or $\{y_{n_k}\}$ such that $\|x_{n_k}\| > n_k\|x_{n_k} + y_{n_k}\|$ or $\|y_{n_k}\| > n_k\|x_{n_k} + y_{n_k}\|$; suppose the former occurs. Now, replacing x_{n_k} by $x_{n_k}/\|x_{n_k}\|$ and y_{n_k} by $y_{n_k}/\|x_{n_k}\| \in Y$, we may assume that $\|x_{n_k}\| = 1$ and $\|x_{n_k} + y_{n_k}\| < 1/n_k$ for all n_k. Note that $|1 - \|y_{n_k}\|| = |\,\|x_{n_k}\| - \|-y_{n_k}\|\,| \leq \|x_{n_k} + y_{n_k}\| < 1/n_k$ and $\|y_{n_k}\| \to 1$. Therefore

$$\|x_{n_k} + y_{n_k}/\|y_{n_k}\|\| \leq \|x_{n_k} + y_{n_k}\| + \|-y_{n_k} + y_{n_k}/\|y_n\|\|$$
$$< 1/n_k + |1 - \|y_{n_k}\|| < 2/n_k.$$

Thus $c < 2/n_k$ for a sequence $n_k \to \infty$, and so $c = 0$.

80. Let $T : c \to c$ be a bounded linear operator with $\|T\| = 2 - \eta < 2$ such that $T(e_n) = e_n$ for all n. With $e = (1, 1, \ldots)$, we have $\|e - 2e_n\|_\infty = 1$, so that $\|T(e) - 2e_n\|_\infty \leq 2 - \eta$. In particular, the n-th term of $T(e)$ must be $\geq \eta$ for all n, and therefore $T(e) \notin c_0$.

Now, $L_\infty(x) = \lim_n x_n$, the linear functional on c defined in Problem 8.111, is bounded with $\|L_\infty\| = 1$. Then the mapping $P : c \to c_0$ given by $P(x) = y$ where $y_n = x_n - L_\infty(x)$ is a bounded projection of norm 2.

82. We prove (b) first. For the sake of argument suppose that P is not bounded and let $\{x_n\} \subset X_2$ be such that $\|P(x_n)\| = 1$, while $x_n \to 0$ in X. Since M is finite dimensional, passing to a subsequence if necessary we may assume that $\lim_n P(x_n) = z$ exists in M, and, therefore, $(I - P)(x_n) \to -z$ in X_1. Since both M and X_1 are closed, $z \in M \cap X_1$, and, therefore, $z = 0$. However, $\|z\| = \lim_n \|P(x_n)\| = 1$, which cannot happen.

(a) Let $\{x_n\} \subset X_2$ be such that $\lim_n x_n = x \in X$. Then, by (b), $\{P(x_n)\}$ is a Cauchy sequence in M and, hence, converges to $z \in M$. Consequently, $\{(I - P)(x_n)\}$ converges to $w \in X_1$. Hence $x_n = P(x_n) + (I - P)(x_n)$ converges to $z + w \in X_2$. Thus $x = z + w \in X_2$, which is therefore closed.

85. For the sake of argument suppose T is one such mapping. Then with ℓ_n the n-th coordinate functional on ℓ^∞, i.e., $\ell_n(x) = x_n$, $x \in \ell^\infty$, $n = 1, 2, \ldots$, observe that, since $x \in K(T)$ iff $\ell_n(T(x)) = 0$ for all n, $K(T) = \bigcap_n K(\ell_n \circ T)$.

Now, we claim that c_0 cannot be expressed as the countable intersection of kernels of continuous linear functionals on ℓ^∞. To see this let $\{N_\alpha\}_{\alpha \in I}$ denote the uncountable family of infinite subsets of $\mathbb{N}$ such that $N_\alpha \cap N_\beta$ is finite whenever $\alpha \neq \beta$ constructed in Problem 1.56, and to each $\alpha \in I$ assign the ℓ^∞ sequence $x^\alpha = \chi_{N_\alpha}$, i.e., $x_n^\alpha = 1$ if $n \in N_\alpha$ and $x_n^\alpha = 0$ otherwise.

We claim that if $c_0 \subset K(L)$ for a bounded linear functional L on ℓ^∞, then $\{\alpha \in I : L(x^\alpha) \neq 0\}$ is countable. To see this let $A_n = \{\alpha \in I : |L(x^\alpha)| \geq 1/n\}$, $n = 1, 2, \ldots$, pick distinct elements $\alpha_1, \ldots, \alpha_k$ of A_n, and let $x = \sum_{j=1}^{k} \mathrm{sgn}(L(x^{\alpha_j})) \, x^{\alpha_j}$; clearly $L(x) \geq k/n$. Now consider $M_j = N_{\alpha_j} \setminus \bigcup_{i \neq j} N_{\alpha_i}$, $1 \leq j \leq k$. Then $N_{\alpha_j} \setminus M_j$ is finite and if $\tilde{x} = \sum_{j=1}^{k} \mathrm{sgn}(L(x^{\alpha_j})) \chi_{M_j}$, $x - \tilde{x} \in c_0$ and $L(x) = L(\tilde{x})$. Moreover, since the M_j are pairwise disjoint, it follows that $\|\tilde{x}\|_\infty = 1$, and thus $k/n \leq L(x) = L(\tilde{x}) \leq \|L\|$. Hence A_n has at most $n\|L\|$ elements and $\{\alpha \in I : L(x^\alpha) \neq 0\}$ is countable.

Now, if $\{L_n\} \subset \ell^{\infty*}$ and $C = \{\alpha \in I : L_n(x^\alpha) \neq 0 \text{ for some } n\} = \bigcup_n \{\alpha \in I : L_n(x^\alpha) \neq 0\}$, then C is countable, $I \setminus C \neq \emptyset$, and, therefore, for some $\alpha \in I$, $x^\alpha \in \bigcap_n K(L_n)$. Hence $c_0 \neq \bigcap_n K(L_n)$.

Note that, in particular, c_0 is not complemented in ℓ^∞.

90. (a) $\|L_n\| = n$.

(b) Since each x in c_{00} has vanishing terms after some $N = N_x$, $|L_n(x)| \leq \sum_{k=1}^{N} |x_k|$ for all n, and, therefore, $\{x \in c_{00} : \sup_n |L_n(x)| < \infty\} = c_{00}$. On the other hand, $\sup_n \|L_n\| = \infty$, and the conclusion of the uniform boundedness principle does not hold. Consequently, c_{00} is of first category in itself.

91. The statement is false: T is linear but not necessarily bounded.

94. (a) Since $T_n(x) \to T(x)$ in X for each $x \in B$, $\|T_n(x)\|_X \leq \|T_n(x) - T(x)\|_X + \|T(x)\|_X \leq c_x$ for some finite constant c_x. Thus $\{\|T_n(x)\|_X\}$ is pointwise bounded in B and by the uniform boundedness principle, normed bounded. Hence $\sup_n \|T_n\| < \infty$.

(b) Since for each $\ell \in X^*$, $\lim_n T_n^* \ell(x) = \lim_n \ell(T_n(x)) = \ell(T(x)) = T^* \ell(x)$ for all $x \in B$, by (a), $\sup_n \|T_n^*(\ell)\|_{X^*} < \infty$. Since this holds for all

$\ell \in X^*$ and X^* is a Banach space, by the uniform boundedness principle it follows that $\sup_n \|T_n^*\| < \infty$, and, therefore, $\sup_n \|T_n\| < \infty$ as well.

(c) First, T is clearly linear. Now, since for any subsequence $\{T_{n_k}(x)\}$ of $\{T_n(x)\}$, $\lim_{n_k} T_{n_k}(x) = T(x)$ in X, picking a subsequence n_k along which the norm assumes the liminf, it follows that $\|T(x)\|_X = \lim_{n_k} \|T_{n_k}(x)\|_X \leq \lim_{n_k} \|T_{n_k}\| \, \|x\|_B = \liminf_n \|T_n\| \, \|x\|_B$, and the conclusion follows.

95. (c) implies (a) For the sake of argument suppose that $\lim \sup_n \|T_n\| = \infty$. Then there is a strictly increasing sequence n_k such that $\|T_{n_k}\| \geq 2^k$ for all k, and so there exist $\{x_{n_k}\} \subset B$ with norm 1 such that $\|T_{n_k}(x_{n_k})\|_X \geq 2^k$, or $\|T_{n_k}(x_{n_k}/2^k)\|_X \geq 1$. Now, since $\sum_k \|x_{n_k}/2^k\|_B < \infty$, by assumption $\|T_{n_k}(x_{n_k}/2^k)\|_X \to 0$, which is not the case.

96. Necessity is clear: Since Y is bounded we have $\sup_{y \in Y} |L(y)| \leq \|L\| \sup_{y \in Y} \|y\| < \infty$ for each $L \in X^*$. Conversely, let $y \in Y$ and put $y^{**} = J_X(y)$. Since $y^{**}(L) = L(y)$ for $L \in X^*$ we have $|y^{**}(L)| = |L(y)| < \infty$ for each $L \in X^*$, and, consequently, $\{y^{**}\} \subset X^{**}$ is pointwise bounded, i.e., bounded at each $L \in X^*$. Now, since X^* is a Banach space, by the uniform boundedness principle $\{y^{**}\}$ is norm bounded and $\|y^{**}\|_{X^{**}} \leq c$ for some constant $c > 0$, for all $y \in Y$. Finally, since $\|y\|_X = \|y^{**}\|_{X^{**}}$, $Y \subset X$ is bounded.

98. For the sake of argument suppose that the statement is true. Define the linear functionals L_n on $\mathcal{P}$ by $L_n(p) = a_0 + a_1 + \cdots + a_n$. Then, if $p(t) = a_0 + a_1 t + \cdots + a_m t^m$, $|L_n(p)| \leq (m+1)\|p\| < \infty$ for all n, and by Problem 97, $\{\|L_n\|\}$ is bounded. However, if $p_n(t) = 1 + t + t^2 + \cdots + t^n$, then $\|p_n\| = 1$ and $L_n(p_n) = (n+1)\|p_n\|$. Hence $n + 1 = |L_n(p_n)|/\|p_n\| \leq \|L_n\| \leq c$ for all n, which is not the case.

99. Only (c) implies (a) offers any difficulty. For fixed $x \in B$, let $J_X(T(x)) = T(x)^{**}$. Then $L(T(x)) = J_X(T(x))(L)$ for each $L \in X^*$, and, consequently, $\sup_{T \in \mathcal{T}} |J_X(T(x))(L)| = \sup_{T \in \mathcal{T}} |L(T(x))| < \infty$ for each $L \in X^*$. Now, by the uniform boundedness principle applied to X^* it follows that $\sup_{T \in \mathcal{T}} \|J_X(T(x))(L)\| < \infty$ and since $\|T(x)\|_X = \|J_X(T(x))\|_{X^{**}}$, also $\sup_{T \in \mathcal{T}} \|T(x)\|_X < \infty$. Now, this estimate is true for each $x \in B$ and by the uniform boundedness principle, $\sup_{T \in \mathcal{T}} \|T\| < \infty$.

101. Regard $\{T_n(x)\} \subset B_1$ as a subset of B_1^{**}. Now, for each $x \in B$ and $\ell \in B_1^*$ the scalar sequence $\{\ell(T_n(x))\}$ converges and so is bounded. The uniform boundedness principle then gives that the norms $\{\|T_n(x)\|\}$ are uniformly bounded for each $x \in B$. We are in a situation where we can apply the uniform boundedness principle again to conclude that $\{\|T_n\|\}$ is bounded.

102. Since (b) implies (a) follows readily, we prove (a) implies (b). Consider the Banach space $c_0(B)$ introduced in Problem 8.41, and let T_n, T

be functionals on $c_0(B)$ given by $T_n(x) = \sum_{m=1}^{n} L_m(x_m)$ for all n, and $T(x) = \sum_m L_m(x_m)$, $x \in B$, respectively; by assumption $\lim_n T_n(x) = T(x)$ exists for every $x \in B$ and $T(x)$ is well-defined. Then, $\|T_n\| \leq \sum_{m=1}^{n} \|L_m\|$ for all n and $|T(x) - T_n(x)| \leq \sum_{m=n+1}^{\infty} |L_m(x_m)| \to 0$ as $n \to \infty$ for all $x \in c_0(B)$. Now, for $x \in c_0(B)$, let $\lambda_m = \operatorname{sgn}(L_m(x_m))$, $1 \leq m \leq n$, and observe that $\sum_{m=1}^{n} |L_m(x_m)| = \sum_{m=1}^{n} \lambda_m L_m(x_m) = \sum_{m=1}^{n} L_m(\lambda_m x_m) = T_n(y)$ where $y_m = \lambda_m x_m$ for $1 \leq m \leq n$ and $y_m = 0$ otherwise. Then $\sum_{m=1}^{n} |L_m(x_m)| \leq \|T_n\| \|y\| \leq \|T_n\|$. By picking the x_m judiciously it follows that $\sum_{m=1}^{n} \|L_m\| \leq \|T_n\|$, and so $\sum_{m=1}^{n} \|L_m\| = \|T_n\|$ for all n. Therefore $T_n(x) \to T(x)$ for all $x \in c_0(B)$, and the mappings are linear and continuous, and, therefore, by the uniform boundedness principle, $\sup_n \|T_n\| < \infty$, which completes the proof.

Observe that it suffices for the assumption to hold for a dense subset X of B. To see this, if $x_n \to 0$ in B, pick $y_n \in X$ such that $\|x_n - y_n\| < 1/2^n$; then $y_n \to 0$ and by assumption $\sum_m |L_m(y_m)| < \infty$. As above $\{\|L_m\|\}$ is bounded, and, therefore, $\sum_m |L_m(y_m)| < \infty$. Moreover, since the norms are uniformly bounded, $|\sum_m (L_m(x_m) - L_m(y_m))| \leq M \sum_m 2^{-m} < \infty$, therefore absolutely convergent, which implies convergent, and (a) follows.

103. For $x \in B$, $|L_A(x)| \leq \int_A |T(x)(t)|\,dt \leq \int_{\mathbb{R}^n} |T(x)(t)|\,dt$ for all $A \in \mathcal{L}(\mathbb{R}^n)$, and so $\sup_{A \in \mathcal{L}(\mathbb{R}^n)} |L_A(x)| < \infty$ for all $x \in B$. Therefore by the uniform boundedness principle, $\|L_A\| \leq c$ for all $A \in \mathcal{L}(\mathbb{R}^n)$. Now, $\operatorname{sgn}(T(x)(t)) = \chi_{A^+}(t) - \chi_{A^-}(t)$, where $A^+ = \{t \in \mathbb{R}^n : T(x)(t) > 0\}$ and $A^- = \{t \in \mathbb{R}^n : T(x)(t) < 0\}$. Then $\int_{\mathbb{R}^n} |T(x)(t)|\,dt = \int_{A^+} T(x)(t)\,dt - \int_{A^-} T(x)(t)\,dt = L_{A^+}(x) - L_{A^-}(x) \leq 2c\,\|x\|_B$.

104. The continuous functions

$$f_t(x) = \frac{f(x+t) - f(x)}{t}$$

converge to $f'(x)$ pointwise, so $M(x) = \sup |f_t(x)| < \infty$ for each x. Now, by the uniform boundedness principle there is an open set U on which $M(x)$ is uniformly bounded, and then $f|_U$, since it has a bounded derivative, is Lipschitz there.

107. Let B, B_1 be Banach spaces and $T : B \to B_1$ a bounded surjective mapping, which we assume to be injective. Then T has a well-defined inverse T^{-1} and its graph $\{(T(x), x) : x \in B\}$, being the image of the graph $\{(x, T(x)) : x \in B\}$ of T under the isometry $\phi : B \times B_1 \to B_1 \times B$ given by $\phi(x, y) = (y, x)$, is closed. By the closed graph theorem, T^{-1} is bounded, and so, T is open.

In general, let T be bounded and onto B_1 and as in Problem 21, write $T = \widetilde{T} \circ \pi$ where π is the canonical map onto $B/K(T)$. Then, $\widetilde{T}$ is bounded,

1-1, and onto, and has a bounded inverse. By the first part of the argument $\widetilde{T}$ is open and by Problem 8.159(d), T is open.

108. Suppose that G is not of first category in B; then G is of second category in B. Let $X_n = \{x \in B : \sup_{T \in \mathcal{F}} \|T(x)\|_X \le n\}$; each X_n is closed and $G = \bigcup_n X_n$. Therefore by the Baire category theorem, there exists $n_0 > 0$ such that X_{n_0} contains a ball $B(x_0, r)$, say. Now, if $x \in B(0, r)$, $\|T(x)\|_X \le \|T(x - x_0)\|_X + \|T(x_0)\|_X \le n_0 + n_0 = 2n_0$, and, if $x \in B$, $rx/2\|x\| \in B(0, r)$ and $\|T(x)\|_X \le (4n_0/r)\|x\|_B$ for all $T \in \mathcal{F}$. Hence $x \in G$ and $G = B$.

This is an equivalent formulation: If $\sup_{T \in \mathcal{F}} \|T\| = \infty$, there exists a "point of resonance", i.e., $x \in B$ such that $\sup_{T \in \mathcal{F}} \|T(x)\|_X = \infty$.

109. Consider the sequence $\{G_m\}$ of (good) subsets of B given by $G_m = \{x \in B : \limsup_n \|T_n^m(x)\|_{X_m} < \infty\}$, $m = 1, \ldots$. As in Problem 108 it readily follows that each G_m is of first category in B. Since B is complete, by the Baire category theorem we get that $A = B \setminus \bigcup_m G_m$ is of second category in B.

110. (a) Since T is not onto, from the open mapping theorem it follows that $R(T)$ is of first category in B_1. Since $R(T)$ is dense in B_1, the interior of $\overline{R(T)}$ is nonempty, and so $R(T)$ is not nowhere dense in B_1.

(b) Let B_X be the closed unit ball in X. Suppose that $T(B_X)$ is nowhere dense. Then $T(X) = \bigcup_n T(nB_X) = \bigcup_n nT(B_X)$ is a countable union of nowhere dense sets and, hence, of first category. On the other hand, if the convex set $F = \overline{T(B_X)}$ has nontrivial interior, then it contains a ball $B_Y(0, r)$ for some $r > 0$. The usual proof of the open mapping theorem gives that $T(X) = Y$ and T is onto.

111. Observe that for $p < q \le \infty$, $L^q(I) \subset L^p(I)$ and the inclusion mapping $J : L^q(I) \to L^p(I)$ is bounded. Moreover, since $L^q(I)$ contains the continuous functions, it is dense in $L^p(I)$. By Problem 110, $L^q(I)$ is of first category in $L^p(I)$, but not nowhere dense there.

113. For the sake of argument suppose that T is onto. Then by the open mapping theorem T^{-1} is bounded, and by Problem 65 this is not the case.

114. Let $T_1(x_1) = T_2(y_1)$ and $T_1(x_2) = T_2(y_2)$. Then $T(x_1) = y_1$, $T(x_2) = y_2$, and for a scalar λ, $T_1(x_1 + \lambda x_2) = T_1(x_1) + \lambda T_1(x_2) = T_2(y_1) + \lambda T_2(y_2) = T_2(y_1 + \lambda y_2)$; hence $T(x_1 + \lambda x_2) = T(x_1) + \lambda T(x_2)$ and T is linear. Next, let $x_n \to x$ in B_1 and $T(x_n) \to y$ in B_2. Since T_1 is bounded, $T_1(x_n) \to T_1(x)$ in X. Let y_n be the unique element in B_2 such that $T_1(x_n) = T_2(y_n)$; then $T(x_n) = y_n$. Now, $y_n = T(x_n) \to y$ in B_2 and since T_2 is bounded, $T_2(y_n) \to T_2(y)$ in X. Since $T_2(y_n) = T_1(x_n)$ for all n, $T_2(y_n) \to T_1(x)$ in X and by the uniqueness of limits $T_1(x) = T_2(y)$. Now, by assumption there is only one y with this property and so $T(x) = y$ and by the closed graph theorem T is continuous.

115. Let $\{x_n\} \subset B$ and $y \in B_1$ be such that $x_n \to x$ in B and $T(x_n) \to y$ in B_1. Then for all $L \in B_1^*$ it follows that $L(y) = \lim_n L(T(x_n)) = \lim_n S(L)(x_n) = S(L)(x) = L(T(x))$. Hence, since B_1^* separates points in B_1, $T(x) = y$ and T is bounded by the closed graph theorem. Finally, since B separates points in B^*, $S = T^*$.

118. First observe that by scaling, in this case multiplying through by 2^{-n}, it follows that $B_2(0, 1/2^n) \subset B_1(0, r/2^n) + B_2(0, 1/2^{n+1})$ for all $n \geq 1$. Now let $\|y\|_2 < 1$, and $y_1 = y$. Then, if $\|y_n\|_2 < 1/2^{n-1}$, we have $y_n = x_n + y_{n+1}$, where $\|x_n\|_1 < r/2^{n-1}$ and $\|y_{n+1}\|_2 < 1/2^n$ for all $n \geq 1$. Now, by completeness, there exists $x \in X$ such that $\lim_n \sum_{k=1}^n x_k = x$ in $(X, \|\cdot\|_1)$ and, therefore, by the continuity of I, $\lim_n \sum_{k=1}^n x_k = x$ in $(X, \|\cdot\|_2)$. Moreover, since $y = \sum_{k=1}^n x_k + y_{n+1}$ and $\lim_n \|y_{n+1}\|_2 = 0$, then $\lim_n \sum_{k=1}^n x_n = y$ in $(X, \|\cdot\|_2)$, and $x = y$. Hence, $\|y\|_1 = \|x\|_1 \leq \sum_n \|x_n\|_1 < 2r$ and $B_2(0, 1) \subset B_1(0, 2r)$. Now, since for any $0 \neq y \in X$, $y/(2\|y\|_2)$ has $\|\cdot\|_2$ norm < 1, then $\|y\|_1 \leq 4r\|y\|_2$. Finally, the continuity of I gives that $\|y\|_2 \leq c\|y\|_1$ for all $y \in X$, and the norms are equivalent.

119. Let $Y^p = (Y, \|\cdot\|_p)$, $Y^q = (Y, \|\cdot\|_q)$; by assumption Y^p, Y^q are complete. We claim that the identity map $I : Y^p \to Y^q$, which is onto and invertible, is continuous. By the closed graph theorem it suffices to verify that if $\{x_n\} \subset Y$ converges in Y^p to y and to y_1 in Y^q, then $y = y_1$ μ-a.e. This follows at once from the fact that norm convergent sequences have μ-a.e. convergent subsequences.

120. (a) First, X is closed in $L^p(I)$. Indeed, let $\{x_n\} \subset X$ be such that $x_n \to x$ in $L^p(I)$. Then by Hölder's inequality $x_n \to x$ in $L^1(I)$, and since X is closed in $L^1(I)$, $x \in X$. Thus X is complete in $L^p(I)$.

Now, by Hölder's inequality, $\|x\|_1 \leq \|x\|_p$ and, therefore, the inclusion mapping $I : (X, \|\cdot\|_p) \to (X, \|\cdot\|_1)$ is continuous, 1-1, and onto. Therefore, since the spaces involved are complete, as in Problem 116 the inverse mapping, which is the inclusion mapping $I : (X, \|\cdot\|_1) \to (X, \|\cdot\|_p)$, is continuous and so $\|x\|_p \leq c\|x\|_1$ for some constant c and all $x \in X$. Thus the L^1 and L^p norms are equivalent in X. Moreover, since by Hölder's inequality, $\|x\|_2 \leq \|x\|_p$ it readily follows that $\|x\|_2 \leq \|x\|_p \leq c\|x\|_1 \leq c\|x\|_2$, and the L^p and L^2 norms are equivalent. Thus the identity mapping gives an isomorphism of X with the Hilbert space $L^2([0, 1])$.

(b) As in (a), $(X, \|\cdot\|_\infty)$ is isomorphic to $(X, \|\cdot\|_p)$ with $p < \infty$, which is separable. Then by Problem 5.67(b), $(X, \|\cdot\|_\infty)$ is finite dimensional.

Similarly, if $(X, \mathcal{M}, \mu)$ is a finite measure space, $1 \leq p < \infty$, and $W \subset L^p(X) \cap L^\infty(X)$ is a closed subspace of $L^p(X)$, then W is finite dimensional.

Indeed, along the lines of (a), W is a closed subspace of $L^\infty(X)$ and by the open mapping theorem there is a constant c such that $\|x\|_\infty \leq c\,\|x\|_p$ for $x \in W$. Thus $(X, \|\cdot\|_\infty)$ is isomorphic to $(X, \|\cdot\|_p)$ with $p < \infty$, which is separable, and as before $(X, \|\cdot\|_\infty)$ is finite dimensional.

121. (g) Observe that there exists $c > 0$ such that $\|x\|_\infty \leq c\,\|x\|_2$ for $x \in X$. First, if $1 \leq p \leq 2$, by (c), $\|x\|_\infty \leq M\|x\|_p \leq M\,\|x\|_2$. And, if $2 < p < \infty$, $\|x\|_p \leq \|x\|_2^{2/p}\|x\|_\infty^{1-2/p} \leq \|x\|_2^{2/p}(M\,\|x\|_p)^{1-2/p}$, and, therefore, $\|x\|_p \leq M^{p/2-1}\|x\|_2$, and so by (c) again, $\|x\|_\infty \leq M^{p/2}\|x\|_2$.

Let $\{e_k\}$, $1 \leq k \leq K$, be an ONS for $X \subset L^2(I)$ and note that for $x = \sum_{k=1}^{K} \lambda_k e_k$, $\|x\|_\infty^2 \leq c^2 \sum_{k=1}^{K} |\lambda_k|^2$. Hence $\left|\sum_{k=1}^{K} \lambda_k e_k(t)\right|^2 \leq c^2 \sum_{k=1}^{K} |\lambda_k|^2$ for a.e. $t \in I$, where the exceptional null set depends on the choice of the λ_k, but by restricting ourselves to rational sequences the set can be made independent of the λ_k. Then, letting $\lambda_k \to e_k(t)$ for each t outside the exceptional set it follows that $\left|\sum_{k=1}^{K} |e_k(t)|^2\right|^2 \leq c^2 \sum_{k=1}^{K} |e_k(t)|^2$, or $\sum_{k=1}^{K} |e_k(t)|^2 \leq c^2$ for a.e. $t \in I$. Whence integrating we get $K \leq c^2$, and X is finite dimensional.

123. Recall that as in Problem 8.44, $\|p\| = \max\{|p(t)| : t = 0, 1, \dots, n\}$ is a norm on $\mathcal{P}_n$. Now, since $\mathcal{P}_n$ is finite dimensional, by Problem 8.16 all norms in $\mathcal{P}_n$ are equivalent, and in particular $\|\cdot\|$ is equivalent to the L^1 norm and, therefore, there exist constants M, M_1 such that $\int_{-1}^{1} |p(t)|\, dt \leq M\,\|p\| \leq M_1 \int_{-1}^{1} |p(t)|\, dt$. Note that by Problem 98, M cannot be chosen to be independent of n.

126. (a) Problem 5.142 is relevant here.

(b) Necessity first. As in (a), taking $y = e^k$, we have that $x_k^n = \sum_j y_j x_j^n \to 0$ as $n \to \infty$. Since $x^n \in \ell^1$, it represents a bounded linear functional on c_0 and we have that $|\sum_k x_k^n y_k| \leq \|x^n\|_1 \|y\|_\infty$. The uniform boundedness principle then gives that $\sup_n \|x^n\|_1 < \infty$.

Sufficiency next. Fix $y \in c_0$ and note that for any $\varepsilon > 0$ there exists K such that $|y_k| < \varepsilon$ for $k \geq K$. Then

$$\sum_k y_k x_k^n = \sum_{k=1}^{K-1} y_k x_k^n + \sum_{k=K}^{\infty} y_k x_k^n = A + B\,,$$

say. Then $B \leq \sum_{k=K}^{\infty} |y_k|\,|x_k^n| \leq \varepsilon\|x_n\|_1 \leq \varepsilon \sup_n \|x_n\|_1$, which can be made arbitrarily small by choosing ε appropriately. Moreover, since $x_k^n \to 0$ for $1 \leq k \leq K-1$, pick N so that $|x_k^n| \leq \varepsilon/K$ for those k and $n \geq N$, and observe that $A \leq \sum_{k=1}^{K-1} |y_k|\,|x_k^n| \leq \|y\|_\infty \sum_{k=1}^{K-1} |x_k^n| \leq \varepsilon\|y\|_\infty$ for those n. The conclusion follows by combining these estimates.

127. For the sake of argument suppose that such a positive sequence a exists and let $T : \ell^\infty \to \ell^1$ be given by $T(x) = y$ where $y_n = a_n x_n$ for all n; T is bounded with $\|T\| \le \sum_n a_n$. Moreover, since a has the slowest rate of decay for convergent sequences, given a convergent sequence y, the sequence x with terms $x_n = y_n/a_n$ for all n is bounded, $T(x) = y$, and T is onto. Thus, T is surjective and, by the open mapping theorem, open.

Next observe that if $S : X \to Y$ is a bounded open mapping between Banach spaces X and Y, then the preimage of a dense subset in Y is dense in X. To see this let $A \subset Y$ be dense. Then for $U \subset X$ open, $S(U)$ is open in Y and hence has nontrivial intersection with A, and since $U \cap S^{-1}(A) \ne \emptyset$ for an arbitrary U, $S^{-1}(A)$ is dense in X.

In particular, the preimage of a dense subset of ℓ^1 under T is dense in ℓ^∞. Now, applying this observation to the dense subspace ℓ^1 consisting of sequences with finitely many nonzero terms, it follows that this subspace is dense in ℓ^∞, which is not the case.

129. (a) The statement is true.

(b) The statement is false. Let $x_n(t) = y_n(t) = n^{1/2}t^{n-1}$, $n \ge 1$. Then $\|x_n\| = \|y_n\| = n^{-1/2}$, and $x_n, y_n \to 0$ in X. So, if B were jointly continuous, it would follow that $B(x_n, y_n) \to B(0,0) = 0$, but this is not the case since $B(x_n, y_n) = n \int_0^1 t^{2n-2}\, dt = n/(2n-1) \to 1/2$ as $n \to \infty$.

130. For $y \in B_Y(0,1)$, let $T_y : X \to Z$ be the linear operator given by $T_y(x) = T(x, y)$. Then by the continuity of T for x fixed, $\|T_y(x)\|_Z \le c(x)\|y\|_Y \le c(x)$, and, therefore, $\sup_{y \in B_Y(0,1)} \|T_y(x)\|_Z \le c(x)$ for each $x \in X$. Hence by the uniform boundedness principle $\|T_y\| \le c < \infty$ for a constant $c > 0$ and all $y \in B_Y(0,1)$. Therefore $\|T(x,y)\|_Z = \|T_y(x)\|_Z \le c\|x\|_X$ for all $x \in X$ and $y \in B_Y(0,1)$. Now, $T_y(x)$ is homogenous in y, i.e., $T_{\lambda y}(x) = \lambda T_y(x)$ for each scalar λ, and, therefore, since $y' = y/\|y\|_Y \in B_Y(0,1)$, $T(x,y) = \|y\|_Y T(x, y')$ and $\|T(x,y)\|_Z \le c\|x\|_X \|y\|_Y$ for all $x \in X, y \in Y$.

Then, by bilinearity $T(x_0, y_0) - T(x, y) = T(x_0, y_0 - y) + T(x_0 - x, y)$, and, therefore, by the triangle inequality and the above estimate,

$$\begin{aligned}
\|T(x_0, y_0) - T(x, y)\|_Z &\le c\|x_0\|_X \|y - y_0\|_Y + c\|x_0 - x\|_X \|y\|_Y \\
&\le c\|x_0\|_X \|y_0 - y\|_Y + c\|x_0 - x\|_X \|y - y_0\|_Y \\
&\quad + c\|x_0 - x\|_X \|y_0\|_Y.
\end{aligned}$$

It follows that we may take $\delta = \min\{1, \varepsilon/c(1 + \|x_0\|_X + \|y_0\|_Y)\}$.

131. The statement is false.

134. For the sake of argument suppose that T^{-1} exists and is bounded. Then by Problem 133(b), $\|T(x)\| \ge c\|x\|$ for some $c > 0$ and all $x \in D(T)$, and in particular, $\|T(x_n)\| \ge c$ for all n, which is not the case since $T(x_n) \to 0$.

135. First, observe that T^{-1} is closed: If $\{y_n\} \subset D(T^{-1})$ is such that $y_n \to y$ in B_1 and $T^{-1}(y_n) \to x$ in B, there exists $\{x_n\} \subset D(T)$ with $T(x_n) = y_n \to y$ in B_1, and, consequently, $T^{-1}(y_n) = x_n \to x$ in B. Since T is closed, $x \in D(T)$ and $T(x) = y$, i.e., $y \in D(T^{-1})$ and $T^{-1}(y) = x$. But if T^{-1} is closed, $D(T^{-1}) = R(T)$ is closed.

Alternatively, since the mapping $f : X \times Y \to Y \times X$ given by $f(x, y) = (y, x)$ is an isometry and $G(T)$ is closed, $G(T^{-1}) = \big\{ (T(x), x) \, : \, x \in D(T) \big\} \subset Y \times X$ is closed.

136. From the given relations it readily follows that $T(S + I) + I = 0$ and $(S+I)T+I = 0$. Thus $I = -T(S+I) = -(S+I)T$ and $T^{-1} = -(S+I)$.

137. The statement is true when X is finite dimensional. Let S denote the inverse of TT_1. Then $TT_1S = I$ and T_1S is the right inverse of T. Now, since X is finite dimensional, the right inverse is the inverse and so $T^{-1} = T_1S$. Moreover, $T_1 = T^{-1}TT_1$ is the product of invertible mappings and so invertible. Note that the statement is true for $T_1 \cdots T_n$, $n \geq 2$.

On the other hand, the statement is not true when X is infinite dimensional. To see this consider $X = \ell^2$, T the backward shift given by $T(x) = y$ where $y_n = x_{n+1}, n \geq 1$, and T_1 the forward shift given by $T_1(x) = y$ where $y_1 = 0$ and $y_n = x_{n-1}$ for $n \geq 1$. In the case of an infinite-dimensional linear space one can only deduce that T is onto and T_1 is 1-1. Indeed, since $(TT_1)(TT_1)^{-1}(y) = T(T_1(TT_1)^{-1})(y) = y$, T is onto. And, $T_1(x) = T_1(y)$ implies $T(T_1(x)) = T(T_1(y))$ and applying inverses, $x = y$, and, consequently, T is 1-1.

Thus, if T is 1-1, T is invertible. Now, $T_1 = T^{-1}(TT_1)$, $T_1^{-1} = (TT_1)^{-1}T$ is continuous, and $T^{-1} = T_1(TT_1)^{-1}$ is bounded.

138. (a) $(TS - I + S)T = T(ST) - T + ST = I$, and so by uniqueness $TS - I + S = S$. Hence $TS = I$ and $T^{-1} = S$.

(b) Since $S(T(x)) = x$, $\|T(x)\| \geq \|x\|/\|S\|$ and T is 1-1. Now, since T is onto, T^{-1} is well-defined and $\|T^{-1}\| \leq \|S\|$. Then $S = S(TT^{-1}) = (ST)T^{-1} = T^{-1}$.

139. (a) First, necessity. Let $S : B_1 \to B$ denote the right inverse of T. Since $T(S(y)) = y$ for $y \in B_1$, T is onto. Next, since T is continuous, $K(T)$ is closed. To see that $R(S)$ complements $K(T)$ in B, let $x \in B$ and put $z = T(x)$ and $w = S(z)$. Now, $x = (x - w) + w$ with $w \in R(S)$, and since $T(x - w) = T(x) - T(S(z)) = z - z = 0$, $x - w \in K(T)$; thus $B = K(T) + R(S)$. Now, if $x \in K(T) \cap R(S)$, then $x = S(y)$, and so $0 = T(x) = T(S(y)) = y$, and, therefore, $x = S(y) = 0$. Finally, let $\{w_k\} \subset R(S)$, $w_n \to w$ in B. Then if $\{y_k\} \subset B_1$ is such that $S(y_k) = w_k$ for all k, since T is continuous, $\lim_k y_k = \lim_k T(S(y_k)) = \lim_k T(w_k) = T(w)$, and since S is continuous, $w = \lim_k S(y_k) = S(T(w)) \in R(S)$; thus $R(S)$ is closed and $B = K(T) \oplus R(S)$.

Sufficiency next; we seek to define the right inverse S of T. Given $y \in B_1$, let $x \in B$ be such that $T(x) = y$. Since $B = K(T) \oplus N$, $x = x_M + x_N$ with $x_M \in K(T)$ and $x_N \in N$, and we define $S(y) = x_N$. Then $T(S(y)) = T(x_N) = T(x - x_M) = T(x) = y$, and S is a right inverse of T. We claim that S is well-defined. Given $y \in B_1$, let $x \neq x' \in B$ be such that $T(x) = T(x') = y$; then $x' - x \in K(T)$. Now, $x' = (x' - x + x_M) + x_N$, and since $(x' - x + x_M) \in K(T)$ and the decomposition is unique, by picking x or x' as the pre-image of y, it follows that $S(y) = x_N$, and S is well-defined. The linearity of S follows from that of T. Finally, continuity. Let $y_n \to y$ in B_1 and $S(y_n) \to z$ in B; note that since $\{S(y_n)\} \subset N$, $z \in N$. Now, since $T(S(y_n)) = y_n$ for all n, $y = \lim_n y_n = T(z)$, and, therefore, since $T(S(y)) = y$, $z - S(y) \in K(T)$. Hence $z - S(y) \in K(T) \cap N = \{0\}$, $S(y) = z$, and the continuity of S follows from the closed graph theorem.

140. (a) Note that $T(I - ST) = (I - TS)T$. Now, if $(I - ST)(x) = 0$, then $T(I - ST)(x) = 0$, and so $(I - TS)T(x) = 0$. Hence, since $I - TS$ is 1-1, $T(x) = 0$, and, consequently, $x = ST(x) = 0$ and $I - ST$ is injective.

(b) Let $y \in X$. Since $I - TS$ is surjective there exists $x \in X$ such that $(I - TS)(x) = T(y)$. Thus with $z = S(x) + y$, $x = T(z)$, and, consequently, $T(I - ST)(z) = T(y)$, which implies that $(I - ST)(z) = y + v$ with $v \in K(T)$. Then setting $w = z - v$ it follows that $(I - ST)(w) = y$ and $I - ST$ is onto.

(c) Follows by combining (a) and (b). Alternatively, if $(I - TS)$ is invertible, $S(I - TS)^{-1}T + I$ gives the inverse of $(I - ST)$.

142. (a) Let $y_n \to y$ in Y and $T^{-1}(y_n) \to x$ in X. Let $\{x_n\} \subset D(T)$ be such that $y_n = T(x_n)$ for all n, and note that $x_n = T^{-1}(y_n) \to x$ and $T(x_n) = y_n \to y$. Then since T is closed it follows that $x \in D(T)$ and $y = T(x)$. Hence $y \in R(T)$, $x = T^{-1}(y)$, and T^{-1} is closed.

(b) By (a), T^{-1} is linear and closed. Now let $y_n \to y$ in Y and $T^{-1}(y_n) \to w$ in X. We rewrite this as $T^{-1}(y_n) \to w$ and $T(T^{-1}(y_n)) \to T(T^{-1}(y))$, which since T is closed gives $T(w) = T(T^{-1}(y))$. Thus $w = T^{-1}(y)$, T^{-1} is closed, and by the closed graph theorem, bounded.

143. By Problem 133(a), T is 1-1 and T^{-1} is bounded. Let $\ell(x) = L(T^{-1}(x))$; ℓ is linear and since $|\ell(x)| \leq \|L\| \|T^{-1}(x)\| \leq \|L\| \|T^{-1}\| \|x\|$, it is also bounded. Then $\ell(T(x)) = L(x)$ for all $x \in B$.

146. $(T - \lambda I)(-\sum_{n=0}^{N-1} \lambda^{-(n+1)}T^n) = -\sum_{n=1}^{N-1} \lambda^{-n}T^n + \sum_{n=0}^{N-1} \lambda^{-n}T^n = I$.

147. The limit is $(I - T)^{-1}$. The conclusion also holds under the assumption that $\lim_n \|T^n\|^{1/n} < 1$, for then there exists $\eta < 1$ such that $\lim_n \|T^n\|^{1/n} < \eta$, and so $\|T^n\| < \eta^n$ for n sufficiently large and $\sum_{n=0}^{\infty} \|T^n\|$

converges. Then if λ is a scalar such that $\lim_n \|T^n\|^{1/n} < |\lambda|^{-1}$, it follows that $\lim_n \|(\lambda T)^n\|^{1/n} < 1$, and, therefore, $I - \lambda T$ is invertible and $(I - \lambda T)^{-1} = \sum_{n=0}^{\infty} \lambda^n T^n$. This result is known as Neumann's formula.

148. $\{P_n\}$ converges to $(I - T)^{-1}$.

150. Let $R = \|T\|$. Then for $|r| > R$,

$$\|rx - T(x)\| \geq |\,\|rx\| - \|T(x)\|\,| \geq (|r| - \|T\|)\,\|x\|\,,$$

and by Problem 133(b), $rI - T$ is invertible. Now, for $x \in B$, let $y = (rI - T)^{-1}(x)$. Then as before if $|r| > \|T\|$, $\|x\| \geq (|r| - \|T\|)\,\|y\|$ and $\|(rI - T)^{-1}\| \leq 1/(|r| - \|T\|)$.

152. (a) Since $\|T_n - T\| < 1$ for n large and T is invertible, by Problem 149, $T_n = T_n - T + T$ is invertible.

(b) Let $T_n, T : L^2(I) \to L^2(I)$ be given by $T_n(x) = \chi_{[1/n,1]}\, x$ and $T(x) = \chi_I\, x$, respectively. Then T is invertible but by Problem 5.25, $\|T_n - T\| = \|\chi_{[0,1/n)}\|_\infty = 1$ for all n, and so the assumption for (a) doesn't hold. Note that no T_n is invertible. However, by Cauchy-Schwarz, $\|T_n(x) - T(x)\|_2 \leq n^{-1/2}\|x\|_2 \to 0$ as $n \to \infty$.

154. (c) The limit is $-T^{-1}ST^{-1}$.

155. Let $T : \ell_0^p \to \ell_0^p$ be given by $T(x) = y$ where $y_1 = 0$ and $y_n = x_{n-1}/2$ for $n \geq 2$; then $\|T(x)\|_p = 2^{-1}\|x\|_p$ and $\|T\| < 1$. Now, if x is such that $x - T(x) = e_1 = (1, 0, \ldots)$, then $x_n = 2^{1-n}$ for all $n \geq 1$. Thus $x \in \ell^p \setminus \ell_0^p$.

157. The equation is solvable for $|\lambda| < 1/2$.

158. (a) If $y = 0$, $x = 0$ is a solution of minimal norm and the estimate in (b) holds for any $c > 0$. Otherwise, if $y \neq 0$ and x_0 is a solution, then every other solution x is of the form $x = x_0 + z$ where $(I - T)(z) = 0$. Let $\phi(z) = \|x_0 + z\|$, $z \in K(I - T)$; since $K(I - T)$ is closed and $x_0 \notin K(I - T)$, ϕ is a continuous real-valued function bounded below by the positive number $a = \inf\{\phi(z) : z \in K(I - T)\}$. Let $\{z_n\} \subset K(I - T)$ be such that $\phi(z_n) = \|x_0 + z_n\| \to a$; note that since $\{x_0 + z_n\}$ converges, $\{z_n\}$ is bounded. Now, since $K(I - T)$ is finite dimensional, passing to a subsequence if necessary we may assume that $\{z_n\}$ converges to some $z_0 \in K(I - T)$, say, and then $\phi(z_n) \to \phi(z_0) = \|x_0 + z_0\| = a$. Thus the equation has the solution $x = x_0 + z_0$, which has minimal norm.

(b) For the sake of argument assume that if x is a solution of minimal norm corresponding to y the quantity $\|x\|/\|y\|$ is unbounded and pick $\{x_n\}$ and $\{y_n\}$ such that $\|x_n\|/\|y_n\| \to \infty$. Since for $\lambda \neq 0$, λy_n corresponds to the minimal solution λx_n, we can assume that $\|x_n\| = 1$ for all n, and in this case $\|y_n\| \to 0$. Since T is compact, passing to a subsequence if necessary we may assume that $\{T(x_n)\}$ converges to x_0', say. Moreover, since $x_n =$

$T(x_n) - y_n$ and $\lim_n y_n = 0$, $x_n \to x'_0$ and, consequently, $T(x_n) \to T(x'_0)$. Thus $x'_0 = T(x'_0)$ and $x'_0 \in K(I - T)$. Then, by the minimality of the norm of the solution x_n, it follows that $\|x_n - x'_0\| \geq \|x_n\| = 1$, and $\{x_n\}$ does not converge to x'_0, which is not the case. Thus $\|x\|/\|y\|$ is bounded and the desired inequality follows with $c = \sup(\|x\|/\|y\|)$.

From this result it follows at once that, if the equation $(I - T)(x) = y$ has a unique solution for every $y \in B_1$, then the solution is stable under perturbations, i.e., there exists a constant C such that, if $\|y_1 - y_2\|_{B_1} \leq \varepsilon$, the corresponding solutions x_1 and x_2 satisfy $\|x_1 - x_2\|_B \leq C\varepsilon$. Indeed, the unique solution is minimal, and so $\|x_1 - x_2\|_B \leq c\|y_1 - y_2\| \leq c\varepsilon$.

The observant reader will note that the fact that $R(I - T)$ is closed follows at once from this. Let $\{y_n\} \subset R(I - T)$ converge to y_0; passing to a subsequence if necessary we may assume that $\|y_n - y_0\| \leq 1/2^{n+1}$ and so $\|y_{n+1} - y_n\| \leq 1/2^n$. Let x_0 be a minimal solution to $(I - T)(x) = y_1$ and x_n a minimal solution to $(I - T)(x) = y_{n+1} - y_n$, $n = 1, 2, \ldots$. Then $\|x_n\| \leq c\|y_{n+1} - y_n\| \leq c/2^n$, and, therefore, $\sum_n \|x_n\|$ converges. Then, if $x' = \sum_{k=0}^{\infty} x_k$, it follows that $(I - T)(x') = \lim_n \sum_{k=0}^{n}(I - T)(x_k) = \lim_n(y_1 + \sum_{k=1}^{n}(y_{k+1} - y_k)) = \lim_n y_{n+1} = y_0$, and $y_0 \in R(I - T)$, which is therefore closed.

160. Note that this result implies that if $x - T(x) = 0$ only has the trivial solution $x = 0$, then $L - T^*(L) = \ell$ is solvable for all ℓ.

163. By Problem 133, $\|T(x)\| \geq c\,\|x\|$. Let $\varepsilon = c/2$. Then

$$\|T(x) + S(x)\| = \|T(x) - (-S(x))\| \geq \|T(x)\| - \|S(x)\|$$
$$\geq c\,\|x\| - (c/2)\,\|x\| = (c/2)\,\|x\|.$$

Then again by Problem 133, $T + S$ has the same properties.

164. If $f \in K(A)$, $A(f)(z) = 0$ for all $z \in D$. Hence $f(z) = 0$ for $z \in D \setminus \{0\}$ and by the continuity of f also $f(z) = 0$. Thus A is $1 - 1$. Take now any function $f_n \in C(D)$ such that $f_n(z) = 0$ if $|z| \geq 1/n$, integer n, and note that $\|Af_n\| = \max_D |z^3 f_n(z)| = \max_{|z| \leq 1/n} n^{-3}|f_n(z)|$. Therefore A does not satisfy $\|Af\| \geq c\|f\|$ and $R(A)$ cannot be closed.

165. (b) The statement is true. First, by Problem 57, T is bounded with $\|T\| \leq (\int_0^1 \int_0^1 k^2(s,t)\,ds\,dt)^{1/2} = 1/\sqrt{6}$. We claim that the solution is $x = (I - 2T)^{-1}(y)$. Since $\|2T\| \leq 2/\sqrt{6} < 1$, by Neumann's formula in Problem 147, $I - 2T$ is invertible with inverse $\sum_{k=0}^{\infty}(2T)^k(y)$, and this gives $x = \sum_{k=1}^{\infty}(2T)^k(y)$ with the series converging in X. Note that since $T(y) > 0$ for $y > 0$, this makes x positive for y positive and continuous for y continuous.

166. (a) iff (b) Necessity first. Since $\widetilde{T}$ is a bounded linear bijection, if $R(T)$ is closed in Y, it is complete and hence $\widetilde{T}$ is an isomorphism by the open mapping theorem. Conversely, if $\widetilde{T}$ is an isomorphism, then $R(T)$ is isomorphic to the Banach space $X/K(T)$ and is hence closed.

(b) iff (c) Necessity first. Let $x \in X$. Since $(\widetilde{T})^{-1}$ is bounded, let M be such that $\|[x]\|_{X/K(T)} \leq M\|\widetilde{T}([x])\|_Y$ for every $x \in X$. Then by the definition of quotient norm there exists $z \in K(T)$ such that $\|x + z\|_X \leq 2\|[x]\|_{X/K(T)}$. Let $x' = x + z$ and observe that, since $\widetilde{T}([x]) = T(x)$, we have $\|x'\|_X \leq 2M\|T(x)\|_Y$.

Sufficiency next. Since $x' - x \in K(T)$, it readily follows that $\|[x]\|_{X/K(T)} \leq \|x'\|_X \leq M\|T(x)\|_Y = \|\widetilde{T}([x])\|_Y$ and, therefore, $(\widetilde{T})^{-1}$ is bounded.

(c) iff (d) Necessity first. Given $y \in R(T)$ with $\|y\|_Y < 1/M$, let $x \in X$ be such that $y = T(x)$ and pick $x' \in X$ with $T(x') = T(x)$ and $\|x'\|_X \leq M\|T(x)\|_Y < 1$. Then $B_{R(T)}(0, 1/M) \subset T(B_X(0,1))$.

Sufficiency next. Given $x \in X$, let $y = T(x)$. Then $(r/\|y\|_Y)y \in B_{R(T)}(0, r)$ and there exists $x'' \in B_X(0,1)$ with $T(x'') = (r/\|y\|_Y)y$. Put $x' = (\|y\|_Y/r)x''$. Then $T(x') = T(x)$ and $\|x'\|_X = (1/r)\|T(x)\|_Y$. Therefore the conclusion holds with $M = 1/r$.

(d) iff (e) Necessity first. By the linearity of T we see that, for any $x \in X$ and $\varepsilon > 0$, $T(B_X(x, \varepsilon)) \supset B_Y(T(x), r\varepsilon)$. It then readily follows that $T(U)$ is open in Y whenever U is open in X. Conversely, $T(B_X(0,1))$ is open and certainly it contains $0 \in Y$, so $T(B_X(0,1))$ contains some open ball centered at 0.

167. Necessity first. Let $\widetilde{T} : X/K(T) \to Y$ be the operator defined in Problem 21; since $R(\widetilde{T}) = R(T)$ is closed and $\|\widetilde{T}\| = \|T\|$, $\widetilde{T}$ is a continuous homeomorphism and by the inverse mapping theorem $\widetilde{T}^{-1}$ is continuous and $\|\widetilde{T}^{-1}(y)\|_{X/K(T)} \leq c\|y\|_Y$, $y \in R(\widetilde{T})$, or equivalently, $\|[x]\|_{X/K(T)} \leq c\|\widetilde{T}([x])\|_Y$ for all $x \in X$. Therefore, since $T(x) = \widetilde{T}([x])$, it follows that $d(x, K(T)) = \|[x]\|_{X/K(T)} \leq c\|\widetilde{T}([x])\|_Y = c\|T(x)\|_Y$ for all $x \in D(T)$.

Sufficiency next. Let $\{y_n\} \subset R(T)$ be such that $y_n \to y$ in Y, and pick $\{x_n\} \subset X$ with $T(x_n) = y_n$ for all n; then $d(x_n - x_m, K(T)) \leq c\|y_n - y_m\|_Y \to 0$ as $m, n \to \infty$. Thus $\{[x_n]\}$ is Cauchy in $X/K(T)$ and, therefore, it converges to $[x] \in X/K(T)$, say. Let $\{z_n\} \subset K(T)$ be such that $\|x_n - x - z_n\|_X \leq 2\|[x_n] - [x]\|_{X/K(T)}$ for all n. Then $\{x_n - z_n\}$ converges to $x \in X$ and by the continuity of T, $T(x_n - z_n) = T(x_n) \to T(x)$. Thus by the uniqueness of limits $y = T(x)$ and $y \in R(T)$, which is therefore closed.

Observe that since for a bounded linear isometry $T : B \to B_1$, $\|T(x)\| = \|x - y\|$ for all $y \in K(T)$, then $R(T)$ is closed in B_1.

170. We claim that the conclusion holds with $R = r/(1 - \varepsilon)$. Let $y \in B_Y(0,1)$. We define recursively sequences $\{x_n\} \subset X$ and $\{y_n\} \subset Y$ with $y_1 = y$ and satisfying the following properties: (i) $\|y_n\|_Y < \varepsilon^{n-1}$,

(ii) $\|x_n\|_X < r\|y_n\|_Y$, (iii) $\|T(x_n) - y_n\|_Y < \varepsilon^n$, and (iv) $y_{n+1} = y_n - T(x_n)$. We proceed as follows. Let $y_1 = y$; since $\|y_1\|_Y < 1$, by assumption there exists x_1 with the required properties. Then let $y_2 = y_1 - T(x_1)$, $\|y_2\|_Y < \varepsilon$. In general, if y_n has been picked with $\|y_n\|_Y < \varepsilon^{n-1}$, let $y' = \varepsilon^{1-n} y_n \in B_Y(0,1)$. Then by assumption there exists $x' \in B_X(0,r)$ such that $\|y' - T(x')\|_Y < \varepsilon$. If we now put $x_n = \varepsilon^{n-1} x'$ and $y_{n+1} = y_n - T(x_n)$, the required properties are satisfied.

Now, by (ii), $\sum_n \|x_n\|_X < r/(1-\varepsilon) = R$ and so $\sum_n x_n$ converges to some $x \in X$ with $\|x\|_X < R$, and by (iii) and (iv), $\|y - \sum_{k=1}^n T(x_k)\|_Y = \|y_{n+1}\|_Y < \varepsilon^n$, so that $\lim_k \sum_k T(x_k) = y$ in Y. Thus, by the continuity of T, $T(x) = y$.

173. (a) Necessity first. Since $K(T) = R(T^*)^{\perp} = \{0\}$, T is injective. Furthermore, since Y^*, X^* are Banach spaces and $T^* : Y^* \to X^*$ is surjective, by the open mapping theorem $B_{X^*}(0,r) \subset T^*(B_{Y^*}(0,1))$ for some $r > 0$ and, therefore, scaling, $B_{X^*}(0,2) \subset T^*(B_{Y^*}(0,2/r))$. Thus if $\|L\|_{X^*} < 2$ there exists $\ell \in Y^*$ such that $\|\ell\|_{Y^*} \leq 2/r$ and $T^*(\ell) = L$. Now, let $0 \neq x \in X$; then by Hahn-Banach there exists $L \in X^*$ such that $\|L\|_{X^*} = 1$ and $L(x) = \|x\|_X$. Hence there is $\ell \in Y^*$ with norm $r/2$ such that $T^*(\ell) = L$. So $\|x\|_X = L(x) = T^*(\ell)(x) = \ell(T(x)) \leq \|\ell\|_{Y^*}\|T(x)\|_Y < (2/r)\|T(x)\|_Y$, and by Problem 133, T^{-1} is bounded.

Conversely, given $L \in X^*$, $L \circ T^{-1}$ is a bounded linear functional on $R(T)$. By Hahn-Banach there exists $\ell \in Y^*$ such that $\ell(y) = L \circ T^{-1}(y)$ for all $y \in R(T)$. Now, for all $x \in X$, $T(x) \in R(T)$ and so $T^*\ell(x) = \ell(Tx) = L(T^{-1}(Tx)) = L(x)$. Thus T^* is surjective.

(b) Since $X^{**} = X$, $(T^*)^*(x)(L) = x(T^*L) = T^*L(x) = L(Tx) = (Tx)(L)$, and so $(T^*)^* = T$. Then, since $A^{\perp} = \overline{A}^{\perp}$, $K(T) = K(T^{**}) = \{0\}$ iff $\overline{R(T^*)} = X^*$.

174. T is onto iff T^* is bounded below with bound c, say. Then, since $\|S\| = \|S^*\|$, if $\varepsilon < c/2$, $\|T^*(\ell) + S^*(\ell)\|_{X^*} \geq \|T^*(\ell)\|_{X^*} - \|S^*(\ell)\|_{X^*} \geq (c/2)\|\ell\|_{X^*}$, and so $\|T^* + S^*\| \geq c/2$ is also bounded below, and, consequently, $T + S$ is onto.

176. Since $(I^*(x^*))(y) = x^*(I(y))$ for all $y \in Y$, it follows that $I^*(x^*) = x^*|_{Y^*}$ for all $x^* \in X^*$, and I^* is the restriction operator to Y^*.

181. Let $U : R(S^*) \to X^*$ be the linear mapping given by $U(S^*(\ell)) = T^*(\ell)$; since $K(S^*) = R(S)^{\perp} \subset R(T)^{\perp} = K(T^*)$, U is well-defined. We claim that U is bounded. For the sake of argument suppose this is not the case and let $\{\ell_n\} \subset Y^*$ be such that $\|S^*(\ell_n)\|_{X^*} = 1$ for all n and $\|T^*(\ell_n)\|_{X^*} \to \infty$. Now take $x \in X$. Then, by assumption there exists $z \in X$ such that $S(z) = T(x)$. So $T^*(\ell_n)(x) = \ell_n(T(x)) = \ell_n(S(z)) =$

$S^*(\ell_n)(z)$ and so $|T^*(\ell_n)(x)| \leq \|z\| < \infty$ for each n. But as in Problem 94, $\sup_n \|T^*(\ell_n)\|_{X^*} < \infty$, which is not the case.

182. (c) Let $C : X \to Y = L^\infty(\Omega)$ be given by $C(x) = S(x)$ for each $x \in X$. Now, if $x_k \to x$ in X and $C(x_k) \to y$ in Y, then $y_k = C(x_k) = S(x_k) \to S(x) = C(x)$ in Y, and with $y_k \to y$ this gives $y = C(x)$. Thus C is continuous, so $x_k \to x$ in X implies uniform convergence (L^∞ norm) for $S(x_k) \to S(x)$, since the continuous functions are dense in X, each $S(x)$ is the uniform limit of continuous functions, and hence continuous.

183. (a) Not every Schauder basis is a Hamel basis: $B = \{e_n\}$ is a Schauder basis and not a Hamel basis for ℓ^p, $1 \leq p < \infty$. Since $x = \sum_n x_n e_n$ for every $x \in \ell^p$, B is a Schauder basis for ℓ^p. Now, if $x \in \ell^p$ has $x_n \neq 0$ for infinitely many n, then $x - \sum_{k=1}^n x_k e_k \neq 0$ for all n, x cannot be expressed as a finite sum of elements in B, and B is not a Hamel basis for ℓ^p.

(b) If $x \in c_0$, then $\|x - \sum_{n=1}^N x_n e_n\| = \sup_{n>N} |x_n| \to 0$ as $N \to \infty$, and, therefore, $x = \sum_n x_n e_n$. Also, if $x = \sum_n x'_n e_n$, then $|x_n - x'_n| \leq \lim_N \|x - \sum_{n=1}^N x'_n e_n\| = 0$. So $x'_n = x_n$ for all n, and the expansion is unique. Hence $\{e_n\}$ is a Schauder basis for c_0.

(c) No.

(d) For $x \in c$, let $x_\infty = \lim_n x_n$ and recall that as in Problem 8.111(c), $x = x_\infty e + \sum_n (x_n - x_\infty) e_n$, where the representation is unique and the series converges in ℓ^∞ and hence in c. Now, since $x \in c_0$ can be written as $(0, x_1, x_2, \ldots)$ in terms of the coordinates for c, this gives an intuitive explanation why $\dim(c/c_0) = 1$.

185. Clearly $\| \cdot \|_1$ is a norm. Now, since $\|x - \sum_{k=1}^n \lambda_k x_k\| \to 0$, we have $\|x\| \leq \|x - \sum_{k=1}^n \lambda_k x_k\| + \sup_n \|\sum_{k=1}^n \lambda_k x_k\|$, and $\|x\| \leq \|x\|_1$. Also note that since $\|P_n(x)\|_1 \leq \sup_m \|\sum_{k=1}^m \lambda_k x_k\| = \|x\|_1$, P_n maps $(X, \| \cdot \|_1)$ into itself with norm ≤ 1 for all n.

Next, completeness. Let $\sum_k \|y_k\|_1 < \infty$. As noted above $\sum_k \|y_k\| < \infty$ and since X is complete there exists $z \in X$ such that $\lim_n \sum_{k=1}^n y_k = z$ in X. Similarly, $\sum_k \|P_n(y_k)\| \leq \sum_k \|P_n(y_k)\|_1 \leq \sum_k \|y_k\|_1 < \infty$, and, consequently, there are $\{z_n\} \subset X$ such that $\lim_N \sum_{k=1}^N P_n(y_k) = z_n$ for all n. And, since $\|z_n - \sum_{k=1}^N P_n(y_k)\| \leq \sum_{k=N+1}^\infty \|P_n(y_k)\| \leq \sum_{k=N+1}^\infty \|P_n(y_k)\|_1 \leq \sum_{k=N+1}^\infty \|y_k\|_1 < \varepsilon$ for N large enough, the convergence is uniform in n. Now, since $R(P_m)$ is finite dimensional along the lines of Problem 20, P_n is continuous on $R(P_m)$ for all n, m, and, therefore, $P_n(z_m) = P_n(\lim_N \sum_{k=1}^N P_m(y_k)) = \lim_N P_n P_m(\sum_{k=1}^N y_k) = \lim_N P_{n \wedge m}(\sum_{k=1}^N y_k) = z_{n \wedge m}$. This implies, in particular, that $P_n(z_m) = z_n$ for all $m \geq n$, and so for some scalar sequence λ, $z_n = \sum_{k=1}^n \lambda_k x_k$ for all n. Then by the definition of z, $P_n(z) = z_n$ for all n.

Finally, $\|P_n(z - \sum_{k=1}^{N} y_k)\| = \|z_n - \sum_{k=1}^{N} P_n(y_k)\| \leq \|\sum_{k=N+1}^{\infty} P_n(y_k)\| \leq \sum_{k=N+1}^{\infty} \|P_n(y_k)\| \leq \sum_{k=N+1}^{\infty} \|P_n(y_k)\|_1 \leq \sum_{k=N+1}^{\infty} \|y_k\|_1 < \varepsilon$ uniformly in n. Thus $\|z - \sum_{k=1}^{N} y_k\|_1 = \sup_n \|P_n(z - \sum_{k=1}^{N} y_k)\| \leq \varepsilon$ for N sufficiently large and $z = \sum_k y_k$ in $(X, \|\cdot\|_1)$, which is therefore complete.

The equivalence of the norms follows from the inverse mapping theorem.

186. The statement is false.

187. Since $L_n(x_n) = 1$, $1 \leq \|L_n\| \|x_n\|$. Next, let $x = \sum_n \lambda_n x_n \in B$. Then since $\lambda_n x_n = \sum_{k=1}^{n} \lambda_k x_k - \sum_{k=1}^{n-1} \lambda_k x_k$ it follows that $|L_n(x)| \|x_n\| = \|\lambda_n x_n\| \leq \|\sum_{k=1}^{n} \lambda_k x_k\| + \|\sum_{k=1}^{n-1} \lambda_k x_k\| \leq 2 \sup_n \|\sum_{k=1}^{n} \lambda_k x_k\| = 2\|x\|_1 \leq 2\,c\,\|x\|$, and, therefore, $\|x_n\| \|L_n\| \leq 2c$.

Alternatively, note that for an integer m, with $P_0 = 0$, $P_m(x) - P_{m-1}(x) = L_m(x)x_m$, and if we pick $\ell_m \in B^*$ with $\ell_m(x_m) = 1$ and $\|\ell_m\| = 1/\|x_m\|$, it follows that $L_m(x) = \ell_m(P_m(x) - P_{m-1}(x))$, and, hence, $L_m \in B^*$ with $\|L_m\| \leq 2 \sup_k(\|P_k\|/\|x_k\|)$.

190. Let $\{x_n\}$ be a Schauder basis for B and $x_0 \in B$ with infinitely many nonvanishing coefficients with respect to the basis. Let $Y_n = \mathrm{sp}\{x_1, \ldots, x_n\}$, $Y = \bigcup_n Y_n$, and $X = \mathrm{sp}\{x_0, Y\}$; note that since Y is dense in H, so is X. Now, decompositions in X are unique: If $x = \lambda x_0 + y_\lambda = \mu x_0 + y_\mu$, then $(\lambda - \mu)x_0 = y_\mu - y_\lambda$, which can only hold if $\lambda = \mu$ and $y_\lambda = y_\mu$. Let T be defined by $T(x) = \lambda x_0$ where $x = \lambda x_0 + y \in X$; T is well-defined on X and obviously linear.

We claim that T is not closable. Indeed, let $y_N = \sum_{n=1}^{N} a_n(x_0)x_n \in Y$; then, $y_N \to x_0$ in B and $T(y_N) = 0$. Thus $\lim_N(x_0 - y_N, T(x_0 - y_N)) = \lim_N(x_0 - y_N, x_0) = (0, x_0) \in \overline{G(T)}$. Finally, (x_0, x_0) and $(x_0, 0)$ both belong to $\overline{G(T)}$, and, consequently, $\overline{G(T)}$ is not the graph of a linear mapping.

192. With $J_B : B \to B^{**}$ the natural map and $J_B^* : B^{***} \to B^*$ its adjoint, let $P = J_{B^*} \circ J_B^*$; we claim that $P : B^{***} \to B^{***}$ is well-defined and bounded. First, for all $L \in B^*$, since for any $x \in B$ we have $J_B^*(L)(x) = J_B(x)(L) = L(x)$, it follows that $J_B^*|_{B^*} = I_{B^*}$. Moreover, since $J_B^*(L) = L$ for all $L \in B^*$, $P|_{B^*} = I_{B^*}$. Note that the argument actually shows that $R(P) \subset B^*$ and $P^2 = P$. Lastly, to compute the norm of P, since natural maps are isometries, $\|P\| \leq \|J_B^*\| \|J_B^*\| = \|J_{B^*}\| \|J_B\| = 1$. On the other hand, $\|P\| \geq \|P|_{B^*}\| = 1$.

193. The statement is false. Let J_B be the natural map from B into B^{**} where B is a nonreflexive Banach space. Then, with $J_{B^*} : B^* \to B^{***}$ the natural map from B^* into B^{***}, restricting ourselves to elements of B^{**} of the form $J_B(x)$, $x \in B$, for $L \in B^*$ we have $L(x) = J_B(x)(L) = J_{B^*}(L)(J_B(x)) = J_B^*(J_{B^*}(L))(x)$, and, consequently, $J_B^* \circ J_{B^*}|_{B^*} = I_{B^*}$ is the identity on B^*. Then, as observed in Problem 139(a) J_B^* is onto. For the sake of argument suppose that J_B^* is 1-1. Then J_{B^*} maps B^* onto B^{***} and B^* is reflexive, contrary to the fact established in Problem 8.152 that it is not.

194. $R(\pi^*) = M^\perp$; more precisely, π^* is bijective from $(X/M)^*$ onto $M^\perp$, with $\|\pi^*(\xi)\|_{X^*} = \|\xi\|_{X/M}$ for all $\xi \in (X/M)^*$. Thus, $(X/M)^*$ is isometrically isomorphic to $M^\perp$.

195. (b) First note that for $\ell \in K(T^*)$, $\ell(T(x)) = T^*(\ell)(x) = 0$ for all $x \in X$. Let $\phi : K(T^*) \to (Y/R(T))^*$ be defined by $\phi(\ell) = \lambda$ where $\lambda([y]) = \ell(y)$ for all $[y] \in Y/R(T)$; since $\ell(y + T(x)) = \ell(y)$ for all $x \in X$, it is always the case that λ is a well-defined linear functional on $Y/R(T)$ with $\|\lambda\| \leq \|\ell\|$.

Now, when $R(T)$ is closed, the canonical projection $\pi : Y \to Y/R(T)$ is bounded, and given a linear functional λ on $Y/R(T)$, let $\ell \in Y^*$ be defined by $\ell = \lambda \circ \pi$; then $\|\ell\| \leq \|\lambda\|$. Then $(T^*\ell)(x) = (\lambda \circ \pi)(T(x)) = 0$ for all $x \in X$ and $\ell \in K(T^*)$. It is readily seen that this inverts the previous construction.

196. Necessity first. Since T is open, $B_Y(0, c) \subset \overline{T(B_X(0, 1))}$ for some $c > 0$, and, consequently, also $\overline{B_Y(0, c)} \subset \overline{T(B_X(0, 1))}$. Then, if $\ell \in Y^*$, it follows that $\sup_{\|y\|_Y \leq c} |\ell(y)| \leq \sup_{\|x\|_X \leq 1} |\ell(T(x))| = \sup_{\|x\|_X \leq 1} |T^*(\ell)(x)|$. This gives that $c\|\ell\|_{Y^*} \leq \|T^*(\ell)\|_{X^*}$.

Conversely, let $Z = \overline{T(B_X(0, 1))} \subset Y$ and $w \in Y \setminus Z$; since $B_X(0, 1)$, and hence $T(B_X(0, 1))$, is balanced and convex, Z is closed, balanced and convex. Then by Hahn-Banach there is a bounded linear functional ℓ on Y such that $\sup_{z \in Z} |\ell(z)| < |\ell(w)|$. Now, since $|\ell(w)| \leq \|\ell\|_{Y^*}\|w\|_Y$ and by the properties of the Minkowski functional $\sup_{z \in Z} |\ell(z)| = \|T^*(\ell)\|_{X^*} \geq c\|\ell\|_{Y^*}$, it readily follows that $c\|\ell\|_{Y^*} < |\ell(w)| \leq \|\ell\|_{Y^*}\|w\|_Y$, and, consequently, $\|w\|_Y > c$ for all $w \in Y \setminus Z$. Thus $\overline{B_Y(0, c)} \subset Z$ and since X is complete it follows that $B_Y(0, c) \subset T(B_X(0, 1))$ and T is open.

The reader should verify that analogous statements with the roles of T and T^* reversed also hold.

197. (a) Necessity first. Let c be such that $\|x\|_X \leq c\|T(x)\|_Y$ for all $x \in X$. Now, any $L \in X^*$ induces a well-defined bounded linear functional ℓ on $Z = R(T) \subset Y$ by $\ell(T(x)) = L(x)$; note that $|\ell(T(x))| \leq \|L\|_{X^*}\|x\|_X \leq c\|L\|_{X^*}\|T(x)\|_Y$, so that $\|\ell\|_{Y^*} \leq c\|L\|_{X^*}$. Then, by Hahn-Banach ℓ extends to $\ell' \in Y^*$ with $\|\ell'\|_{Y^*} = \|\ell\|_{Y^*}$. This gives $\ell' \in Y^*$ with $T^*(\ell') = L$ and $\|\ell'\|_{Y^*} \leq c\|L\|_{X^*}$, and by Problem 166, $\widetilde{T^*}$ is an isomorphism.

Conversely, given $x \in X$, by Problem 8.78 there exists $L \in X^*$ with $\|L\|_{X^*} = 1$ and $L(x) = \|x\|_X$. Then by Problem 166 there exists $\ell \in Y^*$ with $T^*(\ell) = L$ and $\|\ell\|_{Y^*} \leq c$. Thus $\|x\|_X = L(x) = \ell(T(x)) \leq \|\ell\|_{Y^*}\|T(x)\|_Y \leq c\|T(x)\|_Y$, and T is an isomorphic embedding.

Chapter 20

Hilbert Spaces

Solutions

3. The statement is false as the example $x = (1, i), y = (1, -i)$ in $\mathbb{C}^2$ shows. On the other hand, if X is a real inner product space $\|x + y\|^2 = \|x\|^2 + \|y\|^2 + 2\langle x, y \rangle$ and so $2\langle x, y \rangle = 0$.

5. (a) implies (b) Since $\|x \pm \lambda y\|^2 = \|x\|^2 \pm 2\Re(\overline{\lambda}\langle x, y \rangle) + |\lambda|^2\|y\|^2$ the implication follows at once.

(b) implies (c) $\|x + \lambda y\|^2 = (\|x + \lambda y\|^2 + \|x - \lambda y\|^2)/2 = \|x\|^2 + |\lambda|^2\|y\|^2 \geq \|x\|^2$.

(c) implies (a) Note that for all scalars λ, $\|x + \lambda y\|^2 - \|x\|^2 = 2\Re(\overline{\lambda}\langle x, y \rangle) + |\lambda|^2\|y\|^2 \geq 0$. Now, the value $\lambda = -\langle x, y \rangle/\|y\|^2$ gives

$$-2\frac{|\langle x, y \rangle|^2}{\|y\|^2} + \frac{|\langle x, y \rangle|^2}{\|y\|^2} = -\frac{|\langle x, y \rangle|^2}{\|y\|^2} \geq 0,$$

which can only hold if $\langle x, y \rangle = 0$.

6. If $x = 0$ or $y = 0$ there is nothing to prove so assume $x \neq 0, y \neq 0$. First, always $\|x\| \geq d(x, Y)$. Now, if $\langle x, y \rangle = 0$, $\langle x, z \rangle = 0$ for all $z \in Y$. Thus by the Pythagorean theorem $\|x - z\|^2 = \|x\|^2 + \|z\|^2 \geq \|x\|^2$ for all $z \in Y$, and, consequently, $d(x, Y) \geq \|x\|$ and x is orthogonal to Y.

On the other hand, if $\langle x, y \rangle \neq 0$, let $\lambda = \langle x, y \rangle/\langle y, y \rangle$; then $\lambda \neq 0$ and $\langle x - \lambda y, z \rangle = 0$ for all $z \in Y$. Therefore, since $x = (x - \lambda y) + \lambda y$, by the Pythagorean theorem $\|x\|^2 = \|x - \lambda y\|^2 + |\lambda|^2\|y\|^2 > \|x - \lambda y\|^2 \geq d(x, Y)^2$. Hence $\|x\| > d(x, Y)$ and x is not orthogonal to Y.

9. $x = x_0/\|x_0\|$ and the distance is $\|x - x_0\| = |\|x_0\| - 1|$.

10. Since T^{-1} is an isometric linear isomorphism and the parallelogram law holds in H, $\|x - y\|_B^2 + \|x + y\|_B^2 = \|T^{-1}(x - y)\|_H^2 + \|T^{-1}(x + y)\|_H^2 = 2\|T^{-1}(x)\|_H^2 + 2\|T^{-1}(y)\|_H^2 = 2\|x\|_B^2 + 2\|y\|_B^2$. Thus the norm in B satisfies the parallelogram law and B is a Hilbert space.

11. The statement is not true in general. First, $\langle \cdot, \cdot \rangle_1$ is always an inner product and $\|x\|_X^2 = \langle x, x \rangle_1 = \sum_k \lambda_k |x_k|^2$. Suppose that $\sum_k \lambda_k < \infty$ and consider the sequence $\{x^n\} \subset X$ given by $x^n = e_1 + \cdots + e_n$, $n \geq 1$. Given $\varepsilon > 0$, let N be such that $\sum_{k=N}^\infty \lambda_k \leq \varepsilon$. Now, for all $m > n > N$, $\|x^m - x^n\|_X^2 = \sum_{k=n+1}^m \lambda_k \leq \varepsilon$. Then $\{x^n\}$ is a Cauchy sequence in X but since $(1, \ldots, 1, \ldots) \notin \ell^2$, it does not converge in X.

12. Given $x, y \in \overline{X}$, let $\{x_n\}, \{y_n\} \subset X$ be such that $x_n \to x$ and $y_n \to y$ in $\overline{X}$. We claim that $\{\langle x_n, y_n \rangle\}$ is a Cauchy sequence. Indeed, since $\|x_n\|, \|y_n\| \leq c < \infty$, it follows that for sufficiently large n, m,

$$|\langle x_n, y_n \rangle - \langle x_m, y_m \rangle| \leq |\langle x_n - x_m, y_n \rangle| + |\langle x_m, y_n - y_m \rangle|$$
$$\leq c \|x_n - x_m\| + c \|y_n - y_m\| \to 0$$

as $n, m \to \infty$. Therefore $\{\langle x_n, y_n \rangle\}$ converges and we define $\langle x, y \rangle_1 = \lim_n \langle x_n, y_n \rangle$. Now, this expression is well-defined since if $x_n' \to x$ and $y_n' \to y$, then $|\langle x_n, y_n \rangle - \langle x_n', y_n' \rangle| \leq |\langle x_n - x_n', y_n \rangle| + |\langle x_n', y_n - y_n' \rangle| \to 0$ as $n \to \infty$. Next, $(\overline{X}, \langle \cdot, \cdot \rangle_1)$ is a Hilbert space. $\langle \cdot, \cdot \rangle_1$ is bilinear (or sesquilinear if the underlying field is $\mathbb{C}$) because limits and the inner product are linear. It is symmetric (or conjugate symmetric) and positive for the same reason. Finally, $\langle x, x \rangle_1 = 0$ iff $\langle x_n, x_n \rangle \to 0$, so $x_n \to 0$ and $x = 0$; hence it is positive-definite. Thus $\langle \cdot, \cdot \rangle$ extends to $\overline{X}$ and for $x \in \overline{X}$, $\|x\|_1^2 = \lim_n \|x_n\|^2 = \lim_n \langle x_n, x_n \rangle = \langle x, x \rangle_1$.

13. (a) The functions $x(t) = 1/2 - t$, $y(t) = x(t)$ for $0 \leq t \leq 1/2$, and $y(t) = -x(t)$ for $1/2 < t \leq 1$ do not satisfy the parallelogram law and, consequently, $C(I)$ is not a Hilbert space.

14. By Problem 8.160(b), X/M equipped with $\|[x]\| = \inf\{\|x - m\| : m \in M\} = d(x, M)$ is a complete normed space so it only remains to verify that the norm is induced by an inner product. By the projection theorem $\|[x]\| = \|x - x_M\|$ where x_M denotes the unique element in M such that $\|x - x_M\| = d(x, M)$. Given $[x], [y]$ in H/M, let x and y be elements in $[x]$ and $[y]$, respectively, and define $\langle [x], [y] \rangle = \langle x - x_M, y - y_M \rangle$; we claim that this is a well-defined inner product in $H/M \times H/M$. Suppose that x, x' are elements in $[x]$ and y, y' in $[y]$, respectively; then $x - x' \in M$. Observe that for $m \in M$, $x' - m = x - (m + x - x')$ and since $\|x' - m\|$ is minimized when $m = x_M'$ and $\|x - (m + x - x')\|$ when $m + x + x' = x_M$, by uniqueness it follows that $x - x_M = x' - x_M'$. Similarly, $y - y_M = y' - y_M'$, and, consequently, $\langle x - x_M, y - y_M \rangle = \langle x' - x_M', y' - y_M' \rangle$ and $\langle \cdot, \cdot \rangle$ is well-defined. All the other properties of inner product are readily verified.

15. (a) The statement is false in general.

(b) Let $H' = H/K(T)$; by Problem 14, H' is a Hilbert space. Now, T induces a linear mapping $T' : H' \to B$ which in addition to being bounded and onto is also 1-1. By the inverse mapping theorem T' has a continuous inverse and establishes a linear homeomorphism between B and H'.

(c) The statement is false. Let B be a reflexive Banach space which is not a Hilbert space. Then $B \oplus B^*$ equipped with the norm $\max(\|b\|_B, \|b^*\|_{B^*})$ is isometrically isomorphic to its dual, without being a Hilbert space.

16. Expanding, $\|x + e^{it}y\|^2 e^{it} = (\|x\|^2 + \|y\|^2)\, e^{it} + \langle x, y \rangle + \langle y, x \rangle\, e^{i2t}$, which when integrated gives $(2\pi)^{-1} \int_0^{2\pi} \|x + e^{it}y\|^2 e^{it}\, dt = \langle x, y \rangle$.

17. Since $9,900$ ordered pairs can be selected from 100 integers,

$$\Big\| \sum_{n=1}^{100} x^n \Big\|^2 = \sum_{n=1}^{100} \|x^n\|^2 + \Re\Big(\sum_{n \neq m}^{100} \langle x^n, x^m \rangle \Big) \leq 100 + 9,900/10 = 1,090\,,$$

and, consequently, $\| \sum_{n=1}^{100} x^n \| \leq 1,090^{1/2}$. To see that the estimate is sharp consider the sequences $x^n \in \ell^2$ with terms $x_1^n = \sqrt{.1}$, $x_{n+1}^n = \sqrt{.9}$, and the remaining terms equal to 0, for all $n = 1, \ldots, 100$. Then the assumptions are satisfied and $\| \sum_{n=1}^{100} x^n \| = 1,090^{1/2}$.

19. (a) Let $\{e_n\}$ be an ONS in H and set $x_n = e_n/n$, $n = 1, 2, \ldots$. Then $\| \sum_{k=1}^m x_k - \sum_{k=1}^n x_k \|^2 = \sum_{k=n+1}^m 1/k^2 \to 0$ and so $\sum_k x_k \in H$. However $\sum_k \|x_k\| = \sum_k 1/k = \infty$.

(b) Let $y_n = \sum_{k=1}^n x_k \in H$. Then $\|y_{m+n} - y_n\| \leq \sum_{k=n+1}^{n+m} \|x_k\| \to 0$ as $n \to \infty$. Therefore $\{y_n\}$ is Cauchy in H and so converges to some $x \in H$. Also $\|y_n\| \leq \sum_{k=1}^n \|x_k\| \leq \sum_k \|x_k\|$. Thus $\|x\| = \lim_n \|y_n\| \leq \sum_k \|x_k\|$.

(c) Let $\{e_n\}$ be an ONS in an infinite-dimensional Hilbert space H and for $\lambda \in (0,1)$ let $x_n = \lambda^n e_n$. Then $\sum_n \|x_n\| = \sum_n \lambda^n = \lambda/(1-\lambda)$ and $\|x\|^2 = \sum_n \lambda^{2n} = \lambda^2/(1-\lambda^2)$. Hence $(\sum_n \|x_n\|)/\|x\| = ((1+\lambda)/(1-\lambda))^{1/2}$ is arbitrarily large as $\lambda \to 1$.

(d) We need to assume that the x_n are pairwise orthogonal for otherwise we may have $\sum_n \|x_n\| > 0$ with $\|x\| = 0$. Let $n = \dim(H)$, $\{e_1, \ldots, e_n\}$ an ONB for H, and $x_k = a_k e_k$ for $k = 1, \ldots, n$. Then $\sum_{k=1}^n \|x_k\|/\|x\| = \sum_{k=1}^n |a_k|/(\sum_{k=1}^n |a_k|^2)^{1/2} \leq \sqrt{n}$. Equality holds iff $a_1 = \ldots = a_n \neq 0$.

20. Since $\| \sum_{k=n}^m \lambda_k e_k \|^2 = \sum_{k=n}^m |\lambda_k|^2$, $\lim_n \sum_{k=1}^n \lambda_k e_k = x$ exists in H iff $\lambda \in \ell^2$. Let $\sigma : \mathbb{N} \to \mathbb{N}$ be a bijection; we claim that $x_n^\sigma = \sum_{k=1}^n \lambda_{\sigma(k)} e_{\sigma(k)}$ converges to x in H. Pick n so that $\|x - \sum_{k=1}^n \lambda_k e_k\|^2 < \varepsilon^2$ and let $N \geq n$ be large enough so that all the terms $\lambda_1 e_1, \ldots, \lambda_n e_n$ appear in the sum that defines x_N^σ. Then $\|x - x_m^\sigma\|^2 \leq 2 \sum_{k=N+1}^\infty |\lambda_k|^2 < 2\varepsilon^2$ for $m \geq N$ and $\lim_n x_n^\sigma = x$.

21. Since $\{e_n\}$ is a countable ONB for ℓ^2, all ONB's for ℓ^2 are countable. On the other hand, the sequences $x^\eta = (1, \eta, \eta^2, \ldots)$, $0 < \eta < 1$, form a linearly independent uncountable subset of ℓ^2 and, consequently, a Hamel basis of ℓ^2 is uncountable.

22. For the sake of argument suppose that $x \in X$ has $\|x\| = 1$. Now, $x_n = 1/(1 + 1/n)e_n \in X$ and $\|x_n\| \to 1$. Therefore

$$0 \leq \sum_n ((1 + 1/n)^2 - 1)|\langle e_n, x\rangle|^2 \leq 1 - \|x\|^2 = 0,$$

which implies that $\langle e_n, x\rangle = 0$ for all n. Hence $x = 0$, which is not the case since $\|x\| = 1$.

24. (a) Given $x \in H$ and $y \in K$, note that $\lambda x_K + (1 - \lambda)y \in K$ for all $0 \leq \lambda \leq 1$, and since $d(x, K) = \|x_K - x\|$ and $\lambda x_K + (1 - \lambda)y - x = (x_K - x) + (1 - \lambda)(y - x_K)$ we have

$$\|x_K - x\|^2 \leq \|(x_K - x) + (1 - \lambda)(y - x_K)\|^2$$
$$= \|x_K - x\|^2 + (1 - \lambda)^2\|y - x_K\|^2 + 2(1 - \lambda)\Re\langle x_K - x, y - x_K\rangle.$$

Thus $2\Re\langle x - x_K, y - x_K\rangle \leq (1 - \lambda)\|y - x_K\|^2$ and letting $\lambda \to 1$ it follows that $\Re\langle x - x_K, y - x_K\rangle \leq 0$. Conversely, suppose that given $x \in H$, $z \in K$ verifies the inequality for all $y \in K$. Then since $x - y = (x - z) + (z - y)$ and $\Re\langle x - z, y - z\rangle = -\Re\langle x - z, z - y\rangle$ it follows that $\|x - y\|^2 = \|x - z\|^2 + \|z - y\|^2 - 2\Re\langle x - z, z - y\rangle$, and, consequently, $\|x - y\|^2 \geq \|x - z\|^2$. Since $y \in K$ is arbitrary, $z = x_K$.

(b) By (a), $\Re\langle x - x_K, x_K - y_K\rangle \geq 0$ and $\Re\langle y_K - y, x_K - y_K\rangle \geq 0$. Thus adding, $\Re\langle (x - y) + (y_K - x_K), x_K - y_K\rangle \geq 0$, and, consequently, $\Re\langle x - y, x_K - y_K\rangle - \|x_K - y_K\|^2 \geq 0$. Therefore

$$\|x_K - y_K\|^2 \leq \Re\langle x - y, x_K - y_K\rangle \leq \|x - y\|\,\|x_K - y_K\|$$

and, canceling, $\|x_K - y_K\| \leq \|x - y\|$.

(c) This is the unique element x_K of K such that $d(0, K) = \|x_K\|$.

25. The statement is true.

26. We claim that if $\eta = \inf_{x \in K} J(x)$ and $\{x_n\} \subset K$ is such that $J(x_n) \to \eta$, $\{x_n\}$ is Cauchy. Indeed, $(x_n + x_m)/2 \in K$ and

$$\left\|\frac{x_n - x_m}{2}\right\|^2 = J(x_n) + J(x_m) - 2J\left(\frac{x_n + x_m}{2}\right) \leq J(x_n) + J(x_m) - 2\eta.$$

Thus $\|x_n - x_m\| \to 0$ as $n, m \to \infty$, and, therefore, $\{x_k\}$ is Cauchy and converges to $x \in K$, say, where the minimum is attained. Now, if the minimum is attained at x and x_0, say, as above $\|(x - x_0)/2\|^2 \leq J(x) + J(x_0) - 2\eta = 0$.

27. Let $x_K \in K$ denote the closest element of K to x. Then by Problem 24(a), $\Re\langle x - x_K, x_n - x_K\rangle \leq 0$, and, consequently,

$$\Re\langle x - x_K, x_n - x\rangle + \Re\langle x - x_K, x - x_K\rangle$$
$$= \Re\langle x - x_K, x_n - x\rangle + \|x - x_K\|^2 \leq 0.$$

Therefore $\|x-x_K\|^2 \le -\Re\langle x-x_K, x_n-x\rangle$ and since by the weak convergence the right-hand side tends to 0, the left-hand side, or $\|x-x_K\|$, is 0, and $x = x_K \in K$.

28. Fix $r > 0$ and write $H = \bigcup_{y \in H} B_r(y)$; since H is separable there are countably many balls such that $H = \bigcup_k B_r(y_k)$. Since μ is not the trivial measure, $\mu(B_r(y_k)) > 0$ for some k and by translation all the balls of radius r have nonzero measure equal to $\mu(B_r(y_k))$. Let $c = \mu(B_{r_0/30}(y))$ for any (and hence all) $y \in H$ and observe that if $\{e_n\}$ is an ONB for H, then $B_{r_0/30}(e_n/2) \subset B_{r_0}(0)$ for all n. By the Pythagorean theorem the balls $B_{r_0/30}(e_n/2)$ are pairwise disjoint and so $\mu(B_{r_0}(0)) \ge \sum_n c = \infty$ unless μ is identically 0.

29. Let $\{e_n\}$ be an ONB for H and $T : H \to H \times H$ the linear mapping given by $T(x) = \left(\sum_n \langle x, e_{2n-1}\rangle e_n, \sum_n \langle x, e_{2n}\rangle e_n\right)$. We claim that T is 1-1, onto, and $\|T(x)\|_{H \times H} = \|x\|_H$, $x \in H$. First, the convergence of the Fourier series gives that T is well-defined and linear. Injectivity follows since $T(x) = 0$ implies that $\langle x, e_n\rangle = 0$ for all n and, hence, $x = 0$. Surjectivity is clear since $S : H \oplus H \to H$ given by $S(x_1, x_2) = \sum_n \langle x_1, e_n\rangle e_{2n-1} + \sum_n \langle x_2, e_n\rangle e_{2n}$ is a 2-sided inverse of T. Finally, since by Plancherel's equality $\|S(x_1, x_2)\|^2 = \|x_1\|^2 + \|x_2\|^2$, S is an isometry and so is T.

31. Let $\{e_n\}$ be an ONB for H and consider the mapping $J : H \to \ell^2$ given by $J(x) = y$ where $y_n = \langle e_n, x\rangle = \alpha_n$ for all n; J is clearly linear. Bessel's inequality gives that $\{\alpha_n\} \in \ell^2$ and Parseval's equation gives that the inner product is preserved, i.e., $\langle x, x_1\rangle = \langle J(x), J(x_1)\rangle$. Isometries are 1-1 and since J is onto we have finished.

32. Consider the polynomials $\{t^n\}_{n=0}^\infty$ in $[0, 1]$; by the Gram-Schmidt orthogonalization process there are pairwise orthogonal polynomials $\{p_n\}_{n=0}^\infty$ on I where $p_0(t) = 1$ and p_n has positive leading coefficient and is of degree n for each $n \ge 1$. This sequence is unique.

33. If H is finite dimensional, H is the only dense subspace of itself and the result holds since H has an ONB. Suppose then that H is infinite dimensional and let $\{x_n\}$ be a countable dense subset of H, which need not be a subset of M. Since M is dense in H for each n there is a sequence $\{x_m^n\} \subset M$ such that $\lim_m x_m^n = x_n$; $D = \bigcup_{n,m}\{x_m^n\} \subset M$ is then a countable dense subset of H. Let $B = \{x_n\}$ be a maximal linearly independent subset of D. Then the linear span of B contains D, is dense in H, and, therefore, its closure is H. Let $\{e_n\}$ denote the Gram-Schmidt orthonormalization of $\{x_n\}$. Since $x_n \in M$ for all n and each e_n is a finite linear combination of the x_n, $\{e_n\} \subset M$, and since the closed linear span of $\{e_n\}$ is equal to that of B, $\{e_n\}$ is an ONB for H.

34. (a) Clearly $\{x_n\}$ is an ONS. Now, since $\|y_n\|^2 = (1-4^{-n})+(2^{-n})^2 = 1$ and as a simple computation shows $\langle y_m, y_n \rangle = 0$ if $m \neq n$, $\{y_n\}$ is an ONS. Next, suppose that $x = \sum_n \lambda_n e_n \in X \cap Y$. Since $x \in X$ the odd Fourier coefficients of x vanish and since $x = \sum_n \mu_n(\sqrt{1-4^{-n}}\, e_{2n} + 2^{-n} e_{2n-1}) \in Y$ it follows that $\mu_n 2^{-n} = 0$, and so $\mu_n = 0$ for all n, and $x = 0$. Finally, we claim that $\{e_n\} \subset X+Y$ and so $X+Y$ is dense in H. Indeed, $e_{2n} = x_n \in X$ and $e_{2n-1} = 2^n(y_n - (1-4^{-n})^{1/2}x_n) \in X+Y$ for all n.

(b) Since $\sum_n |\langle v, e_n\rangle|^2 < \infty$, $v \in H$. For the sake of argument suppose that $v = \sum_n a_n x_n + \sum_n b_n y_n \in X+Y$. Then $2^{-k} = \langle v, e_{2k-1}\rangle = b_k 2^{-k}$, which is impossible since it implies that $b_k = 1$ for all k, and the series defining y cannot converge. Therefore $X+Y$ is dense but not closed in H.

35. The statement is not always true. Let L be the linear functional on ℓ_0^2 given by $L(x) = \sum_n n^{-1}x_n$. Since $|L(x)| \leq (\pi/\sqrt{6})\,\|x\|$ for all $x \in \ell_0^2$, L is continuous. Now, since $e_n \in \ell_0^2$ and $L(e_n) = 1/n$ for all n, were x_L to exist it would follow that $(x_L)_n = 1/n$ for all n, and so $x_L \in \ell^2 \setminus \ell_0^2$.

36. By Problem 30, H has a countable ONB $\{e_n\}$, say. Let $L(e_n) = \overline{b_n}$. Then $\sum_{k=1}^n |b_k|^2 = \sum_{k=1}^n b_k L(e_k) = L(\sum_{k=1}^n b_k e_k) \leq \|L\|\,\|\sum_{k=1}^n b_k e_k\| = \|L\|\,(\sum_{k=1}^n |b_k|^2)^{1/2}$ and $\sum_{k=1}^n |b_k|^2 \leq \|L\|^2$ for all n. Hence $\sum_k |b_k|^2 < \infty$ and $\sum_{k=1}^n b_k e_k$ converges to $x_L \in H$, say, with $\|x_L\| = (\sum_k |b_k|^2)^{1/2} \leq \|L\|$. Now, given $x \in H$, let $x_n = \langle x, e_n\rangle$; since $\sum_{k=1}^n x_k e_k \to x$ by the linearity and continuity of L it follows that $L(x) = \lim_n L(\sum_{k=1}^n x_k e_k) = \lim_n \sum_{k=1}^n x_k \overline{b_k} = \langle x, x_L\rangle$. Finally, by Cauchy-Schwarz $|L(x)| = |\langle x, x_L\rangle| \leq \|x\|\,\|x_L\|$ and $\|L\| \leq \|x_L\|$. The uniqueness of x_L is clear.

37. Let L be a continuous linear functional on a Hilbert space H. If $L = 0$, $x_L = 0$ will do. Otherwise $K(L)$ is a proper closed subspace of H and by Problem 8.63(a), $K(L)^\perp$ is 1-dimensional; then $K(L)^\perp = \mathrm{sp}\{y\}$, $y \neq 0$, and $K(L) = K(L)^{\perp\perp} = y^\perp$. Thus the linear functional $\ell(x) = \langle x, y\rangle$, $x \in H$, satisfies $K(\ell) = K(L)$ and by Problem 8.65(a) there exists $\lambda \neq 0$ such that $L = \lambda \ell$, i.e., $L(x) = \lambda\langle x, y\rangle = \langle x, \overline{\lambda}y\rangle = \langle x, x_L\rangle$ with $x_L = \overline{\lambda}y$. Clearly $\|L\| = \|x_L\|$.

38. Let $y \in \overline{X}$, the completion of X discussed in Problem 12. Since $L(x) = \langle x, y\rangle_1$ is a continuous linear functional on X by assumption there is $x_L \in X$ such that $\langle x, y\rangle_1 = L(x) = \langle x, x_L\rangle = \langle x, x_L\rangle_1$ for all $x \in X$. Now, by the continuity of the inner product, $\langle z, x_L - y\rangle_1 = 0$ for all $z \in \overline{X}$ and, consequently, $y = x_L \in X$. Therefore $X = \overline{X}$ and X is complete.

39. The following is an example of disjoint convex subsets of a Hilbert space H that cannot necessarily be separated by a continuous linear functional on H: In $H = L^2([-1,1])$ consider $X_\gamma = \{x \in H : x \text{ is continuous and } x(0) = \gamma\}$, $\gamma \in \mathbb{R}$; for $\alpha \neq \beta$, X_α, X_β are disjoint convex subsets of H that cannot be separated by a continuous linear functional on H.

Now, in our case, note that $C = A - B$ is closed, convex, and does not contain 0. Let x_0 denote the projection of 0 on C and $X = x_0/2 + x_0^\perp$. Then X separates C and $\{0\}$ and so A and B are separated.

40. Since X is a Hilbert space in its own right by the Riesz-Fréchet representation theorem there is a unique $x_\ell \in X$ with $\|x_\ell\| = \|\ell\|$ such that $\ell(x) = \langle x, x_\ell \rangle$ for all $x \in X$. Let L be given by $L(x) = \langle x, x_\ell \rangle$ for $x \in H$; then $L|_X = \ell$ and $\|L\| = \|x_\ell\| = \|\ell\|$. Suppose now that L_1 is another linear functional on H such that $L_1|_X = \ell$ and $\|L_1\| = \|\ell\|$. Then for all $x \in H$, $L_1(x) = \langle x, x_1 \rangle$ where $x_1 \in H$ with $\|x_1\| = \|L_1\|$. Since $L_1|_X = \ell$ we have $\langle x, x_1 - x_\ell \rangle = 0$ for all $x \in X$, i.e., $x_1 - x_\ell \in X^\perp$. Hence $\|\ell\|^2 = \|L_1\|^2 = \|x_1\|^2 = \|x_1 - x_\ell\|^2 + \|x_\ell\|^2 = \|x_1 - x_\ell\|^2 + \|\ell\|^2$. Therefore $\|x_1 - x_\ell\|^2 = 0$, $x_1 = x_\ell$, and the extension is unique.

41. Since $H = X \oplus X^\perp$, each $x \in H$ can be written as $x = P(x) + Q(x)$, and so $L(x)$ is linear, $L|_X = \ell$ and $\|L\| \geq \|\ell\|$. On the other hand, $\|L(x)\| \leq \|\ell\| \|P\| \|x\| = \|\ell\| \|x\|$ since $\|P\| = 1$.

Suppose now that L_1 is another bounded linear functional on H such that $L_1|_X = \ell$ and $\|L_1\| = \|\ell\|$. Then by the Riesz-Fréchet representation theorem, $L_1(x) = \langle x, b \rangle$ with $\|b\| = \|L_1\|$. Then since $b = P(b) + Q(b)$, for all $y \in X$, $L(y) = L_1(y) = \langle y, P(b) \rangle + \langle y, Q(b) \rangle = \langle y, P(b) \rangle$. This gives $\|L\| = \|P(b)\|$. Now, since $\|L\| = \|L_1\|$, then $\|L_1\| = \|P(b)\|$. And, since $\|L_1\| = \|b\|$ this implies that $Q(b) = 0$ and so $L_1(x) = \langle x, b \rangle = \langle x, P(b) \rangle = \langle P(x), b \rangle = L(x)$ (for all $x \in X$?)

42. (a) Since $\sum_n n^{-2}$ converges, by Problem 20, $x_0 = \sum_n (1/n) e_n \in H$ and we can define the linear functional $L(x) = \langle x, x_0 \rangle$, $x \in H$. Then $L(e_n) = 1/n$ for all n and $\|L\| = \|x_0\| = \pi/\sqrt{6}$.

43. The statement is true. Let L be a continuous linear functional on $L^2(\mathbb{R}^n)$ that vanishes on V; by Riesz-Fréchet there exists $h \in L^2(\mathbb{R}^n)$ with $\|h\|_2 = \|L\|$ such that $L(g) = \int_{\mathbb{R}^n} g(y) h(y) \, dy$ for all $g \in L^2(\mathbb{R}^n)$ and, in particular, $L(g) = \int_{\mathbb{R}^n} f(x) g(x) \, e^{ix \cdot \xi} \, dx = 0$ for all $\xi \in \mathbb{R}^n$. Now, by the inversion formula for the Fourier transform, $f(x) g(x) = 0$ a.e. and since $g(x) \neq 0$ a.e. this gives $f = 0$ a.e. and so L is the zero functional. Consequently, by Problem 8.95(b), V is dense in $L^2(\mathbb{R}^n)$.

44. The statement is true. Let $V = \mathrm{sp}\{\tau_x f : x \in \mathbb{R}^n\}$ and L a continuous linear functional on $L^2(\mathbb{R}^n)$ that vanishes on V. By the Fréchet-Riesz representation theorem there exists $h \in L^2(\mathbb{R}^n)$ with $\|h\|_2 = \|L\|$ such that $L(g) = \int_{\mathbb{R}^n} g(y) h(y) \, dy$ for all $g \in L^2(\mathbb{R}^n)$ and, in particular, on V, $\int_{\mathbb{R}^n} f(x + y) h(y) \, dy = 0$ for all $x \in \mathbb{R}^n$. Thus the convolution $f * h$ vanishes everywhere on $\mathbb{R}^n$ and so $\widehat{f}(\xi) \widehat{h}(\xi) = 0$ for all $\xi \in \mathbb{R}^n$. But since $\widehat{f}(\xi) \neq 0$ a.e. this implies that $\widehat{h}(\xi) = 0$ a.e. and by the inversion formula for the Fourier transform $h = 0$ a.e. Therefore L is the zero functional and by Problem 8.95(b), V is dense in $L^2(\mathbb{R}^n)$.

45. Let $X = \{g \in L^2(\mathbb{R}^N) : g$ is compactly supported and continuous$\}$ and L the functional given by $L(g) = \int_{\mathbb{R}^N} f(x)g(x)\,dx$, $g \in X$; clearly L is linear and finite for every $g \in X$. Suppose L is bounded in X. Then by Hahn-Banach L can be extended to a bounded linear functional L_1 on $L^2(\mathbb{R}^N)$ and by Riesz-Fréchet there is a unique $h \in L^2(\mathbb{R}^N)$ such that $L_1(g) = \langle g, h \rangle$ for all $g \in L^2(\mathbb{R}^N)$. But this implies that $f = h$ a.e., which since $h \in L^2(\mathbb{R}^N)$ cannot happen. Therefore L is not bounded and by Problem 8.63(b), $K(L)$ is a dense subspace of $L^2(\mathbb{R}^N)$. Finally, since $L^2(\mathbb{R}^N)$ is separable, by Problem 33 there is an ONB for $L^2(\mathbb{R}^N)$ consisting of elements in $K(L)$.

47. The statement is true.

49. Since for $k \neq \ell$ the supports of $e_{n,k}$ and $e_{m,\ell}$ have at most one point in common for all n, m, their product is zero a.e. and $\langle e_{n,k}, e_{m,\ell} \rangle = 0$. On the other hand, if $n = m$, $\langle e_{n,k}, e_{n,\ell} \rangle = \langle e_k, e_\ell \rangle = \delta_{k,\ell}$ and $\{e_{n,k}\}$ is an ONS in $L^2(\mathbb{R})$. To prove completeness it suffices to verify that the $e_{n,k}$ span a dense subset of $L^2(\mathbb{R})$ and since compactly supported functions are dense in $L^2(\mathbb{R})$ this reduces to proving the density in $L^2([-K, K))$ for any positive integer K. Given $\varepsilon > 0$ and $x \in L^2(\mathbb{R})$ supported in $[-K, K)$, let $x = \sum_{k=-K}^{K-1} x_k$ be a decomposition of x into an orthogonal sum of functions where $x_k = x\,\chi_{[k,k+1)} \in L^2([k, k+1))$. Now, by the translation invariance of the Lebesgue integral, for each k there is $y_k \in \mathrm{sp}\{e_{n,k} : n \in \mathbb{N}, k \in \mathbb{Z}\}$ such that $\|x_k - y_k\| \leq \varepsilon/4K$, $-K \leq k \leq K - 1$. Setting $y = \sum_{k=-K}^{K-1} y_k$, by the orthogonality of the respective terms we have

$$\|x - y\|^2 = \sum_{k=-K}^{K-1} \|x_k - y_k\|^2 \leq \sum_{k=-K}^{K-1} (\varepsilon/4K)^2 = (\varepsilon/2)^2$$

as desired.

50. Note that for an ONB $\{e_n\}$ for H, the ONS $\{e_2, \ldots, e_n, \ldots\}$, and $x \in H$, we have $\sum_{n=2}^{\infty} \langle x, e_n \rangle e_n = x - \langle x, e_1 \rangle e_1$.

51. (a) The statement is false since $(1, -1, 0, \ldots) \perp y_n$ for all n.

(b) The statement is true: $x \perp z_n$ iff $x_n = 2^{n-2} x_1$ for all n, and so $x \in \ell^2$ only if $x_1 = 0$ and then $x = 0$.

52. For the sake of argument suppose that $\{f_n\}$ is not complete and let $0 \neq x \in H$ satisfy $\langle x, f_n \rangle = 0$ for all n. Then $\|x\|^2 = \sum_n |\langle x, e_n \rangle|^2 = \sum_n |\langle x, e_n - f_n \rangle|^2 \leq (\sum_n \|e_n - f_n\|^2)\,\|x\|^2 < \|x\|^2$, which cannot happen since $x \neq 0$.

53. Pick N such that $\sum_{n>N}^{\infty} \|e_n - f_n\|^2 < 1$ and let X denote the closed span of $\{e_{N+1}, e_{N+2}, \ldots\}$ in H; by Problem 52, X is the closed span of $\{f_{N+1}, f_{N+2}, \ldots\}$ in H. Now, let $1 \leq k \leq N$. Since $H = X \oplus X^{\perp}$, $f_k \in X^{\perp} = \mathrm{sp}\{e_1, \ldots, e_N\}$ and $f_k = \sum_{n=1}^{N} \lambda_n^k e_n$. Furthermore, since $\{f_1, \ldots, f_N\}$ is an ONS and, in particular, linearly independent, by linear algebra the $N \times N$ matrix $A = (\lambda_n^k)$, $1 \leq k, n \leq N$, is nonsingular. Hence if $\langle f_k, x \rangle = 0$ for some $x \in H$ and all k, $0 = \langle f_k, x \rangle = \sum_{n=1}^{N} \lambda_n^k \langle e_n, x \rangle$, $1 \leq k \leq N$, and, consequently, since the determinant of the above homogenous system is not 0, the system only admits the trivial solution and $\langle e_1, x \rangle = \ldots = \langle e_N, x \rangle = 0$. Finally, suppose that $\langle x, f_n \rangle = 0$ for all n. Combining the above results, by Bessel's inequality $\|x\|^2 = \sum_n |\langle x, e_n \rangle|^2 = \sum_{n=N+1}^{\infty} |\langle x, e_n - f_n \rangle|^2 \leq \|x\|^2 \sum_{n=N+1}^{\infty} \|e_n - f_n\|^2 < \|x\|^2$, and so $x = 0$ and $\{f_n\}$ is complete.

54. If $n < m$, $\langle f_n, f_m \rangle = 0$ and the f_n are pairwise orthogonal. Similarly, $\|f_n\|^2 = (n^2 + n)^{-1}(n + n^2) = 1$ for all n, and $\{f_n\}$ is an ONS.

Finally, we claim that the f_n span H. Let L be a bounded linear functional on H that vanishes on $X = \mathrm{sp}\{f_1, \ldots, f_n, \ldots\}$. If $L(f_1) = 0$, then $L(e_1 - e_2)/\sqrt{2} = 0$ and $L(e_1) = L(e_2)$. Next, we proceed by induction and prove that if $L(e_1) = \ldots = L(e_n)$, then $L(e_{n+1}) = L(e_1)$. Indeed $L(f_n) = (n^2 + n)^{-1/2} L(\sum_{k=1}^{n} e_k - n e_{n+1}) = (n^2 + n)^{-1/2}(n L(e_1) - n L(e_{n+1})) = 0$ and, consequently, $L(e_1) = L(e_{n+1})$. Now, let $x = \sum_n e_n/n \in H$. Since L is continuous, $L(x) = \lim_N L(\sum_{n=1}^{N} e_n/n) = \lim_N (\sum_{n=1}^{N} n^{-1}) L(e_1)$ and this expression diverges unless $L(e_1) = 0$ in which case $L(e_n) = 0$ for all n and $L = 0$. Therefore L is the 0 functional and by Problem 8.95(b), $X = H$.

55. (b) Let $f_1 = e_1$ and

$$f_n = \sqrt{\frac{n}{n+1}}\, e_n + \frac{1}{\sqrt{n+1}}\, e_{n+1}, \qquad n = 2, \ldots .$$

As in Problem 54, $\mathrm{sp}\{f_n\}$ is dense in H. Furthermore

$$T\left(\sum_n \lambda_n e_n\right) = \sum_n \left(\lambda_n \sqrt{\frac{n}{n+1}} - \lambda_{n-1} \frac{1}{\sqrt{n}}\right) e_n.$$

So if we wish to solve $T(x) = y = \sum \mu_n e_n$ we can solve for the λ_n recursively and it is not hard to see that if $\sum_n |\mu_n|^2 < \infty$ the same is not necessarily true for the λ_n.

58. Let $\{e_n\}$ be an ONB for H, $\{q_n\}$ an enumeration of the rationals in $[0,1]$, and $X_s = \{x \in H : \langle e_n, x \rangle = 0 \text{ if } q_n > s\}$. These sets are obviously closed (orthogonal complement of $\{e_n : q_n > s\}$) and if $s < t$, $X_s \subset X_t$. Moreover, since there is always a rational q_n, say, between s and t, $e_n \in X_t \setminus X_s$ and the spaces are not equal.

60. We claim that M is dense in X. Let $y \in X$ be such that $y_n = 0$ for $n > N$ and $\sum_{n=1}^{N} y_n = \eta$, say. For an integer K, let x be given by

$$x_n = \begin{cases} y_n, & 1 \leq n \leq N, \\ -\eta/K, & N+1 \leq n \leq N+K, \\ 0, & n > N+K. \end{cases}$$

Then $x \in X$ and $\|x - y\|^2 = |\eta|^2/K \to 0$ as $K \to \infty$. So M is dense but not closed in X and $M^{\perp} = \{0\}$.

Next, M_1 is the kernel of a bounded linear functional on ℓ^2 and, therefore, is closed in ℓ^2. By taking relative topologies $M_1 \cap X$ is closed but not dense in X. A direct proof also works. Let $x \in X$ be a limit point of M_1 and given $\varepsilon > 0$, pick $m \in M_1$ such that $\|x - m\| < \varepsilon$; then by Cauchy-Schwarz $|\sum_k x_k/k| = |\sum_k x_k/k - \sum_k m_k/k| \leq \sum_k |x_k - m_k|/k \leq (\sum_k k^{-2})^{1/2}\varepsilon$. Since this is true for arbitrary $\varepsilon > 0$, $\sum_k x_k/k = 0$ and $x \in M_1$, which is therefore closed. Now, we claim that $y \in M_1^{\perp}$ iff $y = y_1(1, 1/2, 1/3, \ldots)$, and so $y \in X$ iff $y = 0$. This is readily seen because the sequence $\{e_1 - n\,e_n\}$ is in M_1, and so $y_1 - ny_n = 0$ for all n when $y \in M_1^{\perp}$. Then y is in X iff $y = 0$ and $M_1^{\perp} = \{0\}$. Hence $X \neq M_1 \oplus M_1^{\perp} = M_1$.

Finally, $M_2 = K(L)$ where L is the linear functional given by $L(e_n) = 1/\sqrt{n}$; since $\{1/\sqrt{n}\} \notin \ell^2$, L is not bounded and, consequently, M_2 is dense but not closed in X. A direct proof of the density of M_2 goes as follows. Let $y \in M_2$ and N be such that $y_k = 0$ for all $k \geq N$. Let $C = \sum_{k=1}^{N-1} y_k/\sqrt{k}$, for $n \geq N$ define $A_n = \sum_{k=N}^{n} 1/k$, and let x be given by

$$x_k = \begin{cases} y_k, & k < N, \\ -C/A_n\sqrt{k}, & N \leq k \leq n \\ 0, & n < k. \end{cases}$$

Then $\sum_k x_k/\sqrt{k} = \sum_{k=1}^{N-1} y_k/\sqrt{k} + \sum_{k=N}^{n} -C/(A_n k) = C - (C/A_n)A_n = 0$ and $x \in M_2$. Now, $\|y - x\|^2 = \sum_{k=N}^{n} C^2/(A_n^2 k) = (C^2/A_n^2)A_n = C^2/A_n \to 0$ since $A_n \to \infty$ as $n \to \infty$. Thus M_2 is dense in X.

62. The statement is not true in the case of the inner product space ℓ_0^2 as Problem 60 shows. It is not true for the Hilbert space ℓ^2 either. Since ℓ_0^2 is dense in ℓ^2 by Problem 60 so is $M = \{x \in \ell^2 : \sum_n x_n = 0\}$ and, therefore, $M^{\perp} = \{0\}$. Note that $e_1 \in \overline{M} \setminus M$. Indeed, let $f_k = e_1 - k^{-1}(e_1 + \cdots + e_k)$; $f_k \in M$ and $\|f_k - e_1\|_2 = k^{-1} k^{1/2} = k^{-1/2} \to 0$ as $k \to \infty$. Thus $e_1 \in \overline{M} \setminus M$ and $H \neq M + M^{\perp}$. On the other hand, $H = \overline{M} \oplus M^{\perp}$. Indeed, since by Problem 61(a) $\overline{M}^{\perp} = M^{\perp}$, the assertion follows at once from the projection theorem.

65. M is not closed; for instance, $\cos(t)$ is the limit of polynomials of even degree. We claim that $M^{\perp}$ consists of the odd functions in $L^2([-1, 1])$;

clearly odd functions are in $M^\perp$. Conversely, if $x \in M^\perp$, $\int_{-1}^{1} x(t)y(t)\,dt = \int_0^1 (x(t) + x(-t))\,y(t)\,dt = 0$ for all $y \in L^2([-1,1])$. In other words, $x(t) + x(-t) \in L^2([0,1])^\perp$, and so $x(t) + x(-t) = 0$ a.e. Therefore $M^{\perp\perp} = \{x \in L^2([-1,1]) : x \text{ is even}\}$. As for the projections, they are $P_{M^\perp}(x)(t) = (x(t) - x(-t))/2$ and $P_{M^{\perp\perp}}(x)(t) = (x(t) + x(-t))/2$.

66. Clearly M is a subspace of X and $M^\perp = \{z \in X : z = 0 \text{ a.e.}$ in $[0,1]\}$. For the sake of argument suppose that $x(t) = \chi_{[-,1,1]}(t)$ can be written as $y + z$ with $y \in M$ and $z \in M^\perp$. Since $y \in M$, $y = 0$ in $[0,1]$ and, therefore, $z = 1$ in $[0,1]$. Similarly, $z = 0$ in $[-1,0]$ implies $y = 1$ in $[-1,0]$, and, therefore, $y, z \notin C([-1,1])$.

67. $Y^\perp = \left\{x \in X : \sum_{k=1}^{n} \langle x, x_k \rangle x_k = 0\right\}$.

69. Necessity first. We claim that $\operatorname{sp}\{X\}^\perp = \{0\}$. Let $y \in H$ be such that $\langle y, z \rangle = 0$ for all $z \in \operatorname{sp}\{X\}$. Since $X \subset \operatorname{sp}\{X\}$, $\langle y, x \rangle = 0$ for all $x \in X$ and, since X is complete, $y = 0$. Since $\operatorname{sp}\{X\}$ is a subspace of H by Problem 61(d), $\overline{\operatorname{sp}}\{X\} = H$.

As for sufficiency, let $\langle y, x \rangle = 0$ for all $x \in X$ and pick $y_n \to y$ where each y_n is a finite linear combination of elements in X. Then $\langle y, y_n \rangle = 0$ for all n and by the continuity of the inner product $\|y\|^2 = \lim_n \langle y, y_n \rangle = 0$. Thus $y = 0$ and X is complete.

70. If $M = \{0\}$, $M^\perp = H$ and $x = x_{M^\perp}$. Clearly $c = \|x\|$. Also, if $\|y\| = 1$, $|\langle x, y \rangle| \le \|x\|\,\|y\| \le \|x\|$ and $C \le \|x\|$; to see there is equality just take $y = x/\|x\|$. On the other hand, if $M \ne \{0\}$ by the projection theorem $x = x_M + x_{M^\perp}$ with $x_M \in M$ and $x_{M^\perp} \in M^\perp$. Now, for $y \in M$, $x_M - y \in M$ and $x_M - y \perp x_{M^\perp}$. Thus $\|x_M - y + x_{M^\perp}\|^2 = \|x_M - y\|^2 + \|x_{M^\perp}\|^2$ and taking $y = x_M$ it follows that $c = \|x_{M^\perp}\|$. Now, if $y \in M^\perp$ has norm 1, $|\langle x, y \rangle| = |\langle x_{M^\perp}, y \rangle| \le \|x_{M^\perp}\|$ and so $C \le \|x_{M^\perp}\|$. To see there is equality take $y = x_{M^\perp}/\|x_{M^\perp}\| \in M^\perp$, $\|y\| = 1$. Then $\langle x_{M^\perp}, y \rangle = \|x_{M^\perp}\|$ and $C = \|x_{M^\perp}\|$.

71. Let $x_1(t) = 1, x_2(t) = t, x_3(t) = t^2$, and $Y = \operatorname{sp}\{x_1, x_2, x_3\}$; then $M = Y^\perp$. Y is a finite-dimensional and, hence, closed subspace of $L^2(I)$, and so $L^2(I) = Y \oplus Y^\perp = Y \oplus M$. Now, given $y \in L^2(I)$, y can be written uniquely as $y = y_Y + y_M$ with $y_Y \in Y$ and $y_M \in M$, and by the Pythagorean theorem the answer is y_M. A quick way to obtain y_M explicitly is to consider an ONB $\{e_1, e_2, e_3\}$ of Y and the answer is then $y_M = y - \langle y, e_1 \rangle e_1 - \langle y, e_2 \rangle e_2 - \langle y, e_3 \rangle e_3$. In our case, by the Gram-Schmidt process $e_1 = 1$, $e_2 = x_2 - \langle x_2, e_1 \rangle e_1 = t - 1/2$, and $e_3 = x_3 - \langle x_3, e_1 \rangle e_1 - \langle x_3, e_2 \rangle e_2 = \sqrt{5}(6t^2 - 6t + 1)$.

72. Let $M = \operatorname{sp}\{1, t, t^2\}$; M is finite dimensional and, hence, closed. Then the max in question is equal to $\min\{\int_I (t^3 - (at^2 + bt + c))^2\,dt :$

$a, b, c \in \mathbb{R}\}$ by Problem 70. Expanding, integrating, and minimizing, by calculus it readily follows that the minimum value is attained at $a = 3/2$, $b = -3/5$, $c = 1/20$, and is equal to $1/2800$.

As for the min, since $\int_I (t^5 - a - bt)t^k \, dt = 0$, $k = 0, 1$ it follows that $a = -4/21$ and $b = 5/7$ and, consequently, the minimum is $100/4851$. Alternatively, squaring, a simple integration gives that the function to minimize is a paraboloid opening up, so the critical point corresponds to a minimum. By calculus we obtain the same values of a, b as before.

73. $d(y, M) = |\int_I y(t) \, dt|$.

74. $L(x) = \langle x, y \rangle$ is a continuous linear functional on H with $\|L\| = \|y\|$ and $K(L) = y^\perp$. Then by Problem 70 with $M = y^\perp$ there, $d(x, y^\perp) = |\langle x, y \rangle|/\|y\|$.

75. As in Problem 65, $M^\perp$ consists of the odd functions in $L^2([-1, 1])$. Therefore by Problem 70, $d(y, M) = \max\{|\int_{-1}^1 y(t)x(t) \, dt| : x \text{ odd}, \|x\| = 1\}$. Now, to estimate these integrals, since the even part of y integrates to 0 against x we can replace y by its odd part $y_o(t) = (y(t) - y(-t))/2$ and $|\int_{-1}^1 y(t)x(t) \, dt| \leq \|y_o\|$. On the other hand, let $y = x_o/\|x_o\|$; y is odd, of norm 1. Hence $\max \geq |\int_{-1}^1 y_o(t) \, x(t) \, dt| = \|y_o\|$ and $d(y, M) = \|y_o\|$.

76. (a) Given $x, y \in H$ and a scalar λ, $P(x + \lambda y) - P(x) - \lambda P(y) = P(x + \lambda y) - (x + \lambda y) + (x - P(x)) + (\lambda y - P(\lambda y))$ where the left-hand side is in X and the right-hand side is perpendicular to X. Since the only element that satisfies this property is 0, P is linear.

(c) Let $x_n \to x$ and $P(x_n) \to y$. Then for $z \in H$, $\langle y, z \rangle = \lim_n \langle P(x_n), z \rangle = \lim_n \langle x_n, P(z) \rangle = \langle x, P(z) \rangle = \langle P(x), z \rangle$. Thus $P(x) = y$ and P is continuous by the closed graph theorem.

(d) Let $y \in X$. Then $\langle x - P(x), P(x) - y \rangle = 0$ for every $x \in H$ and so $\|x - y\|^2 = \|x - P(x)\|^2 + \|P(x) - y\|^2 \geq \|x - P(x)\|^2$ with equality iff $y = P(x)$.

77. We consider sufficiency. We claim that $K(P)^\perp = R(P)$. For the sake of argument suppose that $0 \neq x \in K(P)^\perp \setminus R(P)$. Then $P(x) = (P(x) - x) + x$ with $0 \neq P(x) - x \in K(P)$ and so $x \in K(P)^\perp$. Hence by the Pythagorean theorem $\|P(x)\|^2 = \|P(x) - x\|^2 + \|x\|^2 > \|x\|^2$ and $\|P\| > 1$, which is not the case. Thus $K(P)^\perp \subset R(P)$.

To prove that $R(P) \subset K(P)^\perp$ we begin by observing that $P(v) = v$ for $v \in K(P)^\perp$. Indeed, since $v - P(v) \in K(P)$, $\langle v - P(v), v \rangle = 0$ and $\|v\|^2 = \langle P(v), v \rangle$. Then $\|v\|^2 = \langle P(v), v \rangle \leq \|P(v)\| \|v\| \leq \|v\|^2$ and by equality in the Cauchy-Schwarz inequality, $P(v) = \lambda v$ with $|\lambda| = 1$. And since $\|v\|^2 = \langle P(v), v \rangle = \lambda \langle v, v \rangle$, $\lambda = 1$. Now, let $y = P(x) \in R(P)$, $x \in X$. Since $X = K(P) \oplus K(P)^\perp$, $x = x_1 + x_2$ with $x_1 \in K(P)^\perp$ and $x_2 \in K(P)$, and so $y = P(x_1)$. Thus, if $w \in K(P)$, $\langle y, w \rangle = \langle P(x_1), w \rangle = \langle x_1, w \rangle = 0$, $y \in K(P)^\perp$, and $R(P) \subset K(P)^\perp$.

Now that we have $H = R(P) \oplus R(P)^\perp$ the proof proceeds as follows: Given $x, y \in H$, let $x = x_1 + x_2$ and $y = y_1 + y_2$ where $x_1, y_1 \in R(P)$ and $x_2, y_2 \in R(P)^\perp$. Then since $K(P)^\perp = R(P)$, $\langle P(x), y \rangle = \langle P(x), y_1 \rangle = \langle P(x_1), y_1 \rangle = \langle x_1, y_1 \rangle$. Similarly, $\langle x, P(y) \rangle = \langle x_1, y_1 \rangle$ and P is symmetric.

79. (c) implies (a) Note that $P_M P_N = P_N P_M = 0$: Given $x, y \in H$, $P_M(x) \in M, P_N(y) \in N$ and $\langle P_M(x), P_N(y) \rangle = \langle x, P_M P_N(y) \rangle = 0$. Since x, y are arbitrary we have $P_M P_N = 0$; similarly, $P_N P_M = 0$.

Now, since P is a sum of symmetric operators, P is symmetric. Furthermore, $P^2 = (P_M + P_N)^2 = P_M^2 + P_M P_N + P_N P_M + P_N^2 = P_M + P_N = P$ and P is a projection. Finally, $P(x) = P_M(x) + P_N(x) \in M \oplus N$ as x varies over X. Conversely, if $x = x_1 + x_2 \in M \oplus N$, $x_1 \in M, x_2 \in N$, since $PP_M = P_M, PP_N = P_N$ and since $x_1 = P_M(x_1) = PP_M(x_1)$, $x_2 = P_N(x_2) = PP_N(x_2)$, we have $P(x) = P(x_1 + x_2) = PP_M(x_1) + PP_N(x_2) = x_1 + x_2 = x$ and $P = P_{M \oplus N}$.

81. $x \in X$ iff x is orthogonal to $e_1 - e_2$ and $e_2 + e_3$; that is, $X = \mathrm{sp}\{e_1 - e_2, e_2 + e_3\}^\perp$. Since $\mathrm{sp}\{e_1 - e_2, e_2 + e_3\}$ is a finite-dimensional subspace of ℓ^2 and, therefore, closed, $X^\perp = \mathrm{sp}\{e_1 - e_2, e_2 + e_3\}$. Now, $f_1 = e_1 - e_2, f_2 = e_1 + e_2 + 2e_3$ are orthogonal vectors that span $X^\perp$, and so P is given by

$$P(x) = \frac{\langle x, f_1 \rangle}{\langle f_1, f_1 \rangle} f_1 + \frac{\langle x, f_2 \rangle}{\langle f_2, f_2 \rangle} f_2, \quad x \in X,$$

or $P(x) = 4^{-1}(3x_1 - x_2 + 2x_3, -x_1 + 3x_2 + 2x_3, 2x_1 + 2x_2 + 4x_3, 0, \ldots)$.

83. (a) Yes.

(b) First, necessity. Since u, v are linearly independent by Hahn-Banach there is a linear functional L on H such that $L(v) = 1$, $L(u) = 0$, and $\|L\| = 1$. In other words, there is $x \in X$ such that $\langle x, u \rangle = 0$ and $\langle x, v \rangle \neq 0$. Now, if P is a projection, $P^2 = P$, and, consequently, $\langle x, u \rangle u + \langle x, v \rangle v = \langle x, u \rangle \langle u, u \rangle u + \langle x, v \rangle \langle v, u \rangle u + \langle x, u \rangle \langle u, v \rangle v + \langle x, v \rangle \langle v, v \rangle v$ for all $x \in H$. Hence $\langle x, u \rangle = \langle x, u \rangle \langle u, u \rangle + \langle x, v \rangle \langle v, u \rangle$, and, similarly, $\langle x, v \rangle = \langle x, u \rangle \langle u, v \rangle + \langle x, v \rangle \langle v, v \rangle$. Picking x as above, from the first equation we get $|\langle u, v \rangle|^2 = 0$, and so $u \perp v$, and from the second $\langle v, v \rangle = 1$. Putting $x = u$ in the first equation gives $\|u\|^4 = \|u\|^2$ and $\|u\| = 0$ or $= 1$.

Next, sufficiency. Let $P_1(x) = \langle x, u \rangle u$ and $P_2(x) = \langle x, v \rangle v$; by (a), P_1 and P_2 are orthogonal projections on H. Moreover, $P_1 P_2(x) = \langle P_2(x), u \rangle u = \langle \langle x, v \rangle v, u \rangle u = \langle x, v \rangle \langle v, u \rangle u = 0$. Thus $P_1 P_2 = 0$ and, similarly, $P_2 P_1 = 0$. Hence, by Problem 79, $P = P_1 + P_2$ is an orthogonal projection.

84. No. Let $P = -I$; since P^2 is symmetric and $P^4 = (P^2)^2 = I^2 = I$, $P^2 = I$ is an orthogonal projection. However, although P is symmetric

$(\langle P(x), y \rangle = \langle -x, y \rangle = \langle x, -y \rangle = \langle x, P(y) \rangle)$, since $P^2 = I \neq P = -I$, P is not a projection.

85. (a) If $P^3 = P^2$, then $P^4 = P^3 = P^2$ and $(P^2 - P)^2 = P^4 - 2P^3 + P^2 = 0$. Moreover $P^2 - P$ is self-adjoint and so by Problem 142(c), $P^2 - P = 0$ and P is an orthogonal projection.

(b) In this case $P^5 = P^3$ and $P^n = P^3$ for all $n \geq 3$. Therefore $(P^3 - P^2)^2 = P^6 - 2P^5 + P^4 = 0$, and since $P^3 - P^2$ is self-adjoint by Problem 142(c), $P^3 - P^2 = 0$. Then by (a) $P^2 = P$ and P is an orthogonal projection.

(c) No. If $P = -I$, $P^2 = P^4 = I$, yet P is not a projection.

87. By the projection theorem $d(x, K_1) = \|x - P_{K_1}(x)\|$, and similarly for K_2. Now, by the parallelogram law

$$2d(x, K_1)^2 + 2d(x, K_2)^2$$

$$= 4\left\|x - \frac{P_{K_1}(x) + P_{K_2}(x)}{2}\right\|^2 + \|P_{K_1}(x) - P_{K_2}(x)\|^2,$$

and since $K_1 \subset K_2$, $(P_{K_1}(x) + P_{K_2}(x))/2 \in K_2$ and we have

$$\left\|x - \frac{P_{K_1}(x) + P_{K_2}(x)}{2}\right\| \geq d(x, K_2).$$

Therefore $\|P_{K_1}(x) - P_{K_2}(x)\|^2 \leq 2(d(x, K_1)^2 + d(x, K_2)^2) - 4d(x, K_2) = 2(d(x, K_1)^2 - d(x, K_2)^2)$.

88. (a) Since the union of convex sets is convex as is the closure of a convex set, K is closed and convex. And, since $K_n \subset K$ for all n, $d(x, K) \leq d(x, K_n)$ for all n and $x \in H$. On the other hand, given $\varepsilon > 0$, let $y \in \bigcup_n K_n$ be such that $d(P_K(x), y) \leq \varepsilon$ and let N be such that $y \in K_N$. Then for all $n \geq N$ we have $d(x, K) \leq d(x, K_n) \leq d(x, y) \leq d(x, P_K(x)) + d(P_K(x), y) \leq d(x, K) + \varepsilon$, which implies that $\lim_n d(x, K_n) = d(x, K)$. Finally, since $K_n \subset K$ for all n, by Problem 87, $\|P_{K_n}(x) - P_K(x)\|^2 \leq 2(d(x, K_n)^2 - d(x, K)^2)$, and, consequently, $\lim_n P_{K_n}(x) = P_K(x)$.

90. Given $x \in H$, let $x_n = P_{X_n}(x)$ and for each $x_n \neq 0$, let $e_n = x_n / \|x_n\|$. Since $\{e_n\}$ is an ONS by Bessel's inequality $\sum_n |\langle x, e_n \rangle|^2 \leq \|x\|^2$, and, consequently, $\sum_n \langle x, e_n \rangle e_n$ converges in H. Now, since $x - x_n \in X_n^\perp$, $\langle x - x_n, x_n \rangle = 0$ for all n, which gives $\langle x, x_n \rangle = \|x_n\|^2$ and if not zero, $\langle x, e_n \rangle e_n = \|x_n\|^{-2} \langle x, x_n \rangle x_n = x_n$. Thus $\sum_n x_n$ converges; it remains to compute the sum, which we call x_0. Now, for every $k \geq 1$ and every $y \in X_k$, $\langle x_0, y \rangle = \lim_N \sum_{n=1}^N \langle x_n, y \rangle$. Also, by assumption $\langle x_n, y \rangle = 0$ for all $n \neq k$ and so $\langle x_0, y \rangle = \langle x_k, y \rangle$. Now, since $x = x_k + (x - x_k)$ and $x - x_k \in X_k^\perp$, $\langle x, y \rangle = \langle x_k, y \rangle$, and, consequently, $\langle x - x_0, y \rangle = 0$ for all $y \in X_k$ and $x - x_0 \in X_k^\perp$ for all k, which implies that $x - x_0 \in \bigcap_k X_k^\perp = \{0\}$ and $x = x_0$ is the desired sum.

Finally, uniqueness. Let $x \in H$ and suppose that $x = \sum_n y_n$ with $y_n \in X_n$. Then, if $y \in X_k$, as before $\langle x, y \rangle = \langle y_k, y \rangle$ and $\langle x, y \rangle = \langle x_k, y \rangle$. Therefore $\langle x_k - y_k, y \rangle = 0$ and since $x_k - y_k \in X_k$, by picking $y = x_k - y_k$ it follows that $\|x_k - y_k\| = 0$ and $x_k = y_k$.

92. Let $P : H \to N$ denote the orthogonal projection onto N and $T = P|_M$ the restriction of P to M; clearly T is injective. Hence, if $M_1 \subset M$ is an arbitrary subspace of M with $\dim(M_1) = k$, then $\dim (T(M_1)) = k \leq \dim(N)$, which implies that $\dim(M) \leq \dim(N)$.

94. (a) Necessity first. The case when $H_n = \mathrm{sp}\{e_n\}$ where $\{e_n\}$ is an ONB for H is essentially done in Problem 5.139. In the general case, for the sake of argument suppose that $\lambda \notin \ell^2$ and let $x_n \in H_n$ with $\|x_n\| = \lambda_n$; then $\{\sum_{k=1}^n x_k\} \subset A_{\lambda,\mathcal{H}}$. Now, by the Pythagorean theorem $\| \sum_{k=1}^n x_k\|^2 = \sum_{k=1}^n \lambda_k^2 \to \infty$ as $n \to \infty$ contrary to the fact that $A_{\lambda,\mathcal{H}}$, being compact, is bounded.

Sufficiency next. First, $A_{\lambda,\mathcal{H}}$ is clearly closed. Next, we claim that the identity operator I on $A_{\lambda,\mathcal{H}}$ can be approximated by the finite rank operators given by $I_n = \sum_{k \leq n} P_{H_k}$ where P_{H_k} is the orthogonal projection onto H_k for all k. Given $\varepsilon > 0$, let N be large enough so that $\sum_{k=n}^\infty \lambda_k^2 \leq \varepsilon^2$ for $n \geq N$. Now, for $x \in A_{\lambda,\mathcal{H}}$, since $(I - I_n)(x) = \sum_{k=n}^\infty P_{H_k}(x)$ we have $\|(I - I_n)(x)\|^2 = \sum_{k=n}^\infty \|P_{H_k}(x)\|^2 \leq \sum_{k=n}^\infty \lambda_k^2 \leq \varepsilon^2$ for all $n \geq N$, and, consequently, $\|I - I_n\| \to 0$ as $n \to \infty$. Thus $I|_{A_{\lambda,\mathcal{H}}}$ is the limit in the norm of finite rank operators and, hence, compact, and so $A_{\lambda,\mathcal{H}}$ is compact.

(b) Fix an integer $j \geq 1$. Since K is compact K is totally bounded and there are $\{x_1, \ldots, x_m\} \subset K$ such that given $x \in K$, $\|x - x_i\| < 1/8j$ for some $1 \leq i \leq m$; moreover, by taking $N_j > N_{j-1}$ sufficiently large, $x_1, \ldots, x_m \in E_j = \bigoplus_{i=1}^{N_j} H_i$. Note that if P_j denotes the orthogonal projection into E_j, then $\|x - P_j(x)\| \leq \|x - x_i\| + \|P_j(x - x_i)\| \leq 1/4j$. Let $H_j' = \bigoplus_{i=N_{j-1}+1}^{N_j} H_i$; then $H = \bigoplus_j H_j'$ is an orthogonal decomposition. Now, if $x \in K$ we have $x = \sum_j x_j$ where $x_j \in H_j'$. Then $\|x_j\| = \|P_j(x) - P_{j-1}(x)\| \leq \|x - P_j(x)\| + \|x - P_{j-1}(x)\| \leq 1/j$. Hence $K \subset A_{\mu,\mathcal{H}'}$ with $\mu_j = 1/j$ and $\mathcal{H}' = \{H_j'\}$.

95. (a) Let $x \in H$; then $x = \sum_n \lambda_n e_n$ where $\lim_n \lambda_n = \lim_n \langle x, e_n \rangle = 0$. Now, given a linear functional L on X, let $x_L \in H$ such that $L(x) = \langle x, x_L \rangle$ and $\|L\| = \|x_L\|$. Then $\lim_n L(e_n) = \lim_n \langle e_n, x_L \rangle = 0$ and $e_n \rightharpoonup 0$ in H.

(b) Let $x^k = k^{-1} \sum_{n=N+1}^{N+k} e_n$. Then each x^k is a convex combination of the e_n for $n \geq N$ and $\|x^k\|^2 = k/k^2 = 1/k \to 0$.

98. First, $\|x - x_n\|^2 = \|x\|^2 - \langle x, x_n \rangle - \langle x_n, x \rangle + \|x_n\|^2$. Next, by weak convergence $\lim_n \langle x_n, x \rangle = \langle x, x \rangle = \|x\|^2$, and since $\langle x, x_n \rangle = \overline{\langle x_n, x \rangle}$, $\lim_n \langle x_n, x \rangle = \|x\|^2$. Therefore $\liminf_n (\|x\|^2 - \langle x, x_n \rangle - \langle x_n, x \rangle + \|x_n\|^2) = \liminf_n \|x_n\|^2 + \|x\|^2 - 2\|x\| \geq 0$ and so $\liminf_n \|x_n\| \geq \|x\| = 1$. Now, since by assumption $\limsup_n \|x_n\| \leq 1$, $\lim_n \|x_n\| = 1 = \|x\|$, and the conclusion follows from Problem 97.

99. In particular, note that if $D = \mathrm{sp}\{y_\alpha\}$ is dense in H, then $x_n \rightharpoonup x$ in H iff $\|x_n\| \leq c$ and $\lim_n \langle x_n, y_\alpha \rangle = \langle x, y_\alpha \rangle$ for all α.

100. Since $\|x_n\|^2 = \sum_{k=1}^n k^{-1}$, $\{\|x_n\|\}$ is unbounded and $\{x_n\}$ does not converge weakly in H, and, consequently, not in norm. Now, $\|y_n\| = 1$ for all n and since $\langle e_m, y_n \rangle = 1/\sqrt{n}$ for all m, n, $\lim_n \langle e_m, y_n \rangle = 0$ for all m. Hence by Problem 99, $y_n \rightharpoonup 0$ in H. Moreover, since $\|y_n\| \nrightarrow 0$, $\{y_n\}$ does not converge in H.

101. Consider the ℓ^2 sequences y^m defined by

$$y_k^m = \begin{cases} 1, & k = m, \\ 0, & k \neq m. \end{cases}$$

Let L be a linear functional on ℓ^2 and $x_L \in \ell^2$ such that $L(x) = \langle x, x_L \rangle$ and $\|L\| = \|x_L\|$. Then $L(y^m) = x_{L,m} \to 0$ as $m \to \infty$, and, therefore, $y^m \rightharpoonup 0$ in ℓ^2. Now, $x_{m,n} = y^m + m y^n$, so for a fixed m, $x_{m,n} = y^m + m y^n$ converges weakly to y^m in ℓ^2 as $n \to \infty$. Thus y^m is in the weak closure of S and since $y^m \rightharpoonup 0$ in ℓ^2, 0 is in the weak closure of S.

For the sake of argument suppose that $\{x_{m,n}\}$ in S converges weakly to 0. Then $\{x_{m,n}\}$ is bounded and, consequently, m is bounded. Now, if $z \in \ell^2$ is given by $z_n = 1/n$, we have $\langle z, x_{m,n} \rangle = 1/m + m/n \to 0$ as $n \to \infty$. Hence $m \to \infty$, which is not the case.

103. A particular instance of this result is the following: Let $x \in \ell^2$ and put $x^n = (0, \ldots, 0, x_1, x_2, \ldots)$, $n \geq 1$, where the first n terms of x^n are 0. Then $x^n \rightharpoonup 0$ in ℓ^2.

104. Note that with $e_k = e^{ikt}$ the k-th Fourier coefficient of x_n is given by

$$\frac{1}{2\pi} \int_T x(nt) \overline{e_k(t)}\, dt = \frac{1}{2\pi} \frac{1}{n} \int_0^{2\pi n} x(t) \overline{e_k(t/n)}\, dt$$

$$= \frac{1}{2\pi} \frac{1}{n} \sum_{\ell=0}^{n-1} \int_0^{2\pi} x(t) \overline{e_k((t + 2\pi\ell)/n)}\, dt$$

$$= \frac{1}{2\pi} \int_0^{2\pi} x(t) \frac{1}{n} \Big(\sum_{\ell=0}^{n-1} e^{-ik2\pi\ell/n} \Big) \overline{e_k(t/n)}\, dt.$$

Now let $y = \sum_{|k| \leq N} d_k e^{ikt}$ be a trigonometric polynomial of degree N. Then by Plancherel's equality

$$\langle x, y \rangle = \frac{1}{2\pi} \int_0^{2\pi} x(t) \Big(\sum_{|k| \leq N} \overline{d_k} e^{-ikt/n} \frac{1}{n} \Big(\sum_{\ell=0}^{n-1} e^{-ik2\pi\ell/n} \Big) \Big)\, dt.$$

Let $\Phi(n, N, t)$ denote the function multiplying $x(t)$ in the integral above. To compute $\lim_n (2\pi)^{-1} \int_0^{2\pi} x(t)\Phi(n, N, t)\, dt$ we consider two cases. If $k = 0$ the limit is $\overline{d_0}$ and if $k \neq 0$, by Weyl's lemma, the limit is $\int_0^1 e^{-ik2\pi}\, dx = 0$, and so

$$\langle x, y \rangle = \frac{1}{2\pi} \int_0^{2\pi} x(t)\, dt \frac{1}{2\pi} \int_0^{2\pi} \overline{y}(t)\, dt = \frac{1}{2\pi} \int_0^{2\pi} c_0 \overline{y}(t)\, dt = \langle c_0, y \rangle$$

for all $y \in L^2(I)$. Hence $x_n \rightharpoonup c_0$ in $L^2([0, 2\pi])$.

105. Considering $\{x_n - x\}$ if necessary we may assume that $x = 0$. Let $x_{n_1} = x_1$. Having chosen $x_{n_1}, \ldots, x_{n_{k-1}}$, pick n_k so that $|\langle x_{n_{k-1}}, x_n \rangle| \leq 2^{-k}$ for all $n \geq n_k$; this choice is possible by the weak convergence to 0. Recall that also $\|x_n\| \leq c$. Now, a direct computation gives

$$\left\| \frac{1}{N} \sum_{k=1}^N x_{n_k} \right\|^2 = \frac{1}{N^2} \sum_{k=1}^N \|x_{n_k}\|^2 + \frac{1}{N^2} \sum_{1 \leq k \neq m \leq N} \langle x_{n_k}, x_{n_m} \rangle = I_N + J_N ,$$

say. $I_N \leq c^2/N$ and J_N is bounded by

$$2\frac{1}{N^2} \sum_{1 \leq k < m \leq N} |\langle x_{n_k}, x_{n_m} \rangle| \leq 2\frac{c^2}{N^2} \sum_{1 \leq k < m \leq N} \frac{1}{2^{k+1}}$$

$$\leq 2\frac{c^2}{N^2} \sum_{k=1}^N \frac{N - k}{2^k} \leq \frac{c^2}{2N} .$$

Since $I_N, J_N \to 0$ with N, we have finished.

106. (a) It is readily verified that p is a norm on H and that by Cauchy-Schwarz $p(x)^2 \leq (\sum_n 2^{-2n})(\sum_n |\langle x, e_n \rangle|^2) \leq \|x\|^2$. Therefore, were (H, p) complete, by the inverse mapping theorem the norms would be equivalent but this is not the case since $\|e_n\| = 1$ and $p(e_n) = 2^{-n}$ for all n.

(b) Considering $\{x_k - x\}$ if necessary we may assume that $x = 0$. Necessity first. Let $\|x_k\| \leq c$ for all k. Given $\varepsilon > 0$, let m be large enough so that $c \sum_{n=m}^\infty 2^{-n} \leq \varepsilon/2$. Since $x_k \rightharpoonup 0$ there exists k_0 such that $|\langle x_k, e_n \rangle| \leq \varepsilon/2(m - 1)$ for all $1 \leq n < m$ and $k \geq k_0$. Therefore, since $|\langle x_k, e_n \rangle| \leq c$, $p(x_k) \leq \sum_{n=1}^{m-1} |\langle x_k, e_n \rangle| + c \sum_{n=m}^\infty 2^{-n} \leq (m - 1)\varepsilon/2(m - 1) + \varepsilon/2 = \varepsilon$, and so $p(x_k) \to 0$.

Sufficiency next. By Problem 99 it suffices to verify that $\langle x_k, e_n \rangle \to 0$ for all n. Fix n and given $\varepsilon > 0$, let k_0 be large enough so that $p(x_k) \leq 2^{-n}\varepsilon$ for $k \geq k_0$. Then $2^{-n}|\langle x_k, e_n \rangle| \leq p(x_k) \leq 2^{-n}\varepsilon$ and $|\langle x_k, e_n \rangle| \leq \varepsilon$ for $k \geq k_0$.

(c) Let $x_k = ke_k$, $k \geq 1$. Since $\{\|x_k\|\}$ is unbounded $\{x_k\}$ does not converge weakly in H. On the other hand, $p(x_k) = k2^{-k} \to 0$ as $k \to \infty$.

(d) Let $\{x_n\} \subset H$ be bounded. Then by Problem 8.154(a) a subsequence $\{x_{n_k}\}$ of $\{x_n\}$ converges weakly to $x \in H$, say, and, therefore, by (b), $p(x_{n_k} - x) \to 0$. Thus every sequence in the unit ball of $(H, \|\cdot\|)$ has a convergent subsequence in (H, p) and the unit ball is compact there.

107. Let $\{e_n\}$ be an ONS in H. The result is clearly true if $\|x\| = 1$. Otherwise let $b_n = \sqrt{1 - \|x\|^2 + |c_n|^2} - c_n$ where $c_n = \langle x, e_n \rangle$ are the Fourier coefficients of x with respect to the ONS $\{e_n\}$; note that since $|c_n| < 1$, $|b_n| < 1 + \sqrt{2}$. We claim that if $y_n = x + b_n e_n$, then $\|y_n\| = 1$ for all n. Indeed, since $x - c_n e_n \perp e_n$,

$$
\begin{aligned}
\|y_n\|^2 &= \| x - c_n e_n + (\sqrt{1 - \|x\|^2 + |c_n|^2}) e_n \|^2 \\
&= \| x - c_n e_n \|^2 + \| (\sqrt{1 - \|x\|^2 + |c_n|^2}) e_n \|^2 \\
&= (\|x\|^2 - |c_n|^2) + (1 - \|x\|^2 + |c_n|^2) = 1 .
\end{aligned}
$$

Therefore for $z \in H$, $|\langle z, y_n \rangle - \langle z, x \rangle| = |b_n||\langle z, e_n \rangle| \le (1 + \sqrt{2})|\langle z, e_n \rangle|$, which tends to 0 as $n \to \infty$.

In particular, the result gives that $\{x : \|x\| < 1\}$ is not weakly open because each weak ball centered at a point x_0 in the set < 1 contains a point of $S = \{\|x\| = 1\}$.

108. (a) Let $\{x_n\} \subset K$ be such that $x_n \rightharpoonup x$ in H. Then by Problem 105 there is a subsequence $\{x_{n_k}\}$ of $\{x_n\}$ such that $y_N = N^{-1} \sum_{k=1}^{N} x_{n_k} \to x$ in H. Since K is convex, $y_N \in K$ for all N, and since K is closed the limit of the y_N, i.e., x, is in K. Thus K is weakly closed.

A similar argument gives that if K is a convex subset of a Hilbert space H, 0 is in the closure of K iff there is a sequence $\{x_n\} \subset K$ that converges weakly to 0 in X.

(b) The right-hand side above is finite for all $x_0 \in H$ since $K \ne \emptyset$. If $x_0 \in K$, pick $x = x_0$. Otherwise, call the inf η and let $\{x_n\} \subset K$ be such that $\|x_n - x_0\| \to \eta$. Then $\{x_n\}$ is bounded and since H is reflexive, by Problem 8.154(a), passing to a subsequence if necessary we may assume that there exists $x \in H$ such that $x_n \rightharpoonup x$ in H. Then as in (a) above $x \in K$ and $\eta \le \|x - x_0\|$. Moreover, since $x_n - x_0 \rightharpoonup x - x_0$, by Problem 8.137(b) $\|x - x_0\| \le \liminf_n \|x_n - x_0\| = \eta$, and, therefore, $\|x - x_0\| = \eta$.

111. First, by Cauchy-Schwarz $|\langle T(x), y \rangle| \le \|T(x)\| \, \|y\| \le \|T\| \, \|x\| \, \|y\|$, and so $\sup_{\|x\|=\|y\|=1} |\langle T(x), y \rangle| \le \|T\|$. Conversely, the conclusion is obvious if $T = 0$. Otherwise, for an arbitrary $\varepsilon > 0$, let x_0 be such that $\|T(x_0)\| > (1 - \varepsilon)\|T\|$, $\|x_0\| = 1$. Then with $x = x_0$ and $y = T(x_0)/\|T(x_0)\|$ we have $|\langle T(x), y \rangle| = |\langle T(x_0), T(x_0) \rangle|/\|T(x_0)\| = \|T(x_0)\| \ge (1 - \varepsilon)\|T\|$ and the conclusion follows.

113. Let P_n denote the projection onto $\mathrm{sp}\{e_1,\ldots,e_n\}$ and put $T_n = P_n \circ T$; each T_n is of finite rank, bounded, and $T(x) - T_n(x) = (I - P_n) \circ T(x)$. Since $I - P_n = R^n \circ L^n$, by Problem 112 it follows that $\|T(x) - T_n(x)\| \leq \|R^n\|\,\|L^n\|\,\|T(x)\| \to 0$ as $n \to \infty$.

115. By assumption and Cauchy-Schwarz we get $\|T\| = |\langle T(x_0), x_0 \rangle| \leq \|T(x_0)\|\,\|x_0\| \leq \|T\|$, and so $|\langle T(x_0), x_0 \rangle| = \|T(x_0)\|\,\|x_0\|$. The conclusion follows from Problem 7.

116. The statement is true. Let $\{e_n\}$ be the standard ONB for ℓ^2 and Λ an uncountable Hamel basis for ℓ^2 that contains the e_n. Then pick $e \in \Lambda \setminus \{e_n\}$ and define T on the e_λ by

$$T(e_\lambda) = \begin{cases} 1, & e_\lambda = e, \\ 0, & \text{otherwise.} \end{cases}$$

For $x \in H$ write $x = \sum_{\lambda \in \Lambda} x_\lambda\, e_\lambda$, where the sum is actually finite, and let $T(x) = \sum_{\lambda \in \Lambda} x_\lambda T(e_\lambda)$. Then T is linear and $T(e_n) = 0$ for all n. Since $e = \lim_N \sum_{n=1}^{N} \langle e, e_n \rangle e_n$, if T were bounded it would follow that $1 = T(e) = \lim_N \sum_{n=1}^{N} \langle e, e_n \rangle T(e_n) = 0$, which is not the case.

117. Let $T_n(x) = \langle x, e_n \rangle e_n$, $x \in H$.

118. Yes. For an integer k let $T : H \to H$ be given by $T(x) = \langle x, e_1 + \cdots + e_k \rangle e_1$. Then $T(e_n) = e_1$ if $n \leq k$ and $T(e_n) = 0$ otherwise. Thus $\|T(e_n)\| \leq 1$ for all n, and if $x_k = (e_1 + \cdots + e_k)/\sqrt{k}$, $\|x_k\| = 1$, and $\|T\| \geq \|T(x_k)\| = \sqrt{k}$.

119. Let $\mathcal{I} = \{n \in \mathbb{N} : e_n \in X\}$. Since $T|_X = I_X$ it follows that $\sum_{n \in \mathcal{I}} 1 = \sum_{n \in \mathcal{I}} |\langle e_n, T(e_n) \rangle|^2 < \infty$. Therefore X can only contain finitely many e_n, and, consequently, is finite dimensional.

121. Given x we must find t so that $x + tv \perp u$. This is equivalent to $t = -\langle x, u \rangle / \langle v, u \rangle$, which gives $T(x) = x - (\langle x, u \rangle / \langle v, u \rangle)v$. T is clearly linear. Moreover, since $\|T(x)\| \leq (1 + (\|u\|\,\|v\|/|\langle u, v \rangle|))\|x\|$, T is bounded.

We claim that $R(T) = u^\perp$. Indeed, from the definition of T, $R(T) \subset u^\perp$. Moreover, if $y \perp u$, then $y = T(y + 0 \cdot v)$ which gives the opposite inclusion. Finally, if u, v are linearly dependent, T becomes the orthogonal projection onto $u^\perp$.

125. (a) implies (b) Given $x, y \in H$ and a scalar λ, expanding $\langle T(x + \lambda y), T(x + \lambda y) \rangle = \langle x + \lambda y, x + \lambda y \rangle$ it follows that $\|T(x)\|^2 + \Re\bar{\lambda}\langle T(x), T(y) \rangle + \|T(y)\|^2 = \|x\|^2 + \Re\bar{\lambda}\langle x, y \rangle + \|y\|^2$. Since T is an isometry this equation becomes $\Re\bar{\lambda}\langle T(x), T(y) \rangle = \Re\bar{\lambda}\langle x, y \rangle$ for any scalar λ. If H is a real space the choice $\lambda = 1$ will do. If H is a complex space pick $\lambda = 1$ and $\lambda = i$ and verify that $\langle T(x), T(y)\rangle$ and $\langle x, y\rangle$ have the same real and imaginary parts.

(d) implies (a) Let $x = \sum_\alpha \lambda_\alpha e_\alpha \in H$. Then it readily follows that $\langle T(x), T(x)\rangle = \sum_{\alpha,\beta} \lambda_\alpha \overline{\lambda_\beta} \langle T(e_\alpha), T(e_\beta)\rangle = \sum_\alpha |\lambda_\alpha|^2 = \langle x, x\rangle$ and T is an isometry.

126. Necessity first. Let $M = K(T)^\perp$ and $P : H \to M$ the orthogonal projection of H onto M. By Problem 78, $M = \{x \in H : \|P(x)\| = \|x\|\}$, and if $x \in M$, then $\langle T^*T(x), x\rangle = \|T(x)\|^2 = \|x\|^2 = \|P(x)\|^2 = \langle P(x), x\rangle$. And if $x \in M^\perp = K(T)$, $\langle T^*T(x), x\rangle = 0 = \langle P(x), x\rangle$. Thus $\langle (T^*T - P)(x), x\rangle = 0$ for all $x \in H$ and since $T^*T - P$ is self-adjoint, by Problem 142, $T^*T = P$.

Sufficiency next. As above $\|T(x)\|^2 = \|P(x)\|^2$ for all $x \in M$ and, therefore, $\|T(x)\| = \|x\|$ if $x \in M$ and $\|T(x)\| = 0$ if $x \in M^\perp = K(T)$, respectively.

127. (b) and (c) Since $\{e_1, e_2, \ldots\} \subset R(T)$, e is orthogonal to the e_n. Also, since isometries preserve inner products we have $\langle e_n, e_{n+k}\rangle = \langle T^n(e), T^n(e_k)\rangle = \langle e, e_k\rangle = 0$ for $n \geq 0$ and $k > 0$. And since T is an isometry, $\|e_n\| = \|T(e_{n-1})\| = \|e_{n-1}\| = \ldots = \|e\| = 1$. Hence $\{e_n\}$ is an ONS in H.

128. (a) T is readily seen to be linear. Changing variables twice we get $\|T(x)\|^2 = (2\pi)^{-1} \int_0^\alpha |x(2\pi + t - \alpha)|^2 \, dt + (2\pi)^{-1} \int_\alpha^{2\pi} |x(t - \alpha)|^2 \, dt = (2\pi)^{-1} \int_{2\pi-\alpha}^{2\pi} |x(s)|^2 \, ds + (2\pi)^{-1} \int_0^{2\pi-\alpha} |x(s)|^2 \, ds = \|x\|^2$.

(b) For $0 \leq t < \alpha$ we have $T(e_n)(t) = e_n(2\pi + t - \alpha) = e^{in(t-\alpha)} = e^{-in\alpha} e_n(t)$, and, similarly, when $t \geq \alpha$, $T(e_n) = e^{-in\alpha} e_n$ for all $n \in \mathbb{Z}$.

(c) Since T is an isometry, the T^k are isometries for all $k \geq 1$. Thus $\|S_K\| = K^{-1} \|\sum_{k=0}^{K-1} T^k\| \leq K^{-1} \sum_{k=0}^{K-1} \|T^k\| \leq 1$. Now, since by (b) $T^k(e_n) = (e^{-in\alpha})^k e_n$ for all $k \geq 0$, it readily follows that $S_K(e_n) = s_K e_n$ where $s_K \in \mathbb{C}$ is given by

$$s_K = \frac{1}{K} \sum_{k=0}^{K-1} e^{-ink\alpha} = \frac{1}{K} \frac{e^{-inK\alpha} - 1}{e^{-in\alpha} - 1}.$$

Thus $|s_K| \leq 2 |e^{-in\alpha} - 1|^{-1}/K$ which tends to 0 as $K \to \infty$.

(d) Since the constant function $e_0 = 1$ is invariant for T, $S_K(e_0) = e_0$ for all $K \geq 1$. Thus if $y = \sum_{n=-N}^N c_n e_n$ is a trigonometric polynomial, by (c), $\lim_K S_K(y) = c_0 e_0$. Now, if $x \in H$ is arbitrary given $\varepsilon > 0$, let $y = \sum_{n=-N}^N \langle f, e_n\rangle e_n$ be a partial sum of x with N large enough so that $\|x - y\| < \varepsilon/2$. Then, since $\|S_K(x) - S_K(y)\| \leq \|S_K\| \|x - y\| < \varepsilon/2$ and

$\|S_K(y) - c_0 e_0\| < \varepsilon/2$ for all K sufficiently large, we have $\|S_K(x) - c_0 e_0\| < \varepsilon$ for those K. Thus $\{S_K(x)\}$ converges to the constant function $c_0 = \langle x, e_0 \rangle = (1/2\pi) \int_0^{2\pi} x(t)\, dt$, i.e., the average of x.

129. Necessity first. Let $R(T) = \mathrm{sp}\{x_0\}$, $0 \neq x_0 \in X$. Then $T(x) = \ell(x) x_0$ where $\ell(x)$ is the bounded linear functional

$$\ell(x) = \Big\langle \ell(x) x_0, \frac{x_0}{\|x_0\|^2} \Big\rangle = \frac{1}{\|x_0\|^2}\, \langle T(x), x_0 \rangle = \frac{1}{\|x_0\|^2}\, \langle x, T^*(x_0) \rangle$$

and, therefore, $T(x) = \langle x, y \rangle z$ where $z = x_0$ and $y = \|x_0\|^{-2} T^*(x_0)$.

Sufficiency next. Clearly T is a linear mapping and $\dim(R(T)) = 1$. Now, $\|T(x)\| = |\langle x, y \rangle|\, \|z\| \leq \|x\|\, \|y\|\, \|z\|$ and $\|T\| \leq \|y\|\, \|z\|$. Moreover, setting $x = y/\|y\|$ we have $\|T\| \geq |\langle y/\|y\|, y \rangle|\, \|z\| = \|y\|\, \|z\|$, and, consequently, $\|T\| = \|y\|\, \|z\|$.

As for T^*, since $\langle T(x), w \rangle = \langle \langle x, y \rangle z, w \rangle = \langle x, y \rangle \langle z, w \rangle = \langle w, z)(x, y \rangle = \langle x, \langle w, z \rangle y \rangle$ we have $T^*(w) = \langle w, z \rangle y$.

Finally, T is self-adjoint when $T^*(x) = T(x)$ or $y = (\langle x, y \rangle / \langle x, z \rangle)z$, i.e., there exists $\lambda \in \mathbb{C} \setminus \{0\}$ such that $y = \lambda z$. Therefore, $\overline{\lambda}\langle x, z \rangle z = \lambda \langle x, z \rangle z$ for all $x \in H$. In particular, setting $x = z/\|z\|^2$ it follows that $(\overline{\lambda} - \lambda)z = 0$ and so $\lambda \in \mathbb{R} \setminus \{0\}$.

131. Let $\{x_n\}$ be a bounded sequence in H, $\|x_n\| \leq c$ for all n. Then, if $\{y_1, \ldots, y_K\}$ is an ONB for $R(T)$ we have $T(x_n) = \sum_{k=1}^{K} \langle T(x_n), y_k \rangle y_k$ where $|\langle T(x_n), y_k \rangle| \leq \|T(x_n)\|\, \|y_k\| \leq c\|T\|$ for all n, k. Pick first a subsequence $\{x_{n_1}\}$ of $\{x_n\}$ such that $\lim_{n_1} \langle T(x_{n_1}), y_1 \rangle = a_1$ exists, then a further subsequence $\{x_{n_2}\}$ of $\{x_{n_1}\}$ such that $\lim_{n_2} \langle T(x_{n_2}), y_2 \rangle = a_2$ exists and so on until we pick a subsequence $\{x_{n_K}\}$ of $\{x_{n_{K-1}}\}$ such that $\lim_{n_K} \langle T(x_{n_\ell}), y_\ell \rangle = a_\ell$ exists for $1 \leq \ell \leq K$. Now let $y = \sum_{k=1}^{K} a_k y_k$. Then $\|T(x_{n_K}) - y\|^2 = \sum_{k=1}^{K} |\langle T(x_{n_K}), y_k \rangle - a_k|^2 \to 0$ as $n_K \to \infty$ and $\{T(x_n)\}$ has a convergent subsequence

Finally, since $\|e_\alpha - e_\beta\|^2 = 2$ for any distinct elements of an ONB for H, I is not of finite rank unless H is finite dimensional.

132. Necessity first. The answer is immediate if H is separable. Indeed, let $\{e_n\}$ be an ONB for H and $P_n(x) = \sum_{k=1}^{n} \langle x, e_k \rangle e_k$ the orthogonal projection onto $\mathrm{sp}\{e_1, \ldots, e_n\}$. Then $\|P_n(x) - I(x)\| \to 0$ for all $x \in H$ and since $P_n \circ T$ is of finite rank and, hence, compact, by Problem 9.59, $\lim_n \|P_n \circ T - T\| = 0$.

133. (a) $\|T\| = \sup_n |\lambda_n|$.

(b) $T^*(y) = \sum_n \overline{\lambda}_n \langle y, f_n \rangle e_n$. Thus when $e_n = f_n$, T is self-adjoint iff the λ_n are real.

(c) Necessity first. For the sake of argument suppose that $\lambda_n \nrightarrow 0$ and let $\{\lambda_{n_k}\}$ be a subsequence of $\{\lambda_n\}$ and $\eta > 0$ such that $|\lambda_{n_k}| \geq \eta$ for all k. Then $T(e_{n_k}) = \lambda_{n_k} f_{n_k}$ and $\|T(e_{n_k}) - T(e_{n_j})\|^2 = \|\lambda_{n_k} f_{n_k} - \lambda_{n_j} f_{n_j}\|^2 = |\lambda_{n_k}|^2 + |\lambda_{n_j}|^2 \geq 2\eta^2$. Therefore $\{T(e_{n_k})\}$ has no convergent subsequence and T is not compact.

Sufficiency next. Let T_k denote the partial sum of T of order k; T_k is a finite rank operator and, hence, compact. Moreover $\|(T - T_k)(x)\|^2 \leq \sup_{k \geq n} |\lambda_k|^2 \|x\|^2$, $\|T - T_k\| \leq \sup_{k \geq n} |\lambda_k|^2 \to 0$ and T is compact.

136. (a) The statement is false. On $H = \ell^2 \oplus \ell^2$ define $T : H \to H$ by $T(x, y) = (0, x)$. Then $T^2 = 0$ is compact yet T, which is essentially the identity in ℓ^2, is not.

(b) The statement is true. Let $\{x_n\} \subset H$ be bounded, i.e., $\|x_n\| \leq c$ for all n. Passing to a subsequence if necessary we may assume that $\{T^*T(x_n)\}$ converges. Now, since $\|T(x_n) - T(x_m)\|^2 = \langle x_n - x_m, T^*T(x_n) - T^*T(x_m) \rangle \leq 2c \|T^*T(x_n) - T^*T(x_m)\|$, $\{T(x_n)\}$ is Cauchy, hence convergent, and T is compact.

137. Necessity first. Observe that the sup can be taken over $x \in \{e_1, \ldots, e_n\}^{\perp}$ with $\|x\| \leq 1$. Now, $\{\mu_n\}$ is a nonincreasing sequence bounded below and, consequently, it converges to a limit $\mu \geq 0$, say. For all n sufficiently large there exists $x_n \in \{e_1, \ldots, e_n\}^{\perp}$, $\|x_n\| = 1$, such that $\|T(x_n)\| \geq \mu/2$. Since by Problem 103, $x_n \rightharpoonup 0$ and T is compact, by Problem 9.62, $T(x_n) \to 0$ and $\mu = 0$.

As for sufficiency, consider the finite rank operators T_n given by $T_n(x) = \sum_{k=1}^{n} \langle e_k, x \rangle T(e_k)$. Then $T(x) - T_n(x) = \sum_{k=n+1}^{\infty} \langle e_k, x \rangle T(e_k)$, and, consequently, $\|T_n - T\| = \mu_n$. Finally, since $\mu_n \to 0$, T is the uniform limit of finite rank operators and, consequently, compact.

142. (a) The statement is false if H is a real Hilbert space as a ± 90 degree rotation in $\mathbb{R}^2$ shows. In the real case the condition is $\langle T(x), y \rangle = 0$ for all x, y in H for then, given $x \in H$, picking $y = T(x)$ it follows that $\|T(x)\|^2 = \langle T(x), T(x) \rangle = 0$ and so $T(x) = 0$ for all $x \in H$.

On the other hand, the statement is true if the underlying field is $\mathbb{C}$. First, observe that $\langle T(x + \lambda y), x + \lambda y \rangle = \langle T(x), x \rangle + \lambda \langle T(y), x \rangle + \bar{\lambda} \langle T(x), y \rangle + |\lambda|^2 \langle T(y), y \rangle = 0$, and so $\lambda \langle T(y), x \rangle + \bar{\lambda} \langle T(x), y \rangle = 0$ for all $\lambda \in \mathbb{C}$. Picking $\lambda = 1, i$ we get $\langle T(y), x \rangle + \langle T(x), y \rangle = 0$ and $\langle T(y), x \rangle - \langle T(x), y \rangle = 0$, respectively, which together imply that $\langle T(x), y \rangle = 0$ for all $x, y \in H$, and so as before $T = 0$.

(b) No. Consider $T : \mathbb{R}^2 \to \mathbb{R}^2$ given by the matrix

$$M = \begin{pmatrix} 0 & 1 \\ 0 & 0 \end{pmatrix}.$$

(c) Yes. Observe that for $x \in H$, $\|T(x)\|^2 = \langle T(x), T(x) \rangle = \langle x, T^*T(x) \rangle = 0$. Thus $T(x) = 0$ for all $x \in H$ and $T = 0$. In particular, if T is self-adjoint and $T^2 = 0$, since $T^* = T$, $T = 0$.

(d) Since $\langle S^*S(x) + T^*T(x), x \rangle = \langle S(x), S(x) \rangle + \langle T(x), T(x) \rangle = 0$ for all $x \in H$, $S = T = 0$.

143. The statement is false. On ℓ_0^2 let T be given by $T(x) = y$ where $y_n = nx_n$ for all n. As is readily seen $\langle T(x), y \rangle = \sum_n nx_n \overline{y}_n = \langle x, T(y) \rangle$. Since $\|e_n\| = 1$ and $\|T(e_n)\| = n$ for all n, T is unbounded.

144. First, linearity. Let $x, x_1 \in H$ and λ a scalar. Then for all $y \in H$, $\langle T(x + \lambda x_1), y \rangle = \langle x + \lambda x_1, T(y) \rangle = \langle x, T(y) \rangle + \lambda \langle x_1, T(y) \rangle = \langle T(x), y \rangle + \lambda \langle T(x_1), y \rangle = \langle T(x) + \lambda T(x_1), y \rangle$, and, consequently, T is linear.

Next, $R(T)$ is closed. Let $x_n \to x$ and $T(x_n) \to y$. Then for all $z \in H$, $\langle y, z \rangle = \lim_n \langle T(x_n), z \rangle = \lim_n \langle x_n, T(z) \rangle = \langle x, T(z) \rangle = \langle T(x), z \rangle$. Thus $T(x) = y$ and, consequently, T is continuous by the closed graph theorem.

Finally, the norm estimate. Let $\eta = \sup_{\|x\| \leq 1} |\langle T(x), x \rangle|$; since $|\langle T(x), x \rangle| \leq \|T(x)\| \|x\| \leq \|T\| \|x\|^2$, $\eta \leq \|T\|$. Now, since $\langle T(x + y), x + y \rangle - \langle T(x - y), x - y \rangle = 2(\langle T(x), y \rangle + \langle T(y), x \rangle)$ and $|\langle T(z), z \rangle| \leq \eta \|z\|^2$ it readily follows that

$$2|\langle T(x), y \rangle + \langle T(y), x \rangle| \leq \eta(\|x + y\|^2 + \|x - y\|^2)$$
$$= 2\eta(\|x\|^2 + \|y\|^2)$$

for all $x, y \in H$. Now let $\|x\| = 1$ be such that $T(x) \neq 0$ and put $y = \|T(x)\|^{-1} T(x)$. Then $\|y\| = 1$ and $\langle T(x), y \rangle + \langle T(y), x \rangle = \langle T(x), y \rangle + \langle y, T(x) \rangle = 2\|T(x)\|$, and, therefore, $4\|T(x)\| \leq 2\eta(1 + 1)$, or $\|T(x)\| \leq \eta$, which obviously holds when $T(x) = 0$. Hence $\sup_{\|x\|=1} \|T(x)\| = \|T\| \leq \eta$.

145. The statement is false. Let $\{e_n\}$ be an ONB of H and $T_n : H \to H$ be given by $T_n(x) = \langle x, e_n \rangle e_1$; note that $T_n^*(y) = \langle y, e_1 \rangle e_n$ for all n. Then $T_n(x) \to 0$ and $T_n^*(y) \rightharpoonup 0$ for all $x, y \in H$, but $T_n T_n^*(x) = \langle x, e_1 \rangle e_1$ does not tend to 0 weakly in H.

147. Since T is compact and T^* bounded, by Problem 9.58, TT^* is compact. Hence, since $T = T^{**}$, $T^{**}T^*$ is compact and by Problem 136(b), T^* is compact.

149. The only property that offers any difficulty in verifying that $\|\cdot\|_1$ is a norm is $\|T\|_1 = 0$ implies $T = 0$. Now, if $\|T\|_1 = 0$, then $\langle T(x), x \rangle = 0$ for all $x \in H$ with $\|x\| = 1$ and by Problem 142(a), $T = 0$. It is clear that $\|T\|_1 \leq \|T\|$. Moreover, since for a general T with adjoint T^*, $(T + T^*)/2$ and $(T - T^*)/2i$ are self-adjoint and $T = (T + T^*)/2 + i(T - T^*)/2i$, by Problem 144, $\|T\|_1 \leq \|(T + T^*)/2\|_1 + |i| \|(T - T^*)/2i\|_1 \leq 2$.

To see that 2 is the best possible constant consider on $\mathbb{C}^2$, $A = \begin{pmatrix} 0 & 1 \\ 0 & 0 \end{pmatrix}$. Take any $x = (x_1, x_2) \in \mathbb{C}^2$ such that $\|x\| = \sqrt{|x_1|^2 + |x_2|^2} = 1$; then $\|A(x)\| = \|(x_2, 0)\| \le \|x\| = 1$ with equality if $x_1 = 0$. Hence $\|A\| = 1$. On the other hand, $|\langle A(x), x \rangle| = |x_2 \overline{x_1}| \le (|x_1|^2 + |x_2|^2)/2 = 1/2$ with $=$ if $|x_1| = |x_2|$. Hence

$$\|A\| = 1 = 2(1/2) = 2 \sup_{\|x\|=1} |\langle A(x), x \rangle|.$$

150. (a) Note that for $x \in K(T)$, $0 = BT(x) = I(x) - S(x)$, i.e., $S|_{K(T)} = I|_{K(T)}$, and the identity operator is compact on $K(T)$. In particular, the unit ball of $K(T)$ is compact and $K(T)$ is finite dimensional.

(b) Let $\{y_n\} \subset R(T)$ converge to y in H_1 and $\{x_n\} \subset H$ such that $T(x_n) = y_n$ for all n. Now, if $\{\|x_n\|\}$ is bounded let $\{x_{n_k}\}$ be a subsequence of $\{x_n\}$ such that $S(x_{n_k})$ converges to $z \in H$, say. Then $x_{n_k} = I(x_{n_k}) = BT(x_{n_k}) + S(x_{n_k}) \to B(y) + z$, and, consequently, by continuity $y = \lim_{n_k} T(x_{n_k}) = T(\lim_{n_k} x_{n_k}) = T(B(y)+z)$ and $y \in R(T)$. On the other hand, if $\{\|x_n\|\}$ is unbounded, passing to a subsequence if necessary we may assume that $\|x_n\| \to \infty$. Also, since $H = K(T) \oplus K(T)^{\perp}$ we may assume that $x_n \perp K(T)$ for all n. Consider $z_n = x_n/\|x_n\|$. Now, since $\{T(x_n)\}$ converges, it is bounded and, therefore, $T(z_n) = (1/\|x_n\|)T(x_n) \to 0$. Then repeating the argument for bounded sequences, a subsequence $\{z_{n_k}\}$ of $\{z_n\}$ converges to z, say. By the continuity of the norm $\|z\| = 1$, and since $K(T)^{\perp}$ is closed, $z \perp K(T)$. Finally, by the continuity of T, $T(z) = 0$, i.e., $z \in K(T)$, which is not possible since $\|z\| = 1$. Hence $R(T)$ is closed. Alternatively, replacing B by $(I - (S - F))^{-1}B$ as in (a) we may assume that S is of finite rank. Then if $y_n = T(x_n) - y$ in H_1 it follows that $x_n - F(x_n) = B(y_n) \to B(y)$. A similar argument as above then gives that $y \in R(T)$.

(c) First, taking adjoints we have $B^*T^* = I_1 - S_1^*$. Since $R(T)^{\perp} = K(T^*)$ and S_1^* is compact, applying (a) to T^* we get that $R(T)^{\perp} = K(T^*)$ is finite dimensional.

151. (a) Let $y = T(x) \in R(T)$. Then for $z \in K(T^*)$, $\langle y, z \rangle = \langle T(x), z \rangle = \langle x, T^*(z) \rangle = 0$ and so $R(T) \subset K(T^*)^{\perp}$. Furthermore, since $K(T^*)^{\perp}$ is closed, $\overline{R(T)} \subset K(T^*)^{\perp}$. On the other hand, if $y \in R(T)^{\perp}$, $0 = \langle T(x), y \rangle = \langle x, T^*(y) \rangle$ for all $x \in H$, and, consequently, $T^*(y) = 0$. Thus $R(T)^{\perp} \subset K(T^*)$, and, consequently, $K(T^*)^{\perp} \subset R(T)^{\perp\perp} = \overline{R(T)}$, which proves (a).

(b) We apply (a) to T^* in place of T and use that $T^{**} = T$.

(c) Take orthogonal complements in (a) and (b).

This result also implies that if $T : H \to H$ is a self-adjoint linear operator, then $H = \overline{R(T)} \oplus K(T^{\perp})$.

152. (a) Clearly $K(T) \subset K(T^*T)$. Now, if $x \in K(T^*T)$, $\langle x, T^*T(x) \rangle = \langle T(x), T(x) \rangle = 0$, and, therefore, $\|T(x)\| = 0$ and $x \in K(T)$.

(b) By Problem 151(b), $\overline{R(T^*)} = K(T)^{\perp}$, and, therefore, by (a), $\overline{R(T^*)} = K(T^*T)^{\perp}$. Now, since $(T^*T)^* = T^*T$, $\overline{R(T^*)} = R(T^*T)^{\perp\perp}$, which gives the conclusion by Problem 61(c).

153. Since $T^* = T$ the conclusion follows from Problem 152(a).

154. Since the sum defining $T(x)$ is finite for $x \in D(T)$, T is well-defined. Let $y = \sum_n y_n e_n(t) \in D(T^*)$. Then $\langle y, T(x) \rangle = \sum_n y_n x(n)$, where the sum is actually finite, and $|\sum_n y_n x(n)| = |\langle T^*(y), x \rangle| \leq c\|x\|$ for all $x \in D(T)$. Now, fix an integer n and $\eta > 0$, and pick $x_{\eta, n_0} \in D(T)$ symmetric about n so that $x_{\eta, n}(n) = 1$ and $x_{\eta, n}(x) = 0$ for $|x - n| > \eta$ and $\|x_{\eta, n}\| \to 0$ with η. Then as $\eta \to 0$ the right-hand side of the above inequality goes to 0 and the left-hand side is equal to y_n (if $\eta < 1$). It thus follows that $y_n = 0$, and since n is arbitrary, $y_n = 0$ for all n, $y = 0$, and $D(T^*) = \{0\}$

155. Recall that for a densely defined T, $D(T^*)$ is the maximal subspace of H for which T^* is defined, i.e., $D(T^*) = \{y \in H : \text{there exists } T^*(y) \text{ in } H$ such that $\langle T(x), y \rangle = \langle x, T^*(y) \rangle$ whenever $x \in D(T)\}$. First, let $y \in D(T^*)$ and consider the functional ℓ_y on $D(T)$ given by $\ell_y(x) = \langle T(x), y \rangle$. ℓ_y is linear and $|\ell_y(x)| = |\langle T(x), y \rangle| = |\langle x, T^*(y) \rangle| \leq \|x\| \|T^*(y)\|$; hence ℓ_y is bounded on $D(T)$ with $\|\ell_y\| \leq \|T^*(y)\|$ and $D(T^*) \subset D\{y \in H : \ell_y(x) = \langle T(x), y \rangle$ is a bounded linear functional on $D(T)\}$. Conversely, if $y \in D$, since $D(T)$ is dense in H, ℓ_y extends uniquely to a bounded linear functional on H and by the Riesz-Fréchet representation theorem there exists a unique x_ℓ in H such that $\langle T(x), y \rangle = \langle x, x_\ell \rangle$; let $T^*(y) = x_\ell$. Then $D(T^*) \supset D$.

156. (a) implies (b) If $y \in D(T)$ the functional $\ell_y(x) = \langle T(x), y \rangle = \langle x, T(y) \rangle$ satisfies $|\ell_y(x)| \leq \|x\| \|T(y)\|$ and is thus bounded; by Problem 155, $y \in D(T^*)$.

(b) implies (c) Since $\langle T(x), y \rangle = \langle x, T(y) \rangle$ for $x, y \in D(T)$ it follows that T^* extends T. Then T is symmetric, $\langle T(x), x \rangle = \overline{\langle x, T(x) \rangle} = \overline{\langle T(x), x \rangle}$ for $x \in D(T)$, and so $\langle T(x), x \rangle \in \mathbb{R}$.

(c) implies (a) Expanding as in Problem 109(a) it readily follows that $\lambda \langle T(y), x \rangle + \overline{\lambda} \langle T(x), y \rangle$ is real for all $\lambda \in \mathbb{C}$. Picking $\lambda = i$ we get that $\langle T(y), x \rangle - \langle T(x), y \rangle$ is purely imaginary and for $\lambda = 1$ that $\langle T(y), x \rangle + \langle T(x), y \rangle$ is real. From this it readily follows that $\langle T(x), y \rangle = \overline{\langle T(y), x \rangle}$, and, consequently, $\langle T(x), y \rangle = \langle x, T(y) \rangle$.

157. (a) We prove that $R(T - iI)$ is closed, the other assertion being similar. For $x \in D(T)$, by symmetry $\|(T - iI)x\|^2 = \langle T(x) - ix, T(x) - ix \rangle = \|T(x)\|^2 + i\langle T(x), x \rangle - i\langle x, T(x) \rangle + \|x\|^2 = \|T(x)\|^2 + \|x\|^2$. Therefore, if $y_n = (T - iI)(x_n)$ is a sequence in $R(T - iI)$ that converges to $y \in H$, say, it readily follows that $\|T(x_n - x_m)\|^2 + \|x_n - x_m\|^2 \to 0$ as $n, m \to \infty$. Thus $\{x_n\}$

and $\{T(x_n)\}$ are Cauchy sequences in H and, consequently, $\{(x_n, T(x_n))\}$ converges to $(x, y') \in H \times H$, say. Now, since T is closed, $x \in D(T)$ and $y' = T(x)$, and, therefore, $y_n = T(x_n) - ix_n \to y' - ix = (T - iI)(x)$. Thus $y \in R(T - iI)$, which is therefore closed.

(b) Follows from Problem 151.

158. (a) implies (b) If T is self-adjoint, T is closed. Now, if $x \in K(T^* - iI)$, then $T(x) = T^*(x) = ix$, and so $i\|x\|^2 = \langle T(x), x \rangle = \langle x, T(x) \rangle = -i\|x\|^2$ and $x = 0$. A similar computation gives that the same is true for $x \in K(T^* + iI)$.

(b) implies (c) By Problem 157(a), $R(T - iI)$ is closed and by Problem 157(b), $\{0\} = K(T^* + iI) = R(T - iI)^\perp$, i.e., $R(T - iI) = H$.

(c) implies (a) Since T is symmetric, T^* extends T and $D(T) \subset D(T^*)$. Now, if $x \in D(T^*)$, let $x' \in D(T)$ be such that $(T^* + iI)(x) = (T + iI)(x')$. Then $(T^* + iI)(x - x') = 0$ and $x - x' \in K(T^* + iI) = R(T - iI)^\perp = \{0\}$. Thus $x = x' \in D(T)$.

159. (a) Let $x \in D(S)$; since $R(T + iI) = R(S + iI)$ there exists $x' \in D(T)$ with $(T + iI)(x') = (S + iI)(x)$, which, since S extends T, implies $(S + iI)(x - x') = 0$. Hence $S(x - x') = -i(x - x')$, and by Problem 156(c), $\langle S(x - x'), x - x' \rangle = -i\|x - x'\|^2$ is real. Therefore $x - x' = 0$ and $x = x' \in D(T)$.

(b) For the sake of argument suppose that S is a self-adjoint extension of T. Since $H = R(T + iI) = R(S + iI)$, by (a), $S = T$ and T is self-adjoint. However $R(T - iI) \neq H$ implies, by Problem 158(c), that T is not self-adjoint.

160. (a) That M is dense was done in Problem 60. Now, for $x \in M$, let N_x be the least integer such that $\sum_{n=0}^{N_x} x_n = 0$ and $x_n = 0$ if $n > N_x$. Let $x, y \in M$ and assume, as we may, that $N_x \leq N_y$. Now, since $\sum_{m=0}^{n-1} x_m + x_n + \sum_{m=n+1}^{N_x} x_m = 0$ and $\sum_{m=0}^{n} x_m + \sum_{m=n+1}^{N_x} x_m = 0$, we have $(Tx)_n = -i\left(x_n + 2\sum_{m=n+1}^{N_a} x_m\right)$, and, consequently, $\langle T(x), y \rangle = -i\sum_{n=0}^{N_x} x_n y_n - 2i\sum_{n=0}^{N_x}\sum_{m=n+1}^{N_x} x_m y_n = C - 2iD$, say. Now,

$$D = \sum_{m=0}^{N_x}\sum_{n=0}^{m-1} x_m y_n = \sum_{n=0}^{N_x}\sum_{m=0}^{n-1} x_n y_m = \sum_{n=0}^{N_x}\sum_{m=n+1}^{N_y} x_n y_m - \sum_{n=0}^{N_x} x_n y_n$$

$$= i\left(\sum_{n=0}^{N_x} x_n y_n - \sum_{n=0}^{N_x}\sum_{m=n+1}^{N_y} x_n y_m\right)$$

$$= i\sum_{n=0}^{\infty}\left(y_n + 2\sum_{m=n+1}^{N_y} y_m\right) x_n = \langle x, T(y) \rangle.$$

(b) Let $x \in D(T)$. Then it follows that

$$w_n = ((T + iI)x)_n = i \left(\sum_{m=0}^{n-1} x_m - \sum_{m=n+1}^{N_x} x_m \right) + i x_n$$

$$= -2i \sum_{m=n+1}^{N} x_m = 2i \sum_{m=0}^{n} x_m.$$

Let $v \in K(T^* - iI)$ and $a_n \to v$, $a^n \in D(T)$. Since T^* is closed, we have $(T^* - iI)a^n \to (T^* - iI)v = 0$. Let $b(n) = (T^* - iI)a^n$. Hence, $b_k^n = 2i \sum_{m=0}^{k} a_m^n$. Since $2ia_k^n = b_{k+1}^n - b_k^n$, we have $2ia^n = U(b^n) - b^n$, where U is the unilateral left shift. Therefore, $2\|a^n\| \le 2\|b^n\| \to 0$. Hence $v = 0$, thus $0 = K(T^* - iI) = R(T + iI)^\perp$. Therefore, $R(T + iI)$ is dense in ℓ^2.

(c) Let $\tilde{w} = e_1$ and $v \in D(T)$. We have $|\langle T(v), \tilde{w}\rangle|^2 = |iv_0|^2 \le \|v\|^2$. Hence, $\tilde{w} \in D(T^*)$. Furthermore, $\langle (T - iI)v, \tilde{w}\rangle = \langle T(v), \tilde{w}\rangle + \langle -iv, \tilde{w}\rangle = iv_0 - iv_0 = 0$ for all v. Hence, $\langle v, (T^* + iI)\tilde{w}\rangle = 0$ for all v. Therefore $(T^* + iI)\tilde{w} = 0$.

(d) By (b), $R(\overline{T} + iI) = H$. By (a), $\overline{T}$ is symmetric. By (c), $R(\overline{T} - iI)^\perp \mathrm{sp}\{\tilde{w}\}$, hence $R(\overline{T} - iI) \ne H$. Then, by Problem 159, $\overline{T}$ has no self-adjoint extension.

162. (a) Let $x \in H$; since $Q(x) \in K(S + S_1)$, $T(Q(x)) \in K(S + S_1)$ and $(S + S_1)TQ(x) = T(S + S_1)Q(x) = 0$; hence $QTQ(x) = TQ(x)$. Similarly, $QT^*Q(x) = T^*Q(x)$ for all $x \in H$. Therefore $QT = (T^*Q)^* = (QT^*Q)^* = QTQ = TQ$.

(b) Let $x \in K(S)$. Then $\|S_1(x)\|^2 = \langle S_1(x), S_1(x)\rangle = \langle S_1^2(x), x\rangle = \langle S^2(x), x\rangle = \|S(x)\|^2 = 0$ and $x \in K(S_1)$. Since these steps are reversible, $K(S) = K(S_1)$. Thus, if $x \in K(S)$, $x \in K(S + S_1)$ and since Q projects onto $K(S + S_1)$, $Q(x) = x$ for $x \in K(S)$.

(c) Since $S^2 = S_1^2$, $(S + S_1)(S - S_1)(x) = (S^2 - S_1^2)(x) = 0$ for all $x \in H$, hence $(S - S_1)(x) \in K(S + S_1)$, i.e., $Q(S - S_1)(x) = (S - S_1)(x)$ for all $x \in H$. On the other hand, since $Q(S + S_1) = (S + S_1)Q$ and $Q(x) \in K(S + S_1)$ for all $x \in H$, it follows that $Q(S + S_1)(x) = (S + S_1)Q(x) = 0$. Finally, from the relations $QS + QS_1 = 0$ and $QS - QS_1 = S - S_1$ it follows that $2QS_1 = S_1 - S$, and so $S = (I - 2Q)S_1$.

163. (a) Recall that $G(T^*) = \{(v, w) \in H \times H : \langle x, w\rangle = \langle T(x), v\rangle$ for all $x \in D(T)\}$. Thus if $\{(v_n, w_n)\} \subset G(T^*)$ converges to (v, w) in $H \times H$, then $\langle x, w\rangle = \langle T(x), v\rangle$ for all $x \in D(T)\}$ and $(v, w) \in G(T^*)$, which is therefore closed.

(b) Necessity first. Since $D(T^*)$ is dense in H, T^{**} is defined; we claim that $D(T) \subset D(T^{**})$. Indeed, if $v \in D(T)$, $L(w) = \langle T^*(w), v\rangle = \langle w, T(v)\rangle$

is a bounded linear functional on $D(T^*)$ with norm $\leq \|T(v)\|$, and, consequently, by Problem 155, $v \in D(T^{**})$ and $T^{**}(v) = T(v)$ for $v \in D(T)$. Thus T^{**} is a closed extension of T, which is therefore closable.

Sufficiency next. For the sake of argument suppose $0 \neq v_0 \in D(T^*)^\perp$. Then $(0, v_0) \in G(T^*)^\perp$ and from (a) it follows that $G(T^*)^\perp = W^{\perp\perp} = \overline{W}$. Now, it is clear that $\overline{W} = \{(w, -v) : (v, w) \in \overline{G(T)}\}$, and, hence, $(0, v_0) \in \overline{G(T)}$, which implies by Problem 9.1 that $\overline{G(T)}$ is not the graph of a linear operator.

(c) If T is closable, T^* is densely defined and at the beginning of (b) we checked that $\overline{T} \subset T^{**}$. In fact, there is equality, because we can use the computation in (a) applied to T^* instead of T to determine the graph of T^*: $G(T^{**}) = V^\perp$, where $V = \{(T^*(x), -x) \in H \times H : x \in D(T^*)\} \subset H \times H\}$. Thus V is obtained from the graph of T^* as $A(G(T^*))$, where A is the linear isometry $A(v, w) = (w, -v)$. In (b) we saw that $G(T*)^\perp = \overline{W}$ and since $A(\overline{W}) = \overline{G(T)}$, this gives $G(T^{**}) = (A(G(T^*)))^\perp = A(G(T^*))^\perp = A(\overline{W}) = \overline{G(T)}$. Finally, for T closable we have $\overline{T}^* = (T^{**})^* = \overline{T^*} = T^*$ since T^* is closed.

164. Necessity first. For the sake of argument suppose that $m_1 = 0$ and let $\{x_n\}$ be such that $\|x_n\| = 1$ for all n, and $\|T(x_n)\|^2 = \langle T^*T(x_n), x_n \rangle \to 0$. Then $1 = \|x_n\| = \|T^{-1}T(x_n)\| \leq \|T^{-1}\|\|T(x_n)\| \to 0$, which is not the case. Similarly $m_2 > 0$.

Sufficiency next. Pick $c_1, c_2 > 0$ such that $c_1 m_1 < 1, c_2 m_2 < 1$, and define the linear operators $S_1 = c_1 T^*T$, $S_2 = c_2 TT^*$. Then $S_1, S_2, I - S_1, I - S_2$ are self-adjoint and $\|I - S_1\| = \sup_{\|x\|=1}\langle (I - c_1 T^*T)(x), x \rangle = 1 - c_1 \inf_{\|x\|=1}\langle T^*T(x), x \rangle = 1 - c_1 m_1 < 1$; similarly $\|I - S_2\| < 1$. Therefore by Problem 9.149, S_1 and S_2 are invertible and, consequently, $(T^*T)^{-1}$ and $(TT^*)^{-1}$ exist. Hence, since $I = (T^*T)^{-1}T^*T = TT^*(TT^*)^{-1}$, T has a left and right inverse and is therefore invertible. Therefore $T^{-1} = (T^*T)^{-1}T^* = T^*(TT^*)^{-1}$ exists.

165. Note that the result applies to $T = I + SS^*$ where $S : H \to H$ is a bounded linear operator.

166. (a) Since $(S^*TS)^* = S^*TS$ and $\langle S^*TS(x), x \rangle = \langle TS(x), S(x) \rangle \geq 0$ for $x \in H$, S^*TS is positive.

(b) The proof proceeds by induction for T^{2n} and T^{2n+1} simultaneously. First, for $n = 0$, $T^0 = I$ and $T^1 = T$ are positive. Suppose next that T^{2n} and T^{2n+1} are positive. Then by (a), $T^{2(n+1)} = TT^{2n}T$ and $T^{2(n+1)+1} = TT^{2n+1}T$ are positive.

(c) Let b be the functional on $H \times H$ given by $b(x, y) = \langle T(x), y \rangle$. First, for $y \in H$, the mappping $b(\cdot, y)$ is linear and since $T = T^*$, $b(y, x) = \langle T(y), x \rangle = \overline{\langle x, T(y) \rangle} = \overline{\langle T(x), y \rangle}$ for all $y \in H$, and b is hermitian. Also,

$b(x, x) = \langle T(x), x \rangle \geq 0$ for $x \in H$. Therefore, by the Cauchy-Schwarz inequality for b we have $|\langle T(x), y \rangle|^2 \leq \langle Tx, x \rangle \langle T(y), y \rangle$ for all $x, y \in H$. Note that letting $y = T(x)$ it follows that $\langle T(x), T(x) \rangle^2 = \|T(x)\|^4 \leq \langle T(x), x \rangle \langle T^2(x), T(x) \rangle$.

167. (a) If $n = 2k$ is even, $T^n - T^{n+1} = T^k(I - T)T^k$ is positive by Problem 166(a). On the other hand, if $n = 2k + 1$ is odd, $T^n - T^{n+1} = T^k(T - T^2)T^k$ is positive by Problems 166(e) and (a).

(b) By (a), $\{\langle T^n(x), x \rangle\}$ is decreasing and by Problem 166(b) bounded below by 0. Hence, it converges for all $x \in H$.

(c) Let $n, m \in \mathbb{N}$, $n \leq m$. Clearly $(T^n - T^m)^* = T^n - T^m$. Now, as we saw in (b), $\langle T^0(x), x \rangle \geq \langle T^n(x), x \rangle \geq \langle T^m(x), x \rangle \geq 0$, and, therefore, $0 \leq \langle (T^n - T^m)(x), x \rangle \leq \langle T^0(x), x \rangle$. Whence $T^n - T^m$ and $I - (T^n - T^m)$ are positive. Now, by Problem 166(d), $\|T^n(x) - T^m(x)\|^2 \leq \langle T^n(x), x \rangle - \langle T^m(x), x \rangle$ for all $x \in H$. Then since $\{\langle T^n(x), x \rangle\}$ converges, $\{T^n(x)\}$ is Cauchy and, hence, converges to a limit $P(x)$, say, in H for every $x \in H$. Clearly P is linear. Since $\|T^n(x)\| \leq \|T\|^n\|x\| \leq \|x\|$ for all n and $x \in H$, the same is true in the limit, i.e., $\|P(x)\| \leq \|x\|$. Therefore P is continuous and $\{T_n\}$ converges in $\mathcal{B}(H)$ in the strong topology.

(d) By the continuity of the inner product, $\lim_n \langle T_n(x), y \rangle = \langle P(x), y \rangle$ for all $x, y \in H$. Now, since $(T^n)^* = T^n$ for all n, it follows that $\langle P^*(x), y \rangle = \langle x, P(y) \rangle = \lim_n \langle x, T^n(y) \rangle = \lim_n \langle T^n(x), y \rangle = \langle P(x), y \rangle$, and so $P = P^*$. Since T is continuous, $TP(x) = \lim_n TT^n(x) = P(x)$, and, therefore, $TP = P$.

(e) If $T(x) = x$, $T^n(x) = x$ for all n, and so $P(x) = x$. Let $y \in H$ and put $x = P(y)$. By (d), $T(x) = TP(y) = P(y) = x$, and, consequently, $P(x) = x$. This being the case for all $y \in H$ we have $P^2 = P$. By (d), $(T - I)P = 0$. and, therefore, $R(P) \subset K(T - I)$. For $x \in K(T - I)$, $P(x) = x$, and so $x \in R(P)$, and, therefore, $R(P) = K(T - I)$. Finally, if $x \in R(P)^\perp$ for all $y \in H$ we have $\langle P(x), y \rangle = \langle x, P(y) \rangle = 0$, and so $P(x) = 0$. We deduce that $P(x + y) = x$ if $x \in K(T - I)$ and $y \in (K(T - I))^\perp$, and, therefore, P is the orthogonal projection onto $K(T - I)$.

169. Let $x \in D(T^*)$. By assumption $R(T) = R(T^*) = H$ and T and T^* agree on the dense set $D(T)$. So there is $x' \in D(T^*)$ such that $T(x) = T^*(x')$. Hence $\langle T(y), x' \rangle = \langle y, T^*(x') \rangle = \langle y, T(x) \rangle = \langle T^*(y), x \rangle = \langle T(y), x \rangle$ for all $y \in D(T) \subset D(T^*)$. Hence $\langle T(y), x - x' \rangle = 0$. Finally, since there is $y \in D(T)$ such that $T(y) = x - x'$, $\|x - x'\| = 0$.

170. First, since $\overline{R(T)}^\perp = K(T^*) = K(T) = \{0\}$, $R(T)$, which is the domain of T^{-1}, is dense in H. Now, $y \in D(T^{-1*})$ iff $\langle T^{-1}(x), y \rangle = \langle x, T^{-1*}(y) \rangle$ for all $x \in R(T)$, or, equivalently, $\langle z, y \rangle = \langle T(z), T^{-1*}(y) \rangle$ for all $z \in D(T)$. This means that $T^{-1*}(y) \in D(T^*)$ and $T^*T^{-1*} = I$. Similarly, since $R(T^*)$

is dense in H, $y \in D(T^{-1*})$ iff $\langle T^{*-1}(x), y \rangle = \langle x, (T^{*-1})^*(y) \rangle$ and, letting $x = T^*(z)$ we get $\langle z, y \rangle = \langle T^*(z), (T^{*-1})^*(y) \rangle = \langle T^{*-1}T^*(z), y \rangle$. Thus, $T^{*-1}T^* = I$.

172. The statement is false. Let $(x, y) \in X \times Y$ with $\phi(x, y) \neq 0$, $\lambda_n = n + 1/n$, and observe that $\|\lambda_n(x, y) - n(x, y)\| = n^{-1}\|(x, y)\| \to 0$ as $n \to \infty$. However,

$$|\phi(\lambda_n(x, y)) - \phi(n(x, y))| = |(n + 1/n)^2 \phi(x, y) - n^2 \phi(x, y)|$$
$$= (2 + n^{-2})|\phi(x, y)| \to 2|\phi(x, y)| \neq 0$$

as $n \to \infty$.

173. Let $x \in H$. Since $\phi(x, \cdot)$ is conjugate linear in the second variable, $\overline{\phi(x, \cdot)}$ is a bounded linear functional on H with norm $\leq K\|x\|$ where K is the norm of ϕ. Thus, by the Riesz-Fréchet representation theorem, there is a unique $x_\phi \in H$ such that $\overline{\phi(x, y)} = \langle y, x_\phi \rangle$ and $\|\overline{\phi(x, \cdot)}\| = \|x_\phi\|$. Hence, $\phi(x, y) = \langle x_\phi, y \rangle$. Let $T : H \to H$ be given by $T(x) = x_\phi$. First, T is well-defined. Indeed, if $\tilde{T} : H \to H$ is another such mapping, then $\phi(x, y) = \langle T(x), y \rangle = \langle \tilde{T}(x), y \rangle$ for all $x, y \in H$, and so $T(x) = \tilde{T}(x)$ for all $x \in H$. Next, T is linear. Indeed, for all $x, x', y \in H$ and scalars λ,

$$\langle T(x + \lambda x'), y \rangle = \phi(x + \lambda x', y) = \phi(x, y) + \lambda \phi(x', y)$$
$$= \langle T(x), y \rangle + \lambda \langle T(x'), y \rangle = \langle T(x) + \lambda T(x'), y \rangle.$$

Thus, since this relation holds for all $y \in H$, $T(x + \lambda x') = T(x) + \lambda T(x')$.

Finally, T is an isomorphism. First, T is continuous: By Problem 134, $\|T(x)\| \leq K\|x\|$ and T is bounded with $\|T\| \leq K$. Next, T is injective: Let $x \in H$. Then, by Cauchy-Schwarz, $c\|x\|^2 \leq \phi(x, x) = \langle T(x), x \rangle \leq \|T(x)\| \|x\|$. Thus, if $x \neq 0$, $\|T(x)\| \geq c\|x\|$ and, in particular, T is injective. Next, T is onto. To see this, observe that $R(T)$ is closed in H. Indeed, let $\{T(x_n)\}$ be a convergent sequence in $R(T)$. Then $\|T(x_n) - T(x_m)\| \geq c\|x_n - x_m\|$, and, therefore, $\{x_n\}$ is Cauchy, and converges to $x \in H$, say. By the continuity of T, $T(x_n) \to T(x) \in R(T)$, and $R(T)$ is closed. So, to prove that T is surjective it suffices to verify that $R(T)^\perp = \{0\}$. Let $x \in R(T)^\perp$. Then, since $T(x) \in R(T)$, $0 = \langle T(x), x \rangle = \phi(x, x) \geq c\|x\|^2$, and $x = 0$. Finally, by the inverse mapping theorem, T is a homoemorphism.

174. First, by the Riesz-Fréchet representation theorem, there is a unique $y_L \in H$ with $\|y_L\| = \|L\|$ such that $L(x) = \langle y_L, x \rangle$ for all $x \in H$. Let T be the mapping defined in Problem 173, put $u = T^{-1}(y_L)$ and observe that $\phi(u, x) = \langle T(u), x \rangle = \langle y_L, x \rangle = L(x)$ for all $x \in H$. We claim that u is unique. Let u' be such that $\phi(u', x) = L(x)$. Then, $\phi(u - u', x) = 0$ for all $x \in H$ and, in particular, $\phi(u - u', u - u') = 0$. Then, $c\|u - u'\|^2 \leq \phi(u - u', u - u') = 0$ and $u = u'$.

175. (i) For $x \in H$ and $t \in \mathbb{R}$, let $f(t) = \Phi(x + ty)$. Then, f is a real-valued function of t and, from the symmetry of ϕ, it follows that $f(t) = (t^2/2)\phi(y,y) + t(\phi(x,y) - L(y)) + (1/2)\phi(x,x) - L(x) + C$, and $f'(t) = t\phi(y,y) + (\phi(x,y) - L(y))$. Thus, $\Phi(x)$ has a global minimum at $x \in H$ iff $f(t)$ has a global minimum at $t = 0$, i.e., $\Phi(x + ty) = \Phi(x) + tf'(0) + (t^2/2)\phi(x,x) \geq \Phi(x)$ for all $t \in \mathbb{R}$, for all $y \in H$ iff $f'(0) = \phi(x,y) - L(y) = 0$, for all $y \in H$. This establishes the equivalence of (a) and (b), and (i) is done.

176. Necessity first. Let M be a total ONS in H and pick $y_1, y_2 \in T(M)$ and $x_1, x_2 \in M$ such that $y_1 = T(x_1)$ and $y_2 = T(x_2)$. Then $\langle y_1, y_2 \rangle = \langle T(x_1), T(x_2) \rangle = \langle x_1, T^*T(x_2) \rangle = \langle x_1, T^{-1}T(x_2) \rangle = \langle x_1, x_2 \rangle = 1$ if $x_1 = x_2$ and $\langle y_1, y_2 \rangle = 0$ otherwise. Now, since T is a bijection, $x_1 = x_2$ only when $y_1 = y_2$ and so $\langle y_1, y_2 \rangle = 1$ if $y_1 = y_2$ and $\langle y_1, y_2 \rangle = 0$ if $y_1 \neq y_2$ for all $y_1, y_2 \in T(M)$. Thus $T(M)$ is an ONS in H. Next, let $y \in T(M)^{\perp}$. Then $\langle y, T(x) \rangle = \langle T^*(y), x \rangle = 0$ for all $x \in M$ and since M is total, $T^*(y) = 0$. Whence, since T^* is injective when it exists, $y = 0$, and $T(M)$ is total in H.

Sufficiency next. First, T is bounded. Given $x \in H$ with $\|x\| = 1$, let $X = \mathrm{sp}\{x\}$; X is a closed subspace of H and, consequently, $H = X \oplus X^{\perp}$. Now, since $X^{\perp}$ is a closed subspace of H, $X^{\perp}$ is a Hilbert space in its own right and has an ONB M_1, say. Let $M = \{x\} \cup M_1$; since $\|x\| = 1$ this is an ONS in H and M_1 is orthogonal to X and, hence, to x. Moreover, since $\overline{\mathrm{sp}}(M) \supset \overline{\mathrm{sp}}(M_1) \cup \overline{\mathrm{sp}}(\{x\}) = X^{\perp} \cup X = H$, M is a total ONS in H. Then $T(M)$ is a total ONS in H and, in particular, $\|T(x)\| = 1$. Since x is arbitrary, $\|T(x)\| = 1$ whenever $\|x\| = 1$ and T is bounded with norm 1. Now, since $x/\|x\|$ always has norm 1, except when $x = 0$ but then $T(0) = 0$, it follows that $\|T(x)\| = \|x\|$ for all $x \in H$ and, in particular, T is injective.

Next, T is onto. Let M be a total ONS in H; by assumption $\overline{\mathrm{sp}}(T(M)) = H$. Moreover, since $T(M) \subset T(H)$, $\mathrm{sp}(T(M)) \subset T(H)$ and $H = \overline{\mathrm{sp}}(T(M)) \subset \overline{T(H)}$. Therefore $\overline{T(H)} = H$. Then, given $y \in H$, there exist $\{y_n\} \subset T(H)$ and $\{x_n\}$ in H such that $y_n = T(x_n)$ for all n, and $y_n \to y$. Now, since T is an isometry $\|x_m - x_n\| = \|T(x_m - x_n)\| = \|y_m - y_n\|$, and so $\{x_n\}$ is Cauchy and, hence, converges to x in H, say. Finally, since T is bounded, $T(x) = y$ and T is surjective.

Next, by polarization, from the isometry one deduces $\langle Tx, Ty \rangle = \langle x, y \rangle$ for all $x, y \in H$. (This is true for real and complex H; for complex H do the real and imaginary parts separately.) Now, this implies that $\langle T^*T(x), y \rangle = \langle T(x), T(y) \rangle = \langle x, y \rangle$ for all $x, y \in H$. Hence, since y is arbitrary, $T^*T(x) = x$ for all $x \in H$.

Index

Selected Published Titles in This Series

For a complete list of titles in this series, visit the
AMS Bookstore at **www.ams.org/bookstore/gsmseries/**.